W0105940

Nonlinear Model Based Process Control

NATO ASI Series

Advanced Science Institute Series

A Series presenting the results of activities sponsored by the NATO Science Committee, which aims at the dissemination of advanced scientific and technological knowledge, with a view to strengthening links between scientific communities.

The Series is published by an international board of publishers in conjunction with the NATO Scientific Affairs Division

A Life Sciences	Plenum Publishing Corporation
B Physics	London and New York
C Mathematical and Physical Sciences	Kluwer Academic Publishers
D Behavioural and Social Sciences	Dordrecht, Boston and London
E Applied Sciences	
F Computer and Systems Sciences	Springer-Verlag
G Ecological Sciences	Berlin, Heidelberg, New York, London,
H Cell Biology	Paris and Tokyo
I Global Environment Change	

PARTNERSHIP SUB-SERIES

1. Disarmament Technologies	Kluwer Academic Publishers
2. Environment	Springer-Verlag / Kluwer Academic Publishers
3. High Technology	Kluwer Academic Publishers
4. Science and Technology Policy	Kluwer Academic Publishers
5. Computer Networking	Kluwer Academic Publishers

The Partnership Sub-Series incorporates activities undertaken in collaboration with NATO's Cooperation Partners, the countries of the CIS and Central and Eastern Europe, in Priority Areas of concern to those countries.

NATO-PCO-DATA BASE

The electronic index to the NATO ASI Series provides full bibliographical references (with keywords and/or abstracts) to about 50,000 contributions from international scientists published in all sections of the NATO ASI Series. Access to the NATO-PCO-DATA BASE is possible via a CD-ROM "NATO Science and Technology Disk" with user-friendly retrieval software in English, French, and German (©WTV GmbH and DATAWARE Technologies, Inc. 1989). The CD-ROM contains the AGARD Aerospace Database.

The CD-ROM can be ordered through any member of the Board of Publishers or through NATO-PCO, Overijse, Belgium.

Series E: Applied Sciences – Vol. 353

Nonlinear Model Based Process Control

edited by

Rıdvan Berber

Department of Chemical Engineering,
Ankara University,
Tandoğan, Ankara, Turkey

and

Costas Kravaris

Department of Chemical Engineering,
The University of Michigan,
Ann Arbor, Michigan, U.S.A.

Volume I

Springer Science+Business Media, B.V.

Proceedings of the NATO Advanced Study Institute on
Nonlinear Model Based Process Control
Antalya, Turkey
August 10–20, 1997

A C.I.P. Catalogue record for this book is available from the Library of Congress.

ISBN 978-94-010-6140-7   ISBN 978-94-011-5094-1 (eBook)
DOI 10.1007/978-94-011-5094-1

Printed on acid-free paper

All Rights Reserved
© 1998 Springer Science+Business Media Dordrecht
Originally published by Kluwer Academic Publishers in 1998

No part of the material protected by this copyright notice may be reproduced or
utilized in any form or by any means, electronic or mechanical, including photo-
copying, recording or by any information storage and retrieval system, without written
permission from the copyright owner.

CONTENTS

Part V. Industrial Applications

PREFACE

The ASI on Nonlinear Model Based Process Control (August 10-20, 1997; Antalya - Turkey) convened as a continuation of a previous ASI which was held in August 1994 in Antalya on Methods of Model Based Process Control in a more general context. In 1994, the contributions and discussions convincingly showed that industrial process control would increasingly rely on nonlinear model based control systems. Therefore, the idea for organizing this ASI was motivated by the success of the first one, the enthusiasm expressed by the scientific community for continuing contact, and the growing incentive for on-line control algorithms for nonlinear processes. This is due to tighter constraints and constantly changing performance objectives that now force the processes to be operated over a wider range of conditions compared to the past, and the fact that many of industrial operations are nonlinear in nature.

The ASI intended to review in depth and in a global way the state-of-the-art in nonlinear model based control. The list of lecturers consisted of 12 eminent scientists leading the principal developments in the area, as well as industrial specialists experienced in the application of these techniques. Selected out of a large number of applications, there was a high quality, active audience composed of 59 students from 20 countries. Including family members accompanying the participants, the group formed a large body of 92 persons. Out of the 71 participants, 11 were from industry. A good balance was reached in attendance between high-profile presenters, representatives of leading companies and PhD students, which led to very interesting discussions.

20 main lectures (2 from industry), 10 invited contributions (5 from industry) and one special tutorial lecture (upon request from the students) were given. 5 poster contributions with 10 minute flash presentations were made. Two volumes of pre-prints (one for main lectures and invited contributions, and one for poster presentations) were distributed to the participants at registration. This facilitated the understanding of the lectures and the interaction among lecturers and students. A well prepared and frequently updated web-site was maintained for the convenience of participants. It helped the participants to get beforehand information about the technical program, lecturers, presentations, meeting site and social program.

Two panel sessions and one open forum were held during the ASI. The first panel was right at the opening session to have the opinion of a group of prominent lecturers on the status of current control technology, its potential and its limitations. The second panel session was scheduled in the middle of the technical program to discuss if there had been any changes in the views after the presentations of most recent developments in the area. An industrialist's view and conception of interaction between the academia and industry, the list of seven "DON'Ts", and the description of challenging and unsolved problems were particularly welcomed by the audience and resulted in fruitful discussions. Having come to the end of the presentations, a free forum was organized to let everybody, especially those who had not had the chance to speak, express themselves briefly about where they think we should be heading. This session focused on three themes; (i) assessment of the state-of-art, (ii) critique of the current ASI, and (iii) suggestions for a possible future ASI. All participants underlined the high quality of the scientific and administrative organization and appreciated the high level of presentations and active discussions.

The tradition of distributing 'Special Awards', which started at the previous ASI held in 1994, continued. This time, however, every speaker got a 'Special Prize' distributed by Prof. Bequette on a cartoon display regarding the boat tour that was organized within the social program. The role that every contributor played on the cartoon of the sail boat reflected his/her scientific views and discussions. This very special assessment of the meeting is included in the appendix.

The meeting hotel, Falez, again provided a very pleasant atmosphere for working as well as relaxing between sessions. The warm and sunny weather, and the picturesque scenery around the hotel formed a very pleasant setting for the meeting, and in particular, for all non-scientific programs. Exciting excursions to places of cultural and historical interest and social program, including titles such as "Boat Tour Under Full Moon", "Swimming in History" and the like, gave us an opportunity to combine the work with ease and contributed to the good spirit of all lecturers and participants for pursuing discussions in a more relaxed atmosphere.

For the final preparations of this book, the Organizing Committee decided to have all papers reviewed by two reviewers, in order to provide feedback to the authors. This was considered as a positive step towards a high quality book, and was agreed upon by the lecturers and contributors during the ASI. The material in the volume is arranged in five parts each covering the major themes presented at the Institute. The question of how to use linear models to effectively control nonlinear processes is addressed first. Classical gain scheduling techniques are reviewed, followed by two new approaches, which involve multiple models. Part II covers a broad spectrum of nonlinear model based controller synthesis methods. It includes investigations of stability, robustness, constraint handling and non-minimum phase issues on recent and new approaches, as well as new classes of problems, like in distributed parameter differential algebraic equation systems. On line optimization approaches are covered in Part III. Constraint handling and stability issues were also emphasized in this part. Nonlinear state and parameter estimation are significant for the development of a reliable model as well as calculating feedback action from available measurements. These issues are treated in Part IV. A number of contributions came from industry and deal with implementations of nonlinear model based control. These are grouped into Part V. It is hoped that the present book will ultimately help motivate the applications of the developed methods and will contribute to opening new frontiers in research directions. One can also find new research results, ideas and views for future needs and challenges. The book can be used for a graduate level course, and is a comprehensive guide for researchers and industrial control engineers to explore the latest trends in the field.

Clearly, this ASI would not have been possible without the financial resources. The Organizers are indebted to NATO-Scientific Affairs Division for its financial support. Special thanks of appreciation are extended to Dr. L. Veiga da Cunha (NATO-Brussels). The National Administrators of NATO ASIs in Turkey, Greece and Portugal, and the National Science Foundation in the USA provided supplementary funds for participants from these countries. Additional support in different forms were also supplied by Ankara University, Graduate School of Natural and Applied Sciences; AKSA Acrilic Chemical Industries, Co. (Yalova); PAK-GIDA A.Ş. (İzmit) and T. Garanti Bankası (İstanbul). All of these contributions, which came through the help of the individuals; Prof. Dr. Aziz Ekşi, Mustafa Yılmaz, Tuncay Yurdesin and Mazlum İnal are gratefully acknowledged. The efforts by the staff of Falez Hotel, and especially those of Nesrin Şimşek, for making our stay a really enjoyable one are appreciated.

It is clear that the success of a school is ultimately determined by the interest and commitments of the lecturers and participants. They were at a very high level in this ASI. We would like to express our very deep gratitude to all lecturers for their work and contributions, and to the participants for their active involvement and expressions of their great satisfaction in attending the meeting.

R. Berber
Ankara, April 1998

C. Kravaris
Ann Arbor, April 1998

ORGANIZING COMMITTEE

Rıdvan BERBER   (Director)   *Department of Chemical Engineering*
                                    University of Ankara
                                    Tandoğan, 06100 Ankara TURKEY

Costas KRAVARIS (Co-director)   *Department of Chemical Engineering*
                                    The University of Michigan
                                    Ann Arbor MI 48109 USA

Yaman ARKUN   *School of Chemical Engineering*
                                    Georgia Institute of Techology
                                    Atlanta, GA 30332-0100 USA

Wolfgang MARQUARDT   *Department of Process Engineering*
                                    RWTH Aachen, Turmstrasse 46
                                    D-5100 Aachen GERMANY

MAIN LECTURERS
(In addition to the Organizing Committee members)

Jens G. BALCHEN   *Department of Engineering Cybernetics*
                                    Norwegian University of Science and Technology
                                    N-7034 Trondheim NORWAY

B. Wayne BEQUETTE   *Department of Chemical Engineering*
                                    Rensselaer Polytechnic Institute
                                    Troy, NY 12180-3590 USA

Coleman B. BROSILOW   *Department of Chemical Engineering*
                                    Case Western Reserve University
                                    10900 Euclid Av. Cleveland OH 44106-7217 USA

Prodromos DAOUTIDIS   *Dept. of Chemical Eng. & Material Sciences*
                                    University of Minnesota
                                    421 Washinton Ave. SE
                                    Minneapolis MN 55455 USA

Denis DOCHAIN   *Université Catholique de Louvain*
                                    Laboratoire d'Automatique, Dynamique et
                                    Analyse des Systèmes
                                    B-1348 Louvain-la-Neuve BELGIUM

Ferhan KAYIHAN

IETek, Integrated Engineering Technologies
5533 Beverly Ave NE, Tacoma
WA 98422-1402 USA

Ahmet PALAZOĞLU

Department of Chemical Engineering
University of California Davis
Davis, CA 95616 USA

George STEPHANOPOULOS

Department of Chemical Engineering
Massachusetts Institute of Technology
Cambridge, Massachusetts 02139 USA

INVITED CONTRIBUTORS

Frank ALLGÖWER

Automatic Control Laboratory
Swiss Federal Institute of Technology
ETH Zentrum CH-8092 Zurich, Switzerland

Andre J. DAMSLORA

Norsk Hydro ASA, P.O.Box 2560
N-3901, Porsgrunn NORWAY

Aydın KONUK

Aspen Technology - AC&OD
9896 Bissonnet
Houston, Texas 77036 USA

Masoud NIKRAVESH

Earth Sciences Division
Lawrence Berkeley National Laboratory
Berkeley, CA 94720

Simone L. de OLIVEIRA

Technology Center, Cidade University
Blocco G, Ilha do Fundao
Rio de Janeiro, 2194-970, RJ BRAZIL

Ronald K. PEARSON

The DuPont Co.
P.O. Box 80101
Wilmington DE 19880-0101 USA

Masoud SOROUSH

Department of Chemical Engineering
Drexel University
Philadelphia, PA 19104, USA

Alex ZHENG

Department of Chemical Engineering
University of Massachusetts
Amherst, MA 01003-3110, USA

Part I

NONLINEAR CONTROL BASED ON LINEAR MODELS

PRACTICAL APPROACHES TO NONLINEAR CONTROL

A Review of Process Applications

B. W. Bequette
Rensselaer Polytechnic Institute
Troy, NY 12180-3590 USA

Abstract

Control techniques explicitly based on nonlinear process models have received much research attention in recent years, while little emphasis has been given to traditional nonlinear control. This paper reviews traditional nonlinear control techniques, including gain scheduling, measurement transformations and selection of actuator characteristics. Included is a review of nonlinear PI-based approaches that have been applied to linear systems, such as surge drum level control. Although the focus is on traditional applications, recent work combining fuzzy logic control with gain scheduling or nonlinear PI-based control is also surveyed. We also develop a general procedure for gain scheduling of Hammerstein systems and control-affine systems with a specific structure. Low-order tutorial examples are provided to illustrate problems satisfying "linearization conditions" when performing gain scheduling.

1. Motivation

The purpose of this workshop is to present and discuss new approaches to nonlinear model-based control of chemical processes. Before detailed presentations and discussions take place, it is appropriate to consider "traditional" approaches to controlling nonlinear processes. After all, chemical process systems did not become nonlinear overnight. How have these processes been controlled in the past? The literature is actually fairly "lean" on this topic. One reason is that traditional approaches tend to be ad-hoc, developed to solve a specific problem, and may not be easily generalized. Even a recent industrial review of nonlinear control by Ogunnaike and Wright (1997) provides no discussion of traditional techniques (such as gain scheduling) for handling nonlinear systems. A recent monograph on nonlinear process control (Henson and Seborg, 1997a) is devoted solely to model-based approaches to nonlinear control.

The most common way of controlling chemical processes has been to tune a single linear, PI controller (most control systems are designed as SISO) so that it is closed-loop stable to likely disturbances or changes in operating conditions (alternatively, loops can simply be retuned when operating at a new condition for a period of time). Clearly there must be some sacrifice in performance at a specific operating condition. For a distillation application, McDonald et al. (1988) characterize the varying process gains as a function of operating condition as structured uncertainty and compare several

R. Berber and C. Kravaris (eds.), Nonlinear Model Based Process Control, 3-32.
© 1998 *Kluwer Academic Publishers.*

robust control design techniques. Shinnar (1986) discusses the proper selection of a linear model set for linear control system design for highly nonlinear systems. He presents an adiabatic packed bed reactor, with an exothermic reaction, and demonstrates that much of the nonlinearity can be captured by a varying gain. A robust controller can be designed based on the maximum process gain. The characterization of nonlinearities by a varying gain is the basis of gain scheduling control which is covered in section 3.

Many techniques to handle at least mild nonlinearities are now routinely discussed in undergraduate process control textbooks. Cascade control can be successfully used when the nonlinearity is primarily located in the secondary (fast) loop (for example, a control valve nonlinearity). Anti-reset windup is commonly used when manipulated variable saturation occurs.

A criticism of past process control research is that it has tended to focus on algorithms and not structure; algorithms will tend to be the focus of this workshop as well. When actually implementing a control strategy, an algorithm is probably a small part of the total overall effort, although model-based control development can take significant time. Many practical issues involving the hardware and robust implementation of software may dominate the time and economics of a control system installation. Most control researchers (even those in industry) tend to overlook how advanced even relatively common control strategies can be. For example, a standard furnace control system with high/low selectors, fuel/air ratio control, stack gas oxygen composition control (with constraints), etc., can be quite complex.

This paper will focus on three common approaches for nonlinear control. Two approaches are based on modifications to fixed PID algorithms, creating a nonlinear control algorithm, while the other approach is to select actuators or measurements to linearize the system. In section 2 we review methods where PID parameters are a function of the error. Gain scheduling, where controller parameters are a function of an auxiliary variable are covered in section 3. Section 4 reviews measurement/actuator selection, and section 5 presents recent fuzzy logic-based nonlinear PID-type of algorithms. Several different control methods are compared in two example problems in section 6; these results are generalized for specific nonlinear model structures in section 7. Section 8 summarizes the paper.

Basically, there are several topics that we wish to address in this paper:

- What processes have yielded successful applications of traditional nonlinear control?

- When is the performance of traditional nonlinear control equivalent to model-based control?

- What is needed in terms of research effort or presentation of applications results?

We do not review the active research on methods to minimize the effect of integral windup, due to manipulated variable saturation. Nor do we significantly review robust linear control system design.

2. Nonlinear PID Control

The fixed-parameter PID controller is ubiquitous in the process industries. The *ideal* PID controller has the following form

$$u(t) = k_c \left[e(t) + \frac{1}{\tau_I} \int_o^t e(t^*)\, dt^* + \tau_D \frac{de(t)}{dt} \right] \tag{1}$$

where it is understood that any bias associated with steady-state operation is absorbed into the integral term. A review of other forms for PID control (including using the derivative of the measured output rather than error) is provided by Astrom et al. (1993).

Often it is useful to have a nonlinear controller even if the process is linear. A common example is level control of a surge drum where the primary objective is to maintain a reasonably consistent outlet flow under disturbances in the inlet flowrate, subject to high and low drum level constraints. In this situation some small nominal controller gain can be used when the level is close to the setpoint, and as the error increases the gain can increase. A common choice is to make the proportional gain a function of the absolute value of the error

$$k_c = k_{c0}\,(1 + a\,|\,e(t)\,|) \tag{2}$$

Marroquin and Luyben (1972) study four configurations of a temperature cascade control system on a batch reactor, all based on a controller gain that is a function of the absolute error. They find that the case where the slave gain is varied by the master loop error gives better control than a linear controller, particularly under widely varying load conditions.

An alternative to making the controller gain a continuous function of the error is to form a piecewise linear controller (sometimes called a "gap" controller), where (often $k_{cu} = k_{cl}$)

$$
\begin{aligned}
k_c &= k_{cl} & \text{for } e < e_l \\
k_c &= k_{c0} & \text{for } e_l \le e \le e_u \\
k_c &= k_{cu} & \text{for } e > e_u
\end{aligned}
\tag{3}
$$

Marlin (1995) compares a piecewise linear controller with linear PI for a level control problem and shows that the piecewise linear control gives more favorable responses in the outlet flow to sine disturbances in the inlet flow. Indeed, level control appears to be the most common application where tuning parameters are a function of the error.

Shunta and Fehervari (1976) report a "wide range controller" developed by Foxboro to satisfy specific level control objectives; this is a parameter scheduled controller where the output of the proportional part is

$$u(t) = k_{c0}\, 25^{|e(t)|K}\, e(t) \tag{4}$$

therefore the proportional gain is

$$k_c = k_{c0}\,(1 + |\,e(t)\,|\,K\ln 25)\, 25^{|e(t)|K} \tag{5}$$

and the reset time, τ_I, is determined to keep the product $k_c \tau_I$ constant

$$\tau_I = \frac{\tau_{I0}}{(1 + |\,e(t)\,|\,K\ln 25)\, 25^{|e(t)|K}} \tag{6}$$

where τ_{I0} is the reset time at zero deviation (error). Shunta and Fehervari also show how a standard PI controller with low and high level overrides can provide similar responses to the nonlinear PI controller, and note that pH is another application of the wide range controller.

Cheung and Luyben (1980) study the "wide range" controller of Shunta and Fehervari for surge level control and find that the controller is hard to tune and the responses are difficult to predict. They also study several piecewise linear controllers, but recommend a limited output change controller where the rate of change of the controller output was limited. Simulation results are verified experimentally in the thesis by Cheung (1978).

McDonald et al. (1986) derive several controllers that implement optimal level control policies, with the objective of minimizing the rate of change of outlet flow subject to minimum and maximum level constraints. One is a nonlinear proportional feedback control law

$$q(t) = q_s + B[1 - \sqrt{1 - |e|/H}] \tag{7}$$

where $q(t)$ is the outlet flowrate, q_s is the nominal steady-state flowrate, B is the magnitude of the maximum step inlet flow disturbance, $e = h_s - h$, and $H = h_{max} - h_s = h_s - h_{min}$ where h_s is the nominal steady-state height (setpoint). A slightly modified version can be used when the range in tank heights is not symmetric around the setpoint. The proportional feedback law is extended to include integral action. McDonald et al. also formulate an optimal predictive control law and show how it can be reduced to a continuous-time feedforward/feedback controller with the form

$$q(t) = \frac{(q_i - q)^2}{2A(h_m - h)} \tag{8}$$

where q_i is the inlet flowrate and $h_m = h_{max}$ (for $q_i > q$), $h_m = h_{min}$ (for $q_i < q$). Note two effects of the controller in (8):

- as the outlet flow begins to match the inlet flow, the rate of change of outlet flow becomes small

- when the height is not close to a min/max constraint, the rate of change of outlet flow is small, but as it approaches a constraint it can become very large

Since the surge drum is an integrating system any bias in the measurements can lead to drift in the control; to remedy this problem, McDonald et al. modify the control law (8) by appending PI terms

$$\dot{q}(t) = \frac{(q_i - q)^2}{2A(h_m - h)} + k_c[e(t) + \frac{1}{\tau_I} \int_0^t e(t^*) \, dt^*] \tag{9}$$

They suggest "loose tuning" for the PI portion so that the rate of change of outlet flow is not greatly affected by the PI term.

Jutan (1989) develops a nonlinear PI(D) controller where the controller gain is a function of the magnitude of the current error, and a function of the past trends in the process output. A small improvement in ITAE performance over a classical PID controller for a first-order plus time process is shown.

Clearly level control is the most widely-published application of PI-type control based on nonlinear functions of the error (that is, "scheduling" control parameters based on the error). The next section reviews the more general topic of PI-type control based on varying tuning parameters as functions of auxiliary parameters.

3. Controller Parameter (Gain) Scheduling

Perhaps the most widely mentioned technique for controlling nonlinear processes is gain scheduling. In this section we review the following gain scheduling topics: traditional, model-based, and recent theory based on linear parameter-varying systems. We also show examples where the "linearization conditions" may not be satisfied.

3.1 Traditional Gain Scheduling

Let α represent the scheduling variable. A gain-scheduled PI controller is then represented by

$$u(t) \quad = \quad k_c(\alpha) \left[\; e(t) \; + \; \frac{1}{\tau_I} \int_o^t e(t^*) \, dt^* \; \right] \tag{10}$$

Notice that the primary difference between this algorithm and the nonlinear PID controllers of section 2 is that α is a general scheduling variable, whereas the controllers in the previous section were based on error as the "scheduling variable". Although all of the controller parameters can be scheduled, controller gain is commonly scheduled because processes are often characterized as a changing gain with relatively constant dynamics. Typically k_c is varied to keep $k_c k_p$ constant, which then keeps the stability margin constant. Astrom et al. (1993) note that most commercially available single-loop controllers offer a gain scheduling option.

An important step in developing a gain scheduled controller is determining the proper scheduling variable. Rules of thumb such as scheduling on a "slow variable" or using a variable that "captures the nonlinearities" are often used. It is most common, then, to schedule based on either the setpoint or measured output, since the process input (manipulated variable) will vary more rapidly than the output. When there are additional measurements, then a measured or inferred "auxiliary" variable can be used for scheduling. The four basic steps in developing a parameter-scheduled controller are (most publications lack specific details about one or more of these steps):

1. Develop a linear process model for a set (usually a discrete number) of operating conditions

2. Design linear controllers for each operating condition (model)

3. Develop a schedule for the controller parameters

4. Implement the parameter-scheduled controller on the nonlinear plant

8

Note that steps 1 and 2 will often be combined by using some form of closed-loop tuning at discrete operating points. This is particularly true if the "auto tune" feature of commercial controllers (see Astrom et al., 1993) is used; the use of three operating points appears to be common.

There are several different options for the scheduling of the controller parameters

- Switch parameters at discrete values of the scheduling variable

- Interpolate parameters as a function of the scheduling variable

- Vary parameters continuously with the scheduling variable

Clearly, a major advantage of a parameter scheduled controller is that linear control system design procedures can be used, and at least the local control system behavior can be understood. A problem can occur in steps three and four where parameter scheduling is developed and implemented. If the dynamic effect of the changing auxiliary variable is not included in the design process, then even the local closed-loop behavior may not be predicted correctly. This problem of satisfying the linearization conditions is discussed in sections 3.3 and 3.4. Typically controller performance is assessed by extensive simulations of the gain-scheduled controller and nonlinear plant.

A general representation of a gain scheduled controller is shown in Figure 1, where α represents the scheduling variable.

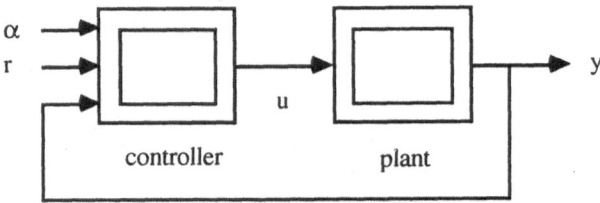

Figure 1. Parameter-Scheduled Control Strategy.

Steadman (1978) presents a number of applications where gain scheduling is successfully used to compensate for nonlinear process behavior. Example calculations are performed for the temperature control of a process furnace, temperature control of an adiabatic CSTR, and furnace excess air control. He also notes that proper use of feedforward and ratio control can reduce nonlinear effects in some systems.

Whatley and Pott (1984) present industrial plant results for scheduled cascade control of temperature in a polymerization reactor. The outer loop (reactor temperature) controller uses a one-sided controller gain schedule, where the controller gain increases if the reactor temperature is too high (this is to protect against reactor runaway). The inner loop (jacket temperature) controller uses a controller gain that is inversely proportional to the difference between the jacket temperature and the make-up jacket feed temperature, in order to compensate for the process gain which varies in the opposite direction. They provide actual plant test results and note that there was no temperature control induced off-spec product after the control system was installed. They expect a 150% return on investment for similar installations on other processes.

Cardello and San (1987, 1988) use an auxiliary variable to apply gain scheduling to the control of dissolved oxygen in batch fermentation reactors. In this simulation study a look-up table is used to select control parameter settings based on an oxygen uptake rate measurement. The settings are based on Cohen-Coon tuning for three different operating regimes (representing three different times in a nominal batch run). A composite controller combining feedforward with gain-scheduled feedback control has better performance than any other control strategy tested. It is not clear, however, how the authors decide on switching points between each of the three regimes.

Engell and Klatt (1993a,b) study temperature and concentration control of a CSTR with the classic van de Vusse reaction scheme. Linear control is used for temperature and continuous gain scheduling is used for concentration control. In (1993a) they assume that the temperature is perfectly controlled, reducing the concentration dynamics to a second-order nonlinear system. They design a gain-scheduled PI controller to maintain a 45° phase margin on the linearized system over the range of operating conditions. The controller gain is a linear function of the steady-state flowrate (manipulated variable). Although not stated explicitly in their paper, it would be undesirable to attempt to use the dynamic value of the manipulated variable as a scheduling variable. They design a nonlinear function generator, based on the output of the time-invariant portion of the controller, which has the same gain relationship as the steady-state flowrate. In (1993b) they assume a constant heat transfer rate (rather than constant temperature), and design a similar gain-scheduled concentration controller.

Knoop and Perez (1994) develop a gain-scheduled PI controller based on continuously linearizing a process model at the current operating point. Their general procedure is for multivariable PI controllers. Although the theory is developed for lumped systems, they show an application to a single input-single output continuous flow furnace (which is a distributed parameter system). Simulation results show the improved closed-loop behavior of the parameter-scheduled PI controller over a fixed-parameter PI controller.

Marini and Georgakis (1984) developed several nonlinear reaction-rate controllers for low-density-polyethylene reactors. One version is a PI controller with the tuning parameters scheduled as a function of the steady-state operating conditions. The scheduled controller is more effective than a constant PI controller, but there are conditions where the full nonlinear controller has better performance. The general approach developed by Marini and Georgakis lead to the idea of extensive variable control (Georgakis, 1986), discussed section 4.

Tsogas and McAvoy (1985) apply "one-way" gain scheduling to control the distillate composition by manipulating the distillate/feed ratio in a distillation column. The term "one-way" means that the controller gain is compensated (using a method suggested by Shinskey, 1977) if the distillate purity is too high, but kept at a nominal value if the purity is too low

$$k_c(x_D) = k_{c0} \frac{1 - x_{D0}}{1 - x_D} \qquad \text{for } x_D > x_{D0}$$

$$k_c(x_D) = k_{c0} \qquad \text{for } x_D < x_{D0}$$

(11)

Since the process gain increases with lower purity ($x_D < x_{D0}$), maintaining a constant controller gain will speed-up the response when the distillate is less pure. Another modification that the authors made is in the integral term. Normally, the integral mode would be represented by

$$\frac{k_c(x_D)}{\tau_I} \int_0^t e(t^*) \, dt^* \qquad (12)$$

however Tsogas and McAvoy find better results if the following representation is used

$$\frac{1}{\tau_I} \int_0^t k_c(x_D) \, e(t^*) \, dt^* \qquad (13)$$

Wong and Seborg (1986a,b) develop a low-order nonlinear model of a distillation column by correlating gains and time constants as continuous functions of the inputs and states. They develop a multivariable control law which can be interpreted as a variable-gain, PI control system. Wong and Seborg (1988) present a parameter-varying PI controller for a first-order nonlinear SISO system. In addition they provide a generalization to the Smith Predictor, based on a first-order nonlinear representation of the process. They also find that the gain scheduled-like approach works better if the process gain is predicted based on the deadtime (i.e. where the process output will be when the control action actually takes effect).

It is common in batch reactor control systems to have a recirculating jacket fluid system that can operate in two modes (using split-range control); see Liptak (1986), for example. The static and dynamic behavior of the two modes can be substantially different, indeed the gain for a single mode can easily vary by a factor of 5. Djavdan (1995) presents an approach for gain scheduling the jacket temperature (slave) control loop in a batch reactor with dual mode control.

pH control has been a common application of gain scheduling. A large part of the problem is estimating the titration curve, which can be used to estimate the process gain. Gray (1981) suggests that five parameters can be selected to approximate a titration curve. This approximate titration curve is then used to create "characterized pH". Linear PID control is then based on the measured (and setpoint) characterized pH's. The net result is the same as gain scheduling.

Gulaian and Lane (1990) develop a technique to construct a titration curve as a mixture of preselected titration curves. The advantage of their approach is that the best fit is a linear estimation problem. The process gain is represented as a continuous function of pH and the controller gain is adjusted so that $k_c k_p$ is constant. Experimental results for caustic soda and phosphoric acid in a pilot plant reactor are shown. Their approach is more related to adaptive control since the titration curve is frequently updated. They show how their adaptive controller outperforms a non-adaptive controller when most of the buffering capacity is removed.

Lin and Yu (1993) modify the Gulaian and Lane method so that the titration curve is a linear combination of two base titration curves. They also use the autotune variation (ATV) method, based on relay feedback (Astrom and Hagglund, 1984) for controller tuning at a nominal operating point. They then schedule the controller gain so that $k_c k_p$ is constant. Simulation results show that the gain scheduled ATV approach outperforms a fixed PI controller (which goes unstable for a decrease in weak acid concentration). Experimental results for a laboratory scale (4.5 liter) reactor are shown by Chan and Yu (1995).

Wright and Kravaris (1996) show that input/output linearization (GLC) results in nonlinear PID controllers for first- and second-order nonlinear systems. An exothermic batch reactor (with a zero-order reaction) with temperature controlled by manipulating the jacket temperature, is shown to result in a two-degree-of-freedom nonlinear PI controller with a constant proportional gain. A batch reactor with jacket dynamics results in an ideal nonlinear PID controller.

Rugh (1987) develops a continuous parameter scheduled control strategy, based on the process setpoint. He applies Ziegler-Nichols tuning for a gain-scheduled PI level controller on the third tank of three tanks in series. The parameter-scheduled controller has much better performance, for large setpoint changes, than a fixed PI controller designed at a nominal operating point. Rugh (1991) also notes that it is not clear that interpolating tuning parameters from values at discrete operating conditions is better than interpolating the process models between discrete points and using a continuously-parameterized controller.

Nystrom et al. (1997) characterize a nonlinear plant by a discrete set of linear models and associated norm-bounded uncertainties. One control problem that they solve is for a single controller which solves for optimal quadratic performance subject to robustness bounds for the set of models describing the plant. Another controller is based on gain scheduling, where the design is performed by scaling all of the models to have the same gain; the controller is then implemented by "unscaling" the resulting control law at each operating point (they also use a high order scheduled controller, but there is no reason that their approach cannot be applied to lower order PID-type controllers). They find that the scheduling must be performed such that the linearization conditions (see sections 3.3 and 3.4) are satisfied. They also find that scheduling the controller outputs yields better results than scheduling the controller gain, but involves a much more complex control system implementation. The application is a pH problem.

3.2 Gain Scheduled Model-Based Control

A number of recent papers have combined model-based control techniques with gain scheduling concepts. Klatt and Engell (1996a) combine exact linearization (Kravaris and Kantor, 1990; Henson and Seborg, 1997b) with gain scheduling to achieve benefits associated with the best features of each. They assume an open-loop stable process and use an open-loop observer for the state feedback used in the exact linearization process. Without any output error compensation, this would be a nonlinear feedforward controller. A gain scheduled controller is used to compensate for model output errors. The manipulated variable applied to the process consists of contributions from the feedback linearizing controller and the gain scheduled controller. Good setpoint tracking and disturbance rejection are shown for pH control in a laboratory scale neutralization of acetic acid by sodium hydroxide. Klatt and Engell (1996b) use a similar control approach to the same van de Vusse reactor studied (using gain scheduling only) by Engell and Klatt (1993a,b).

There has been an effort in developing gain scheduled approaches to model predictive control (see Garcia et al., 1989, for a review of MPC). McDonald and McAvoy (1987) compare linear MPC with four different gain scheduled MPC methods: (i) one-way, (ii) two-way, (iii) one-way gain and time constant, and (iv) two-way gain and time constant; (i) and (ii) are used on a high purity column simulation, while all four approaches were applied to a moderate purity column. Gains are obtained using approximate analytical expressions. They recommend using one-way gain scheduling.

It could be argued that many of the nonlinear model-based predictive control approaches are related to gain scheduling. For example, Garcia (1984) uses a model linearized at the current operating point to predict the effect of future control moves. The linear model is obtained by linearizing the first-principles nonlinear model based on the current value of the inputs and states. The state values are obtained by on-line integration of the nonlinear model (i.e. an open-loop observer is used). The large number of related approaches will not be reviewed here (see Bequette, 1991, for related techniques).

For more advanced model-based approaches such as MPC, it has become more common to use the term "linear parameter varying" (LPV) rather than gain scheduling. Kothare et al. (1997) use 17 models to describe the behavior of a steam level control system; each model is based on steady-state operation at a different power level. They use switching (rather than interpolation) to change the current model/controller as a function of power level. Johansen and Murray-Smith (1997) have noted similarities between gain scheduling approaches and those based on multiple linear models.

3.3 Gain Scheduling Theory and Applications in Other Disciplines

Although gain scheduling (in one form or another) appears to have been implemented for the past thirty years, little theoretical analysis has been performed until recently. Most of this analysis has been based on continuously parameterized systems, while many of the (mostly unpublished) applications have probably used switching methods. A nice overview of recent theoretical work is provided by Rugh (1991), who notes that gain scheduling has suffered from the theory/application gap but in this case the application is ahead of the theory. This is not of particular concern to process control enthusiasts, who should note that model predictive control had many successful industrial applications before academics began to rigorously analyze MPC in the 1980's. The most widely-published application area for gain-scheduling is aircraft control (see Biannic et al., 1997, for example). It should be noted that most of the LPV techniques reviewed in this section are directly applicable to multivariable systems, unlike traditional gain scheduling.

Shamma and Athans (1992) note that the guidelines of scheduling variables that "capture the plant's nonlinearities" and "vary slowly" place fundamental limitations on the achievable performance of current gain scheduling techniques. They also present a simple two-state example of a system that is asymptotically stable for all constant values of parameter within a specified range, yet the system is unstable when the parameter varies with time. The same problem occurs in the closed-loop, where controllers designed based on a constant parameter value are unstable when the parameter varies with time. Shamma and Athans (1991) develop methods to analyze the behavior of gain-scheduled controllers applied to linear parameter-varying processes. "Linearization conditions" which guarantee that certain properties of the fixed operating point designs carry over to the global gain scheduled design are given.

A limit to most gain scheduling approaches is that static scheduling is used, that is, the controller parameters have the same value whenever the scheduling variable is the same. In reality, one can often get better control if the "history" or knowledge of the rate of change of the scheduling variable was used. Kwatra et al. (1996) develop a general method for dynamic gain scheduling for systems that have a relative order one input/output relationship. The controller is synthesized by using input/output linearization and internal model control, then performing algebraic manipulations to convert the control system to gain scheduled form. The gain schedule is implemented as a two-dimensional grid in the output and rate-of-change of the output variable. They show results for an isothermal reactor and a complex polyethylene polymerization reactor.

Sureshbabu and Rugh (1995) develop an approach to gain scheduling based on time derivatives of the scheduling variable. They show an example of a biochemical reactor modeled by Monod kinetics. The substrate feed concentration is used as the scheduling variable. They assume that both state variables are measured and compare their results with a proportional state feedback controller.

Another recent avenue of research is based on H_∞ theory. Packard (1994) develops a linear fractional transformation (LFT) approach to represent a linear parameter-varying plant and controller. The robust control problem is solved using affine matrix inequalities and convex optimization. A related approach is developed by Apkarian and Gahinet (1995). A high performance aircraft application is shown by Biannic et al. (1997).

Kaminer et al. (1995) develop a gain scheduling method that provides integral action at the plant input and uses derivatives of some of the measured outputs in the parameter-scheduled control law. This procedure assures that the linearization conditions are satisfied locally. They use a simple first-order nonlinear (in the state) system to illustrate their main points.

3.4 Local Behavior of Gain-Scheduled Controllers

Here we illustrate the reason that a parameter-scheduled controller, based on controller parameters obtained for locally linear models, will generally have different closed-loop dynamics than the corresponding fixed parameter control laws (that is, the linearization conditions are not satisfied). A block diagram for a locally linearized parameter scheduled controller is shown in Figure 2, where we separate the contributions from the two inputs to the controller, the error and the auxiliary (scheduling) parameter.

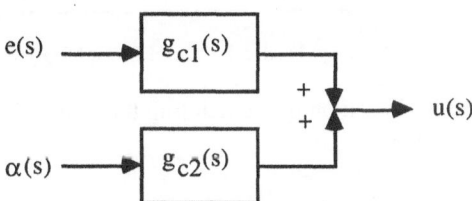

Figure 2. Block Diagram Representation of Linearized Parameter-Scheduled Controller.

A feedback block diagram (based on the linearized system) for the case of scheduling based on the output variable, is shown in Figure 3. The corresponding diagram for scheduling based on the input variable is shown in Figure 4. Either control strategy[1] can be rearranged to form the general feedback strategy shown in Figure 5.

---

[1] Scheduling on the setpoint is left as an exercise for the reader.

14

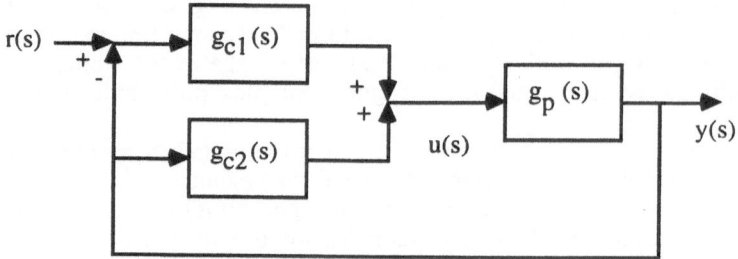

Figure 3. Block Diagram Representation of Output-based Parameter-Scheduled Control Strategy.

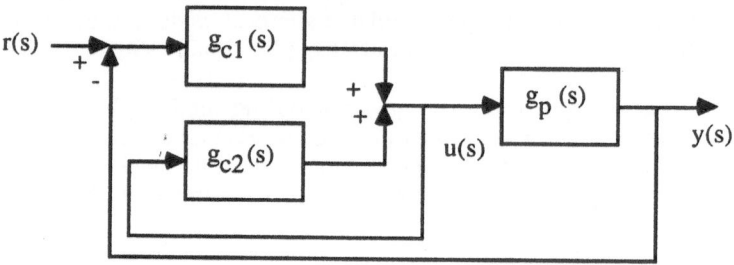

Figure 4. Block Diagram Representation of Input-based Parameter-Scheduled Control Strategy.

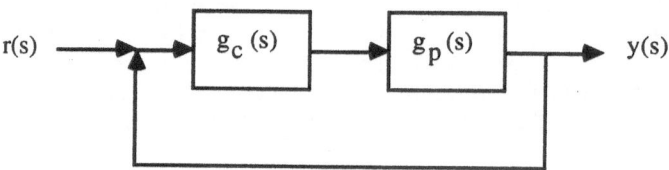

Figure 5. Standard Feedback Control Block Diagram.

For the case of the controller based on the scheduling the output (Figure 3) the feedback controller can be represented as

$$g_c(s) \quad = \quad \frac{g_{c1}(s)}{1 - g_p g_{c2}(s)} \tag{14}$$

with a closed-loop response relationship of

$$y(s) \quad = \quad \frac{g_p(s) g_{c1}(s)}{1 + g_p(s)(g_{c1}(s) - g_{c2}(s))} \; r(s) \tag{15}$$

For the case of the controller based on the scheduling the input (Figure 4) the feedback controller can be represented as

$$g_c(s) \quad = \quad \frac{g_{c1}(s)}{1 - g_{c2}(s)} \tag{16}$$

with a closed-loop response relationship of

$$y(s) \quad = \quad \frac{g_p(s)g_{c1}(s)}{1 + g_p(s)g_{c1}(s) - g_{c2}(s)} \; r(s) \tag{17}$$

In either case we see that there is a dynamic contribution from the transfer function associated with the scheduling variable. Unless appropriately designed, the parameter scheduled controller will not have the same closed-loop response as a fixed-parameter controller designed at the same operating point. Examples are shown in sections 6 and 7.

4. Measurement/Actuator Selection

Perhaps the best way to assure that a nonlinear control system is well-maintained is through design of the physical control devices (actuators and measurements). Shinskey (1962) presents common ways of linearizing control loops through measurement/actuator selection; examples include pH, pressure, temperature and flow control loops. Stout (1956) also shows how to compensate for static nonlinearities. Most process control textbooks have an illustrative example where the installed characteristic of an equal-percentage valve makes the valve position/flowrate relationship linear over a wider range of operating conditions than other valve types, for systems where a significant pressure drop is due to pipe flow resistance at high flowrates.

In distillation control, Ryskamp (1982) notes that logarithms of the product compositions can be used to linearize the input-output relationship. Koung and Harris (1987) use this approach and obtain better closed-loop results than with a partial linearization technique. Skogestad and Morari (1988a) find that the use of logarithms eliminates the effect of nonlinearity at high frequency, while Skogestad and Morari (1988b) show that logarithmic composition-based controllers can effectively operate over a wide range of operating conditions. Longwell (1991) shows that an azeotropic distillation column has better closed-loop performance when the logarithm of the impurity concentration is used as the measured variable, rather than the impurity concentration directly.

Georgiou et al. (1988) also use logarithms to create a more linear input/output relationship for a distillation column, then use the logarithmic compositions in a DMC strategy. They find that this approach yields better results than the parameter scheduled DMC strategy of McDonald and McAvoy (1987), and is simpler to implement. Trelea et al. (1997) use the logarithm of moisture content in a corn dryer to improve control.

Lee et al. (1997) use a single first-order control-affine (input linear) model to characterize nonlinear behavior. They show two different transformations of the output variable that linearize the process input-output behavior. The transformation that they recommend (because it is less sensitive to uncertainty) is shown to be a logarithmic transformation in certain circumstances.

Georgakis (1986) presents an approach based on extensive variables of a process, which are generally related to the total energy or the mass of a component. A linear control system is typically developed based on the extensive process variables. The overall control strategies are inherently nonlinear and multivariable and can maintain desirable closed-loop characteristics over a larger region than simple linear controllers.

5. Fuzzy Logic-Based Controllers

Fuzzy logic-based controllers have received much attention in recent years. Here we review fuzzy logic-based approaches to parameter-compensated PID control. A study of the literature finds two basic approaches. One can be considered related to the nonlinear PI-based approaches presented in section 2 since only functions of the error (and its rate of change) are used to change the controller parameters. The other basic approach is more related to the gain scheduling approaches covered in section 3 since an auxiliary variable is used to schedule the controller.

5.1 Fuzzy logic-based nonlinear PID

Zhao et al. (1993) develop a FL-based nonlinear PID controller based on a desired response to a step setpoint change. Initially, a large control signal is desired to achieve a fast rise time, so the controller gain is large and the derivative time is small. As the setpoint is reached, a small controller gain is desired so that the overshoot is not too great. They develop fuzzy rules to change the controller parameters based on the error and its rate of change at each time step. Several linear transfer functions are used as example processes. The basic objective, then, is to achieve a desired nonlinear response even if the process is linear. No discussion of applications to nonlinear processes is provided.

5.2 Fuzzy logic-based gain scheduled PID

Ling and Edgar (1992) compare three gain scheduling algorithms for a variable gain and time constant first-order plus time-delay process. The three algorithms are (i) gain scheduling (GS), (ii) fuzzy gain scheduling (FGS), and (iii) model-based fuzzy gain scheduling (MFGS). GS is a gain scheduling algorithm with hard switching, while FGS and MFGS are simply different ways of interpolating parameters in a gain scheduled controller; the auxiliary variable used is simply the process output. The MFGS interpolation is linear with the inverse of the controller gain, and linear with both the integral and derivative times. Ling and Edgar (1994) compare the three gain scheduling strategies with fixed PID and nonlinear model predictive control (NMPC) on a laboratory-scale water gas shift reactor. They find that MFGS compares favorably with NMPC.

McMillan et al. (1994) modify the FGS approach of Ling and Edgar (1992) to include deadbands in between the fuzzy regions where controller parameters are linearly interpolated. They apply a three-region design to control of a simulated pH system, and show better results than a standard fixed PI controller and a rigid gain scheduler. Qin and Borders (1993) develop a fuzzy logic-based PI controller for a pH process. They show better results for their three-region design than a single region design in a simulation study.

6. Example Problems

To illustrate principles and issues with gain scheduling we focus on some simple first-order nonlinear systems. In each case we use the IMC-based PI design procedure (Morari and Zafiriou, 1989) for the locally linearized controllers. For the nominal operating point characterized by the scheduling variable (α_0), the controller gain and time constant are represented by

$$k_{c0} = \tau_p(\alpha_0)/(k_p(\alpha_0)\,\lambda) \qquad\qquad \tau_{I0} = \tau_p(\alpha_0) \qquad\qquad (18)$$

where λ is the desired closed-loop time constant. The controller gain and integral time at any other operating point are

$$k_c(\alpha) = \tau_p(\alpha)/(k_p(\alpha)\,\lambda) \qquad\qquad \tau_I = \tau_p(\alpha) \qquad\qquad (19)$$

We compare input/output linearization (globally linearizing control, GLC) with gain scheduling based on both the input (IGS) and output (OGS) variables.

System 1. A Hammerstein (static input nonlinearity, followed by linear dynamics) example.

Consider the following system, which has the steady-state input/output curve and input/gain relationship shown in Figure 6.

$$\dot{x} \quad = \quad -x \quad + \quad u \quad + \quad 0.5\,u^2 \qquad\qquad (\text{system 1}) \qquad (20)$$
$$y \quad = \quad x$$

The process gain, in terms of the steady-state input and output, is

$$k_p \quad = \quad 1 + u_s \quad = \quad \sqrt{1 + 2x_s} \qquad\qquad (21)$$

The process time constant is independent of the operating point

$$\tau_p \quad = \quad 1 \qquad\qquad (22)$$

Note from Figure 6 than this system has input multiplicity. For the purposes of this example we assume that it is desirable to operate in the region where $u > -1$. The open-loop step responses for ± 1 changes in the input are shown in Figure 7.

One would assume that this would be a prime application for gain scheduling, since only the process gain (and not time constant) varies with the operating condition. Consider the nominal operating point of $(u_0, y_0) = (0,0)$, with a desired closed-loop time constant of 1/4 the open-loop time constant ($\lambda = 0.25$). Then

$$k_{c0} = 4 \qquad\qquad \tau_I = 1 \qquad\qquad (23)$$

i. *Input gain scheduling*. Scheduling on the input variable yields

$$k_c \quad = \quad 4/(1 + u) \qquad\qquad (\text{IGS-1}) \qquad (24)$$

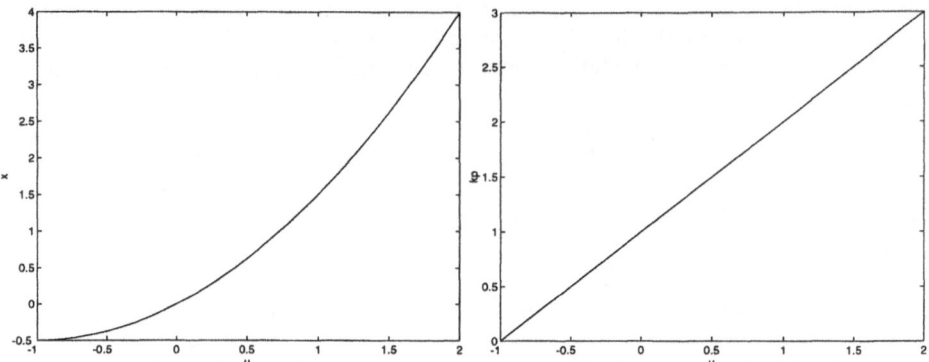

Figure 6. Steady-state input-output and input-gain relationships (only $u > -1$ is shown).

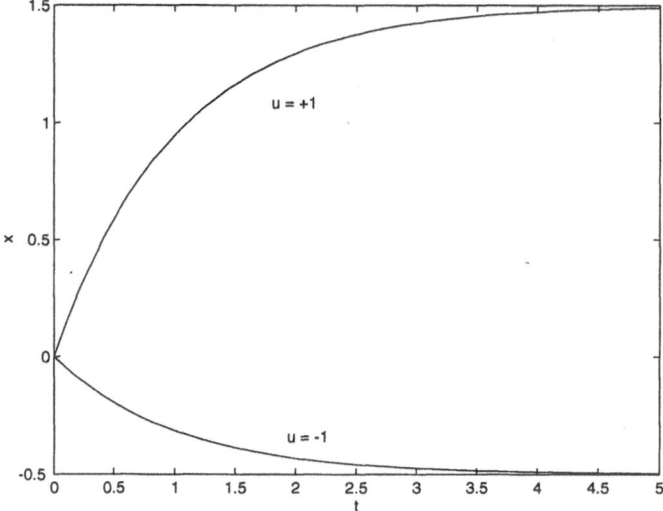

Figure 7. Open-loop step response of system 1.

It turns out this controller does not globally linearize the closed-loop behavior; except at the nominal operating point, the closed-loop time constant varies from 0.25. Later we show an alternative input gain schedule, which does globally linearize the behavior, later (IGS-2).

ii. *Output gain scheduling*. Scheduling on the output variable yields

$$k_C = 4/\sqrt{1 + 2x} \qquad \text{(OGS)} \qquad (25)$$

iii. *Input/output linearization (GLC)*. Let v represent the controller output. Although this system is not in control-affine (input-linear) form, it is easy to find a state feedback solution which linearizes the v-y map using the method detailed in Henson and Seborg

(1997b). Although an explicit solution is not guaranteed in general using this procedure, in this example we find an explicit solution. Choosing the target system

$$y(s) \quad = \quad \frac{1}{s + 1} \; v(s) \tag{26}$$

we find the following state-feedback compensator (*which doesn't actually require state feedback!*)

$$u \quad = \quad -1 + \sqrt{1 + 2v} \tag{27}$$

which, combined with a PI controller with $k_c = 4$ and $\tau_I = 1$, provides a first-order closed-loop response with a time constant of 0.25.

Summary of intermediate results. It turns out that neither (IGS-1) or (OGS) leads to a closed-loop response with a time constant of $\lambda = 0.25$, as shown in Figure 8.

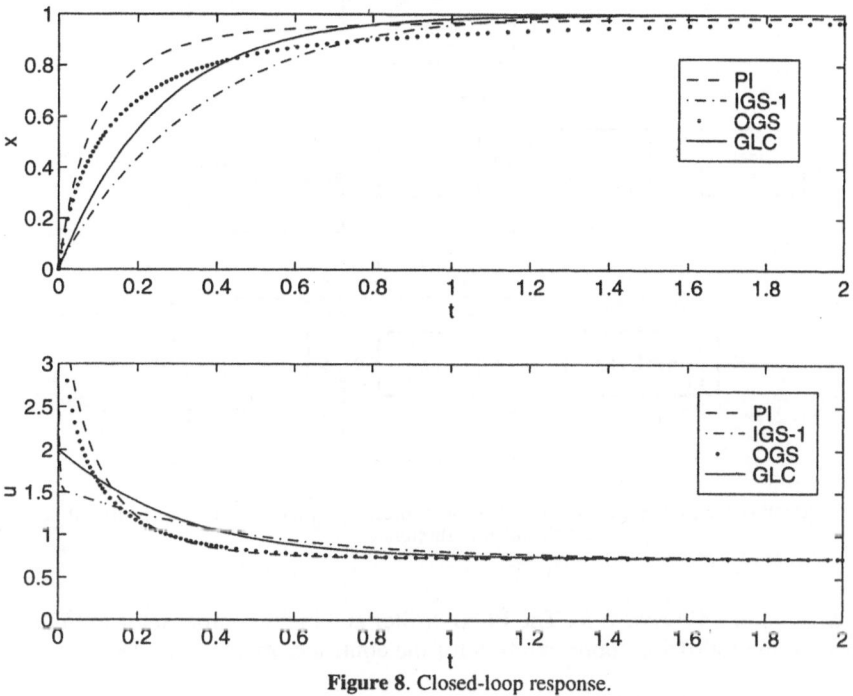

Figure 8. Closed-loop response.

It is easy to show that the closed-loop transfer function of the input gain scheduled controller (IGS-1) is (see exercise 1)

$$y(s) = \frac{1}{\dfrac{1 + 2u_s}{1 + u_s}\lambda\, s + 1}\, r(s) \tag{28}$$

which indicates that the closed-loop response will be slower than the desired response for $u > 0$ and faster for $-1 < u < 0$. Similarly, the input-scheduled controller will be more sensitive to process uncertainty for $u > 0$ and less sensitive for $-1 < u < 0$ (compared to GLC).

In this example it is easy to understand why the controller gain must be scheduled by $1/(1 + u)$, rather than $1/(1 + 0.5u)$, to have a global closed-loop response time of λ. Consider the block diagram shown in Figure 9, where a simple static input compensator linearizes the system.

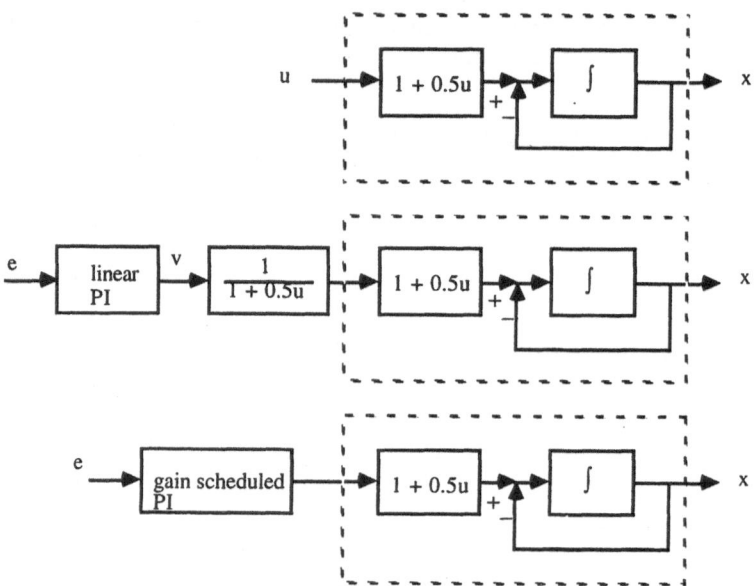

Figure 9. Block Diagram Representation for System 1: process (top), GLC^2 (middle), Alternative Scheduled PI (bottom).

iv. *Alternative input gain schedule.* The PI controller and input compensator in Figure 9 is equivalent to the following representation for the controller gain

$$k_c = 4/(1 + 0.5\, u) \qquad \text{(IGS-2)} \qquad (29)$$

The input/output performance is then equivalent to the GLC controller/compensator pair, with the following closed-loop response

---

[2] Note that $u = \dfrac{v}{1 + 0.5u}$ is equivalent to $u = -1 + \sqrt{1 + 2v}$ (27).

$$y(s) \quad = \quad \frac{1}{0.25\ s + 1}\ r(s) \qquad (30)$$

Although gain scheduling is based on a linear controller, designed at a particular operating point, based on a linear representation of the plant, the linear closed-loop characteristics at that operating point may not be as expected. This is because the change of the controller parameters due to a change in the auxiliary scheduling variable is generally not considered in the design. It should be noted that a scheduled controller based on "switching" does not suffer from this problem. The performance of a switching scheduled controller may suffer, however, because an extreme change in operating condition may be required before the switch to a new set of parameters. The new set of parameters may only provide the desired closed-loop characteristics at a single operating point within the entire range covered by that set of parameters.

Lessons from System 1

There are several lessons to be learned from System 1. One is that the rule of thumb for "scheduling on a slow variable" is not valid for this system. Systems that are Hammerstein in nature (static input nonlinearity followed by linear dynamics) can be globally linearized by static gain scheduling based on the process input; the gain schedule is not, however, the same as "classic gain scheduling" (keeping $k_c k_p$ constant). Input/output linearization (globally linearizing control) and "globally linearizing" input gain scheduling give the same results for these systems.

Although interpolation results are not shown in this example, notice that (for input scheduling) linear interpolation of the inverse controller gains will yield equivalent results to the continuous scheduling.

System 2. First-order bilinear system.

Consider the following system[3], which has the same nominal gain and time constant as system 1

$$\dot{x} \quad = \quad -x \quad + \quad (1-x)\,u \qquad \text{(system 2)} \quad (31)$$
$$y \quad = \quad x$$

The process gain[4], in terms of the steady-state input and output, is

$$k_p \quad = \quad 1/(1+u_s)^2 \quad = \quad (1-x_s)^2 \qquad (32)$$

and process time constant is

$$\tau_p \quad = \quad 1/(1+u_s) \quad = \quad 1-x_s \qquad (33)$$

---

[3] It should be noted that system 2 models an isothermal chemical reactor with a single, first-order irreversible reaction, if time is scaled by the reaction rate constant and the concentration is scaled by the inlet concentration. Kwatra et al. (1996) have applied a dynamic gain scheduling technique to an isothermal CSTR.

[4] A number of publications have used similar models to represent gain-varying systems. Often the general form, $\tau_p\ dy/dt = -y + (k_{p0} + bx)\,u$, is used. Note that for the linearized model, the time constant is not constant and the linearized gain does not vary linearly with the state!

22

The open-loop step response is shown in Figure 10

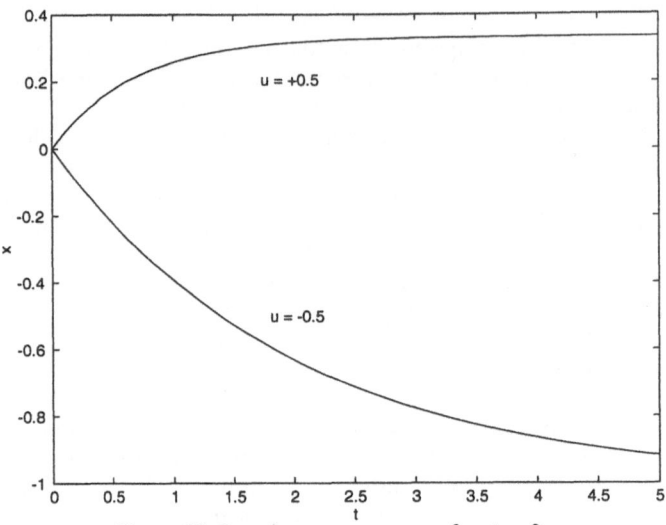

Figure 10. Open-loop step response of system 2.

Consider the nominal operating point of $(u_0,y_0) = (0,0)$, with a desired closed-loop time constant of 1/4 the nominal open-loop time constant ($\lambda = 0.25$). Then

$$k_{c0} = 4 \quad \text{and} \quad \tau_I = 1 \tag{34}$$

Remember that we are using the IMC-based PI procedure, so

$$k_c = \tau_p/(k_p \lambda) \quad \text{and} \quad \tau_I = \tau_p \tag{35}$$

i. *Input parameter scheduling.* Scheduling on the input variable yields

$$k_c = 4(1 + u) \quad \text{and} \quad \tau_I = \frac{1}{1 + u} \tag{36}$$

ii. *Output parameter scheduling.* Scheduling on the output variable yields

$$k_c = 4/(1 - x) \quad \text{and} \quad \tau_I = 1 - x \tag{37}$$

iii. *Input/output linearization (GLC).* Let v represent the controller output. Choosing the target system

$$y(s) = \frac{1}{s + 1} v(s) \tag{38}$$

we find the following state-feedback compensator

$$u \quad = \quad \frac{v}{(1 - x)} \tag{39}$$

which, combined with a PI controller with $k_c = 4$ and $\tau_I = 1$, provides a first-order closed-loop response with a time constant of 0.25

$$y(s) \quad = \quad \frac{1}{0.25 \, s + 1} \ r(s) \tag{40}$$

iv. *Alternative input gain schedule*. The input/output linearizing static compensator is equivalent to the following representation for the scheduled controller gain

$$k_c \quad = \quad 4/(1 - x) \qquad \text{and} \qquad \tau_I \ = \ 1 \tag{41}$$

so that the closed-loop response is equivalent to that of the input/output linearizing (GLC) control.

Lessons from System 2

There are several things that we learned from System 2. This system can be globally linearized by static gain scheduling based on the process output; the gain schedule is not, however, the same as "traditional gain scheduling" (keeping controller gain*process gain constant). Input/output linearization (globally linearizing control) and "globally linearizing" output gain scheduling give the same results for this system. Traditional controller parameter scheduling involves the modification of both the proportional gain and integral time as a function of the process output; it is interesting that "globally linearizing" gain scheduling only requires the controller gain to be changed and not the integral time.

7. Low-order Scheduled Controllers for Specific Nonlinearities

The procedures used to develop gain-scheduled controllers for example systems 1 and 2 allow us to generalize gain-scheduled PID design for some specific low-order nonlinear structures. The gain-scheduled design for Hammerstein systems is shown in 7.1, while design for a specific control-affine structure is presented in 7.2.

7.1 A gain-scheduling procedure for SISO Hammerstein systems

Consider the Hammerstein representation shown in Figure 11, where a static input nonlinearity is followed by linear dynamics (with some abuse of notation). The notation $k_p(u)$ denotes that the gain[5] is a function of the input only. For example, if the static nonlinearity is the form of a polynomial then $k_p(u) = k_{p0}(1 + b_1 u + b_2 u^2 + ...)$.

Our results from system 1 are easily generalized, since only the gain is varying and not the dynamics. The IMC-based PID procedure (Table 6.1-1 from Morari and Zafiriou, 1989) can be used to find the tuning parameters, by simply substituting $k_c(u)k_p(u)$ for

---

[5] Note that, in general, the linearized gain at any operating point will not be $k_p(u)$.

$k_c k_p$. Consider the following common examples.

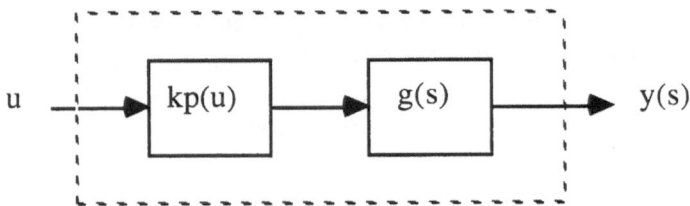

Figure 11. General input-output representation for a Hammerstein system.

First order dynamics. When the dynamics are characterized by

$$g(s) \quad = \quad \frac{1}{\tau_p s + 1} \qquad (41)$$

and the static nonlinearity is represented by

$$k_p(u) \quad = \quad k_{p0}(1 + b_1 u + b_2 u^2 + ...) \qquad (42)$$

A representative differential equation is

$$\tau_p \frac{dy}{dt} \quad = \quad -y + k_p(u) \, u \qquad (43)$$

The gain-scheduled parameters are

$$k_c(u) \quad = \quad \frac{\tau_p}{k_p(u) \, \lambda} \quad = \quad \frac{\tau_p}{k_{p0}(1 + b_1 u + b_2 u^2 + ...) \, \lambda} \qquad (44)$$

$$\tau_I \quad = \quad \tau_p \qquad (45)$$

and the closed-loop response is

$$y(s) \quad = \quad \frac{1}{\lambda s + 1} \, r(s) \qquad (46)$$

Second order dynamics. When the dynamics are second order

$$g(s) \quad = \frac{1}{\tau^2 s + 2\zeta\tau s + 1} \qquad (47)$$

and $k_p(u) = k_{p0}(1 + b_1 u + b_2 u^2 + ...)$, the gain-scheduled parameters

$$k_c(u) \quad = \quad \frac{2\zeta\tau}{k_p(u)\,\lambda} \quad = \quad \frac{2\zeta\tau}{k_{p0}(1 + b_1u + b_2u^2 + ...)\,\lambda} \tag{48}$$

$$\tau_I \quad = \quad 2\zeta\tau \qquad\qquad \tau_D = \frac{\tau}{2\zeta} \tag{49}$$

yield a globally first-order closed-loop response.

Second order nonminimum-phase. When the dynamics are second-order with a right-half-plane zero and $k_p(u) = k_{p0}(1 + b_1u + b_1u^2 + ...)$

$$g(s) \quad = \quad \frac{-\beta s + 1}{\tau^2 s + 2\zeta\tau s + 1} \tag{50}$$

The gain-scheduled parameters (for a PID controller with a first-order filter) are

$$k_c(u) \quad = \quad \frac{2\zeta\tau}{k_p(u)(2\beta + \lambda)} \quad = \quad \frac{2\zeta\tau}{k_{p0}(1 + b_1u + b_2u^2 + ...)(2\beta + \lambda)} \tag{51}$$

$$\tau_I \quad = \quad 2\zeta\tau \qquad \tau_D \quad = \quad \frac{\tau}{2\zeta} \qquad \tau_F \quad = \quad \frac{\beta\lambda}{2\beta + \lambda} \tag{52}$$

which yields the closed-loop response

$$y(s) \quad = \quad \frac{-\beta s + 1}{(\beta s + 1)(\lambda s + 1)}\, r(s) \tag{53}$$

Slightly different results are obtained with a different factorization of the right-half-plane zeros in the IMC design procedure.

7.2 A gain-scheduling procedure for a specific control-affine structure

Consider the representation shown in Figure 12, where a function of the output multiplies the input, followed by linear dynamics (again, with some abuse of notation)

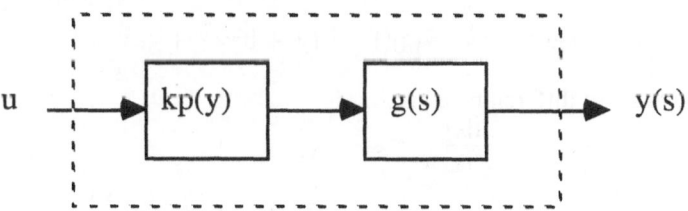

Figure 12. A specific control-affine structure.

The notation $k_p(y)$ denotes that the gain[6] is a function of the output only. For example, if the function multiplying the input is the form of a polynomial then $k_p(y) = k_{p0}(1 + b_1y + b_2y^2 + ...)$.

Our results from system 2 are easily generalized here. The IMC-based PID procedure (Table 6.1-1 from Morari and Zafiriou, 1989) can be used to find the tuning parameters, by simply substituting $k_c(y)k_p(y)$ for k_ck_p. Consider the following common examples.

First order dynamics. When the dynamics are characterized by

$$g(s) = \frac{1}{\tau_ps + 1}$$

and the function multiplying the input is represented by

$$k_p(y) = k_{p0}(1 + b_1y + b_2y^2 + ...) \tag{54}$$

then the gain-scheduled parameters

$$k_c(y) = \frac{\tau_p}{k_p(y)\,\lambda} = \frac{\tau_p}{k_{p0}(1 + b_1y + b_2y^2 + ...)\,\lambda} \tag{55}$$

$$\tau_I = \tau_p$$

yield a globally first-order response

$$y(s) = \frac{1}{\lambda s + 1}\,r(s)$$

Second order dynamics. When the dynamic portion is represented by

$$g(s) = \frac{1}{\tau^2s + 2\zeta\tau s + 1}$$

and the nonlinear function is

$$k_p(y) = k_{p0}(1 + b_1y + b_2y^2 + ...)$$

A representative set of differential equations is

$$\frac{dx_1}{dt} = x_2$$

$$\frac{dx_2}{dt} = -\frac{1}{\tau^2}x_1 - \frac{2\zeta}{\tau}x_2 + \frac{k_p(y)}{\tau^2}u$$

$$y = x_1 \tag{56}$$

---

[6] Note that in general, the gain of the linearized process model at any operating point will not be equal to $k_p(y)$. Also, the open-loop pole(s) will vary with operating condtion.

The gain-scheduled parameters are

$$k_c(y) \quad = \quad \frac{2\zeta\tau}{k_p(y)\,\lambda} \quad = \quad \frac{2\zeta\tau}{k_{p0}(1 + b_1y + b_2y^2 + \ldots)\,\lambda} \tag{57}$$

$$\tau_I \;=\; 2\zeta\tau \qquad\qquad \tau_D \;=\; \frac{\tau}{2\zeta} \tag{58}$$

yield a first-order closed-loop response.

Second order nonminimum-phase. When the open-loop dynamics include a right-half-plane zero

$$g(s) \quad = \frac{-\beta s + 1}{\tau^2 s + 2\zeta\tau s + 1}$$

and

$$k_p(y) \quad = k_{p0}(1 + b_1y + b_2y^2 + \ldots)$$

The gain-scheduled parameters are

$$k_c(y) \quad = \frac{2\zeta\tau}{k_{p0}(1 + b_1y + b_2y^2 + \ldots)(2\beta + \lambda)} \tag{59}$$

$$\tau_I \;=\; 2\zeta\tau \qquad \tau_D \;=\; \frac{\tau}{2\zeta} \qquad \tau_F \;=\; \frac{\beta\lambda}{2\beta + \lambda}$$

which yields the closed-loop response

$$y(s) \quad = \quad \frac{-\beta s + 1}{(\beta s + 1)(\lambda s + 1)}\, r(s)$$

7.3   Implementing integral action

It appears that, in most cases, the integral gain should be scheduled "outside" of the integration. Some publications have suggested the following form for the PI-scheduled control law

$$z \quad - \quad \frac{k_c(\alpha)}{\tau_I(\alpha)}\, e$$

$$u \;=\; k_c(\alpha)\, e \;+\; z \tag{60}$$

which implements

$$u(t) \quad = \quad k_c(\alpha)\, e \;+\; \int_o^t \frac{k_c(\alpha)}{\tau_I(\alpha)}\, e(t^*)\, dt^* \tag{61}$$

28

The actual realization, in general, should be

$$\dot{z} = e$$

$$u = k_C(\alpha)\, e + \frac{k_C(\alpha)}{\tau_I(\alpha)}\, z \qquad (62)$$

which implements

$$u(t) = k_C(\alpha)\, e + \frac{k_C(\alpha)}{\tau_I(\alpha)} \int_0^t e(t^*)\, dt^* \qquad (63)$$

It is not clear which algorithm is used for the gain-scheduling option in most industrial controllers.

8. Summary and Closing Comments

While there has been an explosion of nonlinear model-based control methods published during the past decade (with a focus on differential geometric-based and model predictive control approaches), there has been a relative dearth of publications involving "traditional" nonlinear control methods. The vast majority of applications have involved single-input single-output systems. Table 1 is a concise summary of the papers reviewed here[7]. See Kwatra and Doyle (1997) for a review of some other applications of gain scheduling.

There is certainly a need for industrial practitioners to provide more details of nonlinear PI methods that have been used successfully. It is also appropriate for academic researchers to consider how some of the more advanced model-based control methods can be cast into a form that is more easily implemented in typical single-loop and distributed control system structures. For linear systems the IMC procedure (Morari and Zafiriou, 1989) has been used to design linear PID-type controllers that are equivalent to IMC. Perhaps input/output linearization (GLC) and other nonlinear techniques can be used to design PID-based parameter scheduled controllers for various types of low-order models. In this paper we have shown how to perform IMC-based PID gain scheduling for Hammerstein systems (static input nonlinearity), and for systems with a specific control-affine structure. Wright and Kravaris (1996) derive controller parameter schedules for general relative order one, first and second-order nonlinear systems. There is a need to develop low-order nonlinear model structures and characterize the behavior of common chemical processes. Pearson (1994) reviews the types of static and dynamic behavior that can occur in certain types of nonlinear models.

It should be noted that many nonlinear model-based control techniques attempt to achieve a globally linear closed-loop response. Although this is nice conceptually, since we understand linear system behavior, in practice there is no reason to believe that the closed-loop response at one operating condition should be the same as at another operating condition. Open-loop dynamics and model uncertainty should place different performance constraints on each operating condition. This is certainly recognized in

---

[7] Please note that I have not performed a detailed search of ISA conferences, for example. There is also probably much more experience with gain scheduling than is widely published. For example, I have advised a number of senior projects on gain scheduling, including simulation studies of CSTR's (pH and temperature control) and distillation columns, and experimental studies of laboratory heat exchangers.

29

Table 1. Nonlinear PI-based Process Applications

Biochemical Reactors
Cardello and San (1987,1988) - batch
Sureshbabu and Rugh (1995)

Chemical Reactors (2 or 3 state CSTR)
Engell and Klatt (1993a,b)
Klatt and Engell (1996b) - GS w/I-O Linearization
Kwatra et al. (1996) - GS designed using IOL
    Linearization
Lee et al. (1997)
Steadman (1978)

Chemical Reactors (more complex)
Djavdan (1995) - batch esterification
Kwatra et al. (1996) - polymerizaton, GS using IOL
Ling and Edgar (1994) - fixed bed reactor -
    experimental
Marini and Georgakis (1984) - low density
    polyethylene
Marroquin and Luyben (1972) - batch
Whatley and Pott (1984) - polymerization - *plant
    results*

Distillation
Georgiou et al. (1988) - log compositions - MPC
Koung and Harris (1987) - log compositions
Longwell (1991) - azeotropic, log compositions
McDonald and McAvoy (1987) - GS MPC
Ryskamp (1982)
Shinskey (1977)
Skogestad and Morari (1988a,b) - log compositions
Tsogas and McAvoy (1985)
Wong and Seborg (1986a,b) - low order nonlinear
    model

Level
Cheung and Luyben (1980)
Cheung (1978) - *experimental*
McDonald et al. (1986)
Marlin (1995)
Rugh (1987)
Shunta and Fehervari (1976)

pH
Chan and Yu (1995) - *experimental*
Gray (1981)
Gulaian and Lane (1990)
Klatt and Engell (1996a) - GS w/I-O Linearization
Lin and Yu (1993) - autotune/GS
McMillan et al. (1994) - fuzzy GS
Qin and Borders (1993)

Other Unit Operations
Knoop and Perez (1994) - Furnace
Lee et al. (1997) - Heat exchanger
Steadman (1978) - Furnace
Trelea et al. (1997) - Corn dryer, log moisture
    composition

aircraft control where pilots prefer a faster response at high speeds and a slower response at low speeds (Biannic et al., 1997). A major advantage to a parameter-scheduling type of approach is that it is easy to tune for different responses at each operating point; this was part of the motivation of the work by Klatt and Engell.

Although much of the recent research in gain scheduling has involved continuous parameterizations of nonlinear models, the real power of gain scheduling is in implementing a schedule based on on-line tuning at several operating conditions (usually three points). Commercial controllers generally offer the option to auto-tune at several conditions and schedule parameters between those conditions. The real effort, then, is to determine the best gain schedule (and the best scheduling variable). For example, should controller be directly linearly interpolated, or linearly interpolated with respect to the inverse of the controller gain? Can the schedule based on a local, fixed-parameter, design be easily modified to satisfy linearization conditions?

Acknowledgment

This work has been supported by the National Science Foundation (BES-9522639) and The Petroleum Research Fund, administered by the American Chemical Society.

References

1. Apkarian, P. and Gahinet, P. (1995) A convex characterization of gain-scheduled H_∞ controllers, *IEEE Trans. Auto. Cont.*, **40**, 853-864.
2. Astrom, K.J. and Hagglund, T. (1984) Automatic tuning of simple regulators with specifications on phase and amplitude margins, *Automatica*, **20**, 645.
3. Astrom, K.J., Hagglund, T., Hang, C.C. and Ho, W.K. (1993) Automatic tuning and adaptation for PID controllers - a survey, *Control Eng. Practice*, **1**(4), 694-714.
4. Bequette, B.W. (1991) Nonlinear control of chemical processes: a review, *Ind. Eng. Chem. Res.* **30**(7), 1391-1413.
5. Biannic, J.-M., Apkarian, P. and Garrard. W.L. (1997) Parameter varying control of a high performance aircraft, *J. Guid. Cont. Dyn.*, **20**(2), 225-231.
6. Cardello, R.J. and San, K.-Y. (1987) Application of gain scheduling to the control of batch bioreactors, *Proc. American Control Conf.*, Minneapolis, Mn, 682-686.
7. Cardello, R.J. and San, K.-Y. (1988) The Design of Controllers for Batch Bioreactors, *Biotech. Bioeng.*, **32**, 519-526.
8. Chan, H.-C. and Yu, C.-C. (1995) Autotuning of gain-scheduled pH control: an experimental Study, *Ind. Eng. Chem. Res.*, **34**, 1718-1729.
9. Cheung, T.F. and Luyben, W.L. (1980). Nonlinear and nonconventional liquid level controllers, *Ind. Eng. Chem. Fundam.*, **19**, 93-98.
10. Cheung, T.F. Ph.D. Dissertation, Lehigh University (1978).
11. Djavdan, P. (1995) Temperature control of batch esterification reactor overcoming non-linearities using adaptive gain, in J.B. Rawlings (ed.) *Proc. DYCORD+ '95*, Denmark, 421-426.
12. Engell, S. and Klatt, K.-U. (1993a) Gain scheduling control of a non-minimum-phase CSTR, *Proc. European Control Conference*, Groningen, 2323-2338.
13. Engell, S. and Klatt, K.-U. (1993b) Nonlinear control of a non-minimum-phase CSTR, *Proc. American Control Conference*, San Francisco, 2041-2045.
14. Garcia, C.E. (1984) Quadratic/Dynamic matrix control of nonlinear processes: an application to a batch reaction process, presented at the AIChE Annual Meeting, San Francisco, CA.
15. Garcia, C.E., Prett, D.M. and Morari, M. (1989) Model predictive control: theory and practice - a survey, *Automatica*, **25**, 3355-3348.
16. Georgakis, C. (1986) On the use of extensive variables in process dynamics and control, *Chem. Eng. Sci.*, **41**, 1471-1484.
17. Georgiou, A., Georgakis, C. and Luyben, W.L. (1988) Nonlinear dynamic matrix control for high-purity distillation columns, *AIChE J.*, **34**, 1287-1298.
18. Gray, D.M. (1981) Characterized feedback and feedforward pH control, *ISA Trans.*, **20**(2), 63-66.
19. Gulaian, M. and Lane, J. (1990) Titration curve estimation for adaptive pH control, *Proc. American*

    Control Conference, San Diego, CA, 1414.
20. Henson, M.A. and Seborg, D.E. (1997a) (eds.) *Nonlinear Process Control*, Prentice Hall, Upper Saddle River, NJ.
21. Henson, M.A. and Seborg, D.E. (1997b) Feedback Linearizing Control, Chapter 4 in *Nonlinear Process Control*, Prentice Hall, Upper Saddle River, NJ.
22. Johansen, T.A. and Murray-Smith, R. (1997) The operating regime approach, Chapter 1 in *Multiple Model Approaches to Modeling and Control*, Taylor and Francis, London.
23. Jutan, A. (1989) A Nonlinear PI(D) controller, *Can. J. Chem. Eng.*, **67**, 485-493.
24. Kaminer, I., Pascoal, A. M., Khargonekar, P.P. and Coleman, E.E. (1995) A velocity algorithm for the implementation of gain-scheduled controllers, *Automatica*, **31**(8), 1185-1191.
25. Klatt, K.-U. and Engell, S. (1996a) Nonlinear control of neutralization processes by gain-scheduling trajectory control, *Ind. Eng. Chem. Res.*, **35**(10), 3511-3518.
26. Klatt, K.-U. and Engell, S. (1996b) Gain-scheduled trajectory control of a continuous stirred tank reactor," *Comp. Chem. Engng.*, submitted.
27. Knoop, M. and Perez, J. (1994) Nonlinear PI-controller design for a continuous-flow furnace via continuous gain scheduling, *J. Proc. Cont.*, **4**(3), 143-147.
28. Kothare, M.V., Mettler, B., Morari, M., Bendotti, P. and Falinower, C.-M. (1997) Linear parameter varying model predictive control for steam generator level control, *Comp. Chem. Engng.*, **21**(Suppl), S861-S866.
29. Koung, C.W. and Harris, T.J. (1987) Analysis and control of high-purity distillation columns using nonlinearly transformed composition measurements, Presented at the Canadian Engineering Centennial Conference, Montreal.
30. Kravaris, C. and Kantor, J.C. (1990) Geometric methods for nonlinear process control. I. Background II. Controller synthesis, *Ind. Eng. Chem. Res.*, **29**, 2295-2324.
31. Kwatra, H.S., Doyle III, F.J. and Schwaber, J.S. (1996) Dynamic gain scheduled process control, submitted for publication.
32. Kwatra, H.S. and Doyle III, F.J. (1997) A review of gain scheduled control in chemical process systems, submitted.
33. Lawrence, D.A. and Rugh, W.J. (1995) Gain scheduling dynamic linear controllers for a nonlinear plant, *Automatica*, **31**, 381-390.
34. Lee, J., Cho, W. and Edgar, T.F. (1997) Control system design on a nonlinear first-order plus time delay model, *J. Proc. Cont.*, **7**(1), 65-73.
35. Lin, J.-Y. and Yu, C.-C. (1993) Automatic tuning and gain scheduling for pH control, *Chem. Eng. Sci.*, **48**(18), 3159-3171.
36. Ling, C. and Edgar, T.F. (1992) A New fuzzy gain scheduling algorithm for process control, *Proc. American Control Conference*, 2284-2290.
37. Ling, C. and Edgar, T.F. (1994) Experimental verification of model-based fuzzy gain scheduling technique, *Proc. American Control Conference*, 2475-2480.
38. Liptak, B.G. (1986) Controlling and optimizing chemical reactors, *Chem. Eng.*, 69-8, May 26.
39. Longwell, E.J. (1991) Chemical processes and nonlinear control technology, in CPC-IV (W.H. Ray and Y. Arkun, eds.), Padre Island, TX, pp 445-475, A CACHE Publication.
40. McDonald, K.A., McAvoy, T.J. and Tits, A. (1986) Optimal averaging level control, *AIChE J.*, **32**(1), 75-86.
41. McDonald, K.A. and McAvoy, T.J. (1987) Application of dynamic matrix control to moderate and high-purity distillation towers, *Ind. Eng. Chem. Res.*, **26**(5), 1011-1018.
42. McDonald, K.A., Palazoglu, A. and Bequette, B.W. (1988) Impact of model uncertainty descriptions for high-purity distillation control, *AIChE J.* **34** (12), 1996-2004.
43. McMillan, G., Wojsznis, W. and Borders, G. (1994) Flexible gain scheduler, *ISA Trans.*, **33**, 35-41.
44. Marini, L. and Georgakis, C. (1984) Low-density polyethylene vessel reactors. Part II. A novel controller, *AIChE J.*, **30**, 409-415.
45. Marlin, T.E. (1995) *Process Control. Designing Processes and Control Systems for Dynamic Performance*, McGraw Hill, New York.
46. Marroquin, G. and Luyben, W.L. (1972) Experimental evaluation of nonlinear cascade controllers for batch reactors, *Ind. Eng. Chem. Fundam.*, **11**, 552-556.
47. Morari, M. and Zafiriou, E. (1989) *Robust Process Control*, Prentice Hall.
48. Nystrom, R.H., Sandstrom, K.V., Gustafsson, T.K. and Toivonen, H.T. (1997) Multimodel robust control of nonlinear plants: A case study, Report 97-1, Process Control Laboratory, Abo Akademi University, Abo, Finland.
49. Ogunnaike, B.A. and Wright, R.A. (1997) Industrial Applications of Nonlinear Control, in Chemical Process Control V, AIChE Symposium Series.
50. Packard, A. (1994) Gain scheduling via linear fractional transformations, *Sys. Cont. Letters*, **22**, 79-92.
51. Pearson, R.K. (1994) Nonlinear input/output modeling, in *Proc. ADCHEM '94*, Kyoto, Japan.
52. Qin, S.J. and Borders, G. (1993) A multi-region fuzzy logic controller for controlling processes with nonlinear gains, *Proc. Int. Symp. Intell. Control*, 445-450, Chicago, IL, IEEE.
53. Rugh, W.J. (1987) Design of nonlinear PID controllers, *AIChE J.*, **33**, 1738-1742.
54. Rugh, W.J. (1991) Analytical framework for gain scheduling, *IEEE Cont. Sys. Mag.*, 79-84, Jan.
55. Ryskamp, C. (1982) Explicit vs. implicit decoupling in distillation columns, in Chemical Process Control

2. Seborg, D.E. and Edgar, T.F. (eds), United Engineering Trustees, New York, 361-375.

56. Shamma, J.S. and Athans, M. (1991) Guaranteed properties of gain scheduled control for linear parameter-varying plants *Automatica*, **27**(3), 559-564.

57. Shamma, J.S. and Athans, M. (1992) Gain scheduling: potential hazards and possible remedies, *IEEE Control Systems Magazine*, 101-107. Based on a paper from 1991 ACC, pp 516-21.

58. Shinnar, R. (1986) Impact of model uncertainty and nonlinearities on modern controller design. Present status and future Goals, in M. Morari and T.J. McAvoy (eds.), CPC-III, A CACHE publication, Elsevier.

59. Shinskey, F.G. (1962) Controls for nonlinear processes, *Chem. Eng.*, 155-158, March.

60. Shinskey, F.G. (1977) *Distillation Control*, McGraw-Hill, New York.

61. Shunta, J.P. and Fehervari, W. (1976) Nonlinear control of Liquid level, *Instrum. Technol.*, **23**, 43-48.

62. Skogestad, S. and Morari, M. (1988a) Understanding the dynamic behavior of distillation columns, *Ind. Eng. Chem. Res.*, **27**, 1848-1862.

63. Skogestad, S. and Morari, M. (1988b) LV Control of a high purity distillation column, *Chem. Eng. Sci.*, **43**, 33-48.

64. Steadman, J.F. (1978) Control techniques for nonlinear chemical processes, *ISA Trans.*, 167.

65. Stout, T.M. (1956) Nonlinearities in control systems. 3. Deliberately nonlinear systems, *Cont. Eng.*, 77-85, April.

66. Sureshbabu, N. and Rugh, R.J. (1995) Output regulation with derivative information, *IEEE Trans. Auto. Cont.*, **40**(10), 1755-1766.

67. Trelea, I.-C., Courtois, F. and Trystram, G. (1997) Dynamics analysis and control strategy for a mixed flow corn dryer, *J. Proc. Cont.*, **7**(1), 57-64.

68. Tsogas, A. and McAvoy, T.J. (1985) Gain scheduling for composition control of distillation columns, *Chem. Eng. Commun.*, **37**, 275-291.

69. Whatley, M.J. and Pott, D.C. (1984) Adaptive gain improves reactor control, *Hydrocarbon Processing*, 75-78, May.

70. Wong, S.K.P. and Seborg, D.E. (1986a) Low-order, nonlinear, dynamic models for distillation columns, in *Proc. American Control Conference*, Seattle, 1192-1198.

71. Wong, S.K.P. and Seborg, D.E. (1986b) Control strategies for nonlinear multivariable systems and time delays, in *Proc. American Control Conference*, Seattle, 1023-1024.

72. Wong, S.K.P. and Seborg, D.E. (1988) Control strategies for single-input single-output nonlinear systems with time delays, *Int. J. Control*, **48**(6), 2303-2327.

73. Wright, R.A. and Kravaris, C. (1996) Model-based synthesis of nonlinear PI and PID controllers, *Proceedings 1996 IFAC World Congress*, Vol. M, 133-139.

74. Zhao, Z.-Y., Tomizuka, M. and Isaka, S. (1993) Fuzzy gain scheduling of PID controllers, *IEEE Trans. Sys., Man, and Cybernetics*, **23**(3), 1392-1398 (1993).

MULTIPLE MODEL ADAPTIVE CONTROL (MMAC)

Using a bank of linear models to control a nonlinear system

KEVIN D. SCHOTT and B. WAYNE BEQUETTE
Rensselaer Polytechnic Institute
Troy, NY 12180-3590

Abstract

Multiple-Model Adaptive Control (MMAC) is shown to be an effective strategy for controlling nonlinear chemical processes. A brief review of past work is presented, highlighting the major areas where MMAC has been applied. The structure of the MMAC strategy is shown, including comments on implementation issues. Four basic examples, including a non-linear CSTR with output multiplicities, are presented to discuss decisions and problems faced when designing an MMAC controller. In all cases the MMAC controller provides better performance than a fixed-parameter controller. Directions for future research are discussed.

1. Introduction

For many control specialists the notion of model-based nonlinear process control refers to the explicit use of a first-principles model for feedback control. Although this may often be the preferred approach for controlling processes that are operated over a wide range of operating conditions, clearly there are many challenges to implementing such a strategy. Problems often encountered include trying to estimate parameters that are valid over the entire range, and changing model structure in different operating regimes.

Another active area of research involves the use of multiple models to describe process behaviour over a wide range of operating conditions; the control parallel is to use multiple controllers to achieve good feedback control over the entire range. The basic idea of the multiple model/control approach is shown in Figure 1. There are three basic functionalities that are illustrated in Figure 1: (i) model bank, (ii) controller bank, and (iii) a weighting function to determine the process input (manipulated variable). There are then three immediate issues to a multiple model/control application that come to mind:

1. Modeling. What types of local models will be used (input/output, state space, linear/nonlinear) in the model bank? What number of local models will be used? Will the local models correspond to specific operating regimes, or will they be selected so that parameters span a certain range?

33

R. Berber and C. Kravaris (eds.), Nonlinear Model Based Process Control, 33-57.
© 1998 *Kluwer Academic Publishers.*

2. Control. What types of local controllers will be used (PID, IMC, MPC, LQG) in the controller bank? How will each local controller be tuned (closed-loop performance specification at a particular operating point, robustness specification based on a range of conditions, ad-hoc)?

3. Controller output weighting. Will the controller outputs be "switched" or "interpolated"? Is the weighting "open-loop" based on the operating regime (much like gain scheduling), or "closed-loop" based on plant/model prediction performance (or perhaps the predicted closed-loop performance of a model/controller pair)?

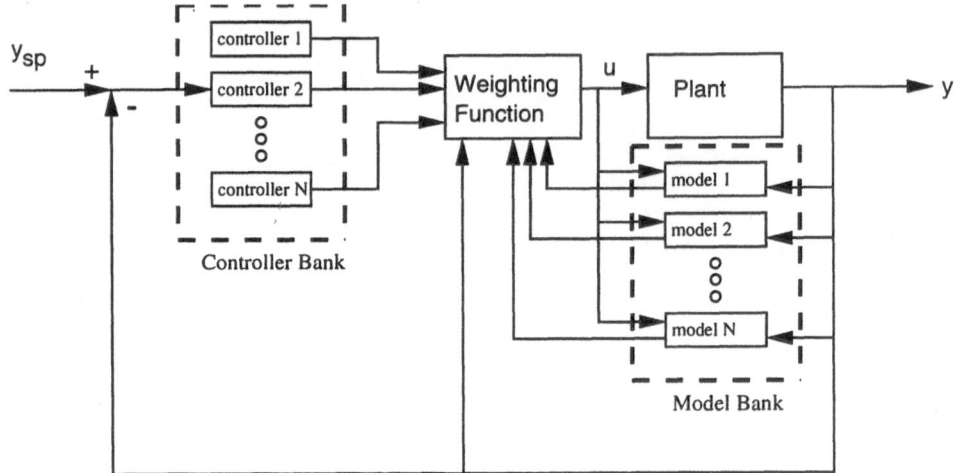

Figure 1. Schematic of multiple-model adaptive control (MMAC) structure

Before we discuss the main topic of this paper, multiple-model adaptive control (MMAC), we provide a brief overview of multiple model-based approaches. A recent monograph edited by Murray-Smith and Johansen (1997) presents a number of different approaches to modeling and/or control based on multiple models with potential process control applications. Johansen and Murray-Smith (1997) note that multiple model approaches have appeared more or less independently in several branches of science and engineering. They provide a very nice summary of the many and varied approaches.

Arkun et al. (1997), in a paper in this volume, present self-scheduled MPC approaches for linear parameter varying (LPV) models. One approach solve linear matrix inequalities (LMIs) on-line, while the other approach solves a QP-based MPC problem, using predictions from the LPV model. Continuous and batch reactor examples are shown.

Kothare et al. (1997) use 17 models to describe the behavior of a steam level control system; each model is based on steady-state operation at a different power level. They use switching (rather than interpolation) to change the current model/controller as a function of power level. Kordon et al. (1997) develop a parallel control system which consists of a bank of virtual control loops and a coordinator to assume good performance in each operating regime and bumpless transfer between regimes. A chemical reactor exhibiting input multiplicity is used as an example.

Qin et al. (1997) apply an interpolating, multiple model-based MPC strategy to a waste treatment system. Gendron et al. (1993) develop a model weighting adaptive control strategy and compare it with MMAC. Their strategy has been successfully operating on a pulp and paper process for over four years.

2. Background

The basic parameter-scheduling philosophy of multiple-model adaptive control can be implemented in many forms. MMAC is a model-based form of scheduled adaption in which a set of paired models and controllers are used to identify and control the plant. The primary difference between this technique and other adaptive control schemes is that the model/controller pairs in the MMAC approach are fixed-parameter, whereas most other adaptive schemes identify specific model parameter-values on-line, then update a single controller to reflect changes in a single process model. By using fixed-parameter models as a basis of identification, nearly any model-based controller design strategy can easily be used. Once the appropriate model or combination of models has been identified, a corresponding controller or combination of controllers is used to supply a control signal to the plant.

The MMAC approach has several advantages over other adaptive and non-linear control strategies. Variations in model structure over time can be handled in a straightforward manner and reliance on parameter estimates of "zero" for unused or unwanted parameters is not necessary. In addition, if linear models and controllers are used, a wealth of well-known design and tuning methodologies for linear controllers can be employed, a clear benefit to the practicing control engineer who does not have complete knowledge of the multitude of nonlinear strategies available in the current literature. MMAC also provides adaption bounds: bad data will only force weighting to an incorrect model; it will not identify a single new model (and corresponding controller) founded on the inferior data.

The MMAC method has some drawbacks too. The most important deficiency is that MMAC is often sub-optimal. For optimal performance the plant needs to be exactly described by one perfect model. MMAC uses a weighting function to decide which model or combination of models best fits the process data, however there is no guarantee that this fit is a good one. The perfect model may not be in the bank, or it may take a long time to converge to the correct model and any control utilizing the "wrong" controller is then sub-optimal; performance is not necessarily poor, simply less than ideal. In addition, some identification and weighting schemes can be sensitive to noise (as discussed in an example in a following section), and deciding on tuning parameters within the weighing function *a priori* can be difficult.

Most of the fundamental background for MMAC is based in stochastic, optimal control, specifically solving a problem with Kalman Filtering and LQG control with the assumption of perfect knowledge of the plant parameters in different operating regions. MMAC was originally developed to avoid the (then) computationally demanding task of identification and adaptive control. Instead of allowing free controller adaption, a set number of cases were developed off-line. These specific cases are then used as the model and controller banks. A typical block diagram for MMAC based on KF/LQG is shown in Figure 2. Theoretical bases are established by Saridis and Dao (1972), Deshpande et al. (1973), Lainiotis (1976a,b), and some basic ideas are presented in Stengel (1986).

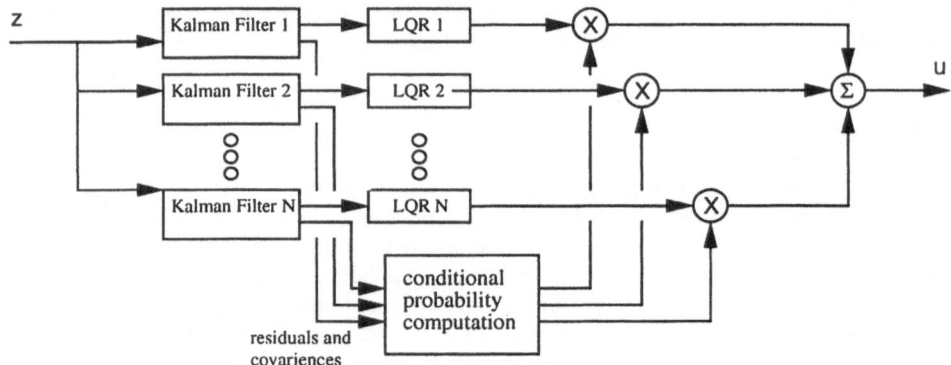

Figure 2. Standard state-space approach to the multiple-model adaptive controller

Greene and Willsky (1980) study a simple 2-state system with 2 models in the model bank; neither model contains the actual plant. They perform a detailed stability analysis and find that not all control gains which stabilize the individual models result in a stable closed-loop for the actual system. This result is very important as it shows that proper design of an MMAC system requires more than simply designing for good nominal performance based on each model. Narendra and Balakrishnan (1997) combine MMAC fixed models with "free running" adaptive models in parallel. Fixed models allow quick responses to changes while adaptive models handle conditions not covered by the fixed models.

Maybeck (1989) notes that a basic problem with the MMAC approach is the potentially large number of models required. For example, if there are two uncertain parameters that can each assume 10 values, then 100 separate filters and controllers must be implemented. Maybeck reviews an approach known as Moving-Bank MMAC that he and collaborators present in a number of applications papers. The basic idea is to include only the models closest to the current operating point in the model bank. He presents five different methods for changing the size or moving the active model bank. His paper also contains a concise review of the standard state-space MMAC approach.

Applications of MMAC have been primarily focused in three different areas:

    (i)   aircraft and spacestructures;
    (ii)   drug delivery; and
    (iii) chemical processes.

Aircraft and spacestructure control
A major challenge in the control of aircraft is to be able to operate under a number of different flight conditions. Although the dynamic behaviour of an aircraft is known to be nonlinear, specifications for control system design are often based on linear models at a number of equilibrium flight conditions. Athans et al. (1977) present an MMAC

design for the F-8C aircraft. The linear models are characterized by nine states, two manipulated inputs and six measured outputs. At several equilibrium conditions a steady-state, discrete, constant-gain Kalman filter is designed. An LQG compensator is designed based on each Kalman filter, as shown in Figure 2. Simulation results are presented using a nonlinear model at a single flight condition. Four models (based on four flight conditions) are used in the model bank. They note that the robustness and sensitivity of the MMAC algorithm appears to rely on careful tuning of the Kalman filters.

Maybeck and coworkers at Wright-Patterson Air Force Base have presented a large number of simulation examples of applications of MMAC to flexible space structure control. Gustafson and Maybeck (1994) apply Moving-Bank MMAC (Maybeck, 1989) to a 352-state model of a SPace Integrated Control Experiment (SPICE) structure. There are 18 manipulated variables and 18 measurements (an additional 36 measurements are used for comparison purposes). They reduce the order of the model to obtain an order tractable for implementation. They find that their moving bank approach provides almost instantaneous tracking of parameters and stabilizing control over the full-range of parameter variations.

Drug infusion control
An area which has had many experimental applications of MMAC is drug infusion control. He et al. (1986) apply MMAC via simulation and animal experiments to the control of blood pressure in dogs by controlling the infusion rate of nitroprusside. Their model bank consists of eight transfer function models and the controller bank consists of eight PI controllers, each based on one of the models. The models all have the same dynamic characteristics with different gains. The model gains are selected so that the corresponding controller would satisfy phase margin requirements for each model interval. Martin et al. (1987) combine pole-placement, a Smith predictor and PI control into an MMAC framework for blood pressure control using nitroprusside. The control strategy is applied to a nonlinear simulation model. MMAC-PI is used by Yu et al. (1987) to control arterial oxygen by adjusting the inspired oxygen fraction in mechanically ventilated dogs.

Yu et al. (1992) use a model predictive controller (MPC) in the MMAC framework to regulate arterial pressure and cardiac output (blood flowrate) by manipulating the infusion rate of two drugs. Thirty-six models are used to span the entire space of expected responses in dogs exhibiting symptoms of cardiac heart failure. To reduce the computational load, only the six highest probability models are used for the control calculation at each time step. Major advantages to using the MPC approach are that constraints are easily taken care of through the solution of a quadratic program, and multivariable systems are handled naturally. A nice review of MPC is provided by Garcia et al. (1989).

MMAC is particularly advantageous in drug delivery applications for a number of reasons. Patient to patient sensitivity to drugs can vary significantly and the response of single patient to a drug can be very time-dependent. Model parameters then vary greatly from patient to patient and in the same patient at different times. There is a major advantage to MMAC during the initial stages of administering a drug to a patient, because no initialization time is required for parameter identification. MMAC begins by controlling with a certain weight distribution among the models, then new control calculations begin as soon as measurements are available, whereas most adaptive control procedures require an initialization time to determine the model parameters. It should be noted that a number of special functions are typically used in the feedback loop to achieve better control of drug delivery. He et al. (1986) and Martin et al. (1987) use a nonlinear unit to freeze the infusion rate of the drug Nitroprusside if the patient's blood pressure drops too low; another nonlinear unit places minimum and maximum limits on the allowable infusion rate. Schott and Bequette (1997) use a simple gain-varying example to illustrate the basic approach to drug infusion control.

Chemical processes control
Chemical reactors are unit operations that exhibit interesting static and dynamic non-linear behaviour. In Schott and Bequette (1995), MMAC is applied in simulations to two different chemical reactors: (I) the van de Vusse reactor and (ii) a classic, exothermic continuous stirred-tank reactor (CSTR). Banerjee et al. (1994, 1997) present a multiple-model "self-scheduling" approach for control of an exothermic CSTR.

3. Implementation Issues

In this section some of the issues which arise when designing an MMAC controller are examined. The state-space, Kalman filter/LQG approach of Figure 2 is discussed in a number of papers (e.g. Athans et al. 1977, Maybeck 1982) and is not reviewed here, however a state-space, KF/LQG example is presented for reference. This paper instead focuses on MMAC implementation outside of the state-space framework, using models in transfer function form and weighting functions which do not involve statistical information about the process. These methods offer performance improvements with comparatively minimal design effort and computational resources.

3.1 IDENTIFICATION

Plant identification in the MMAC controller is done by comparing the plant output to those of the hypothesized models in the model bank. The models can be linear or non-linear, depending upon the detail of information available for the process to be controlled. When linear models are used, the choice of state-space or input/output form is dictated primarily by available process measurements and how well the process noise can be characterized. The task of choosing models for the model bank may be broken into two broad categories: (i) are the models chosen with little knowledge of the actual process? or (ii) are models chosen using much experience or knowledge of specific process operation regions (e.g. linearization of a well-known non-linear process)? If little is known about the process, the size of the model bank may become quite large if many models and parameter combinations are considered. In contrast, when much is known about the process, the size of the model bank is much more defined since the only question is how best to span the parameter space, not how large is the parameter space.

For plant identification, discrete, linear difference models, estimating in closed-loop fashion may be used

$$\hat{y}_{k,n} = (-a_{1,n}y_{k-1} - a_{2,n}y_{k-2} - ... + b_{0,n}u_k + b_{1,n}u_{k-1} + ...) \qquad (1)$$

where $\hat{y}_{k,n}$ is the nth model estimate at the current time step k, and y_k and u_k are process outputs and inputs, respectively. Using a model in this position form permits gain information in the weighting calculations. The velocity form of equation 1 could also be used

$$\Delta\hat{y}_{k,n} = (-a_{1,n}\Delta y_{k-1} - a_{2,n}\Delta y_{k-2} - ... + b_{0,n}\Delta u_k + b_{1,n}\Delta u_{k-1} + ...) \qquad (2)$$

if more emphasis is required on discerning between models based on dynamic differences only. Model form clearly has a direct impact on the tuning of the weighting function. An MMAC approach using equation 2 needs to converge much more rapidly than one which uses equation 1 because model identification can only take place when the system is dynamically active.

3.2 WEIGHTING

For controller weighting, two schemes are used most often: Bayesian probability or a summed square error over a past-error horizon. Within a stochastic, optimal control framework (Maybeck 1982), Bayes theorem can be represented recursively by

$$p_{k,n} = \frac{exp(-r_{k,n}^T\theta^{-1}r_{k,n})p_{k-1,n}}{\sum_{j=1}^{N}[exp(-r_{k,j}^T\theta^{-1}r_{k,j})p_{k-1,j}]} \qquad (3)$$

where $p_{k,n}$ is the probability that the nth model represents the plant at the kth time step, $r_{k,n}$ is residual, N is the number of models in the model bank, and θ is the residual covariance matrix. This probability calculation has the advantage of directly using noise information via the covariance matrix, however it has the drawback of not having any other term to vary the convergence rate.

Alternatively, if a state-space approach is not taken, then a recursive probability can be found using

$$p_{k,n} = \frac{exp(-\varepsilon_{k,n}^T K \varepsilon_{k,n})p_{k-1,n}}{\sum_{j=1}^{N}[exp(-\varepsilon_{k,j}^T K \varepsilon_{k,j})p_{k-1,j}]} \qquad (4)$$

where K is a "convergence" factor used to tune the speed which the probabilities can shift and ε is the residual

$$\varepsilon_{k,n} = y_k - \hat{y}_{k,n} \qquad (5)$$

or the relative residual (He et al. 1986)

$$\varepsilon_{k,n} = \frac{y_k - \hat{y}_{k,n}}{y_{initial} - y_{k,sp}} \tag{6}$$

where $y_{initial}$ is the initial output of the plant and all models, and $y_{k,sp}$ is the reference\ signal which the system is to track.

Because this Bayesian weighting approach was originally developed to find the single correct model amongst a bank of models (Lainiotis, 1971), the recursive weighting function is allowed to use all past measurements in its probability calculation. Over time then, the probability of most models is expected to approach zero, however if a probability were allowed to actually reach zero it would remain zero for all time thereafter. (Theoretically, the probability would not reach zero, but computer machine-precision can cause a small probability to be rounded down to zero.) To avoid this problem the probabilities can be bounded for the next iteration

$$\begin{aligned} p_{k,n} &= \delta & for\ p_{k,n} &\leq \delta \\ p_{k,n} &= p_{k,n} & p_{k,n} &> \delta \end{aligned} \tag{7}$$

and then bounded probabilities are removed from the final weighting

$$\begin{aligned} W_{k,n} &= \frac{p_{k,n}}{\sum_{j=1}^{N} p_{k,j}} & for\ (p_{k,n} > \delta)\ and\ (p_{k,j} > \delta) \\ W_{k,n} &= 0 & for\ p_{k,n} = \delta \end{aligned} \tag{8}$$

$$u_k = \sum_{n=1}^{N} W_{k,n} u_{k,n} \qquad or \qquad \Delta u_k = \sum_{n=1}^{N} W_{k,n} \Delta u_{k,n} \tag{9}$$

The bounding process also limits the number of past observations contained in the current probability estimate. Low δ values require more iterations of equations 3 or 4 before a formerly low-probability model contributes a relevant portion of the overall control signal. Large values of δ yield faster model switching because non-contributing model probabilities are kept artificially high, however large δ values reduce the phase margin of the system (He et al. 1986). Banerjee et al. (1997) solve this "wind-up" problem in a more elegant fashion by recasting the weighting function to only include a past horizon of plant measurements, thereby reducing the history contained in the probability estimate and replacing the physically disconnected tuning parameter δ with a physically pertinent tuning parameter, a time horizon over which the plant might change.

Practically, an unfortunate consequence of using the weighting scheme of (4) is that the control engineer must supply values for tuning parameters K and δ, neither of which have clear a physical connection with the problem and therefore a trial-and-error tuning approach must be taken for the weighting function. In addition, there is no guarantee that the correct weighting will be generated by equations 3 through 8, as shown in an example in the next section.

Weighting schemes may also be based on the past error history, such as

$$E_{k,n} = \sum_{i=1}^{H} (y_{k-i} - \hat{y}_{k-i,n})^2 \tag{10}$$

or

$$E_{k,n} = \sum_{i=1}^{H} \lambda^{H-i} (y_{k-i} - \hat{y}_{k-i,n})^2 \tag{11}$$

where H is a past-time horizon and λ is a forgetting factor. Rule-based weighting schemes (including fuzzy logic) which form combinations of models with low errors could be used, or more commonly full weighting is assigned to the model with the lowest history of error

$$\begin{aligned} W_{k,n} &= 1 \qquad \textit{for } \min(E_{k,n}) \\ W_{k,n} &= 0 \qquad \textit{for all other } E_{k,n} \end{aligned} \tag{12}$$

and the overall controller weighting is given again by equation 9.

3.3 IMPLEMENTATION COMMENTS

Of special note is how well the MMAC approach works when using an adaptive scheme in conjunction with a distributed control system (DCS). Local process computers often do not have the resources to handle very complex control algorithms, and so advanced controller calculations must be done on a supervisory computer connected to a data highway. The MMAC method works nicely within the ÐCS when PID controllers are used in the controller bank because the controller weighting of equation 5 translates into weighting of the individual controller parameters.

$$\Delta u_k = \sum_{n=1}^{N} w_{k,n} \left[K_{c,n} \left\{ (e_k - e_{k-1}) + \frac{\Delta t}{\tau_{1,n}} e_k + \frac{\tau_{D,n}}{\Delta t} (e_k - 2e_{k-1} + e_{k-2}) \right\} \right]$$

$$\Delta u_k = \sum_{n=1}^{N} w_{k,n} K_{c,n} (e_k - e_{k-1}) + \sum_{n=1}^{N} w_{k,n} K_{c,n} \frac{\Delta t}{\tau_{1,n}} e_k + \sum_{n=1}^{N} w_{k,n} K_{c,n} \frac{\tau_{D,n}}{\Delta t} (e_k - 2e_{k-1} + e_{k-2}) \tag{13}$$

$$\Delta u_k = (e_k - e_{k-1}) \sum_{n=1}^{N} [w_{k,n} K_{c,n}] + e_k \Delta t \sum_{n=1}^{N} [w_{k,n} K_{c,n} \frac{1}{\tau_{1,n}}] + \frac{(e_k - 2e_{k-1} + e_{k-2})}{\Delta t} \sum_{n=1}^{N} [w_{k,n} K_{c,n} \tau_{D,n}]$$

and after a few more algebraic steps K_c overall, τ_I overall, and τ_D overall may be defined

$$\Delta u_k = K_{c,overall} \left[(e_k - e_{k-1}) + \frac{\Delta t}{\tau_{I,overall}} e_k + \tau_{D,overall} \frac{(e_k - 2e_{k-1} + e_{k-2})}{\Delta t} \right] \tag{14}$$

Weighting calculations therefore can be done on a supervisory computer and then only tuning parameters to a single PID, not actual changes to the manipulated variable, need to be transmitted across the data highway. If the data highway or supervisory computer is disabled, control at the local process computer continues with the last set of tuning parameters and the integrity of the DCS concept is retained.

Computationally, the additional requirements for an MMAC strategy are generally meager compared to many other adaptive techniques. There are no matrix inversions, optimization routines, or other computationally demanding steps inherent in the MMAC approach. The computation and storage requirements are dictated primarily by the choice of controller. For example, a bank of N PID controllers will not need nearly the same computing power as a bank of N linear QDMC controllers.

4. Examples

Four examples are presented to demonstrate the MMAC approach. The plant non-linearities are kept basic and model/controller banks are kept small so that the mechanics of the problems are clear. The first example uses KF/LQG controllers in a state-space approach as shown in Figure 2. The next three examples use discrete, position form models given in equation 1 and either IMC or IMC-tuned PID controllers. PID controller results are presented because PIDs are well understood and used widely in the chemical industry. Any other model-based controller (e.g. QDMC, etc.) could also be have been used in conjunction with the model/controller banks.

4.1. EXAMPLE 1: A STATE SPACE APPROACH

Consider a second order process whose poles move such that it can be either underdamped (eigenvalues at -0.25 ± 0.97i) or critically damped (two eigenvalues at -1). The plant has a gain of 1.0 in all operating regions and there is noise (Gaussian, $\sigma=0.05$) on the single measured state. For the MMAC controller bank, two LQG controllers are used with Kalman filters as shown in Figure 2, and the Bayesian probability of equation 3 is used as the weighting function. The controller weighting is initially evenly divided between the two controllers. For the first simulation the plant is exactly the same as model 1 (underdamped, gain of 1) and remains so for the entire trial.

Figure 3 shows the results of several setpoint changes on the system. The response to the first setpoint change is predictably sub-optimal because the controller is a combination of the two controllers whereas the plant is actually equivalent to one of the models. Over time, the weighting function moves toward the correct model/controller pair, however the convergence is extremely slow because there are only differences in

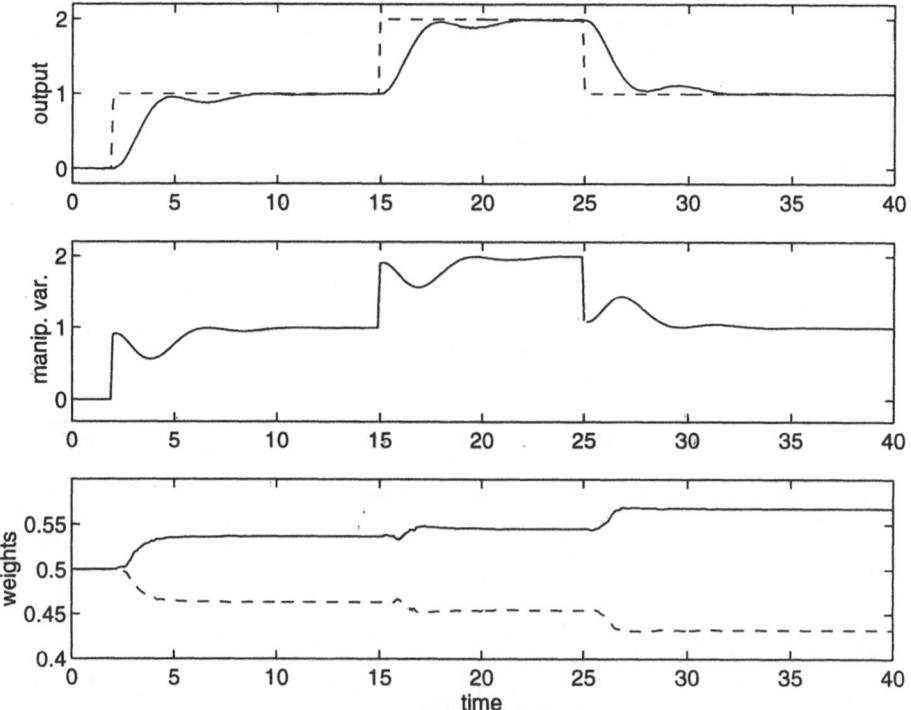

Figure 3. Example 1. Two models used to identify unknown plant dynamics
weight legend: solid = model 1   dashed = model 2

the dynamics of the system, not the gain, and therefore identification can only take place when the system is not at steady-state. After repeated setpoint changes the weighting is eventually driven to full weighing on the correct model. Because the convergence is a function of the noise in the system only, there is no way to increase convergence rate toward the correct model and we can only improve the identification by getting more information (e.g. increasing the sampling rate) during the periods of dynamic activity. For comparison, a vary small change in the gain of one of the models was made in the trial shown in Figure 4. Here model 1 has a gain of 1.05 while the gain for model 2 remains at 1.00. Although the response to first setpoint change is still a combination of the two controllers and is not ideal, identification takes place much more rapidly and the proper controller is selected for the subsequent setpoint change. In this simulation, the plant is switched to equal model 2 at t=25.

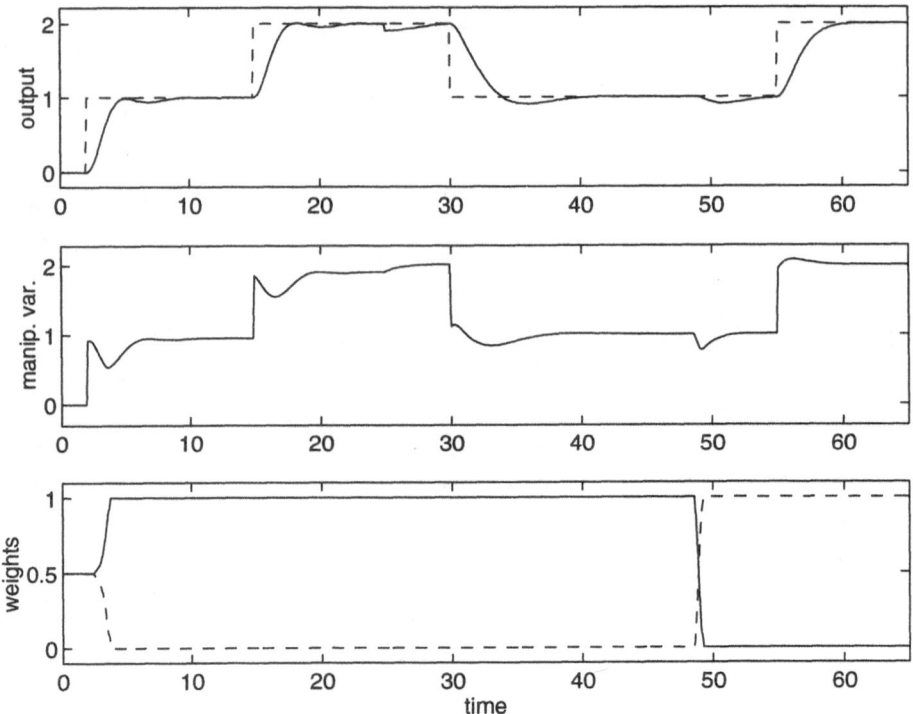

Figure 4. Example 1. Convergence rate difference with 5% difference in model gains
weight legend: solid = model 1   dashed = model 2

Note that the weighing function does not respond to the change until near t=49. This poor identification is due in part to the probabilities "winding up", that is, since the models are not very different and there is little noise, the weighting function requires a good deal of data before it is "convinced" that the plant has indeed changed. Mathematically, the exponential terms in (3) are driven very near 1.0 and 0.0 for the correct and incorrect models, respectively. Figure 5 shows the effect of additional noise on the same system. The convergence when switching models is slower because noise introduces less confidence into each sample, however the change in plant is detected sooner because the weighting function is "not as sure" it had the right model earlier. Mathematically, the exponential terms in (3) are not as dissimilar as the previous, lower noise case.

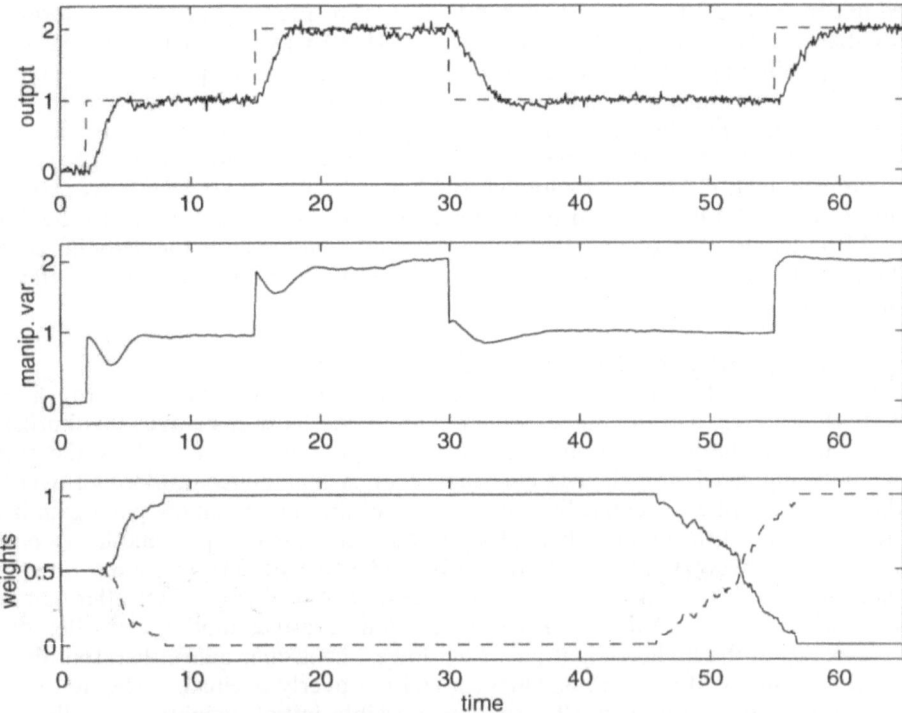

Figure 5. Example 1. Difference in model gains and larger measurement noise
weight legend: solid = model 1   dashed = model 2

4.2. EXAMPLE 2: UNKNOWN PROCESS

In some instances, such as drug delivery to a new patient or start-up of new process
equipment, an adaptive controller is required to identify and control a process with very
little *a priori* information. This problem can be very difficult since most adaptive
strategies require some linear model specification, even if only the order of the plant.
In addition, choosing tuning parameters for the adaptive portion of the algorithm (i.e.
forgetting factors, covariance matrices, etc.) can be difficult when little knowledge of the
process is available. In contrast, an MMAC approach may be taken which requires
only a set of proposed models and a weighting function.

Process Gain Uncertainty
   Consider a first-order process where the gain is not known precisely initially,
though it is known to be positive, greater than or equal to 0.6 and less than or equal to
10.0, and may vary within that range. The time constant and deadtime are known to
be 2 and 1, respectively. For this particular problem where the deadtime does not vary
and the order of the plant does not change, it would be easy to set-up a recursive
estimator and accompanying controller, however the "bare-bones" MMAC
implementation is even more straightforward to construct, yields good performance,
and provides transparent adaption bounds. All of the controllers in the controller bank

are unconstrained PID controllers, implemented in discrete, incremental form, and IMC-based PID tuning (Morari and Zafiriou, 1989) was used with identical tuning criteria (desired closed-loop time constant λ set to one half the process time constant, i.e. $\lambda=1$). The sample time is 0.10 and there is no noise in the system.

Four model/controller pairs were used to span the unknown gain range with model gains of 1,2,4, and 8. The number of models was chosen according to the performance requirement that, if the model with the gain closest to that of the plant's was correctly chosen, then: 1.) there would be no more than 20% overshoot, and 2.) the output would be within 20% of the final value within 5 time units of the setpoint change. The models were implemented in deviation variables, in discrete form as stated in equation 1.

MMAC using Past Error-Horizon Weighting Function

The weighting function from equations 9 and 10 are used with a past error-horizon of H=10 samples, and the model which exhibits the least error over this horizon is given full weighting (equation 12). The three plots in Figure 6 show the system output, manipulated variable, and contribution of each model/controller pair over time when using the MMAC controller. At the start of the simulation the plant gain is 2.7, however no identification can take place because the system input (and hence output) are zero. The weighting remains at the pre-set value of full weighting on model/controller pair number 4 until there is non-zero data available. Controller number 4 was chosen as the initial controller to be conservative; model/controller pair 4 corresponds to the highest plant gain (and lowest controller gain), therefore the first setpoint change will at worst be sluggish and not overly oscillatory should the plant gain initially be lower than 10. Another possible initial weighting distribution is equal weighting on all model/controllers. After the first setpoint change and the process deadtime has passed, the correct model (number 2) is quickly identified.

At $t=25$ the process gain changes abruptly to 7. The weighting function briefly settles on model 3 then moves on correctly to model 4. The process responds very well on the subsequent setpoint change. At $t=55$ the process gain suddenly drops to 1.2 and again the closest model is quickly identified correctly. An additive disturbance of -0.05 is applied to the output at $t=90$. The performance of a single PID controller tuned using a model gain of 4.7, the middle of the unknown gain range, is also shown for comparison.

MMAC using a Bayesian Weighting Function

Because the plant in the previous example has gain uncertainty only, a linear combination of two models should be adequate to cover the entire gain uncertainty region if the weighting system is able to combine the two corresponding controllers. The Bayesian weighting function of equation 4 has the potential to generate this blending. Figure 7 shows the results of using a two model/controller pair MMAC approach with the Bayes weighting on the previous example, except here the plant's gain remains at 5.0 throughout the run. A convergence factor of K=100 was chosen by trial-and-error for fair overall performance for this demonstration and δ was set at 0.05. The two models used in the model bank have gains of 1.0 and 8.0, and again IMC-based PID tuning was used. The results clearly show that the correct blending does not take place; instead of converging on a ratio of 3/7 to 4/7 (model 1 contribution to model 2 contribution) the weighting instead continues to converge to weight completely on the model with the closest gain.

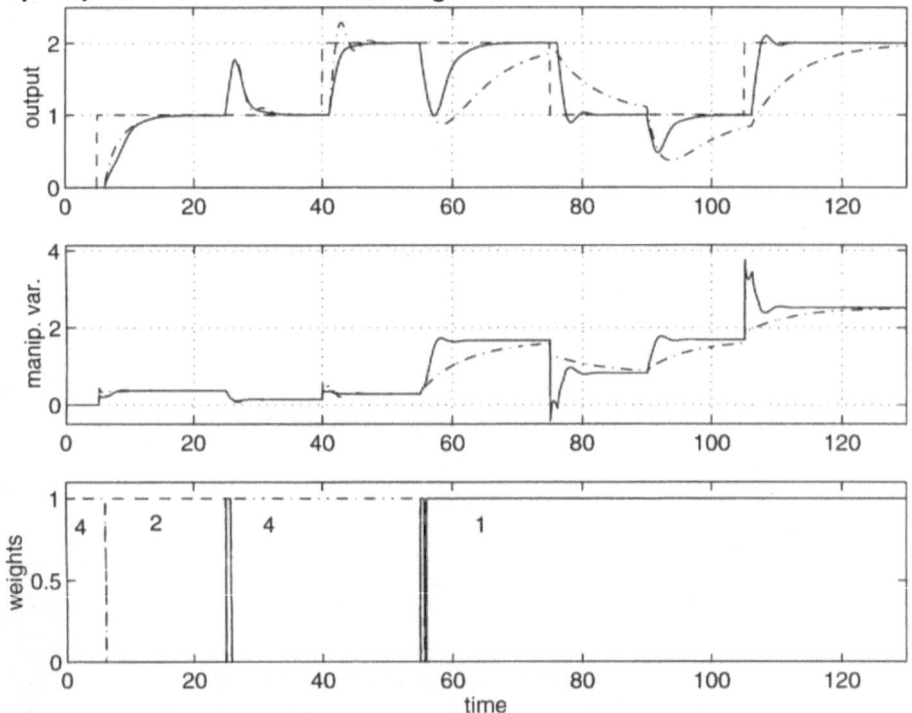

Figure 6. Example 2. Horizon-based weighting; four models
output legend: solid = MMAC   dash-dot = fixed PID

Although equation 4 mathematically allows blending based on probabilities, this version of the Bayes probability equation does not necessarily guarantee that blending will take place in the correct proportions. As a numerical example using this problem, suppose the actual plant gain was 6.25 and the model gains were 1 and 8 for models "1" and "2" respectively. The appropriate linear combination of models to represent the plant is 0.75 of the model "2" and 0.25 of the model "1". Further suppose that equation (4) did reach steady-state with these correct weights, so that probabilities $p_{k-1,n} = p_{k,n} = p_{k+1,n} = \ldots = p_n$ and plant/model differences $\varepsilon_{k-1,n} = \varepsilon_{k,n} = e_{k+1,n} = \ldots = \varepsilon_n$ do not change with time. Therefore, the weighting function would become

$$p_n = \frac{p_n \exp(-\varepsilon_n^T K \varepsilon_n)}{p_n \exp(-\varepsilon_n^T K \varepsilon_n) + p_{n-1} \exp(-\varepsilon_{n-1}^T K \varepsilon_{n-1})}$$

$$0.75 = \frac{0.75 \exp(-\varepsilon_n^T K \varepsilon_n)}{0.75 \exp(-\varepsilon_n^T K \varepsilon_n) + 0.25 \exp(-\varepsilon_{n-1}^T K \varepsilon_{n-1})}$$

(15)

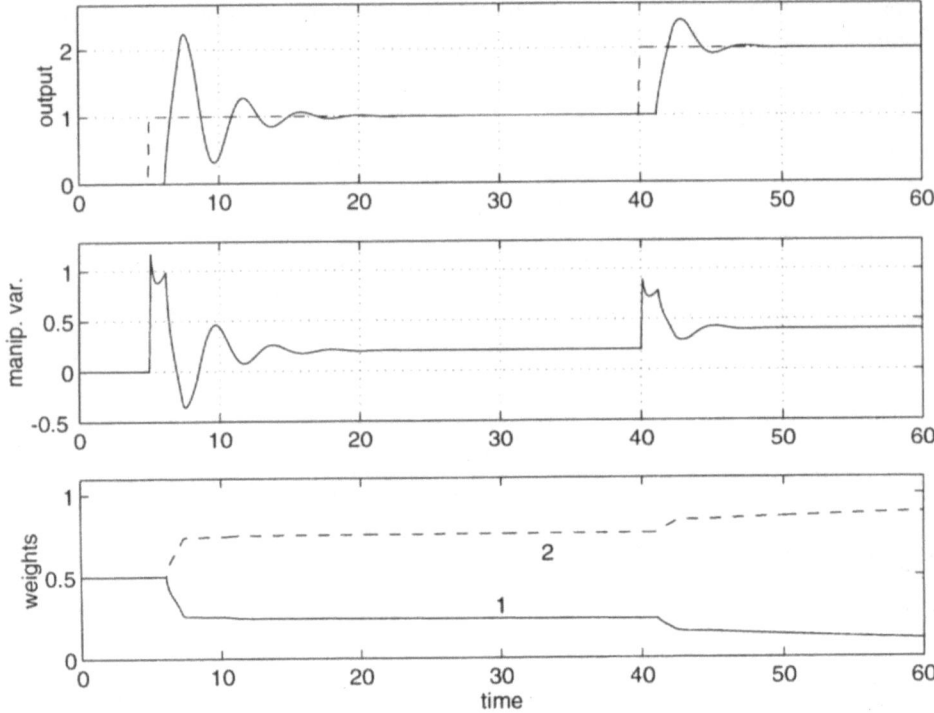

Figure 7. Example 2. Bayesian weighting; two models

and each exponential term can be replaced with a constant

$$0.75 = \frac{0.75\alpha}{0.75\alpha + 0.25\beta} \tag{16}$$

The only α,β which satisfy (16) are $\alpha=\beta$, and because K is constant and $\varepsilon_n \neq \varepsilon_{n-1}$, equation 16 cannot be true. This result does not say that *some* combination will not be found, resulting in better control than either controller would provide individually; it simply shows that the resulting combination will not be the best one. For example, Figure 8 shows the response of a 4-model controller with the same plant gain changes as in the earlier example, however using the Bayes weighting of equation 4, K=50 and δ=0.05. Initial weighting is again set at 100% of model "4", the model with the highest gain and therefore lowest controller gain.

The convergence was deliberately chosen to be slow to show how the controller may settle on combinations of model/controller pairs. The response to the first setpoint change is within the performance specifications, however the weighting function settles

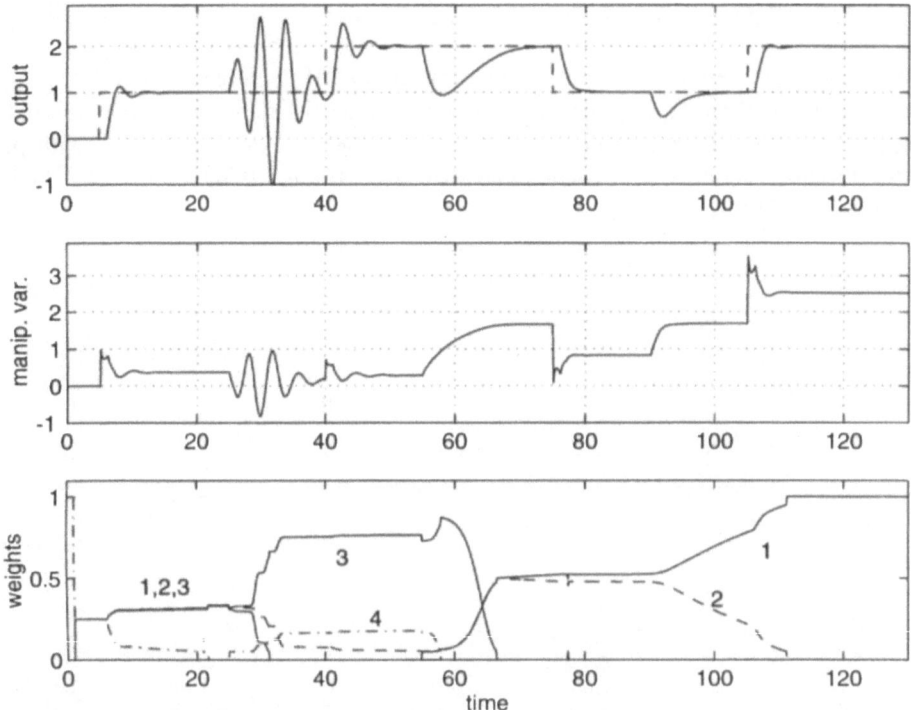

Figure 8. Example 2. Bayesian weighting; four models

incorrectly on a combination of three models even though only two models are needed to describe the uncertain gain. The result is a severe response to the plant's gain change at $t=25$. The response to the next setpoint change does not meet the performance specifications. After the gain change at $t=55$ the subsequent responses to setpoint changes do fall within the performance specifications, even though mathematically the weights do not describe the actual plant.

4.3. NOISE AND MODEL SPACING

When using the type of model identification and weighting function given in equations 1 and 4, noise of sufficient magnitude can mask the differences between model outputs and cause rapid switching between model/controller pairs in the overall controller weighting. The difference, or the "distance" between two model outputs when there is noise on the plant output is

$$\hat{y}_{k,n} - \hat{y}_{k,n-1} = (-a_{1,n}(y_{k-1}+e_{k-1})-...+b_{0,n}u_k+...)-(-a_{1,n-1}(y_{k-1}+e_{k-1})-...+b_{0,n-1}u_k+...)$$
$$\Delta_{n,n-1} = -(a_{1,n}-a_{1,n-1})(y_{k-1}+e_{k-1})-...+(b_{0,n}-b_{0,n-1})u_k$$

$$(17)$$

and the difference between the noisy plant output and the nth model output is then

50

$$\Delta_{y,n} = (y_k + e_k) - \hat{y}_{k,n} = (y_k + e_k) - (-a_{1,n}(y_{k-1} + e_{k-1}) - \ldots + b_{0,n}u_k + \ldots) \qquad (18)$$

where e_k, e_{k-1},... are noise terms at each time step k. If the noise terms cause $\Delta_{y,n}$ to become larger than $\Delta_{n,n-1}$ then the incorrect model is identified at that time step. simulation results for the same system and controller as example 2 above with past-error horizon weighting (H=15), however now there is noise (Gaussian, σ=0.07) added to the plant output.

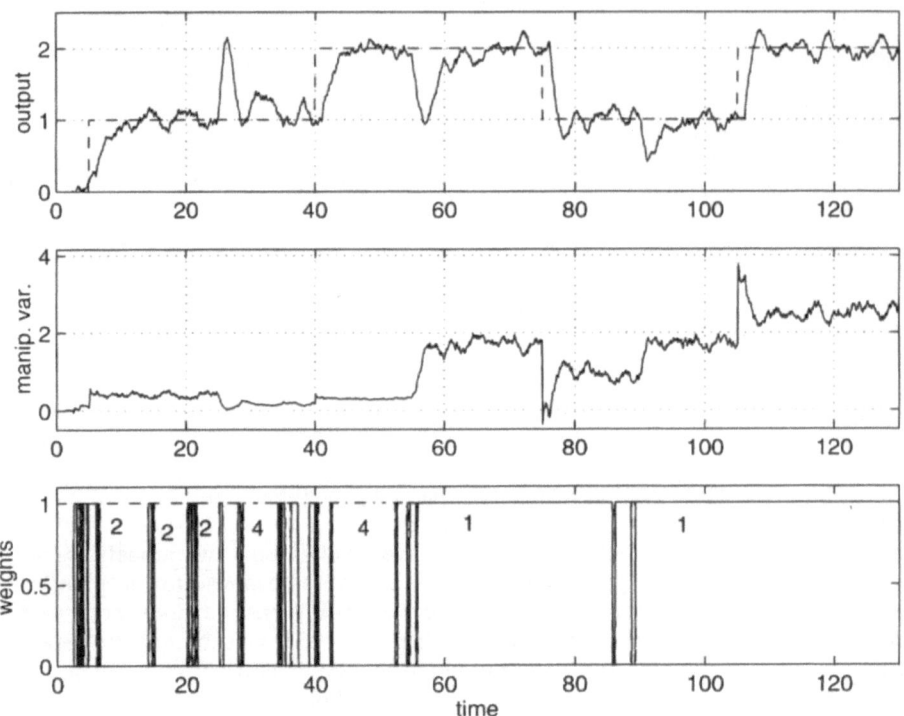

Figure 9. Example 2. Horizon-based weighing; four models and measurement noise

The poor identification and rapid switching in the first part of this simulation demonstrates the need to properly space models in the model bank. Narendra and Balakrishnan (1997) use knowledge of the physical system to disallow weighting changes if sufficient time has not passed since the previous weighting switch. Equation 17 shows that the distance between the models changes not only due to noise, but also as a function of the plant input, further complicating the choice of a past error horizon. Small plant inputs (u_k, u_{k-1}, ...) will cause the difference between adjacent models to shrink. As a numerical example, consider the same plant, 4-model bank, and 4-controller bank as used previously. The models differ in gain only, and the first two models have a gain difference of 1.0, so the output-difference between the two model-pairs is

$$\hat{y}_{k,n} - \hat{y}_{k,n-1} = -(a_{1,n} - a_{1,n-1})y_{k-1} + (b_{1,n} - b_{1,n-1})u_{k-1}$$
$$\hat{y}_{k,2} - \hat{y}_{k,1} = -(0)y_{k-1} + (0.0477)u_{k-1}$$

$$(19)$$

With a manipulated variable of u=1, the distance between these models is only 0.0477 and even a small amount of noise on the plant output will cause identification problems. However, if a higher setpoint forces the manipulated variable to come to steady state with u=2, the distance between models is doubled and more noise (and therefore a smaller error horizon) can be tolerated. A disturbance (input or output) will therefore also effect the weighting performance by shrinking or expanding the distance between models through manipulated variable. Figure 10 illustrates how the model spacing changes depending on the magnitude of the manipulated variable.

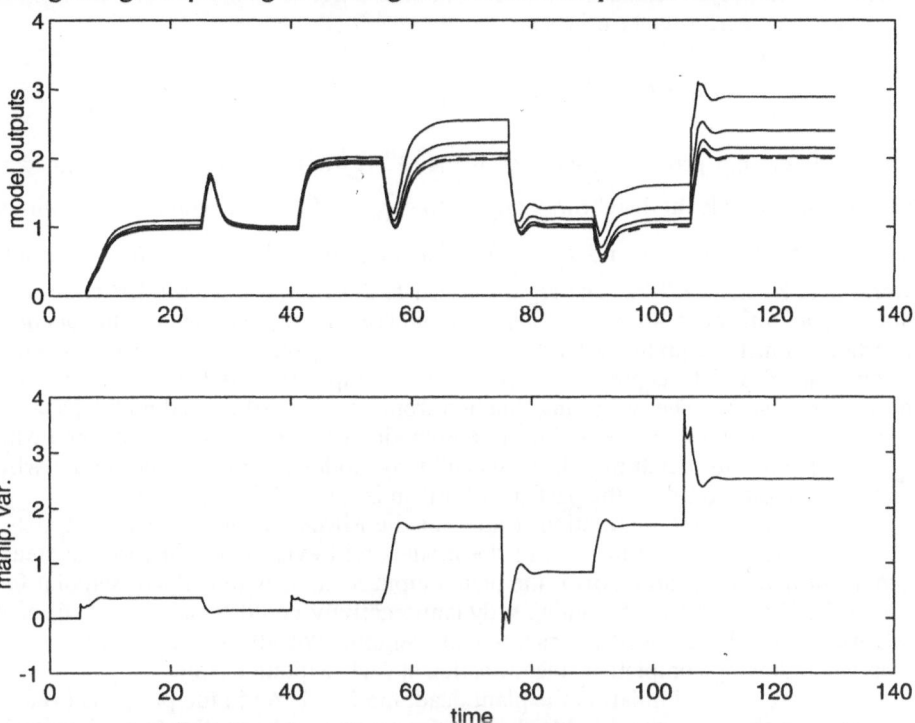

Figure 10. Due to gain differences in the models, the magnitude of the model outputs changes with the magnitude of the manipulated variable
(additive output disturbance of -0.05 applied at t=90)

As formulated in equations 3 through 9, the Bayes weighting scheme suffers the same difficulty as that of the past-error horizon weighting: although changing the noise (or K) varies the rate of convergence, the weighting function is again a function of the manipulated variable via the model output, and therefore the performance of the weighting function will change as the region of plant operation changes. Noise effects do not prevent the use of MMAC outright, but as with any adaptive control system, care must be taken to distinguish how much information is usable on a noisy signal.

4.4. EXAMPLE 3: VARIATION IN PROCESS DEADTIME ONLY

Consider a first-order process where the deadtime is not well known, however it is known to between 1 and 3.4 and may vary within that range. The gain and time constant are known to be 1 and 2, respectively. Variable deadtime problems can be challenging since some adaptive algorithms require accurate knowledge of the system deadtime in their basic formulation. Here five model/controller pairs were used to span the unknown deadtime range with model deadtimes of 1.4, 1.8, 2.2, 2.6, and 3. All of the controllers in the controller bank are unconstrained IMC controllers, implemented in discrete form, and all controllers have identical tuning criteria (desired closed-loop time constant l =1). Since stability and robustness are significant concerns in a variable deadtime system, the five models and their differences in deadtime were chosen using the robust performance criterion

$$\left| \tilde{p}_n \tilde{q}_n f_n \bar{l}_{m,n} \right| + \left| (1 - \tilde{p}_n \tilde{q}_n f) w \right| < 1 \qquad \forall \omega \qquad (20)$$

where $\tilde{p}_n$ is the nth process model, $\tilde{q}_n$ is the nth H_2-optimal controller designed via the IMC method, f is the IMC filter augmenting $\tilde{q}_n$, w is the maximum peak allowed on the sensitivity function, and $\bar{l}_{m,n}$ is the bound on allowed multiplicative uncertainty (Morari and Zafiriou,1989). The region from 1 to 3.4 was spanned by first choosing λ, choosing an initial deadtime value, then increasing or decreasing the deadtime uncertainty until inequality 20 was nearly violated. Some overlap between model regions was allowed keeping (20) under 0.9. Sample time is 0.10 and there is no noise in the system. The weighting function from equation 10 is used with a past-error horizon of 15 samples. Since no linear combination of models can be generated which will fit a plant with deadtime "in-between" two model deadtimes, the model which exhibits the least error over the past error horizon is given full weighting.

Figure 11 shows the simulation results for the MMAC controller using models in the form of equation 1. Since all of the models will eventually converge at steady-state, identification is turned off if the plant output remains within 1% of setpoint for 2 time units. Other criteria for judging dynamic activity could be used as well, or the weighting could be allowed to reach equal weighting of all model/controllers pairs, essentially allowing complete re-identification at each setpoint change.

At the start of the simulation the plant deadtime is 2.1. As in the previous example, the initial weighting was pre-set to the most conservative controller (model/controller pair 5 in this case). At t=25 the deadtime changes to 3.2, but because the system is at steady-state, deadtime identification cannot take place. However, during the dynamic activity of the next setpoint change, the correct model is identified. Note then that at each setpoint change after the plant's deadtime has changed, control moves continue to be based on the previous (i.e. wrong) model/controller until there is sufficient data to identify the new deadtime. At t=55 the deadtime changes to 1.3 and the correct model is identified at the setpoint change at t=75.

The identification process breaks down with the addition of an additive disturbance at t=90 because it forces the output beyond the bounds of all the models. To the model bank, the disturbance is seen as a gain uncertainty (Figure 12) and so none of the models track the process well. One solution to this problem is to use the velocity form of the discrete model (equation 2) or, the size of the model bank could be increased to account for the possibility of this disturbance.

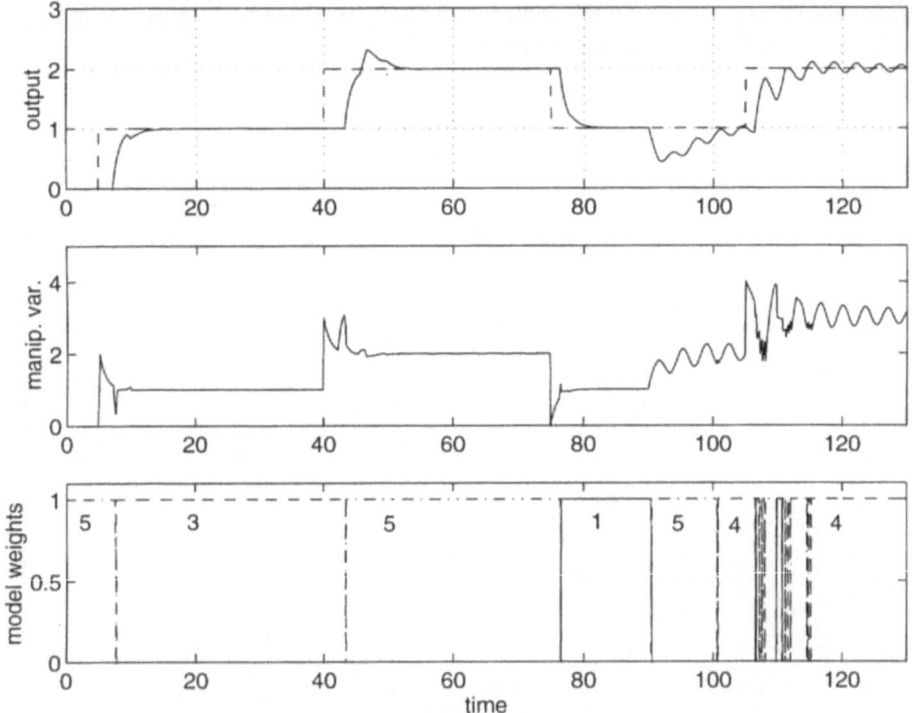

Figure 11. Example 3. Horizon-based weighing; five models

4.5. EXAMPLE 4: PARTIALLY KNOWN PROCESS

In some cases a process is well known in some operating regions, yet perhaps not so well in others. Examples include aircraft (Athens et al, 1977) and industrial chemical processes which run primarily at a few operating points and there is no opportunity for experimentation to expand ones knowledge of the plant. MMAC is an effective approach to improving control in this situation because the well known model/ controller combination can be retained, giving a basis for hypothesized models in nearby operating regions.

The dimensionless modeling equations for a two-state exothermic CSTR are

$$\frac{dx_1}{dt} = -\phi x_1 \kappa(x_2) + q(x_{1f} - x_1)$$

$$\frac{dx_2}{dt} = \beta\phi x_1 \kappa(x_2) - (q + \delta)x_2 + \delta u + qx_{2f} \qquad (21)$$

$$\kappa(x_2) = \exp(\frac{x_2}{1 + x_1}\gamma)$$

where x_1 and x_2 are dimensionless concentration and reactor temperature and u is the dimensionless jacket temperature. Definitions of the dimensionless parameters and the parameters values used for this example are listed in Table 1. Figure 13 shows the steady-state input-output behaviour of the system.

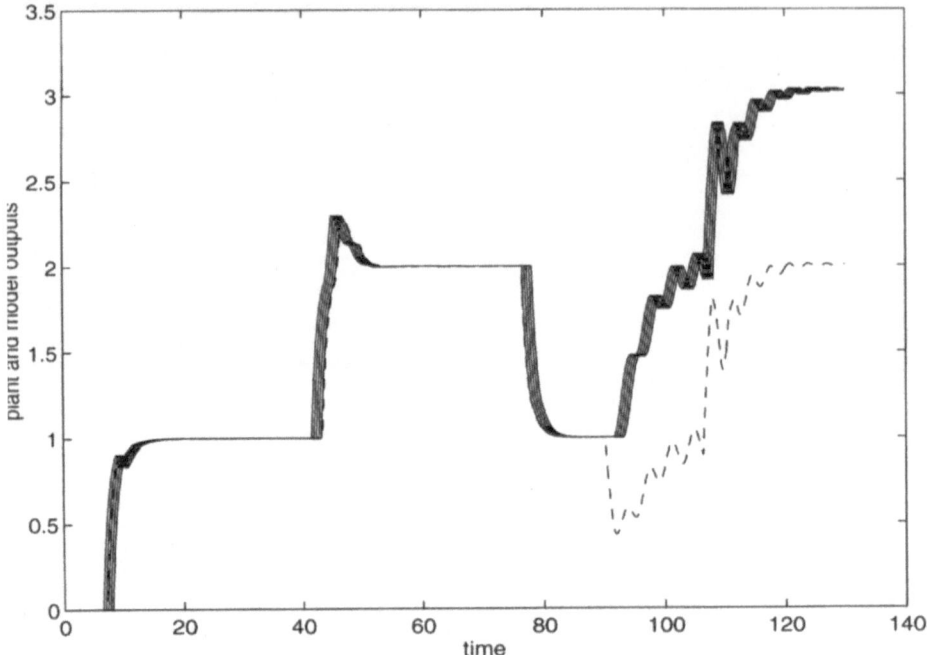

Figure 12. An additive disturbance on the output forces the system output out of the operating space defined by the five models
legend: solid = model outputs   dashed = plant output

Table 1. Dimensionless variables for CSTR model

symbol	variable definition	Parameter value for simulation
ϕ	Nominal Damkohler number	0.072
β	Dimensionless heat of reaction	8.0
δ	Dimensionless heat transfer coefficient	0.3
x_{2f}	Dimensionless reactor feed temperature	0.0
x_{1f}	Dimensionless reactor feed concentration	1.0
q	Dimensionless reactor flowrate	1.0
γ	Dimensionless activation energy	20.0

Although a single controller can be tuned to function in both the open-loop stable and open-loop unstable operating regions, its overall performance may be poor. A single PID controller tuned for operation in an open-loop stable region can perform well for small setpoint changes close to where the controller is designed to operate, yet the same controller exhibits large overshoot when it moves to and from the open-loop unstable region. Using the controller designed for the open-loop unstable region leads to even worse performance.

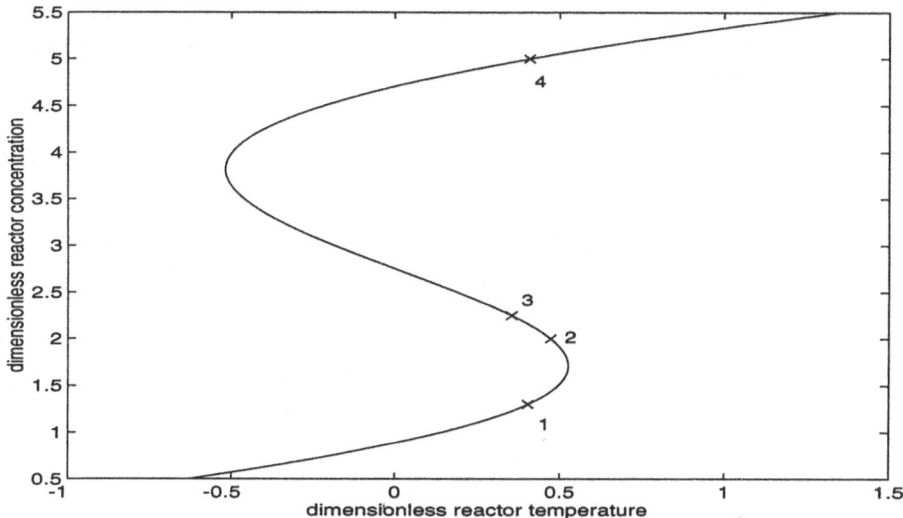

Figure 13. Steady-state input-output curve for exothermic CSTR

For the MMAC controller, four linear models are used, corresponding to dimensionless reactor temperatures of 1.30, 2.00, 2.25, and 5.0. Model "1" is on the lower open-loop stable branch, models "2" and "3" lie on the middle, open-loop unstable branch, and model "4" is on the upper, open-loop stable branch. These models were chosen for demonstration of the MMAC method, representing knowledge in some operating regions but not overall, detailed information about the system's global behaviour. There is no performance measure associated with this model spacing and there was no attempt made to mimic the entire non-linear system's behaviour with a complete set of linear models. Discrete, linear models of the same form as equation 1 are used in the model bank and unconstrained PID-plus-filter controllers in velocity form are used in the controller bank. IMC-based PID tuning rules were used to find the controller's tuning parameters. Identical tuning criteria was used between regions: all regions are tuned for a desired closed-loop time constant of $\lambda=2$. The sample rate is 0.1 and the weighting function of equation 10 is used with a past-error horizon of 15 samples. With the MMAC approach the appropriate controller is used in each operating region, significantly lowering overshoot in the open-loop unstable region. Figure 14 shows that the overall time away from setpoint is reduced when using the MMAC controller.

5. Conclusions and Research Direction

Multiple-Model Adaptive Control is an effective approach for controlling nonlinear systems. It is an intuitively appealing, divide-and-conquer approach which incorporates well-known linear methods to control a non-linear system. In this paper we demonstrate primarily the scheduled-adaption approach, switching fully to a new controller when a new operating region was identified, and that controller "blending" is not a practical expectation with the weighting functions presented. Two interconnected

56

issues remain for future research efforts: (i) an appropriate weighting function for correct blending needs to be developed and, (ii) for a given weighting function, a method for determining the minimum number of models required to span a given parameter space.

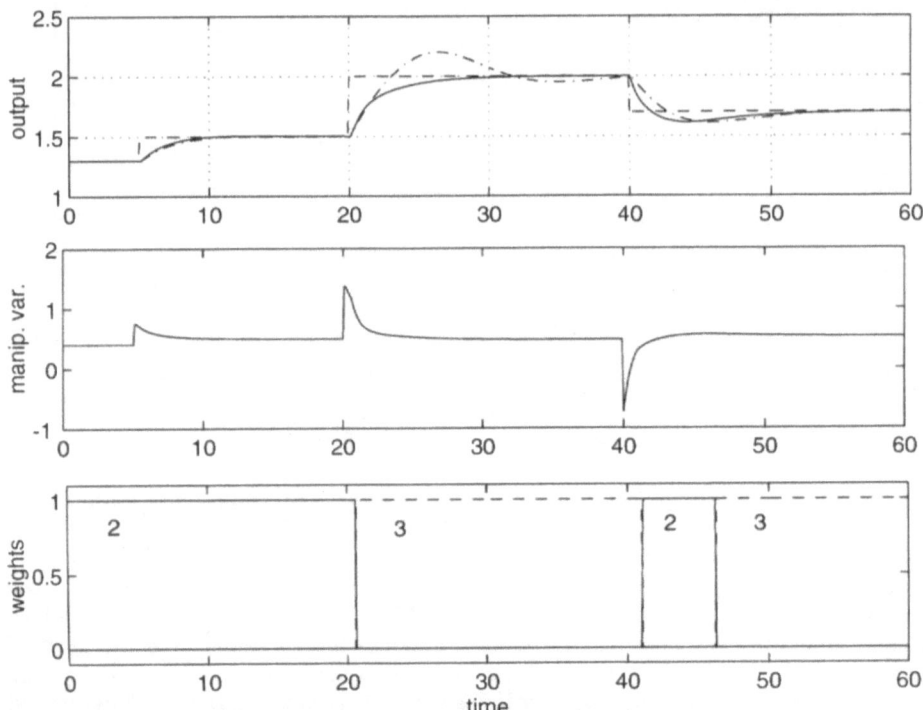

Figure 14. Example 4. Horizon-based weighing; four models
output legend: solid = MMAC dashed = fixed PID

Acknowledgment

This work has been partially supported by the National Science Foundation (BES-9522639) and the Biomedical Engineering Grants program of the Whitaker Foundation.

References

1. Arkun, Y., Banerjee, A. and Lakshmanan, N.M. (1997) Self scheduling MPC using LPV models, in R. Berber and C. Kravaris (eds.), Nonlinear Model Based Process Control, Kluwer, Dordrecht.
2. Athans, M., D. Castanon, K-P. Dunn, C.S. Greene, W.H. Lee, N.R. Sandell, and A.S. Willsky (1977) The Stochastic Control of the F-8C Aircraft Using a Multiple-Model Adaptive Control (MMAC) Method - Part I: Equilibrium Flight, IEEE Trans. Automat. Control AC-22(5), 768-780.
3. Banerjee A., Y. Arkun, B. Ogunnaike, and R. Pearson (1994) Robust Nonlinear Control by Scheduling Multiple Model Based Controllers, 1994 AIChE Annual Meeting, paper 230a, San Francisco, CA.
4. Banerjee A., Y. Arkun, B. Ogunnaike, and R. Pearson (1997) Estimation of Nonlinear Systems Using Linear Multiple Models, AIChE J., 43(5), 1204-1226.

5. Deshpande, J., T.N. Upadhyay, and D.G. Lainiotis (1973) Adaptive Control of Linear Stochastic Systems, Automatica (9), 107-115

6. Garcia, C., D.M. Prett, and M. Morari (1989) Model Predictive Control: Theory and Practice - a Survey, Automatica 23(3), 335-348.

7. Gendron , S., Perrier, M. and Barrette, J. (1993) Deterministic adaptive control of SISO processes using model weighting adaptation, Int. J. Cont., 58(5), 1105-1123.

8. Greene, C. and A.S. Willsky (1980) An Analysis of the Multiple Model Adaptive Control Algorithm, Proceedings Conf. Decision Control, 1142-1145.

9. Gustafson, J. and P.S. Maybeck (1994) Flexible Spacestructure Control via Moving-Bank Multiple-Model Algorithms, IEEE Trans Aero Elec Sys, 30(3)   750-757 .

10. He, W.G., H. Kaufman, and R.J. Roy (1986) Multiple-Model Adaptive Control Procedures for Blood Pressure Control,î IEEE Trans. Biomed. Eng., BME-33(1), 10-19.

11. Johansen, T.A. and R. Murray-Smith, (1997) The Operating Regime Approach, Chapter 1 (pp.1-72) in: Multiple Model Approaches to Modelling and Control, R. Murray-Smith and T.A. Johanson (eds.), Taylor & Francis, London, UK

12. Kordon, A., Fuentes, Y.O., Ogunnaike, B.A. and Dhurjati, P.S. (1997) An intelligent parallel control system structure for plants with multiple operating regimes, Comp. Chem. Engng., 21(Suppl.), S119-S124.

13. Kothare, M.V., Mettler, B. Morari, M., Bendotti, P. and Falinower, C.-M. (1997) Linear parameter varying model predictive control for steam generator level control, Comp. Chem. Engng., 21(Suppl.), S861-S866.

14. Lainiotis, D.G. (1971) Optimal Adaptive Estimation: Structure and Parameter Adaptations, IEEE Trans. Automat Contr., AC-16(2).

15. Lainiotis, D.G. (1976a) Partioning: a unifying framework for adaptive systems. I: Estimation, Proc IEEE 64, 1126-1143.

16. Lainiotis, D.G. (1976b) Partioning: a unifying framework for adaptive systems. II: Control, Proc IEEE 64, 1144-1161.

17. Martin, J. A.M. Schneider and N.T. Smith (1987) Multiple Model Adaptive Control of Blood Pressure Using Sodium Nitroprusside, IEEE Trans Biomed Eng BME-4(8), 603-611.

18. Maybeck, P. (1982) Stochastic Models, Estimation, and Control vol 2 Academic Press, New York NY

19. Maybeck, P. (1989) Moving-Bank Multiple Model Adaptive Estimation and Control Algorithms: An Evaluation, vol 31 of Control and Dynamic Systems, Academic Press, New York, NY

20. Morari M and E. Zafiriou (1989) Robust Process Control, Prentice Hall Englewood Cliffs, NJ

21. Murray-Smith, R. and T.A. Johanson (eds.), (1997) Multiple Model Approaches to Modelling and Control, Taylor & Francis, London, UK

22. Narendra, K.S. and Balakrishnan, J. (1997) Adaptive control using multiple models, IEEE Trans. Auto Cont. 42(2), 171-187.

23. Qin, S.J., Martinez, V.M. and Foss, B.A. (1997) An interpolating model predictive control strategy with application to a waste treatment plant, Comp. Chem. Engng., 21(Suppl.), S881-S886.

24. Saridis G. and T.K. Dao (1972) Learning Approach to the Parameter-Adaptive Self-Organizing Control Problem, Automatic (8), 589-597.

25. Schott, K.D. and B.W. Bequette (1995) Control of Chemical Reactors Using Multiple-Model Adaptive Control (MMAC), 4th IFAC Symposium on Dynamics and Control of Chemical Reactors, Distillation Columns, and Batch Reactors, Dycord+, Helsingor, Denmark 345-350.

26. Schott, K.D. and B.W. Bequette (1997) Multiple Model Adaptive Control Chapter 11 (pp.269-291) in: Multiple Model Approaches to Modelling and Control, R. Murray-Smith and T.A. Johanson (eds.), Taylor & Francis, London, UK

27. Stengel R. (1986) Stochastic Optimal Control: Theory and Application, John Wiley, New York, NY

28. Yu C., R.J. Roy, H. Kaufman and B.W. Bequette (1992) Multiple-Model Adaptive Predictive Control of Mean Arterial Pressure and Cardiac Output, IEEE Trnas Biomed Eng 39(8), 765-788.

29. Yu C., G. He, J.M. So, R. Roy, H. Kaufman and J.C. Newell (1987) Improvement in Arterial Oxygen Control Using Multiple Model Adaptive Control Procedures, IEEE Trans Biomed Eng BME-34(8), 567-574.

SELF-SCHEDULING MPC USING LPV MODELS

Y. ARKUN[1], A. BANERJEE AND N. M. LAKSHMANAN
School of Chemical Engineering
Georgia Institute of Technology
Atlanta, GA 30332-0100
U.S.A.

Abstract This paper presents a self-scheduling MPC framework for plants described by LPV (Linear Parameter Varying) models. Such a controller adjusts to variations in plant dynamics by using measured values of the parameters in the control law. We apply the method to control of nonlinear plants approximated by LPV models constructed from multiple local linear models. In this context the parameters constitute model validity functions which are estimated on-line and used for scheduling MPC. Both quadratic programming based finite horizon MPC and min-max type LMI based MPC algorithms are discussed and applied to continuous and batch systems.

1   Introduction

Linear Parameter Varying (LPV) plants are parameter-dependent linear systems of the form:

$$\begin{aligned} \dot{x} &= A\left(p\right)x + B\left(p\right)u \\ y &= C\left(p\right)x + D\left(p\right)u \end{aligned} \tag{1}$$

where $p\left(t\right)$ is the vector of time-varying parameters belonging to a compact set. This modeling framework is general enough to encompass many

---

[1]To whom all correspondence should be addressed, `yaman.arkun@che.gatech.edu`

R. Berber and C. Kravaris (eds.), Nonlinear Model Based Process Control, 59-84.
© 1998 Kluwer Academic Publishers.

types of problems. For example, equation (1) could represent linear time-invariant systems with time-varying parameters, or parametric linearizations of a nonlinear system along its parameter trajectories. In the literature the LPV models are often used in the context of gain-scheduling. In fact the LPV description has allowed the development of a sound theoretical foundation for gain scheduling which is usually practiced as an art without a priori guarantees for stability or robust performance. For details on gain scheduling using LPV models, the reader is referred to the following original papers: Shamma and Athans [8]; Packard [6] and Apkarian et al. [1].

The main design idea in these papers is to make the controller also parameter-dependent so that when time-varying parameters are measured in real-time, the controller becomes self-scheduling and offers potential performance improvement over fixed robust controllers. In our past work Banerjee et al. [2] and [3] we have also used this key idea but with some important differences. Due to our interest in controlling nonlinear plants during transition from one operating condition to another, we have treated the parameters as validity functions of some local models. Basically a family of linear models for a nonlinear plant are obtained at different operating conditions either by identification or linearization, and a global LPV model is interpolated between them using the elements of $p(t)$ as interpolating or model validity functions. The states and the parameters of the global LPV model are then estimated on-line to match the nonlinear plant dynamics during the transition period. Banerjee et al. [2] presents both a Bayesian estimator and an optimization based moving horizon estimator for that purpose. The controller in Banerjee et al. [3] is a self-scheduling H_∞ controller with affine dependence on the estimated parameters which serve as the "scheduling variables" as shown in figure 1. The controller is designed to be robust against norm-bounded time-varying uncertainty which is added to the LPV model to account for model-plant mismatch.

In this paper instead of Linear Fractional Transformation (LFT) or H_∞ design employed in previous works, the LPV model is incorporated in a Model Predictive Control framework including constraints. In the spirit of MPC the method is a natural extension (at least conceptually) of gain-scheduling to model-scheduling. In case a first-principles nonlinear model is not available, the multiple linear models can be identified from local input/output data and used in the global LPV model. If a first-principles model exists, an MPC controller scheduled by the local multiple models obtained from jacobian linearization constitutes a viable option to MPC

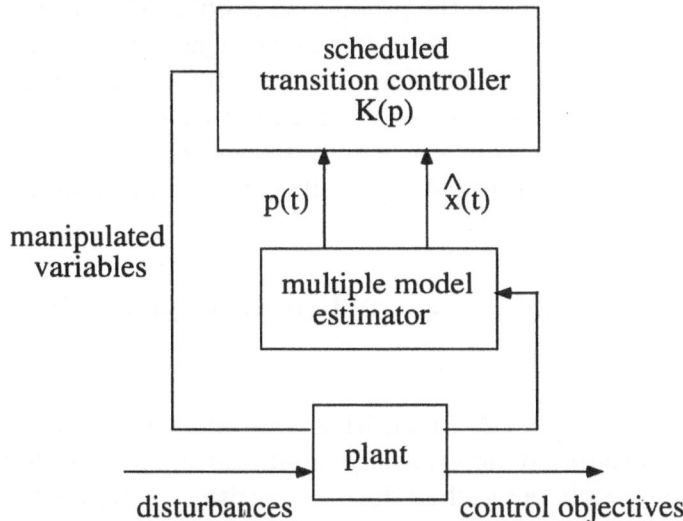

Figure 1: Transition control using a multiple model estimator

based on a more rigorous nonlinear optimization performed on-line.

The rest of the paper is organized as follows. We first present the non-linear transition control problem to motivate the development of the theory. Since construction of the LPV model and estimation of the validity functions will be common to all the MPC algorithms, this background is briefly reviewed next. This is followed by two different MPC formulations. The first one solves certain linear matrix inequalities (LMIs) on-line; the second one is similar to the traditional QP based finite-horizon MPC, but uses the LPV model's predictions. Finally examples are given to demonstrate the application of the method to continuous and batch reactors.

2 Motivation

In recent years it is becoming increasingly common for chemical plants to operate at a number of different operating points in response to changing product specifications (e.g. polymer product and grade changeovers) or persistent plant disturbances (e.g. variations in the plant feed conditions). This wide variety of products and disturbances usually results in long periods of transition from one operating region to another. Poor control during these transitions can lead to the production of off-specification material and unsafe operating conditions. Therefore an efficient strategy for carrying out transition control in order to reduce the time taken for transitions, and to perform them safely, can be expected to have considerable economic impact on industrial practice. Our aim here is to develop feedback strategies by which a plant can be controlled well while in transition as well as in isolated operating regimes.

3   Approximating a nonlinear system by an LPV model

In order to design a good transition controller, it is necessary to obtain a model that accurately describes the nonlinear plant at each of the different operating points, as well as during transition. Often a first principles model that includes the underlying physics and chemistry of the process is difficult to obtain. Even if such a model is available, it may be inappropriate for state estimation and controller design. One alternative is to identify an empirical model using plant input/output data. However since chemical processes are non-linear, unmodeled dynamics which are negligible at one operating point may be dominant at another. Therefore in order to uncover all the necessary plant dynamics, the identification algorithm might require inputs with large amplitude and/or large frequency which may not be practically implementable. This would make it difficult to identify an empirical model that is valid for all operating conditions of interest.

The approach followed here is to use a local model for each of the different plant operating regimes. Each model should satisfactorily describe the plant in a region around the point that it was obtained, but may perform poorly when the plant moves out of that domain. Therefore this section describes ways of combining the information contained in the local models into a global description of the plant in the form of an LPV model, and then carrying out state estimation by tracking transitions on-line.

Let the plant be represented by an unknown nonlinear discrete relationship without loss of generality :

$$
\begin{aligned}
x(k+1) &= f(x(k), u(k)) \\
y(k) &= g(x(k))
\end{aligned}
\tag{2}
$$

where x is the state, y is the output and u is the input. This can be approximated by a time-varying linear system as follows :

$$
x(k+1) = f(x(k), u(k))
\tag{3}
$$

$$\approx \quad f(x(k-1), u(k-1)) +$$

$$\left.\frac{\partial f}{\partial x}\right|_{x(k-1), u(k-1)} (x(k) - x(k-1)) +$$

$$\left.\frac{\partial f}{\partial u}\right|_{x(k-1), u(k-1)} (u(k) - u(k-1)) \tag{4}$$

$$= \quad x(k) + \left.\frac{\partial f}{\partial x}\right|_{x(k-1), u(k-1)} (x(k) - x(k-1)) +$$

$$\left.\frac{\partial f}{\partial u}\right|_{x(k-1), u(k-1)} (u(k) - u(k-1)) \tag{5}$$

Next define new states and inputs in terms of deviations :

$$\hat{x}(k) = x(k) - x(k-1) \qquad \hat{u}(k) = u(k) - u(k-1)$$

Substituting into equation (5) gives :

$$\hat{x}(k+1) \approx \left.\frac{\partial f}{\partial x}\right|_{x(k-1), u(k-1)} \hat{x}(k) + \left.\frac{\partial f}{\partial u}\right|_{x(k-1), u(k-1)} \hat{u}(k) \tag{6}$$

Similarly for the output :

$$\hat{y}(k) \approx \left.\frac{\partial g}{\partial x}\right|_{x(k-1)} \hat{x}(k) \tag{7}$$

Let there be N different steady state operating points :

$$(x_{s,i}, u_{s,i}) \qquad (i = 1, \ldots, N)$$

and assume that a linear state space model $[A_i, B_i, C_i]$ has been obtained at each point. These local models may have been obtained either through identification, or by linearizing a first principles model. So during transition from the i'th regime to the j'th one, the jacobians change from $[A_i, B_i, C_i]$ to $[A_j, B_j, C_j]$, and the local steady state changes from $(x_{s,i}, u_{s,i})$ to $(x_{s,j}, u_{s,j})$. In order to approximate the plant's dynamics during transition, the jacobians given in Equations (6) and (7) need to be estimated on-line as they change. One way of doing this would be to keep interpolating between the state space matrices of the local models at each time step. This is similar in principle to linearizing the plant at discrete time steps during transition, except that linearization is replaced by interpolation between a finite number of models. Here the interpolation is done using model validity functions assigned to each of the local models. These functions make up a vector

$$p(t) = [p_1(t), \ldots, p_N(t)]^T \in R^N$$

and may be defined so that $p_i(t)$ maps the plant's outputs and inputs on to a measure of the validity of the i'th model :

$$p_i(t) : (y(t), u(t)) \rightarrow [0, 1] \tag{8}$$

where

$$
\begin{aligned}
p_i(t) &\rightarrow 1 \quad \textit{when the i'th model is valid,} \tag{9}\\
p_i(t) &\rightarrow 0 \quad \textit{otherwise}
\end{aligned}
$$

and

$$\sum_i^N p_i(t) = 1 \tag{10}$$

Equation (10) implies that as the plant moves into a region where one of the models becomes more trustworthy than the others, the other models lose their validity.

Now equations (6) and (7) can be approximated by a global LPV model interpolated between local models :

$$
\begin{aligned}
\hat{x}(k+1) &= \sum_i p_i(k) A_i \hat{x}(k) + \sum_i p_i(k) B_i \hat{u}(k) \\
\hat{y}(k) &= \sum_i p_i(k) C_i \hat{x}(k) \tag{11}
\end{aligned}
$$

The model validity functions $p_i(t)$ are estimated on-line using Bayesian techniques or on-line optimization (Banerjee et al. 1997a) in such a way that the outputs from equation (11) match the measurements from the true unknown nonlinear plant given by equation (2). The LPV model satisfies the linearization property, i.e. each local model is recovered exactly from the global model at the operating point at which it was obtained, provided the $p_i(t)$ are estimated properly.

4 MPC using LPV models

MPC will be carried out in two ways using LPV models. These are :

1. Linear Matrix Inequalities (LMI) approach. Stability guarantees exist for this approach, but treatment of constraints can be conservative.

2. Finite horizon constrained optimization at each time step, using the time-varying state space matrices given by equation (11). No stability guarantees are given, but treatment of constraints is not conservative.

4.1   LMI approach

Assume that the state matrix is parametrized by the time-varying vector $p(t)$, as given by the following relation :

$$A(p(k)) = \sum_{i=1}^{N} p_i(k) A_i \qquad (12)$$

This implies that $A(p(k))$ always belongs to the polytope:

$$A \in Co\{A_1, \ldots, A_N\} \qquad (13)$$

where Co denotes the convex hull. The subsequent controller design requires that all the local models satisfy the following conditions :

1. The models are stabilizable and detectable.

2. All models have the same B_i, C_i matrices. This is not a serious restriction as shown in Apkarian [1]. The condition will be met if the models include sensor and actuator dynamics. If not, then this condition can be satisfied by filtering the inputs and outputs. This will not significantly change the problem if the filter bandwidths are chosen to be sufficiently high.

Now we show step by step how to cast the LPV model (11) into a form suitable for MPC design. This requires the introduction of additional states as outlined below. For simplicity and without loss of generality, we will assume that all the models have the same C_i matrices.

1. Augment outputs. First the original state vector $\hat{x}(k)$ is augmented by the outputs $y(k)$, and the new local models are redefined accordingly:

$$A_{1,i} = \begin{bmatrix} A_i & 0 \\ CA_i & I \end{bmatrix} \qquad B_{1,i} - \begin{bmatrix} B_i \\ CB_i \end{bmatrix}$$

2. Augment reference trajectory dynamics for controller. Consider a single output without loss of generality and generate its reference trajectory by:

$$\begin{aligned}
x_r(k+1) &= \alpha x_r(k) \\
r(k+1) &= \alpha x_r(k) + r(k) \qquad (14) \\
x_r(0) &= x_{r0} \quad r(0) = 0 \qquad (15)
\end{aligned}$$

Output $y(k)$ is expected to track $r(k)$.

This gives the following local models for the state equations :

$$A_{2,i} = \begin{bmatrix} A_{1,i} & 0 & 0 \\ 0 & \alpha & 0 \\ 0 & \alpha & I \end{bmatrix} \qquad B_{2,i} = \begin{bmatrix} B_{1,i} \\ 0 \end{bmatrix}$$

3. Convert B_i to common B.

Let the input filter dynamics be given by (see Apkarian [1]):

$$\begin{aligned} z(k+1) &= A_z z(k) + B_z \tilde{u}(k) \\ \hat{u}(k) &= z(k) \end{aligned} \qquad (16)$$

The augmented state matrices of the local models are now :

$$A_{3,i} = \begin{bmatrix} A_{2,i} & B_{2,i} \\ 0 & A_z \end{bmatrix} \qquad B_{3,i} = \begin{bmatrix} 0 \\ B_z \end{bmatrix}$$

Here if $A_z = 0$ and $B_z = I$ then the pre-filter simply delays the inputs by one time step. Note that this step is necessary only in order for the local models to have common B matrices, and need not be carried out if this is already the case.

After the above steps the states of the LPV system are given by :

$$x(k+1) = \left[\sum_{i=1}^{N} p_i(k) A_{3,i} \right] x(k) + B\tilde{u}(k) \qquad (17)$$

$$(18)$$

where the augmented state is $x = [\hat{x} \; y \; x_r \; r \; z]$; and :

$$A_{3,i} \equiv \begin{bmatrix} A_i & 0 & 0 & 0 & B_i \\ CA_i & I & 0 & 0 & CB_i \\ 0 & 0 & \alpha & 0 & 0 \\ 0 & 0 & \alpha & I & 0 \\ 0 & 0 & 0 & 0 & A_z \end{bmatrix} \qquad B \equiv \begin{bmatrix} 0 \\ 0 \\ 0 \\ 0 \\ B_z \end{bmatrix} \qquad (19)$$

Now consider the following infinite horizon quadratic performance objective :

$$J_\infty(k) = \sum_{i=0}^{\infty} \left(x(k+i|k)^T Q_1 x(k+i|k) + \tilde{u}(k+i|k)^T R\tilde{u}(k+i|k) \right) \qquad (20)$$

where $Q_1 > 0$ and $R > 0$ are symmetric weighting matrices. At every sampling time k MPC performs the following optimization :

$$\min_{\tilde{u}(k+i|k)i=0,1....M} \quad \max_{A(k+i)\in Co[A_1,A_2.....A_N]} \quad J_\infty(k) \qquad (21)$$

This min-max optimization finds the optimal future inputs for the worst-case performance J_∞ among all the plants in the given polytope. Controller synthesis procedure is a modification of the one given by Kothare et al [5] taking into account that the parameter vector $p(k)$ is measured (or estimated) on-line and used as a scheduling-variable in the control law.

The methods seeks a quadratic function $V(x) = x^T P x, \quad P > 0$ such that at every sampling time k, it constitutes an upper bound on the objective function :

$$J_\infty(k) \le V(x(k|k)) \qquad (22)$$

Hence the MPC synthesis reduces to finding a time-varying controller gain

$$\tilde{u}(k+i|k) \quad = \quad Fx(k+i|k) \qquad (23)$$

$$= \quad \left[\sum_{i=1}^{N} p_i F_i\right] x(k+i|k) \qquad (24)$$

which minimizes the upper bound V $(x(k|k))$ at every sampling time k. Note that the gain is made explicitly dependent on the parameters p_i.

Theorem 1. Suppose that the state x$(k|k)$ and the parameter vector p$(k|k)$ of the LPV system (17) are measured at every sampling time k. Consider the unconstrained MPC. The controller that minimizes the upper bound V $(x(k|k))$ at sampling time k is given by

$$F(p) \quad = \quad \sum_j p_j F_j$$

$$= \quad \sum_j p_j Y_j Q^{-1} \qquad (25)$$

where $Q > 0$ and Y_j are obtained from the solution to the following set of LMIs when it exists :

$$\min_{\gamma,Q,Y_1,...,Y_N} \gamma \qquad (26)$$

subject to

$$\begin{bmatrix} 1 & x(k|k)^T \\ x(k|k)^T & Q \end{bmatrix} \ge 0 \qquad (27)$$

and

$$
\begin{bmatrix}
Q & QA_i^T + Y_i^T B^T & QQ_1^{\frac{1}{2}} & Y_i^T R^{\frac{1}{2}} \\
A_iQ + BY_i & Q & 0 & 0 \\
Q_1^{\frac{1}{2}}Q & 0 & \gamma I & 0 \\
R^{\frac{1}{2}}Y_i & 0 & 0 & \gamma I
\end{bmatrix} \geq 0, i = 1, 2, \ldots, N \quad (28)
$$

Proof. It follows similar steps to those given in [5].

4.1.1 Constraints

The input constraints can be either on $u(k) - u(k-1)$ or on $u(k)$. For now consider only those constraints on $u(k)$. However $u(k)$ does not appear as an input in (17) because the input to our state apace model (17) is $\tilde{u}(k)$. To get around this problem, we must use an extra state to keep track of $u(k)$, make this state an output, and then constrain this additional output. In other words the input constraint must be converted into an equivalent output constraint.

By definition $u(k) - u(k-1) = \hat{u}(k)$. Therefore $u(k+1) = u(k) + \hat{u}(k+1)$. Combining this equation with equation (16), we get :

$$
\begin{aligned}
u(k+1) &= u(k) + z(k+1) \\
&= u(k) + A_z z(k) + B_z \tilde{u}(k) \quad (29)
\end{aligned}
$$

Therefore the LPV model is augmented with this additional state to give the new state vector: $x = [\hat{x}\ y\ x_r\ r\ z\ u]$. The state space matrices are augmented accordingly:

$$
A_{3,i} \equiv
\begin{bmatrix}
A_i & 0 & 0 & 0 & B_i & 0 \\
CA_i & I & 0 & 0 & CB_i & 0 \\
0 & 0 & \alpha & 0 & 0 & 0 \\
0 & 0 & \alpha & I & 0 & 0 \\
0 & 0 & 0 & 0 & A_z & 0 \\
0 & 0 & 0 & 0 & A_z & I
\end{bmatrix}
\qquad
B \equiv
\begin{bmatrix}
0 \\ 0 \\ 0 \\ 0 \\ B_z \\ B_z
\end{bmatrix}
\quad (30)
$$

Similarly the input rate is given by $u(k+1) - u(k) = A_z z(k) + B_z \tilde{u}(k)$. In our simulations we have chosen $A_z = 0$ and $B_z = I$; then, the pre-filter simply delays the inputs by one time. In that case $u(k+1) - u(k) = \tilde{u}(k)$ which is the input of our state space model (17). Both input and output constraints can now be expressed in terms of certain LMIs generalizing the results in Boyd et al. [4] and Kothare et al. [5] to our case. Conditions are given below.

1. Constraints on the rate $u(k) - u(k-1) = \tilde{u}(k-1)$. Let us consider peak bounds on each of the inputs :

$$|\tilde{u}_j(k+i|k)| \leq \tilde{u}_{j,max}, \quad i \geq 0$$

A sufficient condition for satisfying this constraint is given by the existence of a symmetric matrix X such that:

$$\begin{bmatrix} X & Y_k \\ Y_k^T & Q \end{bmatrix} \geq 0, \quad (k = 1, \ldots, N)$$

with

$$X_{jj} \leq \tilde{u}_{j,max}^2, \quad j = 1, \ldots, n_u$$

2. Constraints on $u(k)$ and outputs. Since $u(k)$ is a state of our model, it is treated as an output constraint. A sufficient condition for a particular output $h_j = C_j x$ to be bounded i.e.

$$\max_{A(k+i) \in Co[A_1, A_2 \ldots A_n]} \|h_j(k+i)\|_2 \leq h_{j,max}^2 \quad i \geq 1$$

is given by the following LMIs (again one for each model): For the j'th output :

$$\begin{bmatrix} Q & (A_k Q + BY_k)^T C_j^T \\ C_j(A_k Q + BY_k) & h_{j,max}^2 I \end{bmatrix} \geq 0, \quad (k = 1, \ldots, N)$$

Theorem 2. The feasible receding horizon control law obtained from the solution of the LMIs stated in *Theorem 1* and the LMIs for input and output constraints asymptotically stabilizes the closed-loop LPV system.
Proof. It follows similar steps to those given in [5].

4.2 Finite Horizon Quadratic Programming Based MPC

The second MPC algorithm is similar to the traditional state space MPC. At every sampling time k an objective function similar to (20) where the states and inputs are defined by the LPV model (11) is minimized over a finite horizon subject to input and output constraints, and the control law is implemented as a receding horizon controller. At every sampling time k the estimator based on the LPV model gives the best state estimate, and the future predictions are obtained recursively from the same LPV model assuming that the parameter vector p remains constant and equal to its best estimate at time k. At the next sampling time using new plant measurements the parameter vector along with the LPV model states are updated and the new control input is computed. No maximization over the polytope of possible plants is performed. Therefore unlike the LMI approach which minimizes the worst-case objective value on-line, the method does not provide any stability guarantees although it offers several tuning parameters.

Since the theoretical requirements of the LMI approach are not a consideration here, there is no need to force the local models to have common B and C matrices. Therefore input filtering is not required. However the setpoint dynamics still need to be added to the LPV model.

5   Examples

We consider three physical examples. The first one is an LPV plant, thus no approximation is involved. The second example is a nonlinear CSTR approximated by an LPV model. The last example is a batch reactor showing how the method can be applied to batch plants.

Example 1

The first example is an LPV plant representing an angular positioning system taken from [5]. A motor drives a rotating antenna which is supposed to point in the direction of a moving object. This is to be achieved by manipulating the voltage input to the motor (u). The state equations are given by :

$$
\begin{aligned}
x(k+1) &= \begin{bmatrix} \theta(k+1) \\ \dot{\theta}(k+1) \end{bmatrix} \\
&= \begin{bmatrix} 1 & 0.1 \\ 0 & 1 - 0.1\alpha(k) \end{bmatrix} x(k) + \begin{bmatrix} 0 \\ 0.0787 \end{bmatrix} u(k) \\
&= A(\alpha)x(k) + Bu(k)
\end{aligned}
$$

where θ and $\dot{\theta}$ are the angular position and angular velocity of the antenna, respectively, both of which are neasured. The parameter $\alpha(k)$ is proportional to the coefficient of friction and is time-varying in the range 0.1 to 10. Therefore $A(\alpha(k))$ always belongs to the polytope :

$$
A \in Co\{A_1, A_2\} \tag{31}
$$

where :

$$
A_1 = \begin{bmatrix} 1 & 0.1 \\ 0 & 0.99 \end{bmatrix} \qquad A_2 = \begin{bmatrix} 1 & 0.1 \\ 0 & 0 \end{bmatrix}
$$

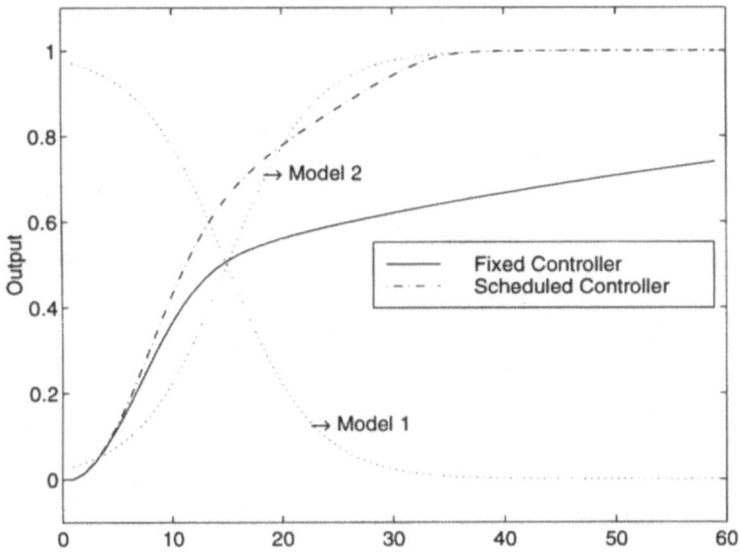

Figure 2: Set point tracking for scheduled and fixed controller

We assume that we have exact knowledge of the parameter vector so that we can test the theory under the conditions it has been derived. We consider a setpoint change in the angular position from 0 rad to 1 rad. The input voltage is constrained as $|u(k)| \leq 2V$. During this transition it is assumed that α increases smoothly between 0.1 to 10, the plant dynamics change from $[A_1, B]$ to $[A_2, B]$. The closed loop results are shown in figures 2, 3 and 4. The constraints are within their limits in a nonconservative way. It is shown that if the controller is not scheduled, the output response becomes very sluggish as the plant response enters into the validity region of the second model. This is due to the fact the the fixed controller gain is based on the first model $[A_1, B]$ and never adjusts to the dynamics of the second model $[A_2, B]$ which includes an integrator. On the other hand as shown in figure 4 the scheduled controller's gain is time-varying to meet the performance specifications and the input limits as the plant varies inside its polytope which results in superior performance.

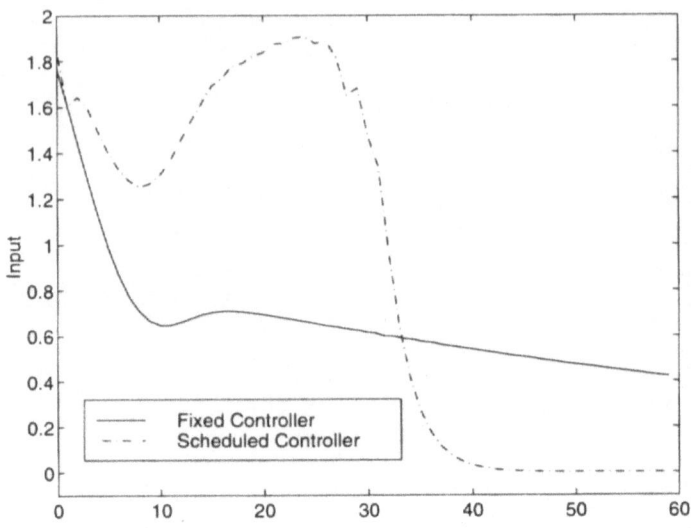

Figure 3: Inputs for scheduled and fixed controller

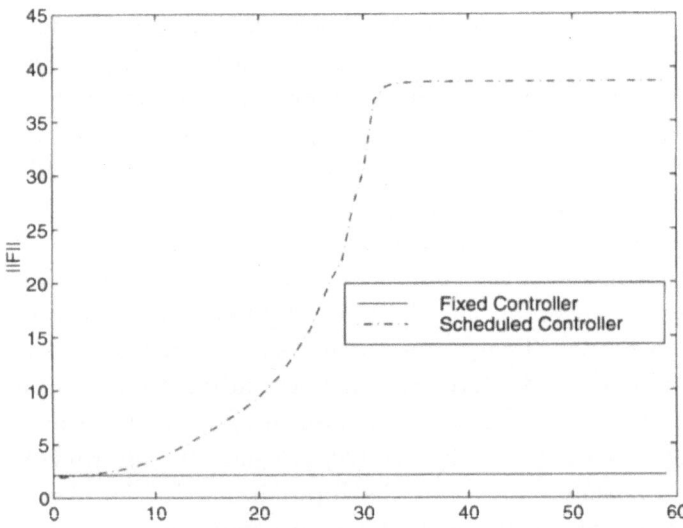

Figure 4: Norm of ontroller gain for scheduled and fixed controller

Example 2

The second test system to be studied is a continuously stirred tank reactor (CSTR) with a first-order exothermic reaction. The dynamic behavior is described by the following equations [9]:

$$\frac{dx_1}{dt} = -x_1 + D_a (1 - x_1) \exp\left(\frac{x_2}{1 + x_2/\gamma}\right)$$

$$\frac{dx_2}{dt} = -x_2 + BD_a (1 - x_1) \exp\left(\frac{x_2}{1 + x_2/\gamma}\right)$$

$$+ \beta (u - x_2)$$

$$y = x_2$$

where the states x_1 and x_2 are the dimensionless concentration and reactor temperature respectively, and t is dimensionless time. The input u is the dimensionless temperature of the cooling jacket surrounding the reactor. The constants are $D_a = 0.072$, $\gamma = 20$, $B = 8$ and $\beta = 0.3$. This system was chosen because it exhibits output multiplicity, as can be seen from the steady state curve given in figure 5. Furthermore the middle branch of the steady state curve is unstable, though the upper and lower branches are stable. Therefore the system has different dynamics in the regions

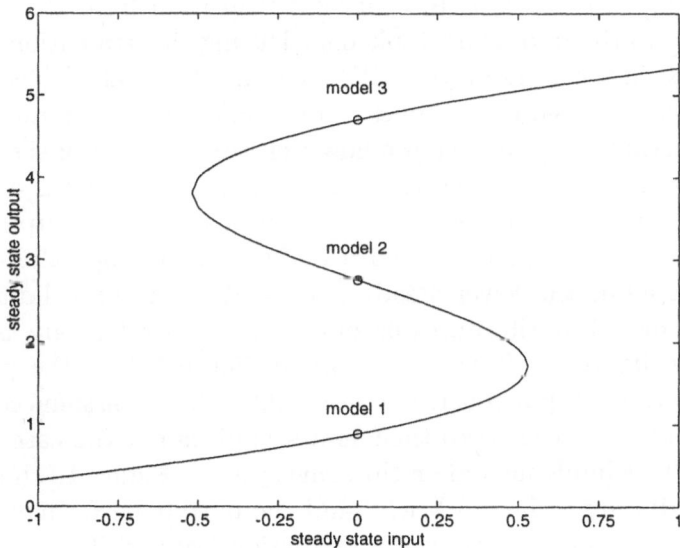

Figure 5: Steady state curve for CSTR, showing the model locations

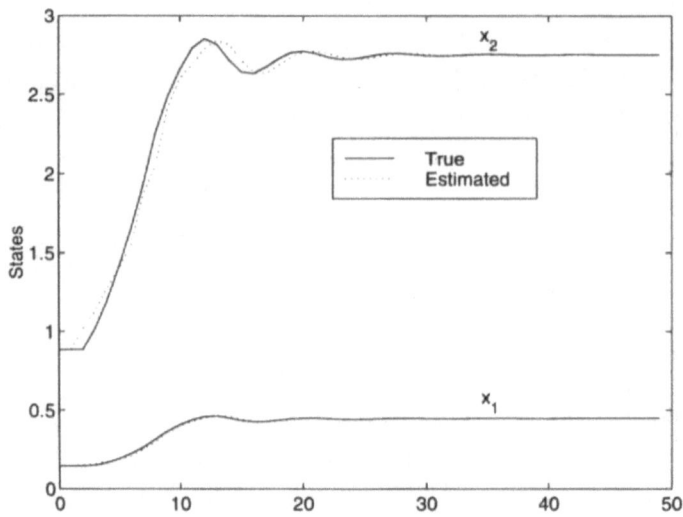

Figure 6: Estimated and actual values of the states

surrounding the three branches, and so is well suited for application of the theory.

The control objective is to achieve a transition from the lower stable steady-state to the middle unstable one. During this transition the moving horizon Bayesian estimator (MHBE) in Banerjee et al. [2] is used (with a window size of 5 sampling times) to estimate the states and the model validity functions. Figure 6 shows how well this is accomplished. The true states are simulated from the rigorous model; the estimates are based on the LPV model. The model validity functions and the output responses for different cases are given in figure 7. As we leave the validity region of the first model at the lower steady state and enter the validity region of the second model at the unstable point, the controller gain is scheduled as shown in figure 8. It is also shown in figure 9 how the performance deteriorates as the input constraint gets tighter. There is some conservatism as the inputs are not close to their limits. This is not the case when finite horizon MPC is implemented on the same system. Figure 10 shows that the input is on its upper bound briefly and the output performance improves as depicted in figure 11. Here the prediction horizon P=5; move horizon M=1; input weight uwt=0.5 and output weight ywt=0.5. The same MHBE is used with a window size of five sampling times.

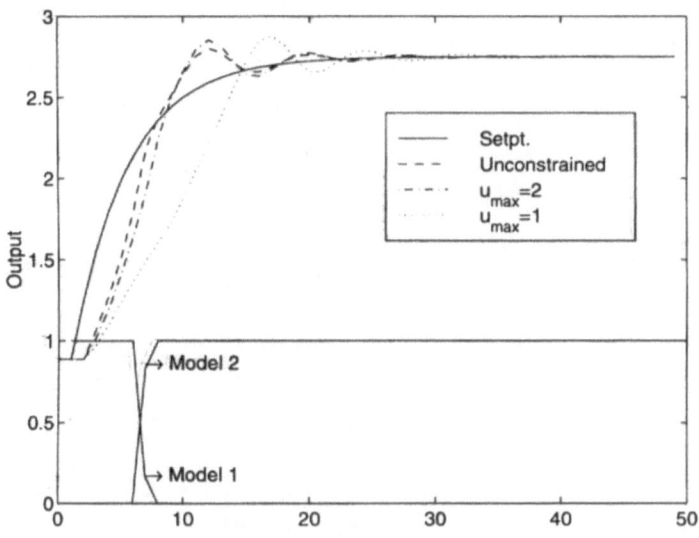

Figure 7: Outputs for different scheduled controllers

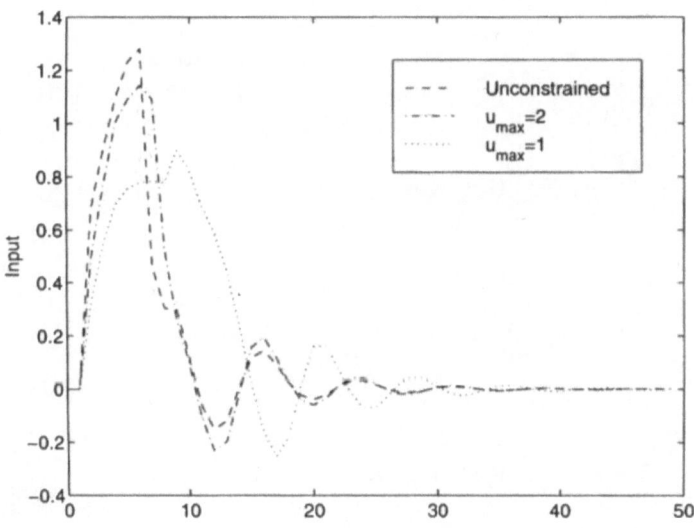

Figure 8: Inputs for different scheduled controllers

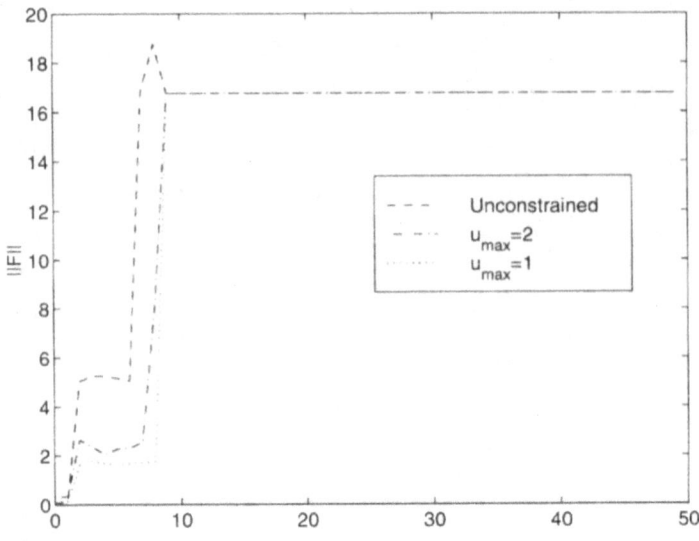

Figure 9: Norm of the controller gain for different scheduled controllers

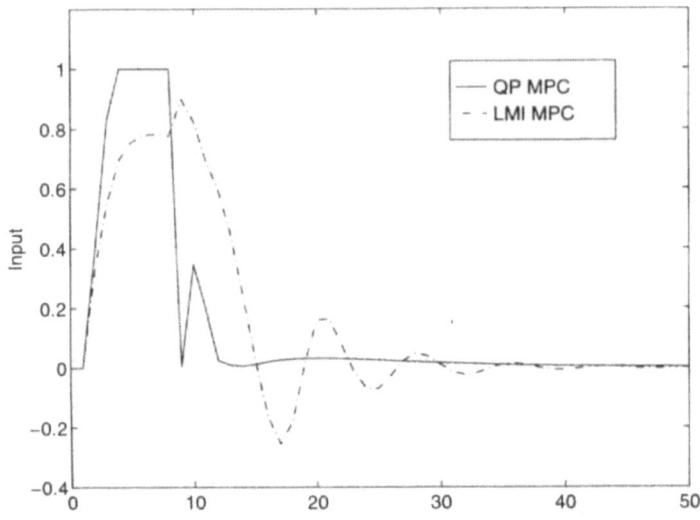

Figure 10: Inputs for different QP and LMI-based MPC

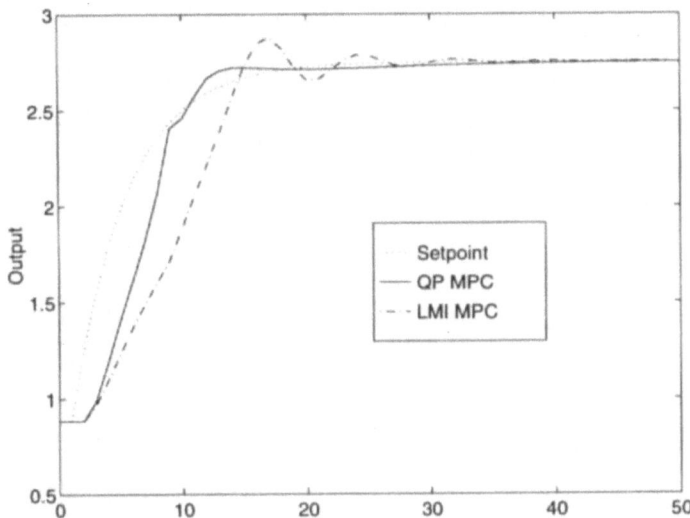

Figure 11: Outputs for different QP and LMI-based MPC

Finally it should be noted that the LPV based controller will work fine if the true plant's jacobian is inside the polytope of local models. In figure 12 we show the region in the state space for which this is true. It is seen that all the closed loop trajectories remain inside this region except some isolated excursions around the unstable point.

Example 3

The last example is a semibatch reactor ([7]) for free-radical polymerization of polymethyl methacrylate. A LPV formulation similar to that proposed in Section 3 is used to approximate the non-linear model. Unlike continuous systems, batch systems don't operate around steady-states. Therefore, the local models are obtained around several operating points on a given optimal profile. This profile could have either been obtained from a rigorous off-line optimization or given empirically based on information from previous batches. The time-varying batch system is then formulated in terms of deviations from this profile as follows:

$$
\begin{aligned}
x\left(k+1\right) &= f(x\left(k\right), u\left(k\right)) \\
&\approx f(x_o\left(k\right), u_o\left(k\right)) + \\
&\quad \left.\frac{\partial f}{\partial x}\right|_{x_o(k), u_o(k)} \left(x\left(k\right) - x_o\left(k\right)\right) +
\end{aligned}
\tag{32}
$$

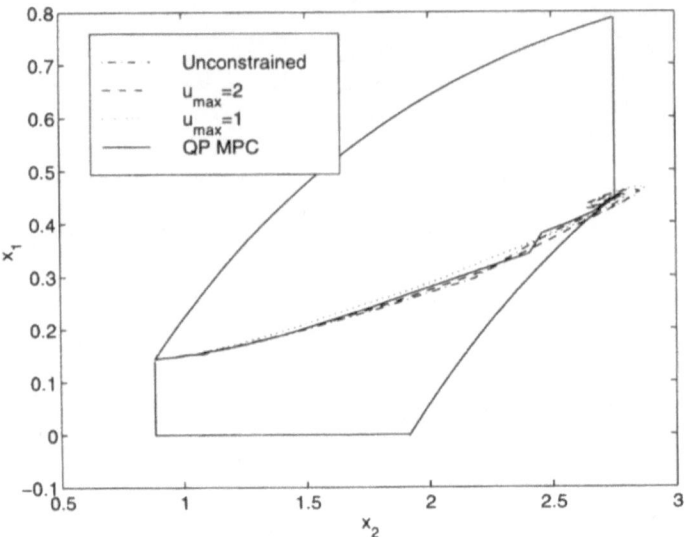

Figure 12: The region in which the plant jacobian is covered by the polytope

$$\frac{\partial f}{\partial u}\bigg|_{x_o(k),u_o(k)} (u(k) - u_o(k)) \tag{33}$$

$$= x_o(k+1) + \frac{\partial f}{\partial x}\bigg|_{x_o(k),u_o(k)} (x(k) - x_o(k)) +$$

$$\frac{\partial f}{\partial u}\bigg|_{x_o(k),u_o(k)} (u(k) - u_o(k))$$

$$y(k) = y_o(k) + \frac{\partial g}{\partial x}\bigg|_{x_o(k),u_o(k)} (x(k) - x_o(k)) +$$

$$\frac{\partial g}{\partial u}\bigg|_{x_o(k),u_o(k)} (u(k) - u_o(k)) \tag{34}$$

The system consists of a jacketed reactor with a mixture of initiator and solvent fed during the batch. The initiator is benzoyl peroxide and the solvent used is ethyl acetate. The monomer is methyl methacrylate which is charged only initially and not added during the batch. The manipulated variables are the feed (initiator plus solvent) flow rate (u_1) and the cooling jacket temperature (u_2). Measurements of monomer and solvent concentrations, reactor temperature and the number average molecular weight are assumed to be available every twenty seconds. The system is fully described by seven non-linear state equations comprising of the component mass balances, the energy balance and the equations of the first two moments for the molecular weight distribution. This system was chosen due to the strong nonlinearities posed by the gel effect and the exponential dependence of reaction rates on temperature.

The optimal profile is obtained off-line as follows:

$$\min_{u_1, u_2} \quad t_f \tag{35}$$

subject to

$$
\begin{aligned}
Conversion(t_f) &> 85\% \\
|NAMW(t_f) - 150000| &\leq 500 \\
330K < Reactor\ Temperature &< 340K \qquad 0 \leq t \leq t_f
\end{aligned}
$$

where t_f is the batch duration and $NAMW$ is the number average molecular weight of the polymer.

In the case of continuous systems, the control aim is regulation of outputs around a setpoint trajectory. In batch control targeting the product specifications at the end of the batch is more important.

For the given polymerization reactor, the control objective is to meet the desired $NAMW$ of 150000 at the end of the batch. One of the most common disturbances is that of a variation in the feed concentration. This is simulated here as a drop in the monomer concentration by 5% from its nominal value. If no feedback control action is taken and the optimal inputs computed from (Eq. 35) are fed to this reactor, the NAMW drops by almost 20% at the end of the batch to 120000 (figure 14) which violates the product specifications.

The off-line computed optimal trajectory for number average molecular weight ($NAMW$) is shown in figure 13. The circles on this trajectory show the times at which the model jacobians were obtained by linearizing the non-linear model. The first two models were obtained where the gel effect correlation is almost linear and the later models are obtained at points corresponding to times after 1.5 hr into the batch when the gel coefficient decreases more drastically and causes the reaction to accelerate.

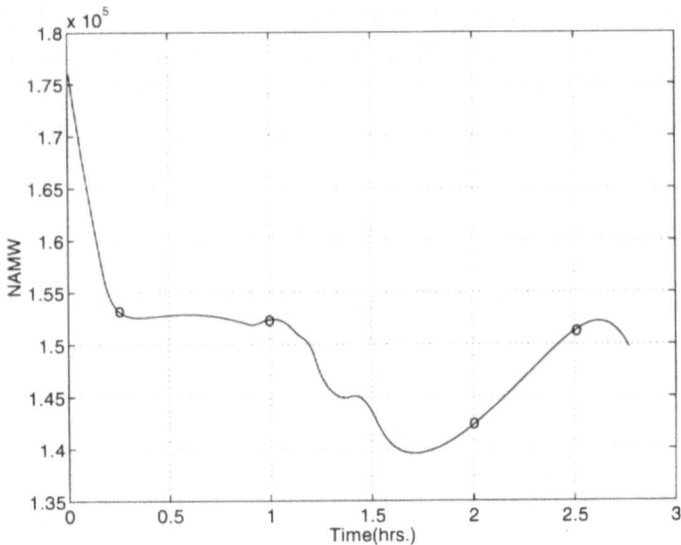

Figure 13: Off-line computed optimal profile for the semibatch polymerization reactor showing the model locations

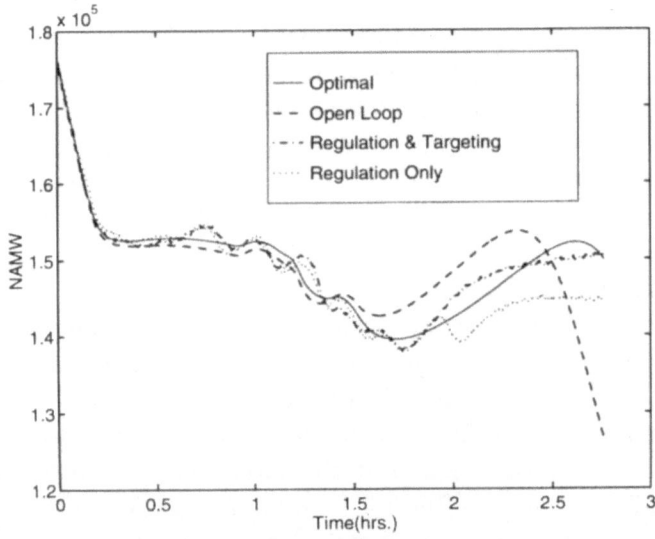

Figure 14: Different NAMW profiles for the semibatch polymerization reactor

Since the control objective here is targeting, a finite horizon MPC as described in Section 5 should predict till the end of the batch. During the initial phases of the batch, the dynamics are governed by the first two models. However since the validity regions of these models do not extend till the end of the batch, using them for targeting can produce erroneous results. Therefore these models can be used only for regulation purposes during the early part of batch.

Therefore, we suggest switching the MPC control objective between regulation and targeting in the following fashion. Initially MPC tries to bring the perturbed output trajectory back to the optimal trajectory and regulate it there until the validity regions of the last two models are entered, as detected by the values of the estimated validity functions. Since the end-point predictions of these models have been found to be acceptable after off-line validation studies, MPC switches to targeting. This is achieved by tuning Q, the matrix weighting the output predictions into the future, in a time-varying manner. By assigning more weight to the predictions towards the end of the batch, the MPC optimization aims at matching the polymer $NAMW$ with the desired value of 150000.

The initial condition disturbance is assumed to be measured and the model validity functions are estimated using the moving horizon bayesian estimator (MHBE) such that the LPV model dynamics match the nonlinear batch dynamics. The past data window size is equal to 20 sampling times for the estimator. The finite horizon MPC is then implemented with M=20, P=40, ywt=1 and uwt=[0.1 0.1] when the model is in the region of validity of the first two models. For the later models, the prediction horizon P is equal to the number of sampling times till the end of the batch. Also, the ywt is changed to 0.01 for all the output predictions except for the last prediction for which the ywt is maintained at 1. The feed flow rate is constrained to a maximum of 0.06 m^3/hr and the jacket temperature is maintained within ± 10% of 340K.

The model validity functions from the MHBE are shown in figure 15. As expected, it picks up the first three models in the order in which they were obtained. But the estimator picks up model 4 for a while before switching back to model 3. This is due to the control action which causes the states towards the end of the batch to be closer in magnitude to the states for which model 3 was obtained and hence, the nonlinear dynamics are better matched by that model.

Regulating the batch in the validity region of the first two models and switching to targeting in the validity region of the third and fourth models

82

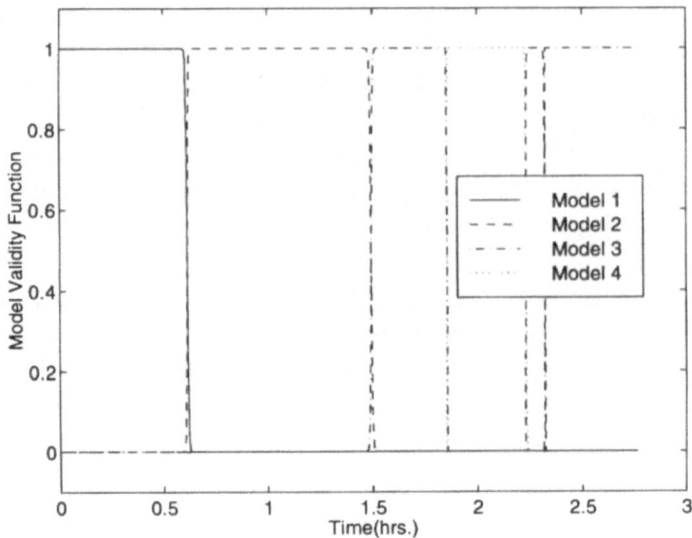

Figure 15: Model validity functions for the semibatch polymerization reactor for the case of regulation and targeting

helps in achieving the desired $NAMW$ of 150000 (figure 14). It's also seen that only regulation would have fallen short of achieving this target value. The optimal inputs and closed loop inputs from this MPC formulation are shown in figure 16. In the initial time period , it's seen that the inputs deviate just enough from the optimal profile such that the NAMW tracks the optimal profile. At a time of about 1.5 hr., when the batch enters the region of validity of the third model (figure 15), the MPC switches to targeting. This is reflected in the inputs that deviate considerably from the optimal profile and drop to their lower bounds. This results in lower initiator concentrations and lower reaction rates. As a consequence, fewer new chains are formed and the NAMW of the polymer increases. Towards the end of the batch, the inputs again start increasing as the control action earlier was too aggressive and if continued, would have resulted in a NAMW above the desired value of 150000.

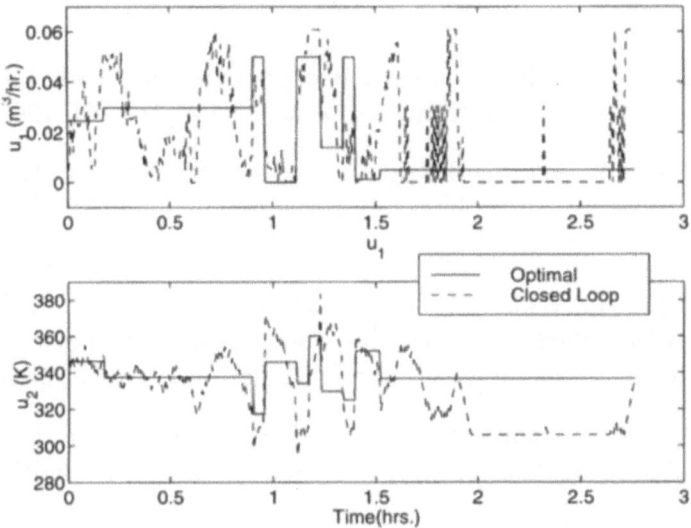

Figure 16: Inputs to the semibatch polymerization reactor for the case of regulation and targeting

6 Conclusion

This paper has considered control of nonlinear plants that undergo transitions between different operating conditions. The control strategy is based on on-line scheduling of multiple linear models using MPC. The underlying MPC model is an LPV model constructed by interpolating between the linear local models using time-varying validity functions. A stabilizing constrained receding horizon controller is developed for such systems which schedules on-line the local models. On-line optimization requires the solution of certain Linear Matrix Inequalities. Applications demonstrate both the new LMI-based method and the more traditional QP-based method using LPV models.

7 Acknowledgements

The authors gratefully acknowledge the financial support of E. I. DuPont de Nemours & Co., Inc. and the National Science Foundation.

References

[1] P. Apkarian, P. Gahinet, and G. Becker. Self-scheduled H_∞ control of linear parameter-varying systems. *Automatica*, 31:1251–1261, 1995.

[2] A. Banerjee, Y. Arkun, B. Ogunnaike, and R. Pearson. Estimation of nonlinear systems using linear multiple models. *AIChE*, 43:1204–1226, 1997.

[3] A. Banerjee, Y. Arkun, B. Ogunnaike, and R. Pearson. H_∞ control of nonlinear processes using multiple linear models. In Local Approaches to Nonlinear Modeling and Control, Taylor and Francis, London, 1997.

[4] S. Boyd, L. El. Ghaoui, E. Feron, and V. Balakrishnan. *Linear Matrix Inequalities in System and Control Theory*. SIAM, Philadelphia, 1994.

[5] M. V. Kothare, V. Balakarishnan, and M. Morari. Robust constrained model predictive control using linear matrix inequalities. *Automatica*, 32:1361–1379, 1996.

[6] A. Packard. Gain scheduling via linear fractional transformations. *Syst. Contr. Lett.*, 19:271–280, 1992.

[7] T. Petersen, E. Hernandez, Y. Arkun, and F. J. Schork. A nonlinear dmc algorithm and its application to a semibatch polymerization reactor. *Chemical Engineering Science*, 47(4):737–753, 1992.

[8] J. P. Shamma and M. Athans. Gain scheduling : Potential hazards and possible remedies. In *Proceedings of the American Control Conference*, pages 516–521, January 1991.

[9] A. Uppal, W. H. Ray, and A. B. Poore. On the dynamic behavior of continuous stirred tank reactors. *Chemical Engineering Science*, 29(967), 1974.

Part II

NONLINEAR MODEL BASED CONTROLLER SYNTHESIS

INSIGHTS INTO THE RELATIONSHIPS BETWEEN LINEAR AND NONLINEAR MODEL BASED CONTROL AND ISSUES FOR FURTHER RESEARCH

R. BERBER
Department of Chemical Engineering
University of Ankara, Tandoğan
06100 Ankara, Turkey

C. BROSILOW
Department of Chemical Engineering
Case Western Reserve University
Cleveland, OH 44106-7217 USA

Abstract
We show that the control laws for linear and nonlinear SISO IMC and their various implementations can all be obtained by the same approach. This insight leads to computationally simple control laws for linear and nonlinear systems in state space realizations. The proposed linearizing model state feedback controller makes the nonlinear system track a linear filter with only one tuning parameter, eliminating the need for an external linear controller. This derivation of the control law also suggests new strategies for systems with right half plane transmission zeros, and other ill behaved zeros. Simulated applications of the proposed controller to some literature problems illustrate the approach.

1. Introduction

A Chemical Engineer encountering the field of nonlinear control for the first time is quite likely to be overwhelmed by the now extensive literature and multitude of various approaches. This is likely to be true even if said engineer has a good background in linear model based control methods. The aim of this paper is to help such an engineer by showing that many, if not most, of the control laws for both linear and nonlinear internal model control (IMC) systems can be obtained in exactly the same way. Further, while the linear and nonlinear model state feedback implementations of such control system have very nearly the same structures, the differences between the two are also instructive. Such differences lead us to pose several questions which we hope will lead to interesting and useful discussions at the NATO Advanced Study Institute on Nonlinear Model Based Control at which this paper is being presented. Our paper concludes with several simulation examples which hopefully illustrate some of the ideas presented in the paper.

87

R. Berber and C. Kravaris (eds.), Nonlinear Model Based Process Control, 87-114.
© 1998 *Kluwer Academic Publishers.*

88

We do not claim that the insight that the control laws for linear and nonlinear systems can be obtained using the same approach is unique to the authors. Indeed, when the authors bounced this idea off of Professors Kravaris and Soroush, they were already familiar with it, and Professor Soroush kindly provided us with a copy of a recent, as yet unpublished paper [14] where the idea is presented for model state feedback systems. Nonetheless, we have troubled to write this paper in hope that it will help in disseminating an important unifying concept.

2. Derivation of Feedback Control Laws

In this section we show that the control laws for linear and non-linear SISO IMC [3, 14] and their various implementations can all be obtained by the same approach. This approach is to: (1) Choose the control effort to force the pth derivative of the process model to track the highest (i.e. the pth) derivative of differential equation which defines the desired closed loop behavior[1], and (2) Set the states of the differential equation for the desired output to the model output and its first p-1 derivatives. We start by showing that the above approach yields any desired IMC control law.

2.1. DEVELOPMENT OF THE IMC CONTROL LAW

The standard IMC block diagram is given in Figure 1a.

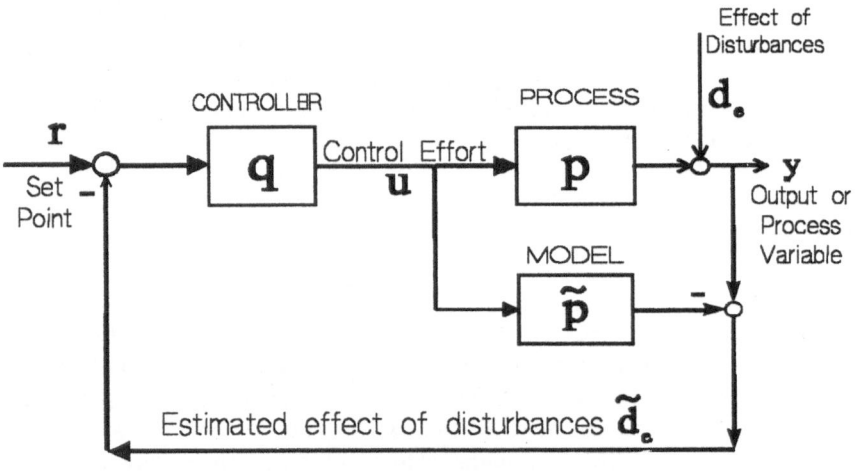

Figure 1a. IMC Block Diagram

For a process model of the form

$$\tilde{y}(s) = \tilde{p}(s)u(s) \tag{1a}$$

---

[1] We will define the pth order more carefully when we actually need it.

$$\tilde{p}(s) = \frac{N(s, e^{-\alpha_j s})}{D(s)} e^{-Ts} \qquad (1b)$$

where: N(s) = A (n-r)th order polynomial in s possibly containing terms of the form

$s^j e^{-\alpha_j s}$ except that (n-r)th order term in s does not contain a dead time.

D(s) = A nth order polynomial in s whose roots are all in the left half plane

r = The relative order of p(s)

the IMC controller, q(s,ε), is formed by inverting part of p(s) and multiplying this inverse by a filter of the form 1/(ε s+1)^m, where m is chosen to make the controller strictly proper (i.e. the relative order of the controller is zero). That is,

$$q(s, \varepsilon) = \frac{D_\mu(s)}{N_m(s, e^{-\alpha_j s})(\varepsilon s + 1)^m} \qquad (2a)$$

where; $D_\mu(s)$ is the portion of the denominator of the model that the IMC controller inverts. That is,

$$D(s) \equiv D_\mu(s) D_v(s) \qquad (3a)$$

In general, $N_m(s, e^{-\alpha_j s})$ has no right half plane zeros, may cancel some or all of the left half plane zeros of $N(s, e^{-\alpha_j s})$, and may have other zeros which are the reflection of the right half plane zeros of $N(s, e^{-\alpha_j s})$ about the imaginary axis (e.g. the reflection of the right half plane zero (-s +1) about the imaginary axis is (s+1)). To deal with all these possibilities we let,

$$N_m(s, e^{-\alpha_j s}) \equiv N_\mu(s, e^{-\alpha_j s}) N_\eta(s, e^{-\alpha_j s}) \qquad (3b)$$

$$N(s, e^{-\alpha_j s}) \equiv N_\mu(s, e^{-\alpha_j s}) N_v(s, e^{-\alpha_j s}) \qquad (3c)$$

where; N_μ contains the zeros of N that are to be cancelled by the IMC controller.

N_η contains the reflected right half plane zeros of N.

N_v contains the right half plane zeros of N, and/or zeros of N that we do not wish to cancel.

Finally, in order that the IMC control system not have an offset, the gain of $q(s)$ must be the inverse of the model gain, or

$$D_\mu(0)/N_m(0) = N(0)/D(0) \qquad (3d)$$

The output to set point response of the control system for a perfect model is therefore given by:

$$y_d(s) = \tilde{p}(s)q(s) \tag{4a}$$

or:

$$y_d(s) = \frac{N_v(s, e^{-\alpha_j s})e^{-Ts}}{N_\eta(s, e^{-\alpha_j s})D_v(s,)(\varepsilon s + 1)^m} y_{sp} \tag{4b}$$

where: $y_d(s)$ is the desired, or perfect model response.

The IMC control law given by (2a) can also be obtained by equating the model output, $y(s)$, given by (1), to the desired response, $y_d(s)$, given by (4). Such an operation, however, requires cancellation of term $N_v(s, e^{-\alpha_j s})e^{-Ts}$ with the same factors in (1). If $N_v(s, e^{-\alpha_j s})$ contains any right half plane zeros, and the cancellation is not exact, then the resulting controller will be unstable. A better method is to equate a modified output to a modified desired output, both formed by dropping the undesirable zeros. That is, let

$$y^*(s) \equiv \frac{N_\mu(s, e^{-\alpha s})}{D(s)} u(s) \tag{5}$$

$$y_d^*(s) \equiv \frac{1}{N_\eta(s, e^{-\alpha s})D_v(s)(\varepsilon s + 1)^m} y_{sp} \tag{6}$$

The relative order of (5) is the same as the relative order of (6), which is the same as its order. Let p be the order of (6) (i.e. $p = m + \dim D_v + \dim N_\eta$). The rule that we wish to demonstrate is that equating the pth derivative of $y^*(t)$, defined by (5) to the pth derivative of $y_d^*(t)$ defined by (6) gives the IMC control law given by (2a). This can be done relatively easily in the Laplace domain with the aid of some tedious algebra.

Let

$$N_\eta(s, e^{-\alpha_j s})D_v(s)(\varepsilon s + 1)^m \equiv \beta_p s^p + f(s, e^{-\alpha_j s}) \tag{7}$$

where $f(s, e^{-\alpha_j s})$ = a (p-1) order polynomial in s containing terms of the form $s^j e^{-\alpha_j s}$.

The Laplace transform of the pth derivative of desired response, $y_d^*(s)$, from (6) and (7) is:

$$s^p y_d^*(s) = (-f(s, e^{-\alpha_j s})y_d^*(s) + N_\eta(s, e^{-\alpha_j s})e^{-Ts}y_{sp})/\beta_p \tag{8}$$

We can substitute y*(s) from (5) for $y_d^*(s)$, in (8) because the term $f(s,e^{-\alpha_j s})$ will differentiate $y_d^*(t)$ at most p-1 times. This gives:

$$s^p y_d^*(s) = (-f(s,e^{-\alpha_j s})y^*(s) + N_\eta(s,e^{-\alpha_j s})e^{-Ts}y_{sp})/\beta_p \qquad (9)$$

Now substituting for y*(s) from (5) into (9) gives:

$$s^p y_d^*(s) = (-\frac{f(s,e^{-\alpha_j s}N_\mu(s,e^{-\alpha_j s})u(s)}{D(s)} + N_\eta(s,e^{-\alpha_j s})y_{sp})/\beta_p \qquad (10)$$

Equating (10) to the pth derivative of the model output, y*(s) from (5) gives

$$\frac{s^p N_\mu(s,e^{-\alpha_j s})u(s)}{D(s)} = \frac{-f(s,e^{-\alpha_j s})N_\mu(s,e^{-\alpha_j s})u(s)}{\beta_p D(s)} + N_\eta(s,e^{-\alpha_j s})y_{sp} \qquad (11)$$

Collecting terms in u(s) in (11) yields:

$$\frac{(\beta_p s^p + f(s,e^{-\alpha_j s})N_\mu(s,e^{-\alpha s})}{D(s)}u(s) = N_\eta(s,e^{-\alpha s})y_{sp} \qquad (12)$$

Using the definitions given by (3) and (7) and solving for u(s) gives:

$$u(s) = \frac{D_\mu(s)y_{sp}}{N_m(s,e^{-\alpha s})(\varepsilon s+1)^m} \qquad (13)$$

$$\equiv q(s)y_{sp} \qquad \text{from (2a)}$$

2.2. MODEL STATE FEEDBACK

The model state feedback structure for constrained linear processes demonstrating attractive constraint handling capabilities was proposed by Coulibaly et al. [2]. The same structure for unconstrained nonlinear processes was first identified by Kravaris and Daoutidis [9]. The properties of this structure was later studied in Kravaris et al. [10]. The model state feedback controller of Figure 1b forms the IMC control effort from a linear combination of the states of the model rather than from the output of lead-lag controller as in a standard implementation [3]. The advantage of such an implementation is that it makes use of the model's knowledge of past controls which is contained in its state. Thus, a model state feedback controller automatically compensates for control effort saturation, unlike the standard lead-lag IMC controller implementation which computes its output undeterred by whether or not the computed controls are actually applied.

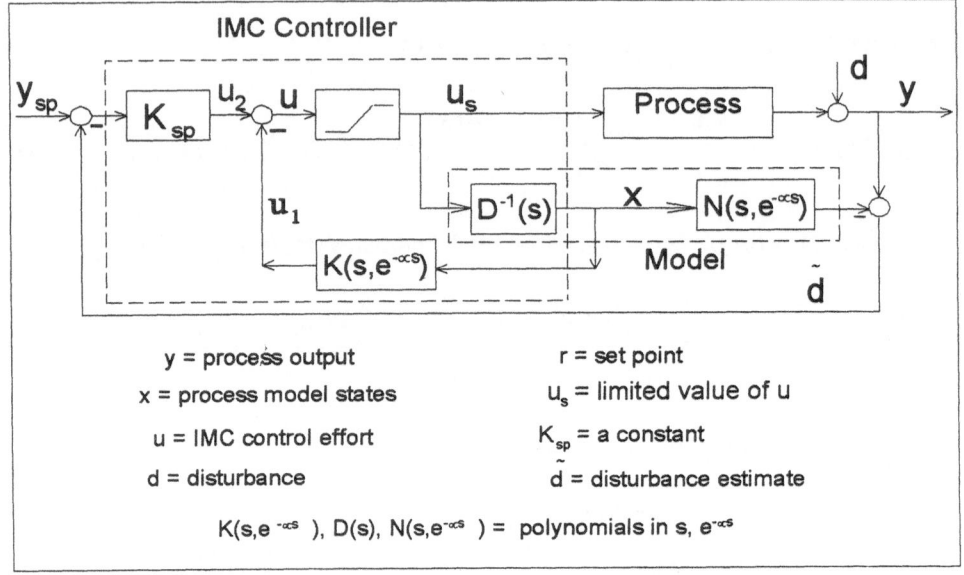

Figure 1b. The Model State Feedback diagram

The model state feedback control law from Figure 1b is given by:

$$u(s) = -K(s, e^{\alpha s})x(s) + K_{sp}(y_{sp}(s)-d(s)) \qquad (14)$$

The inverse Laplace transform of the term $K(s, e^{-\alpha s})x(s)$ is a linear combination of the output $x(t)$ and its n-1 derivatives delayed as necessary, with $K(s, e^{-\alpha s})$ given by:

$$K(s, e^{-\alpha s}) = K_{sp}D_v(s)N_m(s, e^{-\alpha s})(\varepsilon s+1)^m - D(s) \qquad (15)$$

where K_{sp} is chosen so that the nth order terms in $K(s, e^{-\alpha s})$ cancel.

The control law given by (14) and (15) yields exactly the same control effort as that given by (2a) in the absence of constraints. Also, just like (2a), $K(s, e^{-\alpha s})$ can be obtained by choosing the control effort, $u(t)$, to force the pth derivative of the process model output   (p = m + dim D_v + dim N_η) to track the pth derivative of $y_d(t)$, and setting the states of the differential equation for $y_d(t)$ to the states of the model output.

Figure 2 is the state space model state feedback diagram equivalent to Figure 1b, except that the dependence of the elements of the row vectors c and K_x on delays has been suppressed to simplify notation. The process model in Figure 2 is

$$\dot{x}(t) = A x(t) + b u(t)$$
$$y(t) = c x(t)$$

(16)

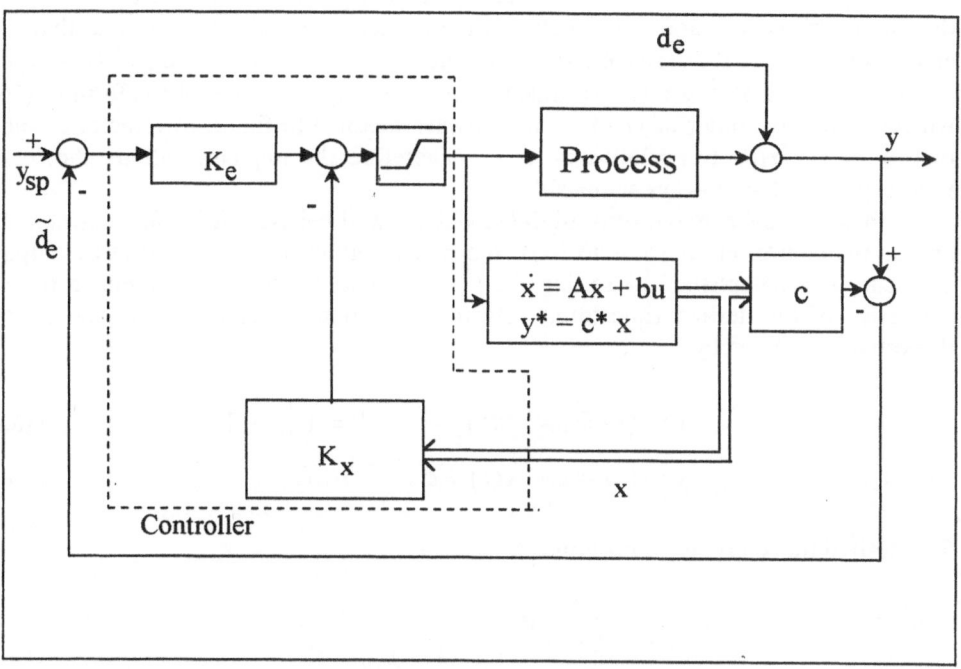

Figure 2. Model State Feedback diagram in state space form (without dead times)

If the model state, $x(t)$, in Figure 2 is defined so that $x_1(t) \equiv y^*(t)$ and $x_{j+1}(t) \equiv \dfrac{d^j}{dt^j} y^*(t), j = 1..n-1$ then the Laplace transform of $y^*(t)$ is given by $y^*(s) = u(s)/D(s)$ just as in Figure 1b. In this case the matrices A,   b, and c* are in the form:

$$A = \begin{pmatrix} 0 & 1 & 0 & . & . \\ 0 & 0 & 1 & 0 & . \\ . & & & & \\ . & & & & \\ -\alpha_0 & -\alpha_1 & . & .. & -\alpha_{n-1} \end{pmatrix} \qquad b = \begin{pmatrix} 0 \\ 0 \\ . \\ . \\ b_n \end{pmatrix} \qquad c^* = \begin{pmatrix} 1 & 0 & . & 0 \end{pmatrix} \qquad (17)$$

and $b_n/\alpha_0 = D^{-1}(0)$

More generally, however, all that is required for the equivalence of Figures 1b and 2 is that the matrices A, b and c^* be such that the relative order of $y^*(t)$ is n, and that the eigenvalues of A be the zeros of $D(s)$. The relative order of $y^*(t)$ is n if $c^*A^j b = 0$ for $j = 0 .. n-2$ and $c^*A^{(n-1)} b \neq 0$. For linear systems, given any model of the form of (16) where the relative order of $y(t)$ is r, it is always possible to find a row vector c^* such that $y^*(t) = c^*x(t)$ has relative order n. Therefore the implementations shown in Figures 1b and 2 are always achievable.

In those cases where the zeros of the model are well behaved, it is often desirable to choose the control effort so as to force the model output to track the desired output. This can be accomplished by setting the rth derivative of the model output to the rth derivative of the desired trajectory.[2] By the definition of relative order the model derivatives are given by:

$$y^{(k)}(t) = cA^k x(t) \qquad k = 1..r-1 \qquad (18a)$$
$$y^{(r)}(t) = cA^r x(t) + cA^{(r-1)}bu(t) \qquad (18b)$$

The desired trajectory, $y_d(t)$, is defined by

$$\left(\varepsilon \frac{d}{dt} + 1\right)^r y_d(t) = \dot{y}_{sp} - \tilde{d}_e \qquad (19a)$$

Therefore

$$y_d^{(r)}(t) = \left(\dot{y}_{sp} - \hat{d}_e\right)/\varepsilon^r - \frac{1}{\varepsilon^r} \sum_{j=1}^{r} \binom{r}{j} \varepsilon^{r-j} y_d^{(r-j)}(t) \qquad (19b)$$

where

$$\binom{r}{j} \equiv \frac{r!}{(j!)(r-j)!} \qquad and \ 0! \equiv 1 \qquad (19c)$$

---

[2] Under the foregoing conditions, dim D_v = dim N_v = 0, and p = m = r.

Equating the rth derivatives of $y(t)$ and $y_d(t)$, and solving for $u(t)$ gives (after recalling that $y_d^{(j)}(t) = y^{(j)}(t)$ for $j=1...r-1$):

$$u(t) = \frac{\left(y_{sp} - \tilde{d}_e\right) - \left(\sum_{j=0}^{r} \binom{r}{j} \varepsilon^{r-j} c A^{r-j}\right) x(t)}{\varepsilon^r c A^{r-1} b} \tag{20}$$

Therefore in Figure 2

$$K_e = \frac{1}{\varepsilon^r c A^{r-1} b} \tag{21a}$$

and

$$K_x = \left(\sum_{j=0}^{r} \binom{r}{j} \varepsilon^{r-j} c A^{r-j}\right) / \left(\varepsilon^r c A^{r-1} b\right) \tag{21b}$$

2.3. NONLINEAR SYSTEMS

A delay free nonlinear process is generally described by the following;

$$x'(t) = f(x) + g(x)u(t) \tag{22a}$$
$$y(t) = h(x) \tag{22b}$$

where $x(t)$ is the state vector of dimension n, $f(x)$ and $g(x)$ are vectors of dimension n, and $u(t)$, $h(x)$ and $y(t)$ are scalar fields. The process output is $y(t)$, and the control effort is $u(t)$.

The derivatives of the output of a nonlinear system can be obtained in terms of the Lie derivatives. For a detailed mathematical background, the reader is referred to the book by Isidori [6] or excellent review papers by Kravaris and Kantor [11, 12]. For a process of relative order r, the derivatives of the output are:

$$y'(t) = L_f h(x)$$
$$y''(t) = L_f^2 h(x)$$
$$....$$
$$y^{(r-1)}(t) = L_f^{r-1} h(x) \tag{23}$$
$$y^{(r)}(t) = L_f^r h(x) + (L_g L_f^{r-1} h(x))u(t)$$

and

$$L_g L_f^k h(x) = 0 \quad for \; 0 \le k \le r - 2 \tag{24}$$

where $\quad L_f h(x) \equiv \left\langle f, \dfrac{\partial h}{\partial x} \right\rangle = \displaystyle\sum_{i=1}^{n} f_i \dfrac{\partial h}{\partial x_i}$

$$L_f^2 h(x) \equiv L_f(L_f h(x))$$

Assuming that the process given by (22) is invertible (i.e. that the zeros of the process output linearized about the trajectory defined by $u(t)$ are not in the right half plane, or near the imaginary axis in the left half plane); then the control, $u(t)$, that forces the output, $y(t)$ to track the trajectory defined by (19) is given by:

$$u(t) = \frac{\left(y_{sp} - \tilde{d}_e\right) - \left(\displaystyle\sum_{j=0}^{r} \binom{r}{j} \varepsilon^{r-j} L_f^{r-j} h(x)\right)}{\varepsilon^r L_g L_f^{r-1} h(x)} \tag{25}$$

The term 'invertible systems' here is to be interpreted as 'systems with stable inverses'. The control law given by (25) was obtained as before by equating the rth derivatives of the output and the desired trajectory[3], and substituting the derivatives of the output for the derivatives of the desired trajectory for all derivatives of order less than r. Figure 3 shows the block diagram implementation of the above control law. The operators in Figure 3 are:

$$K_e = \frac{1}{\varepsilon^r L_g L_f^{r-1} h(\cdot)} \tag{26a}$$

and

$$K_x = \left(\sum_{j=0}^{r} \binom{r}{j} \varepsilon^{r-j} L_f^{r-j} h(\cdot)\right) \Big/ \left(\varepsilon^r L_g L_f^{r-1} h(\cdot)\right) \tag{26b}$$

---

[3] Again, the foregoing assumptions lead to p = r.

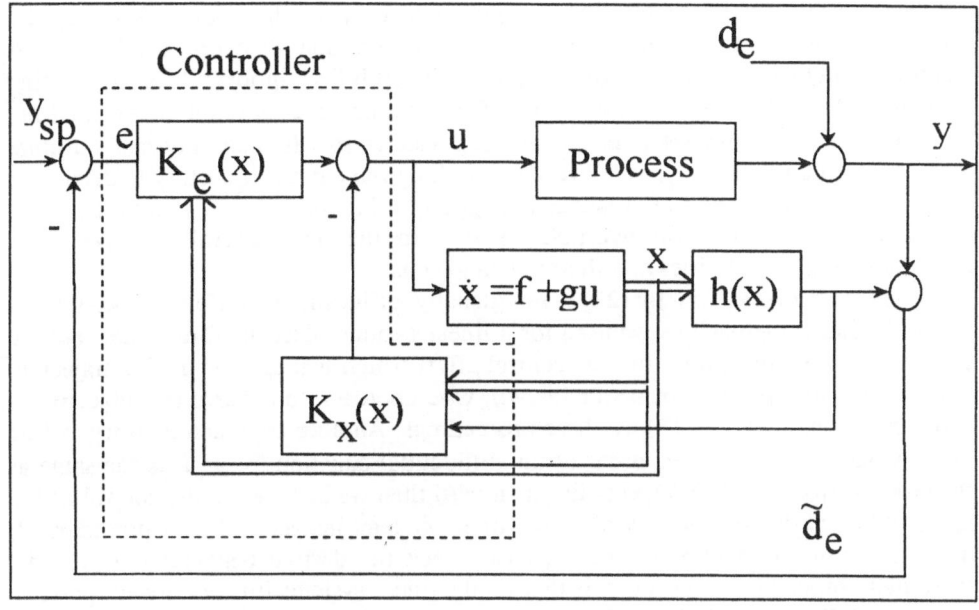

Figure 3. Linearizing Model State Feedback for invertible nonlinear processes

3. Issues Requiring Further Research

Up to this time, it has been common practice to compute the Lie derivatives required in (25) analytically. However, this is convenient only for relatively simple nonlinear systems such as pH control and simple reactor control. There is another possible approach. Compute the required derivatives by automatic differentiation [1]. Quoting from the foregoing reference, "Automatic differentiation techniques rely on the fact that every function, no matter how complicated, is executed on a computer as a (potentially very long) sequence of elementary operations such as additions, [and] multiplications, and elementary functions such as sin and cos. By applying the chain rule

$$\frac{\partial}{\partial t} f(g(t))\Big|_{t=t_o} = (\frac{\partial}{\partial s} f(s)\Big|_{s=g(t_o)})\left(\frac{\partial}{\partial t} g(t)\Big|_{t=t_o}\right)$$

over and over again to the composition of those elementary operations, one can compute, in a completely mechanical fashion, derivatives of f that are correct up to machine precision." The foregoing reference also describes a code, ADIFOR2.0, for the automatic differentiation of Fortran 77 programs. Thus far it appears that the code has been used only to generate Jacobians. Could it and should it be extended to generate Lie derivatives? Assuming that such an extension is possible how practical will it be for on-line controllers?

The control law given by (25) is applicable only to those processes that are invertible. That is, the control effort must not be unstable or overly oscillatory. A significant problem is how does one determine invertibility a-priori? It is easy to find examples [15] where the process is invertible for a positive set point change and not invertible for a negative set point change and vice-versa. Is there any method other than computing the local zeros at every point along the current trajectory to determine invertibility? We have no alternatives to suggest at this time. Further, what does one do when encountering non-invertible zeros? In the next several paragraphs we consider several possibilities and their pros and cons.

When the process given by (22) is not globally, or locally, invertible, we would like to use the same approach as we used for a linear system. That is, form a new output, $y^*(t)$, that is invertible, and find the control effort which makes $y^*(t)$ track a trajectory that leads to the specified behavior for $y(t)$. One approach to achieve this objective is given by Kravaris *et al.* [13] in these proceedings. Another approach is to try to find an $h^*(x)$ so that the relative order of the differential equation for $y^*(t)$ is the same as the order of $x(t)$ (*i.e.* n). If $h^*(x)$ is linear in $y^*(t)$ then we can choose the control $u(t)$ to make $y(t)$ track the trajectory of a desired linear system, just as for linear processes. If $h^*(x)$ is nonlinear in $y^*(x(t))$, then $y(t)$ can track the desired behavior only locally. Even more problematic, however, is that unlike linear systems it is not always possible to find a $y^*(t)$ whose relative order is n. In order to find a $y^*(t)$ that is invertible, the matrix $\left[g(x^0) \quad ad_f g(x^0) \quad \ldots \quad ad_f^{n-1} g(x^0) \right]$ must be nonsingular (*i.e.* have rank n) and the span $\left\{ g(x) \quad ad_f g(x) \ldots ad_f^{n-2} g(x) \right\}$ must be involutive in a neighborhood of x^0, where;

$$ad_f g(x) \equiv [f, g](x)$$

(27a)

$$\equiv \frac{\partial g}{\partial x} f(x) - \frac{\partial f}{\partial x} g(x)$$

$$ad_f^k g(x) \equiv \left[f, ad_f^{k-1} g \right](x)$$

(27b)

The span of $\left\{ g(x) \quad ad_f g(x) \ldots ad_f^{n-2} g(x) \right\}$ is involutive [6] if and only if the Lie brackets of any two vector fields $ad_f^j g(x)$, $ad_f^k g(x)$, $k, j = 0 \ldots n-2$ lie in the span of $\left\{ g(x) \quad ad_f g(x) \ldots ad_f^{n-2} g(x) \right\}$.

The above conditions guarantee the existence of a scalar function $h^*(x)$ that satisfies

$$L_g h^*(x) = L_g L_f h^*(x) = \ldots = L_g L_f^{n-2} h^*(x) = 0 \text{ for all } x \text{ near some } x^0 \quad (28a)$$

and

$$L_g L_f^{n-1} h^*(x^0) \neq 0 \tag{28b}$$

Isidori [6] shows that the equations given by (28) are equivalent to the following partial differential equations:

$$L_g h^*(x) = L_{ad_f g} h^*(x) = \ldots = L_{ad_f^{n-2} g} h^*(x) = 0 \tag{29a}$$

$$L_{ad_f^{n-1} g} h^*(x) \neq 0 \tag{29b}$$

Rather than view (29) as a set of PDE's in $h^*(x)$, it is also possible to view (29) as a matrix vector set of equations in $\dfrac{\partial h^*(x)}{\partial x}$ as follows:

$$\begin{bmatrix} \\ g(x^0) & ad_f g(x^0) & \ldots & ad_f^{n-1} g(x^0) \\ \\ \end{bmatrix} \begin{bmatrix} \dfrac{\partial h^*(x^0)}{\partial x_1} \\ \dfrac{\partial h^*(x^0)}{\partial x_2} \\ \vdots \\ \dfrac{\partial h^*(x^0)}{\partial x_n} \end{bmatrix} = \begin{bmatrix} 0 \\ 0 \\ \vdots \\ 0 \\ N \end{bmatrix} \tag{30}$$

where N = any number.

The matrix $\begin{bmatrix} g(x^0) & ad_f g(x^0) & \ldots & ad_f^{n-1} g(x^0) \end{bmatrix}$ is nonsingular by our first condition. This matrix corresponds to the controllability matrix $\begin{bmatrix} b & Ab & A^2 b & \cdots & A^{n-1} b \end{bmatrix}$ for linear systems, and its non-singularity generally doesn't present a problem. However, in order for the vector computed from (30) to be the gradient of a scalar, we need the involutivity condition. If it is satisfied, then we can get $h^*(x)$ by numerically integrating $dh^*(x)$. That is,

$$h^*(x) = \int_0^t \left(\frac{\partial h^*(x)}{\partial x} \right) \dot{x}(t) dt \tag{31}$$

Alternately, and more simply, we can simply set $h^*(t) = h(t)$ at every time step since N in (30) is arbitrary. Example 5 illustrates this approach.

Unfortunately, however, the involutivity condition is usually not satisfied. What then? Kravaris et al. [13], in this volume, recommend forming an $h^*(x)$ which has a relative order of one and where the local zeros that are placed so as to be invertible. He shows that this can always be done. The difficulty here is that since the relative order of his $h^*(x)$ is less than n, the state $x(t)$ will not be the state of a linear system, even if $h^*(x)$ is made to be the output of a specified linear system. Therefore, the process output, $y(t)$, will not generally follow the specified linear behavior, even locally. However, the control effort will be stable. Also, Kavaris's idea leads to the following (interesting?) speculations. Using his approach, the local zeros of $h^*(x)$ can be placed anywhere, including approaching minus infinity. Such a placement is undoubtedly not a good idea from the point of view of computing a control effort because of numerical ill conditioning. However, a system with local zeros near minus infinity is in some sense close to a system with no zeros. Is it possible then, that the vector computed from (30) will be the gradient of a nonlinear system which is in some sense close to original system given by (22a)? At least in concept, we can compute the Lie brackets of all pairs of vector fields $ad_f^j g(x^0)$, $ad_f^k g(x^0)$, $k, j = 0.... n-2$ and can develop a measure of how far these vectors lie outside the span of $\{g(x^0)\ ad_f g(x^0) ... ad_f^{n-2} g(x^0)\}$. In this manner one may be able to determine a-priori whether $h^*(x^0)$ computed from (31) is useful for control.

4. Examples

In this section, we give examples for simulated applications of the proposed model state feedback control law. First two examples are arbitrarily chosen linear processes of high dimensions. Following three nonlinear systems are representative processes extracted from often-cited case studies in nonlinear control literature. For nonlinear systems, the controller of (20) and for nonlinear systems the controller of (25) was implemented. All simulations were done in Matlab-Simulink® environment.

EXAMPLE 1 (Linear system).
To illustrate the linear case, we simply pick up a second order system whose relative order is one. As is often the case in many applications, it is convenient to represent the process in parametric transfer function form:

$$G(s) = \frac{s+3}{s^2+3s+2} \qquad (32)$$

Since the system has no right-half plane transmission zeros to yield unstable internal dynamics, the inverse is possible by implementing the controller of (20) with a first order filter. Selection of the state space form is a matter of convenience, since the particular form selected for implementation does not have any effect on the design or performance of the controller. The controllable canonical form of this transfer function given by

$$\begin{pmatrix} \dot{x}_1 \\ \dot{x}_2 \end{pmatrix} = \begin{pmatrix} 0 & 1 \\ -2 & -3 \end{pmatrix} \begin{pmatrix} x_1 \\ x_2 \end{pmatrix} + \begin{pmatrix} 0 \\ 1 \end{pmatrix} u$$

$$y = (3 \quad 1) \begin{pmatrix} x_1 \\ x_2 \end{pmatrix}$$

(33)

is chosen here for this purpose. Steady state gain of this system is $K = -c A^{-1} b = 1.5$
However, for illustrative purposes, we have also implemented the case where relative
order was made two ($r^* = n = 2$). In this context, solving the equation $c^* b = 0$ gives
$c^* = (1 \ 0)$. This vector can be multiplied by any number. On the other hand, the
condition for static equivalence requires that $c^* A^{-1} b = c A^{-1} b$ which results in $c^* = (3 \ 0)$.
Therefore, we multiplied $c^* = (1 \ 0)$ by 3 and the auxiliary output is

$$y^* = (3 \quad 0) \begin{pmatrix} x_1 \\ x_2 \end{pmatrix}$$

(34)

The servo behavior of the original system as well as the modified system are shown
in Figure 4 for set point changes. The required change in the manipulated input (or
the control effort) are also indicated in all performance figures presented. As the
system had no RHP transmission zeros, implementation of the modified system with
relative order being equal to the system order slows the response only because the same
filter time constant ($\varepsilon = 1$) was used for both the first and second order filters.

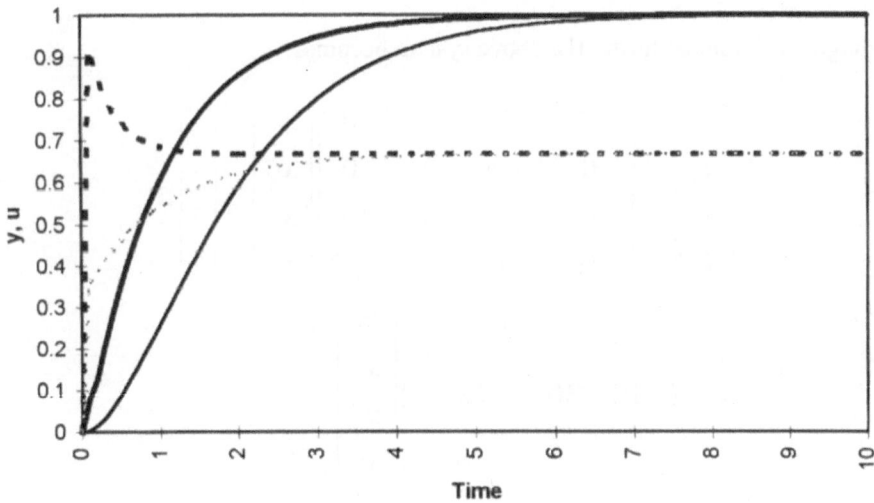

Figure 4. Step set point response for Example 1
(—— y , - - - u , ———— y* — — u*)

Figure 5 depicts the performance of the controlled system for a step disturbance.

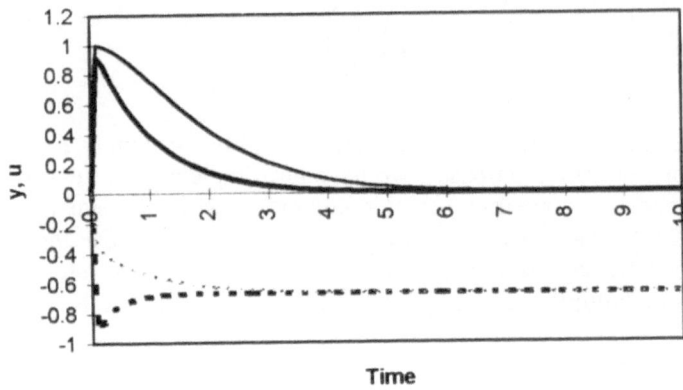

Figure 5. Step disturbance response for Example 1
(— y , - - - u , ——— y* _ _ u*)

EXAMPLE 2 (Linear system).
In this example, we consider a higher order system with RHP transmission zeros.

$$G(s) = \frac{(s+2.5)(s-4)}{(s+0.5)(s+1)(s+1.5)(s+2)} \qquad (35)$$

In diagonal canonical form, the above system becomes:

$$\begin{pmatrix} \dot{x}_1 \\ \dot{x}_2 \\ \dot{x}_3 \\ \dot{x}_4 \end{pmatrix} = \begin{pmatrix} -0.5 & 0 & 0 & 0 \\ 0 & -1 & 0 & 0 \\ 0 & 0 & -1.5 & 0 \\ 0 & 0 & 0 & -2 \end{pmatrix} \begin{pmatrix} x_1 \\ x_2 \\ x_3 \\ x_4 \end{pmatrix} + \begin{pmatrix} 1 \\ 1 \\ 1 \\ 1 \end{pmatrix} u$$

$$\qquad (36)$$

$$y = (-12 \quad 30 \quad -22 \quad 4) \begin{pmatrix} x_1 \\ x_2 \\ x_3 \\ x_4 \end{pmatrix}$$

If the controller is implemented for this process as it is, the system blows up at approximate time of 10 units. This is because of the instability of the zero dynamics. The system was then modified by redefining the output map, i.e. replacing c with c* which was found from the solution of the following system of equations:

$$c*b= 0$$

$$c*Ab= 0$$

$$c*A^2b= 0 \tag{37}$$

$$c*A^{-1}b = cA^{-1}b$$

Solving (37) for c* gives:

$$c* = (-13.3334 \quad 40.0002 \quad -40.0002 \quad 13.3334) \tag{38}$$

When the modified system is implemented with the output map of the system shows nice behavior with this choice of c* leading to a well-behaved and stable inverse as seen in Figure 6. A filter time constant of $\varepsilon=1$ was used for simulations.

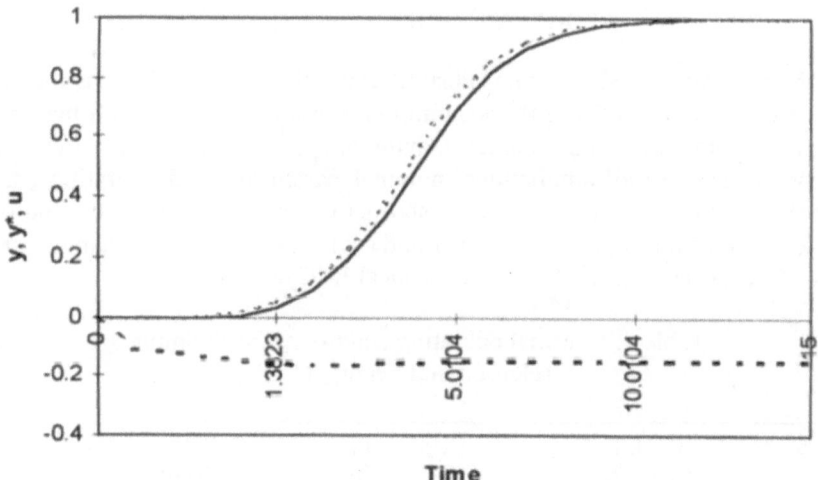

Figure 6. Step set point response for Example 2
(—— y , --- y* , --- u)

104

EXAMPLE 3 (Nonlinear system).

This example is the continuous fermentor studied by Henson and Seborg [5]. Process state variables are composed of biomass concentration (X), substrate concentration (S) and product concentration (P). Input to the process are the feed substrate concentration (S_f) and the dilution rate (D), latter being the manipulated variable. The fermentor is, therefore, described by the following set of differential equations

$$
\begin{aligned}
x_1' &= \mu(x_2, x_3)\, x_1 - x_1\, u \\
x_2' &= -(1/\, Y_{x/s})\, \mu(x_2, x_3)\, x_1 + (S_f - x_2)\, u \\
x_3' &= [\alpha\, \mu(x_2, x_3) + \beta]\, x_1 - x_3\, u
\end{aligned}
\tag{39a}
$$

and the output function

$$
y = x_1 \tag{39b}
$$

which indicates that the biomass concentration is chosen for the controlled output. The specific growth rate, μ, is given by

$$
\mu = \frac{\mu_m(1 - x_3/\, P_m)\, x_2}{K_m + x_2 + (x_2{}^2/K_i)} \tag{40}
$$

exhibiting both substrate and product inhibition, or by a Monod expression as follows:

$$
\mu = \mu_m x_2 /(K_m + x_2) \tag{41}
$$

In the above equations; K_m is the substrate saturation constant, K_i is the substrate inhibition constant, P_m is the product saturation constant, $Y_{x/s}$ is the cell-mass yield, α, β are kinetic parameters and μ_m is the maximum specific growth rate. Relative order of the system is 1. In all simulations, nominal parameters and operating conditions presented in Table 1 were used. Figure 7 shows the servo behavior of the controller for two different filter time constants, $\varepsilon = 0.1$ and $\varepsilon = 1$ under a set point change introduced at t=50. As expected, the controller shows ideal IMC behavior.

Table 1. Nominal operating conditions for Example 3
(Henson and Seborg [5])

$Y_{x/s}$	0.4 g/g	μ_m	0.48
α	2.2 g/g	S_f	20 g/L
β	0.2 h^{-1}	D	0.202 h^{-1}
P_m	50 g/L	X	6.0 g/L
K_m	1.2 g/L	S	5.0 g/L
K_i	22 g?l	P	19.14 g/L

Figure 8 depicts the regulatory behavior of the controller under an unmeasured step disturbance of -12.5 % in the maximum growth rate. The results show good set point tracking and disturbance rejection capabilities of the proposed controller.

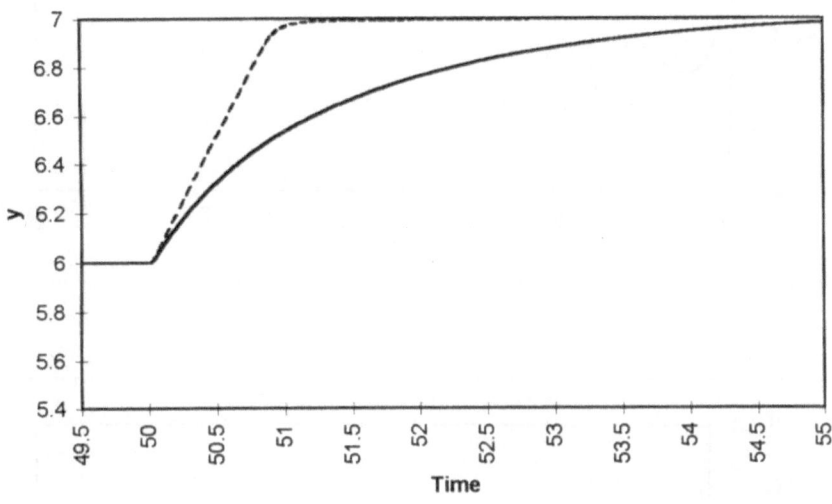

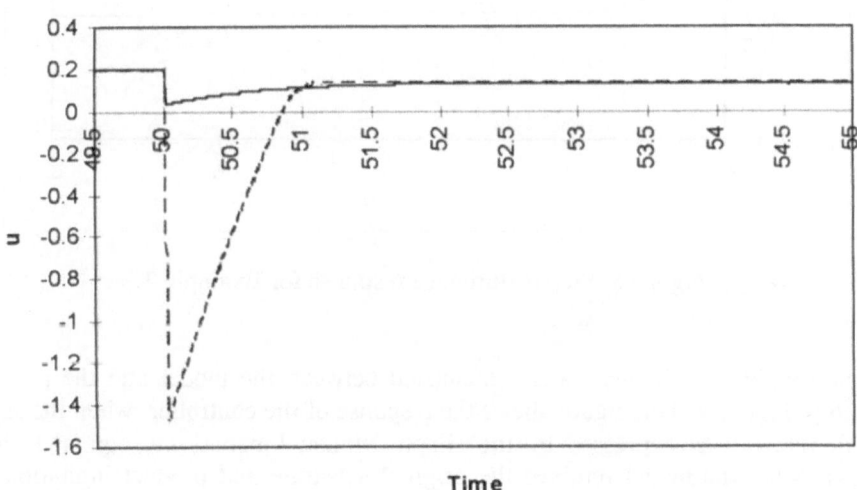

Figure 7. Step set point response for Example 3
($\longrightarrow \varepsilon = 1$, $- - - \varepsilon = 0.1$)

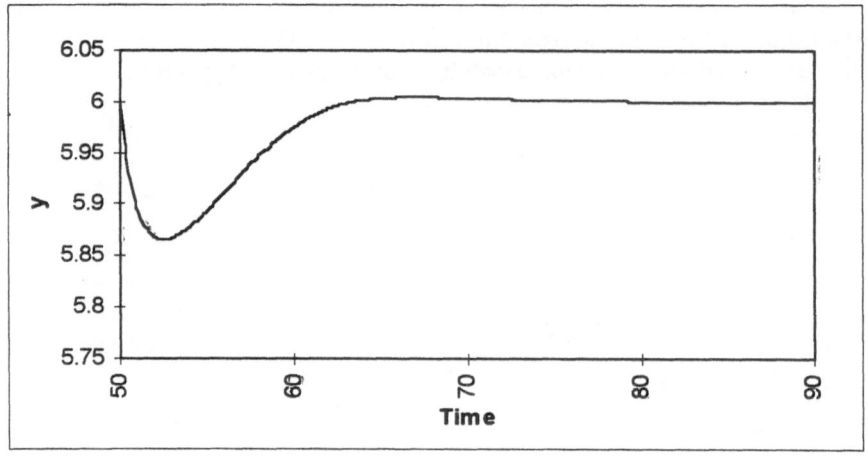

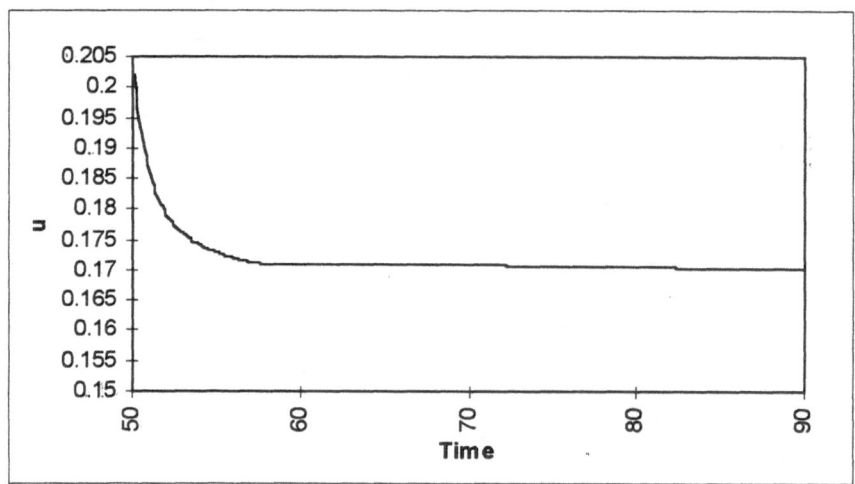

Figure 8. Step disturbance response for Example 3

The case in which there was a mismatch between the model and the process is studied in Figure 9. This figure shows the response of the controller when the specific growth rate μ was expressed by the simple Monod kinetics, *i.e.* eqn. (41) in the process, while the model retained the original substrate and product inhibition, eqn. (40). Results indicate that set point is tracked, unmeasured disturbance in μ rejected and process-model mismatch compensated for delay free process. The controller of (25) results in identical behavior, when compared to the nonlinear IMC controller of

Henson and Seborg [5] but a lot easier to understand and implement because of the fact that it has only two elements, unlike the four component controller of Henson and Seborg [5].

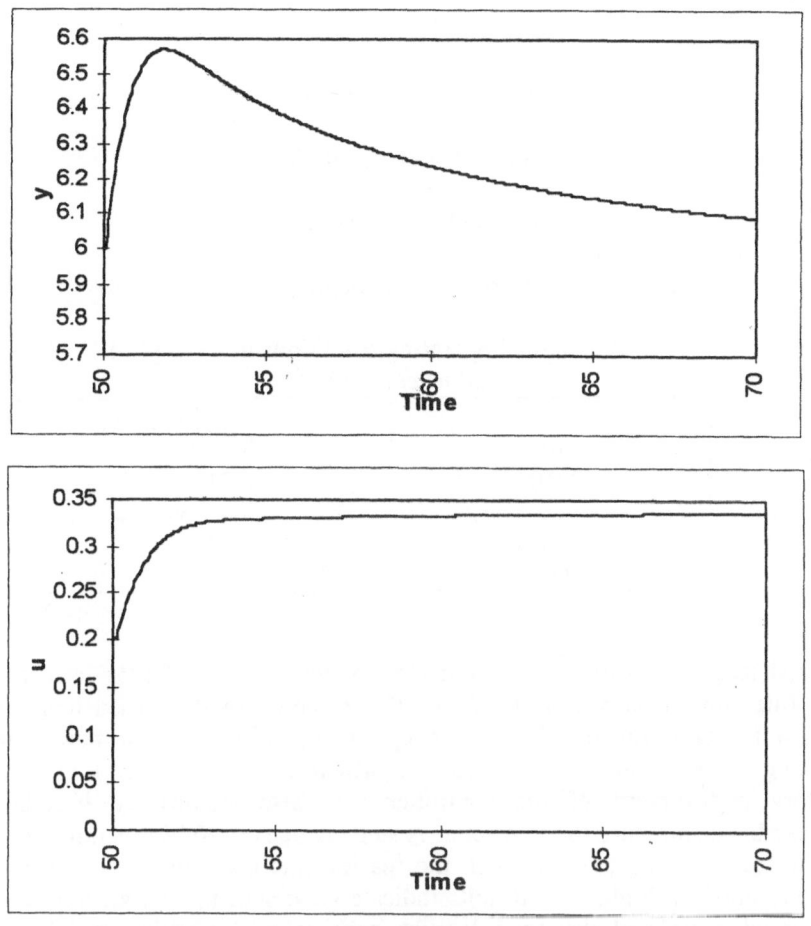

Figure 9. Step disturbance in μ for example 3 using Monod kinetics in the process

EXAMPLE 4 (Nonlinear system).
As a second case study for nonlinear system application, we considered the continuous stirred bioreactor described by Dibiasio *et al.* [4] in which the substrate methanol is used for the growth of *Methylomonas* strain. The same reactor was also studied by Valluri *et al.* [14]. The state vector $x=(S, X)^T$ is composed of substrate and biomass concentrations and the control is the dilution rate, i.e. u–D. Relative order is 1. The controlled output is the substrate concentration in the reactor. The system is described by the following equations

$$x_1^{'} = -\sigma(x_1) x_2 + (S_f - x_1) u$$

$$x_2' = \mu(x_1) x_2 - x_2 u \tag{42}$$
$$y = x_1$$

where the specific growth rate, μ, and the substrate consumption rate, ρ, are given by

$$\mu = \frac{0.504 \, x_1 \, (1-0.204 \, x_1)}{0.000849 + x_1 + 0.406 \, x_1^2} \tag{43}$$

$$\sigma = \frac{x_1 \, (1.32 + 3.86 \, x_1 - 0.661 \, x_1^2)}{0.000849 + x_1 + 0.406 \, x_1^2} \tag{44}$$

The operating conditions and reactor parameters are given in Table 2.

Table 2, Nominal operating conditions in Example 4
(Valluri *et al.* [14])

S_f	1.8 W/V %
$S(0)$	0 W/V %
$X(0)$	0.01 W/V %
S_{ss}	0.4 W/V %
X_{ss}	0.24 W/V %
D_{ss}	0.4 h^{-1}

Closed loop simulations with controller implementation of (25) were conducted with a filter time constant of 0.175 h. The response of the closed loop system is simulated for two situations. The start-up profile of the reactor under suggested linearizing model state feedback control structure is shown in Figure 10. The regulatory performance of the controller was also studied by introducing an unmeasurable disturbance to the reactor. A step change of -20 % was introduced in the substrate feed concentration S_f when the reactor was operating at steady state. The results are given in Figure 11. Results indicate good start-up and settling at unstable middle steady state, and also good disturbance rejection for perfect model around the steady state.

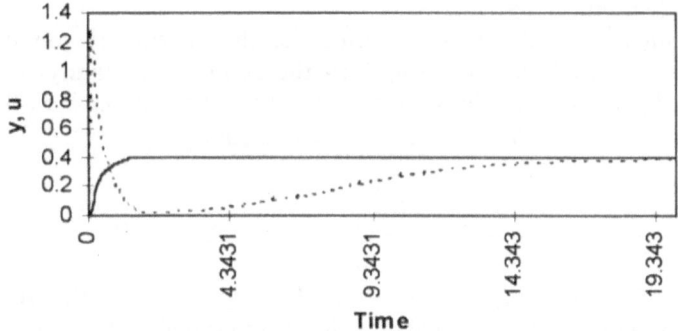

Figure 10. Start up of the reactor in Example 4
(—— y , - - - u)

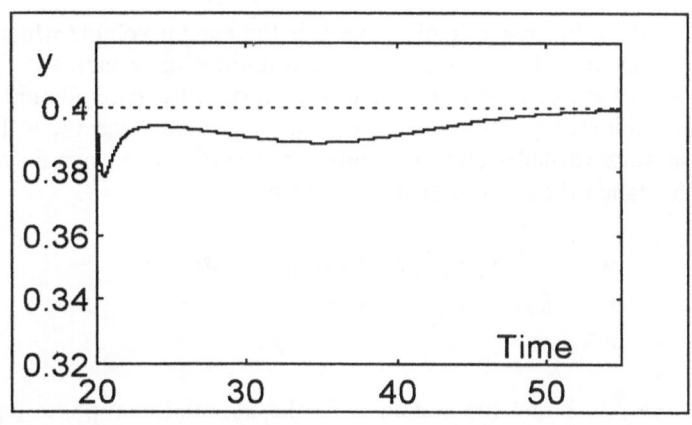

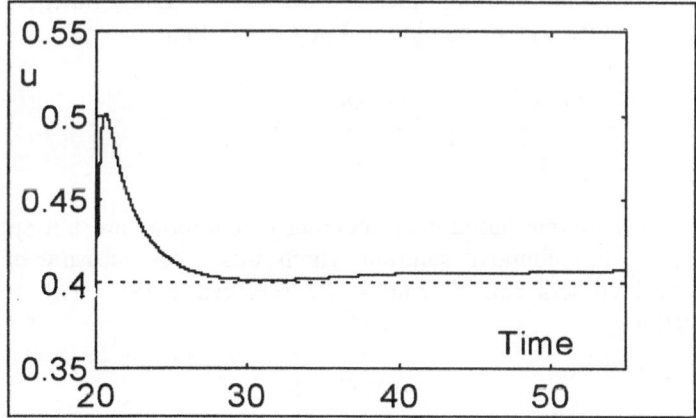

Figure 11. Disturbance rejection of the control system in Example 4

EXAMPLE 5 (Nonlinear system).
The last example studied intends to illustrate the control of nonminimum-phase systems. This is a well-known example in the control community which has been tackled before by many researchers [7, 8, 15]. The system concerns a CSTR where the isothermal series/parallel Van de Vusse reaction is taking place

$$A \xrightarrow{k_1} B \xrightarrow{k_2} C \qquad 2A \xrightarrow{k_3} D$$

where the reaction rate constants are $k_1 = 50 \text{ h}^{-1}$, $k_2 = 100 \text{ h}^{-1}$, $k_3 = 10 \text{ L/mol.h}$. The feed consists of pure A. The mass balances for A and B define the system as follows

$$V(dC_A/dt) = F(C_{A0} - C_A) - k_1 C_A V - k_3 C_A^2 V \qquad (45a)$$

$$V(dC_B/dt) = F(-C_B) + k_1 C_A V - k_2 C_B V \qquad (45b)$$

where F denotes the inlet flow rate of A and F is the reactor volume which is constant in the course of reaction. C_A and C_B are concentrations of reactants A and B in the reactor. The control objective is to maintain C_B constant by manipulating the dilution rate F/V. The initial steady state is given by $C_{As} = 3.0 \text{ mol/L}$ and $C_{Bs} = 1.117 \text{ mol/L}$. By defining the state variables and the control as $x = (C_A, C_B)^T$, $u = F/V$, the system may be put into standard nonlinear state space form:

$$\dot{x}_1 = -k_1 x_1 - k_3 x_1^2 + (C_{A0} - x_1)u \qquad (46a)$$
$$\dot{x}_2 = k_1 x_1 - k_2 x_2 - x_2 u \qquad (46b)$$
$$y = x_2 \qquad (46c)$$

Relative order of y is 1 and the system is in the nonminimum-phase region for the parameters and steady state values used by Wright and Kravaris [15].

To make the system minimum-phase, we seek a scalar function h*(x) as a nontrivial solution to the to the equation $L_g L_f h^*(x) = 0$, that is,

$$g_1 \frac{\partial h^*(x)}{\partial x_1} + g_2 \frac{\partial h^*(x)}{\partial x_2} = 0 \qquad (47)$$

The solution is not unique because the boundary conditions are not specified. For example, Kantor [7] found a solution which was a log measure of the reactor conversion. While others can be found, we replaced h*(x) with the following analytical function:

$$y^* = h^*(x) = \frac{\alpha(t) C_{A0} x_2}{C_{A0} - x_1} \qquad (48)$$

The relative order of $y*$ is two.

In order that $y*$ have the same steady state, and track the same trajectory as the original process output, $y(t)$, we have inserted a function, $\alpha(t)$, into our solution of (47). $\alpha(t)$ is chosen so that

$$\frac{y*(t)}{y(t)} = 1 \quad \Rightarrow \quad \alpha(t)\frac{C_{A0}\, x_2}{C_{A0} - x_1} = x_2$$

this gives:

$$\alpha(t) = \frac{C_{A0} - x_1}{C_{A0}}$$

The performance of the controller for set point tracking is depicted in Figure 12 for a filter time constant of $\varepsilon=0.01$ h. In order to study the regulatory performance, the reactor was subject to a step disturbance of -10% in the inlet concentration of reactant A. The disturbance was considered to be unmeasurable. Figure 13 shows the profiles of the controlled output and manipulated input for this case.

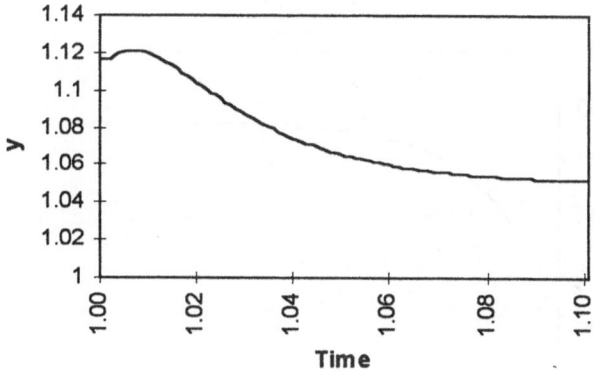

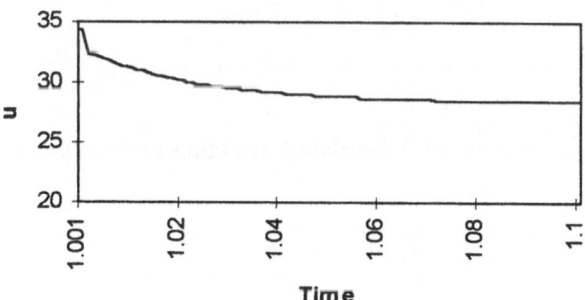

Figure 12. Set point tracking in Example 5

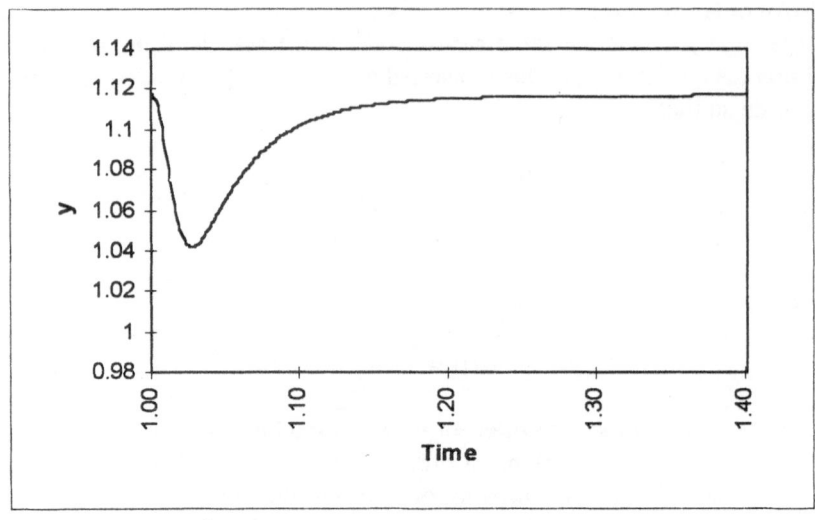

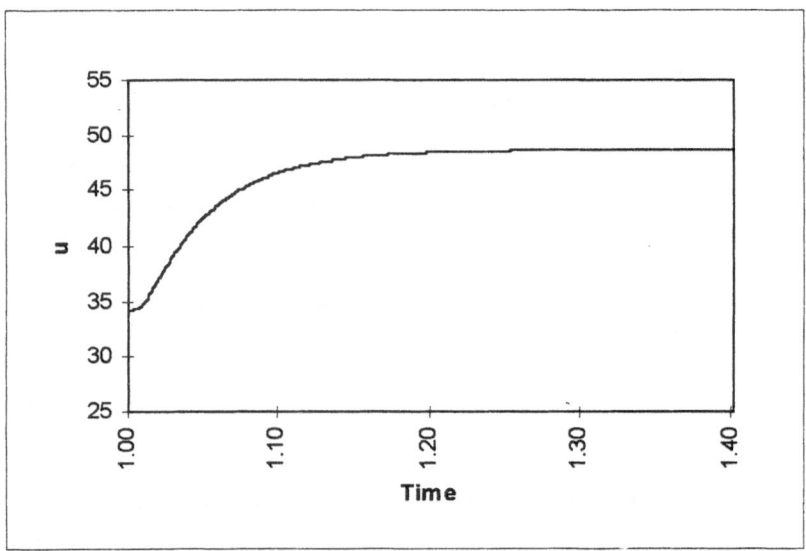

Figure 13. Disturbance rejection in Example 5

5. Conclusions

The model state feedback implementation of IMC for linear and nonlinear systems has been addressed directly in a comprehensive and unifying framework. The key features of the developed control laws are
- Easy to implement, and brings new insights into understanding
- Only one tuning parameter (compared to geometric controllers)
- For nonlinear systems, eliminates coordinate transformations for controller synthesis and also the need for an external controller to drive the transformed input
- Suggests new strategies for controlling non-invertible systems.
- Provides a conceptually unifying framework for linear and nonlinear model state feedback control in state space form.

This work concerned only with open loop stable processes. Stabilization of unstable processes will be presented in a subsequent paper.

Acknowledgment
The authors would like to acknowledge the Fulbright Scholarship that allowed R. Berber to stay at Case Western Reserve University in the course of this work. We also want to thank Prof. Kravaris, for being so willing to patiently explain, and hopefully eliminate, many of our confusions.

References

1. Bischof, C., Carle, A., Khademi, P. and Mauer, A. (1994) "The ADIFOR 2.0 system for the Automatic Differentiation of Fortran 77 Programs", Argonne Preprint ANL-MCS-P481-1194.
2. Coulibaly, E.S., Maiti, S. and Brosilow, C. (1992) Internal model predictive control (IMPC), Paper No. 123j, AIChE Annual Meeting, Miami Beach, FL.
3. Coulibably,E., Maiti,S., and Brosilow, C. (1995) Internal model predictive control (IMPC) *Automatica* **31**, 1471-1482.
4. Dibiasio, D., Lim, H.C. and Weigand, W.A. (1981) An experimental investigation of stability and multiplicity of steady states in a biological reactor, *AIChE J.* **27**, 284.
5. Henson, M. A. and Seborg, D. A. (1991) An internal model control strategy for nonlinear systems, *AIChE J.* **37**, 1065-1081.
6. Isidori, A. (1995) Nonlinear Control Systems: An Introduction, 3rd ed. Springer Verlag, Berlin.
7. Kantor, J. C. (1986) Stability of state feedback transformations for nonlinear systems- Some practical considerations, Proceedings 1986 American Control Conference, Seattle, WA, Vol 2, 1014-1016.
8. Kravaris, C., and Daoutidis, P. (1990) Nonlinear state feedback control of second order nonminimum-phase systems, *Comp. Chem. Eng.* **14**, 439-449.

9. Kravaris, C. and Daoutidis, P. (1992) Output feedback controller realizations for open-loop stable nonlinear processes, Proceedings of 1992 American Control Conference, Chicago, IL, 2596-2601.

10. Kravaris, C., Daoutidis, P. and Wright, R.A. (1994) Output feedback control of nonminimum-phase nonlinear processes, *Chem. Eng. Sci.* **49**, 2107-2122.

11. Kravaris, C. and Kantor, J.C. (1990a) Geometric methods for nonlinear process control. 1. Background, *Ind. Eng. Chem. Research* **29**, 2295-2310.

12. Kravaris, C. and Kantor, J.C. (1990b) Geometric methods for nonlinear process control. 2. Controller synthesis, *Ind. Eng. Chem. Research* **29**, 2311-2323.

13. Kravaris, C., Niemiec, M., Berber, R. and Brosilow, C. (1998) Nonlinear model-based control of nonminimum-phase processes; in R. Berber and C. Kravaris (eds.), *Nonlinear Model Based Process Control*, NATO ASI Series, Kluwer Academic Publishers, Dordrecht, pp. 115-141.

14. Valluri, S., Soroush, M. and Nikravesh, M. (1998) Shortest prediction horizon nonlinear model predictive control, *Chem. Eng. Sci.* **53**, 273-292.

15. Wright, R.A. and Kravaris, C. (1992) Nonminimum-phase compensation for nonlinear processes, *AIChE J.* **38**, 26-40.

NONLINEAR MODEL-BASED CONTROL OF NONMINIMUM-PHASE PROCESSES

COSTAS KRAVARIS and MICHAEL NIEMIEC
Department of Chemical Engineering
The University of Michigan
Ann Arbor, Michigan 48109, USA

RIDVAN BERBER
Department of Chemical Engineering
The University of Ankara
Tandogan, 06100 Ankara, Turkey

COLEMAN B. BROSILOW
Department of Chemical Engineering
Case Western Reserve University
Cleveland, Ohio 44106, USA

Abstract. This paper reviews the properties of model-state feedback and input/output linearizing state feedback for the synthesis of nonlinear controllers. These concepts are extended to nonminimum-phase systems, where a synthetic output, which is statically equivalent to the actual process output, is used to design the model-state feedback controller. Systematic procedures are outlined for the construction of the synthetic output and the assignment of local zeros. The proposed controller is illustrated with a nonminimum-phase van de Vusse reactor through simulations.

1. Introduction

Although there has been considerable progress in the field of nonlinear systems theory and applications, the control of nonlinear nonminimum-phase processes still remains one of the difficult and challenging problems in the field of process control. The challenge lies in the construction of a stable approximation of the process inverse. For linear systems, the system is factored into minimum-phase and nonminimum-phase parts, with only the former being inverted in the controller design. For nonlinear systems, the decomposition into minimum-phase and nonminimum-phase parts is an extremely difficult problem. In the special case of second-order systems, this decomposition problem was solved in Kravaris and Daoutidis [9], and it was shown that it gives rise to ISE-optimal control laws with respect to step changes in the set point. Alternative approaches were developed by Doyle et al. [3],[4]; including approximate

R. Berber and C. Kravaris (eds.), Nonlinear Model Based Process Control, 115-142.
© 1998 *Kluwer Academic Publishers.*

stable/anti-stable factorization of the zero dynamics, an inner-outer based approximation, and a multiple-input approach. These approaches are also applicable to limited classes of nonlinear systems.

A nonminimum-phase compensation structure for nonlinear systems was developed by Wright and Kravaris [21], based on a synthetic output, which is minimum-phase and statically equivalent to the original output. The synthetic output is calculated on-line and controlled to set point via an inverse-based controller. With this control structure, the decomposition problem is completely bypassed and the control problem reduces to constructing an appropriate synthetic output. To this end, Wright and Kravaris [21] proposed an ISE-optimization formulation, which led to analytical results only for a limited class of nonlinear systems. Reduced-order controller realizations, arising form this nonlinear nonminimum-phase compensation structure, were derived in Kravaris et al. [11] and interpreted in the context of the model-state feedback structure.

The present paper starts with a review of previous results and concepts on model-state feedback, input/output linearization, nonminimum-phase compensation, static equivalence of outputs, and then formulates the controller design problem in a more general context. Instead of searching for an optimal synthetic output, the emphasis is placed on characterizing the class of statically equivalent outputs to the given process output. Out of this class, a particular synthetic output with prescribed zeros is then selected in a systematic way. The selected synthetic output is used to construct an input/output linearizing state feedback, which in turn generates a model-state feedback controller. The proposed control strategy is illustrated in a simulation case study for a nonisothermal CSTR where a series/parallel van de Vusse reaction takes place.

This paper is organized as follows: Section 2 reviews the model-state feedback structure and its properties. Section 3 reviews notions and properties of input/output linearizing state feedback. Section 4 discusses the use of input/output linearizing state feedback and the resulting model-state feedback controller on the basis of a synthetic output. Section 5 proposes a systematic procedure for the construction of a synthetic output with prescribed zeros. Finally, in Section 6, the proposed method is evaluated via numerical simulation in a control problem for a nonisothermal CSTR, which exhibits nonminimum-phase behavior.

2. The Model-State Feedback Structure

The model-state feedback control structure is depicted in Figure 1. The process model is simulated on-line to generate the model state. The model state is used to generate a disturbance estimate, which corrects the set point, and is also fed back according to a static state feedback control law. Two different versions of the model-state feedback control structure are shown in Figure 1, unconstrained and constrained. The key feature of the constrained version is that the saturation function is applied on the calculated control action (from the static state feedback) before it goes through the model simulation. In this way, the simulated process states are not corrupted by the presence of input constraints.

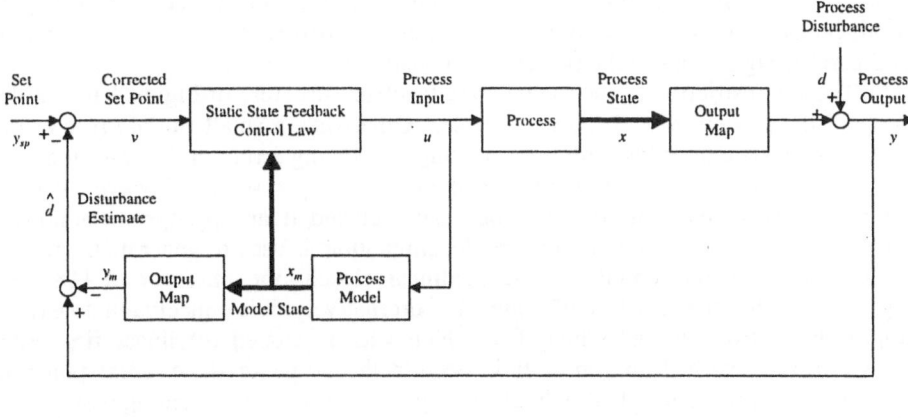

(a)

(b)

FIGURE 1. Model-State Feedback Control Structure (a) Unconstrained (b) Constrained

The model-state feedback structure for unconstrained nonlinear processes was first proposed in Kravaris and Daoutidis [10]. The model-state feedback for constrained linear processes was first proposed in Coulibaly et al. [2], using a companion-form realization of a transfer-function model, where the attractive constraint-handling capabilities of model-state feedback have been demonstrated. The properties of the unconstrained nonlinear model-state feedback were studied in Kravaris et al. [11], including specific conditions for the overall control law to posses integral action. When an input/output linearizing static state feedback is used in the model-state feedback control structure, the resulting controller can be interpreted as a model-predictive controller, both in the unconstrained and constrained cases [13]-[15]. Further results on the constrained model-state feedback control structure have been reported in Mhatre and Brosilow [12] for linear multivariable systems and in Soroush and Nikravesh [17] for nonlinear systems, which study the anti-windup properties of the control structure and addresses design issues in the presence of constraints.

The key restriction of the model-state feedback structure of Figure 1 is that the process and the process model must be stable. Otherwise, a small initial error in the model state may keep growing with time, leading to growing errors in the control action calculation and potential loss of stability of the closed-loop system. However, it must be pointed out that this restriction can be easily relaxed if an appropriate nonlinear closed-loop observer, instead of straight model simulation, is used to generate the model state. For example, one could use the nonlinear closed-loop observer of [18] and appropriately modify the structure of Figure 1. Alternatively, it is conceivable to extend the approach of Brosilow and Cheng [1], which was developed for linear first-order unstable processes with dead time. It is beyond the scope of the present paper to consider unstable processes. Throughout the paper, there will be a standing assumption that both the process and the process model are locally asymptotically stable in the vicinity of the operating steady-state.

In what follows, some key theoretical properties of the model-state feedback structure will be outlined, which will be needed in subsequent developments.

2.1.   MODEL-STATE FEEDBACK CONTROL OF A NONLINEAR PROCESS

Consider a process whose dynamics can be described by a nonlinear state-space model the form:

$$\dot{x} = f(x) + g(x)u \tag{1}$$

where $u \in \Re$ is the manipulated input, $x \in \Re^n$ is the n-vector of process states, $f(x)$ and $g(x)$ are vector functions from $\Re^n$ into $\Re^n$, and the process output $y \in \Re$ is defined via:

$$y = h(x) + d \tag{2}$$

where $h(x)$ is a scalar function from $\Re^n$ into $\Re$, and $d \in \Re$ is a disturbance input.

To generate a model-state feedback controller for the process, one must know the dynamics (1), the output map $h(x)$ in (2), and, in addition, one must be given a static state feedback control law:

$$u = \Psi(x, v) \tag{3}$$

where $v \in \Re$ is a reference input and $\Psi(x, v)$ is a function from $\Re^n \times \Re$ into $\Re$.

For the process dynamics, output map, and static state feedback defined in (1), (2), and (3), respectively, the model-state feedback control structure of Figure 1 takes the more concrete form of Figure 2.

Collecting all the controller blocks together, one can write the model-state feedback controller as a nonlinear dynamic system:

$$\begin{aligned} \dot{x}_m &= f(x_m) + g(x_m)\Psi(x_m, h(x_m) + y_{sp} - y) \\ u &= \Psi(x_m, h(x_m) + y_{sp} - y) \end{aligned} \tag{4}$$

in the unconstrained case, and:

$$\begin{aligned} \dot{x}_m &= f(x_m) + g(x_m)S(\Psi(x_m, h(x_m) + y_{sp} - y)) \\ u &= S(\Psi(x_m, h(x_m) + y_{sp} - y)) \end{aligned} \tag{5}$$

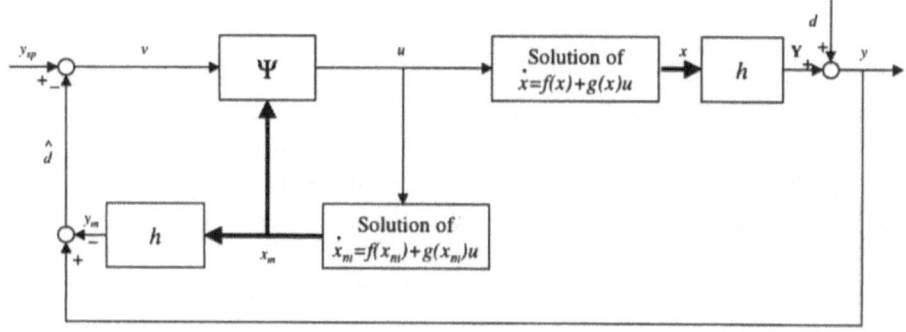

(a)

(b)

FIGURE 2. Model-State Feedback Control of a Nonlinear Process (a) Unconstrained (b) Constrained

in the constrained case. In (5), $S(\bullet)$ represents the standard saturation function defined by:

$$S(u_c) = \begin{cases} u_H, & \text{if } u_c \geq u_H \\ u_c, & \text{if } u_L \leq u_c \leq u_H \\ u_L, & \text{if } u_c \leq u_L \end{cases} \qquad (6)$$

where u_L and u_H are the lower and upper bounds of u, respectively. From Figure 2, one can observe that the input u drives both the process dynamics and the process model. *If there are no modeling errors* (i.e. the model is an exact representation of the process dynamics) *and the process dynamics is stable, then* $x = x_m$. (This property will hold true

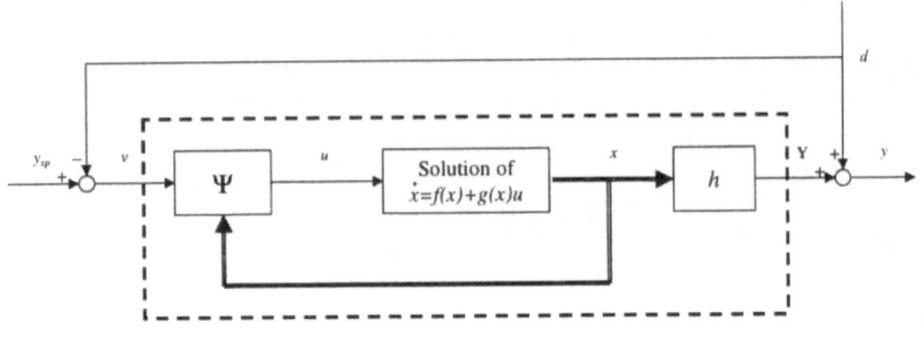

(a)

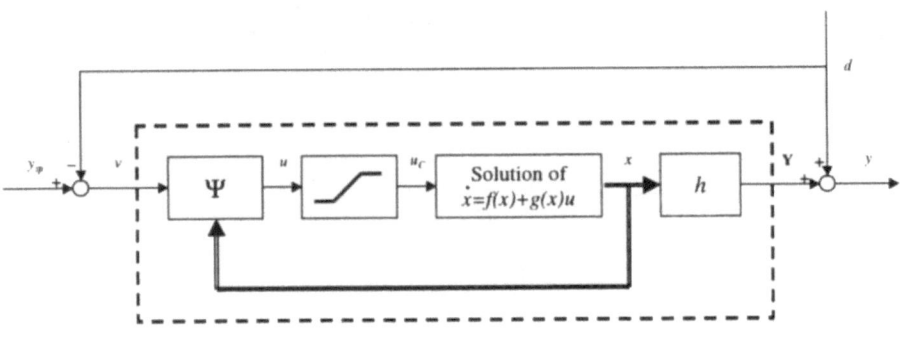

(b)

FIGURE 3. Reduced Closed-Loop System after Cancellation at the Process Modes (a) Unconstrained (b) Constrained

for all times if x and x_m match at $t = 0$; it will be true for large times if there is an initial mismatch between x and x_m).

As a result of the above property, feeding back the model state x_m will have the same effect as feeding back the actual process state. Also, because $x_m = x$, it follows that $y_m = Y$ and therefore $\hat{d} = d$. Consequently, for an open-loop stable process, which is modeled without error, the closed-loop dynamics simplifies as shown in Figure 3. This simplification of the closed-loop dynamics arises from zero-pole cancellations at the process poles, whenever the process dynamics is linear. In a nonlinear context, one could say that the closed-loop dynamics is reduced to the form of Figure 3, after cancellations at the process modes.

The reduced realization of the closed-loop system shown in Figure 3 provides guidance on how the static state feedback law Ψ could be synthesized. If the inner loop in Figure 3 (represented by the dashed lines) can be made stable and have the prescribed input/output behavior, this will also shape the input/output behavior of the overall system:

- The y_{sp} – to – y input/output behavior is exactly the same as the v – to – Y input/output behavior.
- The d – to – y input/output behavior is equal to 1 minus the v – to – Y input/output behavior.

In summary, *the main advantage of the model-state feedback structure is that it reduces the problem of dynamic output feedback synthesis to a problem of static state feedback synthesis, as long as the process is open-loop stable.* The presence of modeling errors will not alter the synthesis formula for Ψ, but it must be accounted for in the choice of tunable parameters in Ψ, in view of the well-known trade-off between performance and robustness.

2.2. PROPERTIES OF THE NONLINEAR MODEL-STATE FEEDBACK CONTROLLER

The properties of the model-state feedback controller, which allow the reduction to a static state feedback controller synthesis problem, will now be formulated in a more precise manner. This will be done for the unconstrained case, although with very minor modifications, the same properties hold for the constrained case as well.

Suppose that the model-state feedback controller (4) is placed around a process modeled by (1) and (2). Then, the resulting closed-loop dynamics is governed by:

$$\dot{x}_m = f(x_m) + g(x_m)\Psi(x_m, h(x_m) - h(x) + y_{sp} - d)$$
$$\dot{x} = f(x) + g(x)\Psi(x_m, h(x_m) - h(x) + y_{sp} - d) \tag{7}$$

As an output of the system, one may consider the process output (2), or, alternatively, the error $e = y_{sp} - y$, i.e.,

$$e = -h(x) + y_{sp} - d \tag{8}$$

The latter is directly related to common performance measures like IAE and ISE.

Property 1: Closed-loop stability under static state feedback implies closed-loop stability under model-state feedback.

Assume that the open-loop process (1) is locally asymptotically stable. If the dynamic system:

$$\dot{x} = f(x) + g(x)\Psi(x, v)$$

is locally asymptotically stable, then the closed-loop system (7) is locally asymptotically stable.

122

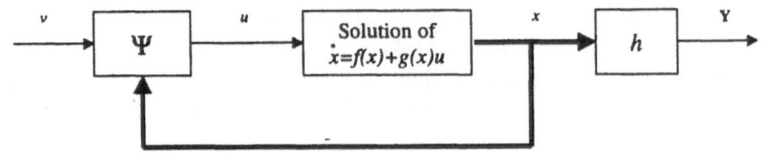

FIGURE 4. Static State Feedback Loop for the Synthesis of the Model-State Feedback Controller

Property 2: **Unity steady-state gain under static state feedback implies zero steady-state error under model-state feedback.**

If the dynamic system:

$$\dot{x} = f(x) + g(x)\Psi(x,v)$$
$$Y = h(x)$$

has unity steady-state gain, i.e. $Y_s = v_s$ at every steady state, then the closed-loop system (7) with output (8) has the property that $e_s = 0$ at every steady-state.

Property 3: **Input/output behavior under static state feedback determines input/output behavior under model-state feedback.**

The dynamic system:

$$\dot{x} = f(x) + g(x)\Psi(x,v)$$
$$e = -h(x) + v$$

driven by $v = y_{sp} - d$, has the same input/output behavior as the closed-loop system (7) with output (8). Equivalently, if one considers the dynamic system:

$$\dot{x} = f(x) + g(x)\Psi(x,v)$$
$$Y = h(x)$$

and denote by R the input/output operator that relates v to y:

$$Y = Rv$$

then the input/output behavior of the closed-loop system (7) with output (2) will be:

$$y = R(y_{sp} - d) + d$$

The theoretical limit of perfect control will be obtained in the limit as R tends to the identity operator.

The proofs of the above three properties are very straightforward; they are omitted for brevity. As a conclusion from the three properties, the synthesis problem reduces to synthesizing a static state feedback Ψ as shown in Figure 4, for which the following design conditions must be imposed:

(i)   Stability
(ii)   Unity steady-state gain
(iii)   Desirable input/output behavior

3. Controller Synthesis on the Basis of Input/Output Linearization

3.1. PRELIMINARIES

Consider a nonlinear system of the form:

$$\dot{x} = f(x) + g(x)u$$
$$Y = h(x)$$

(9)

From Hirschorn [6], the <u>relative order</u> of a system of the form (9) is defined as the smallest integer r for which:

$$L_g L_f^{r-1} h(x) \neq 0$$

In the above condition, L_g and L_f are Lie derivative operators:

$$L_g = g_1(x) \frac{\partial}{\partial x_1} + \cdots + g_n(x) \frac{\partial}{\partial x_n}$$

$$L_f = f_1(x) \frac{\partial}{\partial x_1} + \cdots + f_n(x) \frac{\partial}{\partial x_n}$$

As an immediate consequence of the definition of relative order, the derivatives of the output Y up to order $(r-1)$ can be expressed as functions of the states only:

$$\frac{d^k Y}{dt^k} = L_f^k h(x), \quad k = 1, \ldots, r-1$$

whereas the r-th derivative of the output explicitly depends on u:

$$\frac{d^r Y}{dt^r} = L_f^r h(x) + L_g L_f^{r-1} h(x) u$$

Furthermore, it can be shown that the scalar fields $h(x), L_f h(x), \ldots, L_f^{r-1} h(x)$ are linearly independent and, therefore, can be used as a part of a coordinate transformation. In particular, one can apply a coordinate transformation:

$$z = T(x)$$

(10)

defined by:

$$z_1 = t_1(x)$$
$$\vdots$$
$$z_{n-r} = t_{n-r}(x)$$
$$z_{n-r+1} = h(x)$$
$$\vdots$$
$$z_n = L_f^{r-1} h(x)$$

(11)

where $t_1(x), \ldots, t_{n-r}(x)$ are scalar fields which are linearly independent to each other and linearly independent to $h(x), \ldots, L_f^{r-1} h(x)$. Then, the system (9) can be expressed in terms of the new coordinates (11) as follows:

$$\dot{z}_1 = F_1(z) + G_1(z)u$$
$$\vdots$$
$$\dot{z}_{n-r} = F_{n-r}(z) + G_{n-r}(z)u$$
$$\dot{z}_{n-r+1} = z_{n-r+2}$$
$$\vdots$$
$$\dot{z}_{n-1} = z_n$$
$$\dot{z}_n = F_n(z) + G_n(z)u$$
$$Y = z_{n-r+1}$$

(12)

where:

$$F_j(z) = [L_f t_j(x)]_{x=T^{-1}(z)}, \quad j = 1, \ldots, n-r$$
$$G_j(z) = [L_g t_j(x)]_{x=T^{-1}(z)}, \quad j = 1, \ldots, n-r$$
$$F_n(z) = [L_f^r h(x)]_{x=T^{-1}(z)}$$
$$G_n(z) = [L_g L_f^{r-1} h(x)]_{x=T^{-1}(z)}$$

A realization of the form (12) has the property that the output and its derivatives up to order $(r-1)$ are used as states. This property allows defining of a notion of nonlinear zeros as follows.

As in Isidori [7], the zero dynamics of a nonlinear system of the form (9) is the $(n-r)$-th order unforced dynamic system defined by:

$$\dot{z}_1 = \left[F_1(z) - \frac{F_n(z)}{G_n(z)} G_1(z) \right]_{\substack{z_{n-r+1}=Y_s \\ z_{n-r+2}=0 \\ \vdots \\ z_n=0}}$$

$$\vdots$$

(13)

$$\dot{z}_{n-r} = \left[F_{n-r}(z) - \frac{F_n(z)}{G_n(z)} G_{n-r}(z) \right]_{\substack{z_{n-r+1}=Y_s \\ z_{n-r+2}=0 \\ \vdots \\ z_n=0}}$$

where Y_s is a reference steady-state value for the output.

For a linear system (9) and a linear coordinate transformation (11), the zero dynamics (13) is a linear system whose eigenvalues are exactly the zeros of the transfer function of (9). For nonlinear systems, the eigenvalues of the linear approximation of (13) around a reference steady-state are exactly the zeros of the transfer function of the linear approximation of (9) around the same steady-state.

The notion of zero dynamics enables the characterization of minimum-phase and nonminimum-phase behavior in nonlinear systems. By definition [7], a dynamic system of the form (9) is called locally <u>minimum-phase</u> at Y_s if its zero dynamics (13) is locally asymptotically stable. Otherwise, it is called <u>nonminimum-phase</u>.

In order to test whether a nonlinear system is locally minimum-phase, one can calculate the eigenvalues of the linear approximation of (13) or, equivalently, the zeros of the linear approximation of (9).

3.2.   INPUT/OUTPUT LINEARIZING STATE FEEDBACK AND THE RESULTING MODEL-STATE FEEDBACK CONTROLLER

For a nonlinear system of the form (9) with relative order r, it is always possible to calculate a static state feedback which forces the output y to follow an r-th order linear exponential trajectory defined by:

$$Y + \gamma_1 \frac{dY}{dt} + \cdots + \gamma_r \frac{d^r Y}{dt^r} = v \tag{14}$$

where $\gamma_1, \ldots, \gamma_r$ are constant parameters. Using the expressions for the output derivatives given earlier, it is straightforward to find that the necessary state feedback is given by:

$$u = \Psi(x, v) = \frac{v - h(x) - \sum_{k=1}^{r} \gamma_k L_f^k h(x)}{\gamma_r L_g L_f^{r-1} h(x)} \tag{15}$$

When the above static state feedback is substituted into the general form of model-state feedback controller of (4), it gives rise to the following specific model-state feedback controller:

$$\dot{x}_m = f(x_m) + g(x_m) \frac{(y_{sp} - y) - \sum_{k=1}^{r} \gamma_k L_f^k h(x_m)}{\gamma_r L_g L_f^{r-1} h(x_m)}$$

$$u = \frac{(y_{sp} - y) - \sum_{k=1}^{r} \gamma_k L_f^k h(x_m)}{\gamma_r L_g L_f^{r-1} h(x_m)} \tag{16}$$

As explained in the previous section, the properties of the closed-loop system under model-state feedback (Figure 2), can be extracted from the properties of the closed-loop system under static state feedback (Figure 4).

For the particular state feedback under consideration (15), it can be shown that the resulting closed-loop system:

$$\dot{x} = f(x) + g(x) \frac{v - h(x) - \sum_{k=1}^{r} \gamma_k L_f^k h(x)}{\gamma_r L_g L_f^{r-1} h(x)} \tag{17}$$

$$Y = h(x)$$

(i)   will be locally asymptotically stable if the output trajectory (14) is stable (i.e. all roots of $1 + \gamma_1 s + \ldots + \gamma_r s^r$ have negative real parts) and the system (9) is locally minimum-phase

(ii)   will always have unity steady-state gain

(iii)   will have linear input/output behavior with transfer function:

$$Y = \frac{1}{1 + \gamma_1 s + \cdots + \gamma_r s^r} v$$

Consequently, for the model-state feedback controller of (16), the resulting closed-loop system:

(i)   will be locally asymptotically stable if the output trajectory (14) is stable and the system (9) is locally minimum-phase and locally asymptotically stable

(ii)   will always have zero steady-state error

(iii)   will have linear input/output behavior with transfer function description:

$$y = \frac{1}{1+\gamma_1 s + \cdots + \gamma_r s^r} y_{sp} + \left(1 - \frac{1}{1+\gamma_1 s + \cdots + \gamma_r s^r}\right) d$$

4.   Controller Synthesis for Nonminimum-Phase Systems Using a Synthetic Output

The input/output linearization approach outlined in the previous section will not work for nonminimum-phase systems of the form (9). If, however, instead of the given output map $h(x)$, one uses a different output map $h^*(x)$ in (15), such that the dynamic system:

$$\dot{x} = f(x) + g(x)u$$
$$Y^* = h^*(x) \tag{18}$$

is locally minimum-phase, then the static state feedback:

$$u = \Psi(x,v) = \frac{v - h^*(x) - \sum_{k=1}^{r^*} \gamma_k L_f^k h^*(x)}{\gamma_{r^*} L_g L_f^{r^*-1} h^*(x)} \tag{19}$$

where r^* is the relative order of (18), will guarantee local asymptotic stability of the resulting closed-loop system:

$$\dot{x} = f(x) + g(x)\frac{v - h^*(x) - \sum_{k=1}^{r^*} \gamma_k L_f^k h^*(x)}{\gamma_{r^*} L_g L_f^{r^*-1} h^*(x)} \tag{20}$$

$$Y = h(x)$$

as long as the roots of $1 + \gamma_1 s + \cdots + \gamma_{r^*} s^{r^*}$ have negative real parts.

Note that, under (19), the output trajectory defined by (14) will not be followed in closed-loop; instead, the closed-loop system will follow:

$$Y^* + \gamma_1 \frac{dY^*}{dt} + \cdots + \gamma_{r^*} \frac{d^{r^*}Y^*}{dt^{r^*}} = v \tag{21}$$

where $Y^* = h^*(x)$. Consequently, the unity steady-state gain property will not necessarily hold true and the input/output behavior of the closed-loop system (20) will not be related to controller performance measures in a transparent manner.

When a model-state feedback controller is generated out of the static state feedback (19):

$$\dot{x}_m = f(x_m) + g(x_m)\frac{(y_{sp} - y) + h(x_m) - h^*(x_m) - \sum_{k=1}^{r^*} \gamma_k L_f^k h^*(x_m)}{\gamma_{r^*} L_g L_f^{r^*-1} h^*(x_m)}$$

$$u = \frac{(y_{sp} - y) + h(x_m) - h^*(x_m) - \sum_{k=1}^{r^*} \gamma_k L_f^k h^*(x_m)}{\gamma_{r^*} L_g L_f^{r^*-1} h^*(x_m)} \tag{22}$$

the closed-loop stability property will carry over as long as the dynamics of (9) is locally asymptotically stable. However, this model-state feedback controller will not necessarily induce zero steady-state error in the closed-loop system and closed-loop performance will not be easy to quantify.

The foregoing considerations indicate that special conditions must be imposed on the auxiliary output $h^*(x)$ so that the resulting controller would be useful.

Definition 1 (Wright and Kravaris [21]):
Consider a nonlinear system of the form:
$$\dot{x} = f(x) + g(x)u$$
and denote by E its equilibrium curve:
$$E = \{x \in \Re^n \mid \exists\, u \in \Re \text{ satisfying } f(x) + g(x)u = 0\}$$
The outputs:
$$Y = h(x)$$
and:
$$Y^* = h^*(x)$$
are called statically equivalent if:
$$h(x) = h^*(x) \quad \forall\, x \in E$$
An immediate consequence of the definition is that, if $h^*(x)$ is statically equivalent to $h(x)$, then the static state feedback (19) will induce unity steady-state gain between v and Y in closed-loop. Therefore, the model-state feedback controller (22) will induce zero steady-state error in closed-loop.

Definition 2 (Wright and Kravaris [21]):
Consider a nonlinear system of the form:
$$\dot{x} = f(x) + g(x)u$$
with output:
$$Y = h(x).$$
The auxiliary output:
$$Y^* = h^*(x)$$
is called ISE-optimal if $h^*(x) = v$ is exactly the manifold of optimal state trajectories for the singular optimal control problem:

$$\text{Min} \quad \text{ISE} = \int_0^{\infty} [v - h(x)]^2 dt$$

$$\text{subject to} \quad \dot{x} = f(x) + g(x)u$$

If an ISE-optimal output $h^*(x)$ can be found, this would mean that perfect control of $Y^* = h^*(x)$ would correspond to ISE-optimal control of $Y = h(x)$. Consequently, the speed of the trajectory (21) of Y^* would be a measure of closeness to ISE-optimality, and this would provide a meaning to the tunable parameters $\gamma_1, \ldots, \gamma_{r^*}$. Furthermore, an ISE-optimal $h^*(x)$ will necessarily be minimum-phase and statically equivalent to $h(x)$.

Earlier work of one of the authors focused on the problem of finding an ISE-optimal output as the point of departure for the design of controllers for nonminimum-phase nonlinear systems. In the special case of a second-order system, it was shown in Kravaris and Daoutidis [9] that the ISE-optimal output is:

$$h_{opt}^*(x) = h(x)$$

if $h(x)$ is minimum-phase, whereas:

$$h_{opt}^*(x) = h(x) - 2\frac{L_g h(x)}{\det[[f,g](x)\vdots g(x)]}\det[f(x)\vdots g(x)] \tag{23}$$

if $h(x)$ is nonminimum-phase, where:

$$[f,g](x) = \frac{\partial g}{\partial x}(x)f(x) - \frac{\partial f}{\partial x}(x)g(x)$$

is the Lie bracket of $f(x)$ and $g(x)$. Unfortunately, the ISE-optimization problem is extremely difficult for systems of order higher than 2. In Wright and Kravaris [21], analytical results on ISE-optimal outputs were derived for a very limited class of higher-order systems. Other nonminimum-phase compensation approaches (not directly related to ISE-optimization), which have been proposed in the literature [3],[4] are also limited to very restrictive classes of dynamic systems.

The approach that will be proposed in the present paper will give up the search for an optimal $h^*(x)$. Instead, it will develop a general methodology for generating a large class of statically equivalent outputs, from which it will be possible to select one with prescribed zeros. In this way, instead of looking for optimality, the objective will be to assign zeros.

5. Construction of a Statically Equivalent Synthetic Output with Pre-Assigned Zeros

In this section, a procedure will be proposed for the construction of an output $Y^* = h^*(x)$ which is statically equivalent to a given output $Y = h(x)$ and has pre-assigned zeros. The first step in the procedure involves the construction of $(n-1)$ independent outputs which vanish along the equilibrium curve. With the aid of these $(n-1)$ outputs, a statically equivalent output will be constructed, with adjustable weights. These weights will be subsequently selected to place the zeros at pre-assigned locations. Guidelines for the tuning of the zeros will be suggested at the end of the section.

5.1. CONSTRUCTION OF $(n-1)$ INDEPENDENT OUTPUTS WHICH VANISH ON THE EQUILIBRIUM CURVE

Consider a dynamic system of the form:
$$\dot{x} = f(x) + g(x)u$$
or written more explicitly:

$$\dot{x}_1 = f_1(x) + g_1(x)u$$
$$\vdots$$
$$\dot{x}_{n-1} = f_{n-1}(x) + g_{n-1}(x)u \tag{24}$$
$$\dot{x}_n = f_n(x) + g_n(x)u$$

Whenever this system is at equilibrium (steady-state), the following algebraic equations will be satisfied:

$$0 = f_1(x_s) + g_1(x_s)u_s$$
$$\vdots$$
$$0 = f_{n-1}(x_s) + g_{n-1}(x_s)u_s \qquad (25)$$
$$0 = f_n(x_s) + g_n(x_s)u_s$$

If the vector function $g(x)$ is nonzero, at least one of the components of $g(x)$ will be nonzero. Suppose that $g_n(x)$ is nonzero. Then, one can solve the last equation of (25) for u:

$$u_s = -\frac{f_n(x_s)}{g_n(x_s)}$$

Substituting the result into the first $(n-1)$ equations, leads to $(n-1)$ conditions which must satisfied at equilibrium:

$$0 = f_1(x_s) - \frac{f_n(x_s)}{g_n(x_s)} g_1(x_s)$$
$$\vdots$$
$$0 = f_{n-1}(x_s) - \frac{f_n(x_s)}{g_n(x_s)} g_{n-1}(x_s)$$

The above conditions imply that the functions:

$$n_j(x) = f_j(x) - \frac{f_n(x)}{g_n(x)} g_j(x), \quad j = 1,\ldots,n-1 \qquad (26)$$

must vanish at equilibrium.

Proposition 1:
Consider a dynamic system of the form:
$$\dot{x} = f(x) + g(x)u$$
and denote by E its equilibrium curve:
$$E = \{x \in \mathfrak{R}^n \mid \exists\, u \in \mathfrak{R} \text{ satisfying } f(x) + g(x)u = 0\}$$
Suppose that the n-th component of $g(x)$ is nonzero:
$$g_n(x) \neq 0 \quad \forall\, x \in \mathfrak{R}^n$$
Then the functions:
$$n_j(x) = f_j(x) - \frac{f_n(x)}{g_n(x)} g_j(x), \quad j = 1,\ldots,n-1$$
are well-defined and have the following properties:
(i) $n_j(x) = 0 \quad \forall\, x \in E$
(ii) $n_j(x),\ j = 1,\ldots,n-1$ are linearly independent scalar fields in the vicinity of any equilibrium point such that $\det\left[\dfrac{\partial f}{\partial x}(x_s) + u_s \dfrac{\partial g}{\partial x}(x_s)\right] \neq 0$

Proof: The first property of the functions $n_j(x)$ is an immediate consequence of (25). To prove the second property, observe that:

$$\frac{\partial n_j}{\partial x_k}(x) = \left[\frac{\partial f_j}{\partial x_k}(x) - \frac{f_n(x)}{g_n(x)}\frac{\partial g_j}{\partial x_k}(x)\right] - \frac{g_j(x)}{g_n(x)}\left[\frac{\partial f_n}{\partial x_k}(x) - \frac{f_n(x)}{g_n(x)}\frac{\partial g_n}{\partial x_k}(x)\right] \quad \begin{array}{l} j=1,\ldots,n-1 \\ k=1,\ldots,n \end{array}$$

from which:

$$\begin{bmatrix} \frac{\partial n_1}{\partial x_1}(x) & \frac{\partial n_1}{\partial x_2}(x) & \cdots & \frac{\partial n_1}{\partial x_n}(x) \\ \vdots & \vdots & & \vdots \\ \frac{\partial n_{n-1}}{\partial x_1}(x) & \frac{\partial n_{n-1}}{\partial x_2}(x) & \cdots & \frac{\partial n_{n-1}}{\partial x_n}(x) \end{bmatrix} = \begin{bmatrix} & \vdots & -\frac{g_1(x)}{g_n(x)} \\ I_{n-1} & \vdots & \vdots \\ & \vdots & -\frac{g_{n-1}(x)}{g_n(x)} \end{bmatrix}\left[\frac{\partial f}{\partial x}(x) - \frac{f_n(x)}{g_n(x)}\frac{\partial g}{\partial x}(x)\right]$$

The first factor in the right-hand side has rank $(n-1)$. The second factor in the right-hand side equals $\left[\frac{\partial f}{\partial x}(x_s)+u_s\frac{\partial g}{\partial x}(x_s)\right]$ at equilibrium, and is nonsingular by assumption. Therefore, the right hand side has rank $(n-1)$ at equilibrium.

It is important to emphasize that the functions $n_j(x)$ of Proposition 1 are not the only ones that vanish on the equilibrium curve. Any $(n-1)$ independent eliminant relations from (25) will generate $(n-1)$ functions that vanish at equilibrium. For example, instead of the n_j's given by (26), one may use:

$$n_j(x) = \det\begin{bmatrix} f_j(x) & g_j(x) \\ f_n(x) & g_n(x) \end{bmatrix}, \quad j=1,\ldots,n-1 \qquad (27)$$

These will also vanish on the equilibrium curve and will also be linearly independent scalar fields.

5.2.   A CLASS OF STATICALLY EQUIVALENT OUTPUTS

Proposition 2:
Let $Y = h(x)$ be an output to the dynamic system $\dot{x} = f(x) + g(x)u$. Then, any output of the form:

$$h*(x) = h(x) + \sum_{j=1}^{n-1}\lambda_j(x)\,n_j(x) \qquad (28)$$

where $n_j(x)$, $j=1,\ldots,n-1$, are functions which vanish on the equilibrium curve (e.g. the ones defined by (26) or (27)) and $\lambda_j(x)$, $j=1,\ldots,n-1$, are arbitrary functions, will be statically equivalent to $h(x)$.

Proof: Obvious consequence of Definition 1 and the property of the functions $n_j(x)$.

Remark 1: Comparing with the ISE-optimal $h*(x)$ for second order systems given by (23), we see that it is of the form (28) with:

$$n_1(x) = \det[f(x) \vdots g(x)]$$

$$\lambda_1(x) = -2\frac{L_g h(x)}{\det[[f,g](x) \vdots g(x)]}$$

Remark 2: It is possible to construct a more general class of statically equivalent outputs, where the functions $n_j(x)$ enter nonlinearly. However, this would not be very meaningful from a practical point of view because already (28) defines a very large class of statically equivalent outputs, since the functions $\lambda_j(x)$ are completely arbitrary.

5.3.   ASSIGNMENT OF ZEROS

This subsection will outline a procedure for the selection of an output function $h^*(x)$ out of the class (28) such that the linear approximation of:

$$\dot{x} = f(x) + g(x)u$$

$$Y^* = h^*(x)$$

around a reference equilibrium point has prescribed zeros. To this end, it is sufficient to consider the subclass of (28) with constant weights λ_j, i.e.

$$h^*(x) = h(x) + \sum_{j=1}^{n-1} \lambda_j n_j(x)$$

Let (x_s, u_s) be a reference equilibrium and:

$$P(s) = \frac{\partial h}{\partial x}(x_s) Adj\left[sI - \left(\frac{\partial f}{\partial x}(x_s) + u_s \frac{\partial g}{\partial x}(x_s) \right) \right] g(x_s)$$

$$= p_0 + p_1 s + \cdots + p_{n-1} s^{n-1}$$

$$Q_j(s) = \frac{\partial n_j}{\partial x}(x_s) Adj\left[sI - \left(\frac{\partial f}{\partial x}(x_s) + u_s \frac{\partial g}{\partial x}(x_s) \right) \right] g(x_s)$$

$$= q_{j,1} s + \cdots + q_{j,n-1} s^{n-1}, \quad j = 1,\ldots,n-1$$

be the zeros polynomials corresponding to $h(x)$ and $n_j(x)$, $j = 1,\ldots,n-1$, respectively. Note that the constant term in $Q_j(s)$ will have to equal to zero because $n_j(x)$ vanishes at every steady-state.

Furthermore, let z_j^d, $j = 1,\ldots,n-1$, be the desirable zeros for $h^*(x)$ at the reference equilibrium. The given values of z_j^d and the requirement of static equivalence with $h(x)$ completely specifies the desirable zeros polynomial for $h^*(x)$:

$$P^d(s) = p_0 \prod_{j=1}^{n-1}\left(1 - \frac{s}{z_j^d} \right)$$

$$= p_0 + p_1^d s + \cdots + p_{n-1}^d s^{n-1}$$

The necessary values of the adjustable parameters λ_j that will make the zeros polynomial of $h^*(x)$ equal to $P^d(s)$ can be obtained from the condition:

$$P(s) + \sum_{j=1}^{n-1} \lambda_j Q_j(s) = P^d(s)$$

i.e. by solving the linear equation:

$$\begin{bmatrix} q_{11} & q_{21} & \cdots & q_{n-1,1} \\ \vdots & \vdots & & \vdots \\ q_{1,n-1} & q_{2,n-1} & \cdots & q_{n-1,n-1} \end{bmatrix} \begin{bmatrix} \lambda_1 \\ \vdots \\ \lambda_{n-1} \end{bmatrix} = \begin{bmatrix} p_1^d - p_1 \\ \vdots \\ p_{n-1}^d - p_{n-1} \end{bmatrix}$$

Because $\dfrac{\partial n_j}{\partial x}(x_s), j = 1,\ldots, n-1$ are linearly independent, the columns

$[q_{j1} \quad \cdots \quad q_{j,n-1}]^T$ will be linear independent as long as $\left[\dfrac{\partial f}{\partial x}(x_s) + u_s \dfrac{\partial g}{\partial x}(x_s) \right]$ and $g(x_s)$

form a controllable pair; under this condition, the above linear equation will admit a unique solution.

5.4. DISCUSSION

(a) Under an input/output linearizing state feedback with respect to $Y^* = h^*(x)$, (19), the resulting closed-loop dynamics consists of two parts: a linear part whose poles are exactly the roots of $1 + \gamma_1 s + \cdots + \gamma_{r*} s^{r*}$ and a nonlinear part which represents the zero dynamics of $Y^* = h^*(x)$. When Y^* is viewed as the output of the closed-loop system, the nonlinear part of the closed-loop system becomes unobservable and the input/output behavior becomes linear, following the output trajectory (21). However, when $Y = h(x)$ is viewed as the output of the closed-loop system, the nonlinear part of the closed-loop system is observable and it influences the output performance characteristics. By adjusting the weights λ_j of the function $h^*(x)$, one assigns the $(n - r^*)$ eigenvalues of the linear approximation of the zero dynamics of Y^* and, therefore, $(n - r^*)$ poles of the linear approximation of the closed-loop system. In other words, the proposed controller design approach is essentially a local pole-placement approach, where a subset of the closed-loop poles are assigned implicitly in terms of the zeros of the synthetic output function.

(b) Because of the local nature of the pole placement for the $(n - r^*)$-th-order nonlinear part of the closed-loop dynamics, one may wish to try to maximize r^*, in order to maximize the linear part of the closed-loop dynamics. The ultimate maximum would be $r^* = n$, which would linearize the entire closed-loop dynamics. The case $r^* = n$ corresponds to $h^*(x)$ having no zeros and therefore to $P^d(s)$ being constant. The design procedure outlined in the previous section poses absolutely no restrictions on using $P^d(s) = p_o$, i.e. selecting $p_1^d = \cdots p_{n-1}^d = 0$. One can always use the derived condition and uniquely determine the necessary values of the weights λ_j. It is well known, however, that an output of relative order n does not exist, unless the process dynamics satisfy some very restrictive involutivity conditions [19]. And even if these involutivity conditions are satisfied, the class of outputs of relative order n may not contain a member that is statically equivalent to the given output.

The resolution of this apparent contradiction is the following: the output $h^*(x)$ calculated from the condition $P^d(s) = p_o$ will not, in general, have relative order n, but instead, it will have a relative-order singularity at the design steady-state x_s. Because relative-order singularity is, in general, undesirable, the use of $P^d(s) = p_o$ should be avoided.

(c) With complete linearity of the closed-loop system being infeasible, one may wish to try at least to preserve the relative order of the original output, i.e. try to enforce $r^* = r$. Suppose, for example, that $r = 2$ and one wishes to construct a statically equivalent output with $r^* = 2$. If one tries to enforce this condition by simply using a $(n - 2)$-order polynomial for $P^d(s)$, i.e. setting $p_{n-1}^d = 0$, weights will be uniquely determined but the resulting function $h^*(x)$ could have relative-order singularity at x_s. However, this singularity problem could be prevented if one could choose $n_1(x),\ldots,n_{n-2}(x)$ to have relative order larger or equal to 2. Then, the weights $\lambda_1,\ldots,\lambda_{n-2}$ can be safely adjusted to match the $(n - 2)$ prescribed zeros, setting $\lambda_{n-1} = 0$. It turns out that one can always find $(n - 2)$ functions of relative order 2 that vanish on the equilibrium curve. These will be solutions of the partial differential equation:

$$\sum_{m=1}^{n} \frac{\partial w}{\partial x_m} g_m(x) = 0$$

under the additional condition that

$$w(x) = 0 \quad \forall x \in E$$

but the numerical calculation of these $(n - 2)$ solutions is far from being trivial.

If the original output has relative order r larger than 2 and one wishes to have $r^* = r$, the situation is even more complicated. In order to prevent relative-order singularity, one will seek for $(n - r^*)$ functions which simultaneously satisfy $(r^* - 1)$ partial differential equations and also vanish on E. But then, involutivity restrictions will arise from the theorem of Frobenius and, even when they are met, the numerical calculation will be, for all practical purposes, intractable.

(d) The previously mentioned difficulties in imposing a high relative order r^* suggest that $r^* = 1$ must be used in the application of the proposed method. When a $(n - 1)$-th-degree polynomial is used for $P^d(s)$ (i.e. $p_{n-1}^d \neq 0$), then the output $Y^* = h^*(x)$ is guaranteed to have relative order $r^* = 1$. Of course, one must check for possible relative-order singularities away from the design steady-state x_s. If it turns out that x_s is too close to the singular manifold $L_g h^*(x) = 0$, $P^d(s)$ must be changed, in order to derive a different output $h^*(x)$.

(e) The following design rules are recommended:
- Choose $z_1^d,\ldots,z_{n-r}^d$ to be at the LHP zeros of the linear approximation of the process and at the reflections of the RHP zeros with respect to the imaginary axis.

- Tune the excess zeros $z^d_{n-r+1}, \ldots, z^d_{n-1}$ and the output trajectory pole $-1/\gamma_1$ similarly to linear IMC filter poles, as compromise parameters between performance and robustness.

The above design rules will lead to ISE-optimal response for <u>small</u> step changes in the vicinity of the design steady-state, in the limit as $z^d_{n-r+1}, \ldots, z^d_{n-1}, -1/\gamma_1$ tend to negative infinity. Alternative optimality criteria could be used, like for example a weighted sum of ISE with a quadratic input penalty, leading to different design rules.

(f) The controller design approach outlined in this section gives rise to a nonlinear-model-based controller, but the design specifications are local, around a given steady-state. In order to impose more global design specifications, state-dependent weights $\lambda_j(x)$ must be used. The use of state-dependent weights will also allow constructing statically equivalent outputs of higher relative order. The design of state-dependent weights $\lambda_j(x)$ is beyond the scope of the present paper.

6. Application: Control of a Nonisothermal van de Vusse Reactor

To illustrate the application of the proposed control strategy, a continuous stirred tank reactor (CSTR) is considered, where the series/parallel van de Vusse reaction [8],[20] is taking place:

$$A \xrightarrow{k_1} B$$
$$B \xrightarrow{k_2} C$$
$$2A \xrightarrow{k_3} D$$

In the above mechanism, A is the reactant, B the desired product, C and D are unwanted by-products. An example is the production of cyclopentenol (B) from cyclopentadiene (A) by acid-catalyzed electrophilic addition of water in dilute solution, where cyclopentanediol (C) and dicyclopentadiene (D) are also produced as side products [5].

The control of an isothermal CSTR with a van de Vusse reaction has been investigated in the past by many researchers (see e.g. Kantor [8], Kravaris and Daoutidis [9], Wright and Kravaris [21], Engell and Klatt [5]). This system has unstable zero dynamics and it has been used to test nonlinear control algorithms in terms of their ability to handle nonminimum-phase behavior. In Doyle et al. [3], nonisothermal operation was considered but the system was made minimum-phase through the use of feedback with an additional manipulated input. In this work, nonisothermal operation will be considered and control will be achieved via a single manipulated input; in this way, the process will be nonminimum-phase and third-order, making the control problem more challenging than the isothermal case which was nonminimum-phase and second-order.

The rates of formation of A and B are assumed to be as follows:

$$r_A = -k_1(T)C_A - k_3(T)C_A^2$$
$$r_B = k_1(T)C_A - k_2(T)C_B$$

where the rate coefficients $k_i(T)$ are given by Arrhenius expressions:

$$k_i(T) = k_{i0}\exp(-E_i/RT)$$

and C_A and C_B are the concentrations of the species A and B inside the reactor, respectively. The volume of the CSTR, V, is assumed to be constant during the operation. The feed stream consisting of pure A is fed to the reactor at an inlet flow rate F. The concentration of A in the feed stream is $C_{A0} = 5$ gmol·L^{-1} and the feed temperature is $T_O = 403.15$ K. The dynamics of the CSTR can be described in terms of the material balances for species A and B and an energy balance for the reactor as follows:

$$\frac{dC_A}{dt} = -k_1(T)C_A - k_3(T)C_A^2 + (C_{A0} - C_A)u$$

$$\frac{dC_B}{dt} = k_1(T)C_A - k_2(T)C_B - C_B u$$

$$\frac{dT}{dt} = \frac{(-\Delta H_1)k_1(T)C_A + (-\Delta H_2)k_2(T)C_B + (-\Delta H_3)k_3(T)C_A^2 + Q}{\rho C_p} + (T_0 - T)u$$

where $u = F/V$ is the dilution rate, T is the reactor temperature, $-Q$ is the cooling rate per unit volume (assumed constant), ρ and C_p are the density and specific heat of the reaction mixture, respectively, and ΔH_i are the heats of reaction. These parameters were assumed to have the following values (from Engell and Klatt [5]):

$k_{10} = 1.287 \cdot 10^{12}$ h^{-1}	$k_{20} = 1.287 \cdot 10^{12}$ h^{-1}	$k_{30} = 9.043 \cdot 10^9$ L(mol·h)$^{-1}$
$E_1/R = -9758.3$ K	$E_2/R = -9758.3$ K	$E_3/R = -8560$ K
$\Delta H_1 = 4.2$ kJ·mol^{-1}	$\Delta H_2 = -11$ kJ·mol^{-1}	$\Delta H_3 = -41.85$ kJ·mol^{-1}
$\rho = 0.9342$ kg L^{-1}	$C_p = 3.01$ kJ(kg· K)$^{-1}$	$Q = -451.509$ kJ(L·h)$^{-1}$

The control objective is to maintain the output:

$$y = C_B$$

at set point, by manipulating the dilution rate, $u = F/V$.

The steady-state characteristics of the process can be directly obtained form the model. Solving the steady-state equations for the manipulated input, one obtains the following three conditions:

$$u_s = \frac{k_1(T_s)C_{As} + k_3(T_s)C_{As}^2}{C_{A0} - C_{As}}$$

$$= \frac{k_1(T_s)C_{As} - k_2(T_s)C_{Bs}}{C_{Bs}}$$

$$= \frac{(-\Delta H_1)k_1(T_s)C_{As} + (-\Delta H_2)k_2(T_s)C_{Bs} + (-\Delta H_3)k_3(T_s)C_{As}^2 + Q}{\rho C_p(T_s - T_0)}$$

From these conditions and the given parameter values, one can derive the equilibrium curve of the process; this is shown in Figure 5. The equilibrium curve consists of two branches: a high temperature - high production rate branch, and a low temperature - low production rate branch. The plane of constant output, $y = C_B$, crosses the equilibrium curve at two points, each representing a different steady-state. Considering the fact that there are always limits on the flow rate and the operating temperature, the former of

136

these two branches seems is more desirable for the operation of the reactor. Therefore, for the closed loop simulation studies, which will be presented towards the end of this section, operation along the high temperature - high production rate branch of the equilibrium curve was chosen.

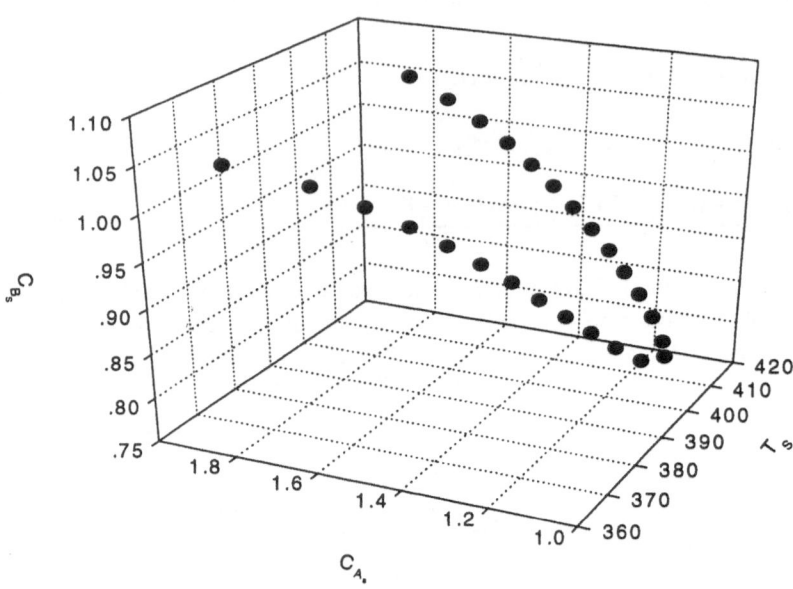

FIGURE 5. Equilibrium Curve of the Process

From the steady-state conditions presented earlier, one can immediately see that the functions:

$$n_1(C_A, C_B, T) = -[k_1(T)C_A + k_3(T)C_A^2]C_B + (C_{AO} - C_A)[k_1(T)C_A - k_2(T)C_B]$$

$$n_2(C_A, C_B, T) = [(-\Delta H_1)k_1(T)C_A + (-\Delta H_2)k_2(T)C_B + (-\Delta H_3)k_3(T)C_A^2 + Q]C_B +$$

$$\rho C_p(T_0 - T)[k_1(T)C_A - k_2(T)C_B]$$

vanish at every point of the equilibrium curve. These are of the form of (27) and will be used in the construction of a statically equivalent output.

The zero dynamics of the process can now be calculated. Observing that the process has relative order $r = 1$ (since the time derivative of the output $y = C_B$ explicitly depends on the input u), and that it is already in the form of (12), the zero dynamics can be directly obtained from (13):

$$\frac{dC_A}{dt} = -[k_1(T)C_A + k_3(T)C_A^2] + (C_{AO} - C_A)\frac{k_1(T)C_A - k_2(T)y_s}{y_s}$$

$$\frac{dT}{dt} = \frac{(-\Delta H_1)k_1(T)C_A + (-\Delta H_2)k_2(T)y_s + (-\Delta H_3)k_3(T)C_A^2 + Q}{\rho C_p} +$$

$$(T_0 - T)\frac{k_1(T)C_A - k_2(T)y_s}{y_s}$$

where y_s is a reference steady-state value of the output $y = C_B$. To test for local minimum-phase behavior, one can calculate the eigenvalues of the linear approximation of the zero dynamics, which will coincide with the zeros of the transfer function of the linear approximation of the process dynamics.

Suppose, for example, that the reference steady-state is $C_{As} = 1.25$ mol·L^{-1}, $C_{Bs} = 0.90$ mol·L^{-1}, $T_s = 407.15$ K, which corresponds to $u_s = 19.5218$ hr^{-1}. Around this steady-state, the process is locally asymptotically stable; the eigenvalues of the linear approximation of the process dynamics are at -96.4652 and $-33.1538 \pm 9.81523\ i$. The zeros polynomial for the output $y = C_B$ (characteristic polynomial of the linear approximation of the zero dynamics, or equivalently, the numerator polynomial of the transfer function of the linear approximation of the process dynamics) is:

$$P(s) = 1233.01 + 100.364s - 0.9s^2$$

with roots at -11.1671 and $+122.683$. This shows that the process is locally nonminimum-phase around the given steady-state, because of the presence of a RHP zero. Similarly, one can calculate the zeros polynomials corresponding to $n_1(C_A, C_B, T)$ and $n_2(C_A, C_B, T)$ given earlier, for the particular steady-state under consideration. These are given by:

$$Q_1(s) = 7227.05s + 649.185s^2$$

$$Q_2(s) = 95163.3s - 1505.32s^2$$

Thus, when one forms the statically equivalent output:

$$y^* = C_B + \lambda_1 n_1(C_A, C_B, T) + \lambda_2 n_2(C_A, C_B, T)$$

its zeros polynomial around the given steady-state:

$$P(s) + \lambda_1 Q_1(s) + \lambda_2 Q_2(s)$$

can be made equal to a given polynomial $P^d(s)$ by appropriately adjusting the weights λ_1 and λ_2. For example, choosing

$$P^d(s) = 1233.01\left(1 + \frac{s}{11.1671}\right)\left(1 + \frac{s}{122.683}\right)$$
$$= 0.9(s + 11.1671)(s + 122.683)$$

so that the zeros of y^* are at the LHP zero of $P(s)$ and at the reflection of the RHP zero, the necessary weights are found to be:

$$\lambda_1 = 0.002774$$

$$\lambda_2 = 5.58478 \cdot 10^{-7} \approx 0$$

Under this choice of weights, the statically equivalent output y^* will be ISE-optimal for small step changes in the set point. Also, it will behave well for larger changes in the

138

set point, although the response will not be ISE-optimal. In the closed-loop simulations, the value of the weight λ_2 was set equal to zero, and therefore,

$$y^* = h^*(C_A, C_B, T) = C_B + 0.002774n_1(C_A, C_B, T)$$

was used in the control law. This output has relative order $r^* = 1$ and therefore the state-feedback:

$$\Psi(C_A, C_B, T, v) = \frac{v - h^*(C_A, C_B, T) - \gamma_1 L_f h^*(C_A, C_B, T)}{\gamma_1 L_g h^*(C_A, C_B, T)}$$

was used to generate the model-state feedback controller. The value $\gamma_1 = 0.1$ was used in all simulations.

Figure 6 shows the closed-loop response under model-state feedback control for a positive step change of 0.1 mol·L^{-1} in the set point. Both the original output $y = C_B$ and the synthetic minimum-phase output $y^* = h^*(C_A, C_B, T)$ are depicted in the figure.

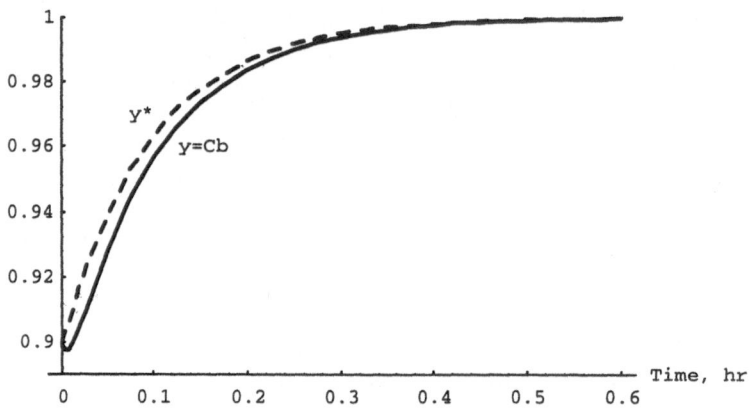

FIGURE 6. Closed-Loop Response for a Step Change in the Set Point

One can observe from this graph that the process output shows an initial inverse response of due to the unstable zero dynamics, while the minimum-phase statically equivalent output follows the linear first-order exponential trajectory on which the controller was based. The nonminimum-phase characteristics of the original output are better reflected in Figure 7, where the time scale is shortened to cover the details of the initial inverse response.

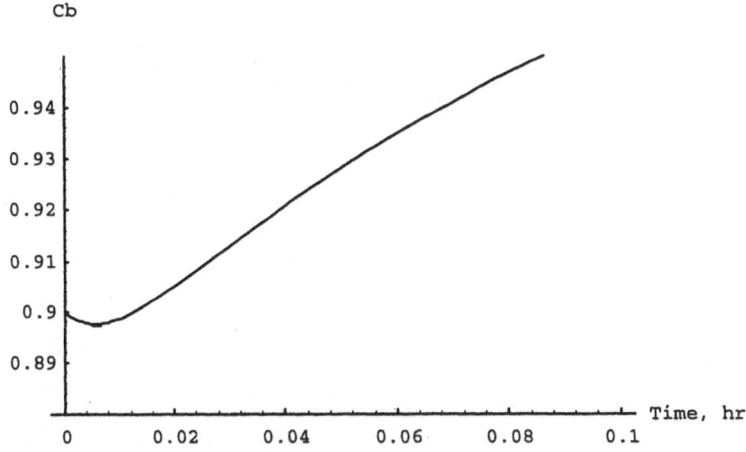

FIGURE 7. Nonminimum-Phase Behavior of C_B for a Step Change in the Set Point

The profiles of the control effort and the reactor temperature are presented in Figures 8 and 9, respectively, under the same set point change.

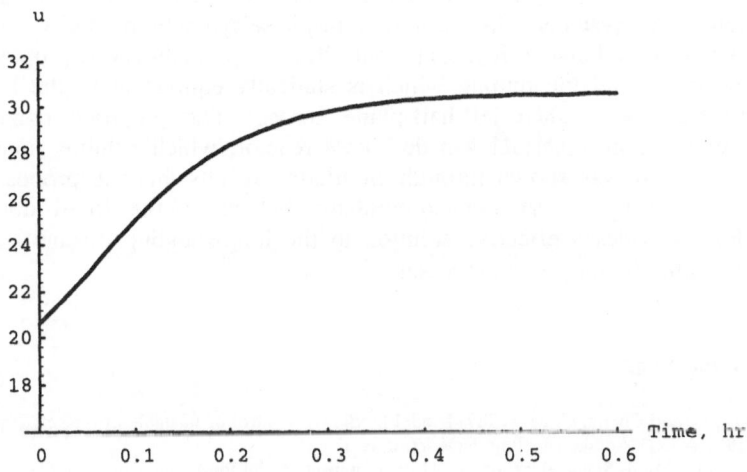

FIGURE 8. Control Effort for a Step Change in the Set Point

140

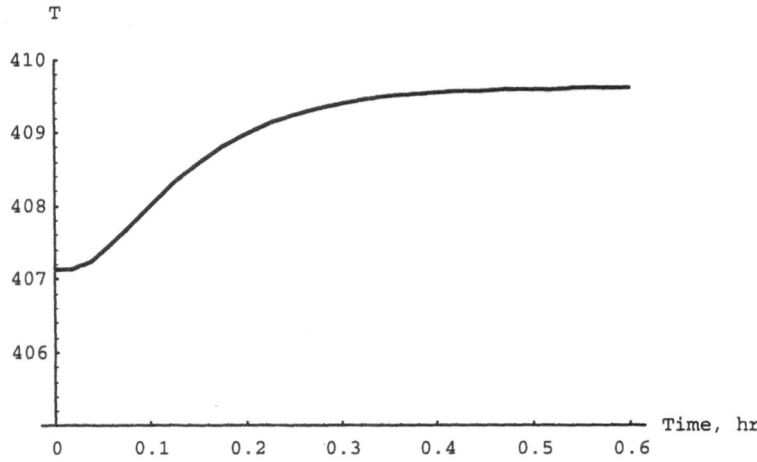

FIGURE 9. Response of Reactor Temperature for a Step Change in the Set Point

7.   Conclusion

This paper first provided a review of the properties of model-state feedback and input/output linearizing state feedback. These concepts were then extended to nonminimum-phase systems, where a minimum-phase synthetic output was utilized for the synthesis of a model-state feedback controller. A procedure was proposed for the construction of a synthetic output, which is statically equivalent to the actual plant output and has pre-assigned left-half-plane zeros.   The proposed concepts were illustrated with a nonisothermal van de Vusse reactor, which exhibits nonminimum-phase behavior. It was shown through simulation results that the proposed scheme leads to excellent set point tracking and regulatory behavior in the closed-loop response, and therefore provides a effective solution to the long-standing difficult problem of controlling nonminimum-phase processes.

8.   References

1.   Brosilow, C. and Cheng, C. M. (1987) Model predictive control of unstable processes, Paper presented at the 1987 AIChE Annual Meeting, New York, N.Y.
2.   Coulibaly, E., Maiti, S. and Brosilow, C. (1995) Internal Model Predictive Control (IMPC), *Automatica*, **31**, 1471-1482; also, presented at the 1992 AIChE Annual Meeting, Miami Beach, Fl.
3.   Doyle III, F. J., Allgower, F., Oliveira, S., Gilles, E. and Morari, M. (1992) On nonlinear systems with poorly behaved zero dynamics, Proceedings 1992 American Control Conference, Chicago, IL, 2571-2575.
4.   Doyle III, F.J., Allgower, F., and Morari, M. (1996) A normal form approach to approximate input-output linearization for maximum phase nonlinear SISO systems, *IEEE Trans. Aut. Cont.*, **41**, 305-309.
5.   Engell, S. and Klatt, K.U. (1993) Nonlinear control of a nonminimum-phase CSTR, Proceedings 1993 American Control Conference, San Fransico, CA, 2341-2945.
6.   Hirschorn, R.M. (1979) Invertibility of nonlinear control systems, *SIAM J. Cont. Opt.*, **17**, 289-297.
7.   Isidori, A. (1989) Nonlinear Control Systems, 2nd ed. Springer, Berlin.

8. Kantor, J.C. (1986) Stability of state feedback transformations for nonlinear systems- Some practical considerations, Proceedings 1986 American Control Conference, Seattle, WA, 1014-1016.

9. Kravaris, C. and Daoutidis, P. (1990) Nonlinear state feedback control of 2nd-order nonminimum-phase nonlinear systems, *Comp. Chem. Engng.*, **14**, 439-449.

10. Kravaris, C. and Daoutidis, P. (1992) Output feedback controller realization for open-loop stable nonlinear processes, Proceedings 1992 American Control Conference, Chicago, IL, 2596-2601.

11. Kravaris, C., Daoutidis, P. and Wright, R.A. (1994) Output feedback control of nonminimum-phase nonlinear processes, *Chem. Eng. Sci.*, **49**, 2107-2122.

12. Mhatre, S. and Brosilow, C. (1996) Multivariable model state feedback, Proceedings IFAC World Congress, San Fransico, CA, Vol. M, 139-144.

13. Soroush, M. and Kravaris, C. (1992) A continuous-time formulation of nonlinear model-predictive control, Proceedings 1992 American Control Conference, Chicago, IL, 1561-1567.

14. Soroush, M. and Kravaris, C. (1992) Discrete time nonlinear controller synthesis by input/output linearization, *AIChE J.*, **38**, 1923-1945.

15. Soroush, M. and Kravaris, C. (1996) A Continuous time formulation of nonlinear model predictive control, *Int. J. Cont.*, **63**, 121-146.

16. Soroush, M. and Kravaris, C. (1996) MPC formulation of GLC, *AIChE J.*, **42**, 2377-2381.

17. Soroush, M. and Nikravesh, M. (1996) Shortest-prediction-horizon nonlinear model predictive control, Proceedings IFAC World Congress, San Fransico, CA, Vol. M, 19-24.

18. Soroush, M. (1997) Nonlinear state-observer design with application to reactors, *Che. Eng. Sci.*, **52**, 387-404.

19. Su, R. (1982) On the linear equivalents of nonlinear systems, *Syst. Control Lett.*, **2**, 48-52.

20. van de Vusse, J.G. (1964), Plug-flow-type reactor versus tank reactor, *Chem. Eng. Sci.*, **19**, 994-998.

21. Wright, R.A. and Kravaris, C. (1992) Nonminimum-phase compensation for nonlinear processes, *AIChE J.* **38**, 26-40.

NONLINEAR MODEL-ALGORITHMIC CONTROL:

A REVIEW AND NEW DEVELOPMENTS

MICHAEL NIEMIEC and COSTAS KRAVARIS
Department of Chemical Engineering
The University of Michigan
Ann Arbor, Michigan 48109-2136
U.S.A.

Abstract. This paper presents a review of Nonlinear Model-Algorithmic Control (MAC), and introduces a new generalized formulation, which incorporates nonminimum-phase compensation. In addition, new nonlinear multirate Model-Algorithmic Control methods are developed. The handling of input constraints within a MAC framework is also discussed. Several chemical engineering examples illustrate the methods throughout the paper.

1. Introduction

The chemical engineering field is dominated by processes that exhibit strong nonlinearities, creating major challenges in design and control. Since the control algorithm is usually based upon a linear approximation of the process dynamics around an operating steady state, performance can be unacceptably poor in the presence of strong nonlinearities. Processes with strong nonlinearities include polymerization reactors, high-purity distillation columns, bioprocesses, and pH processes. The significance of nonlinearities and the insufficiency of traditional control strategies has shifted academic and industrial efforts to develop new nonlinear control schemes. These schemes include model predictive process control and geometric process control. As a direct result, the area of nonlinear process control has evolved as a structurally solid research area with practical applicability, enabling the solution of important practical problems. One class of nonlinear controllers, which combines the theoretical rigor of geometric control methods with the intuitive appeal of model-predictive control methods, is the class of Model-Algorithmic Controllers (MAC). These will be the focus of the present paper. Furthermore, the basis of any nonlinear, digital, computer-based control law is the time-discretization of a nonlinear continuous-time dynamic system. Discretization with fast sampling rates can cause the system to become nonminimum-phase. This creates complications in the design of controllers, which are based on an inverse of the process model. New research efforts to develop nonminimum-phase compensation methods for nonlinear discrete-time systems will be introduced and demonstrated with a chemical engineering example.

143

R. Berber and C. Kravaris (eds.), Nonlinear Model Based Process Control, 143-172.
© 1998 *Kluwer Academic Publishers.*

Another major industrial process control problem results from process measurements that are available at different rates. One common scenario arises when the process outputs, or controlled variables, are measurable only with large sampling periods and delays. These slow sampling rates are primarily imposed by hardware limitations and/or measurement cost, and are associated with distillation columns, packed-bed reactors, and polymerization reactors, where compositions and product properties are measured with various forms of analytical instruments such as chromatography columns. Typically, temperature and pressure measurements, which are available at fast sampling rates, are used in industry to control these important, but slowly sampled, process variables which are hastily incorporated into the control of the process when they are available. A new nonlinear multirate Model-Algorithmic Control (MAC) strategy will be presented that effectively incorporates the slowly sampled measurements while maintaining proper performance in the control of the measurements with fast sampling rates.

2.   Model-Algorithmic Control (MAC)

In its original form, Model-Algorithmic Control is a model-predictive controller, which uses a linear impulse response model to predict the future behavior of the system. MAC was developed in France in the late sixties within the chemical process industry. It originally appeared as a heuristic algorithm under the name of Model Heuristic Control in the late seventies [14], with advancements in the theory occurring in the early eighties [11],[12]. In a comprehensive review of MAC, the controller was shown to be essentially identical to the IMC structure when compared with other theoretical controller synthesis methods [5]. Although the original development of MAC uses a linear input/output model, a generalized state-space approach for both linear and nonlinear systems has been developed following the conceptual steps of the original methodology [15],[18].

2.1.   SISO NONLINEAR MODEL-ALGORITHMIC CONTROL

2.1.1.   *Preliminaries*

Time-Discretization of a Nonlinear Dynamic Model
Most chemical processes can be adequately modeled with unsteady-state material and energy balances in the form of a system of ordinary differential equations (ODE's) as follows:

$$\dot{x} = f(x) + ug(x) \tag{1}$$

where u is the manipulated input, x is the n-vector of states, and $f(x)$, $g(x)$ are nonlinear vector functions. The output of the process is expressed as a nonlinear function of the

states:

$$y = h(x) \tag{2}$$

Given an equidistant grid of points in time:

$$t_k = k\Delta t, \quad k = 0,1,2,\ldots$$

where Δt is the sampling period, and assuming zero-order hold of the input in between samples, the nonlinear system of ODE's (1) can be discretized according to [7]:

$$x(k+1) = x(k) + \Delta t \left[I_n + \sum_{l=1}^{\infty} \frac{\Delta t^l}{(l+1)!} M_l[x(k),u(k)] \right] \left[f[x(k)] + u(k)g[x(k)] \right] \tag{3}$$

where:

$$M_1(x,u) = \frac{\partial f}{\partial x}(x) + u\frac{\partial g}{\partial x}(x)$$

$$M_{l+1}(x,u) = \frac{\partial}{\partial x}\left[M_l(x,u)[f(x) + ug(x)] \right] \quad l = 1,2,3,\ldots$$

The output map of the process is:

$$y(k) = h[x(k)] \tag{4}$$

In general, the infinite series (3) cannot be evaluated in closed form, and therefore, can only approximately discretize the system of ODE's. For example, using a relatively small sampling period Δt, an accurate approximation can be obtained by truncating the series (3), keeping only a few leading terms. The crudest approximation, known as Euler's method, is the zeroth-order truncation:

$$x(k+1) = x(k) + \Delta t[f[x(k)] + u(k)g[x(k)]]$$

Depending on the size of the sampling period, Euler's method may or may not be adequate. Alternative approximate discretization approaches, which are not directly based upon a truncation of the series, include Runge-Kutta, Heun, Adams, etc., and can be found in numerical analysis books.

Two important remarks must be made here. First, in all approximate discretization methods, the continuous-time and time-discretized systems have exactly the same steady-state characteristics. For this reason, steady-state analysis can be performed using the original continuous-time description, which is usually easier. Second, unless Euler's method is used, the discretized system has the manipulated input appearing nonlinearly in the right hand side.

In this paper, we will consider a general discrete-time state-space model of the form:

$$\begin{aligned} x(k+1) &= \Phi[x(k),u(k)] \\ y(k) &= h[x(k)] \end{aligned} \tag{5}$$

where x denotes the vector of state variables, u denotes the manipulated input, and y represents the controlled output. It is assumed that $x \in X \subset \Re^n$, $u \in U \subset \Re$, $\Phi(x, u)$ is an analytic vector function on $X \times U$, and $h(x)$ is an analytic scalar function on X.

Relative Order

For a system with the form of (5), the relative order of the output y with respect to the manipulated input u is the smallest integer r for which:

$$\left[\frac{\partial h(x)}{\partial x}\right]\left[\frac{\partial \Phi(x,u)}{\partial x}\right]^{r-1}\left[\frac{\partial \Phi(x,u)}{\partial u}\right] \not\equiv 0 \tag{6}$$

If such an integer does not exist, then $r = \infty$. Equivalently, r can be viewed as the smallest number of sampling periods after which the manipulated input move $u(k)$ affects the output y. With these ideas, the following notation can be defined:

$$h^0(x) \overset{\Delta}{=} h(x)$$

$$h^l(x) \overset{\Delta}{=} h^{l-1}[\Phi(x,u)], \quad l = 1,\dots,r-1$$

In this notation:

$$\frac{\partial}{\partial u}h^{r-1}[\Phi(x,u)] = \left[\frac{\partial h(x)}{\partial x}\right]\left[\frac{\partial \Phi(x,u)}{\partial x}\right]^{r-1}\left[\frac{\partial \Phi(x,u)}{\partial u}\right] \not\equiv 0 \tag{7}$$

Furthermore, the following relations will hold:

$$y(k+l) = h^l[x(k)], \qquad l = 0,\dots,r-1$$
$$y(k+r) = h^{r-1}[\Phi[x(k), u(k)]]$$

If the system output y does not have a finite relative order ($r = \infty$), the manipulated input u never affects the output y. In a well formulated control problem, the output y must possess a finite relative order r. If a discrete-time process has $r > 1$, then the plant dead-time is equal to $(r-1)\Delta t$, where Δt is the sampling period of the process.

For a process of the form of (5) and finite relative order r, (7) implies that the nonlinear algebraic equation:

$$h^{r-1}[\Phi(x,u)] = \tilde{y} \tag{8}$$

is locally solvable in u via the implicit function theorem. The implicit function will be represented by:

$$u = \Psi(x, \tilde{y}) \tag{9}$$

which is assumed to be well defined and unique on $X \times h(X)$.

Minimum-Phase Behavior and Zero Dynamics

The notion of relative order motivates the decomposition of the process into two subsystems in series. The subsystems are: (i) a delay-free subsystem, and (ii) a pure delay subsystem:

$$\left.\begin{aligned} x(k+1) &= \Phi[x(k), u(k)] \\ \tilde{y}(k) &= h^{r-1}[\Phi[x(k), u(k)]] \end{aligned}\right\} \text{Delay - Free}$$

$$y(k) = \tilde{y}(k-r) \qquad \qquad \} \text{Pure Delay}$$

(10)

This decomposition is a generalization of the factorization of linear discrete-time systems into an invertible part and a pure delay. The inverse of the delay-free subsystem can be constructed due to the solvability of (8):

$$\begin{aligned} x(k+1) &= \Phi[x(k), \Psi[x(k), \tilde{y}(k)]] \\ u(k) &= \Psi[x(k), \tilde{y}(k)] \end{aligned}$$

(11)

If the system is linear, the inverse system of (11) has $n-r$ poles at the process zeros and r poles at the origin. For the inverse system to be stable, the finite process zeros must be inside the unit circle. Accordingly, stability of the nonlinear inverse system can be defined.

Definition 1. Given a discrete-time nonlinear system of the form of (5), the delay-free part is called minimum-phase if the dynamics:

$$x(k+1) = \Phi[x(k), \Psi[x(k), \tilde{y}(k)]]$$

(12)

are locally asymptotically stable. If the dynamics are not stable, then the system will be called nonminimum-phase.

The local asymptotic stability of (12) can be checked, via Lyapunov's first method, by calculating the eigenvalues of the Jacobian evaluated at an equilibrium point. Using (8) and (9), along with the implicit function theorem, the Jacobian of the system is:

$$\mathcal{J}(x, u) = \left[\frac{\partial \Phi(x, u)}{\partial x}\right] - \left[\frac{\partial \Phi(x, u)}{\partial u}\right]\left[\frac{\partial}{\partial x} h^{r-1}[\Phi(x, u)]\right]\left[\frac{\partial}{\partial u} h^{r-1}[\Phi(x, u)]\right]^{-1}$$

(13)

which is evaluated at the reference equilibrium point $(x, u) = (x_{ss}, u_{ss})$. If all the eigenvalues of the Jacobian are in the interior of the unit circle, the dynamics of (12) are guaranteed to be locally asymptotically stable around the reference equilibrium point.

Definition 2. Given a discrete-time nonlinear system of the form of (5), the delay-free part is called hyperbolically minimum-phase if all the eigenvalues of the Jacobian matrix, evaluated at a reference equilibrium point $(x, u) = (x_{ss}, u_{ss})$, are in the interior of the unit circle.

It is important to note that if a system of the form of (5) is hyperbolically minimum-phase, then it will also be minimum-phase. However, the converse of the statement may not hold. The dynamics of (12) can have some eigenvalues on the unit circle and still be asymptotically stable.

2.1.2. *SISO Nonlinear Model-Algorithmic Control for Unconstrained Processes*

The idea of relative order is key for the understanding of the model predictive framework. Starting with a system of the form of (5) and finite relative order r, future changes of the output y can be predicted by simulating the model on-line:

$$y_M(k+1) - y_M(k) = h^1[x_M(k)] - h[x_M(k)]$$
$$y_M(k+2) - y_M(k) = h^2[x_M(k)] - h[x_M(k)]$$
$$\vdots \qquad\qquad\qquad (14)$$
$$y_M(k+r-1) - y_M(k) = h^{r-1}[x_M(k)] - h[x_M(k)]$$
$$y_M(k+r) - y_M(k) = h^{r-1}[\Phi[x_M(k), u(k)]] - h[x_M(k)]$$

where the subscript M denotes variables from the on-line model simulation. These predicted changes can then be added to the measurement output signal $y(k)$ to obtain the following "closed-loop" predictions of the output:

$$\hat{y}(k+1) = y(k) + h^1[x_M(k)] - h[x_M(k)]$$
$$\hat{y}(k+2) = y(k) + h^2[x_M(k)] - h[x_M(k)]$$
$$\vdots \qquad\qquad\qquad (15)$$
$$\hat{y}(k+r-1) = y(k) + h^{r-1}[x_M(k)] - h[x_M(k)]$$
$$\hat{y}(k+r) = y(k) + h^{r-1}[\Phi[x_M(k), u(k)]] - h[x_M(k)]$$

A desired linear reference trajectory can be defined for the output:

$$\hat{y}(k+r) = (1-\alpha)y_{sp} + \alpha\hat{y}(k+r-1) \qquad (16)$$

where the subscript sp denotes the output set-point, and α is a tunable scalar parameter. Combining (15) and (16), such that the output prediction matches the reference trajectory, one can derive the nonlinear SISO Model-Algorithmic Controller:

$$h^{r-1}[\Phi[x_M(k), u(k)]] = (1-\alpha)[e(k) + h[x_M(k)]] + \alpha h^{r-1}[x_M(k)] \qquad (17)$$

Using (8) and (9), the manipulated input can be calculated, with the corresponding control law given by:

$$u(k) = \Psi\big(x_M(k), (1-\alpha)[e(k) + h[x_M(k)]] + \alpha h^{r-1}[x_M(k)]\big) \qquad (18)$$

With the model states x_M obtained by simulating $x_M(k+1) = \Phi[x_M(k), u(k)]$ on-line, the closed-loop system becomes:

$$x(k+1) = \Phi\big(x(k), \Psi\big(x_M(k), (1-\alpha)[y_{sp} - h[x(k)] + h[x_M(k)]] + \alpha h^{r-1}[x_M(k)]\big)\big)$$

$$x_M(k+1) = \Phi\big(x_M(k), \Psi\big(x_M(k), (1-\alpha)[y_{sp} - h[x(k)] + h[x_M(k)]] + \alpha h^{r-1}[x_M(k)]\big)\big) \quad (19)$$

$$y(k) = h[x(k)]$$

It can be seen from (19) that the Model-Algorithmic Controller possesses the structure of Model-State Feedback [9]; this is shown in Figure 1.

FIGURE 1: Model-Algorithmic Controller Structure

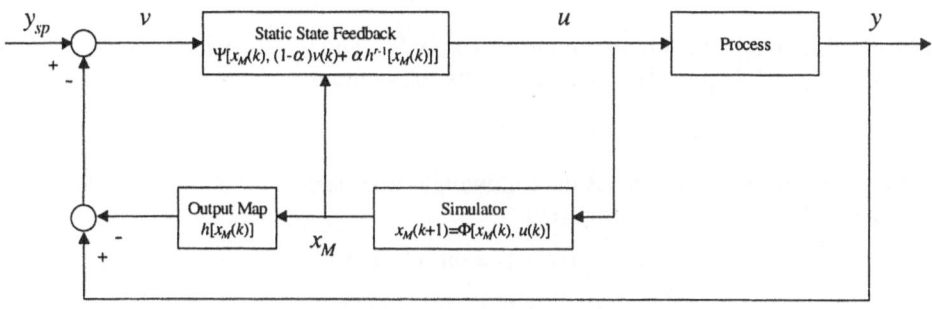

Closed-Loop Stability

As shown in Soroush and Kravaris [15], the following conditions guarantee local asymptotic stability of the closed-loop system:
(i)   The reference trajectory is stable. This is guaranteed if $0 < \alpha < 1$.
(ii) The delay-free part of the system is locally hyperbolically minimum-phase.
(iii) The open-loop process is locally asymptotically stable.
Out of the three conditions, condition (ii) is the most restrictive and will be relaxed in the following section.

2.1.3.   *SISO Nonlinear Model-Algorithmic Control with Nonminimum-Phase Compensation*

The problem of constructing control algorithms for nonminimum-phase processes is of considerable significance in nonlinear control. Research in this area is still at early stages and confined to continuous-time nonlinear systems. In linear systems, the process transfer function is factored into a minimum-phase element and a nonminimum-phase all-pass element. The design of the controller is based on the minimum-phase element, which effectively leaves the nonminimum-phase element in the open-loop. For continuous time nonlinear systems, such a decomposition is currently only available for second order systems [8]. For general nonlinear systems, the issue of decomposition is an open problem. Recently, the minimum-phase output predictor control structure for continuous-time nonminimum-phase compensation has been developed [20]. The advantage of this formulation is that it reduces the control problem to one of calculating a minimum-phase output map, thus eliminating the need for decomposition of the process dynamics into minimum-phase and nonminimum-phase elements.

In what follows, a Model-Algorithmic Controller formulation will be proposed, which performs nonminimum-phase compensation using an auxiliary minimum-phase output in a similar fashion as in Wright and Kravaris [20] for continuous-time systems. Consider a nonlinear system with the form:

$$x(k+1) = \Phi[x(k), u(k)]$$
$$y(k) = h[x(k)]$$

(20)

whose delay-free part is nonminimum-phase, $x \in \Re^n$, and y has relative order r. Together with the original output, also consider an auxiliary output:

$$y^*(k) = h^*[x(k)]$$

(21)

with relative order r^*, which has the following properties:
(i) The delay-free part of the system:

$$x(k+1) = \Phi[x(k), u(k)]$$
$$y^*(k) = h^*[x(k)]$$

is hyperbolically minimum-phase.
(ii) The two outputs y and y^* are statically equivalent in the sense that their values agree at every steady-state.

$$h^*[x_{ss}] = h[x_{ss}]$$

If an output y^* with the above properties can be found, controlling the output y to a constant set point can be accomplished by controlling the output y^* to the same constant set point.

Following the SISO MAC framework, future changes of the output y^* can be predicted by simulating the model on-line:

$$y_M^*(k+1) - y_M^*(k) = h^{*^1}[x_M(k)] - h^*[x_M(k)]$$
$$y_M^*(k+2) - y_M^*(k) = h^{*^2}[x_M(k)] - h^*[x_M(k)]$$
$$\vdots$$

(22)

$$y_M^*(k+r^*-1) - y_M^*(k) = h^{*^{r^*-1}}[x_M(k)] - h^*[x_M(k)]$$
$$y_M^*(k+r^*) - y_M^*(k) = h^{*^{r^*-1}}[\Phi[x_M(k), u(k)]] - h^*[x_M(k)]$$

Then, the "closed-loop" predictions of the auxiliary output can be obtained by adding the changes to the on-line estimate $y^*(k) = y(k) + h^*[x_M(k)] - h[x_M(k)]$:

$$\hat{y}^*(k+1) = y(k) + h^{*^1}[x_M(k)] - h[x_M(k)]$$
$$\hat{y}^*(k+2) = y(k) + h^{*^2}[x_M(k)] - h[x_M(k)]$$
$$\vdots \qquad\qquad (23)$$
$$\hat{y}^*(k+r^*-1) = y(k) + h^{*^{r^*-1}}[x_M(k)] - h[x_M(k)]$$
$$\hat{y}^*(k+r^*) = y(k) + h^{*^{r^*-1}}[\Phi[x_M(k),u(k)]] - h[x_M(k)]$$

A desired linear reference trajectory can be defined for the auxiliary output:

$$\hat{y}^*(k+r^*) = (1-\alpha)y_{sp} + \alpha\hat{y}^*(k+r^*-1) \qquad\qquad (24)$$

where the subscript sp denotes the nonminimum-phase output set-point. It should be noted that the auxiliary output is requested to follow the reference trajectory of (24), and in general, the original nonminimum-phase output will not follow the same trajectory. However, at every steady-state, both the auxiliary and the original output will equal the same set point. Combining (23) and (24), such that the output prediction matches the reference trajectory, one can derive a SISO nonlinear MAC with nonminimum-phase compensation:

$$h^{*^{r^*-1}}[\Phi[x_M(k),u(k)]] = (1-\alpha)[e(k) + h[x_M(k)]] + \alpha h^{*^{r^*-1}}[x_M(k)] \qquad\qquad (25)$$

The manipulated input can be locally calculated, with the corresponding control law given by:

$$u(k) = \Psi\Big(x_M(k),(1-\alpha)[e(k) + h[x_M(k)]] + \alpha h^{*^{r^*-1}}[x_M(k)]\Big) \qquad\qquad (26)$$

With the model states x_M obtained by simulating $x_M(k+1)=\Phi[x_M(k),u(k)]$ on-line, the closed-loop system becomes:

$$x(k+1) = \Phi\Big(x(k),\Psi\Big(x_M(k),(1-\alpha)[y_{sp} - h[x(k)] + h[x_M(k)]] + \alpha h^{*^{r^*-1}}[x_M(k)]\Big)\Big)$$
$$x_M(k+1) = \Phi\Big(x_M(k),\Psi\Big(x_M(k),(1-\alpha)[y_{sp} - h[x(k)] + h[x_M(k)]] + \alpha h^{*^{r^*-1}}[x_M(k)]\Big)\Big) \qquad (27)$$
$$y(k) = h[x(k)]$$

Closed-Loop Stability

Because the output $y^* = h^*[x(k)]$ is locally hyperbolically minimum-phase, the same local stability analysis as in Soroush and Kravaris [15] is applicable to the closed-loop system (27). Local asymptotic stability of the closed-loop system is guaranteed under the following conditions:

(i)   The reference trajectory is stable. This is guaranteed if $0 < \alpha < 1$.
(ii) The open-loop process is locally asymptotically stable.

Statically Equivalent Outputs

Following Kravaris et al. [9], a minimum-phase output can be constructed which is statically equivalent to the original output. Consider the nonlinear auxiliary output of the form:

$$y^* = h(x) + \lambda_1 n_1(x) + \lambda_2 n_2(x) + \cdots + \lambda_{n-1} n_{n-1}(x) \qquad (28)$$

where λ_j is a constant, and $n_j(x)$ is a function that vanishes at every steady-state. Then, this output will necessarily equal y at steady-state. The functions $n_j(x)$, $j=1,\ldots,n-1$, can be calculated by eliminating u from the steady-state equations of the model. This is particularly easy from the original continuous-time model (1):

$$0 = f(x) + ug(x) \qquad (29)$$

Using the n-th equation to solve for u, and substituting the result into the other $(n-1)$ equations, one obtains $(n-1)$ conditions that must hold at every steady-state. The functions:

$$n_j(x) = f_j(x) - \frac{g_j(x)}{g_n(x)} f_n(x), \quad j = 1,\ldots, n-1 \qquad (30)$$

will necessarily vanish at every steady-state. It is possible to show that the functions of (30) are linearly independent:

$$rank \begin{bmatrix} \dfrac{\partial n_1}{\partial x}[x_{ss}] \\ \vdots \\ \dfrac{\partial n_{n-1}}{\partial x}[x_{ss}] \end{bmatrix} = n-1 \qquad (31)$$

The parameters λ_j's are chosen so that the linear approximation of the system has desired zeros, $z_1^d, \ldots, z_{n-1}^d$, which lie inside the unit circle. Using the definition of linear system zeros, coefficients of the zero polynomials can be matched to find $\lambda_1, \ldots, \lambda_{n-1}$:

$$\det \begin{bmatrix} zI - A & -b \\ c^* & 0 \end{bmatrix} = \det \begin{bmatrix} I - A & -b \\ c & 0 \end{bmatrix} \prod_{j=1}^{n-1} \left(\frac{z - z_j^d}{1 - z_j^d} \right) \qquad (32)$$

where:

$$A = \frac{\partial \Phi}{\partial x}(x_{ss}, u_{ss}); \quad b = \frac{\partial \Phi}{\partial u}(x_{ss}, u_{ss}); \quad c = \frac{\partial h}{\partial x}(x_{ss}); \quad c^* = \frac{\partial h}{\partial x}(x_{ss}) + \sum_{j=1}^{n-1} \lambda_j \frac{\partial n_j}{\partial x}(x_{ss}) \quad (33)$$

The λ_j's are then used in (28) to create the nonlinear minimum-phase output map for use with the nonlinear MAC with nonminimum-phase compensation, and leads to a compensation method that is simple and completely general.

Example 1

The nonlinear MAC with nonminimum-phase compensation will now be illustrated with a chemical engineering example. Consider a nonisothermal continuous stirred-tank reactor (CSTR) where Cyclopentenol is being produced through a van de Vusse reaction [4],[19]:

$$C_5H_6 \xrightarrow{\ k_1\ /(+H_2O)\ } C_5H_7OH \xrightarrow{\ k_2\ /(+H_2O)\ } C_5H_8(OH)_2$$

Cyclopentadiene   Cyclopentenol   Cyclopentanediol

$$2\ C_5H_6 \xrightarrow{\ k_3\ } C_{10}H_{12}$$

Cyclopentadiene   Dicyclopentadiene

The reactant is Cyclopentadiene (A), the desired product is Cyclopentenol (B), and the unwanted by-products are Cyclopentanediol (C) and Dicyclopentadiene (D). Using a sufficiently small Δt, the system equations were discretized using Euler's method (see §2.1.1.), giving the following mole and energy balances:

$$C_A(k+1) = C_A(k) + \left(\frac{F}{V}(C_{AO} - C_A(k)) - k_1[T(k)]C_A(k) - k_3[T(k)]C_A^2(k) \right) \Delta t$$

$$C_B(k+1) = C_B(k) + \left(-\frac{F}{V}C_B(k) + k_1[T(k)]C_A(k) - k_2[T(k)]C_B(k) \right) \Delta t \qquad (34)$$

$$T(k+1) = T(k) - \frac{1}{\rho C_P}\left[k_1[T(k)]C_A(k)\Delta H_{R1} + k_2[T(k)]C_B(k)\Delta H_{R2} + k_3[T(k)]C_A^2(k)\Delta H_{R3} \right]\Delta t$$

$$+ \left(\frac{F}{V}(T_o - T(k)) + \frac{Q}{\rho C_P} \right)\Delta t$$

where F/V is the dilution rate and Q is the rate of heat per unit volume that is added or removed from the reactor. C_{AO} is the concentration of Cyclopentadiene in the feed stream, T_O is the temperature of the feed stream, and $(-\Delta H)_{Ri}$ is the heat of reaction. In addition, ρ is the density of the reacting mixture, C_p is the heat capacity of the reacting mixture, and Δt (0.01 hr) is the sampling period. The rate coefficients are dependent on the reactor temperature via the Arrhenius equation:

$$k_i = k_{oi}\exp\left(\frac{E_{Ai}}{RT} \right) \qquad i = 1,2,3$$

The concentration of Cyclopentenol $(y=C_B)$ is maintained at a constant value by manipulating the dilution rate $(u=F/V)$ with a constant Q. Initially, the reactor is operating under the steady-state conditions of Table 1.

TABLE 1. System Parameters

C_{Ass}	1.25 mol/l	Q	-451.51 kJ/l·hr	ΔH_{R1}	4.20 kJ/mol A
C_{Bss}	0.90 mol/l	C_p	3.01 kJ/kg·K	ΔH_{R2}	-11.00 kJ/mol B
T_{ss}	407.15 K	ρ	0.9342 kg/l	ΔH_{R3}	-41.85 kJ/mol A
u_{ss}	19.52 hr^{-1}	k_{o1}	1.287×10^{12} hr^{-1}	E_{A1}	-81.13 kJ/mol
C_{AO}	5.00 mol/l	k_{o2}	1.287×10^{12} hr^{-1}	E_{A2}	-81.13 kJ/mol
T_O	403.15 K	k_{o3}	9.043×10^{9} l/mol·hr	E_{A3}	-71.17 kJ/mol

It is easy to verify that the relative order of $y=C_B$ is equal to one. Therefore, it is necessary to construct two functions that vanish at steady-state. These can be easily constructed by eliminating the input $u=F/V$ from the steady-state equations corresponding to (34):

$$n_1 = -(k_1[T(k)]C_A(k) + k_3[T(k)]C_A^2(k))C_B(k) + (C_{A0} - C_A)(k_1[T(k)]C_A(k) - k_2[T(k)]C_B(k))$$

$$n_2 = (Q - [k_1[T(k)]C_A(k)\Delta H_{R1} + k_2[T(k)]C_B(k)\Delta H_{R2} + k_3[T(k)]C_A^2(k)\Delta H_{R3}])C_B(k)$$
$$+ \rho C_P(T_0 - T(k))(k_1[T(k)]C_A(k) - k_2[T(k)]C_B(k)) \tag{35}$$

It can be easily checked that the above functions are indeed linearly independent by computing the gradients of the functions in (35) at the operating steady-state:

$$rank\begin{bmatrix} 110.44 & -261.45 & 0.0681 \\ -124.15 & 1281.18 & -28.33 \end{bmatrix} = 2$$

The linearized minimum-phase system is chosen to have double zeros at $z^d=0.888$. This places two zeros at the stable zero of the original linearized system zeros. The original zeros are found to be $z=0.888$ and $z=2.227$. λ_1 and λ_2 are calculated using (32), where:

$$A = \begin{bmatrix} 0.135 & 0 & -0.042 \\ 0.502 & 0.303 & 0.010 \\ 1.741 & 1.964 & 0.934 \end{bmatrix}; \quad b = \begin{bmatrix} 0.038 \\ -0.009 \\ -0.040 \end{bmatrix}; \quad c = \begin{bmatrix} 0 & 1 & 0 \end{bmatrix}$$

$$c^* = [\lambda_1 110.44 - \lambda_2 124.15 \quad 1 - \lambda_1 261.45 + \lambda_2 1281.18 \quad \lambda_1 0.0681 - \lambda_2 28.33]$$

The resulting values are $\lambda_1 = 0.016625$ and $\lambda_2 = 3.34495 \times 10^{-6}$. It is seen that λ_2 does not significantly contribute to the placement of the zeros and will not be included in the auxiliary minimum-phase output map. The nonlinear minimum-phase output map takes the form:

$$y^* = C_B + 0.016625 n_1[x]$$

where n_1 is given in (35). This minimum-phase output map serves as the basis for the design of the SISO nonlinear MAC with nonminimum-phase compensation used in the simulations, and is guaranteed to be equal to C_B at steady-state.

Simulation Results

The response of the closed-loop system was simulated for a step change in the set point value of the output. The system was requested to move C_B from a steady-state value of 0.9 mol/l to 1.0 mol/l with the controller parameter $\alpha = 0.7$. The responses for the system variables are given in Figure 2. It is seen in Figure 2A that the proposed MAC with nonminimum-phase compensation is able to track step changes in the set point. Figure 2B shows the response of the manipulated input and the reactor temperature.

FIGURE 2. Closed-Loop Profiles for a Step Change in the Set Point

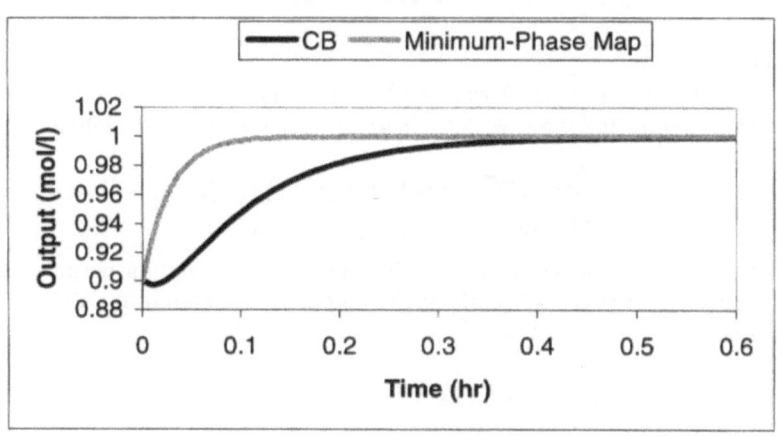

(A)   Output (C_B) and $y*$ Profiles

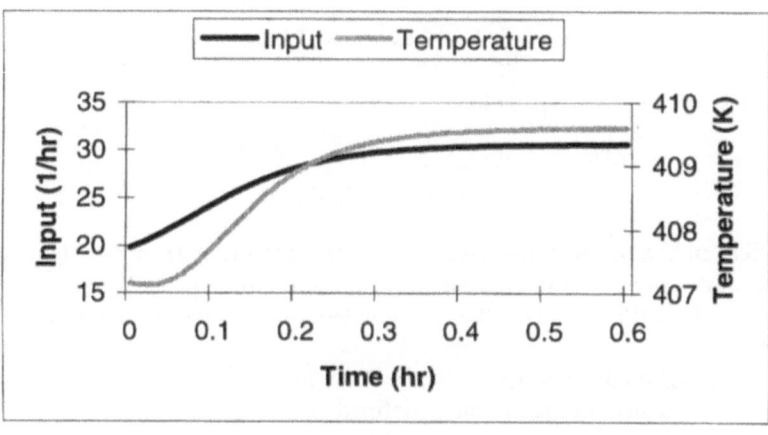

(B)   Manipulated Input (F/V) and Reactor Temperature Profiles

2.2.   MIMO NONLINEAR MODEL-ALGORITHMIC CONTROL

2.2.1.   *Preliminaries*

Consider a MIMO nonlinear process described by a discrete-time state-space model of the form:

$$x(k+1) = \Phi[x(k),u(k)]$$
$$y_i(k) = h_i[x(k)], \qquad i = 1,...,m$$

(36)

where $x = [x_1 ... x_n]^T$ denotes the vector of state variables, $u = [u_1 ... u_m]^T$ denotes the manipulated input vector, and $y = [y_1 ... y_m]^T$ represents the controlled output vector. It is assumed that $x \in X \subset \Re^n$, $u \in U \subset \Re^m$, $\Phi(x, u)$ is an analytic vector function on $X \times U$, and $h(x) = [h_1(x) ... h_m(x)]^T$ is an analytic vector function on X.

Relative Order

For a system with the form of (36), the relative order of the output y_i with respect to the manipulated input vector u is the smallest integer r_i for which:

$$\left[\frac{\partial h_i(x)}{\partial x}\right]\left[\frac{\partial \Phi(x,u)}{\partial x}\right]^{r_i-1}\left[\frac{\partial \Phi(x,u)}{\partial u}\right] \neq [0 \cdots 0] \tag{37}$$

Equivalently, r_i can be viewed as the smallest number of sampling periods after which the manipulated input move $u_j(k)$ affects the output y_i. Extending the SISO notation to MIMO systems:

$$h_i^0(x) \overset{\Delta}{=} h_i(x)$$

$$h_i^l(x) \overset{\Delta}{=} h_i^{l-1}[\Phi(x,u)], \quad l = 1, ..., r_i - 1$$

In this notation:

$$\frac{\partial}{\partial u} h_i^{r_i-1}[\Phi(x,u)] \neq [0 \cdots 0] \tag{38}$$

and the following relations will hold:

$$y_i(k+l) = h_i^l[x(k)], \qquad l = 0, ..., r_i - 1$$
$$y_i(k+r_i) = h_i^{r_i-1}[\Phi[x(k),u(k)]]$$

If the system output y_i does not have a finite relative order ($r_i = \infty$), the manipulated input vector does not affect the output y_i. If a discrete-time process has $r_i > 1$, then the *smallest* plant dead-time between the manipulated input vector and the output y_i is equal to $(r_i-1)\Delta t$.

For a MIMO discrete-time system of the form of (36) and finite relative orders, the characteristic matrix of the system is defined as:

$$\mathcal{C}(x,u) \overset{\Delta}{=} \begin{bmatrix} \dfrac{\partial}{\partial u} h_1^{r_1-1}[\Phi[x(k),u(k)]] \\ \vdots \\ \dfrac{\partial}{\partial u} h_m^{r_m-1}[\Phi[x(k),u(k)]] \end{bmatrix} \tag{39}$$

The characteristic matrix will be assumed to be nonsingular throughout the remaining MIMO formulations.

For a MIMO system with a nonsingular characteristic matrix and finite relative orders, (38) implies that the nonlinear algebraic equation:

$$h_i^{r_i-1}[\Phi(x,u)] = \tilde{y}_i \tag{40}$$

is locally solvable for the manipulated input vector u. The implicit function will be represented by:

$$u = \Psi(x, \tilde{y}) \tag{41}$$

where $\tilde{y} = [\tilde{y}_1 \cdots \tilde{y}_m]^T$, and will be assumed to be well-defined on $X \times h(X)$.

Minimum-Phase Behavior and Zero Dynamics

Following from the SISO system formulation, the process can be decomposed into two subsystems in series:

$$
\left.
\begin{aligned}
x(k+1) &= \Phi[x(k), u(k)] \\
\tilde{y}_i(k) &= h_i^{r_i-1}[\Phi[x(k), u(k)]], \quad i = 1, \dots, m
\end{aligned}
\right\} \text{Delay - Free}
$$
$$
y_i(k) = \tilde{y}_i(k-r), \qquad\qquad i = 1, \dots, m \left.\right\} \text{Pure Delay}
\tag{42}
$$

The inverse of the delay-free subsystem can be constructed due to the solvability of (40):

$$
\begin{aligned}
x(k+1) &= \Phi[x(k), \Psi[x(k), \tilde{y}(k)]] \\
u(k) &= \Psi[x(k), \tilde{y}(k)]
\end{aligned}
\tag{43}
$$

If the system is linear, the inverse system (43) has $n - \sum_{i=1}^{m} r_i$ poles at the transmission zeros and $\sum_{i=1}^{m} r_i$ poles at the origin. For the inverse system to be stable, the finite transmission zeros must be inside the unit circle. Accordingly, stability of the MIMO nonlinear inverse system can be defined.

Definition 3. Given a MIMO discrete-time nonlinear system of the form of (36), its delay-free part is called minimum-phase if the dynamics,

$$x(k+1) = \Phi[x(k), \Psi[x(k), \tilde{y}(k)]] \tag{44}$$

are locally asymptotically stable. Otherwise, the system will be called nonminimum-phase.

The local asymptotic stability of (44) can be checked, via Lyapunov's first method, by calculating the eigenvalues of the Jacobian evaluated at an equilibrium point. Using (40) and (41), along with the implicit function theorem, the Jacobian of the system is:

$$
\mathcal{J}(x, u) = \left[\frac{\partial \Phi(x, u)}{\partial x} \right] - \left[\frac{\partial \Phi(x, u)}{\partial u} \right] [\mathcal{C}(x, u)]^{-1}
\begin{bmatrix}
\dfrac{\partial}{\partial x} h_i^{r_i-1}[\Phi(x, u)] \\
\vdots \\
\dfrac{\partial}{\partial x} h_m^{r_m-1}[\Phi(x, u)]
\end{bmatrix}
\tag{45}
$$

which is evaluated at the reference equilibrium point $(x,u)=(x_{ss},u_{ss})$. If all the eigenvalues of the Jacobian are in the interior of the unit circle, the dynamics of (44) will be locally asymptotically stable around the reference equilibrium point.

Definition 4. Given a discrete-time nonlinear system of the form of (36), its delay-free part is called locally hyperbolically minimum-phase if all the eigenvalues of the Jacobian matrix, evaluated at a reference equilibrium point $(x,u)=(x_{ss},u_{ss})$, are in the interior of the unit circle.

Analogous to the SISO case, a locally hyperbolic minimum-phase delay-free part is also locally minimum-phase, but the converse may not hold.

2.2.2. *MIMO Nonlinear Model-Algorithmic Control for Unconstrained Processes*
The MIMO formulation of model-algorithmic control is a direct extension of the previous SISO MAC formulation. Using the process model to predict the future changes of the output, the "closed-loop" predictions for each output can be constructed for a system of the form of (36) and finite relative orders $r_1...r_m$ as follows:

$$\hat{y}_i(k+1) = y_i(k) + h_i^1[x_M(k)] - h_i[x_M(k)]$$
$$\hat{y}_i(k+2) = y_i(k) + h_i^2[x_M(k)] - h_i[x_M(k)]$$
$$\vdots \tag{46}$$
$$\hat{y}_i(k+r_i-1) = y_i(k) + h_i^{r_i-1}[x_M(k)] - h_i[x_M(k)]$$
$$\hat{y}_i(k+r_i) = y_i(k) + h_i^{r_i-1}[\Phi[x_M(k),u(k)]] - h_i[x_M(k)]$$

For each of the outputs, a desired linear reference trajectory can be defined:

$$\hat{y}_i(k+r_i) = (1-\alpha_i)y_{isp} + \alpha_i\hat{y}_i(k+r_i-1) \tag{47}$$

where the subscript sp denotes the output set-point, and α_i is a tunable scalar parameter between 0 and 1. Combing (46) and (47), such that the output prediction matches the reference trajectory, one can derive the MIMO nonlinear MAC controller.

$$h_1^{r_i-1}[\Phi[x_M(k),u(k)]] = (1-\alpha_1)[e_1(k) + h_1[x_M(k)]] + \alpha_1 h_1^{r_i-1}[x_M(k)]$$
$$\vdots \tag{48}$$
$$h_m^{r_i-1}[\Phi[x_M(k),u(k)]] = (1-\alpha_m)[e_m(k) + h_m[x_M(k)]] + \alpha_m h_m^{r_i-1}[x_M(k)]$$

If the process has the form of (36) with finite relative orders and a nonsingular characteristic matrix, (38) implies that the nonlinear algebraic equations (48) are locally solvable via the implicit function theorem. The corresponding control law is given by:

$$u(k) = \Psi\left(x_M(k), (I_m - \alpha)[e(k) + h[x_M(k)]] + \alpha \begin{bmatrix} h_1^{r_i-1}[x_M(k)] \\ \vdots \\ h_m^{r_i-1}[x_M(k)] \end{bmatrix}\right) \tag{49}$$

where $\alpha \equiv diag\{\alpha_1 \cdots \alpha_m\}$, $e(k) \equiv y_{sp} - y(k)$, and I_m is the $m \times m$ identity matrix. With the model states, x_M, obtained by simulating $x_M(k+1) = \Phi[x_M(k), u(k)]$ on-line, the closed-loop system becomes:

$$x(k+1) = \Phi\left(x(k), \Psi\left(x_M(k), (I_m - \alpha)[y_{sp} - h[x(k)] + h[x_M(k)]] + \alpha\begin{bmatrix} h_1^{r_1-1}[x_M(k)] \\ \vdots \\ h_m^{r_m-1}[x_M(k)] \end{bmatrix}\right)\right)$$

$$x_M(k+1) = \Phi\left(x_M(k), \Psi\left(x_M(k), (I_m - \alpha)[y_{sp} - h[x(k)] + h[x_M(k)]] + \alpha\begin{bmatrix} h_1^{r_1-1}[x_M(k)] \\ \vdots \\ h_m^{r_m-1}[x_M(k)] \end{bmatrix}\right)\right) \quad (50)$$

$$y(k) = h[x(k)]$$

Closed-Loop Stability

As in the SISO formulation, the following conditions guarantee local asymptotic stability of the closed-loop system.
(i)   The reference trajectories are stable. This is guaranteed if $0 < \alpha_i < 1$.
(ii) The delay-free part of the system is locally hyperbolically minimum-phase.
(iii) The open-loop process is locally asymptotically stable.

2.2.3.   *Input Constraints*

With the previously described MAC controller, the outputs will follow the linear reference trajectories of (47) as long as the manipulated inputs do not reach an upper or lower bound. However, this will no longer hold if the inputs do reach an upper or lower bound. These bounds are usually due to the physical limitations of the process such as pump size, flow valve limitations, and heating power constraints. Since the input constraints significantly affect the behavior of the system, a proper design of the controller must account for the presence of the constraints. Intuitively, driving the open-loop state simulator with the actual input that drives the process (as opposed to the calculated input which might exceed the bounds) will give the most accurate model states, and accordingly, the most effective control of the process within the input constraints. The actual input that drives the simulator and process can be defined as:

$$u_{iA} = \begin{cases} u_{iL}, & \text{if } u_{iC} < u_{iL} \\ u_{iC}, & \text{if } u_{iL} \leq u_{iC} < u_{iH} \\ u_{iH}, & \text{if } u_{iC} \geq u_{iH} \end{cases} \quad i = 1, \ldots, m \quad (51)$$

where u_{iA} is the physically realizable input to the process and simulator, u_{iC} is the calculated input from the MAC algorithm, u_{iL} is the lower bound of the input, and u_{iH} is the upper bound of the input.

Example 2

The handling of input constraints within a MIMO MAC setting will now be examined with a second chemical engineering example. Consider the free-radical homopolymerization of methyl methacrylate (MMA) in a CSTR, with azo-bis-isobutyronitrile (AIBN) as the initiator and toluene as the solvent [16]. The model was discretized using Euler's method and is composed of eight state equations:

$$C_m(k+1) = C_m(k) + \left(R_m[x(k)] + \frac{(C_{mms} - (1+\varepsilon X_p(k))C_m(k))F_m}{V} \right)\Delta t$$

$$C_i(k+1) = C_i(k) + \left(R_i[x(k)] + \frac{F_i(k)C_{iis} - (1+\varepsilon X_p(k))C_i(k)F_m}{V} \right)\Delta t$$

$$C_s(k+1) = C_s(k) + \left(\frac{F_m C_{sms} + F_i(k)C_{sis} - (1+\varepsilon X_p(k))C_s(k)F_m}{V} \right)\Delta t$$

$$X_p(k+1) = X_p(k) + \left(R_{X_p}[x(k)] - \frac{(F_m C_{mms} X_p(k)(1-X_p(k))}{VC_m(k)} \right)\Delta t$$

$$T(k+1) = T(k) + \left(\frac{R_h[x(k)]V}{mc} + \frac{(T_{ms} - (1+\varepsilon X_p(k))T(k))F_m}{V} \right)\Delta t + AU_{R1}(T_j(k) - T(k))\Delta t$$
$$+ AU_{R2}(T_\infty - T(k))\Delta t$$

$$Tj(k+1) = T_j(k) + \left(AU_{j1}(T(k) - T_j(k)) + AU_{j2}(T_\infty - T_j(k)) + AU_{j3}Q(k) \right)\Delta t$$

$$\mu_0(k+1) = \mu_0(k) + \left(R\mu_0[x(k)] - \frac{(1+\varepsilon X_p(k))\mu_0(k)F_m}{V} \right)\Delta t \qquad (52)$$

$$\mu_1(k+1) = \mu_1(k) + \left(R\mu_1[x(k)] - \frac{(1+\varepsilon X_p(k))\mu_1(k)F_m}{V} \right)\Delta t$$

where C_m, C_i, C_s are the molar concentrations of the monomer, initiator, and solvent, respectively. X_p is the conversion, T and T_j are the temperatures of the reactor and jacket, μ_0 and μ_1 the mole and mass concentration of the dead polymer chains, and Δt (30 s) is the sampling period. V is the reactor volume, C_{iis} and C_{sis} are the concentrations of the initiator and solvent in the initiator stream, C_{mms} and C_{sms} are the concentrations of the monomer and solvent in the monomer stream, and ε is the volume expansion factor. Additionally, T_∞ and T_{ms} are the ambient and monomer stream temperatures, AU_{kl} are the heat transfer coefficients for the reactor and jacket, m is the mass of reactor contents, and c is the heat capacity of the contents. The rate equations are given by:

$$R_m[x(k)] = -C_m(k)\lambda_0(k)(k_{fm}(k) + k_p(k))$$

$$R_i[x(k)] = -k_i(k)C_i(k)$$

$$R_{X_p}[x(k)] = (1 - X_p(k))\lambda_0(k)(k_{fm}(k) + k_p(k))$$

$$R_h[x(k)] = (-\Delta H_p)k_p(k)\lambda_0(k)C_m(k)$$

$$R_{\mu_0}[x(k)] = (k_{td}(k)\lambda_0(k) + k_{fm}(k)C_m(k) + k_{ts}(k)C_s(k))\lambda_0(k) + 0.5k_{tc}(k)\lambda_0^2(k)$$

$$R_{\mu_1}[x(k)] = (k_{fm}(k)C_m(k) + k_{ts}(k)C_s(k))\lambda_1(k) + k_t(k)\lambda_0(k)\lambda_1(k)$$

with the following equations completing the model:

$$\lambda_0(k) = \left(\frac{2fk_i(k)C_i(k)}{k_t(k)} \right)^{1/2}$$

$$\lambda_1(k) = \left(\frac{2fk_i(k)C_i(k) + (k_p(k) + k_{fm}(k))\lambda_0(k)C_m(k) + k_{ts}(k)\lambda_0(k)C_s(k)}{k_{fm}(k)C_m(k) + k_{ts}(k)C_s(k) + k_t(k)\lambda_0(k)} \right) M_m$$

$$k_l(k) = Z_l Exp[-E_l / RT(k)], \quad l = t, p, i.$$

$$\frac{k_l(k)}{k_p(k)} = Z_l Exp[-E_l / RT(k)], \quad l = fm, ts$$

$$k_t(k) = k_{tc}(k) + k_{td}(k)$$

$$\frac{k_{tc}(k)}{k_{td}(k)} = Z_{tc} Exp[-E_{tc} / RT(k)]$$

In this notation, M_m is the molecular weight of the monomer and f is the initiator efficiency. k_t, k_p, and k_i are the rate constants for the overall termination, propagation, and initiation. k_{fm} and k_{ts} are the rate constants for chain transfer to the monomer and solvent, with k_{td} and k_{tc} designating the rate constants for termination by disproportionation and combination.

The controlled outputs of the system are the conversion ($y_1 = X_p$) and the reactor temperature ($y_2 = T$). It is easy to verify that $r_1 = r_2 = 2$. Inputs to the system are F_m, the constant monomer flow rate, F_i, the initiator flow rate (u_1), and Q, the rate of heat input or removal (u_2). The system parameters are given in Table 2.

TABLE 2. System Parameters

F_m	2.78×10^{-7} m³/s	T_∞	293.2 K	Z_p	4.917×10^5 m³/kmol·s
V	1.20×10^{-3} m³	$-\Delta H_p$	57800 kJ/kmol	E_p	1.820×10^4 kJ/kmol
ε	-0.108	AU_{RI}	0.019 s⁻¹	Z_i	1.053×10^{15} s⁻¹
f	0.58	AU_{R2}	0.0038 s⁻¹	E_i	1.284×10^5 kJ/kmol
c	1.815 kJ/kg·K	AU_{J1}	0.0008 s⁻¹	Z_{fm}	4.661×10^9 m³/kmol·s
C_{iis}	0.263 kmol/m³	AU_{J2}	0.00037 s⁻¹	E_{fm}	7.418×10^4 kJ/kmol
C_{sis}	8.986 kmol/m³	AU_{J3}	0.0664 K/kJ	Z_{ts}	1.010×10^3 m³/kmol·s
C_{mms}	3.984 kmol/m³	M_m	100.12 kg/kmol	E_{ts}	4.769×10^4 kJ/kmol
C_{sms}	5.881 kmol/m³	Z_t	9.800×10^7 m³/kmol·s	Z_{tc}	3.956×10^{-4} m³/kmol·s
T_{ms}	293.2 K	E_t	2.933×10^3 kJ/kmol	E_{tc}	-1.711×10^4 kJ/kmol

Simulation Results

The response of the closed-loop system was simulated for a step change in the set point value of the first output X_p. The system was requested to move the conversion from a steady-state value of 0.4 to 0.5 at a constant reactor temperature of 350 K with $\alpha_1 = \alpha_2 = 0.7$. The initiator flow rate, u_1, is constrained between the upper bound of 5.0×10^{-8} m³/s and the lower bound of 0 m³/s, while the heating rate, u_2, remains unconstrained. Two cases were investigated to examine the proper procedure for constraint handling:

162

<u>Case 1</u>: The calculated input, u_{iC}, from the MAC algorithm is used to drive the open-loop state estimator, while the actual constrained input, u_{iA}, drives the process.

<u>Case 2</u>: The actual constrained input, u_{iA}, drives the open-loop state estimator and the process.

The schematic of the closed-loop system is given in Figure 3.

FIGURE 3. Closed-Loop Process Representation

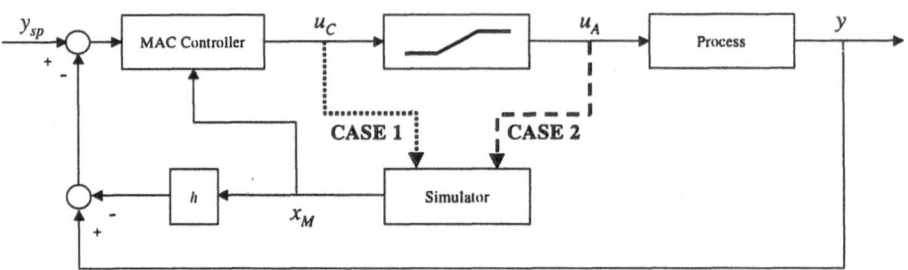

Figures 4 and 5 depict the closed-loop profiles for the step change in conversion. As seen in the two figures, driving the simulator with the actual constrained input, Case 2, gives superior performance to driving the simulator with the calculated input values, Case 1. Case 2 has a considerably faster rise time and does not exhibit the temperature fluctuations that are seen in Case 1. In addition, the first manipulated input, u_1, of Case 2 does not oscillate between the upper and lower bounds as does Case 1.

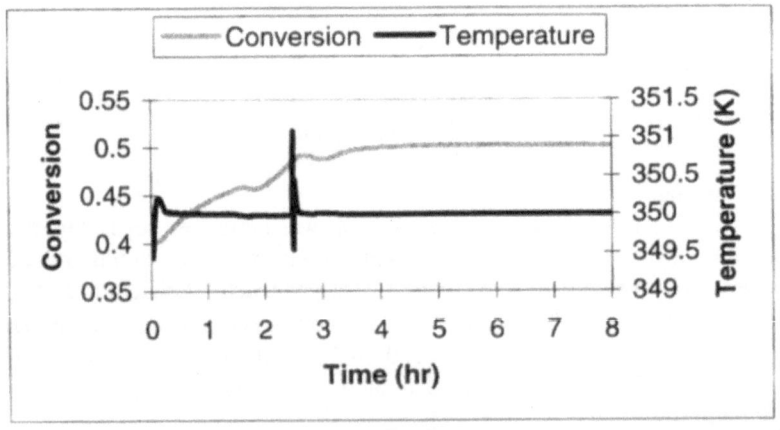

(A)   Controlled Output Profiles

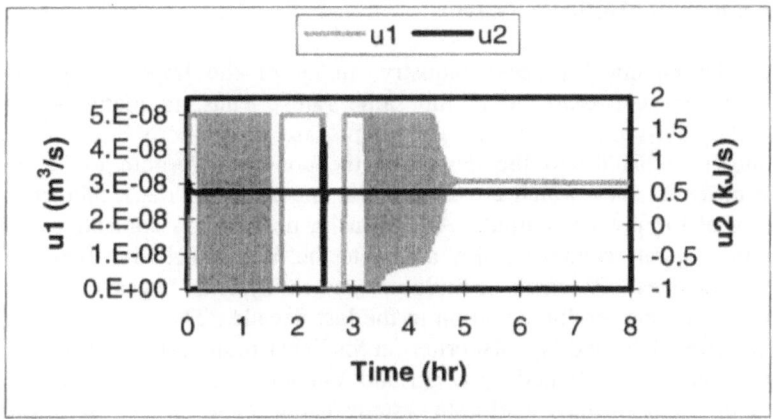

(B)   Manipulated Input Profiles

FIGURE 5.   Closed-Loop Profiles for Case 2

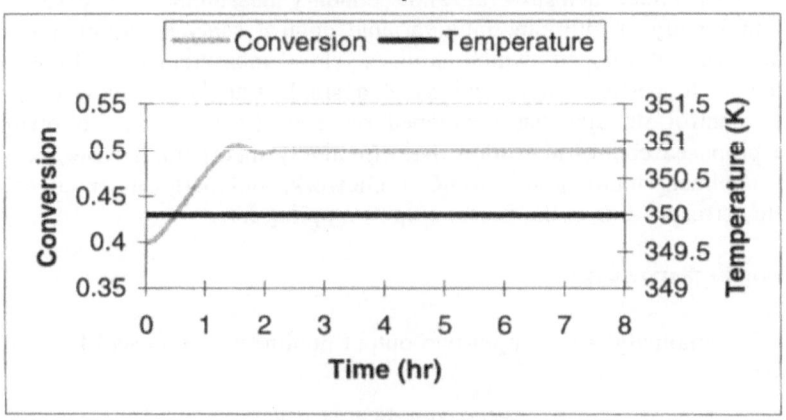

(A)   Controlled Output Profiles

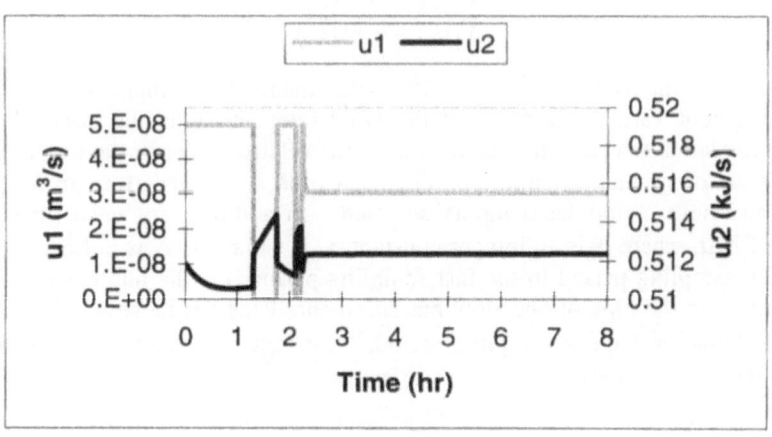

(B)   Manipulated Input Profiles

3.   Control of Multirate Systems

In the chemical process industry, many of the important process output measurements are not available at the same rate. Thus, in many current control strategies, the low-sampling-rate measurements are not directly used to control the process output. This forces the design of the control algorithm to be based upon secondary measurements which are sampled at much higher rates, and often leads to poor control of the process output. Analogously, designing a controller based on the slowest sampling rate cannot quickly attenuate the disturbances in other outputs with faster sampling rates. Research of multirate control systems began in the late 1950's, but has received considerable attention in the last decade. The state-space description, transfer characteristics, and Nyquist criterion for linear multivariable multirate sampled-data systems have been studied by Araki and Yamamoto [2]. More recently, the pole assignment problem for linear multirate systems using a state space description has been looked at extensively [1],[6]. A linear state-space model-predictive technique that uses a sub-optimal cascade filter for dual-rate systems was developed, which utilizes primary measurements available at a slow rate and secondary measurements available at fast rate [10]. Additionally, a multirate multivariable control scheme within a linear MAC framework was developed which utilizes slow measurements along with fast measurements to perform input actions at a single rate [13]. Finally, a nonlinear predictive control strategy was developed by Bequette [3]. The following section details a proposed control algorithm that effectively incorporates slow measurements within a nonlinear multivariable MAC framework, and performs input actions at a single rate corresponding to the fastest output sampling rate.

3.1.   PRELIMINARIES

For simplicity, a two-input-two-output nonlinear system will be considered of the form:

$$x(k+1) = \Phi[x(k), u(k)]$$
$$y_s(k) = h_s[x(k)] \qquad\qquad (53)$$
$$y_f(k) = h_f[x(k)]$$

where $x \in \mathfrak{R}^n$ is the state vector, $u \in \mathfrak{R}^2$ is the manipulated input vector, y_s is the controlled output variable sampled at the slow rate, and y_f is the controlled output variable sampled at the fast rate. Denote by r_s the relative order of the output y_s and by r_f the relative order of the output y_f. For each time $k\Delta t$, where k is an integer, y_f is sampled and both manipulated inputs actuated. In addition, for each time that is a multiple of $N\Delta t$, where N is an integer constant, y_s is sampled. N is defined as the ratio of the slow sampling period to the fast sampling period, i.e. the number of fast output measurements in the time of one slow measurement. This can be seen schematically in Figure 6. Hereafter, the fast sampling period, Δt, is regarded as a constant, and will not appear in the nomenclature.

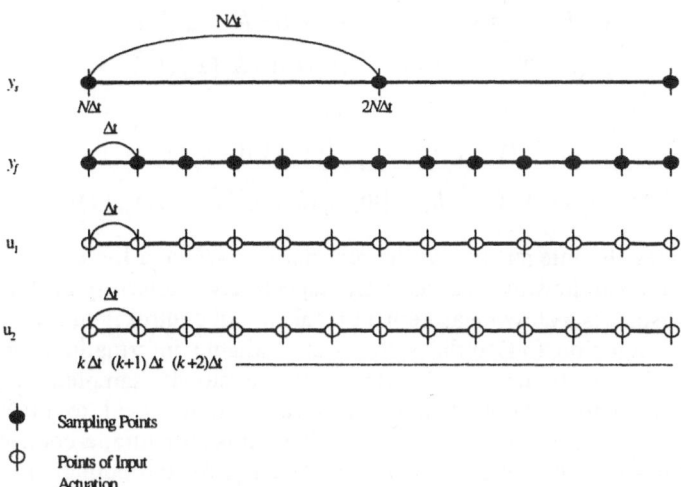

FIGURE 6. Multirate Sampled-Output Process

3.2. NONLINEAR MULTIRATE MODEL-ALGORITHMIC CONTROL

The first scenario to be considered consists of slow measurements with negligible time delay. This occurs in gas phase reactors, where compositions are analyzed quickly with gas chromatography, but due to measurement cost and/or instrument down time can not be implemented at a meaningfully fast rate. Starting within a nonlinear MAC framework, the closed-loop predictions can be subdivided into the output predictions sampled at the slow and fast rate, which take the form:

$$\hat{y}_s(k+1) = y_s\left(\left[\frac{k}{N}\right]N\right) + h_s^{\ 1}[x_M(k)] - h_s[x_M(\left[\frac{k}{N}\right]N)]$$

$$\hat{y}_s(k+2) = y_s\left(\left[\frac{k}{N}\right]N\right) + h_s^{\ 2}[x_M(k)] - h_s[x_M(\left[\frac{k}{N}\right]N)]$$

$$\vdots \tag{54}$$

$$\hat{y}_s(k+r_s-1) = y_s\left(\left[\frac{k}{N}\right]N\right) + h_s^{\ r_s-1}[x_M(k)] - h_s[x_M(\left[\frac{k}{N}\right]N)]$$

$$\hat{y}_s(k+r_s) = y_s\left(\left[\frac{k}{N}\right]N\right) + h_s^{\ r_s-1}[\Phi[x_M(k),u(k)]] - h_s[x_M(\left[\frac{k}{N}\right]N)]$$

166

and:

$$\hat{y}_f(k+1) = y_f(k) + h_f^{-1}[x_M(k)] - h_f[x_M(k)]$$
$$\hat{y}_f(k+2) = y_f(k) + h_f^{-2}[x_M(k)] - h_f[x_M(k)]$$
$$\vdots \qquad (55)$$
$$\hat{y}_f(k+r_f-1) = y_f(k) + h_f^{r_f-1}[x_M(k)] - h_f[x_M(k)]$$
$$\hat{y}_f(k+r_f) = y_f(k) + h_f^{r_f-1}[\Phi[x_M(k),u(k)]] - h_f[x_M(k)]$$

where $[k/N]$ denotes the integer part of the real number k/N and the subscripts s and f refer to outputs sampled the slow and fast rate, respectively. Referring back to Figure 6, at every $[k/N]N$ instances a slow measurement is taken, but control inputs are actuated at every k instant. Equation (54) utilizes the best available information, which is the measurement at $[k/N]N$, to predict the future of the slowly sampled output. The predictions for the output sampled at the fast rate are identical to (46) since the measurements are available at every k instant. Using this structure, a controller can be constructed that not only has good disturbance rejection for the output sampled at the fast rate, but also includes the necessary information for the good control of the output sampled at the slow rate. For each of the outputs, a reference trajectory can be defined:

$$\hat{y}_s(k+r_s) = (1-\alpha_s)y_{ssp} + \alpha_s \hat{y}_s(k+r_s-1)$$
$$\hat{y}_f(k+r_f) = (1-\alpha_f)y_{fsp} + \alpha_f \hat{y}_f(k+r_f-1) \qquad (56)$$

Matching the predictions with the reference trajectories, one can calculate the necessary control actions by simulating (53) and numerically solving (57) on-line.

$$h_s^{r_s-1}[\Phi[x_M(k),u(k)]] = (1-\alpha_s)[e_s\left(\left[\frac{k}{N}\right]N\right) + h_s[x_M\left(\left[\frac{k}{N}\right]N\right)]] + \alpha_s h_s^{r_s-1}[x_M(k)]$$
$$h_f^{r_f-1}[\Phi[x_M(k),u(k)]] = (1-\alpha_f)[e_f(k) + h_f[x_M(k)]] + \alpha_f h_f^{r_f-1}[x_M(k)] \qquad (57)$$

3.3. NONLINEAR MULTIRATE MODEL-ALGORITHMIC CONTROL WITH TIME DELAYS

The second scenario accounts for the significant time delays often associated with the slowly sampled output due to the analytical nature of the measurements. The molecular weight measurement of polymers can have a 10-40 minute time delay from when the sample is first injected into the column to when the measurement is received. Therefore, it is necessary to incorporate this delay into the development of the multirate algorithm. With time delays in the slowly sampled outputs, the system now takes the form:

$$x(k+1) = \Phi[x(k),u(k)]$$
$$y_f(k) = h_f[x(k)] \qquad (58)$$
$$y_s(k) = h_s[x(k-\phi)]$$

FIGURE 7. Multirate Sampled-Output Process with Time Delays

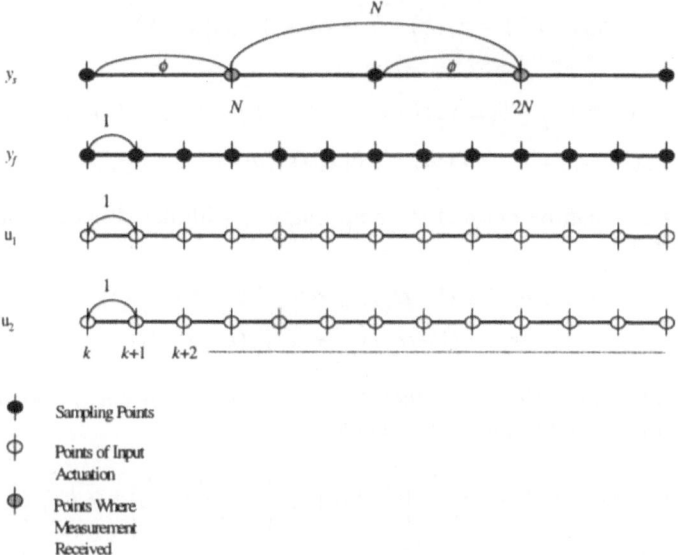

● Sampling Points

Φ Points of Input
Actuation

Φ Points Where
Measurement
Received

where y_s is the time-delayed output with the slow sampling rate, and ϕ is the time delay. A schematic for the multirate system with time delays is given in Figure 7.

Similar to the previous case, the best "closed-loop" predictions, based on the measurements received at $[k/N]N$, are as follows:

$$\hat{y}_s(k+\phi+1) = y_s\left(\left[\frac{k}{N}\right]N\right) + h_s^{\,1}[x_M(k)] - h_s[x_M\left(\left[\frac{k}{N}\right]N - \phi\right)]$$

$$\hat{y}_s(k+\phi+2) = y_s\left(\left[\frac{k}{N}\right]N\right) + h_s^{\,2}[x_M(k)] - h_s[x_M\left(\left[\frac{k}{N}\right]N - \phi\right)]$$

$$\vdots$$

$$\hat{y}_s(k+\phi+r_s-1) = y_s\left(\left[\frac{k}{N}\right]N\right) + h_s^{\,r_s-1}[x_M(k)] - h_s[x_M\left(\left[\frac{k}{N}\right]N - \phi\right)]$$

$$\hat{y}_s(k+\phi+r_s) = y_s\left(\left[\frac{k}{N}\right]N\right) + h_s^{\,r_s-1}[\Phi[x_M(k),u(k)]] - h_s[x_M\left(\left[\frac{k}{N}\right]N - \phi\right)]$$

(59)

where r_s is the relative order of the output h_s without time delay. The predictions for the output with the fast sampling rate are identically (55):

$$\hat{y}_f(k+1) = y_f(k) + h_f^{\ 1}[x_M(k)] - h_f[x_M(k)]$$

$$\hat{y}_f(k+2) = y_f(k) + h_f^{\ 2}[x_M(k)] - h_f[x_M(k)]$$

$$\vdots$$ (60)

$$\hat{y}_f(k+r_f-1) = y_f(k) + h_f^{\ r_f-1}[x_M(k)] - h_f[x_M(k)]$$

$$\hat{y}_f(k+r_f) = y_f(k) + h_f^{\ r_f-1}[\Phi[x_M(k), u(k)]] - h_f[x_M(k)]$$

Reference trajectories can be defined for both outputs, with the slowly sampled output now time-delayed:

$$\hat{y}_s(k+\phi+r_s) = (1-\alpha_s)y_{ssp} + \alpha_s \hat{y}_s(k+\phi+r_s-1)$$

$$\hat{y}_f(k+r_f) = (1-\alpha_f)y_{fsp} + \alpha_f \hat{y}_f(k+r_f-1)$$ (61)

Matching the predictions to the reference trajectories for both cases and numerically solving (62), the input vector can be calculated.

$$h_s^{\ r_s-1}[\Phi[x_M(k), u(k)]] = (1-\alpha_s)[e_s(\left[\frac{k}{N}\right]N) + h_s[x_M(\left[\frac{k}{N}\right]N - \phi)]] + \alpha_s h_s^{\ r_s-1}[x_M(k)]$$

$$h_f^{\ r_f-1}[\Phi[x_M(k), u(k)]] = (1-\alpha_f)[e_f(k) + h_f[x_M(k)]] + \alpha_f h_f^{\ r_f-1}[x_M(k)]$$ (62)

Example 3

Consider the polymerization reactor of Example 2, but now the controlled outputs of the system are the number averaged molecular weight ($y_1=\mu_1/\mu_0$; $r_1=2$) and the reactor temperature ($y_2=T$). The manipulated inputs of the system remain the initiator flow rate ($u_1=F_i$), which is bounded by the constraints of Example 2, and the unbounded rate of heat input or removal ($u_2=Q$). The number averaged molecular weight (NAMW) is measured on-line using a size exclusion chromatography system, with samples taken every 30 minutes. Due to the analytic nature of the measurement, the results are delayed 30 minutes. Since the measurement of the reactor temperature occurs every 0.5 minutes, N is equal to ϕ, which is equal to 60.

Simulation Results

The closed-loop responses of the system utilizing the nonlinear multirate MAC with time delays was simulated for step changes in both controlled outputs. Figure 8 shows the profiles for a step in the NAMW set point from 30000 kg/kmol to 50000 kg/kmol. The reactor temperature set point was held constant at 350.05 K for this case. Figure 9 shows the profiles for a step in the reactor temperature set point from 350.05 K to 345.05 K while holding the NAMW set point constant at 30000 kg/kmol. For each case, $\alpha_s = \alpha_f = 0.993$. As seen Figures 8 and 9, the nonlinear multirate MAC algorithm exhibits good performance in tracking the set point changes for each output while maintaining the other output at its set point value. Figure 8 shows that the algorithm is able to effectively track a slowly sampled output with active input constraints.

FIGURE 8. Closed-Loop Profiles for a Step Change in the NAMW Set Point

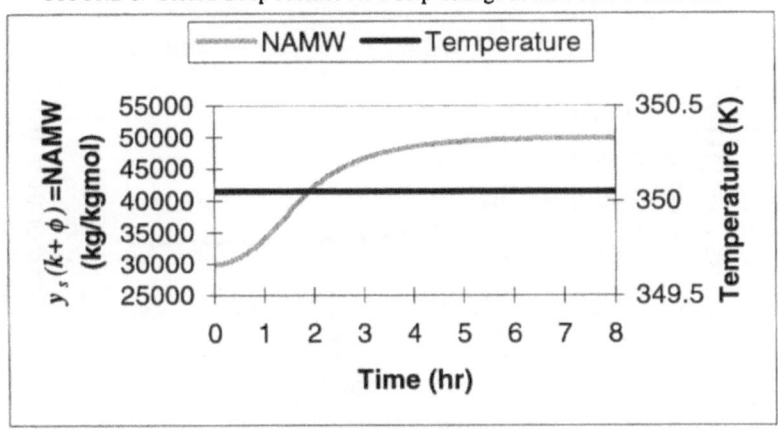

(A) Controlled Output Profiles

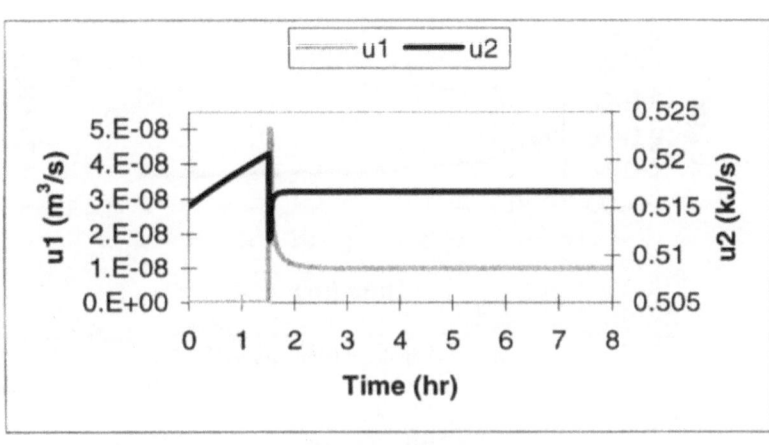

(B) Manipulated Input Profiles

FIGURE 9. Closed-Loop Profiles for a Step Change in the Reactor Temperature Set Point

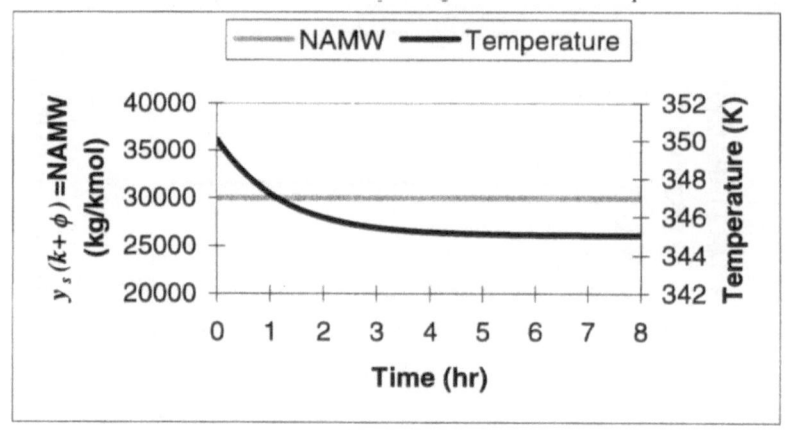

(A)   Controlled Output Profiles

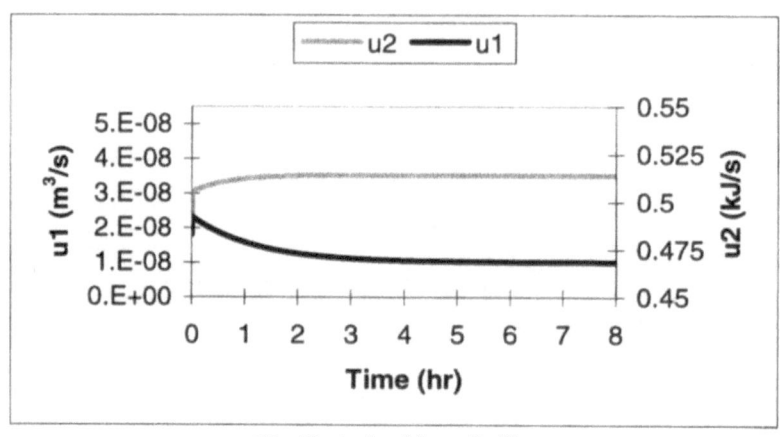

(B)   Manipulated Input Profiles

4.   Conclusion

This paper reviews SISO and MIMO nonlinear model-algorithmic control, with a discussion on input constraint handling. In addition, two new nonlinear model-algorithmic controllers are proposed to deal with significant chemical engineering control problems.   A SISO nonlinear MAC algorithm with nonminimum-phase compensation was derived and applied to a van de Vusse chemical reactor, where the controller was shown to effectively track set point changes for a nonminimum-phase system.   In addition, a MIMO nonlinear multirate model-algorithmic controller was derived and applied to a polymerization reactor. It was shown that the controller was able to effectively track set point changes in a multirate system with active input constraints, both for a slowly sampled output and a more rapidly sampled output. The proposed controllers represent effective methods to deal with the respective control problems.

Acknowledgement

Financial support from the National Science Foundation through Grant CTS–9403432 is gratefully acknowledged.

5.   References

1. Araki, M., and T. Hagiwara (1986) Pole Assignment by Multirate Sampled-Data Output Feedback, *Int. J. Control*, **44**, 1661-1673.
2. Araki, M., and K. Yamamoto (1986) Multivariable Multirate Sampled-Data Systems: State-Space Description, Transfer Characteristics, and Nyquist Criterion, *IEEE Trans. Autom. Contr.*, **AC-31**, 145-154.
3. Bequette, B. W. (1991) Nonlinear Predictive Control Using Multi-Rate Sampling, *Can. J. of Chem. Eng.*, **69**, 136-143.
4. Engell, S., and K. U. Klatt (1993) Nonlinear Control of a Non-Minimum-Phase CSTR, *Proc. ACC*, 2341-2945.
5. Garcia, C. E., and M. Morari (1982) Internal Model Control: 1. A Unifying Review and Some New Results, *Ind. Eng. Chem. Process Des. Dev.*, **21**, 308-323.
6. Hagiwara, T., and M. Araki (1988) Design of a Stable State Feedback Controller Based on the Multirate Sampling of the Plant Output, *IEEE Trans. Autom. Contr.*, **33**, 812-819.
7. Kazantzis, N., and C. Kravaris (1997) System-Theoretic Properties of Sampled-Data Representations of Nonlinear Systems Obtained via Taylor-Lie Series, *Int. J. Control*, **67**, 997-1020.
8. Kravaris, C., and P. Daoutidis (1990) Nonlinear State Feedback Control of Second-Order Nonminimum-Phase Nonlinear Systems, *Comput. Chem. Engng.*, **14**, 439-449.
9. Kravaris, C., M. Niemiec, R. Berber, and C. Brosilow (1998) Nonlinear Model-Based Control of Nonminimum-Phase Processes, in R. Berber and C. Kravaris (eds.), Nonlinear Model Based Process Control, Kluwer Academic Publishers, Dordecht.
10. Lee, J. H., M. Gelormino, and M. Morari (1992) Model Predictive Control of Multi-Rate Sampled-Data Sytems: A State-Space Approach, *Int. J. Control*, **55**, 153-191.
11. Mehra, R. K., and R. Rouhani (1980) Theoretical Considerations on Model Algorithmic Control for Nonminimum Phase Systems, *Proc. ACC*, TA8-B.
12. Mehra, R. K., R. Rouhani, and R. Praly (1980) New Theoretical Developments in Multivariable Predictive Algorithmic Control, *Proc. ACC*, FA9-B.
13. Ohshima, M., I. Hashimoto, H. Ohno, M. Takeda, T. Yoneyama, and F. Gotoh (1994) Multirate Multivariable Model Predictive Control and its Application to a Polymerization Reactor, *Int. J. Control*, **59**, 731-742.
14. Richalet, J., A. Rault, J. L. Testud, and J. Papon (1978) Model Predictive Heuristic Control: Application to Industrial Processes, *Automatica*, **14**, 413-428.
15. Soroush, M., and C. Kravaris (1992) Discrete-Time Nonlinear Controller Synthesis by Input/Output Linearization, *AIChE J.*, **38**, 1923-1945.
16. Soroush, M., and C. Kravaris (1993) Multivariable Nonlinear Control of a Continuous Polymerization Reactor: An Experimental Study, *AIChE J.*, **39**, 1920-1931.
17. Soroush, M., and C. Kravaris (1996) Discrete-Time Nonlinear Feedback Control of Multivariable Processes, *AIChE J.*, **42**, 187-203.
18. Soroush, M., and C. Kravaris (1996) MPC Formulation of GLC, *AIChE J.*, **42**, 2377-2381.
19. van de Vusse, J. G. (1964) Plug-flow type Reactor Versus Tank Reactor, *Chem. Eng. Sci.*, **19**, 994-998.
20. Wright, A. R., and C. Kravaris (1992) Nonminimum-Phase Compensation for Nonlinear Processes, *AIChE J.*, **38**, 26-40.

WINDUP AND DIRECTIONALITY COMPENSATION IN NONLINEAR MODEL-BASED CONTROL

MASOUD SOROUSH

Department of Chemical Engineering
Drexel University
Philadelphia, PA 19104
USA

Abstract

When a process with actuator saturation nonlinearities is controlled with an analytical (non-model predictive) controller, the closed-loop performance may be of low quality due to process directionality and/or windup. This work characterizes these two phenomena and presents (a) an optimal directionality compensator and (b) nonlinear model-based control laws that optimally compensate for process directionality and windup.

Given a controller output, the process directionality compensator calculates an optimal feasible (constrained) plant input that results in a process response as close as possible to the response of the same process to the controller output. The compensator can be used for both linear and nonlinear processes, irrespective of the type of controller being used. The notion of process directionality is defined precisely, and the class of processes that do not exhibit the process directionality are characterized. The performance of the optimal directionality compensator is shown and compared with those of clipping and direction preservation, by linear and nonlinear examples.

The nonlinear model-based control laws include two distinct components: (i) an input-output linearizing controller that inherently include an optimal integral windup compensator and (ii) the optimal directionality compensator. The connections between (a) the derived control laws and (b) model state feedback control and modified internal model control are established. When one of the derived control laws is applied to time-invariant linear processes with a diagonal characteristic matrix, the resulting linear controller is exactly a reduced-order modified internal model controller.

R. Berber and C. Kravaris (eds.), Nonlinear Model Based Process Control, 173-208.
© 1998 *Kluwer Academic Publishers. Printed in the Netherlands.*

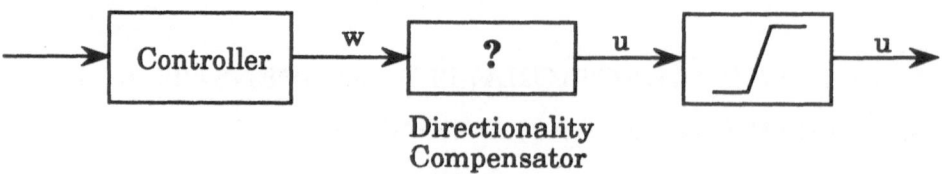

Figure 1. Directionality compensation.

The application and performance of one of the derived control laws are demonstrated by a chemical reactor example.

1. Introduction

When a multi-input multi-output (MIMO) process with actuator saturation nonlinearities is controlled by an analytical dynamic controller, the closed-loop response may be considerably poorer than an integral-of-squared-error (ISE) optimal response. Examples of analytical controllers are PID controllers, internal model controllers and input-output linearizing controllers. This poorer closed-loop performance can be due to windup and/or process directionality. The latter phenomenon is specific to MIMO processes. If the process directionality is not compensated for, then the process actuators naturally render the controller output feasible by clipping (limiting) the controller output components. The problem of process directionality compensation is that of calculating a feasible plant input on the basis of a given unconstrained controller output (see Figure 1). It is worth noting that these two problems are not present in model predictive control, in which constraints are explicitly accounted for and the controller action is solution to a constrained optimization problem.

1.1. PROCESS DIRECTIONALITY

The phenomenon of process directionality is specific to MIMO processes. In single-input single-output (SISO) processes with actuator constraints, the boundary of plant input feasible set (which is naturally closed and convex) consists of only two single isolated points, and when controller output is infeasible, one of the two points which is closest to the unconstrained controller output usually yields an optimal response, i.e. one that is closest to the response of the same process to the unconstrained controller output. In other words, in SISO processes clipping (limiting) an unconstrained controller output usually leads to an optimal feasible plant input. In MIMO processes with actuator constraints, however, the boundary of plant-input feasible set consists of infinite number of points, and when controller output is infeasible, the feasible point which is closest (in the plant input space) to the unconstrained controller output may not yield an optimal (in the sense described above) response. In other words, in MIMO processes clipping the components of an unconstrained controller output may not lead to an optimal feasible plant input.

Compared to integral windup, the phenomenon of process directionality has received less attention. In purely-analytical control methods, a feasible plant input, u, has been obtained by one of the following methods:

- 'Clipping' [10, 26]:

$$u_\ell = sat_\ell[w] \triangleq \begin{cases} u_{l_\ell}, & w_\ell < u_{l_\ell} \\ w_\ell, & u_{l_\ell} \le w_\ell \le u_{h_\ell}, \\ u_{h_\ell}, & w_\ell > u_{h_\ell} \end{cases} \quad \ell = 1, \cdots, m$$

where w is the controller output, and u_{l_ℓ} and u_{h_ℓ} are respectively the lower and upper limits on a plant input u_ℓ.

- Direction preservation [5, 8, 12, 17]:

$$u_\ell = w_\ell \min \left\{ \frac{sat_1(w)}{w_1}, \cdots, \frac{sat_m(w)}{w_m} \right\}, \quad \ell = 1, \cdots, m$$

In [5], the direction preservation approach has been suggested for directionality compensation in processes with ill-conditioned steady state gain matrix.

- Optimization formulation of the conditioning technique [9, 25]. In this approach, when a controller output is infeasible, a feasible controller output is obtained by calculating (via optimization) a new setpoint value which is closest (in the setpoint space) to the original setpoint value and which yields a feasible controller output.

- Controller detuning [15]. When controller output is infeasible, an optimization problem is solved to obtain the values of controller tunable parameters that result in a feasible controller output.

Unlike in SISO processes, in MIMO processes clipping and direction preservation may lead to completely different feasible plant inputs and may steer the process in wrong 'directions', leading to very poor closed-loop performances. As we will see, depending on the structural properties of the process under consideration, one or none of these approaches can yield an optimal feasible controller action. The directionality compensation via optimization formulation of the conditioning technique may also lead to poor performance, because it is based on the controller being used but not on the process being controlled. For example, when completely decentralized control is used, this method is identical to clipping, irrespective of the nature of the process under control.

In Section 2, the notion of process directionality is defined, and the class of processes that do not exhibit the process directionality are characterized. An optimal directionality compensator is presented. The compensator is applicable to processes irrespective of the type of controller being used, and involves solving a simple quadratic program on-line. Given an unconstrained controller output, say w, and the characteristic (decoupling) matrix of the process to which the controller action to be applied, the compensator calculates an optimal feasible plant input u that yields a closed-loop response as close as possible to the response of the same process to the controller output w (see Figure 1).

1.2. WINDUP

Windup is another controller-performance-degradation phenomenon that is associated with actuator saturation. Although this phenomenon has been studied extensively, only a few attempts have been made to define it precisely. Furthermore, while closed-loop-response quality indices such as response time and overshoot have been used to indicate the presence of windup, at the present time there is no specific measure to quantify windup. To characterize the classes of controllers that do not exhibit windup, we adopt the following definitions:

- *Definition 1* [5]: A dynamic controller does not exhibit windup, if and only if the states of the controller are not driven by the error when the actuator is in saturation.
- *Definition 2* [10]: A dynamic controller does not exhibit windup, if and only if when the actuator is in saturation the closed-loop behavior under the controller is identical to that under a static state feedback. This definition is based on the realization that windup is not associated with static feedback controllers.
- *Definition 3:* A dynamic controller does not exhibit windup, if and only if the controller action is solution to a moving-horizon constrained optimization problem.

According to Definition 1 any controller whose states are not driven by the error when the actuator is in saturation, does not exhibit windup. However, according to Definition 3 only model predictive controllers have a windup-free performance. Definition 2 is neither as broad as Definition 1 nor as strict as Definition 3. In other words, if $\mathcal{A}$, $\mathcal{B}$ and $\mathcal{C}$ respectively represent the sets of the controllers that do not exhibit windup according to Definitions 1, 2 and 3, then $\mathcal{C} \subset \mathcal{B} \subset \mathcal{A}$. In the strictest sense, only an ISE-optimal response is windup free.

In linear analytical control, the issues of windup and constraint handling as well as closed-loop stability in the presence of input constraints have been studied extensively (e.g. [1, 3, 9, 10, 13, 21, 25]). In particular, in linear analytical model-based control, powerful results are available in the frameworks of internal model control [26] and model state feedback control [6, 15].

In nonlinear analytical model-based control, the issues of input constraint handling and windup have received considerable attention in recent years. More specifically, there have been several approaches to the problem of integral windup in input-output linearizing control methods. These include:

- Conditional integration (i.e. turning off integration when a constraint is active). This approach was employed in real-time nonlinear control of pilot-scale polymerization reactors (e.g. [20]).
- MPC formulation of input-output linearization [19, 24].
- Input constraint mapping of Calvet and Arkun [4], together with an input-output linearizing state feedback and a linear controller. The input constraint mapping maps the constraints on manipulated inputs to 'state-dependent' constraints on the reference input to the input-output linearizing state feedback. This mapping, together with the input-output linearizing state feedback, converts the nonlinear system with input constraints to a 'linear' system with the state-dependent input constraints. To regulate the 'linearized' system with the state-dependent input constraints, Oliveira et al. [16] and Kurtz and Henson [14] have used linear MPC, and Kendi and Doyle [12] has employed the modified linear internal model control of Zheng et al. [26].
- An observer-based anti-windup approach with a nonlinear gain for nonlinear processes [10]. This approach prevents windup in the sense of Definitions 1 and 2.

In Section 3, two analytical nonlinear model-based control laws for processes with input constraints are presented. They include two distinct components: (i) an input-output linearizing controller that inherently includes an optimal integral windup compensator and (ii) the optimal directionality compensator. Furthermore, they can minimize the mismatch between the constrained and unconstrained process output responses over a short horizon into the future.

1.3. ORGANIZATION OF THIS CHAPTER

Section 2 presents a precise definition of process directionality and the optimal directionality compensator. It also describes the scope of the work and the application and performance of the directionality compensator via numerical simulations. In Section 3, dynamic input-output linearizing control laws that can handle input constraints and constant disturbances and model errors are derived. The connections between (a) the derived control laws and (b) the modified internal model control (IMC) and the model state feedback control (MSFC) are established, followed by the application of one of the derived nonlinear control laws to a chemical reactor example.

2. Optimal Directionality Compensation

2.1. SCOPE

We consider the class of general but affine-in-control, nonlinear multivariable processes described by a state-space model of the form

$$\begin{cases} \dot{x}(t) &= f(x(t)) + g(x(t))u(t) \\ y(t) &= h(x(t)) \end{cases} \tag{1}$$

where $x = [x_1 \cdots x_n]^T \in \mathbb{R}^{n \times 1}$, $u = [u_1 \cdots u_m]^T \in U \subset \mathbb{R}^{m \times 1}$ and $y = [y_1 \cdots y_m]^T \in \mathbb{R}^{m \times 1}$ are the vectors of state variables, plant inputs (manipulated inputs), and controlled outputs respectively. Here $U = \{u | u_{l_\ell} \le u_\ell \le u_{h_\ell},\ \ell = 1, \cdots, m\}$, where u_{l_ℓ}, u_{h_ℓ}, $\ell = 1, \cdots, m$, are scalar constant quantities. A controller output w is said to be feasible, if and only if $w \in U$. It is assumed that: $g_1(x)$, $\cdots$, $g_m(x)$, $h(x)$ and $f(x)$ are smooth vector functions, where $g_j(x)$ represents the jth column of the matrix $g(x)$; the process is minimum phase (has asymptotically stable zero dynamics); each controlled output y_i has a finite relative order r_i, which is the smallest integer for which locally $\left[L_{g_1} L_f^{r_i-1} h_i(x) \cdots L_{g_m} L_f^{r_i-1} h_i(x) \right] \ne 0$; the characteristic matrix of process is locally nonsingular. The characteristic matrix is an $m \times m$ matrix whose ijth entry is $L_{g_j} L_f^{r_i-1} h_i(x)$; it will be denoted by $\mathcal{C}(x)$. Here L_f and L_{g_j} are Lie derivative (in the directions of the vectors f and g_j respectively) operators.

2.1.1. *Characteristic Matrix: Short-Horizon Gain Matrix of Process*
The characteristic matrix of a process characterizes the sensitivity of the process to input changes over a short time horizon, while the steady-state gain matrix of a process characterizes the sensitivity of the process to input changes over an infinite horizon.

Consider a general nonlinear process of the form of (1) with finite relative orders $r_1, \cdots, r_m$, $h(0) = 0$, and initial conditions $x(0) = 0$, where the origin is the nominal equilibrium point of the process corresponding to $u = 0$. For such a process, the output response after a very small time horizon into the future, Δ, is given by [7]

$$y(\Delta) \approx diag \left\{ \frac{\Delta^{r_\ell}}{r_\ell!} \right\} \mathcal{C}(x(0))u(0)$$

This indicates that over a very short horizon the effect of an input char ;e on the controlled outputs depends on the structure of the characteristic matrix. In other words, over a short time horizon, it is the characteristic matrix that primarily determines (i) the direction in which the process response to an plant input change will evolve and (ii) the strength of the effect of a plant input change on the process response. For this reason, as we will see, the characteristic matrix, but not the steady state gain matrix, plays a pivotal role in process directionality compensation.

2.2. PROCESS DIRECTIONALITY

Whether a process is SISO or MIMO, clipping a controller output w, $sat(w)$, yields the feasible plant input which is closest (in the plant input space) to the unconstrained controller output w. In mathematical terms, $u = sat(w)$ is the solution to the quadratic program

$$\min_{u} ||u - w||^2$$

subject to

$$u_{l_\ell} \leq u_\ell \leq u_{h_\ell}, \qquad \ell = 1, \cdots, m$$

where $||\eta||$ denotes the Euclidean norm of a vector η. This feasible plant input, which is closest (in the plant input space) to w, may not lead to an optimal response (i.e. one that is closest to the response of the same process to the unconstrained controller output w). We will refer to this performance degradation as the process directionality, a precise definition of which is given here.

Definition 4: A process of the form of (1) does not exhibit process directionality, if and only if for every plant input $w \in \mathbb{R}^m$ the response of the process to $sat(w)$ is closest (in the output space) to the response of the same process to w.

It is worth noting that in several aspects this notion of process directionality is different than the one known as the dependence of process gain on the direction of plant input vector. For example, the latter does not exist when the condition number of process steady-state gain matrix is one, while as we will see, the former is not present when process characteristic matrix is diagonal.

2.3. OPTIMAL DIRECTIONALITY COMPENSATION

For a process of the form of (1), let $\hat{y}_\ell^*(\tau)$ and $\hat{y}_\ell(\tau)$, $\tau \geq t$, represent the predicted values of a controlled output y_ℓ when the process is subjected to a given unconstrained controller output w and to a feasible plant input u respectively. The objective is to calculate a feasible plant input, u, that renders the predicted value of every output y_ℓ, $\hat{y}_\ell$, as close as possible to $\hat{y}_\ell^*$. In mathematical terms, we seek a feasible controller action u that is solution to the constrained minimization problem:

$$\min_{u(t)} \left\{ \sum_{\ell=1}^{m} q_\ell \, ||\hat{y}_\ell(\tau) - \hat{y}_\ell^*(\tau)||_{p_\ell}^2 \right\} \qquad (2)$$

subject to the input constraints

$$u_{i_\ell} \leq u_\ell(t) \leq u_{h_\ell}, \qquad \ell = 1, \cdots, m \qquad (3)$$

where t represents the present time, $||\xi(\tau)||_{p_\ell}$ is the p_ℓ-function norm of a scalar function $\xi(\tau)$ over a sufficiently short time interval of the form $[t, \, t + T_{h_\ell}]$ with $T_{h_\ell} > 0$:

$$||\xi(\tau)||_{p_\ell} \triangleq \left[\int_t^{t+T_{h_\ell}} |\xi(\tau)|^{p_\ell} d\tau \right]^{(1/p_\ell)}, \quad p_\ell \geq 1,$$

and $q_1, \cdots, q_m$, are adjustable positive scalar parameters whose values are set according to the relative importance of the controlled outputs: the higher the value of a q_ℓ, the smaller the mismatch between the constrained and unconstrained process responses in y_ℓ (the lesser the effect of the constraints on the y_ℓ response).

Theorem 1: For a process of the form of (1), at each time instant t given an unconstrained controller output w, the optimal feasible plant input, denoted by u^+, that minimizes the performance index in (2) subject to the constraints of (3), is solution to the m-dimensional quadratic program:

$$\min_u ||Q\mathcal{C}(x)u - Q\mathcal{C}(x)w||^2 \qquad (4)$$

subject to

$$u_{l_\ell} \leq u_\ell \leq u_{h_\ell}, \qquad \ell = 1, \cdots, m \qquad (5)$$

where Q is a constant $m \times m$ diagonal matrix given by

$$Q = \text{diag} \left\{ \frac{\sqrt{q_\ell} \, ||(\tau - t)^{r_\ell}||_{p_\ell}}{r_\ell!} \right\}$$

The proof is given in [18].

The quadratic program of (4) and (5) is trivially solvable. For example, one can use the computationally efficient, simple method described by the following theorem. This method is based on the work of Barnard [2].

TABLE 1. Converging sequences corresponding to Example 1 with different $w = u^0$.

ℓ	u_1^ℓ	u_2^ℓ	u_1^ℓ	u_2^ℓ	u_1^ℓ	u_2^ℓ	u_1^ℓ	u_2^ℓ
0	2.00	3.0	2.0	-3.0	0.50	3.0	0.5	0.5
1	1.00	2.0	1.0	-2.0	0.50	2.0	0.5	0.5
2	0.92	2.0	1.0	-2.0	0.41	2.0		
$\vdots$	$\vdots$	$\vdots$	$\vdots$		$\vdots$	$\vdots$		
19					-1.0	2.0		
20					-1.0	2.0		
$\vdots$								
28	-1.00	2.0						
29	-1.00	2.0						

Theorem 2: Let u^+ denote the solution to the quadratic program of (4) and (5), $P = [p_{ij}] \triangleq C^T Q^T Q C$,

$$\delta = \|P\|_F \triangleq \sqrt{\sum_{i=1}^{m}\sum_{j=1}^{m} p_{ij}^2} \qquad \text{(Frobenius norm of } P)$$

and $\phi(u) \triangleq sat\,[\delta^{-1}P(w - u) + u]$. Then,
(i) $u^+ = \phi(u^+)$, and
(ii) the fixed-point iteration sequence $\{u^\ell\}$ generated by

$$u^{\ell+1} = \phi\left(u^\ell\right), \quad \ell = 0, 1, 2, 3, \cdots, \quad \forall u^0 \in \mathbb{R}^m$$

converges to u^+ [$\phi(u)$ is a global contraction mapping in $\mathbb{R}^m$].

Example 1: Consider a quadratic program of the form of (4) and (5) with $Q = I$,

$$C = \begin{bmatrix} 1.0 & -10.0 \\ 0.1 & 2.0 \end{bmatrix},$$

$u_{l_1} = -1$, $u_{h_1} = +1$, $u_{l_2} = -2$, and $u_{h_2} = +2$. To obtain the solution to this quadratic program for several values of w, we use the method of Theorem 2 which yields the converging sequences given in Table 1.

Theorem 1 indicates that at each time instant, the optimal feasible plant input, u^+, is calculated on the basis of $C(x)$ and a given unconstrained controller output w. Let $\mathcal{F}[C(x), w]$ denote the solution to the quadratic program of (4) and (5). Then, $u^+ = \mathcal{F}[C(x), w]$ represents the optimal

182

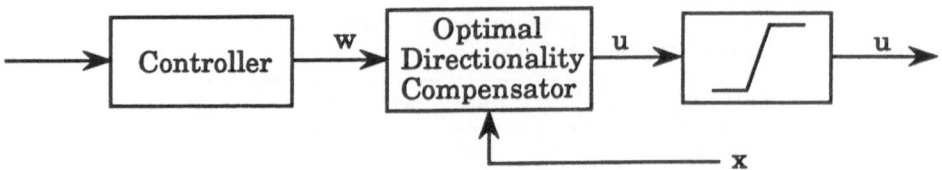

Figure 2. Optimal directionality compensation in nonlinear systems.

directionality compensator. Thus, the characteristic matrix plays a key role in the optimal directionality compensation: to calculate an optimal feasible plant input in a nonlinear process, given a controller output, one needs to know the characteristic matrix and measurements of the state variables of the process (see also Figure 2). It is the nature of characteristic matrix not that of steady state gain matrix that determines when it is optimal to use the clipping approach for directionality compensation. It is noteworthy that characteristic matrix and steady state gain matrix characterize two different aspects of process behavior; the former characterizes the sensitivity of process to input changes over a very short horizon and the latter over an infinite horizon. Structural properties such as singular values and condition number of steady state gain matrix and relative gain array also characterize the process response over an infinite horizon. Using steady state structural properties as a basis for selecting either of the approaches may lead to a very poor closed-loop performance, unless one uses a steady state controller.

Remark 1: For the class of processes with diagonal characteristic matrix, the optimal directionality compensator is identical to m limiters (clippers), i.e.

$$u_\ell = sat_\ell(w), \quad \ell = 1, \cdots, m$$

Thus, for this class of processes the feasible plant input which is closest (in the plant input space) to the unconstrained controller output w, yields to an optimal process response (i.e. one that is closest to the response of the same process to the unconstrained controller output w). In other words, this class of processes do not exhibit process directionality, and thus in the presence of input constraints their closed-loop performance is not degraded by process directionality.

Remark 2: In the case that the weights $q_1, \cdots, q_m$ are chosen such that

$$\frac{q_\ell \, ||(\tau - t)^{r_\ell}||^2_{p_\ell}}{(r_\ell!)^2} = 1, \quad \ell = 1, \cdots, m$$

that is, when the controlled outputs are of equal importance irrespective of the values of their relative orders $(r_1, \cdots, r_m)$, the quadratic program of

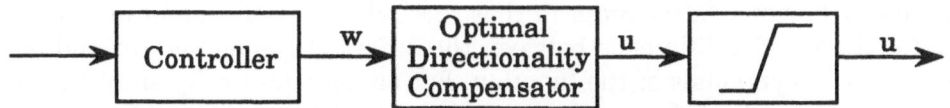

Figure 3. Optimal directionality compensation in linear systems.

(4) and (5) takes the simple form:

$$\min_u \|C(x)u - C(x)w\|^2 \tag{6}$$

subject to $u_{l_\ell} \leq u_\ell \leq u_{h_\ell}$, $\ell = 1, \cdots, m$.

2.4. APPLICATION TO TIME-INVARIANT LINEAR SYSTEMS

Consider multivariable time-invariant linear systems described by a state-space model of the form

$$\begin{cases} \dot{x}(t) &= Ax(t) + Bu(t), \quad x(0) = 0 \\ y(t) &= Cx(t) \end{cases} \tag{7}$$

where A, B and C are $n \times n$, $n \times m$ and $m \times n$ matrices respectively. This class of systems is a special case of (1): $f(x(t)) = Ax(t)$, $g(x(t)) = B$, and $h(x(t)) = Cx(t)$. The input-output behavior of the system of (7) can be represented by the s-domain matrix transfer function

$$\bar{y}(s) = P(s)\bar{u}(s) \tag{8}$$

where

$$P(s) = C (sI - A)^{-1} B \tag{9}$$

In these processes, the relative order r_i is the smallest integer for which $c_i A^{r_i - 1} B \neq [0 \cdots 0]$, where c_i is the ith row of the matrix C, and the characteristic matrix

$$C = \begin{bmatrix} c_1 A^{r_1 - 1} B \\ \vdots \\ c_m A^{r_m - 1} B \end{bmatrix} = \lim_{s \to \infty} [diag\,\{s^{r_\ell}\}\, P(s)],$$

which is independent of x. Thus, as shown in Figure 3, in time-invariant linear processes the optimal directionality compensator does not require information on the state of process: $u = \mathcal{F}[C, w]$.

In linear processes, the characteristic matrix will be diagonal, if and only if the diagonal element of every row of the transfer function matrix has the absolutely lowest relative order in that row, where relative order of a rational function is the difference between the orders of the numerator and denominator polynomials of the function. For an asymptotically stable linear process, the steady state process gain is given by $P(0) = -CA^{-1}B$, which is obviously different than the characteristic matrix $C = \lim_{s \to \infty} [diag \{s^{r_\ell}\} P(s)]$.

Example 2: Consider the following 2×2 linear process:

$$P(s) = \begin{bmatrix} \dfrac{1}{s+1} & \dfrac{1000s}{(s+1)^2} \\ \dfrac{s}{(s+1)^2} & \dfrac{2}{s+1} \end{bmatrix}$$

This process is 'statically', completely decoupled, because its steady state gain matrix and relative gain array are diagonal:

$$K_p = \begin{bmatrix} 1 & 0 \\ 0 & 2 \end{bmatrix}, \quad RGA = \begin{bmatrix} 1 & 0 \\ 0 & 1 \end{bmatrix}$$

However, the process is dynamically, strongly coupled, since its characteristic matrix

$$C = \begin{bmatrix} 1 & 1000 \\ 1 & 2 \end{bmatrix}$$

is not diagonal. Thus, the process with input constraints exhibits the process directionality; for this process clipping is not optimal.

Example 3: Consider the following 2×2 linear process:

$$P(s) = \begin{bmatrix} \dfrac{1}{s+1} & \dfrac{1000}{(s+1)^2} \\ \dfrac{1}{(s+1)^2} & \dfrac{2}{s+1} \end{bmatrix}$$

Checking the steady state gain matrix and relative gain array of the process:

$$K_p = \begin{bmatrix} 1 & 1000 \\ 1 & 2 \end{bmatrix}, \quad RGA = \begin{bmatrix} -0.002 & 1.002 \\ 1.002 & -0.002 \end{bmatrix}$$

we see that this process is 'statically', highly coupled. However, it is dynamically, weakly coupled:

$$C = \begin{bmatrix} 1 & 0 \\ 0 & 2 \end{bmatrix}$$

Thus, this process with input constraints do not exhibit the process directionality; for this process clipping is optimal.

2.5. APPLICATION TO THREE PROCESSES

2.5.1. *Decentralized PI Control of a Linear Process*
Consider the linear two-input two-output process:

$$P(s) = \frac{1}{100s + 1} \begin{bmatrix} 40 & -300 \\ -1 & 40 \end{bmatrix} \tag{10}$$

with $|u_i| \leq 1$, $i = 1$, 2. For this example, $r_1 = 1$, $r_2 = 1$, and

$$C = \begin{bmatrix} 0.4 & -3.0 \\ -0.01 & 0.4 \end{bmatrix}$$

Two completely decentralized PI controllers with $k_{c_1} = 2.1$, $k_{c_2} = 0.36$, $\tau_{I_1} = 174.0$ s and $\tau_{I_2} = 22.1$ s, and with conditional integration (to prevent integral windup) are used to track asymptotically the set point changes $y_{sp1} = 8$ and $y_{sp2} = 3$. The conditional integration involves turning off the integrator of the ith loop when the input u_i saturates.

Figure 4 depicts the closed-loop output response under the same two PI controllers but three different directionality compensators; it shows that while clipping and direction preservation approaches lead to poor responses in y_1, the optimal direction compensator provides a significantly better closed-loop performance.

2.5.2. *Model-Based Control of a Linear Process*
Consider the linear two-input two-output process:

$$\begin{cases} \dot{x}_1 &= -0.01x_1 - 0.0002x_2 + 0.25u_1 \\ \dot{x}_2 &= -0.5x_1 - 0.03x_2 + 4u_2 \\ y_1 &= x_1 \\ y_2 &= x_2 \end{cases} \tag{11}$$

with $|u_1| \leq 0.12$ and $|u_2| \leq 0.12$.

For this example, $r_1 = 1$, $r_2 = 1$,

$$C = \begin{bmatrix} 0.25 & 0 \\ 0 & 4 \end{bmatrix}$$

We use the following mixed error- and state-feedback controller that induces the unconstrained, completely decoupled, closed-loop response: $5\dot{y}_1 + y_1 = y_{sp1}$, $2\dot{y}_2 + y_2 = y_{sp2}$, and that includes an optimal integral windup com-

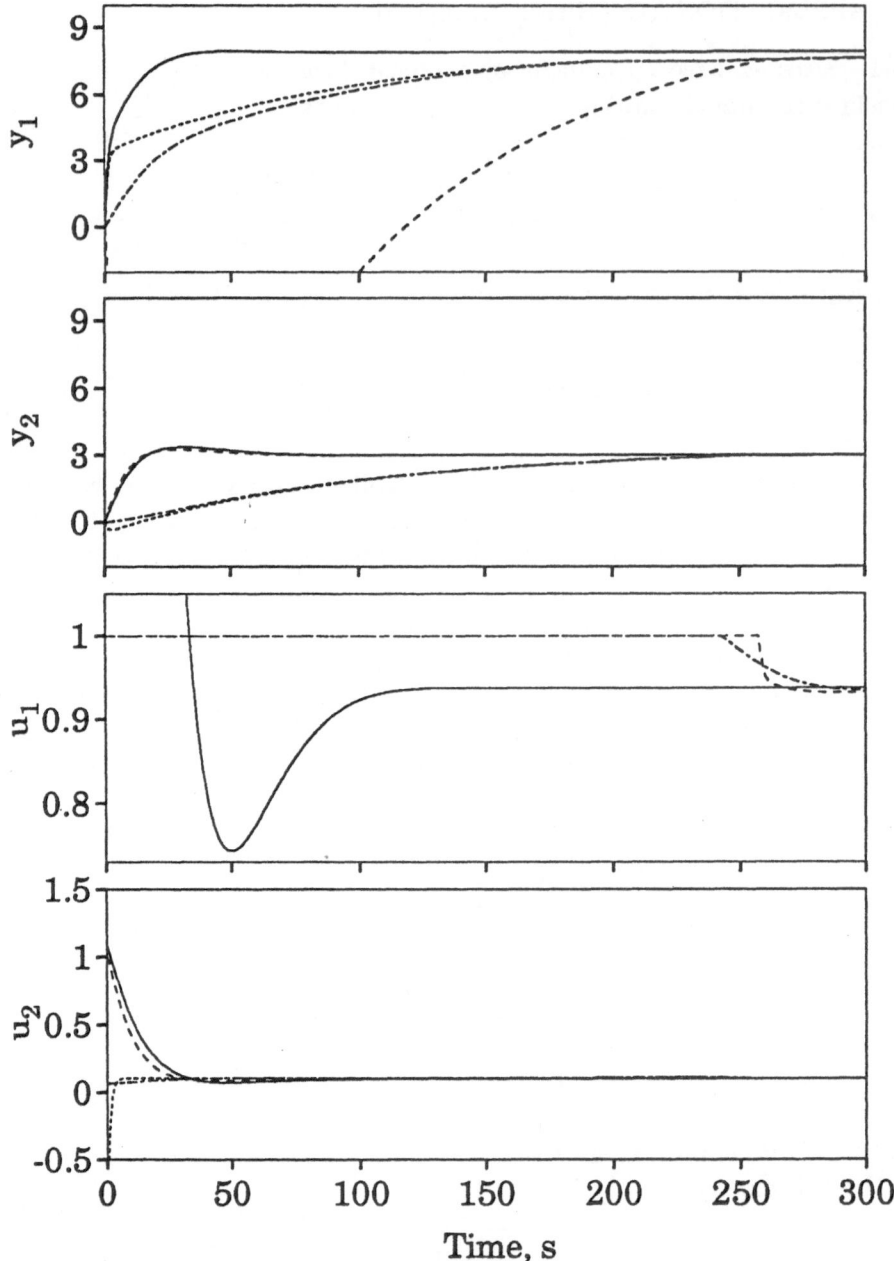

Figure 4. Profiles of the controlled outputs and plant inputs of the linear example of (10): solid = no bounds on the inputs; dashed = clipping, dotted-dashed = direction preservation, and dotted = optimal directionality compensation, when $|u_j| \leq 1$, $j = 1, 2$.

pensator [22]:

$$\begin{cases} \dot{\eta}_1^{(1)} = 10(-\eta_1^{(1)} + 0.999x_1 - 0.00002x_2 + 0.025u_1), & \eta_1^{(1)}(0) = 0 \\ \dot{\eta}_1^{(2)} = 10(-\eta_1^{(2)} - 0.05x_1 - 0.997x_2 + 0.4u_2), & \eta_1^{(2)}(0) = 0 \\ \dot{\xi}_1^{(1)} = 0.2(-\xi_1^{(1)} + \eta_1^{(1)} + e_1), & \xi_1^{(1)}(0) = 0 \\ \dot{\xi}_1^{(2)} = 0.5(-\xi_1^{(2)} + \eta_1^{(2)} + e_2), & \xi_1^{(2)}(0) = 0 \\ w_1 = 0.8\left(\eta_1^{(1)} + e_1 + 49\xi_1^{(1)} + 50.05x_1 + 0.001x_2\right) \\ w_2 = 0.125\left(\eta_1^{(2)} + e_2 + 19\xi_1^{(2)} + x_1 + 20.06x_2\right) \end{cases}$$

Figure 5 depicts the closed-loop output response under the same model-based controller but different directionality compensators; the optimal directionality compensator and clipping exhibit the same performance which is significantly better than that of direction preservation.

2.5.3. 'Input-Output Linearizing' Control of a Nonlinear Bioreactor

Consider a continuous stirred-tank bioreactor described by a mathematical model of the form

$$\begin{cases} \dfrac{dS_1}{dt} = -\dfrac{1}{Y_{X/S_1}}\mu(S_1, S_2) + [S_{f_1} - S_1]D_1 - S_1 D_2 \\ \dfrac{dS_2}{dt} = -\dfrac{1}{Y_{X/S_2}}\mu(S_1, S_2) - S_2 D_1 + [S_{f_2} - S_2]D_2 \qquad (12) \\ \dfrac{dX}{dt} = \mu(S_1, S_2)X - X(D_1 + D_2) \end{cases}$$

where the specific growth rate

$$\mu(S_1, S_2) = \frac{\mu_m S_1}{K_{S_1} + S_1}\frac{S_2}{K_{S_2} + S_2},$$

S_1 and S_2 denote the outlet concentrations of substrates 1 and 2 respectively, S_{f_1} and S_{f_2} respectively represent the concentration of substrate 1 in feed 1 and the concentration of substrate 2 in feed 2, and D_1 and D_2 are respectively the dilution rates of feed streams 1 and 2. The values of the model parameters are the same as in [22].

The controlled outputs and manipulated inputs are as follows: $y_1 = S_1$, $y_2 = S_2$, $u_1 = D_1$, and $u_2 = D_2$ with the bounds $0 \leq D_i \leq 0.4\ h^{-1}$, $i = 1, 2$. The control objective is to operate the reactor at the set points $y_{sp_1} = 2.0\ kg.m^{-3}$ and $y_{sp_2} = 4.9\ kg.m^{-3}$, by an input-output linearizing controller with an optimal integral windup compensator.

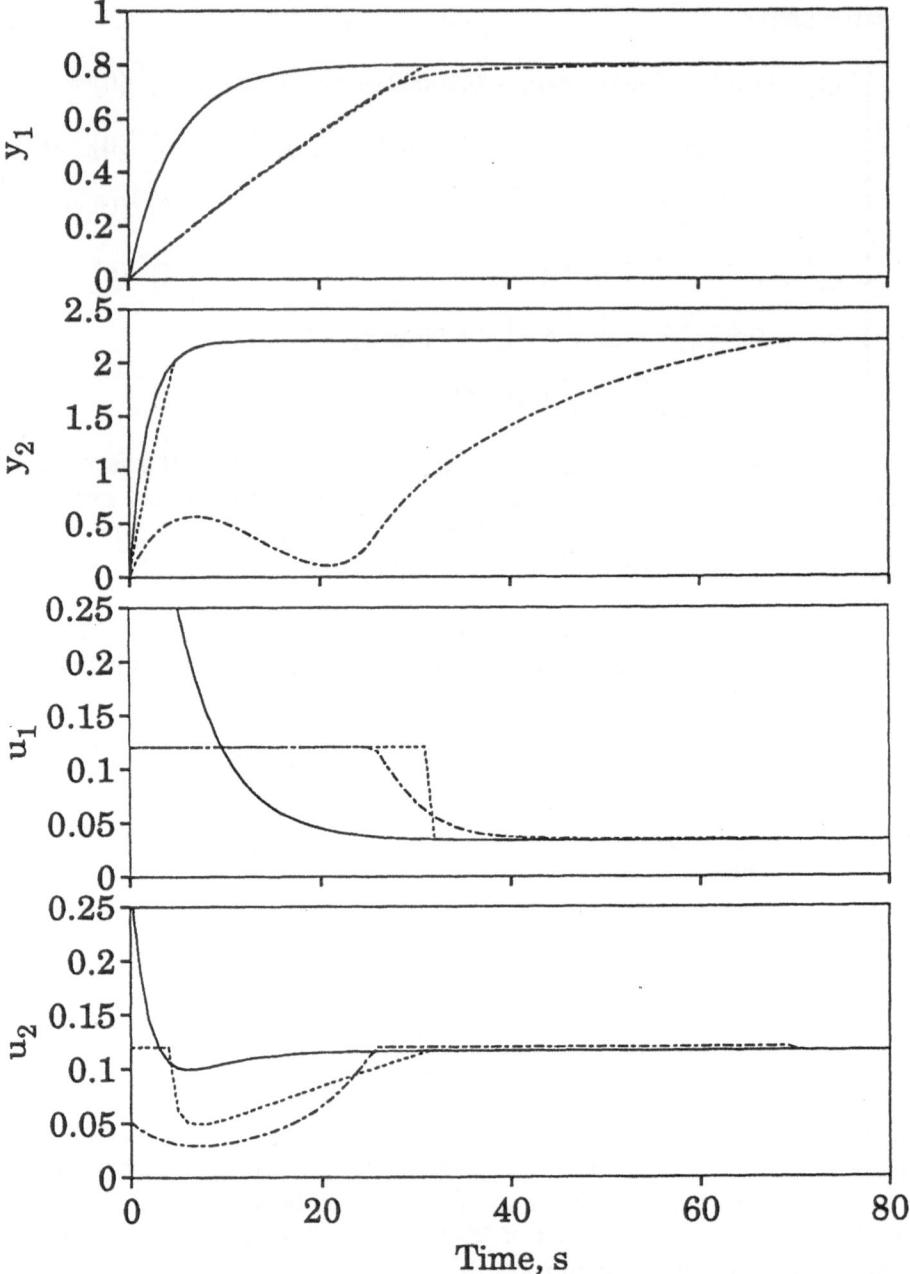

Figure 5. Profiles of the controlled outputs and plant inputs of the linear example of (11): solid = no bounds on the inputs; dotted-dashed = direction preservation, and dotted = optimal directionality compensation and clipping, when $|u_j| \leq 0.12$, $j = 1, 2$.

For this process, $r_1 = 1$, $r_2 = 1$, and

$$C = \begin{bmatrix} S_{f_1} - S_1 & -S_1 \\ -S_2 & S_{f_2} - S_2 \end{bmatrix}$$

Figure 6 depicts the startup profiles of the controlled outputs and plant inputs of the bioreactor under the same nonlinear controller but three different directionality compensators. In the presence of the input constraints, clipping (dashed line) cannot operate the process at the steady state (leads a very poor closed-loop response), and direction preservation (dotted-dashed line) results in a relatively better performance compared to that of clipping. However, the closed-loop performance under the optimal directionality compensator is of higher quality; it is the closest response to that represented by the solid line (obtained in the absence of the plant input bounds).

3. Optimal Windup Compensation

3.1. SCOPE

We consider the broader class of multi-input multi-output, continuous-time, nonlinear processes described by a state-space model of the form

$$\begin{cases} \dot{x}(t) & = & f(x(t)) + g(x(t))u(t), & x(0) = 0 \\ y_i(t) & = & h_i(x(t - \theta_i)), & i = 1, \cdots, m \end{cases} \tag{13}$$

where $x = [x_1 \cdots x_n]^T$ is the vector of state variables, $u = [u_1 \cdots u_m]^T$ is the vector of manipulated inputs, $y = [y_1 \cdots y_m]^T$ is the vector of controlled outputs, and θ_i is the deadtime in the ith output. We make the following assumptions: (a1) every variable of the system of (13) is in the form of deviation from its nominal steady-state value, and thus the origin is an equilibrium point; (a2) $x \in X \subset \mathbb{R}^{n \times 1}$, where X is an open connected set which contains the equilibrium point; (a3) $u \in U = \{u | \ u_{l_\ell} \leq u_\ell \leq u_{h_\ell}, \quad \ell = 1, \cdots, m\} \subset \mathbb{R}^{m \times 1}$, where every pair u_{l_ℓ} and u_{h_ℓ} are scalar constants which satisfy $u_{l_\ell} < 0 < u_{h_\ell}$; (a4) $g(x)$ is a smooth matrix function on X; (a5) $f(x)$ and $h(x)$ are smooth vector functions on X; (a6) output set-point, denoted by y_{sp}, is *achievable* at steady state in the sense that there exits a $u_o \in \text{interior}(U)$, which satisfies $f(\zeta) + g(\zeta)u_0 = 0$, where $\zeta \in X$ and $h(\zeta) = y_{sp}$; (a7) $det[\mathcal{C}(x)] \neq 0$, $\forall x \in X$; and (a8) the delay-free part of process (system of (13) with $\theta_1 = \cdots = \theta_m = 0$) is minimum phase (has asymptotically stable zero dynamics) on X. The measured (actual) value of the vector of the controlled variables will be denoted by $\bar{y}$:

$$\bar{y}_i(t) = h_i(\bar{x}(t - \theta_i)) + d_i(t), \quad i = 1, \cdots, m,$$

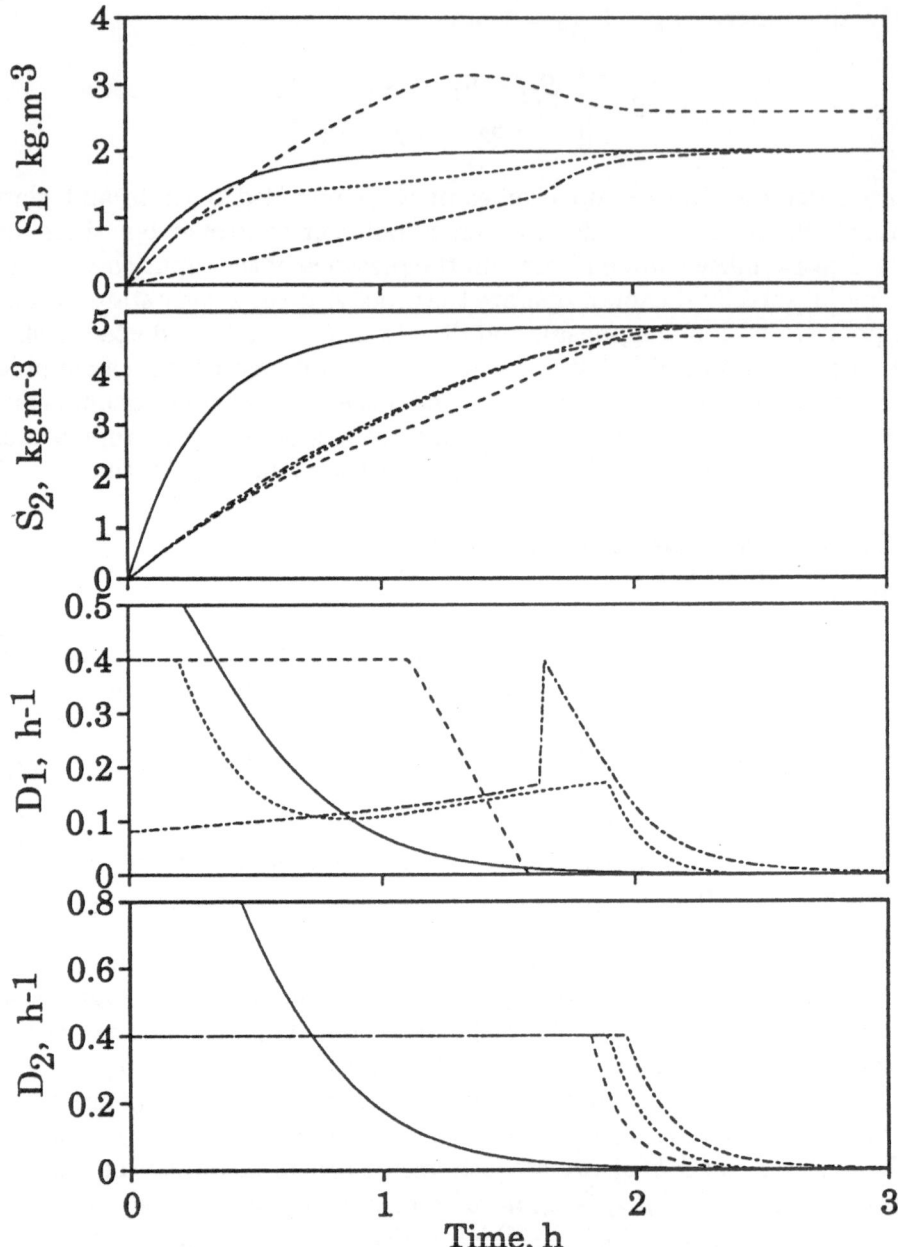

Figure 6. Profiles of the controlled outputs and plant inputs of the bioreactor example of (12): solid = no bounds on the inputs; dashed = clipping, dotted-dashed = direction preservation, and dotted = optimal directionality compensation, when $0 < D_i < 0.4\ h^{-1}$, $i = 1, 2$.

where $\bar{x}$ represents the measured (actual) value of the vector of the state variables and d is the vector of *constant* unmeasurable disturbances. The model-predicted values of the vectors of the controlled variables and state variables will be represented by y and x respectively.

3.2. OVERVIEW OF THE SOLUTION

We seek nonlinear control laws that: (a) in the absence of input constraints induce an adjustable, linear, input-output, closed-loop response; (b) force the mismatch between the linear unconstrained and the constrained responses to decay to zero asymptotically at an adjustable rate; (c) allow one to adjust the time period during which the input constraints are in saturation, without changing the nominal, linear, unconstrained, input-output, closed-loop response; and (d) can induce a closed-loop response with a very low ISE or the lowest ISE, in the presence of the input constraints.

In mathematical terms, we seek analytical nonlinear control laws that:

(a) in the absence of input constraints induce a linear input-output closed-loop response of the form of

$$\begin{bmatrix} \bar{y}_1(t+\theta_1) \\ \vdots \\ \bar{y}_m(t+\theta_m) \end{bmatrix} + \sum_{i=1}^{m} \sum_{\ell=1}^{r_i} \gamma_{i\ell} \frac{d^\ell \bar{y}_i(t+\theta_i)}{dt^\ell} = y_{sp}(t) \qquad (14)$$

where $\gamma_{i\ell} = [\gamma_{i\ell}^1 \cdots \gamma_{i\ell}^m]^T$, $\ell = 1, \cdots, r_i$, $i = 1, \cdots, m$, are m-vector adjustable parameters with $det\,[\gamma_{1r_1} \cdots \gamma_{mr_m}] \neq 0$.

(b) can minimize the quadratic performance index:

$$\sum_{\ell=1}^{m} q_\ell \, \|\hat{y}_\ell(\tau) - y_\ell^*(\tau)\|_{p_\ell}^2 \qquad (15)$$

subject to the input constraints

$$u_{l_\ell} \leq u_\ell(t) \leq u_{h_\ell}, \qquad \ell = 1, \cdots, m \qquad (16)$$

where y^* is the requested, linear, unconstrained, input-output, response described by (14); $\hat{y}_\ell(\tau)$ is the predicted value of an output y_ℓ; t represents the present time; $\|\xi(\tau)\|_{p_\ell}$ is the p_ℓ-function norm of a scalar function $\xi(\tau)$ over a sufficiently short time interval of the form $[t+\theta_\ell,\ t+\theta_\ell+T_{h_\ell}]$ with $T_{h_\ell} > 0$:

$$\|\xi(\tau)\|_{p_\ell} \triangleq \left[\int_{t+\theta_\ell}^{t+\theta_\ell+T_{h_\ell}} |\xi(\tau)|^{p_\ell} d\tau \right]^{(1/p_\ell)}, \qquad p_\ell \geq 1,$$

and $q_1, \cdots, q_m$, are adjustable positive scalar parameters whose values are set according to the relative importance of the controlled outputs; and $T_{h_1}, \cdots, T_{h_m}$ are sufficiently short time horizons into the future.

(c) when the input constraints are not active, eliminate the mismatch between $\bar{y}$ and y^* asymptotically at an adjustable rate. In particular, we request the mismatch to be governed by:

$$[\bar{y}(t) - y^*(t)] + \sum_{i=1}^{m}\sum_{\ell=1}^{r_i}\beta_{i\ell}\frac{d^\ell[\bar{y}_i(t) - y_i^*(t)]}{dt^\ell} = 0 \qquad (17)$$

where $\beta_{i\ell} = [\beta_{i\ell}^1 \cdots \beta_{i\ell}^m]^T$, $\ell = 1, \cdots, r_i$, $i = 1, \cdots, m$, are m-vector adjustable parameters with $det\,[\beta_{1r_1} \cdots \beta_{mr_m}] \neq 0$. This allows one to adjust the rate at which the constrained response approaches the linear unconstrained response, without changing the latter governed by (14).

To derive control laws with the aforementioned theoretical properties, we solve the following constrained minimization problem:

$$\sum_{\ell=1}^{m}q_\ell\,\|\hat{y}_\ell(\tau) - \tilde{y}_\ell^*(\tau)\|_{p_\ell}^2 \qquad (18)$$

subject to the input constraints $u_{l_\ell} \leq u_\ell(t) \leq u_{h_\ell}$, $\ell = 1, \cdots, m$. Here $\tilde{y}^*(t)$ describes the path that the output response will follow to reach the unconstrained response y^* when the constraints are no longer active. It is the solution to the linear system

$$\tilde{y}^*(\tau) + \sum_{i=1}^{m}\sum_{\ell=1}^{r_i}\beta_{i\ell}\frac{d^\ell\tilde{y}_i^*(\tau)}{d\tau^\ell} = \tilde{y}(t) \qquad (19)$$

subject to the initial conditions

$$\tilde{y}_i^*(t + \theta_i) = h_i(x(t)) + \bar{y}_i(t) - h_i(x(t - \theta_i)),$$
$$i = 1, \cdots, m$$

$$\frac{d^\ell\tilde{y}_i^*(t + \theta_i)}{dt^\ell} = L_f^\ell h_i(x(t)),$$
$$\ell = 1, \cdots, r_i - 1, \; i = 1, \cdots, m$$

where the forcing function $\tilde{y}$ is related to the output setpoint, y_{sp}, according to

$$y^* + \sum_{i=1}^{m}\sum_{\ell=1}^{r_i}\gamma_{i\ell}\frac{d^\ell y_i^*}{dt^\ell} = y_{sp}$$

$$\tilde{y} = y^* + \sum_{i=1}^{m}\sum_{\ell=1}^{r_i}\beta_{i\ell}\frac{d^\ell y_i^*}{dt^\ell} \qquad (20)$$

In the limit that $\beta_{i\ell} \to 0$, $i = 1, \cdots, m$, $\ell = 1, \cdots, r_i$, according to (19) and (20), $\tilde{y}^* \to y^*$. Thus, in that limit the solution to the constrained minimization problem of (18) will also be the solution to the constrained minimization problem of (16).

Notations: The following notations will be used in the subsequent parts of this section:

(i) the time-invariant linear system

$$\left\{ \begin{array}{rcl} \dot{\eta} & = & A_c\eta + B_c\tilde{y} \\ \tilde{y}^* & = & C_c\eta \end{array} \right. \tag{21}$$

where $\eta \in \mathbb{R}^{(r_1 + \cdots + r_m) \times 1}$, and A_c, B_c and C_c are constant matrices of appropriate dimensions, will denote a minimal-order state-space realization of the linear system of (19).

(ii) the time-invariant linear system

$$\left\{ \begin{array}{rcl} \dot{\xi} & = & A_c^* \xi + B_c^* y_{sp} \\ \tilde{y} & = & \beta\gamma^{-1} [C_c^* \xi + y_{sp}] \end{array} \right. \tag{22}$$

where $\xi \in \mathbb{R}^{(r_1 + \cdots + r_m) \times 1}$, $\gamma = [\gamma_{1r_1} \cdots \gamma_{mr_m}]$, $\beta = [\beta_{1r_1} \cdots \beta_{mr_m}]$, and A_c^*, B_c^* and C_c^* are constant matrices of appropriate dimensions, will represent a minimal-order state-space realization of the linear system of (20);

(iii)

$$\Phi(x, u) \stackrel{\triangle}{=} h(x) + \sum_{i=1}^{m} \sum_{\ell=1}^{r_i} \beta_{i\ell} L_f^\ell h_i(x) + \beta C(x)u;$$

(iv)

$$\Psi_0(x, \varsigma) \stackrel{\triangle}{=} [\gamma C(x)]^{-1} \left[\varsigma - \gamma\beta^{-1} \left\{ h(x) + \sum_{i=1}^{m} \sum_{\ell=1}^{r_i} \beta_{i\ell} L_f^\ell h_i(x) \right\} \right]$$

3.3. NONLINEAR CONTROLLERS SYNTHESIS

This section includes two theorems that concisely describe the theoretical properties of two nonlinear control laws, one for nonlinear processes with full state measurements and one for nonlinear processes with incomplete state measurements and deadtimes.

3.3.1. *Processes with Full State Measurements*

Theorem 3: For a process of the form of (13) with complete state measurements ($\bar{x}$) and no deadtimes, the dynamic mixed error- and state-feedback

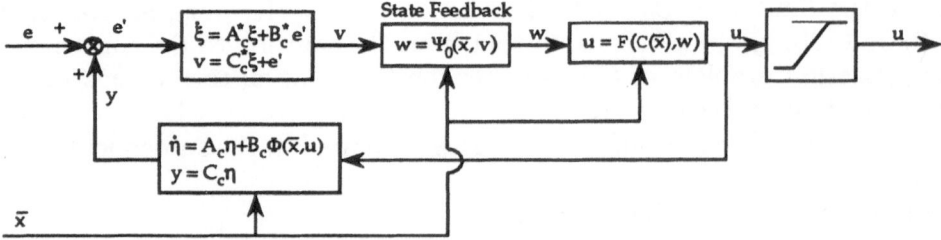

Figure 7. Generalized, mixed error- and state-feedback control structure.

control law:

$$\begin{cases} \dot{\eta} = A_c\eta + B_c\Phi(\bar{x}, u), & \eta(0) = 0 \\ \dot{\xi} = A_c^*\xi + B_c^*[C_c\eta + e], & \xi(0) = 0 \\ u = \mathcal{F}\{\mathcal{C}(\bar{x}), \Psi_0(\bar{x}, C_c\eta + e + C_c^*\xi)\} \end{cases} \tag{23}$$

where $e \stackrel{\triangle}{=} y_{sp} - \bar{y}$,

(a) minimizes the constrained performance index of (18). It minimizes the performance index of (15) in the limit that all the roots of the characteristic equation

$$det\left\{I_m + \left[\left(\sum_{\ell=1}^{r_1}\beta_{1\ell}s^\ell\right)\cdots\left(\sum_{\ell=1}^{r_m}\beta_{m\ell}s^\ell\right)\right]\right\} = 0 \tag{24}$$

are placed far left in the complex plane (i.e. in the limit that $\beta_{i\ell} \to 0$, $i = 1, \cdots, m$, $\ell = 1, \cdots, r_i$).

(b) in the absence of the constraints induces the linear input-output closed-loop response of (14) with $\theta_1 = \cdots = \theta_m = 0$, irrespective of the values of $\beta_{i\ell}$, $\ell = 1, \cdots, r_i$, $i = 1, \cdots, m$.

(c) has integral action: in the presence of constant disturbances and model errors, induces an offsetless closed-loop response.

The proof is given in [23].

The block diagram of the control law of Theorem 3 is depicted in Figure 7. This control structure will be referred to as the generalized, mixed error- and state-feedback control structure. The ξ subsystem provides an extra flexibility for achieving a desirable response in the presence of input constraints.

In the special case that the tunable parameters are chosen such that $\beta_{i\ell} = \gamma_{i\ell}$, $i = 1, \cdots, m$, $\ell = 1, \cdots r_i$, the matrix $C_c^* = 0$, and $\gamma\beta^{-1}\beta_{i\ell} = \gamma_{i\ell}$, $i = 1, \cdots, m$, $\ell = 1, \ldots r_i$. In such a case,

- the control law of (23) will be of order $(r_1 + \cdots + r_m)$.

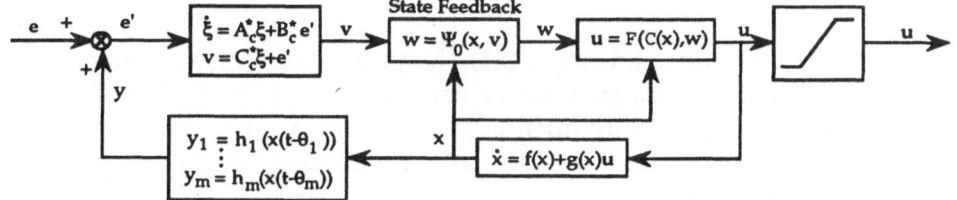

Figure 8. Generalized, error-feedback control structure.

- when the characteristic matrix is diagonal, the control law of (23) belongs to the class of nonlinear control laws developed by Kapoor and Daoutidis [10] using an observer-based approach to integral windup.

3.3.2. *Processes with Incomplete State Measurements and Deadtimes*

Theorem 4: For a process of the form of (13) with incomplete state measurements, the dynamic error-feedback control law:

$$\begin{cases} \dot{\xi} = A_c^*\xi + B_c^*e', & \xi(0) = 0 \\ \dot{x} = f(x) + g(x)\mathcal{F}\left\{\mathcal{C}(x), \Psi_0(x, e' + C_c^*\xi)\right\}, & x(0) = 0 \qquad (25) \\ u = \mathcal{F}\left\{\mathcal{C}(x), \Psi_0(x, e' + C_c^*\xi)\right\} \end{cases}$$

where

$$e_i' = e_i(t) + h_i(x(t - \theta_i)), \quad i = 1, \cdots, m$$

(a) minimizes the constrained performance index of (18). It minimizes the performance index of (15) in the limit that all the roots of (24) are placed far left in the complex plane (i.e. in the limit that $\beta_{i\ell} \to 0$, $i = 1, \cdots, m$, $\ell = 1, \cdots, r_i$).

(b) in the absence of the constraints, induces the linear input-output closed-loop response of (14), irrespective of the values of $\beta_{i\ell}$, $\ell = 1, \cdots, r_i$, $i = 1, \cdots, m$.

(c) has integral action: in the presence of constant disturbances and model errors, induces an offsetless closed-loop response.

The proof is given in [23].

The block diagram of the control law of Theorem 4 is depicted in Figure 8. This control structure will be referred to as the generalized, error-feedback control structure. Again, here, the ξ subsystem provides an extra flexibility for obtaining better closed-loop performance in the presence of input constraints.

In the special case that the tunable parameters are chosen such that $\beta_{i\ell} = \gamma_{i\ell}$, $i = 1, \cdots, m$, $\ell = 1, \cdots r_i$, the controller action will be independent of ξ, and thus the control law of (25) will be of order n.

3.4. CLOSED-LOOP STABILITY

Consider the following conditions: (c1) the process of (13) with $\theta_1 = \cdots = \theta_m = 0$ is minimum-phase (has stable zero dynamics) on X; (c2) the process of (13) is asymptotically open-loop stable on X; (c3) the parameters $\gamma_{i\ell}$, $i = 1, \cdots, m$, $\ell = 1, \cdots r_i$, are chosen such that all the roots of the characteristic equation

$$det\left\{I_m + \left[\left(\sum_{\ell=1}^{r_1}\gamma_{1\ell}s^\ell\right)\cdots\left(\sum_{\ell=1}^{r_m}\gamma_{m\ell}s^\ell\right)\right]\right\} = 0 \qquad (26)$$

lie in the left half of the complex plane; and (c4) the parameters $\beta_{i\ell}$, $i = 1, \cdots, m$, $\ell = 1, \cdots r_i$, are chosen such that all the roots of the characteristic equation of (24) lie in the left half of the complex plane.

Definition 5: For a process of the form of (13), an initial condition x_0 is said to be feasible, if and only if there exists a feasible smooth manipulated input trajectory $u(t)$ (i.e. a smooth $u(t)$ such that $u(t) \in U$, $\forall t \in [0, t_f]$), that can drive the process from x_0 to the origin (nominal equilibrium point) within a finite time t_f. The set of all the feasible initial conditions will be denoted by $\mathcal{X}$.

Definition 6: A process of the form of (13) is said to be asymptotically stable on X, if and only if for every initial condition $x_0 \in X$, the system $\dot{x}(t) = f[x(t)]$ evolves such that $\lim_{t\to\infty}||x(t)|| \to 0$.

Theorem 5: For a process of the form of (13) with complete state measurements ($\bar{x}$) and no deadtimes, the closed-loop system under the dynamic mixed error- and state-feedback control law of (23)

(a) in the absence of input constraints will be asymptotically stable on X, if the conditions c1, c3 and c4 hold.

(b) in the presence of input constraints will be asymptotically stable on $X \cap \mathcal{X}$, if the conditions c1, c3 and c4 hold and the tunable parameters $\beta_{i\ell}$, $i = 1, \cdots, m$, $\ell = 1, \cdots r_i$, are chosen such that the roots of (24) are placed sufficiently close to the origin.

The proof is given in [23].

Theorem 6: For process of the form of (13) with incomplete state measurements, the closed-loop system under the dynamic error-feedback control law of (25)

(a) in the absence of input constraints will be asymptotically stable on X, if the conditions c1, c2, c3 and c4 hold.

(b) in the presence of input constraints will be asymptotically stable on $X \cap \mathcal{X}$, if the conditions c1, c2, c3 and c4 hold and the tunable parameters $\beta_{i\ell}$, $i = 1, \cdots, m$, $\ell = 1, \cdots r_i$, are chosen such that the roots of (24) are placed sufficiently close to the origin.

The proof is given in [23].

Theorems 5 and 6 indicate that for a constrained nonlinear process of the form of (13), the derived control laws offer great flexibility to ensure closed-loop stability, without changing the nominal closed-loop response that the same control laws induce in the absence of the constraints.

Remark 3: The tunable parameters $\gamma_{i\ell}$, $i = 1, \cdots, m$, $\ell = 1, \cdots r_i$, determine the shape of the nominal linear closed-loop response, described by (14), that is induced when there are no constraints. The tunable parameters $\beta_{i\ell}$, $i = 1, \cdots, m$, $\ell = 1, \cdots r_i$, do not affect the shape of the nominal linear closed-loop response, but they influence the shape of the mismatch between the constrained and the linear unconstrained responses. They also affect the time period during which manipulated inputs are in saturation; the further left the locations of the roots of the polynomial of (24) in the complex plane, the longer the time period during which the manipulated inputs saturate. Once the manipulated inputs are no longer in saturation, the control laws of (23) and (25) both induce the closed-loop response of (17). In the case that complete input-output decoupling is desirable, it is achieved simply by setting $\gamma_{i\ell}^j = \beta_{i\ell}^j = 0$, $i \neq j$, $i = 1, \cdots, m$, $j = 1, \cdots, m$, $\ell = 1, \cdots, r_i$.

3.5. MODIFIED IMC AND MSFC PARAMETERIZATIONS OF THE CONTROL LAWS

Modified IMC and MSFC parameterizations of the nonlinear control laws of (23) and (25) are depicted in Figures 9, 10 and 11, and the corresponding controller components are given in Tables 2, 3 and 4. Unlike in the linear modified IMC structure [26] and in the MSFC structure [6, 15], in this nonlinear case, Q_1 and Q_1' depend on the state x. As we will see in the next subsection, when the control law of (25) is applied to time-invariant linear systems with diagonal characteristic matrix, the resulting linear controller is a minimal-order state-space realization of a modified internal model controller. The preceding parameterizations indicate that the implementation of the control law of (25) according to the modified IMC structure will increase the order of the controller by $2n$. Thus, the control law of (25) is a minimal-order state-space realization of a nonlinear modified IMC controller.

3.6. APPLICATION TO LINEAR SYSTEMS

Consider the class of time-invariant, linear processes described by a state-space model of the form

$$\begin{cases} \dot{x}(t) &= Ax(t) + Bu(t), \quad x(0) = 0 \\ y_i(t) &= c_i x(t - \theta_i), \quad i = 1, \cdots, m \end{cases} \tag{27}$$

TABLE 2. Modified IMC parameterization of the nonlinear controller of (25).

| P | $\dot{x}$ | $=$ | $f(x) + g(x)u$ |
| | $y_i(t)$ | $=$ | $h_i(x(t - \theta_i))$, $\quad i = 1, \cdots, m$ |

Q_1	$\dot{x}$	$=$	$f(x) + g(x)u$
	$\dot{\xi}$	$=$	$A_c^* \xi + B_c^* e'$
	w_1	$=$	$[\gamma C(x)]^{-1} [e' + C_c^* \xi]$

| Q_2 | $\dot{x}$ | $=$ | $f(x) + g(x)u$ |
| | w_2 | $=$ | $[\beta C(x)]^{-1} \left[h(x) + \sum_{i=1}^{m} \sum_{\ell=1}^{r_i} \beta_{i\ell} L_f^\ell h_i(x) \right]$ |

| Q_3 | $\dot{x}$ | $=$ | $f(x) + g(x)\mathcal{F}\{C(x), w\}$ |
| | u | $=$ | $\mathcal{F}\{C(x), w\}$ |

TABLE 3. MSFC parameterization of the nonlinear controller of (25).

| P | $\dot{x}$ | $=$ | $f(x) + g(x)u$ |
| | $y_i(t)$ | $=$ | $h_i(x(t - \theta_i))$, $\quad i = 1, \cdots, m$ |

| Q_1' | $\dot{\xi}$ | $=$ | $A_c^* \xi + B_c^* e'$ |
| | w_1 | $=$ | $[\gamma C(x)]^{-1} [e' + C_c^* \xi]$ |

| Q_2' | w_2 | $=$ | $[\beta C(x)]^{-1} \left[h(x) + \sum_{i=1}^{m} \sum_{\ell=1}^{r_i} \beta_{i\ell} L_f^\ell h_i(x) \right]$ |

| Q_3' | u | $=$ | $\mathcal{F}\{C(x), w\}$ |

TABLE 4. MSFC parameterization of the nonlinear controller of (23).

| Q_0'' | $\dot{\eta}$ | $=$ | $A_c \eta + B_c \Phi(\bar{x}, u)$ |
| | y | $=$ | $C_c \eta$ |

| Q_1'' | $\dot{\xi}$ | $=$ | $A_c^* \xi + B_c^* (e + y)$ |
| | w_1 | $=$ | $[\gamma C(\bar{x})]^{-1} [e + y + C_c^* \xi]$ |

| Q_2'' | w_2 | $=$ | $[\beta C(\bar{x})]^{-1} \left[h(\bar{x}) + \sum_{i=1}^{m} \sum_{\ell=1}^{r_i} \beta_{i\ell} L_f^\ell h_i(\bar{x}) \right]$ |

| Q_3'' | u | $=$ | $\mathcal{F}\{C(\bar{x}), w\}$ |

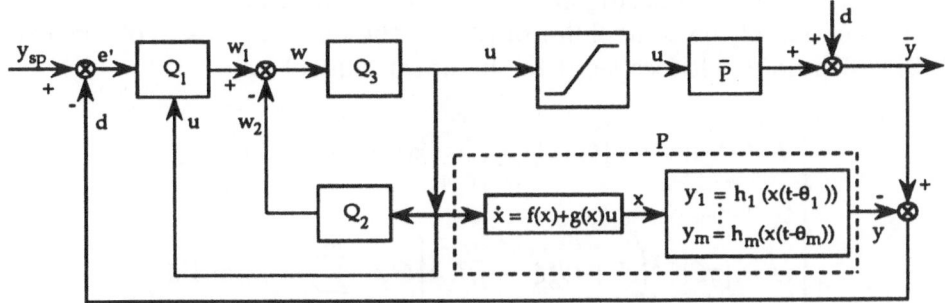

Figure 9. Modified IMC parameterization of the nonlinear controller of (25).

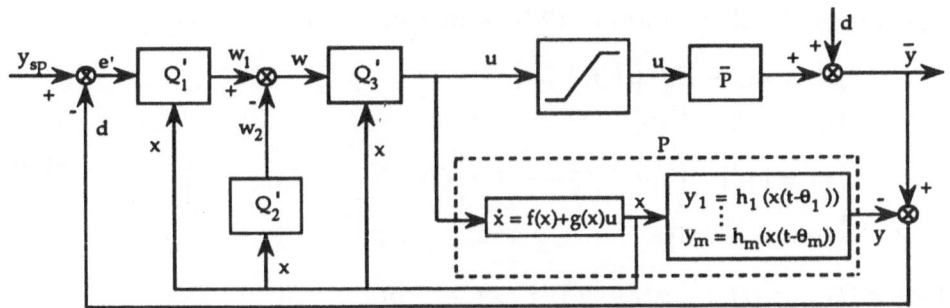

Figure 10. MSFC parameterization of the nonlinear controller of (25).

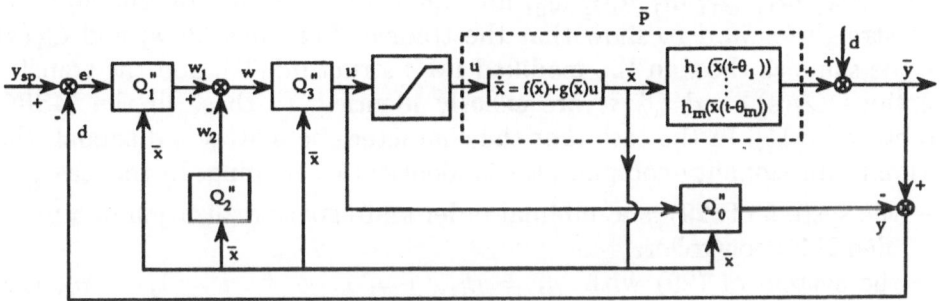

Figure 11. MSFC parameterization of the nonlinear controller of (23).

where A, B and C are $n \times n$, $n \times m$ and $m \times n$ constant matrices respectively. This class of systems is a special case of (13) for $f(x(t)) = Ax(t)$, $g(x(t)) = B$, $h_i(x(t - \theta_i)) = c_i x(t - \theta_i)$, $i = 1, \cdots, m$. It is assumed that the delay-free part of the system of (27) is minimum-phase, has finite relative orders

$r_1, \cdots, r_m$, and has a nonsingular characteristic matrix.

Application of the control law of (23) to the linear processes of the form of (27) with no deadtimes leads to the linear control law:

$$\begin{cases} \dot{\eta} = A_c\xi + B_c\left[C\bar{x} + \sum_{i=1}^{m}\sum_{\ell=1}^{r_i}\beta_{i\ell}c_iA^\ell\bar{x} + \beta Cu\right] \\ \dot{\xi} = A_c^*\xi + B_c^*e' \\ w = [\gamma C]^{-1}\left(e' + C_c^*\xi - \gamma\beta^{-1}\left[C\bar{x} - \sum_{i=1}^{m}\sum_{\ell=1}^{r_i}\beta_{i\ell}c_iA^\ell\bar{x}\right]\right) \\ u = \mathcal{F}\{C, w\} \end{cases} \qquad (28)$$

where $e' = e + C_c\eta$, and the application of the control law of (25) to the linear processes of the form of (27) results in the linear control law:

$$\begin{cases} \dot{x} = Ax + B\mathcal{F}\{C, w\} \\ \dot{\xi} = A_c^*\xi + B_c^*e' \\ w = [\gamma C]^{-1}\left(e' + C_c^*\xi - \gamma\beta^{-1}\left[Cx - \sum_{i=1}^{m}\sum_{\ell=1}^{r_i}\beta_{i\ell}c_iA^\ell x\right]\right) \\ u = \mathcal{F}\{C, w\} \end{cases} \qquad (29)$$

where $e_i'(t) = e_i(t) + c_ix(t - \theta_i)$, $i = 1, \cdots, m$.

Modified IMC and MSFC parameterizations of the linear control laws of (28) and (29) are depicted in Figures 12, 13 and 14, and the corresponding controller components are given in Tables 5, 6 and 7. Note that in this linear case, Q_1, Q_1', Q_1'', Q_3, Q_3', and Q_3'' do not depend on the state x. It is straightforward to show that the transfer functions $Q_1(s)$ and $Q_2(s)$ are the same as those in the modified IMC structure [26], and the transfer functions $Q_1'(s)$, and $Q_2'(s)$ are exactly identical to those in the MSFC structure [6, 15]. In the case that the characteristic matrix is diagonal, the optimal directionality compensator is identical to clipping. In this case,

- the system of (29) is a minimal-order state-space realization of a modified IMC controller.
- the system of (29) with $\gamma_{ij} = \beta_{ij}$, $j = 1, \cdots, r_i$. $i = 1, \cdots, m$, is a state-space realization of a model state feedback controller

Example 4: Consider the same linear example and the unconstrained desired closed-loop response given in [26]:

$$y = \begin{bmatrix} 4 & -5 \\ -3 & 4 \end{bmatrix}\frac{10}{100s + 1}u, \quad -1 \leq u_i \leq +1, \ i = 1, 2$$

$$y_i = \frac{1}{20s + 1}y_{spi}, \quad i = 1, 2 \qquad (30)$$

TABLE 5. Modified IMC parameterization of the linear controller of (29).

P	$\dot{x}$	$=$	$Ax + Bu$
	$y_i(t)$	$=$	$c_i x(t - \theta_i), \quad i = 1, \cdots, m$
Q_1	$\dot{\xi}$	$=$	$A_c^* \xi + B_c^* e'$
	w_1	$=$	$[\gamma C]^{-1} [e' + C_c^* \xi]$
Q_2	$\dot{x}$	$=$	$Ax + Bu$
	w_2	$=$	$[\beta C]^{-1} \left[Cx + \sum_{i=1}^{m} \sum_{\ell=1}^{r_i} \beta_{i\ell} c_i A^\ell x \right]$
Q_3	u	$=$	$\mathcal{F}\{C, w\}$

TABLE 6. MSFC parameterization of the linear controller of (29).

P	$\dot{x}$	$=$	$Ax + Bu$
	$y_i(t)$	$=$	$c_i(x(t - \theta_i)), \quad i = 1, \cdots, m$
Q_1'	$\dot{\xi}$	$=$	$A_c^* \xi + B_c^* e'$
	w_1	$=$	$[\gamma C]^{-1} [e' + C_c^* \xi]$
Q_2'	w_2	$=$	$[\beta C]^{-1} \left[Cx + \sum_{i=1}^{m} \sum_{\ell=1}^{r_i} \beta_{i\ell} c_i A^\ell x \right]$
Q_3'	u	$=$	$\mathcal{F}\{C, w\}$

TABLE 7. MSFC parameterization of the linear controller of (28).

Q_0''	$\dot{\eta}$	$=$	$A_c \eta + B_c \left[C\bar{x} + \sum_{i=1}^{m} \sum_{\ell=1}^{r_i} \beta_{i\ell} c_i A^\ell \bar{x} + \beta C u \right]$
	y	$=$	$C_c \eta$
Q_1''	$\dot{\xi}$	$=$	$A_c^* \xi + B_c^* e'$
	w_1	$=$	$[\gamma C]^{-1} [e' + C_c^* \xi]$
Q_2''	w_2	$=$	$[\beta C]^{-1} \left[C\bar{x} + \sum_{i=1}^{m} \sum_{\ell=1}^{r_i} \beta_{i\ell} c_i A^\ell \bar{x} \right]$
Q_3''	u	$=$	$\mathcal{F}\{C, w\}$

202

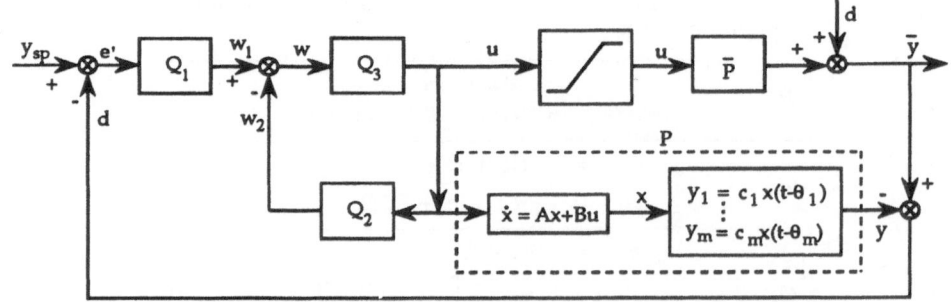

Figure 12. Modified IMC parameterization of the linear controller of (29).

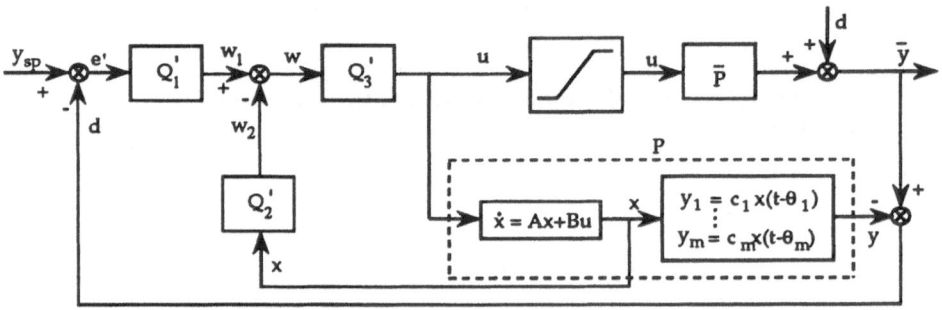

Figure 13. MSFC parameterization of the linear controller of (29).

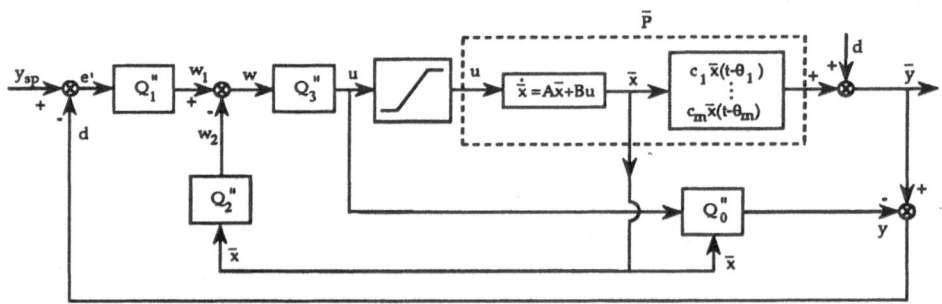

Figure 14. MSFC parameterization of the linear controller of (28).

Here, $r_1 = r_2 = 1$, $\gamma_{11}^1 = \gamma_{21}^2 = 20$, $\gamma_{11}^2 = \gamma_{21}^1 = 0$, and

$$
C = \begin{bmatrix} 0.4 & -0.5 \\ -0.3 & 0.4 \end{bmatrix}
$$

TABLE 8. Modified IMC parameterization of the linear controller of (31).

P	$y \;=\;$	$\begin{bmatrix} 4 & -5 \\ -3 & 4 \end{bmatrix}$	$\dfrac{10}{100s+1}u$
Q_1	$w_1 \;=\;$	$\begin{bmatrix} 4 & 5 \\ 3 & 4 \end{bmatrix}$	$\dfrac{10(s+1)}{20s+1}e'$
Q_2	$w_2 \;=\;$	$\begin{bmatrix} 1 & 0 \\ 0 & 1 \end{bmatrix}$	$\dfrac{99}{100s+1}u$
Q_3	$u \;=\;$	$\mathcal{F}\{\mathcal{C}, w\}$	

For this process, the linear control law of (29) with $\beta_{11}^2 = \beta_{21}^1 = 0$ and $\beta_{11}^1 = \beta_{21}^2 = 1$, takes the form

$$
\begin{cases}
\dot{\xi}_1^{(1)} = -0.05\xi_1^{(1)} + 0.05e_1', & \xi_1^{(1)}(0) = 0 \\
\dot{\xi}_1^{(2)} = -0.05\xi_1^{(2)} + 0.05e_2', & \xi_1^{(2)}(0) = 0 \\
\dot{x}_1 = -0.01x_1 + 0.4u_1 - 0.5u_2, & x_1(0) = 0 \\
\dot{x}_2 = -0.01x_2 - 0.3u_1 + 0.4u_2, & x_2(0) = 0 \\
w = \begin{bmatrix} 2.0 & 2.5 \\ 1.5 & 2.0 \end{bmatrix} \begin{bmatrix} e_1' + 19\xi_1^{(1)} - 19.8x_1 \\ e_2' + 19\xi_1^{(2)} - 19.8x_2 \end{bmatrix} \\
u = \mathcal{F}(\mathcal{C}, w)
\end{cases}
\tag{31}
$$

Note that in the absence of constraints, the linear controller of (31) induces the same closed-loop response of (29), irrespective of the values of β_{ij}^k, $i, j, k = 1, 2$. The modified IMC parameterization of the controller of (31) leads to the controller components given in Table 8. The matrix transfer functions $P(s)$, $Q_1(s)$, and $Q_2(s)$ given in Table 8 are the same as those in [26].

3.7. APPLICATION TO A NONLINEAR CHEMICAL REACTOR

We consider the same chemical reactor described in [20]:

$$
\begin{aligned}
\dot{C}_A &= f_1(C_A, T) + \frac{1}{\tau}C_{A_i} \\
\dot{T} &= f_2(C_A, T) + \frac{1}{\rho c V}Q
\end{aligned}
\tag{32}
$$

where C_A and C_{A_i} are respectively the outlet and inlet concentrations of the reactant, T is the outlet stream temperature, and Q is the rate of heat input to the reactor. The controlled outputs and manipulated inputs are: $y_1 = C_A$, $y_2 = T$, $u_1 = C_{A_i}$, and $u_2 = Q$ with the bounds $5 \leq u_1 \leq 15 \ kmol.m^{-3}$ and $-10 \leq u_2 \leq 10 \ kJ.s^{-1}$. Here $r_1 = r_2 = 1$ and $\mathcal{C} = diag\{1/\tau, \ 1/(\rho cV)\}$. Application of the control law of Theorem 3 with $\gamma_{11}^1 = \gamma_{21}^2 = 100$ and $\gamma_{11}^2 = \gamma_{21}^1 = \beta_{11}^2 = \beta_{21}^1 = 0$, to this chemical reactor leads to the following generalized, mixed error- and state-feedback controller:

$$
\begin{cases}
\dot{\eta}_1^{(1)} = \dfrac{1}{\beta_{11}^1}\left[-\eta_1^{(1)} + \bar{C}_A + \beta_{11}^1 f_1(\bar{C}_A, \bar{T}) + \dfrac{u_1}{\tau}\right] \\[2ex]
\dot{\eta}_1^{(2)} = \dfrac{1}{\beta_{21}^2}\left[-\eta_1^{(2)} + \bar{T} + \beta_{21}^2 f_2(\bar{C}_A, \bar{T}) + \dfrac{u_2}{\rho cV}\right] \\[2ex]
\dot{\xi}_1^{(1)} = 0.01(\eta_1^{(1)} + e_1 - \xi_1^{(1)}) \\[1ex]
\dot{\xi}_1^{(2)} = 0.01(\eta_1^{(2)} + e_2 - \xi_1^{(2)}) \\[1ex]
u = \mathcal{F}(\mathcal{C}, w)
\end{cases}
\tag{33}
$$

where

$$
w_1 = \frac{\eta_1^{(1)} + e_1 - \xi_1^{(1)} + \dfrac{100}{\beta_{11}^1}\left(\xi_1^{(1)} - \bar{C}_A\right) - 100 f_1(\bar{C}_A, \bar{T})}{100\tau}
$$

$$
w_2 = \frac{\eta_1^{(2)} + e_2 - \xi_1^{(2)} + \dfrac{100}{\beta_{21}^2}\left(\xi_1^{(2)} - \bar{T}\right) - 100 f_2(\bar{C}_A, \bar{T})}{100\rho cV}
$$

Figure 15 depicts the startup profiles of the controlled outputs and manipulated inputs under the nonlinear controller of (33) for two values of the pair $(\beta_{11}^1, \beta_{21}^2)$. In the absence of the input constraints, the closed-loop response for both values of the pair (solid line) are exactly the requested, completely input-output decoupled, first-order response. With $\beta_{11}^1 = \beta_{21}^2 = 2\ s$, the mismatch between the unconstrained and constrained closed-loop responses is less and the heat input rate and the inlet concentration stay in saturation for a longer period of time.

4. Conclusions

An optimal directionality compensator was presented. Given a controller output, the compensator calculates an optimal feasible plant input that yields a constrained process response as close as possible to the response of the same process to the controller output. This optimization is over sufficiently short prediction horizons, and its solution is obtained simply by solving a quadratic program. This study revealed that selection among different process directionality compensators should be made on the basis

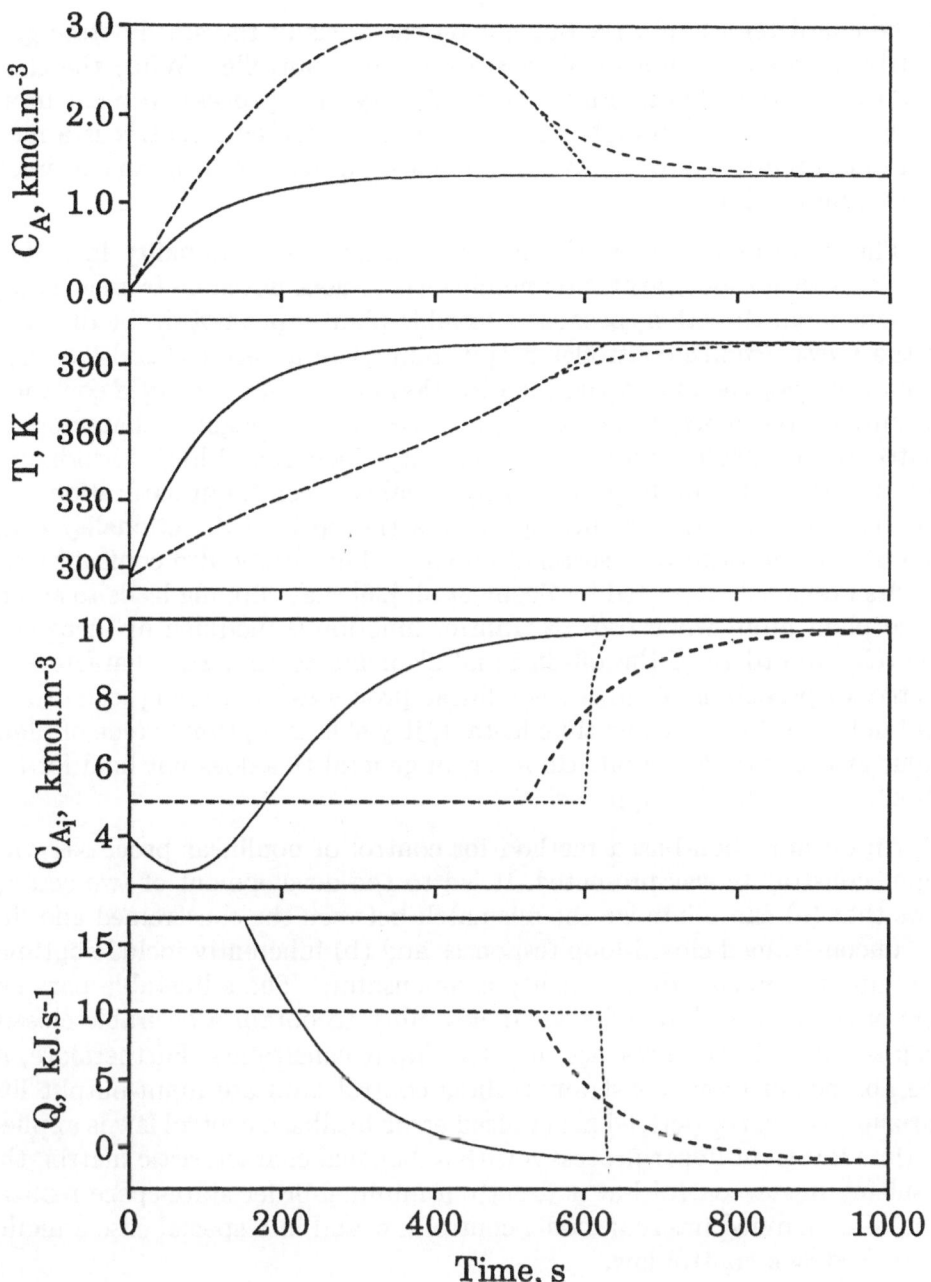

Figure 15. Profiles of the controlled outputs and plant inputs of the reactor example of (32): solid = no bounds on the inputs, $\beta_{11}^1 = \beta_{21}^2 = 100\ s$ and 2 s; dashed = when $\beta_{11}^1 = \beta_{21}^2 = 100\ s$, dotted = when $\beta_{11}^1 = \beta_{21}^2 = 2\ s$, with $5 \leq u_1 \leq 15\ kmol.m^{-3}$ and $-10 \leq u_2 \leq 10\ kJ.s^{-1}$.

of the characteristic matrix but not on the basis of the steady state gain matrix, unless one implements a steady state controller. While the characteristic matrix characterizes the sensitivity of a process to plant input changes over a very short horizon, the steady state gain matrix is a measure of the sensitivity of a process to plant input changes over an extremely large (infinite) horizon.

The performance degradation due to process directionality in MIMO processes with input saturation nonlinearities does not stem from a change in plant input direction, because a feasible plant input not in the direction of the *unconstrained* controller output may yield a better closed-loop performance than the one in the the direction of the *unconstrained* controller output. Furthermore, the class of processes with a diagonal characteristic matrix do not exhibit process directionality; their closed-loop performance does not degrade due to process directionality when input saturation constraints are present. For these processes the optimal directionality compensator is identical to a series of clippers. This study also confirmed the results previously reported by Zheng et al. [26] that clipping leads to an optimal plant input when process transfer function is modified by increasing the relative orders of the off-diagonal elements so that the characteristic matrix of process is diagonal. For linear processes, if clipping leads to an optimal plant input at one time instant, it yields an optimal value of plant input at every time instant. However, in general this does not hold for the direction preservation approach.

An optimization-based method for control of nonlinear processes with input constraints was presented. It led to the development of two control laws that (a) can minimize the mismatch between the constrained and linear unconstrained closed-loop responses and (b) inherently include optimal windup and process directionality compensators. The adjustable parameters of the control laws offer great flexibility to obtain a desirable closed-loop response in the presence of active input constraints. Furthermore, in the absence of input constraints, these control laws are input-output linearizing. When the derived generalized error-feedback control law is applied to time-invariant linear processes with a diagonal characteristic matrix, the resulting linear control law is exactly a minimal-order state-space realization of a modified internal model control law and in a special case a model state feedback control law.

Acknowledgment

Financial support from the National Science Foundation through the grant CTS-970-3278 is gratefully acknowledged.

References

1. Åström, K.J. and Rundqwist, L. (1989) Integrator Windup and How to Avoid It, *Proc. of ACC*, Pittsburgh, PA, 1693.
2. Barnard, R.D. (1976) Continuous-Time Implementation of Optimal-Aim Controls, *IEEE Trans. Auto. Contr.*, **21**, 432.
3. Bernstein, D.S. and Michel, A.N. (1995) A Chronological Bibliography on Saturating Actuators, *International J. of Robust and Nonlinear Contr.*, **5** 375.
4. Calvet, J.P. and Arkun, Y. (1988) Feedforward and Feedback Linearization of Nonlinear Systems and its Implementation Using Internal Model Control, *Ind. Eng. Chem. Res.*, **27**, 1822.
5. Campo, P.J. and Morari, M. (1990) Robust Control of Processes Subject to Saturation Nonlinearities, *Comput. & Chem. Eng.*, **14**, 343.
6. Coulibaly, E., Maiti, S. and Brosilow, C. (1995) Internal Model Predictive Control (IMPC), *Automatica*, **31**, 1471.
7. Daoutidis, P. and Kravaris, C. (1992) Structural Evaluation of Control Configurations for Multivariable Nonlinear Processes, *Chem. Eng. Sci.*, **47**, 1091.
8. Doyle, J.C., Smith, R.S. and Enns, D.F. (1987) Control of Plants with Input Saturation Nonlinearities, *Proc. of ACC*, 1034.
9. Hanus, R. and Kinnaert, M. (1989) Control of Constrained Multivariable System Using Conditioning Technique, *Proc. of ACC*, 1712.
10. Kapoor, N. and Daoutidis, P. (1996) An Observer-based Anti-Windup Scheme for Nonlinear Systems with Input Constraints, *International J. of Contr.*, submitted.
11. Kapoor, N., and Daoutidis, P. (1997) Stabilization of Systems with Input Constraints," *International J. of Contr.*, **66(5)**, 653.
12. Kendi, T.A. and Doyle, F.J. (1997) An Anti-Windup Scheme for Multivariable Nonlinear Systems, *J. of Process Contr.*, **7(5)**, 329.
13. Kothare, M.V., Campo, P.J., Morari, M. and Nett, C.N. (1994) A Unified Framework for the Study of Antiwindup Designs, *Automatica*, **30**, 1869.
14. Kurtz, J.M., and Henson, M.A. (1997) Input-Output Linearizing Control of Constrained Nonlinear Processes, *J. of Process Contr.*, **7(1)**, 3.
15. Mhatre, S. and Brosilow, C. (1996) Multivariable Model State Feedback, *Proc. of IFAC World Congress*, San Francisco, M, 139.
16. Oliveira, S.L., Nevistic, V. and Morari, M. (1995) Control of Nonlinear Systems Subject to Input Constraints, *Preprints of Nonlinear Control Systems Design Symposium*, 15.
17. Park, J.-K. and Choi, C.-H. (1995) Dynamic Compensation Method For Multivariable Control Systems with Saturating Actuators, *IEEE Trans. Auto. Contr.*, **40(9)**, 1635.
18. Soroush, M. and Valluri, S. (1997) Calculation of Optimal Feasible Controller Output in Multivariable Processes with Input Constraints, *Proc. of ACC*, 3475.
19. Soroush, M. and Kravaris, C. (1992) A Continuous-Time Formulation of Nonlinear Model Predictive Control, *Proc. of ACC*, Chicago, 1561.
20. Soroush, M. and Kravaris, C. (1992) Nonlinear Control of a Batch Polymerization Reactor: an Experimental Study, *AIChE J.*, **38**, 1429.
21. Teel, A.R. (1996) A Nonlinear Small Gain Theorem for the Analysis of Control Systems with Saturation, *IEEE Trans. Automat. Contr.*, **41(9)**, 1256.
22. Valluri, S. (1997) *Nonlinear Control of Processes with Actuator Saturations*, Ph.D. Thesis, Drexel University, Philadelphia (1997).
23. Valluri, S. and Soroush, M. (1997) Nonlinear Control of Processes with Actuator Saturation Nonlinearities, submitted to *Automatica*.
24. Valluri, S., Soroush, M. and Nikravesh, M. (1997) Shortest-Prediction-Horizon Nonlinear Model Predictive Control, *Chem. Eng. Sci.*, in press.
25. Walgama, K.S. and Sternby, J. (1993) Conditioning Technique for Multi-Input Multi-Output Processes with Input Saturation, *Proc. IEE Part D*, **140**, 231.

26. Zheng, A., Kothare, M.V. and Morari, M. (1994) Anti-Windup Design for Internal Model Control, *International J. Contr.*, **60**, 1015.

INTERNALLY STABLE LINEAR AND NONLINEAR ALGORITHMIC INTERNAL MODEL CONTROL OF UNSTABLE SYSTEMS

R. BERBER
Department of Chemical Engineering
University of Ankara, Tandogan, 06100 Ankara, Turkey

C. BROSILOW
Department of Chemical Engineering
Case Western Reserve University, Cleveland OH44106 USA

Abstract
An internally stable Algorithmic Internal Model Control (AIMC) strategy which uses linear or nonlinear model state feedback is proposed for unstable systems. The closed loop responses are those that would be obtained from a two degree of freedom IMC control system, if it were stable. Results of several simulations demonstrate the validity of the approach.

1. Introduction

The theory and practice of the control of inherently stable processes have spurred very significant advances within the framework of internal model type control structures [7, 8, 11, 15]. Such control structures incorporate an online or "internal" model of the controlled process to directly infer how disturbances have influenced the controlled variable and take appropriate control action to counter the disturbances. A major practical advantage of such control systems is that they readily accommodate constraints on both the control efforts and controlled variables [7]. Control effort constraints are even more important in the control of inherently unstable processes.

Unfortunately, however, the common internal model type control structure [11] cannot be used with inherently unstable processes due to internal instability in the control structure. This internal instability arises from two sources: (1) in an unstable process, the effect on the controlled variable of disturbances which enter through the process grows without bound, and (2) differences between the model and process states grow without bound. It is relatively easy to rearrange the structure of an internal model controller to estimate the disturbances, which are bounded, rather than the effects of the disturbances which are unbounded and thus eliminate the first problem. However, a more profound restructuring is needed to insure that errors in state estimates do not grow without bound. Algorithmic Internal Model Control (AIMC) described in the next section provides a simple, but effective, solution to the problem.

R. Berber and C. Kravaris (eds.), Nonlinear Model Based Process Control, 209-233.
© 1998 *Kluwer Academic Publishers.*

There is probably substantial similarity between Algorithmic IMC (AIMC) as described in this work and the generalized predictive control used by Clarke *et al.* [6] as a basis for an adaptive controller. Clarke mentioned that his algorithm worked for unstable systems, but did not further pursue the subject. AIMC is also probably similar to some implementations of IDCOM [10]. Richalet [10] marked at CPC-III that a version of IDCOM was used successfully with an integrating process.

Below we define AIMC in terms of a sequence of tasks. The tasks define structure just as block diagrams define the structure of Internal Model Control (IMC) and Inferential Control (IC) [14]. Methods for accomplishing the tasks in AIMC and for designing the controller and estimator blocks in IMC and IC depend on performance specifications, the process and disturbance models and the accuracy of the models. The AIMC structure is defined in terms of tasks rather than block diagrams because the task description seems simpler and more natural.

To explore the properties of stabilizing AIMC we first study two linear processes (first and second order). For the first order plus dead time process, a block diagram description of AIMC is obtained and its equivalence to a two degree of freedom IMC system is shown. The suggested AIMC strategy is then extended to nonlinear systems, and it is shown that it applies equally well to rather general nonlinear processes. Simulation examples are provided for both linear and nonlinear cases.

2. Algorithmic Internal Model Control

AIMC can be best described as a sequence of tasks, which is more comprehensive than a block diagram notation. The approach uses a process model and proceeds through the following tasks:

(1) Describe a desired trajectory between the set point and the controlled variable. This trajectory is usually taken as the output of the IMC filter cascaded with the portion of the process model that the IMC controller does not invert, and that is driven by the set point.

(2) Estimate and predict the disturbances entering the process. If there is a process dead time then the disturbance estimates are extrapolated forward in time by the amount of the dead time. Different estimation strategies are proposed for linear and nonlinear systems as explained in the following section.

(3) Predict the process output one dead time into the future based on current measurements, current and pass controls, and the extrapolated disturbances. This step also evolves the projections of the process states as calculated from the model.

(4) Compute the controls which force the model to track the desired filter states by equating the first r derivatives of the process output to the first r derivatives of the desired trajectory.

(5) Go to the next sampling instant (or increase integration time in simulation) and process from the task 2 above.

The above strategy is shown in Figure 1 as a computing flow diagram for clarity.

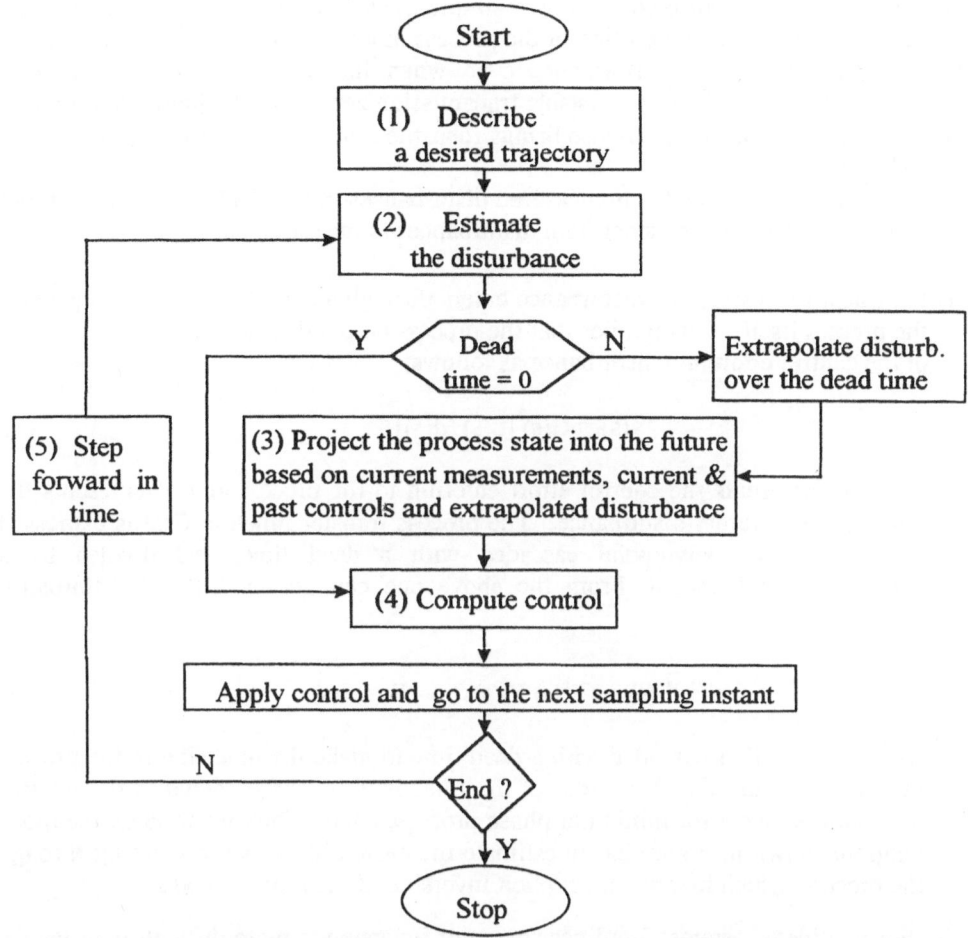

Figure 1. General flow diagram for stabilizing AIMC

Process models to be used in the proposed approach should have unique trajectories for unique inputs. Further, the process outputs and states should depend continuously on the process inputs at each instant of time (i.e. infinitesimal changes in input functions should not cause jumps in the process outputs or states.

Definition of a filter in task 1 above allows the response speed of the dynamical system to be adjustable by manipulating one parameter, namely the filter time constant. It provides a means of tuning the control system to achieve a desirable response. If the model is a perfect representation of the process, then the response of the process will be the same as the response of the dynamic system described by this filter equation. Most commonly, the filter will be chosen as a simple linear system with the same order as the relative order of the lag portion of the process model. A linear dynamic system can be used without loss of performance even when the process model is nonlinear provided that the model has no unstable transmission zeros [3]. Including the filter as the reference dynamical system also brings robustness in the IMC structure in the case of modeling errors.

The ability to estimate the unmeasured disturbances in task 2 above depends on the invertibility of the process model from disturbances to measurements.

(i) For linear processes, the disturbance enters through a disturbance lag or in general the process itself. This implies that the process output depends on the combination of the control effort and disturbance as follows;

$$y(s) = G(s) [u(s)+d(s)] \qquad (1)$$

where u represents the control effort entering to the process and d represents the unmeasured external disturbance. The process transfer function $G(s)$ is expressed as a numerator polynomial cascaded with a dead time, and divided by a denominator polynomial. From the above one can get the filtered disturbance estimate as:

$$e^{-s\theta} \tilde{d}(s) = G^{-1}(s) F_d(s)y(s) - F_d(s)u(s) \qquad (2)$$

where F_d is a filter cascaded with a dead time to make the inversion of the process model $G(s)$ realizable. The order of the filter is to be chosen equal to the relative order of the model for minimum phase processes. Therefore, we propose the block diagram shown in Figure 2a to estimate the current disturbance entering through the process, which involves an explicit inversion of the process model.

(ii) For nonlinear systems, nonlinear operator inversion is more difficult than for the linear case. The only available and mathematically rigorous analytical construction technique is Hirschorn inverse [13], which is very sensitive to noise and/or numerical errors and fails to produce a satisfactory approximation in practical cases [9]. Computationally, however, one can use an iterative algorithm to find the disturbances which cause the model outputs to match observations rather than attempt to actually construct the model inverse. We used a Newton type convergence algorithm for constructing the disturbance that makes the model output match the process output at each sampling instant. Figure 3 shows this strategy as a computing flow diagram.

In case if the disturbance enters through a disturbance transfer function $D(s)$ rather than through the process, as in Figure 2b, then the output will be:

$$y(s) = G(s) u(s) + D(s) d(s)$$

Solving this for $d(s)$ gives the following for the filtered disturbance estimate:

$$d(s)= D^{-1}(s) \; F(s) \; y(s) - D^{-1}(s) \; F(s) \; G(s)u(s) = D^{-1}(s) \; F(s) \; [y(s)- \breve{G}(s)u(s)]$$

This scheme represented in the block diagram in Figure 2b, becomes the same as our original diagram, i.e. Figure 2a, if $D=\breve{G}=G$.

Another possibility for higher order linear systems and nonlinear systems is to estimate the disturbance with a Newton strategy similar to the one explained in the foregoing paragraph. Disturbance estimation under these circumstances, as indicated in Figure 2c, would be straightforward as follows:

Assume the disturbance model (D) is the same as the real disturbance transfer function. The iterative procedure for the disturbance estimate d is:

a) Integrate model and D with current d. This will give model output, y_m.

b) Increase d by a small amount, Δd, and re-integrate model and D (starting from the same initial conditions as above) to get the perturbed model output

c) Calculate "Derivative"

d) Calculate disturbance estimate $d(i+1)= d(i)-\alpha \, [(y_p-y_m)/\text{Derivative}]$

e) Check convergence and re-iterate if necessary.

When the disturbance transfer function is unknown, then disturbance effect d_e (instead of the disturbance itself, d) needs to be estimated. In this case, the estimate of the effect of disturbance on the output will be just an additive signal, i.e. constant, at each sampling time

$$d_e = y_p - y_m$$

which corresponds exactly to the estimation of unmodelled effects in DMC algorithm [12].

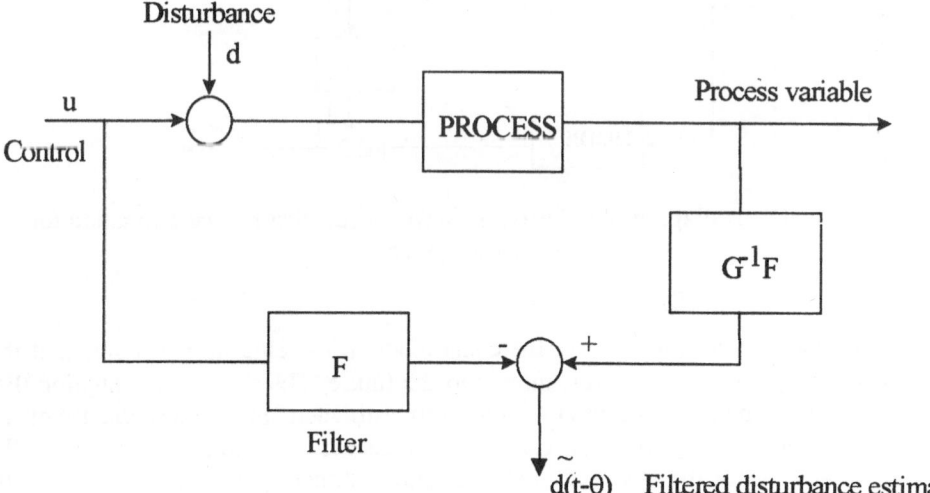

Figure 2a. Disturbance estimation for linear processes

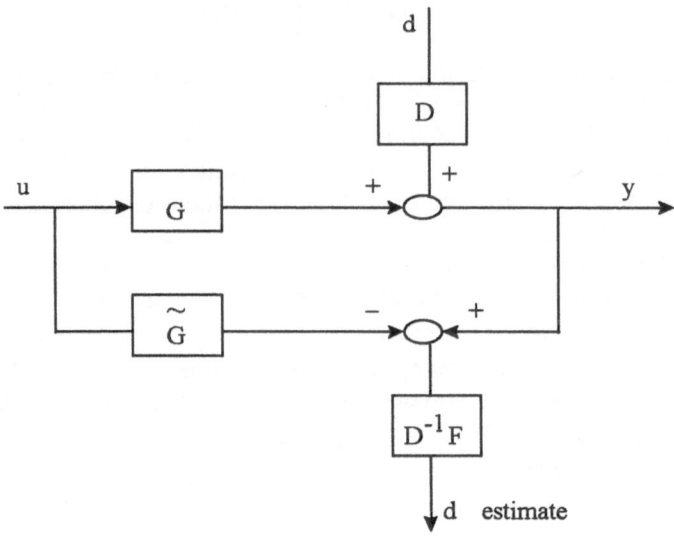

Figure 2b. Disturbance estimation when the disturbance enters through a D(s)

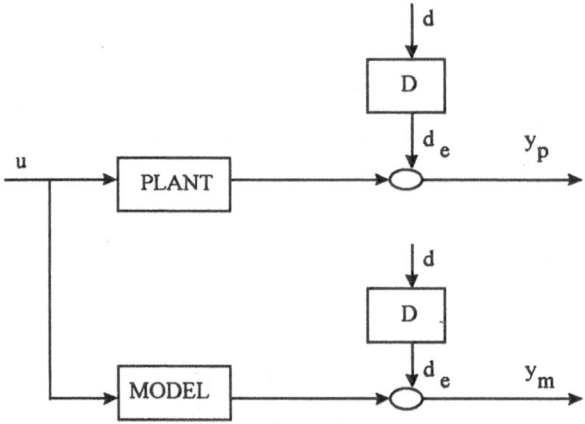

Figure 2c. Block diagram for clarifying Newton-type disturbance estimation for nonlinear systems

If the processes has a dead time, θ, the consistency in the algorithm requires that the disturbance be predicted for θ time units into the future. The simplest assumption that the disturbance remains constant at the currently estimated value allows the use of the simple Newton-type algorithm to compute the disturbance values which force the model output to track the observed process output. Estimating a constant value for the disturbance over a sampling interval will not assure that the time derivatives of the

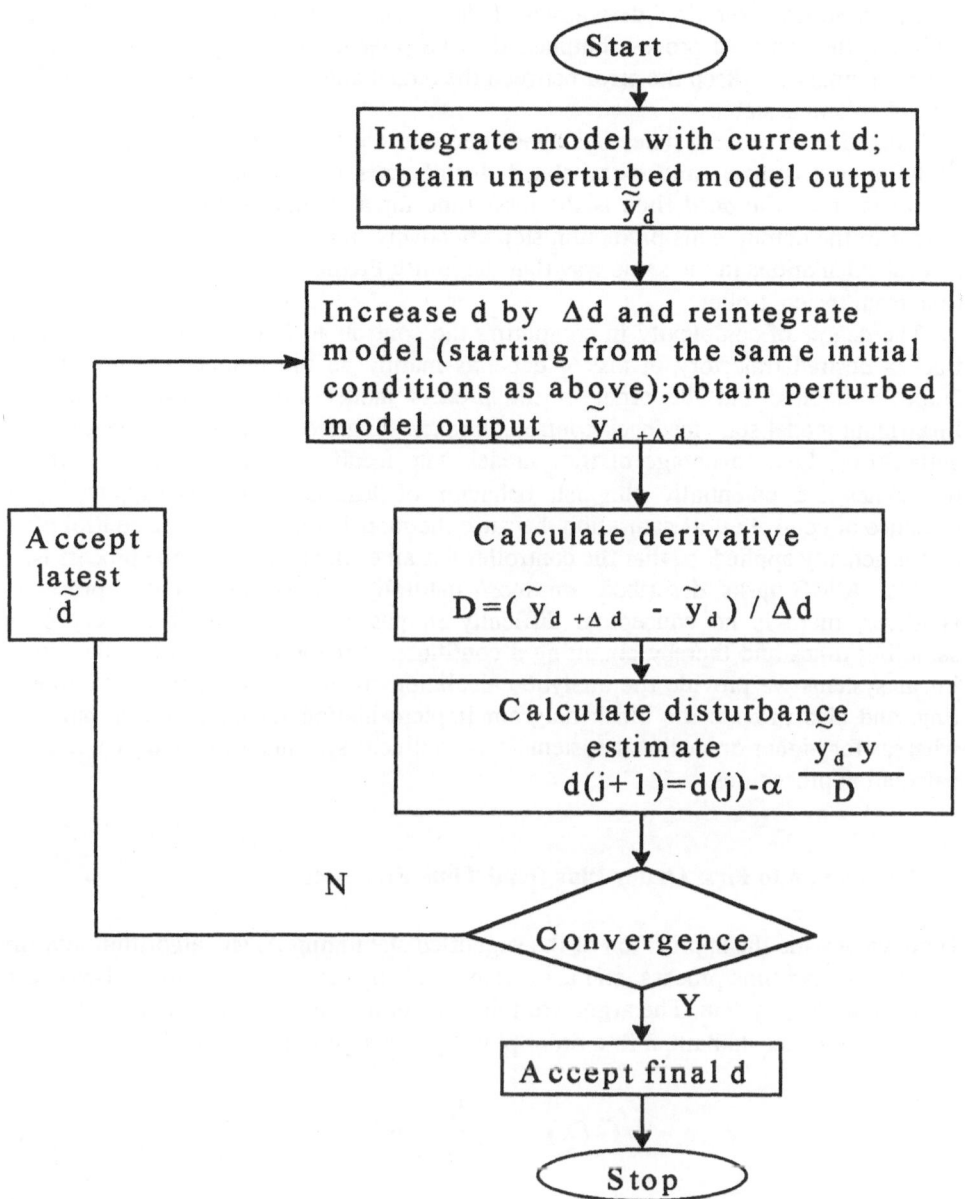

Figure 3. Disturbance estimation for nonlinear processes
(where α is a convergence factor, j is iteration counter)

model output track the time derivatives of the process. However, if the unstable states influence the observed process outputs, then keeping the model outputs tracking the process outputs will keep the error between the model and process states bounded if the control system is stable.

Prediction of process states one dead time into the future (task 3) simply involves applying past controls and extrapolated disturbances to the model. If the model is nonlinear, then the dead time is the least time for a change in the control effort to appear in the output. This prediction step effectively 'removes' the dead time from the control calculations in the same way that the Smith Predictor [16] 'removes' the dead time from the controller.

The degree of complexity in computing the controls to force the process model to track a desired trajectory in task 4 depends mainly on the number of controls and output variables, and the order of the process model. For this task, we use the linearizing model state feedback control approach suggested in a previous paper by the authors [1]. One advantage of the model state feedback implementation is that it overcomes the potentially sluggish behavior of lead-lag IMC controllers in the presence of control effort saturation, because the model states reflect the control effort that is actually applied, so that the controller has an estimate of where the process is.

The AIMC tasks described are most naturally applied to sampled processes. However, there is no conceptual difficulty in passing to the limit of very small sampling times and thereby obtaining a continuous version of AIMC. For the simple linear systems we provide the analytic calculations of the AIMC tasks in continuous time and also provide the block diagram implementation for comparative purposes, whereas for higher order linear systems and nonlinear systems we give only numerical calculation procedures.

3. Application to First Order Plus Dead Time Processes

Here we provide the application of the suggested stabilizing AIMC algorithm to a first order plus dead time process, and show that this is indeed equivalent to the two degree of freedom IMC system. The argument follows that in Brosilow and Cheng [2].

In the Laplace domain, a first order plus dead time process is given by

$$G(s) = \frac{K}{\tau s + 1} e^{-s\theta} \tag{3}$$

where y is the controlled process output, K is the process gain, τ is the process time constant (the process is unstable if $\tau < 0$) and θ is the process dead time.

For a clearer understanding of the control strategy, let us now consider the analytic calculation of the control effort for such simple processes. Since the disturbance enters through the process, the equation (1) holds;

$$y = G(s)\big[u(s) + d(s)\big]$$

For the first task of AIMC, we choose a first order filter as the dynamic system to be followed. The disturbance estimate, therefore, can be obtained from these two equations as follows:

$$e^{-s\theta}\tilde{d}(s) = G^{-1}(s)F(s)y(s) - F(s)u(s)$$

where F is a first order filter cascaded with a dead time whose existence serves the purpose of making the inverse of the process realizable;

$$F(s) = \frac{K}{\varepsilon s + 1}e^{-s\theta} \tag{4}$$

The filter used for disturbance estimation is chosen, generally, to be the same as the dynamic system to be followed. ε is the filter time constant, or tuning parameter for the speed of response. A prediction of the disturbance for θ time units into the future is also needed. The simplest possible prediction is to assume that the disturbance remains constant at the currently estimated value. That is

$$d(t-\theta+\sigma) = d(t-\theta) \quad \text{for } 0\leq \sigma \leq \theta \tag{5}$$

Prediction of the process state at t+θ involves the effect of past controls. The output and the state of a first order system are the same, so estimation of the current state is simply the output or some suitable filtered version thereof. Prediction of the state one dead time ahead requires projection of the current state, past controls, and the projected disturbances. For the continuous representation,

$$y(t+\theta) = e^{-\theta/\tau}y(t) + K\int_t^{t+\theta} e^{-(t+\theta-\sigma)/\tau}u(\sigma-\theta)d\sigma + K(1-e^{-\theta/\tau})d(t-\theta) \tag{6}$$

The integral on the right is the convolution integral evaluated over a period θ, starting from time t (note that when σ=t the integrand is e-$^{\tau/\theta}$u(t-θ), while when σ=t+θ the integrand is u(t)). Because the integral is evaluated over a finite period, it remains bounded even when τ is negative.

The control u(t) is chosen to make the process model follow a desired linear system as it moves toward the set point. The model is given by (1) and (2) and can be expressed in the time domain as

$$\tau\frac{d}{dt}y(t+\theta) + y(t+\theta) = K(u(t)+d(t)) \tag{7}$$

We choose u(t) in this equation to make the response like that of the following linear system:

$$\varepsilon \frac{d}{dt} y(t+\theta) + y(t+\theta) = y_{sp} \tag{8}$$

thus we require that

$$K(u(t)+dt) - y(t+\theta) = -\frac{\tau}{\varepsilon} y(t+\theta) + \frac{\tau}{\varepsilon} y_{sp} \tag{9}$$

Evaluating d(t) from eqn. (5) as d(t-θ) and solving for u(t) gives

$$u(t) = -d(t+\theta) + \frac{\tau}{\varepsilon K} y_{sp} + \frac{1-\tau/\varepsilon}{K} y(t+\theta) \tag{10}$$

Figure 4 gives the block diagram for the continuous version of the AIMC algorithm for a first order plus dead time process.. This figure comes from equation (3) and the transform of equations (6) and (10). The transform of the integral in (6), as derived in Appendix A, is:

$$\mathcal{L}\left\{ K \int_t^{t+\theta} e^{-(t+\theta-\sigma)/\tau} u(\sigma - \theta) d\sigma \right\} = \frac{K(1 - e^{-\theta/\tau} e^{-s\tau})}{\tau s + 1} u(s) \tag{11}$$

The right hand side of this equation implies that integral in Eq. (6) can be evaluated as

$$K \int_t^{t+\theta} e^{-(t+\theta-\sigma)/\tau} u(\sigma - \theta) d\sigma = x(t) - e^{-\theta/\tau} x(t-\theta) \tag{12}$$

where x(t) is given by the solution of

$$\tau \frac{d}{dt} x(t) + x(t) = Ku(t) \tag{13}$$

The last equation is a valid method of computation only for stable x(t). If x(t) is unstable, the right hand side of the equation (12) is the difference between two numbers growing without bound and infinite precision is required to get a finite result. This difficulty is the reason why the two degree of freedom IMC control system given in Figure 5 is computationally equivalent to the AIMC structure only for stable systems. For step disturbances and step set point changes the optimal designs of $G_I(s)$ and $G_d(s)$ in Figure 5 are

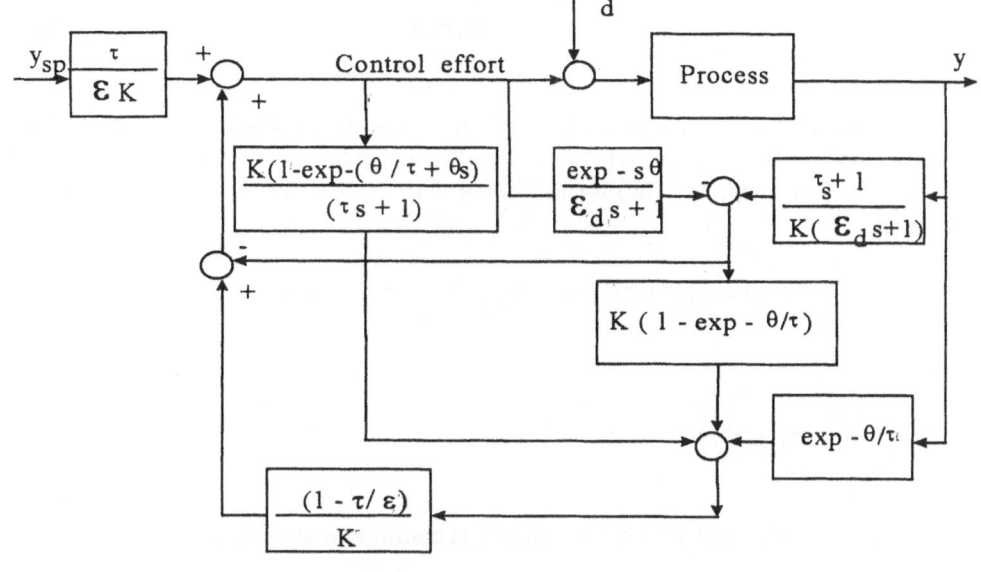

Figure 4. AIMC block diagram for FOPDT process

$$G_I(s) = \frac{\tau s + 1}{(\varepsilon s + 1) K} \tag{14}$$

$$G_d(s) = a + \frac{(1-a)(\tau s + 1)}{(\varepsilon_d s + 1)} \tag{15}$$

where $\qquad a \equiv (1 - \varepsilon / \tau) e^{-\theta/\tau}$

The above choices for $G_I(s)$ and $G_d(s)$ give the best possible response to step set point changes and disturbances as ε and G_d approach zero and the model is a perfect representation of the process.

For perfect models the response of the system in Figure 5 to disturbances and set point changes is

$$y(s) = \frac{y_s(s)}{\varepsilon s + 1} + (1 - \frac{a e^{-s\theta}}{\varepsilon s + 1}) \frac{K e^{-s\theta} d(s)}{\tau s + 1} - \frac{(1-a) e^{-s\theta} K}{(\varepsilon s + 1)(\varepsilon_d s + 1)} d(s) \tag{16}$$

The time domain response for a step input in set point is an exponential rise which approaches a step as the time constant, ε, approaches zero. The time domain response to a step disturbance for $\varepsilon = \varepsilon_d = 0$ is

$$y(t) = 0 \qquad \text{for} \qquad 0 \leq t \leq \theta$$
$$\quad\ = (1-e^{-t/\tau})K \qquad\qquad \theta \leq t \leq 2\theta \qquad\qquad (16a)$$
$$\quad\ = 0 \qquad\qquad\qquad\qquad t > 2\theta$$

No controller with a priori knowledge of when the step disturbance will occur can give a better response than eqn. (16a). When ε and ε_d are greater than zero, the response is the same as that given by (16a) for $t \leq 2\theta$, and is the sum of exponentials given below for $t \geq 2\theta$.

$$y(t) = \alpha e^{-(t-2\theta)/\varepsilon_d} + \beta e^{-(t-2\theta)/\varepsilon} \;\; ; \;\; t > 2\theta \qquad\qquad (16b)$$

where
$$\alpha \equiv (1-a)c$$
$$\beta \equiv (1-a)(1-c)+a-e^{-(\theta/\tau)}$$
$$c \equiv (1-\varepsilon/\varepsilon_d)$$

Notice that in (16b), $y(2\theta)=1- e^{-(\theta/\tau)}$ so that y is continuous at $t=2\theta$.

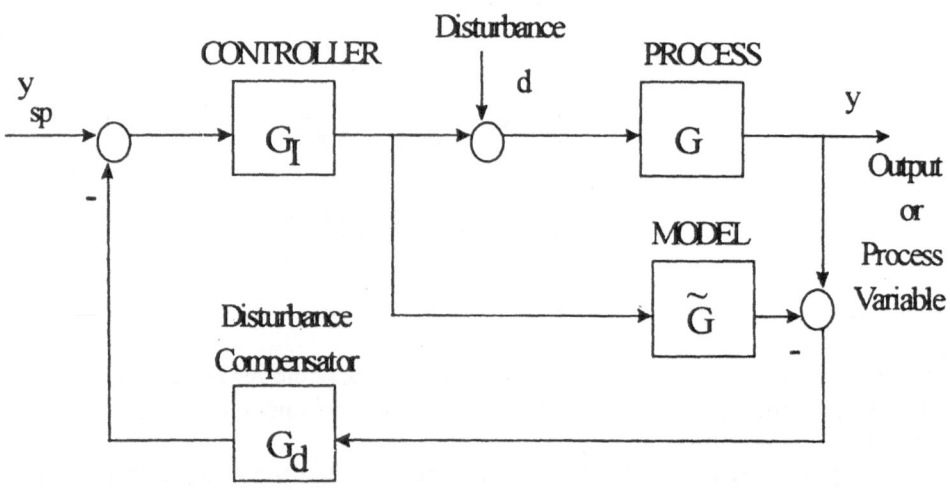

Figure 5. Two degree of freedom IMC control system

It is shown in Appendix B that (3), (6) and (10), and equivalently Figure 4, reduce exactly to the structure given in Figure 5 with $G_I(s)$ and $G_d(s)$ as expressed by equations (14) and (15). Thus, the two figures are analytically equivalent for both positive and negative τ (i.e. stable and unstable processes) but they are computationally equivalent only for positive τ.

Equivalence can also be established between AIMC algorithm and the two degree of freedom IMC structure of Figure 5 for very general linear systems. The key issue in establishing the desired equivalence will be how the model state is estimated in the AIMC algorithm. Since in the IMC structure the model state evolves in an open loop fashion (i.e. without regard to output observations), it is likely that the equivalent AIMC algorithm will update only the model output based on observations, the balance of the model state will evolve in an open loop fashion from one sampling instant to the next. There may, however, be a significant advantage in terms of robustness and performance to use state estimation in the AIMC algorithm when it is applied to processes higher than first order.

The application of the suggested strategy to an unstable process where K=1, τ=-1, θ=.8 are presented in Figures 6-8 for disturbance rejection, set point tracking and two degree of freedom behavior respectively. Simulations were done in MATLAB-SIMULINK® environment via numerical computation. Computing the future desired value of the process state and the control which force the model to track the desired trajectory is accomplished according to the model state feedback law [1, 7]. Figure 6 depicts the disturbance rejection capability of the controller when a disturbance of d=1 is introduced through the process. In these simulations, controller filter and disturbance estimation filter are chosen to be ε =1.5, ε_d =6.

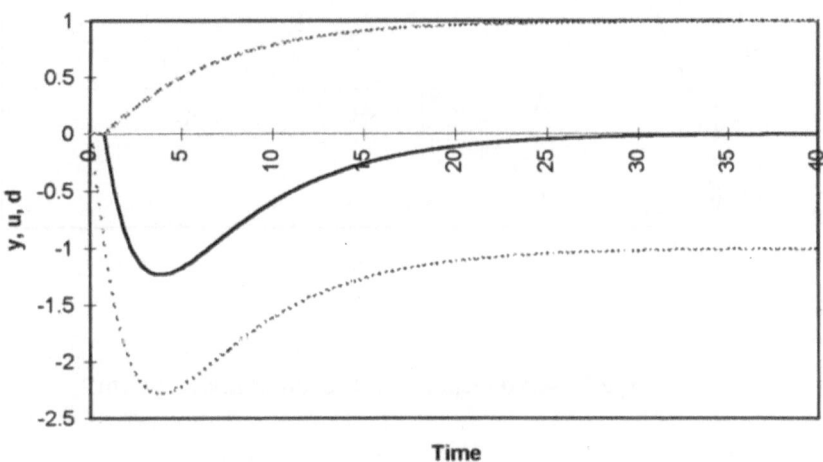

Figure 6. Regulatory response of first order linear system
(— y : Process variable, - - - u : Control, _ _ _ d : Disturbance estimate)

Figure 7 shows the controller under a step disturbance of 1 introduced at time t=0. Figure 8 was obtained under a step change in set point as well as a disturbance of d=0.5 entering through the process where the selection of tuning parameters was

222

$\epsilon=\epsilon_d=1.5$. As reflected in these figures, good set point tracking and disturbance rejection is observed with an unstable process.

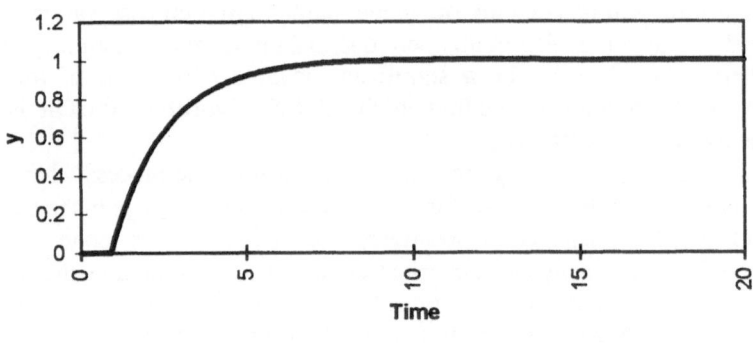

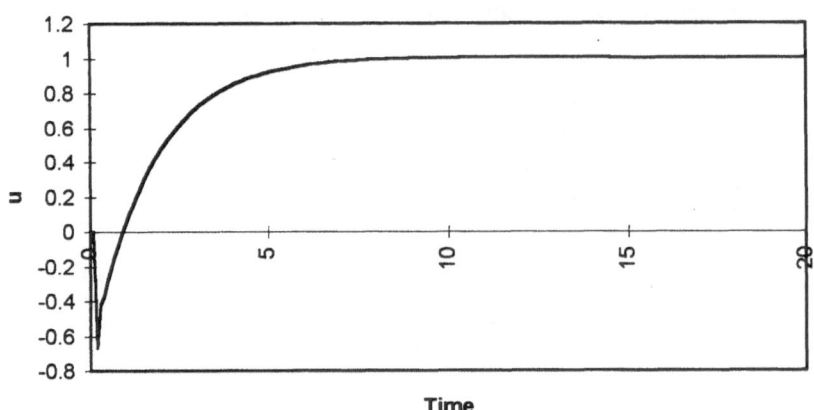

Figure 7. Servo response of first order linear system

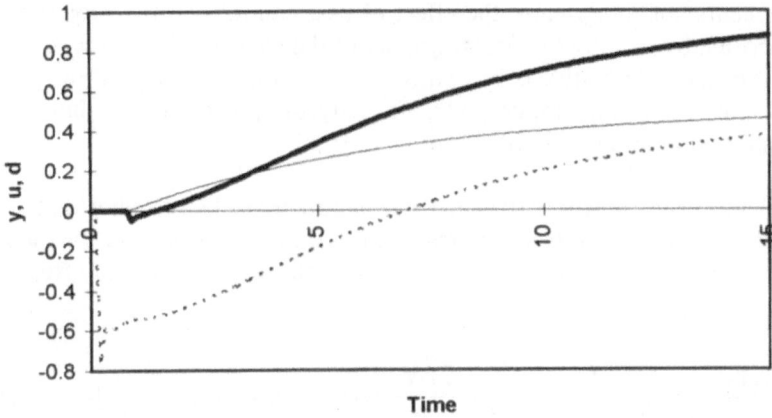

Figure 8. Behavior of first order system under set point changes and step disturbance
(— y : Process variable, - - - u : Control, _____ d : Disturbance estimate)

4. Application to Higher Order Linear Systems

Application of the stabilizing AIMC algorithm to second or higher linear systems is
performed numerically by a main program written in MATLAB® and using
SIMULINK® with separate block diagrams representing the process and the model.
The simulated process is

$$G(s) = \frac{K}{(\tau_1 s + 1)(\tau_2 s + 1)} e^{-s\theta} \qquad (17)$$

where the process parameters are: $\tau_1 = 1$, $\tau_2 = -1.1$ and $\theta = 1$. A perfect model was
assumed in all simulations. The AIMC control algorithm of Section 2 is implemented
according to the specified tasks as follows:

- The desired trajectory is set as a second order filter;

$$\varepsilon^2 \ddot{y}_f + 2\varepsilon\dot{y}_f + y_f = y_{sp} \qquad (18)$$

- Disturbance is estimated as described in Figure 3.

- Process output is projected into the future by integrating the model equation (17)
from the current time, t_k, to $t_k+\theta$ based on current measurement, current and past
controls, and the extrapolated disturbances. In the time domain, (17) becomes:

$$\tau_1\tau_2\ddot{y} + (\tau_1 + \tau_2)\dot{y} + y = K(u+d) \qquad (19)$$

In the course of integration, the effect of past controls on the projection is taken into account by shifting the past trajectory of the control effort forward in time by θ units. Since the derivative of the process output constitutes one of the states of this model, this projection implies that, not only the process output but also its first derivative is projected into the future.

- To calculate the control so that the model output tracks the filter output given by (18), the AIMC control law requires that the second derivative of the model output, from (19) be equal to the second derivative of the desired trajectory from (18). This gives

$$u(t_k) = \frac{\tau_1 \tau_2}{K\varepsilon^2} y_{sp} - d + \frac{1}{K}\left(\tau_1 + \tau_2 - \frac{2\tau_1 \tau_2}{\varepsilon}\right)\dot{y}(t_k + \theta) + \frac{1}{K}\left(1 - \frac{\tau_1 \tau_2}{\varepsilon^2}\right)y(t_k + \theta) \quad (20)$$

In the derivation of the above equation, the first derivative of the filter output was made equal to the first derivative of the model output. Since the system has a dead time, and therefore all model states are projected one dead time into the future, the control effort should be calculated from this equation with output and its first derivative being evaluated at time $t_k+\theta$.

Results from the application of the stabilizing AIMC strategy to the unstable process, given by (17) are shown in Figures 9-11. Figures 9 and 10 depict the performance of the controller for set point tracking and disturbance rejection (under an introduced disturbance of d=1 through the process). Figure 11 shows the transient behavior of the system under a set point change of r=1 and a disturbance of d= - 0.3. All these simulations were done with $\varepsilon=\varepsilon_d=1$. These figures reveal that the algorithm is capable of maintaining an unstable process at the desired set point despite disturbances.

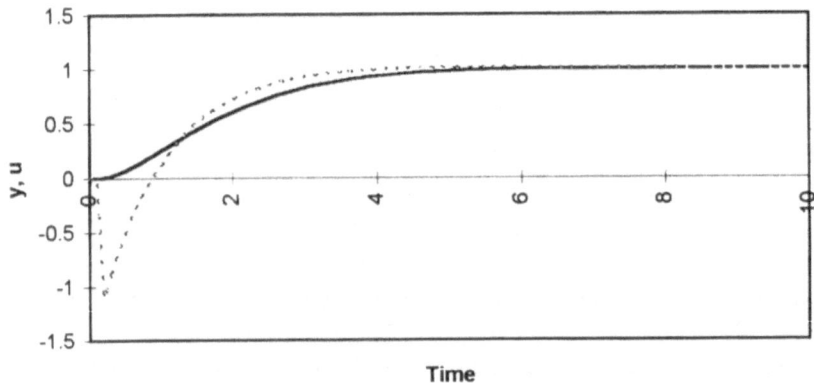

Figure 9. Second order linear system, servo behavior
(—— y : Process variable, - - - u : Control)

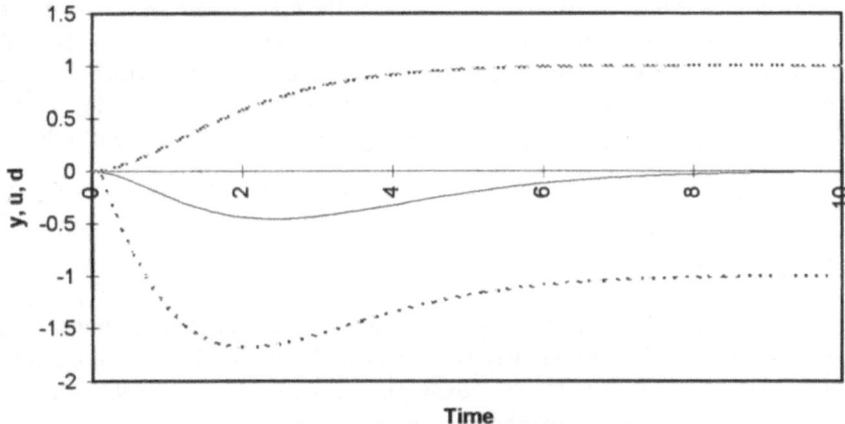

Figure 10. Second order linear system, regulatory behavior
(—— y : Process variable, - - - u : Control, _ _ _ _ d : Disturbance estimate)

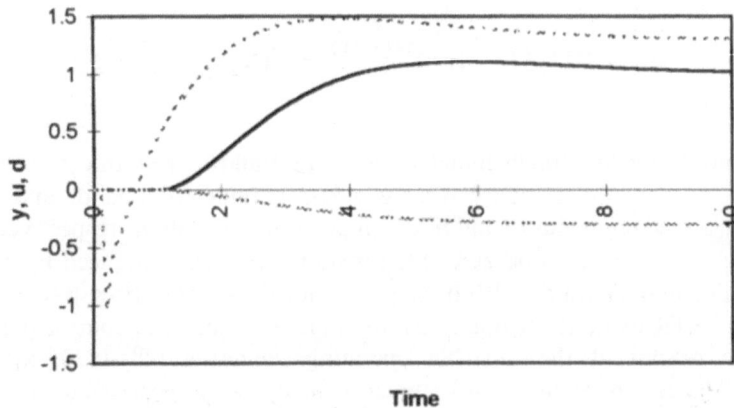

Figure 11. Second order system, under set point change and step disturbance
(—— y : Process variable, - - - u : Control, _ _ _ _ d : Disturbance estimate)

Remarks: In the present work we estimated the disturbance in such a way that the model output be the same as the process output at every instant of time. Then, the control was calculated to force the model output to follow the desired trajectory, by using the model output and its derivative. In this procedure, the model output is the same as the process output, but the derivative of the model output is not necessarily the derivative of the process output. An alternative strategy would be such that the disturbance could be estimated to force the model output, and its derivative to track the process output and its derivative. In this case, the disturbance would be described by two parameters rather than just one. Also, an estimate of the derivative of the process output could have been used in the calculation of control effort. We suspect that this approach may make considerable difference, particularly for imperfect models. These issues require further investigation, and our current efforts are concentrated in this direction.

5. Application to Nonlinear Systems

To illustrate the applicability of the algorithm to unstable nonlinear systems, we have chosen the example studied previously by Calvet and Arkun [4, 5] originating from the work of Uppal et al. [17]. This example considers a continuous stirred tank reactor in which a first order, exothermic, irreversible reaction is taking place. The dimensionless model equations are given by:

$$\dot{x}_1 = -x_1 + D_a(1-x_1)e^{\frac{x_x}{1+(x_2/\nu)}} - d_2 \tag{21a}$$

$$\dot{x}_2 = -x_2 + BD_a(1-x_1)e^{\frac{x_x}{1+(x_2/\nu)}} - \beta(x_2 - x_{20}) + \beta u + d_1 \tag{21b}$$

where x_1 and x_2 are the dimensionless composition and temperature respectively. D_a, B, ν, β and x_{20} are the standard dimensionless parameters, and d_1 and d_2 are the dimensionless feed temperature and feed composition fluctuations respectively [4]. In this work, d_2 is assumed to be zero. Step disturbance was introduced by assigning a specified value to the term d_1. With the parameters D_a=0.072, B=8, β=0.3, ν=20, and x_{20}=0, the CSTR exhibits an ignition/extinction behavior. The control problem is to operate the reactor at the unstable operating condition u^{op}=-0.20, x_1^{op}=0.5 and x_2^{op}=3.03. The reactor is initially assumed to be operating at a stable steady state of x_1=0.2 and x_2=1.33.

Simulation results when the suggested stabilizing AIMC structure is implemented are given in Figures 12-14. The two plots in Figure 12 show the response of the dimensionless temperature to a step change in set point for two different values of filter time constant, as well as the change in the control effort. Tuning parameters used were ε=1 and ε=2. The transition of the process variable reflects a linear behavior, and is smooth allowing the reactor to be operated at an unstable operating point by the AIMC algorithm.

Figures 13a and 13b present the transient behavior of the process variable and the control effort when an unmeasured step disturbance of d_1 =.3 is introduced to the process. Figure 13c shows the estimated step disturbance that makes the model output equal to the process output if it were introduced to the model as an additive signal to the control effort similar to the linear case. It is to be noted that the estimated additive step disturbance of d_e=1 is equivalent to the specified unmeasured disturbance of d=.3 at the steady state. The disturbance estimation given in Figure 13c also makes the model and process output match at each sampling time throughout the transient. It is seen that this disturbance is successfully compensated by the AIMC and the reactor is kept at the unstable operating point. The difference between a small and relatively large filter time constant is as expected, i.e. the larger the filter time constant, the more sluggish the response is. To track the two-degree of freedom capability of the suggested control algorithm, a run is made with an introduced set point change as well as an unmeasured disturbance of $d_1 = 0.3$. The results, as shown in Figure 14, indicate that the controller is capable of tracking the set point successfully under the influence of a disturbance.

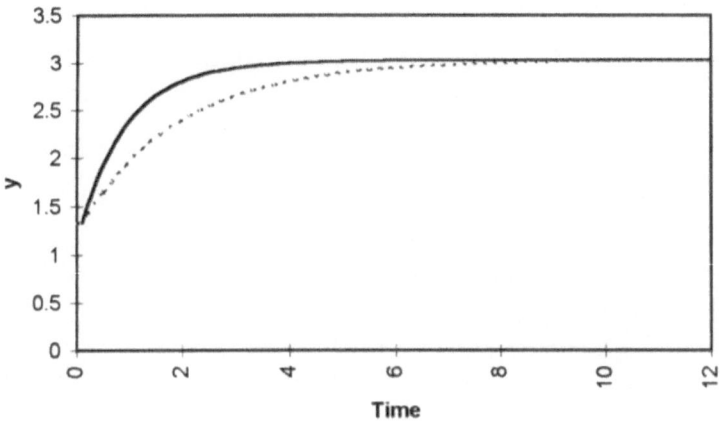

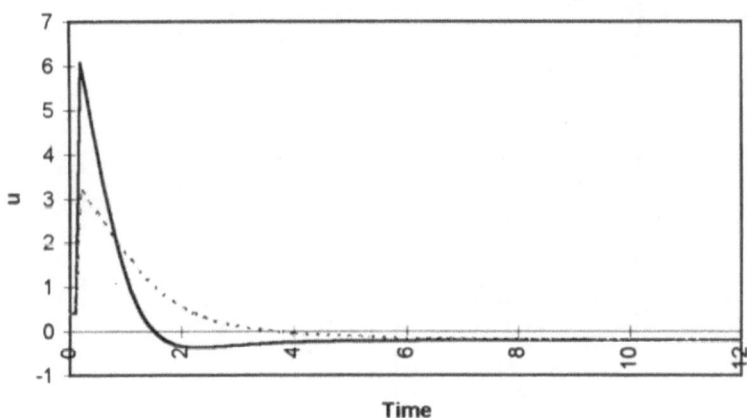

Figure 12. Nonlinear system under a step change (——— $\varepsilon = 1$, - - - $\varepsilon = 2$)

228

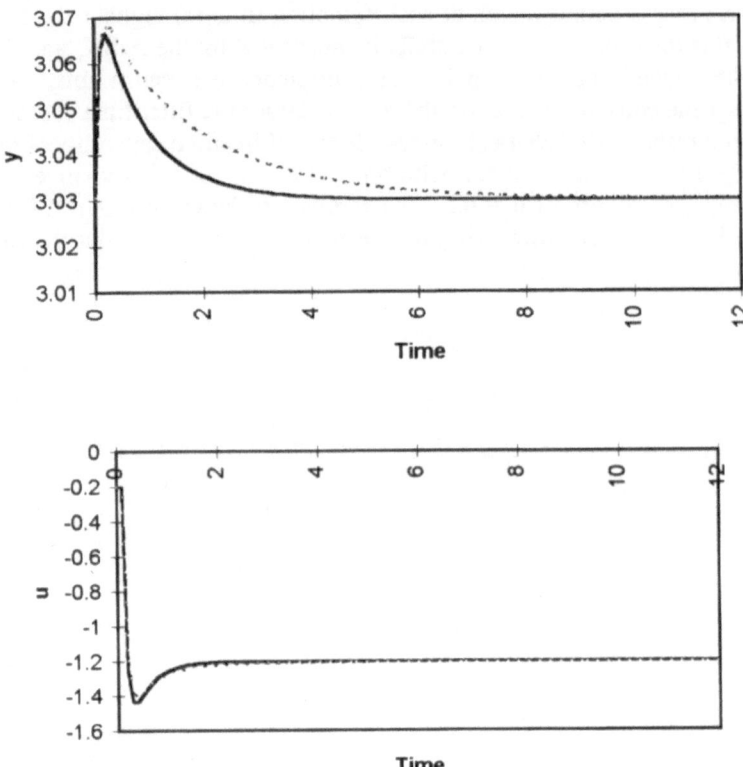

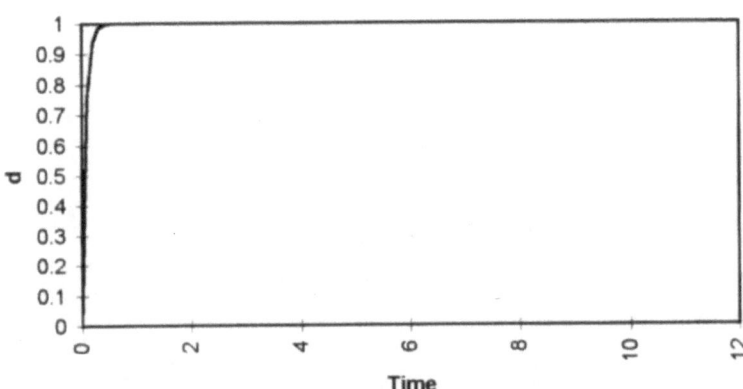

Figure 13. Nonlinear system under an unmeasured disturbance
(—— $\varepsilon = 1$, - - - $\varepsilon = 2$)

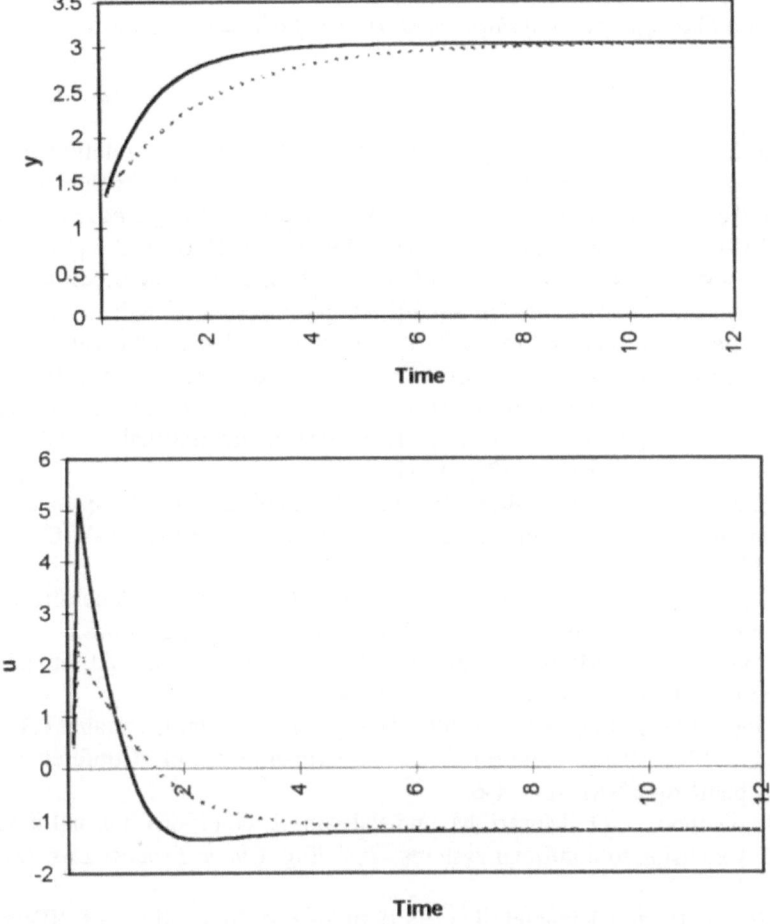

Figure 14. Nonlinear system under step change and disturbance
(—— ε =1, - - - ε =2)

6. Conclusion

The problem of the control of unstable systems has been tackled, and a solution that is applicable to linear as well as nonlinear processes is provided. An algorithmic IMC algorithm, applicable to a very broad range of process models, is presented to stabilize the unstable processes. For a first order plus dead time process, AIMC is analytically equivalent to the two degree of freedom IMC. AIMC is, however, internally stable for unstable models whereas the classical implementation of IMC is not. The suggested AIMC structure is shown via extensive simulations that it can be effectively used for stable operation of unstable processes.

230

Acknowledgment: This work was carried out when one of the authors (R.B.) was at CWRU on a Fulbright Scholarship, which is gratefully acknowledged.

References

1. Berber, R. and Brosilow, C. (1998) Insights into the relationships between linear and nonlinear model-based control and issues for further research, in R. Berber and C. Kravaris (eds.), *Nonlinear Model Based Process Control*, NATO ASI Series, Kluwer Academic Publishers, Dordrecht, pp. 87-114.
2. Brosilow, C. and Cheng, C. M. (1987) Model predictive control of unstable systems, Paper presented at AIChE Annual meeting, New York, N.Y.
3. Byrnes, C. I. and Isidori, A. (1985) Global feedback stabilization of nonlinear systems, Proceedings of IEEE CDC, Fort Lauderdale, FL 1031-1037.
4. Calvet, J. P. and Arkun, Y. (1988) Feedforward and feedback linearization of nonlinear systems and its implementation using internal model control, *Ind. Engng. Chem. Res.* **27**, 1822-1831.
5. Calvet, J. P. and Arkun, Y. (1989) Stabilization of feedback linearized nonlinear processes under bounded perturbations, Proc. of ACC-1989, Vol. 1. 747-752.
6. Clarke, D. W., Mohtadi, C. and Tuffs, P. S. (1987) Generalized predictive control - Part I. The basic algorithm, *Automatica*, **23**, 137-148.
7. Coulibaly, E., Maiti, S. and Brosilow, C. (1995) Internal model predictive control (IMPC), *Automatica*, **31**, 1471-1482.
8. Dong, J. and Brosilow, C. (1997) Design of robust multivariable PID controllers via IMC, Paper presented at American Control Conference, ACC'97, Albuquerque NM, June 4-6.
9. Ecomonou, C. G., Morari, M. and Palsson, B. O. (1986) Internal model control. 5. Extension to nonlinear systems, *Ind. Eng. Chem. Process Des. Dev.* **25**, 403-411.
10. Froisy, B. and Richalet, J. (1986) Industrial applications of IDCOM, in M. Morari and T.J. McAvoy (eds.) *Chemical Process Control CPC-III*, Elsevier, pp. 233-244.
11. Garcia, C. E. and Morari, M. (1982) Internal model control. 1. A unifying review and some new results, *Ind. Eng. Chem. Proc. Des. Dev.* **21**, 308-323.
12. Garcia, C.E. and Morshedi, A.M. (1986) Quadratic programming solution of dynamic matrix control (QDMC) *Chem. Eng. Commun.* **46**, 73-87.
13. Hirschorn, R. M. (1979) Invertibility of nonlinear control systems, *SIAM J. Cont. Opt.* **17**, 289-297.
14. Joseph, B. and Brosilow, C. B. (1978) Inferential control of processes, *AIChE J.* **24**. 485-491.
15. Morari, M. and Zafiriou, E. (1989) Robust Process Control, Prentice-Hall, Inc. Englewood Cliffs, NJ.
16. Smith, O. J. M. (1957) Closer control of loops with dead time, *Chem. Eng. Prog.* **53**, 217-219.
17. Uppal, A., Ray, W.H. and Poore, A. B. (1974) On the dynamic behavior of continuous stirred tank reactors, *Chem. Eng. Sci.* **29**, 967-985.

Appendix A.   Laplace Transform of the Effect of Past Controls on the Future Process Output

The influence of past controls on the predicted output $y(t + \theta)$, represented by y_m here, can be evaluated through the following convolution integral:

$$y_m \equiv K \int_t^{t+\theta} e^{-(t-\theta-\sigma)/\tau} u(\sigma - \theta) d\sigma \qquad (A1)$$

To obtain the transform of (A1), we first express y_m as the difference between two integrals, and then change variables so as to get each of the integrals in standard convolution form as follows:

$$y_m / K = \int_0^{t+\theta} e^{-(t-\theta-\sigma)/\tau} u(\sigma - \theta) d\sigma - \int_0^t e^{-(t-\theta-\sigma)/\tau} u(\sigma - \theta) d\sigma$$

$$= \int_\theta^{t+\theta} e^{-(t-\theta-\sigma)/\tau} u(\sigma - \theta) d\sigma - \int_0^t e^{-(t-\theta-\sigma)/\tau} u(\sigma - \theta) d\sigma \qquad (A2)$$

since $m(\sigma - \theta)$ is zero for $\sigma < \theta$.

Now let $\phi = \sigma - \theta$ in the first integral and factor $e^{-\theta/\tau}$ from the second integral. This gives

$$y_m / K = \int_0^t e^{-(t-\phi)/\tau} u(\phi) d\phi - e^{-\theta/\tau} \int_0^t e^{-(t-\sigma)/\tau} u(\sigma - \theta) d\sigma \qquad (A3)$$

Transforming (A3) gives

$$\mathcal{L}\{y_m / K\} = \frac{u(s)}{\tau s + 1} - \frac{(e^{-\theta/\tau})(e^{-s\theta} u(s))}{\tau s + 1}$$

$$= \frac{1}{\tau s + 1} u(s) - \frac{e^{-\theta/\tau}}{\tau s + 1} e^{-\theta s} u(s)$$

Appendix B. The Analyical Equivalence of AIMC and Two Degree of Freedom IMC

The Laplace transforms of eqns. (10) and (6) in the text are, respectively,

$$u(s) = - e^{-s\theta} d(s) + \frac{\tau}{\varepsilon K} y_{SP}(s) + \frac{(1 - \tau/\varepsilon)}{K} (e^{s\theta} y(s)) \qquad (B1)$$

$$[e^{s\theta} y(s)] = e^{-\theta/\tau} y(s) + \frac{K(1 - e^{-\theta/\tau} e^{-s\theta})}{\tau s + 1} u(s) + K(1 - e^{-\theta/\tau}) e^{-s\theta} d(s) \qquad (B2)$$

Letting $e^{-\theta/\tau} \equiv b$ and rearranging terms gives

$$[e^{s\theta} y(s)] = b [y(s) - \frac{Ke^{-s\theta}}{\tau s + 1} u(s)] + \frac{K}{\tau s + 1} u(s) + K(1-b)e^{-s\theta} d(s) \qquad (B3)$$

The term in parentheses in (B3) is measured output less the effect of all past controls on the output as predicted by the model and as shown in Figure 5.
Let

$$d_e(s) \equiv y(s) - \frac{Ke^{-s\theta}}{\tau s + 1} u(s) \qquad (B4)$$

Substituting (B3) and (B4) into (B1) gives

$$u(s) = \frac{(1 - \tau/\varepsilon)}{K} [bd_e(s) + \frac{K}{\tau s + 1} u(s) + K(1-b)e^{-s\theta} d(s)] + \frac{\tau}{\varepsilon K} y_{SP}(s) - e^{-s\theta} d(s) \qquad (B5)$$

Rearranging terms to solve for u(s) from (B5) gives

$$\left[\frac{\varepsilon s + 1}{\tau s + 1} \right] u(s) = - a\, d_e(s) + (a-1)Ke^{-s\theta} d(s) + y_{SP}(s) \qquad (B5a)$$

where $a \equiv (1 - \varepsilon/\tau)b$

From the manner in which $e^{-s\theta} d(s)$ is computed, c.f. eqn. (2) in the text,

$$e^{-s\theta} d(s) = \frac{\tau s + 1}{(\varepsilon_d s + 1)K} d_e(s) \qquad (B6)$$

Substituting (B6) into (B5a) gives

$$u(s) = \frac{\tau s + 1}{K(\varepsilon s + 1)} \left[\; y_{SP}(s) - [a + (1-a)\frac{(\tau s + 1)}{\varepsilon_d s + 1}] \, d_e(s) \; \right] \qquad \text{(B7)}$$

Letting

$$G_d(s) \equiv a + (1-a)\frac{(\tau s + 1)}{\varepsilon_d s + 1}$$

and

$$G_I(s) \equiv \frac{\tau s + 1}{K(\varepsilon s + 1)}$$

gives exactly the relationships shown diagrammatically in Figure 5.

APPROXIMATE I/O–LINEARIZATION OF NONLINEAR SYSTEMS

F. ALLGÖWER[1]
Institut für Automatik, ETH Zürich
CH-8092 Zürich, Switzerland
email: allgower@aut.ee.ethz.ch

F.J. DOYLE III
Dept. of Chem. Eng., University of Delaware
Newark, DE 19716, USA
email: doyle@che.udel.edu

1   Introduction

Nonlinear controller design by exact input-output linearization (I/O-linearization) has become a popular approach for controlling nonlinear systems during the past decade. In contrast to the common Jacobi-linearization approach, where higher order terms in the Taylor series are just neglected, these nonlinear terms of higher order are exactly compensated in the I/O-linearization approach and no approximation error is made. Hence, I/O-linearized dynamical systems exhibit exactly linear I/O behavior. Subsequently to the I/O-linearization step a linear controller can be designed for this exactly linear system to achieve the desired performance using the well-developed methods of linear control theory. Only the first step, namely the I/O-linearization, requires a large design effort. However, this step has to be performed only once for a given plant. Necessary design iterations for achieving satisfactory performance and a satisfying compromise between conflicting design specifications can then be dealt with exclusively in the second, linear step, where these iterations can be performed efficiently and without significant cost. For a summary on a one-step design using I/O-linearization see the papers of Kravaris in this volume.

The development of the theoretical framework for exact I/O-linearization can be traced back to work in the mid 70s (Porter, 1970; Freund, 1975; Singh and Rugh, 1972; Isidori et al., 1981), but has only become popular with the books of Isidori and Nijmeijer/van der Schaft (Isidori, 1989; Nijmeijer and van der Schaft,

---

[1]Correspondence author: Tel. +41-1-6323557, Fax. +41-1-6321211

R. Berber and C. Kravaris (eds.), Nonlinear Model Based Process Control, 235-274.
© 1998 *Kluwer Academic Publishers.*

1990) in the late 80's. Even though there are several approaches to the exact I/O-linearization problem (Isidori and Ruberti, 1984; Lee and Marcus, 1987; Claude *et al.*, 1983; Monaco and Normand-Cyrot, 1983), the geometric approach using state feedback has become the most popular one (Isidori, 1989; Nijmeijer and van der Schaft, 1990). We refer to (Isidori, 1989; Nijmeijer and van der Schaft, 1990) for a detailed description of the method.

Controller design by exact I/O-linearization has, however, a number of drawbacks and shortcomings. First of all, not all nonlinear systems can be *exactly* I/O-linearized while maintaining internal stability. Standard theory (Isidori, 1989) requires the nonlinear system to have stable zero-dynamics (minimum phase property) and a well-defined relative degree (further restrictions apply in the MIMO case (Isidori, 1989)). Furthermore, issues like input constraints, robustness, the output feedback case, and disturbance attenuation properties cannot be directly addressed in the framework of *exact* I/O-linearization.

These drawbacks can be partially overcome when relaxing the requirement for an *exact* I/O-linearization to that of an *approximate* I/O-linearization. There are many approaches in the literature that treat the approximate I/O-linearization problem considering, for example, the unstable zero-dynamics case, input constraints, systems with a poorly defined relative degree *etc.*. See, for example, (Allgöwer and Doyle III, 1997) for a list of references. However, the vast majority of these approaches is only concerned with stabilizing the closed loop. Only in very few cases is an attempt made to systematically address the issue of degree of linearity achieved. For most approaches it is easy to construct examples where the approximately I/O-linearized system displays more pronounced nonlinearity than the original system. The main difficulty responsible for the small number of approaches that directly address the degree of linearity in connection with approximate I/O-linearization lies in the term "approximate linearity" itself. By definition, a system is called linear if its I/O-behavior satisfies the superposition and homogeneity principle. All other systems are called nonlinear, independent of their "degree of nonlinearity". By making the term "approximate linearity" more precise, One can develop methods to analyze and synthesize appropriate I/O-linearizing control schemes. One approach to define the degree of linearity is to characterize it by the order of the Taylor series expansion of the compensated system up to which no nonlinear term exists. A system will thus be called for example "approximately linear of order five" if its Taylor expansion does not contain any quadratic, cubic, forth, or fifth order terms but arbitrary terms of order strictly higher than five. This is a meaningful definition in many cases. However, the drawback of this definition is obvious: Even if we assume that the Taylor series converges, the size of the factors multiplying each term of different order does not enter the definition, neither does the "place" in the right-hand side of the differential equation where these terms appear and thus the relative importance for the I/O-dynamics. Moreover, such a definition is strongly dependent on the operating region without allowing to consider this in the definition. It is therefore very easy to construct examples for which an approximate linearization of very high order exhibits much stronger nonlinear behavior than that of an approximate linearization of very low order.

In many cases a different definition based on the notion of so-called nonlinearity measures (Allgöwer and Gilles, 1992; Allgöwer, 1995c; Nikolaou, 1994; Guay *et al.*, 1995; Stack and Doyle III, 1997a; Stack and Doyle III, 1997b) is more meaningful. The central idea behind nonlinearity measures is a direct quantification of the degree of nonlinearity in the I/O-behavior. The nonlinearity measure in (Allgöwer, 1995c), that is used in this paper, assigns to each system a number in the interval $[0, 1]$, that directly relates to the strength of the nonlinearity: A linear system has a value of zero, while the strongest possible nonlinearity is characterized by a value of one. A short review of this concept is given in Appendix A.

In this paper we restrict ourselves to the discussion of four different approximate I/O-linearization schemes, that directly address the issue of degree of linearity achieved, even though there exists by now a number of further approaches in the literature (see for example (Paolini *et al.*, 1991; Castillo, 1991; Hauser *et al.*, 1989; Kappos, 1989; Kang, 1994; Grizzle and Di Benedetto, 1992; Xia, 1995)). The schemes we discuss are based on a direct minimization of the nonlinearity measure by numerical optimization (Sect. 2), a consideration of the problem in the framework of disturbance decoupling and disturbance attenuation (Sect. 3), an approximate I/O-linearization using a partitioned inverse (Sect. 4), and an extension of standard I/O-linearization to also include the case of non-minimum phase systems (Sect. 5).

2   Approximate I/O-linearization by minimization of nonlinearity measures

In order to quantify the degree of nonlinearity in the I/O-behavior of nonlinear systems so called nonlinearity measures can be used (see Appendix A for the definition and a short discussion of a nonlinearity measure). Nonlinearity measures have various applications related to approximate I/O-linearization: Firstly, they can be used in the analysis of the linearization quality of approximately linearized systems. Along the same lines the robustness of the approximate linearization quality with respect to parameter uncertainties can be examined or the necessary complexity of I/O-linearizing controllers can be investigated (Allgöwer, 1996b). Secondly, approximately I/O-linearizing feedback can be synthesized via a direct minimization of the degree of nonlinearity (quantified by the nonlinearity measure). This approach is outlined in the sequel. The problem is defined as follows: Find a compensator R

$$u = R[y_m, v] \tag{1}$$

for nonlinear system N

$$y = N[u], \tag{2}$$

such that the nonlinearity measure of the compensated system N_g (depicted in Figure 1)

$$y = N_g[v], \tag{3}$$

238

with new input v and output y is as small as possible. Here, R is a causal non-linear operator and y_m is the vector of measured outputs available for feedback. Alternatively, instead of minimizing the nonlinearity measure, we can also require the nonlinearity measure of the compensated system to be smaller than some given bound β, that represents the maximally tolerable degree of nonlinearity.

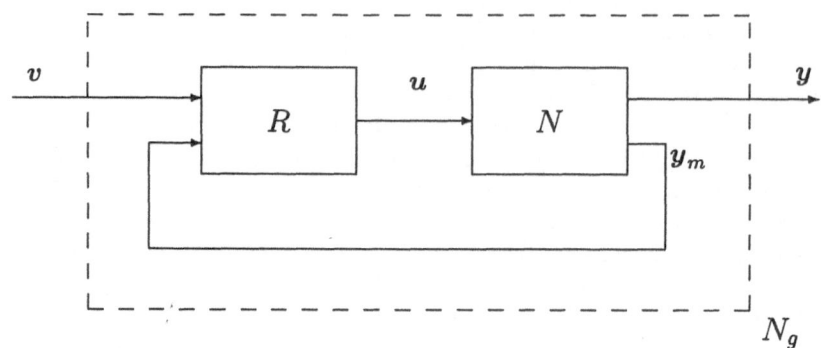

Figure 1: Compensated system N_g with compensator R and plant N.

An analytical computation of an optimal compensator R is in general not easily possible. The problem can however be greatly simplified if we assume that the *structure* of R is fixed and only a limited number of parameters in R are left open. This approach will be explained with the following example.

Example:
We consider the problem of approximately linearizing the I/O-behavior of the non-linear SISO system

$$
\begin{aligned}
y &= x_1 \\
\dot{x}_1 &= -3x_1 + 4x_2 + 2x_1^2 + u \\
\dot{x}_2 &= -x_1 + x_2 - 0.5x_1^3,
\end{aligned}
\tag{4}
$$

with scalar input u, scalar output y, and state $x = [x_1, x_2]^T$. We are interested in the behavior of system (4) in a neighborhood around the steady state $x_s = [0,0]^T$ and assume that the input is constrained by

$$
|u| \le 0.6 \ .
\tag{5}
$$

System (4) displays strongly nonlinear I/O-behavior as demonstrated by the step responses shown in Figure 2: doubling of the input step size does not lead to an output of double the size; a step with opposite sign leads to a qualitatively different output function. The system equations are given in input-normalized Byrnes-Isidori normal form. It can thus be readily seen that (4) has unstable zero dynamics, and hence application of standard I/O-linearization is not possible. As structure for the approximately I/O-linearizing feedback law we consider a static state-feedback

$$
u = \alpha(x, v),
\tag{6}
$$

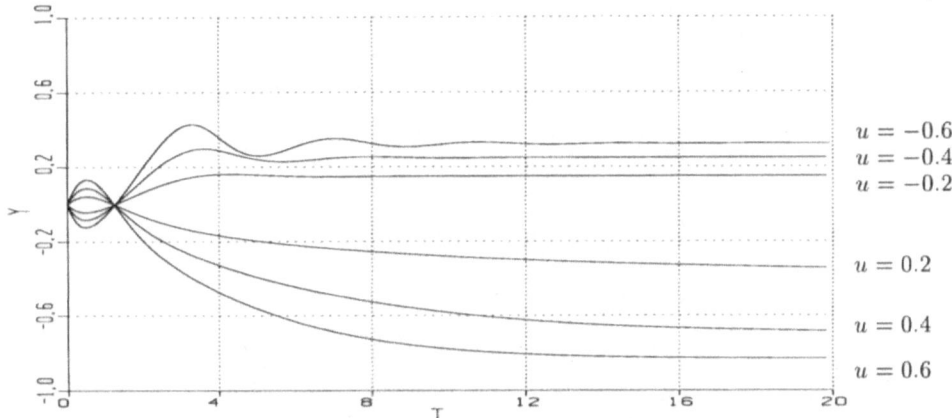

Figure 2: Step responses of uncompensated system (4).

where for simplicity we assume that the structure of α is determined by a (homogeneous) polynomial of degree two

$$\alpha(x,v) = a_1 x_1^2 + a_2 x_2^2 + a_3 x_1 x_2 + v, \qquad (7)$$

with parameters a_i, $i = 1,2,3$ to be determined by the optimization. Compensation structure (6), (7) does not include any linear terms, as we only want to compensate for nonlinearities without changing the Jacobi-linearized system. The control problem defined above is therefore reduced to a parameter optimization problem that can be solved numerically using standard optimization tools. For this example computation of the nonlinearity measure via the Fourier coefficient based lower bound (Allgöwer, 1996a) is used. We denote this lower bound by $\chi_{N_g}^{U_s}$. We initialize the optimization with

$$a_i = 0, \quad i = 1,2,3, \qquad (8)$$

which corresponds to the system without feedback. The value of the nonlinearity measure for the uncompensated system is $\chi_{N_g}^{U_s} = 0.43$, which reflects the pronounced nonlinearities. The solution for the optimization problem can be easily obtained using a standard SQP-Algorithm (Powell, 1982) and is given by

$$\begin{aligned} a_1 &= -1.90 \\ a_2 &= 0.05 \\ a_3 &= -0.14 \ . \end{aligned} \qquad (9)$$

With this feedback the nonlinearity measure of the compensated system is

$$\chi_{N_g}^{U_s} = 0.04 \ . \qquad (10)$$

Figure 3 shows step responses of the compensated system (4) with feedback (6), (7) and parameter values (9).

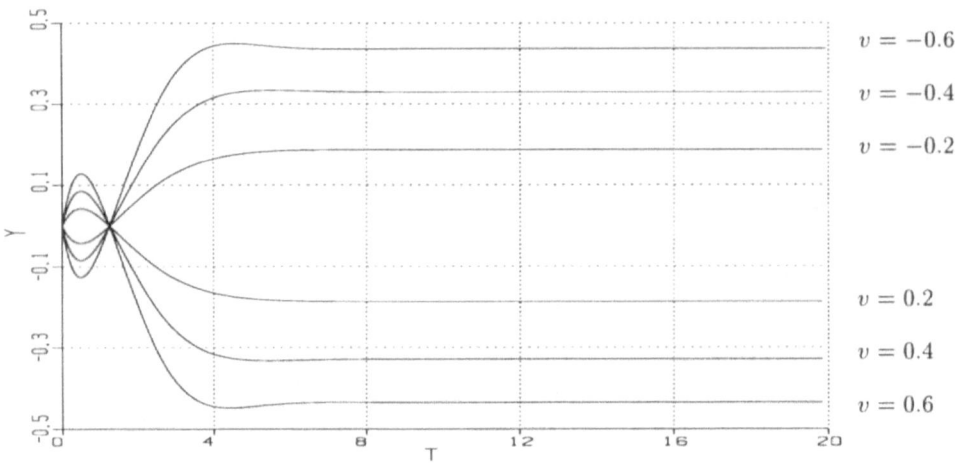

Figure 3: Step responses of system (4) with compensation (6), (7), (9).

As expected by the large reduction of the value of the nonlinearity measure, the compensated system displays considerably "more linear" behavior than the uncompensated system. However, despite the very low value of $\chi_{N_g}^{U_s} = 0.04$ the step responses still display some nonlinear effects (e.g. slight "overshoot" for large input amplitudes). This can only be attributed partly to the fact that no exact I/O-linearization can be expected due to the unstable zero-dynamics. The main reason must be seen in the use of a lower bound to compute the nonlinearity measure (see (Allgöwer, 1996b) for a more detailed discussion).

There is a simple interpretation for the optimal feedback law in this example: The term $-1.9x_1^2$ compensates the nonlinearity in the direct I/O-channel $(+2x_1^2)$. The weakly nonlinear zero dynamics remains in the loop and causes only minor nonlinear effects in the I/O-behavior. Inclusion of further terms of higher order in (7) does not lead to a further reduction of the nonlinearity in this example.

During optimization, stability of the compensated system has to be guaranteed. This can be accomplished either by a parameterization of all stabilizing controllers (e.g. (Paice and van der Schaft, 1996; Desoer and Kabuli, 1987)) or by detecting and penalizing unstable parameter regions during optimization (Allgöwer and Gilles, 1992).

The biggest problem in connection with this approach to approximate I/O-linearization is the determination of an appropriate structure for the approximately linearizing feedback law. In many practical cases a suitable structure can be found using physical insight into the problem. If this is not possible, a general compensation structure, like for instance a polynomial feedback function, can be used. However, in this case the number of parameters to be determined will grow combinatorically both with the state dimension and the desired approximation order. If many parameters have to be determined, the resulting (non-convex) optimization problem will be rather expensive to solve. Our experience with a large number of example systems shows, however, that in general a significant reduction of the size of nonlinearity is possible with relatively small expenditure.

The main advantage of this approach to approximate I/O-linearization is the large class of systems that can be considered. In particular, systems that cannot be linearized exactly, can be approximately linearized this way.

The standard theory of exact I/O-linearization requires that the whole state must be available for feedback. Another big advantage of the approach presented here is that in addition to state feedback also output feedback can be considered in the same framework. In many cases sufficient linearity can even be achieved without the need for a feedback loop at all. Instead, the use of a static precompensator

$$u = \gamma(v) \tag{11}$$

as in Figure 4 is sufficient. We refer to I/O-linearization with a structure like in Fig. 4 as *I/O-linearization by precompensation.*

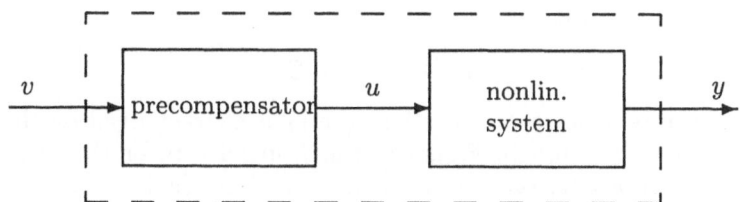

Figure 4: Structure of I/O-linearization by precompensation.

This structure constitutes an interesting special case of the general compensation structure (1) where the output of compensator R does not depend on y_m. There are two main advantages of the precompensator structure when compared to the case with feedback: First, it can be shown (Allgöwer, 1995b) that an approximate I/O-linearization by precompensation is structurally characterized by very good robustness properties both with respect to stability and with respect to the quality of the I/O-linearity. Secondly, the combinatorial explosion of the number of open feedback parameters with growing state dimension is avoided, as the precompensator γ is only a function of one scalar variable v for SISO systems. Our experience is that in most cases a simple sum of selected "basis"-functions for the scalar-valued function $\gamma(v)$ is absolutely sufficient. An application of approximate I/O-linearization by precompensation and minimization of the nonlinearity measure to a process control problem is given below.

The approach presented can furthermore be used to determine *robust* minima of the nonlinearity measure. If for example the system to be linearized contains uncertain parameters, it is possible to find a feedback law that achieves a low value of the nonlinearity measure for all systems within the uncertainty description (Allgöwer and Gilles, 1992). If such a robustly linearizing compensation does not exist, it is advisable not to apply I/O-linearization techniques at all.

2.1 Approximate I/O-linearization of a membrane reactor

We consider production of a substance C in a stirred membrane reactor (cf. Fig. 5) where initial reactant A is fed to the reactor with flow rate q and concentration c_{A0}.

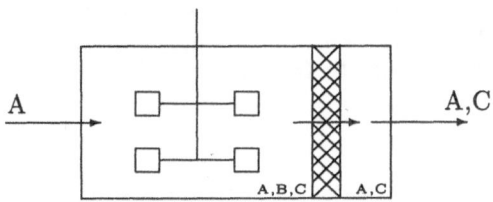

Figure 5: Schematic representation of the membrane reactor for production of substance C from substance A.

The following isothermal reaction

$$A \xrightarrow{r_1} B \xrightarrow{r_2} C \tag{12}$$

takes place in the reactor, where reaction rate r_1 is characterized by the strongly nonlinear relationship shown in Figure 6 (see Appendix B for the corresponding equation) and reaction rate r_2 is assumed to be proportional to the concentration of substance B. Only substances A and C can pass the semi-permeable membrane (symbolized by the hatched area in Figure 5). Intermediate product B cannot pass the membrane and hence cannot exit the reactor. Therefore no costly separation is needed in order to eliminate the unwanted intermediate product B and product yield and selectivity are greatly improved when compared to a production in a conventional CSTR.

The reactor dynamics are described by the following nonlinear differential equations that are derived from component balances for substances A and B:

$$\dot{c}_A = q\,(c_{A0} - c_A) - r_1(c_A)$$
$$\dot{c}_B = r_1(c_A) - r_2 \cdot c_B \,. \tag{13}$$

Concentration c_{A0} of the initial reactant fed to the reactor is considered as manipulated input, that is constrained by $6.5\frac{mol}{l} \leq c_{A0} \leq 9.5\frac{mol}{l}$. As control objective we want to operate the reactor such that concentration c_B is kept constant, and thus substance B is not accumulated in the reactor. Concentration c_B can only be measured with some dead-time t_d, which we consider in the dynamical model via a first order Padé approximation:

$$\dot{x}_m = -\frac{2}{t_d}x_m + \frac{4}{t_d}\left(c_B - c_{Bs} + \frac{x_m}{2}\right)$$

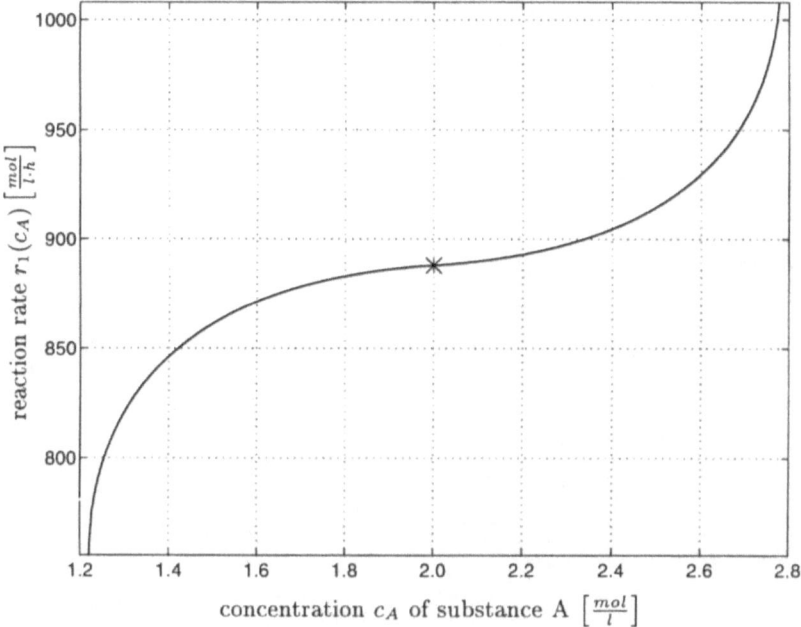

Figure 6: Reaction rate r_1 as function of the concentration of substance A. The asterisk ("'*'") indicates the operating point.

$$(14)$$

$$y = x_m - c_B + c_{Bs} - \frac{x_{ms}}{2} \, .$$

We assume that output y is the only available measurement and to-be-controlled variable. Together, eqs. (13) and (14) describe the dynamical behavior of the reactor. The physico-chemical parameters and the steady state operating point are summarized in Appendix B.

The reactor displays fairly nonlinear I/O-behavior as shown in Figure 7.

The value of the nonlinearity measure for the open loop reactor is

$$\chi_N^{\mathcal{U}_s} = 0.17 \, ,$$

where the computation is performed via the Fourier coefficient based lower bound (Allgöwer, 1996u). This value sustains the assessment that the reactor shows clear, but not excessively strong nonlinear behavior.

Standard I/O-linearization cannot be applied to this reactor because of the measurement delay that gives rise to unstable zero-dynamics and because not the whole state can be measured. Instead, we attempt an approximate I/O-linearization by precompensation based on minimization of the nonlinearity measure in a neighborhood of the operating point. We parameterize the precompensator function γ via a look-up table with eleven entries t_i (cf. Table 1).

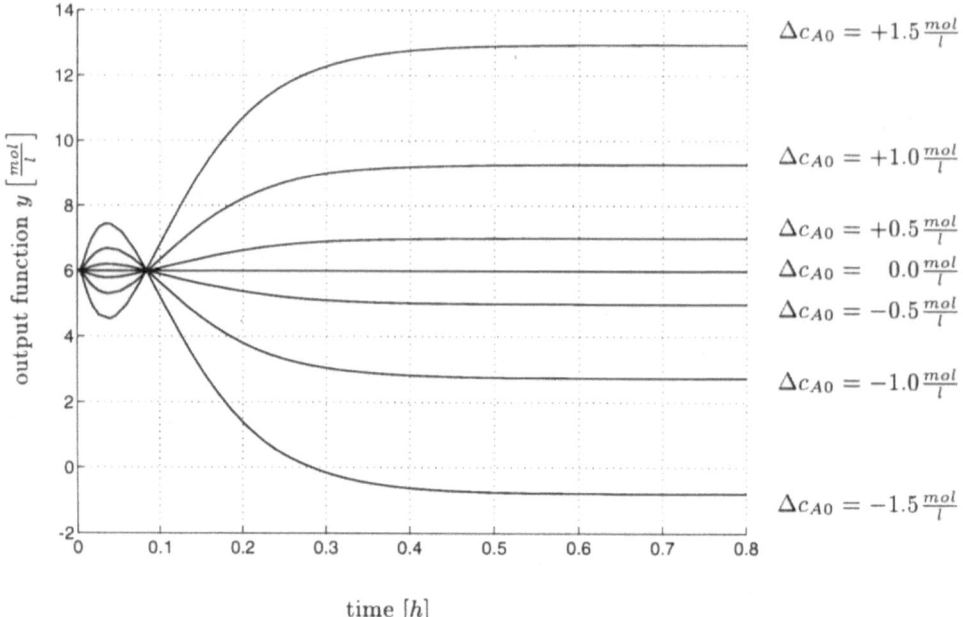

Figure 7: Dynamic response of the open-loop membrane reactor to inputs of different step size Δc_{A0}.

c_{a0}	t_1	t_2	t_3	t_4	t_5	t_6	t_7	t_8	t_9	t_{10}	t_{11}
v	-1.5	-1.0	-0.5	-0.2	-0.1	0.0	$+0.1$	$+0.2$	$+0.5$	$+1.0$	$+1.5$

Table 1: Look-up table for the precompensator with unknown (i.e. to be determined) entries t_i.

Linear interpolation is used between these values and extrapolation for values below or above the given v-range. At the operating point we require $v_s = 0$, which directly implies $t_6 = c_{A0s}$, where c_{A0s} denotes the steady state input value. The remaining ten values are determined via a minimization of the nonlinearity measure of the compensated reactor. The resulting optimization problem is non-convex. Hence, the optimizer might get stuck in local optima. To circumvent this, repeated optimization runs with 100 randomly chosen initial values for the table entries t_i are performed (Monte-Carlo optimization) and the best solution (the lowest value of the nonlinearity measure) is selected. This approach is of course neither theoretically satisfying nor optimal with respect to the computation cost involved. However, for this example a satisfying solution can be found using this approach with little implementation effort. The total CPU time needed to solve this problem was a little more than eight hours on a relatively slow VAX 4090. As optimal solution the precompensation function shown in Figure 8 is found.

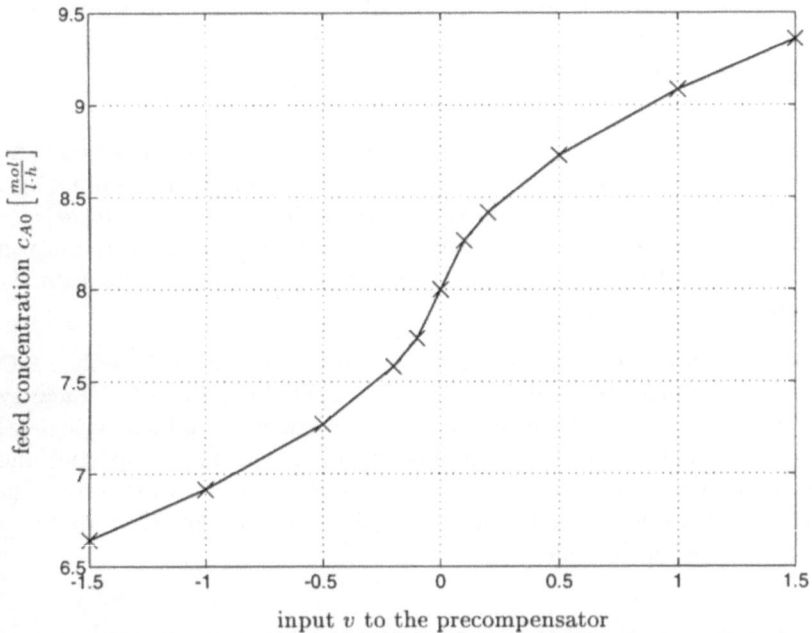

input v to the precompensator

Figure 8: Optimal precompensation function γ. Cross marks ("'×'") indicate the table entries t_i found by the optimization. Between entries linear interpolation is used.

The value of the nonlinearity measure at the optimum is $\chi_N^{u_s} = 0.06$. Hence, the precompensated reactor shows virtually linear I/O-behavior, as can also be seen from the step responses in Figure 9.

Remark 1 *The described precompensation achieves an almost complete linearity of the compensated reactor and a subsequent design of a linear output feedback controller is easily possible. Using for example the root locus method, a simple P-controller can be found that guarantees stability of the closed loop robustly.*

If, however, the nonlinearities of the reactor are not compensated, a design of linear controllers using linear methods (on the basis of the Jacobi-linearized reactor) can lead to problems. We will demonstrate this with the design of a P-controller

$$c_{A0} = k_p(y - y_s) + c_{A0_s}. \tag{15}$$

The parameter range of k_p for which stability is expected on the basis of the Jacobi-linearized reactor is easily shown to be

$$-1.325 = -0.08 \cdot r_2 \leq k_p \leq 0.08 \cdot r_2 + \frac{0.17}{t_d} = 2.65. \tag{16}$$

A P-controller with a gain of $k_p = 1.325$, that is far away from the theoretical (linear) stability boundaries, should thus also lead to stability for the nonlinear system. This is however not the case! Figure 10 shows the closed-loop behavior of the reactor with feedback (15) and a gain of $k_p = 1.325$ if no disturbances are considered, but the initial condition for concentration c_B varies from nominal value by the very small amount of 0.5%, i.e.

$$c_B(0) = 1.005 \cdot c_{BS}. \tag{17}$$

The closed loop is unstable with finite escape time. An analysis of the nonlinear closed loop reveals that there is an unstable limit cycle around the nominal operating point of the closed loop with a very small amplitude. If, however, a comparable linear P-controller is designed for the approximately I/O-linearized reactor, the region of attraction computed based on the Jacobi-linearized system coincides with the true region of attraction.

With the membrane reactor example we wanted to demonstrate that I/O-linearization by precompensation can indeed lead to a satisfying degree of linearity. But we also wanted to show that linear controller design for nonlinear systems based exclusively on the Jacobi-linearization may suggest misleading properties and may often lead to bad controllers. As is well-known (but often forgotten or disregarded) linear theory is only admissible for linear plants, or for systems that are previously linearized using appropriate methods.

3 Approximate I/O-linearization via a related disturbance decoupling problem

In this section we reformulate the classical I/O-linearization problem and cast it as an appropriate disturbance decoupling problem, that can be addressed with various techniques. The underlying idea is to consider the model matching problem depicted in Figure 11: We want to find a compensator K such that the compensated nonlinear system N has the same I/O-behavior as the linear reference system G. This means that the output $e = \bar{y} - y$ has to be zero for all possible inputs w, or in other words, compensator K has to be chosen such that input w is decoupled from output e. The problem of exact disturbance decoupling is by now well understood and the corresponding techniques can be found in standard text books (Isidori, 1989; Nijmeijer and van der Schaft, 1990). Exact decoupling is equivalent to an exact I/O-linearization. If we aim at an approximate I/O-linearization, the problem can be either cast as an *almost* disturbance decoupling problem, where an approximate solution with an arbitrary degree of accuracy is sought, or as a disturbance attenuation problem. For both the almost disturbance decoupling problem and the disturbance attenuation problem, several approaches are known. Here we restrict ourselves to the nonlinear H_∞ formulation (see for example (Isidori and Astolfi, 1992; van der Schaft, 1992; Allgöwer et al., 1994; Allgöwer and Doyle III, 1997) for an introduction and discussion), where a feedback is sought that attenuates, or almost decouples, the effect of L_2-inputs w on the output e (Marino

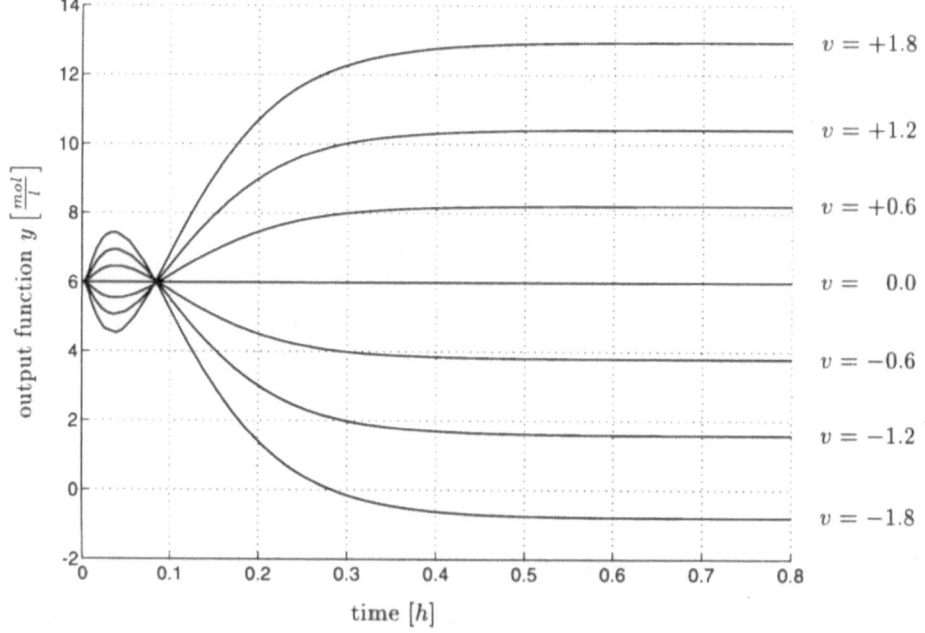

Figure 9: Step responses of the precompensated membrane reactor with precompensation function γ according to Figure 8 and input steps of different size.

et al., 1994; Isidori, 1996; van der Schaft, 1992) and only discuss the disturbance attenuation case. We make two further simplifications in order to be able to apply standard nonlinear H_∞ theory: We only require that the L_2-gain be smaller than some bound γ (representing the maximally tolerable degree of nonlinearity), plus we include an additional term related to the control effort in the criterion. The problem we want to solve is thus: *Find a state-feedback law*

$$u = \alpha(x, \xi, w) \tag{18}$$

where x and ξ are the states of the nonlinear system N and the linear reference system G, so that the following inequality is satisfied for all $w(t) \in L_2$:

$$\int_0^T \left(\|e\|^2 + \beta\|u\|^2 \right) dt \le \gamma^2 \int_0^T \|w\|^2 dt, \quad \forall w(t), T. \tag{19}$$

Through inclusion of the term $\beta\|u\|^2$ we try to formulate a compromise between linearity and control effort. From a practical point of view this is a very sensible and necessary thing to do because we do not want to pay for slightly increased linearity by significantly amplified input moves. The case $\beta = 0$ gives rise to a singular H_∞ problem for which only preliminary results are known (Maas and van der Schaft, 1996; Santosuosso, 1995). A further advantage of considering the

248

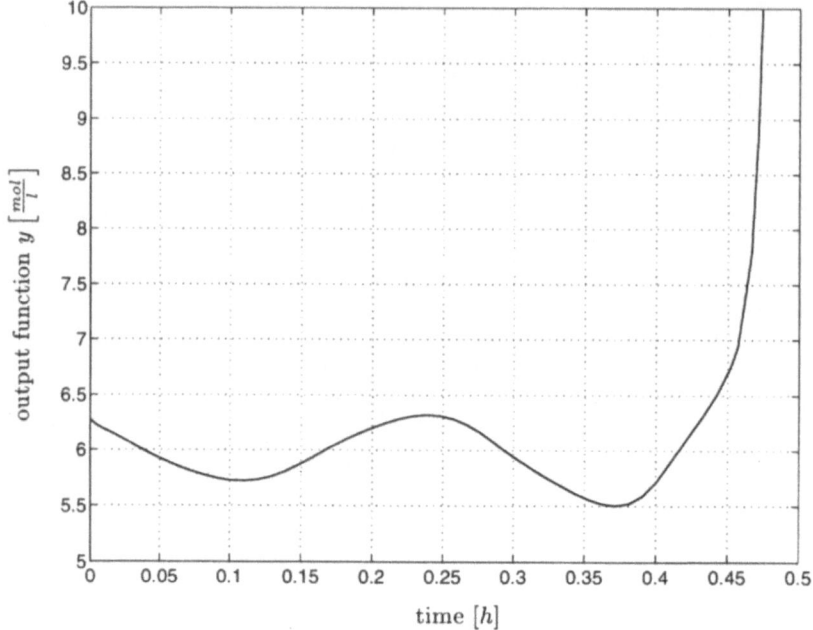

Figure 10: Closed-loop behavior of the membrane reactor with P-controller and $k_p = 1.325$ and initial condition $c_B(0) = 1.005 \cdot c_{BS}$.

case $\beta \neq 0$ is that it is possible to approximately linearize systems, for which an exact I/O-linearization is not possible. This is also demonstrated with the example below.

Example:
We consider the same simple example as in Sect. 2 and determine compensator K such that (19) holds with $\beta = 0.01$ and $\gamma = 0.01$. The nonlinear H_∞ control problem is solved using Lukes' method (Lukes, 1969). Figure 12 shows the I/O-behavior of the compensated system when terms up to order four are considered. As can be seen, the resulting compensated system behaves rather linearly and the behavior is comparable in quality to the results obtained by numerical minimization of nonlinearity measures described in Sect. 2.

An application of this approach to the more challenging approximate I/O-linearization of a CSTR for cyclopentenol synthesis can be found in (Allgöwer *et al.*, 1994). An interesting relationship between approximate I/O-linearization via nonlinear disturbance decoupling and approximate I/O-linearization by minimization of nonlinearity measures is pointed out in (Allgöwer, 1995a).

The main advantage of considering I/O-linearization via a related disturbance decoupling problem is the larger class of systems that can be addressed by looking at the problem from a different perspective. This includes the derivation of new

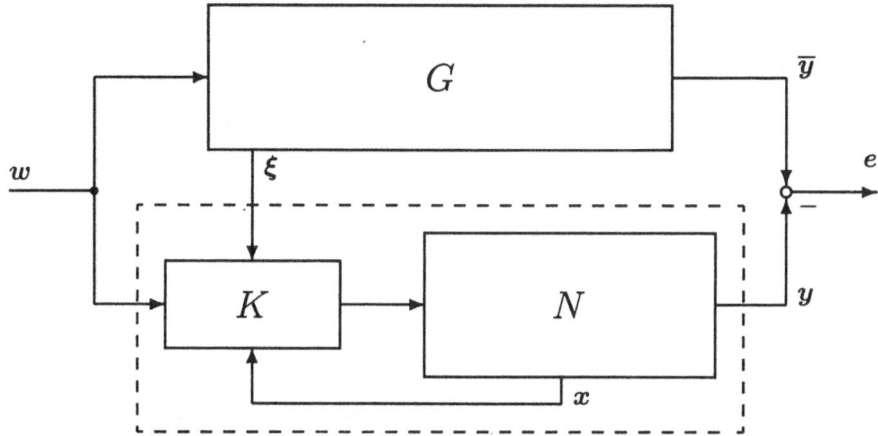

Figure 11: Configuration for approximate I/O-linearization via a related disturbance decoupling problem.

solvability conditions that go beyond the minimum phase and well-defined relative
degree conditions that are only sufficient. Also, a new problem, the *almost exact
I/O-linearization problem*, can be introduced in analogy to the almost disturbance
decoupling problem, that allows to approximately linearize a larger class of systems
with an *arbitrary* degree of linearity. And last but not least, a compromise between
desired degree of linearity and necessary input energy can be formulated. On the
negative side, the computation of the necessary compensators does not become
simpler. In fact, exactly the opposite holds for example for the special case of a
nonlinear H_∞ approach as exemplary discussed above.

4 Partitioned Approach to Approximate I/O-Linearization

In this section, the design of an approximate feedback linearizing controller for a
nonlinear system is derived using a partitioned inverse (Kendi and Doyle III, 1997).
It is assumed that the nonlinear system can be represented as the following general
dynamic input-output model:

$$y = P[u, d] \tag{20}$$

where d are the measured disturbances. For systems without measured disturbances, it has been shown in (Economou and Morari, 1986) that a nonlinear IMC
controller is composed of two elements; (i) an inverse of the nonlinear system, and
(ii) a linear filter for setpoint tracking. In the next theorem, a partitioned nonlinear
inverse is obtained for the nonlinear system with measured disturbances.

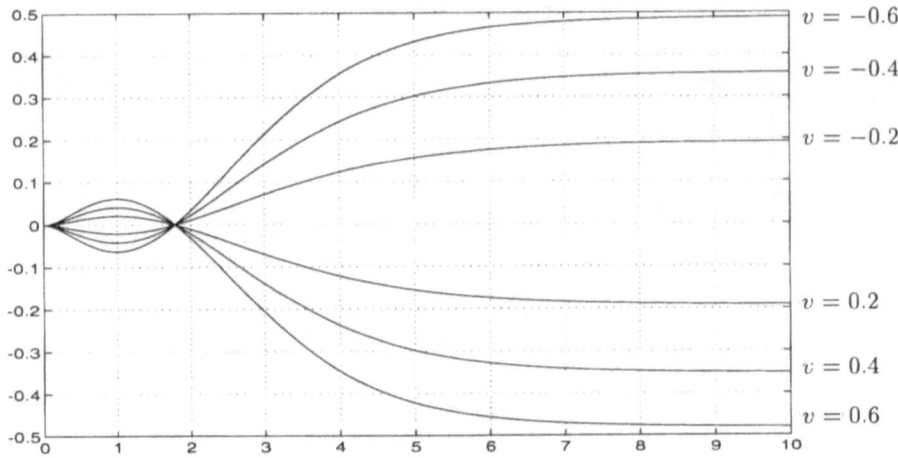

Figure 12: Step responses of system (4) with compensation computed using nonlinear H_∞-methods.

Theorem 4.1 *Assume that the nonlinear operator can be partitioned as follows:*

$$P[u, d] = P_L[u] + P_D[d] + P_{NL}(d)[u] \tag{21}$$

where P_L and P_D are linear operators and P_{NL} is a nonlinear operator which operates on u and is updated using the measured disturbance parameter, d. If the inverse of the linear operator P_L exists, the left inverse of P, denoted as $H[\xi, d]$, is equal to the following:

$$H[\xi, d] = (I + P_L^{-1} P_{NL}(d))^{-1} P_L^{-1}[\xi - P_D[d]] \tag{22}$$

which is shown in block diagram form in Figure 13.

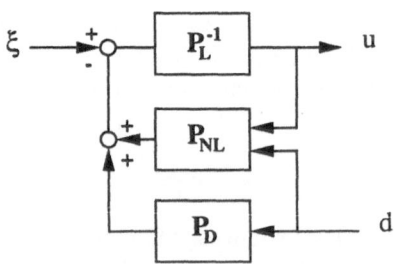

Figure 13: Block diagram realization of the partitioned nonlinear inverse.

Proof:

It is necessary to find an input which drives the output, y, to an arbitrary value, ξ, which is demonstrated mathematically below:

$$y = P_L[u] + P_D[d] + P_{NL}(d)[u] = \xi$$

Subtracting the linear contribution of the disturbance isolates the operators on the manipulated input:

$$P_L[u] + P_{NL}(d)[u] = \xi - P_D[d]$$

This transforms the system into an equation that resembles the traditional partitioned nonlinear system and the inverse is calculated as follows:

$$P_L(I + P_L^{-1} P_{NL}(d))[u] = \xi - P_D[d]$$
$$(I + P_L^{-1} P_{NL}(d))[u] = P_L^{-1}[\xi - P_D[d]]$$
$$u = (I + P_L^{-1} P_{NL}(d))^{-1} P_L^{-1}[\xi - P_D[d]] = H[\xi, d]$$

$\square$

Remark 2 *In performing the decomposition shown in Equation (21) no assumptions are made about the structure of the operators P_L, P_D, and P_{NL}. Therefore, this partition is a general result and not a severe limitation for the proposed framework.*

A linearizing controller which cancels the effects of measured disturbances can be obtained by adding a dynamical filter f to the error signal ξ, within the partitioned nonlinear inverse. A schematic of the linearizing controller with feedforward compensation is shown in Figure 14. Observe that the controller is composed of three sub-controllers; the linear IMC controller (feedback controller) $\bar{Q}$, the linear feedforward compensator $\bar{Q}^*$, and a nonlinear compensator Q^*. In (Doyle III *et al.*, 1995a) it has been observed that a similar decomposition holds for linearizing controllers using Volterra models. For stable minimum phase systems,

$$\bar{Q} = P_L^{-1} f$$
$$\bar{Q}^* = f^{-1} P_D$$
$$Q^* = f^{-1} P_{NL}$$

Remark 3 *The I/O-linearizing controller with feedforward compensation is consistent with previous results in the limiting cases.*

- *In the absence of measured disturbances, the proposed design method yields a nonlinear controller which is consistent with (Economou and Morari, 1986).*

- *In the absence of nonlinear dynamics, the proposed design method yields a linear IMC controller with feedforward compensation which is consistent with (Morari and Zafiriou, 1989).*

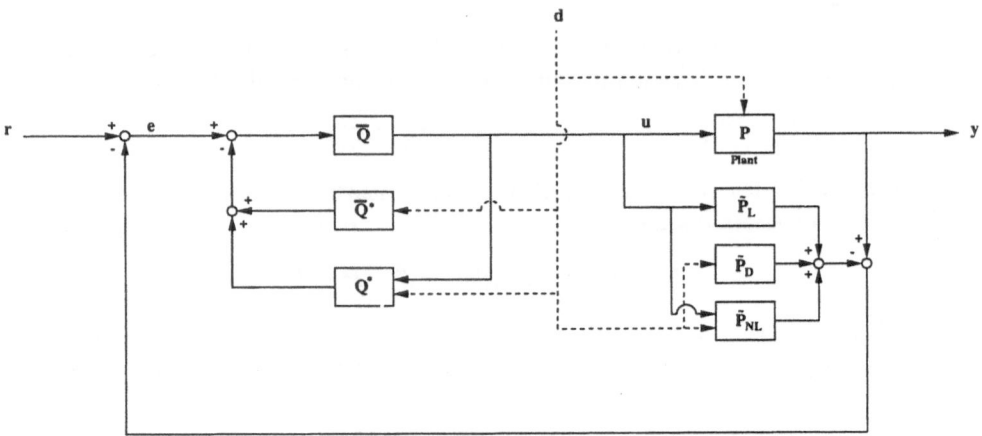

Figure 14: Block diagram of the approximate I/O-linearizing controller with feed-forward architecture.

4.1 Nonminimum Phase Systems

To demonstrate this approximate linearization approach, a class of nonminimum phase systems is considered. Nonminimum phase systems constitute a challenging class of systems for control design since the exact system inverse is unstable. For linear systems, the linear IMC design procedure employs an allpass factorization to partition the process model into a minimum phase part and a nonminimum phase part. It has been shown that a controller which utilizes the inverse of the minimum phase part is ISE-optimal for setpoint changes. It has been shown in (Doyle III *et al.*, 1995*a*) that a controller which employs the allpass factorization for calculation of the linear inverse within the partitioned nonlinear inverse yields superior closed-loop performance over the corresponding linear controller for nonminimum phase systems. Similar efforts to obtain a pseudo-inverse with input-output linearization (Doyle III *et al.*, 1995*b*; Kravaris and Daoutidis, 1990) have yielded results for specific classes (maximally nonminimum phase and second order systems, respectively); however, a general result has not yet been obtained. In this section, we present results which are applicable to a large class of nonlinear systems, including some systems for which input-output linearization is not applicable.

A given nonminimum phase system can be represented as a partitioned nonlinear plant using Equation (21). Furthermore, $P_L(s)$ can be factored into an allpass and a minimum phase contribution as outlined for example in (Morari and Zafiriou, 1989)

$$P_L = P_L^A P_L^M \tag{23}$$

where P_L^A denotes the allpass and P_L^M denotes the minimum phase portion. For stable linear systems, an H_2 optimal pseudo-inverse can be obtained by inverting the minimum phase portion of the plant.

One proceeds as before for the I/O-linearization control synthesis, except that in the design of the linear portion of the control structure, $\bar{Q}$, the minimum phase factor is employed.

All of the results in this section can be extended in a straightforward manner to handle anti-windup synthesis using generalizations of the linear theory and the partitioned control structure (Kendi and Doyle III, 1997).

4.1.1 CSTR Example

The example considered in this section is an isothermal CSTR with the following well studied sequence of reactions (where component B is the desired product):

$$A \longrightarrow B \longrightarrow C$$
$$2A \longrightarrow D$$

A model for the process can be obtained by performing mass balances on components A and B as follows:

$$\dot{C}_A = -k_1 C_A - k_3 C_A^2 + \frac{F}{V}(C_{Af} - C_A)$$
$$\dot{C}_B = k_1 C_A - k_2 C_B - \frac{F}{V} C_B$$
$$y = C_B$$

The control objective of this problem is to regulate the concentration of component B by manipulating the inlet flow rate. This example exhibits the well known phenomena that the gain of the system changes sign at the point of maximum conversion (this can be observed from a plot of the gain locus). In addition, the dynamics of the system also change from nonminimum phase to minimum phase at the point of maximum conversion.

The specific values of the parameters for the CSTR corresponding to an operating point in the nonminimum phase regime are shown in Table 2. The values of the states and input at the operating point are:

$$C_{A0} = 3.0$$
$$C_{B0} = 1.117$$
$$\frac{F_0}{V} = 0.034286$$

Substituting the values for the constants and converting the variables to deviation form, results in the following state-space process description:

$$\dot{x}_1 = -0.05x_1 - 0.1x_1^2 + u(0.01 - 0.001x_1)$$
$$\dot{x}_2 = 0.05x_1 - 0.1x_2 + u(-0.001x_2)$$
$$y = x_2$$
$$-0.03 \leq u \leq 0.03 \tag{24}$$

Table 2: Parameters for the isothermal CSTR.

k_1	0.05 min^{-1}
k_2	0.10 min^{-1}
k_3	0.01ℓ mol^{-1} min^{-1}
C_{Af}	10 mol ℓ^{-1}
V	1000ℓ

where x_1 is the deviation variable for the concentration of component A, x_2 is the deviation variable for the concentration of the component B, and u is the deviation variable for inlet flow.

A linear approximation of the nonlinear model is obtained from a first-order Taylor series expansion:

$$\dot{x}_1 = -0.05x_1 + 0.01u$$
$$\dot{x}_2 = 0.05x_1 - 0.1x_2$$
$$y_L = x_2 \tag{25}$$

The linear model, shown above, is a state-space version of the $\tilde{\boldsymbol{P}}_L$ operator with an equivalent Laplace domain representation shown below:

$$\tilde{\boldsymbol{P}}_L(s) = \frac{-s + 0.17}{0.9s^2 + 0.25s + 0.017} \tag{26}$$

A state-space representation of the nonlinear partition, $\boldsymbol{P}_{NL}$, is as follows:

$$\dot{x}_1 = -0.05x_1 - 0.1x_1^2 + u(0.01 - 0.001x_1)$$
$$\dot{x}_2 = 0.05x_1 - 0.1x_2 + u(-0.001x_2)$$
$$\dot{x}_3 = -0.05x_3 + 0.01u$$
$$\dot{x}_4 = 0.05x_3 - 0.1x_4$$
$$y_{NL} = x_2 - x_4 \tag{27}$$

Observe that the first two state equations are the fundamental modeling equations and the second two state equations are the linearized fundamental modeling equations. The nonlinear contribution is isolated by the difference operation which is performed on the output map, effectively $\tilde{\boldsymbol{P}}_{NL} = \tilde{\boldsymbol{P}} - \tilde{\boldsymbol{P}}_L$.

To obtain the unconstrained nonlinear controller, one applies the allpass factorization to $\tilde{\boldsymbol{P}}_L$:

$$\tilde{\boldsymbol{P}}_L(s) = P_L^A P_L^M$$
$$P_L^A = \frac{s - 0.17}{-s - 0.17}$$
$$P_L^M = \frac{-s - 0.17}{-(0.9s^2 + 0.25s + 0.017)}$$

The unconstrained linear IMC controller, $\bar{Q}$, is obtained by inverting the minimum phase portion of the plant and adding a first order filter to make to controller realizable.

$$\bar{Q} = \frac{0.9s^2 + 0.25s + 0.017}{(-s - 0.17)(\lambda s + 1)}$$

For the subsequent closed-loop simulations, the IMC filter constant is $\lambda = 12$ min. The nonlinear control design is completed by adding the nonlinear feedback loop, where $Q^* = f^{-1}P_{NL}$, around the optimal linear controller as depicted in Figure 14.

The simulation results shown in Figure 15 indicate the performance improvements that are possible from using an approximate linearization scheme for this nonminimum phase system. An additional aspect of the simulation, not detailed here, is that a linear anti-windup design has been employed with the linear IMC partition of each control structure (Kendi and Doyle III, 1997). Clearly, the approximate I/O-linearized structure is able to emerge from the constraint on the input at the "optimal time".

5 Approximate I/O-linearization of non-minimum phase nonlinear systems

Many technical processes of practical importance are characterized by unstable zero-dynamics. In analogy to the linear case such systems are often called "non-minimum phase" systems. Control of non-minimum phase systems is usually significantly more difficult than control of systems with stable zero-dynamics as the unstable zero-dynamics restricts the achievable closed-loop performance. This also manifests itself when trying to I/O-linearize nonlinear non-minimum phase systems: Standard I/O-linearization (Isidori, 1989) always leads to internal instability and is thus not suited for practical controller design. Nonlinear controller design by I/O-linearization does however have many favorable aspects and thus a number of attempts have been made to extend I/O-linearization to non-minimum phase systems. Most of these methods evade the problems associated with non-minimum phase behavior by introducing new outputs, with respect to which the system has minimum phase characteristics. In order to be able to assess the level of linearity achieved with respect to the true outputs, the new outputs have to be chosen such that a meaningful connection between the new and the true outputs exists. In (Kravaris and Wright, 1992) the outputs are related through the notion of *statical equivalence*. This method guarantees that the compensated system is at least linear with respect to the static I/O-behavior. However, this method does not guarantee any linearity in the dynamic transient. Moreover, this method can only be applied if the underlying system is feedback linearizable and thus a restricting involutivity condition has to hold.

The outputs can also be altered by adding another dynamical system in parallel to the original one if the new output is defined as the sum of both outputs. Using a process application it was shown in (Svaricek, 1995) that the zero-dynamics can

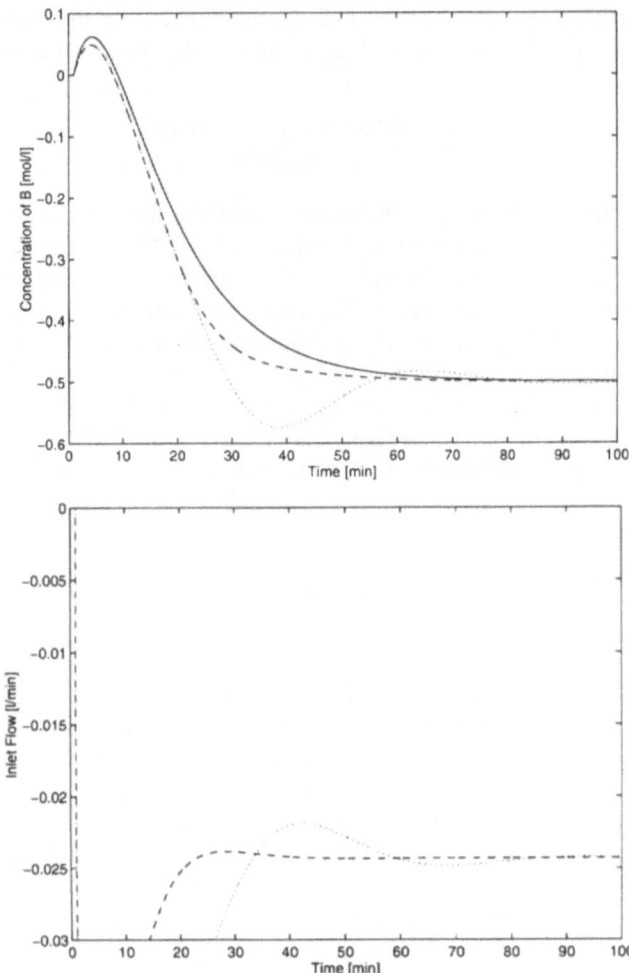

Figure 15: Closed-loop simulation of anti-windup controllers for a step setpoint change of -0.50 mol ℓ^{-1}. Solid line: unconstrained reference, dotted line: linear IMC-based anti-windup, dashed line: approximate linearizing-based anti-windup.

be "stabilized" when adding simple linear systems in parallel. The results obtained for the example process in (Svaricek, 1995) do appear very promising. However, general results that allow to relate the new output to the true one are not available at present.

The method described in (Doyle III *et al.*, 1995*b*) can also be viewed as belonging to the above class of methods where new outputs are introduced. There, the nonlinear system is approximately factored into a minimum phase factor (outer-factor) and an all-pass factor (inner factor). The I/O-behavior of the outer factor is linearized using a dynamic state feedback. The inner factor remains as a limiting element in the loop. The same design "philosophy" leads to optimal ISE performance when applied to linear systems. The approximate inner-outer factorization can however only be derived for maximum phase nonlinear systems, i.e. systems with antistable zero-dynamics and is thus not generally applicable.

In this section we describe a novel method for approximate I/O-linearization of non-minimum phase nonlinear systems (Allgöwer, 1997) that does not require the introduction of new outputs. First the system is transformed into a suitable (approximate) normal form, that allows the (approximate) compensation of the nonlinearities in the I/O-behavior. This normal form is a special case of the Byrnes-Isidori normal form where the zero dynamics is factored into a stable and into an unstable part where the latter one is approximately linearized and decoupled from the stable part. Under certain conditions this method leads to an exact I/O-linearization of nonlinear non-minimum phase systems. However in most cases only an approximate linearization will be achieved.

For the definition of the achieved "quality" of approximation we adopt a standard notion that is characterized by the order of the Taylor series expansion of the compensated system up to which no nonlinear terms exist (Krener, 1984; Hauser, 1991; Xia, 1995). Thus a system will for example be called 'approximately linear of order four' if its Taylor expansion doesn't contain any quadratic, cubic or fourth order terms.

We restrict our attention to nonlinear SISO systems

$$\dot{x} = f(x) + g(x) \cdot u$$
$$y = h(x) \tag{28}$$

with well-defined relative degree $r \leq n$, $x \in \mathbb{R}^n$ and $u, y \in \mathbb{R}$. Without loss of generality we assume $f(0) = 0$ and $h(0) = 0$. Under the assumptions made, system (28) can be transformed to Byrnes-Isidori normal form (B/I-normal form)

$$y = z_1 \tag{29a}$$
$$\dot{\xi} = J\xi + e_r \cdot a(\xi, \eta) + e_r \cdot b(\xi, \eta) \cdot u \tag{29b}$$
$$\dot{\eta} = q(\xi, \eta) \tag{29c}$$

using a nonlinear state transformation (Isidori, 1989)

$$x = T(z) \tag{30}$$

with $z = [\xi, \eta]^T, \xi = [z_1, \ldots, z_r]^T \in \mathbb{R}^r$, $\eta = [z_{r+1}, \ldots, z_n]^T \in \mathbb{R}^{n-r}$ and $e_r = [0, \ldots, 0, 1]^T \in \mathbb{R}^r$. J denotes a matrix of suitable dimension with all zero elements except for ones on the first upper secondary diagonal. We refer to normal form (29) as being *input-normalized* in that eq. (29c) does not explicitly depend on the input vector u. If system (29) is minimum phase then I/O- linearity with internal stability can be achieved using "Standard I/O-linearization" (Isidori, 1989) with

$$u = \frac{1}{b(z)}\Big(-a(z) + v\Big) \tag{31}$$

resulting in

$$y = z_1 \tag{32a}$$
$$\dot{\xi} = J\xi + e_r \cdot v \tag{32b}$$
$$\dot{\eta} = q(\xi, \eta) . \tag{32c}$$

I/O-linearity is thus achieved using two effects: Firstly, feedback (31) compensates the nonlinearities $a(z)$ and $b(z)$ in the direct I/O-channel and secondly, it decouples the nonlinear driven zero-dynamics from the I/O-behavior.

In this section we consider nonlinear systems with *unstable* zero-dynamics. For these systems feedback (31) leads to internal instability as the zero-dynamics (32c) that is unobservable in the compensated system, is unstable by definition. If the zero-dynamics (32c) is factored into its stable and antistable part and if the linearizing feedback is chosen to only decouple the stable part, then the resulting system can be internally stabilized using a proper extension to the decoupling feedback. This approach will also partially linearize the I/O-behavior: Nonlinearities $a(z)$ and $b(z)$ in the direct I/O-channel can still be compensated. Furthermore nonlinearities in the stable part of the zero-dynamics will not contribute to the I/O-behavior. What is left are nonlinearities in the antistable part of the zero-dynamics. Their effect on the I/O-behavior can however also be reduced using an appropriate feedback as we will show. This idea will be made precise in the following and a systematic method to achieve approximate linearity will be introduced.

5.1 An approximate I/O-linearization result

The core of the suggested method is an appropriate transformation of the driven zero-dynamics into a form in which the antistable part is not influenced by the stable part and in which the antistable part is given in (approximately) linear form. As we will see later, it is useful that the transformed system is still represented in input-normalized B/I-normal form, i.e. that the input u does not enter the differential equations that describe the driven zero-dynamics of the transformed system. Therefore transformation T has to be of the special form

$$\begin{bmatrix} \xi \\ \eta \end{bmatrix} = T(\xi', \eta_1, \eta_2) = \begin{bmatrix} \xi' \\ T_\eta(\bar{\xi}', \eta_1, \eta_2) \end{bmatrix} \tag{33}$$

with $\bar{\xi}' = [z_1', \ldots, z_{r-1}']^T \in \mathbb{R}^{r-1}$. Coordinates $\eta_1 \in \mathbb{R}^i$ are used to describe the unstable part of the zero-dynamics in the transformed system, where i denotes the dimension of the unstable invariant manifold of the zero-dynamics. The stable part is described by $\eta_2 \in \mathbb{R}^{n-r-i}$. Transformation (33) leaves coordinates ξ unchanged.

Definition 1 *The driven zero-dynamics of system (28) with input-normalized B/I-normal form (29) is called* approximately antistable decouplable of order k, *if there exists a transformation (33), that transform system (29) into*

$$
\begin{aligned}
y &= z_1' \\
\dot{\xi}' &= J\xi' + e_r \cdot \bar{a}(\xi', \eta_1, \eta_2) + e_r \cdot \bar{b}(\xi', \eta_1, \eta_2) \cdot u \\
\dot{\eta}_1 &= \gamma_1(\xi', \eta_1) + \mathcal{O}^{[k+]}(\xi', \eta_1, \eta_2) \\
\dot{\eta}_2 &= \gamma_2(\xi', \eta_1, \eta_2)
\end{aligned}
\tag{34}
$$

with

$$
\sigma \left(\left. \frac{\partial \gamma_1(\xi', \eta_1)}{\partial \eta_1} \right|_{\substack{\xi'=0 \\ \eta_1=0}} \right) > 0
$$

$$
\sigma \left(\left. \frac{\partial \gamma_2(\xi', \eta_1, \eta_2)}{\partial \eta_2} \right|_{\substack{\xi'=0 \\ \eta_1=0 \\ \eta_2=0}} \right) < 0 .
$$

$\sigma(\cdot)$ is used to denote the spectrum of a matrix. Superscripts in square brackets describe the (homogeneous) order of the respective term. If a plus-sign ("+") is added to the superscript, then the expression contains terms of order strictly larger than the value of the superscript. The expression $\mathcal{O}^{[k+]}(\xi', \eta_1, \eta_2)$ represents arbitrary terms in ξ', η_1 and η_2 with order strictly larger than k.

Note that the transformed system is still given in input-normalized B/I-normal form and that the η_2-coordinate does only enter the $\dot{\eta}_1$ equation in terms that are of order larger than k. In the following we refer to a system given in form (34) as being *approximately antistable decoupled* or describe the system as being given in *approximately antistable decoupled input-normalized B/I-normal form*.

Definition 2 *The driven zero-dynamics of system (28) with B/I-normal form (29) is called* approximately linearizable of order k, *if there exists a transformation (33), such that system (29) is transformed into*

$$
\begin{aligned}
y &= z_1' \\
\dot{\xi}' &= J\xi' + e_r \cdot \bar{a}(\xi', \eta') + e_r \cdot \bar{b}(\xi', \eta') \cdot u \\
\dot{\eta}' &= M\xi' + N\eta' + \mathcal{O}^{[k+]}(\xi', \eta')
\end{aligned}
\tag{35}
$$

with $\eta' = [\eta_1, \eta_2]^T$.

Note that the *driven* zero-dynamics has to be linear, which is a much stronger requirement than the linearity of the undriven zero-dynamics.

Analogous definitions can be made for the linearizability of the antistable or stable part of the driven zero-dynamics.

These definitions allow us to state the following theorem that gives sufficient conditions for exact linearizability of non-minimum phase nonlinear systems, together with the structure of the linearizing feedback. The respective results for the approximate I/O-linearization of order k is added in parenthesis.

Theorem 5.1 *Assume that the Jacobi linearization of the zero-dynamics of system (28) with input-normalized B/I-normal form (29) does not have any eigenvalues with zero real part and i eigenvalues $(i \leq n - r)$ with positive real part. The I/O-behavior of system (28) can be linearized (approximately up to order k) using a static state feedback, if*

(i) system (28) has a well-defined relative degree r,

(ii) the Jacobi linearization of (28) is controllable,

(iii) there exists a transformation

$$z = T^{-1}(x) = \left[t_1^{-1}(x), \ldots, t_n^{-1}(x)\right]^T \tag{36}$$

such that system (28) can be transformed into a B/I-normal form with (approximately) antistable decoupled zero-dynamics (of order k), such that simultaneously the antistable part of the zero-dynamics is (approximately) linear (up to order k).

The (approximately) linearizing state feedback is given by

$$u = \frac{1}{L_g L_f^{r-1} h(x)} \left\{ -L_f^r h(x) + \sum_{l=1}^{r} c_{1l} L_f^{l-1} h(x) + \sum_{m=1}^{i} c_{2m} t_{r+m}^{-1}(x) + v \right\}, \tag{37}$$

where

$$\boldsymbol{\eta}_1 = \left[t_{r+1}^{-1}(x), \ldots, t_{r+i}^{-1}(x)\right]^T$$

denotes the coordinate of the antistable part of the zero-dynamics

$$\dot{\boldsymbol{\eta}}_1 = M_1 \boldsymbol{\xi}' + N_1 \boldsymbol{\eta}_1 \quad \left(+ \mathcal{O}^{[k+]}(\boldsymbol{\xi}', \boldsymbol{\eta}_1, \boldsymbol{\eta}_2)\right).$$

The I/O-behavior of the compensated system is then described by the linear system (A, B, C) of order $r + i$ and

$$A = \begin{bmatrix} J + e_r \cdot c_1^T & e_r \cdot c_2^T \\ M_1 & N_1 \end{bmatrix}, \quad b = \begin{bmatrix} 0_{r-1} \\ 1 \\ 0_i \end{bmatrix}, \quad c^T = \begin{bmatrix} 1 \\ 0_{r+i-1} \end{bmatrix}$$

and [2]

$$c_1 = \left[c_{11}, \ldots, c_{1r}\right]^T, \quad c_2 = \left[c_{21}, \ldots, c_{2i}\right]^T.$$

System (28) with feedback (37) is internally stable if and only if

$$\sigma(A) < 0. \tag{38}$$

Furthermore vectors c_1 and c_2 can always be chosen such that (38) holds.

Proof: From condition (iii) we know that there exists a transformation

$$x = T(z) \tag{39}$$

with inverse transformation (36) such that (28) is transformed into

$$y = z_1'$$
$$\dot{\xi}' = J\xi' + e_r \cdot \bar{a}(\xi', \eta_1, \eta_2) + e_r \cdot \bar{b}(\xi', \eta_1, \eta_2) \cdot u$$
$$\dot{\eta}_1 = M_1\xi' + N_1\eta_1 \quad \left(+ \mathcal{O}^{[k+]}(\xi', \eta_1, \eta_2)\right)$$
$$\dot{\eta}_2 = \gamma_2(\xi', \eta_1, \eta_2)$$

with

$$\sigma(N_1) > 0$$

and

$$\sigma\left(\left.\frac{\partial \gamma_2(\xi', \eta_1, \eta_2)}{\partial \eta_2}\right|_{\substack{\xi'=0 \\ \eta_1=0 \\ \eta_2=0}}\right) < 0. \tag{40}$$

Applying feedback

$$u = \frac{1}{\bar{b}(\xi', \eta_1, \eta_2)}\left(-\bar{a}(\xi', \eta_1, \eta_2) + c_1^T\xi' + c_2^T\eta_1 + v\right) \tag{41}$$

results in

$$y = z_1' \tag{42a}$$
$$\dot{\xi}' = J\xi' + e_r \cdot c_1^T\xi' + e_r \cdot c_2^T\eta_1 + v \tag{42b}$$
$$\dot{\eta}_1 = M_1\xi' + N_1\eta_1 \quad \left(+ \mathcal{O}^{[k+]}(\xi', \eta_1, \eta_2)\right) \tag{42c}$$
$$\dot{\eta}_2 = \gamma_2(\xi', \eta_1, \eta_2). \tag{42d}$$

Condition (i) implies $\bar{b}(0,0,0) \neq 0$ and feedback (41) is thus locally nonsingular. As η_2 only enters eqs. (42a)-(42c) through terms of order larger than k, the I/O-behavior is decoupled (up to order k) from eq. (42d), and hence (42) shows

---

[2]Subscripts $r-1, i$ and $r+i-1$ denote the respective dimension of the zero vector $\mathbf{0}$.

(approximately) linear I/O-behavior (up to order k). This proofs the sufficiency of conditions (i)-(iii) for achieving (approximate) I/O-linearity (up to order k).

Eq. (36) can also be written as

$$\xi = t_\xi^{-1}(x) = \left[t_1^{-1}(x),\ldots,t_r^{-1}(x)\right]^T$$
$$\eta_1 = t_{\eta_1}^{-1}(x) = \left[t_{r+1}^{-1}(x),\ldots,t_{r+i}^{-1}(x)\right]^T$$
$$\eta_2 = t_{\eta_2}^{-1}(x) = \left[t_{r+i+1}^{-1}(x),\ldots,t_n^{-1}(x)\right]^T .$$

Transformation $t_1^{-1}(x)$ to $t_r^{-1}(x)$ represent the usual transformations into B/I-normal form (Isidori, 1989)

$$t_1^{-1}(x) = h(x)$$
$$t_r^{-1}(x) \overset{\cdots}{=} L_f^{r-1}h(x) .$$

Functions $\bar{a}$ and $\bar{b}$ in (41) can therefore be written in original coordinates as

$$a(x) = \bar{a}(t_\xi^{-1}, t_{\eta_1}^{-1}, t_{\eta_1}^{-1}) = L_f^r h(x)$$
$$b(x) = \bar{b}(t_\xi^{-1}, t_{\eta_1}^{-1}, t_{\eta_1}^{-1}) = L_g L_f^{r-1} h(x) .$$

Hence feedback (41), given in original coordinates, reads

$$u = \frac{1}{L_g L_f^{r-1} h(x)} \left\{ -L_f^r h(x) + c_1 t_\xi^{-1} + c_2 t_{\eta_1}^{-1} + v \right\} ,$$

which proofs (37).

It remains to proof internal stability of the transformed and compensated system. For this we consider eq. (42). Stability of eqs. (42a)-(42c), that are independent on η_2, is determined by the dynamic matrix

$$A = \begin{bmatrix} J + e_r \cdot c_1^T & e_r \cdot c_2^T \\ M_1 & N_1 \end{bmatrix} .$$

A necessary and sufficient condition for exponential stability is that the eigenvalues of A do only have negative real parts. This is equivalent to condition (38). Condition (40) guarantees local exponential stability of the uncbservable part (42d). Transformation (39) and nonsingular feedback (41) do not change the controllability of the system. Therefore condition (ii) directly implies controllability of the Jacobian linearization of (42). If the Jacobian linearization of (42) is controllable, then there exists vectors c_1 and c_2, such that the eigenvalues of A can be assigned arbitrary values. This completes the proof. □

It is immediately clear that conditions (i)-(iii) are only sufficient and not necessary for (approximate) I/O-linearizability with internal stability. For minimum phase systems the number of unstable eigenvalues of the Jacobi linearized zero-dynamics is zero, condition (iii) is trivially satisfied and feedback (37) reduces to the standard I/O-linearizing feedback. Hence, Theorem 5.1 contains the standard I/O-linearization result for minimum phase nonlinear systems as a special case. Condition (iii) is certainly less restrictive than the requirement for minimum phaseness. However, in most practical cases no *exactly* antistable decoupling and simultaneously antistable linearizing transformation will exist for the driven zero-dynamics. Hence, an *exact* linearization of non-minimum phase systems will be possible only rarely. The main strength of this result is that it allows to treat the *approximate* I/O-linearization case for which condition (iii) has to be satisfied only up to order k. It is immediately clear that the higher the order k is, the more restricting condition (iii) will be.

In order to compute the (approximately) linearizing feedback law (37) two successive transformations have to be determined: First, system (28) has to be brought to input-normalized B/I-normal form. Then, the zero-dynamics have to be transformed into (approximately) antistable decoupled and antistable linear form. Different from the standard (minimum phase) case both transformations have to be determined explicitly, as both transformation relations do explicitly enter feedback law (37). For SISO systems a transformation to input-normalized B/I-normal form always exists, even though it might be difficult to derive. This is however not the case for MIMO systems any more, where a restricting involutivity condition has to be satisfied. An extension of the approach presented to the MIMO case will thus require an additional involutivity condition to hold.

Remark 4 *Theorem 5.1 only guarantees that the Jacobi linearization of the compensated system is asymptotically stable. Thus the (approximately) I/O-linearized system is* locally *exponentially stable. However, a much stronger statement with respect to closed loop stability can be given. If we assume that feedback (37) exactly linearizes the I/O-behavior of system (28), then the compensated system consists of a series connection of a (globally) asymptotically stable linear part (42a)-(42c) and a nonlinear part (42d). In order to examine the stability conditions for the nonlinear part (42d) we first consider the undriven case with $\xi' = 0$ and $\eta_1 = 0$. Due to the stable/antistable factorization, the dynamics of this system are restricted to the locally stable invariant manifold $W_{lok}^s(0)$ of the undriven zero-dynamics with*

$$W_{lok}^s(0) = \left\{ [\xi', \eta_1, \eta_2]^T \in \mathcal{V} : \xi' = 0, \eta_1 = 0 \right\} . \tag{43}$$

$\mathcal{V} \in \mathbb{R}^n$ *is the neighborhood of the origin where $W_{lok}^s(0)$ can be defined. The stable part of the undriven zero-dynamics that remains unobservable in the loop after compensation, is thus not only stable in the first approximation, but has, in a certain sense, a maximal region of attraction. The true region of attraction of the driven nonlinear part (42d) may however be small even though $W_{lok}^s(0)$ may be defined globally. This restriction is due to the so-called* peaking phenomenon *(Sussmann and Kokotovic, 1991). Of course this assessment also holds for the well-known standard I/O-linearization theory for minimum phase systems. If only an approximate I/O-linearization of order k is achieved, then the local stable invariant manifold is correctly represented up to order k. In general, feedback (37) thus achieves a larger* *region of attraction than a feedback that only places the poles of the Jacobian linearization. Therefore the I/O-linearization method described can also be seen as a method for nonlinear system stabilization.*

5.2 Transformation of the driven zero-dynamics

The method for exact or approximate I/O-linearization of non-minimum phase nonlinear systems proposed in Section 5 crucially depends on the existence and computability of suitable coordinates for the driven zero-dynamics. As discussed, the requirement for (approximate) I/O-linearizability (of order k) is that the driven zero-dynamics can be transformed into a form, where the antistable part of the driven zero-dynamics (when expanded in a Taylor series up to order k) is firstly linear and secondly not influenced by the stable part of the driven zero-dynamics. A systematic and general method for the transformation of the driven zero-dynamics into various forms can be found in (Allgöwer, 1996b). The method described in (Allgöwer, 1996b) is motivated by corresponding results from normal form theory of dynamical systems and contains the transformation into (approximately) antistable decoupled and (approximately) antistable linear form (of order k) as a special case. In the algorithm described in (Allgöwer, 1996b), terms of higher order are successively transformed into the desired form starting with the linear terms, quadratic terms etc. At each step the unknown parameters of the transformation are determined by solving a system of *linear* equations making use of the results from the previous step. Even though the method is computationally very attractive, the number of transformation parameters that needs to be computed at each step grows combinatorially with both the dynamic order n and the approximation order k. In (Löffler *et al.*, 1993) an implementation of this algorithm for the computer-algebra program MATHEMATICA is described, that allows to compute the transformation for systems of relatively high order.

It is obvious that not all systems can be transformed into an input-normalized B/I-normal form with arbitrary structure of the driven zero-dynamics. Some necessary and sufficient conditions for the existence of transformations into the desired special input-normalize B/I-normal form are also given in (Allgöwer, 1996b).

5.3 Summary of the procedure

The previous sections describe a systematic method to exactly or approximately linearize the I/O-behavior of non-minimum phase nonlinear systems. Nonlinearities that influence the I/O-behavior are exactly or approximately compensated using a static state feedback law while maintaining internal stability. The procedure can be summarized as follows:

Step 1: Transformation of the considered system into an arbitrary input-normalized B/I-normal form.

Step 2: (linearization) Compensation of nonlinearities in the direct I/O-channel and decoupling of the driven zero-dynamics by static state feedback. This feedback is equivalent to the standard I/O-linearizing feedback for minimum phase systems.

Step 3: (transformation) Transformation of the driven zero-dynamics into antistable decoupled and antistable linear form.

In many cases a transformation into exactly antistable decoupled and exactly antistable linear form will not be possible or will be too involved. Using the algorithm described in (Allgöwer, 1996b) a successive transformation to approximately antistable decoupled and approximately antistable linear form can be achieved:

Step 3a: Transformation of linear terms
Step 3b: Transformation of second order terms
Step 3c: Transformation of third order terms
• • •

Successively, terms of higher order will be transformed until either the desired approximation order is achieved or until a transformation of terms of higher order will not exist any more.

Step 4: (stabilization) Linear controller synthesis for the transformed system such that the (approximately) linear part is stabilized while the stable zero-dynamics part remains decoupled.

5.4 Example

In order to demonstrate the described procedure we will linearize the I/O- behavior of a simple example system.

We consider the following system in input-normalized B/I-normal form

$$
\begin{aligned}
y &= x_1 \\
\dot{x}_1 &= -x_1^3 + x_2 x_3 + u \\
\dot{x}_2 &= x_2 + \frac{x_2 x_3 + x_1}{1 + x_3} - x_1 x_2^2 \qquad (x_3 \neq -1) \\
\dot{x}_3 &= -x_3 + x_1 x_2 (1 + x_3) ,
\end{aligned}
\tag{44}
$$

for which we want to derive an I/O-linearizing feedback. The system is of order three, has relative degree one, and it is easy to see that the two dimensional zero-dynamics consists of a stable and an unstable part of dimension one each.

Step 1: As system (44) is already given in input-normalized B/I-normal form this step is not needed.

Step 2 (linearization): Feedback

$$
u = x_1^3 - x_2 x_3 + \tilde{v}
$$

compensates the nonlinearities in the direct I/O-channel and decouples the zero-dynamics from the I/O-behavior completely. This feedback is equivalent to the linearizing feedback that would be obtained when applying standard I/O-linearization and leads to

$$
\begin{aligned}
y &= x_1 \\
\dot{x}_1 &= \tilde{v} \\
\dot{x}_2 &= x_2 + \frac{x_2 x_3 + x_1}{1 + x_3} - x_1 x_2^2 \\
\dot{x}_3 &= -x_3 + x_1 x_2 (1 + x_3) \ .
\end{aligned}
\tag{45}
$$

Step 3 (transformation): For this simple third order system a direct transformation $x = T(z)$ of the driven zero-dynamics into exactly antistable decoupled and exactly antistable linear form can be given:

$$
T : \quad
\begin{array}{ccc}
x_1 & = & z_1 \\
x_2 & = & \frac{z_2}{1 + z_3} \\
x_3 & = & z_3 \ .
\end{array}
\tag{46}
$$

A straightforward computation shows that the transformed system is described by

$$
\begin{aligned}
y &= z_1 & \text{(47a)} \\
\dot{z}_1 &= \tilde{v} & \text{(47b)} \\
\dot{z}_2 &= z_2 + z_1 & \text{(47c)} \\
\dot{z}_3 &= -z_3 + z_1 z_2 \ . & \text{(47d)}
\end{aligned}
$$

The third equation represents the antistable part of the driven zero-dynamics, which is indeed linear and independent of the coordinate z_3 that describes the stable part of the zero-dynamics. The stable part of the zero-dynamics is nonlinear and depends on all states. System (47) has linear I/O-behavior but is internally unstable due to the unobservable unstable zero-dynamics. Therefore a stabilization is required in the following step using a linear controller.

Step 4 (stabilization): The linear controller is restricted to only feedback the coordinates z_1 and z_2. As detailed in Sect. 5.1, coordinate z_3, that is associated with the nonlinear stable zero-dynamics part, must not enter the feedback. Linear feedback

$$
\tilde{v} = -3z_1 - 4z_2 + v
$$

places the two poles of subsystem (47b)-(47c) to -1. Thus we obtain the following compensated system that is internally stable and has exactly linear I/O- behavior:

$$
\begin{aligned}
y &= z_1 \\
\dot{z}_1 &= -3z_1 - 4z_2 + v \\
\dot{z}_2 &= z_2 + z_1 \\
\dot{z}_3 &= -z_3 + z_1 z_2 \ .
\end{aligned}
\tag{48}
$$

Combining steps 1 to 4 and using the inverse of transformation (46)

$$T^{-1} : \begin{array}{rcl} z_1 & = & x_1 \\ z_2 & = & x_2(1 + x_3) \\ z_3 & = & x_3 \end{array}$$

we obtain the linearizing feedback in original coordinates:

$$u = x_1^3 - x_2 x_3 - 3x_1 - 4\Big(x_2(1 + x_3)\Big) + v \ . \tag{49}$$

This example is very simple. In (Allgöwer, 1996b) we show an application of this method to the I/O-linearization of a realistic, unstable and non-minimum phase CSTR with complex reaction scheme. As it turns out, the I/O-behavior of this reactor can even be linearized exactly, despite its unstable zero-dynamics.

5.5 Discussion

An exact I/O-linearization with the approach described is only possible if the zero-dynamics can be exactly transformed into this form, and only few practically relevant non-minimum phase systems can thus be exactly I/O-linearized. This requirement is however less restrictive than the assumption needed for standard I/O-linearization, namely the requirement that the zero-dynamics is stable. The main advantage of the presented scheme must be seen in its ability to systematically allow an *approximate* I/O-linearization, which will be possible for a large class of non-minimum phase systems.

The crucial step in the derivation of the linearizing feedback law is the transformation of the driven zero-dynamics. Here we draw from recent results presented in (Allgöwer, 1996b) that present a kind of structure algorithm for the transformation of the zero-dynamics.

Summarizing, the proposed method for approximate I/O-linearization of non-minimum phase nonlinear systems allows to significantly enlargen the class of nonlinear non-minimum phase systems that can be I/O-linearized when compared to other known approaches. The main drawback must be seen in the higher computational effort necessary.

6 Conclusions

Four different approaches to the approximate I/O-linearization problem were presented in this paper. The common feature of these methods is that the degree of linearity achieved after approximate I/O-linearization is address explicitly. The first method aims at a direct minimization of the degree of nonlinearity and differs therefore from most other schemes. The key to this approach is the definition of a suitable nonlinearity measure. In the second approach approximate I/O-linearization is considered via a corresponding disturbance decoupling or disturbance attenuation problem. In the third approach, a suitable partition of the nonlinear I/O-operator allows one to find an approximate inverse which is stable. In the fourth approach standard I/O-linearization is extended to relax the requirement for stable zero-dynamics. All of these methods have different advantages and drawbacks. The first approach via the minimization of nonlinearity measures is probably the most general method possible in that arbitrary static or dynamic compensation structures can be considered. An analytical derivation of these structures is, however, very difficult, and even the presented technique that assumes a fixed structure and merely minimizes over a finite number of parameters, leads to a non-convex, expensive to solve optimization problem. New insights into the problem structure can be gained with the second approach, but computation of the compensators is still a very involved problem. The third approach exploits the close link between system inversion and I/O-linearization and proposes a practical solution based on this. As a drawback the degree of linearity of the resulting compensated system is only influenced indirectly. The fourth approach is directly related to standard I/O-linearization which is a special case of the presented method. The price for relaxing the minimum phase requirement is again, as with other methods, an increased computational complexity.

Summarizing, changing the goal from an exact I/O-linearization to an approximate linearization allows firstly to enlarge the class of systems that can be considered, and secondly brings advantages with respect to application relevant properties of the closed loop. Our experience shows that a sufficient degree of linearity can be achieved in most cases even if additional requirements like robustness or input constraints are considered for the approximate linearization step.

References

Allgöwer, F. (1995a). Definition and computation of a nonlinearity measure and application to approximate I/O-linearization. Internal Report 95-1. Institut für Systemdynamik und Regelungstechnik, Universität Stuttgart.

Allgöwer, F. (1995b). Ein Verfahren zur robusten E/A-Linearisierung durch Vorkompensation. Internal Report 95-13. Institut für Systemdynamik und Regelungstechnik, Universität Stuttgart.

Allgöwer, F. (1995c). Nichtlinearitätsmaße — Ein Werkzeug zur Analyse und Synthese nichtlinearer Regelkreise. In: *Entwurf Nichtlinearer Regelungen* (S. Engell, Ed.). pp. 309–331. Oldenbourg Verlag. München.

Allgöwer, F. (1996a). Definition and computation of a nonlinearity measure. In: *Nonlinear Control System Design 1995* (A.J. Krener and Mayne D.Q., Eds.). Vol. 1. pp. 257–62. Pergamon. Oxford, UK.

Allgöwer, F. (1996b). *Näherungsweise Ein-/Ausgangs-Linearisierung nichtlinearer Systeme*. Fortschritt-Berichte, Reihe 8, Bd. 582. VDI Verlag. Düsseldorf.

Allgöwer, F. (1997). Approximate input-output linearization of nonminimum phase nonlinear systems. In: *Proc. 4rd European Control Conference ECC'97*. Brussels, Belgium.

Allgöwer, F., A. Rehm and E. D. Gilles (1994). An engineering perspective on nonlinear H_∞ control. In: *Proc. 33rd IEEE Conf. Decision Contr.*. Orlando, FL. pp. 2537–2542.

Allgöwer, F. and E. D. Gilles (1992). Approximate input/output-linearization of nonlinear systems. AIChE Annual Meeting, Miami, FL, paper no. 126f.

Allgöwer, F. and F.J. Doyle III (1997). Nonlinear process control: Which way to the promised land?. In: *Chemical Process Control – Assessment and New Directions for Research* (J.C. Kantor, C.E. Garcia and B. Carnahan, Eds.). pp. 24–45. AIChE Symposium Series, Vol. 93, No. 316.

Castillo, B. (1991). Output tracking through singular points for a class of nonlinear systems. In: *Proc. 1st European Control Conference ECC'91*. Grenoble. pp. 1496–1501.

Claude, D., M. Fliess and A. Isidori (1983). Immersion, directe et par bouclage d'un système non linéaire dans un linéaire. *C. R. Hebd. Séanc. Acad. Sci. Paris* **296**, 237–240.

Desoer, C.A. and G. Kabuli (1987). Nonlinear plants, factorizations and stable feedback systems. In: *Proc. 26th IEEE Conf. Decision Contr.*. Los Angeles, CA. pp. 155–156.

Doyle III, F. J., B. A. Ogunnaike and R. K. Pearson (1995a). Nonlinear model-based control using second-order Volterra models. *Automatica* **31**(5), 697–714.

Doyle III, F. J., F. Allgöwer and M. Morari (1995b). A normal form approach to approximate input-output linearization for maximum phase nonlinear systems. *IEEE Trans. Automat. Contr.* **41**(2), 305–309.

Economou, C. G. and M. Morari (1986). Internal model control. 5. Extension to nonlinear systems. *Ind. Eng. Chem. Process Des. Dev.* **25**, 403–411.

Freund, E. (1975). The structure of decoupled nonlinear systems. *Int. J. Contr.* **21**(3), 443–450.

Grizzle, J.W. and M.D. Di Benedetto (1992). Approximation by regular input-output maps. *IEEE Trans. Automat. Contr.* **AC-37**, 1052–1055.

Guay, M., P.J. McLellan and D.W. Bacon (1995). Measurement of nonlinearity in chemical process control systems: The steady state map. *Can. J. Chem. Eng.* **73**, 868–882.

Harris, K. R., M. C. Colantonio and A. Palazoglu (1997). Nonlinearity measure via functional expansion. Preprint.

270

Hauser, J. (1991). Nonlinear control via uniform system approximation. *Syst. Contr. Lett.* **17**(2), 145–154.

Hauser, J., S. Sastry and P. Kokotović (1989). Nonlinear control via approximate input-output linearization: The ball and beam example. In: *Proc. 28th IEEE Conf. Decision Contr.*. Tampa, FL. pp. 1987–1993.

Isidori, A. (1989). *Nonlinear Control Systems: An Introduction*. 2. ed.. Springer-Verlag. Berlin.

Isidori, A. (1996). Global almost disturbance decoupling with stability for non minimum–phase single–input single–output nonlinear systems. *Syst. Contr. Lett.* **28**(2), 115–22.

Isidori, A., A.J. Krener, C. Gori Giorgi and S. Monaco (1981). Nonlinear decoupling via feedback: A differential geometric approach. *IEEE Trans. Automat. Contr.* **AC-26**, 331–345.

Isidori, A. and A. Astolfi (1992). Disturbance attenuation and H_∞-control via measurement feedback in nonlinear systems. *IEEE Trans. Automat. Contr.* **AC-37**(9), 1283–1293.

Isidori, A. and A. Ruberti (1984). On the synthesis of linear input-output responses for nonlinear systems. *Syst. Contr. Lett.* **4**(1), 17–22.

Kang, W. (1994). Approximate linearization of nonlinear control systems. *Syst. Contr. Lett.* **23**, 43–52.

Kappos, E. (1989). A geometrical linearization theory. In: *Proc. 28th IEEE Conf. Decision Contr.*. Tampa, FL. pp. 77–81.

Kendi, T.A. and F.J. Doyle III (1997). Nonlinear internal model control for systems with measured disturbances and input constraints. submitted.

Kravaris, C. and C.-B. Chung (1987). Nonlinear state feedback synthesis by global input/output linearization. *AIChE J.* **33**(4), 592–603.

Kravaris, C. and P. Daoutidis (1990). Nonlinear state feedback control of second-order nonminimum-phase nonlinear systems. *Comp. & Chem. Eng.* **14**, 439–449.

Kravaris, C. and R.A. Wright (1992). Nonminimum-phase compensation for nonlinear processes. *AIChE J.* **38**, 26–40.

Krener, A.J. (1984). Approximate linearization by state feedback and coordinate change. *Syst. Contr. Lett.* **5**, 181–185.

Lee, H.G. and S.I. Marcus (1987). On input-output linearization of discrete-time nonlinear systems. *Syst. Contr. Lett.* **8**(3), 249–259.

Löffler, J., F. Allgöwer and R. Rothfuß (1993). Die Mathematica Near Identity Transformation Toolbox. Internal report. Institut für Systemdynamik und Regelungstechnik, Universität Stuttgart.

Lukes, D. L. (1969). Optimal regulation of nonlinear dynamical systems. *SIAM J. Contr.* **7**(1), 75–100.

Maas, W.C.A and A.J. van der Schaft (1996). Singular nonlinear H_∞ optimal control problem. *Int. J. of Robust and Nonlinear Control* **6**(7), 669–89.

Marino, R., W. Respondek, A. J. van der Schaft and P. Tomei (1994). Nonlinear H_∞ almost disturbance decoupling. *Syst. Contr. Lett.* **23**, 159–168.

Monaco, S. and D. Normand-Cyrot (1983). Formal power series and and input-output linearization of nonlinear discrete-time systems. In: *Proc. 22nd IEEE Conf. Decision Contr.*. San Antonio, TX. pp. 655–660.

Morari, M. and E. Zafiriou (1989). *Robust Process Control.* Prentice-Hall. Englewood Cliffs, NJ.

Nijmeijer, H. and A. J. van der Schaft (1990). *Nonlinear Dynamical Control Systems.* Springer Verlag. New York.

Nikolaou, M. (1994). How nonlinear is a "nonlinear" system? Old and new results under a unifying theory. preprint.

Paice, A.D.B. and A.J. van der Schaft (1996). The class of stabilizing nonlinear plant controller pairs. *IEEE Trans. Automat. Contr.* **AC-41**(5), 634–645.

Paolini, E., J.A. Romagnoli, A.C. Desages and A. Palazoglu (1991). Approximate models for control of nonlinear systems. Technical Report UCD-EECS-SCR-91/1. University of California, Davis, CA.

Porter, W.M. (1970). Diagonalization and inverses for nonlinear systems. *Int. J. Contr.* **10**, 252–264.

Powell, M.J.D. (1982). A fast algorithm for nonlinearly constrained optimization calculations. In: *Biennial Conference on Numerical Analysis, Dundee* (G.A. Watson, Ed.). pp. 144–157. Springer-Verlag. Berlin.

Santosuosso, G.L. (1995). Semiglobal singular H_∞ control problem for nonlinear planar systems. In: *Proc. 3rd European Control Conference ECC'95*. Vol. 2. Rome,. Italy. pp. 1497–502.

Singh, S.N. and W.J. Rugh (1972). Decoupling in a class of nonlinear systems by state variable feedback. *ASME J. Dyn. Syst. Meas. Contr.* **94**, 323–329.

Stack, A.J. and F.J. Doyle III (1997). Application of a Control-Law Nonlinearity Measure to the Chemical Reactor Analysis. *AIChE J.* **43**, 425-439.

Stack, A.J. and F.J. Doyle III (1997). The optimal control structure: an approach to measuring control-law nonlinearity. *Comp. Chem. Eng.* **21**, 1009-1019.

Sussmann, H. J. and P. V. Kokotovic (1991). The peaking phenomenon and the global stabilization of nonlinear systems. *IEEE Trans. Automat. Contr.* **AC-36**(4), 424–440.

Svaricek, F. (1995). Nonlinear dynamic control of a non-minimum-phase CSTR. In: *3rd IFAC Nonlinear Control Systems Design Symposium*. Lake Tahoe, CA. pp. 214–219.

van der Schaft, A. J. (1992). L_2-gain analysis of nonlinear systems and nonlinear state feedback H_∞ control. *IEEE Trans. Automat. Contr.* **AC-37**(6), 770–784.

Xia, X. (1995). Approximate input-output linearization by static state feedback. preprint.

Appendix
A Definition of a nonlinearity measure

In this appendix we briefly review the definition of the nonlinearity measure introduced in (Allgöwer, 1996a). We consider general nonlinear (i.e. not necessarily linear), multivariable I/O-systems with input u and output y. The input signals u are assumed to belong to the space $\mathcal{U}_a$ of admissible input signals. The output signals y are elements of the normed space of output signals $\mathcal{Y}$. The I/O-behavior is described by the nonlinear operator N that maps input signals $u \in \mathcal{U}_a$ into output signals $y \in \mathcal{Y}$:

$$y = N[u]. \tag{50}$$

Functions without an explicit time argument are used to denote trajectories, while the value of a function at some specific time instance is always written with such an argument, e.g. $u(t)$. We consider static as well as dynamic operators N, but restrict our attention to causal, BIBO-stable systems. This setup describes a very general class of systems that comprise finite dimensional systems in continuous and discrete time as well as infinite dimensional systems. Without loss of generality, we assume

$$\lim_{t \to \infty} y(t) = 0 \qquad \text{for} \quad u(t) \equiv 0 \; \forall t. \tag{51}$$

Definition 3 *The nonlinearity measure $\phi_N^{\mathcal{U}}$ of a BIBO-stable, causal I/O-system $N : \mathcal{U}_a \to \mathcal{Y}$ for inputs signals $u \in \mathcal{U} \subseteq \mathcal{U}_a$ is defined by the nonnegative number*

$$\phi_N^{\mathcal{U}} = \inf_{G \in \mathcal{G}} \sup_{u \in \mathcal{U}} \frac{\|G[u] - N[u]\|_{p_{\mathcal{Y}}}}{\|N[u]\|_{p_{\mathcal{Y}}}}, \tag{52}$$

with $G : \mathcal{U}_a \to \mathcal{Y}$ being a linear I/O-operator belonging to the space of linear operators $\mathcal{G}$, $\| \cdot \|_{p_{\mathcal{Y}}}$ a norm in $\mathcal{Y}$, and $\mathcal{U}$ being the set of inputs considered.

The "size" of the nonlinearity of an I/O-system is thus defined as the normalized largest difference between the output of nonlinear system N and a linear system G measured by the norm $\| \cdot \|_{p_{\mathcal{Y}}}$ in $\mathcal{Y}$. The difference is taken with respect to the worst-case input in $\mathcal{U}$ and with respect to the linear system G^* that approximates N best. The set $\mathcal{U}$ contains the subset of inputs that we consider relevant in the controlled loop. An example for a specific choice of nonlinearity measure is given by restricting $\mathcal{U}$ to consist of L_2-signals with bounded energy $\|u\|_{L_2} \leq \hat{u}$ and the norm $\| \cdot \|_{p_{\mathcal{Y}}}$ being the L_2-norm.

It is easy to see that the nonlinearity measure $\phi_N^{\mathcal{U}}$ has the value zero for N being a linear operator. It is however not possible to deduce linearity of N from $\phi_N^{\mathcal{U}} = 0$, as the set $\mathcal{U}$ may not contain all admissible input signals $\mathcal{U}_a$. Yet from $\phi_N^{\mathcal{U}} = 0$ we can conclude the existence of a linear operator G^* such that

$$\|G^*[u] - N[u]\|_{p_{\mathcal{Y}}} = 0 \qquad \forall u \in \mathcal{U}. \tag{53}$$

The behavior of nonlinear system N for the input signals considered is thus the same as that of the linear system G^* and therefore N behaves linear in a certain sense. We refer to this as: "N is linear in $\mathcal{U}$". The nonlinearity measure $\phi_N^{\mathcal{U}}$ always takes values in the interval $[0, 1]$, because of the normalization with $\|N[u]\|_{p_y}$ (Allgöwer, 1996b). Thus, a value of one indicates the strongest possible degree of nonlinearity. It must be stressed, however, that the value of the nonlinearity measure depends of course on the choice of input signals $\mathcal{U}$ considered, which is also indicated by the superscript $\mathcal{U}$ in the notation for the nonlinearity measure $\phi_N^{\mathcal{U}}$.

Computation of the value of the nonlinearity measure according to the above definition requires the solution of a infinite dimensional, nonlinear minimax problem of high complexity. In order to circumvent this difficulty several approximation schemes were developed. A lower bound based on the computation of Fourier coefficients and an approximate solution scheme involving a convex optimization were presented in (Allgöwer, 1996a). A computationally attractive approach via functional expansions is given in (Harris et al., 1997).

B   Data for the membrane reactor

Parameters of the membrane reactor		
manipulated input controlled output	c_{a0} y	
operating point	$c_{As} = 2 \frac{mol}{l}$ $c_{Bs} = 55.85 \frac{mol}{l}$ $x_{ms} = 12 \frac{mol}{l}$	$c_{A0s} = 8 \frac{mol}{l}$ $y_s = 6 \frac{mol}{l}$ $q_s = 148 \frac{1}{h}$
parameters	$\alpha_{11} = 1.46 \cdot 10^3 \frac{mol}{l \cdot h}$ $\alpha_{12} = 148 \frac{1}{h}$ $\alpha_{13} = -3.2 \cdot 10^5 \frac{mol}{l \cdot h^2}$ $\alpha_{14} = 9.0 \cdot 10^5 \frac{mol^2}{l^2 \cdot h^2}$ $t_d = 0.126 \, h$	$\alpha_{21} = 0.93 \cdot 10^3 \frac{mol}{l \cdot h}$ $\alpha_{22} = 148 \frac{1}{h}$ $\alpha_{23} = 3.2 \cdot 10^5 \frac{mol}{l \cdot h^2}$ $\alpha_{24} = -3.9 \cdot 10^5 \frac{mol^2}{l^2 \cdot h^2}$ $r_2 = 15.9 \frac{1}{h}$

Reaction rate r_1 is defined by:

$$r_1(c_A) = \begin{cases} \alpha_{11} - \alpha_{12}c_A - \frac{1}{2}\sqrt{\alpha_{13}c_A + \alpha_{14}} & \text{for } c_A \geq c_{As} \\ \alpha_{21} - \alpha_{22}c_A + \frac{1}{2}\sqrt{\alpha_{23}c_A + \alpha_{24}} & \text{for } c_A < c_{As} . \end{cases} \tag{54}$$

ELEMENTARY NONLINEAR DECOUPLING (END), A GENERAL APPROACH TO MODEL BASED CONTROL OF NONLINEAR MULTIVARIABLE PROCESSES

JENS G. BALCHEN
Professor Emeritus
Department of Engineering Cybernetics
Norwegian University of Science and Technology
7034 Trondheim
NORWAY

Keywords: Multivariable control, nonlinear decoupling, constraint handling

Abstract: The paper gives a survey of a strategy based on Elementary Nonlinear Decoupling (END) used for the control of nonlinear multivariable processes. It describes the structure of the solution and the different optimization procedures leading to the final implementable control strategy. Based on the END strategy it is possible to develop a convenient systematic method for handling constraints both in the control variables (inputs) and state variables. Application of the END strategy has been demonstrated by the simulation of a number of realistic processes such as a distillation column (Balchen and Sandrib 1995a), a fluidized catalytic cracker (FCC) (Balchen and Sandrib 1995b) and a steam boiler (Balchen and Larsen 1998).

1. INTRODUCTION. THE END METHOD

There is a long-standing need for a systematic procedure for the design of a control strategy for a general nonlinear multivariable process and a number of proposals have been presented in the literature. Some of these proposals are variations of the LQG strategy (Balchen 1993) whereas others are variations of MPC (Model based Predictive Control) in which constraint handling is an integral part through some kind of nonlinear programming (Qin and Badgwell 1996).

R. Berber and C. Kravaris (eds.), Nonlinear Model Based Process Control, 275-310.
© 1998 *Kluwer Academic Publishers.*

276

An algorithm based on Elementary Nonlinear Decoupling (END) is related to methods from differential geometry techniques (Isidori 1989, Henson and Seborg 1997) but it solves the invertibility problem differently. In END, a property transformation is introduced defining a set of property variables which becomes the object of control. The property space has a dimension equal to that of the independent control vector.

The END algorithm is illustrated by a block diagram in Figure 1.

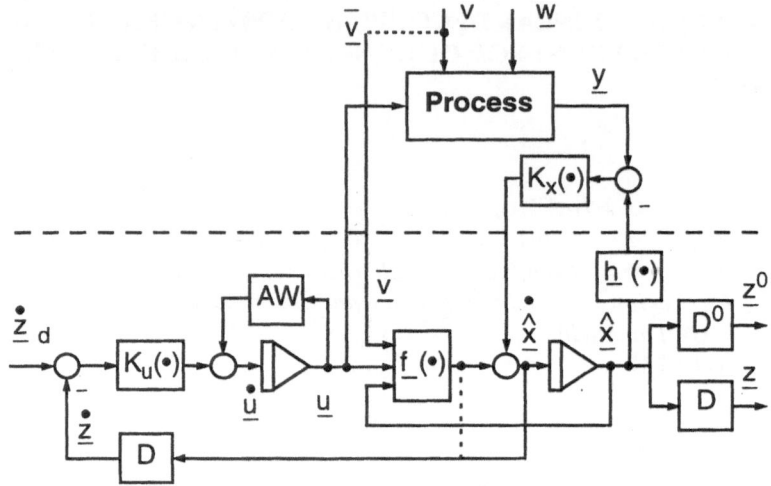

Figure 1 Block diagram of the END-algorithm.

The process to be controlled is in general described by a nonlinear model where

$$\dot{\mathbf{x}} = \mathbf{f}(\mathbf{x}, \mathbf{u}, \mathbf{v}) \tag{1}$$

$\mathbf{x}$: state vector (dim $\mathbf{x}$ = n)

$\mathbf{u}$: control vector (dim $\mathbf{u}$ = r)

$\mathbf{v}$: disturbance vector (dim $\mathbf{v}$ = s)

$\mathbf{f}(\cdot)$: vector of nonlinear functions (dim $\mathbf{f}$ = n)

Furthermore, it is assumed that the measurements from the process are modelled by the relationship

$$y = h(x) + w \tag{2}$$

where

y : measurement vector (dim $y = m$)

w : measurement noise vector (dim $w = m$)

In the following it is assumed that a state estimation scheme in the form of an Extended (Augmented) Kalman Filter or something equivalent is implemented so that the whole state vector (x) is available in the development of the control system.

The END algorithm belongs to the family of *inverse control* algorithms. In END, the invertibility problem is resolved by designing a property transformation $(z = d(x))$ which reflects the actual (desirable) properties (z_0) and a set of corrective variables (Δz)

$$z = z^0 + \Delta z = d(x) = d^0(x) + \Delta d(x) \tag{3}$$

The actual (desirable) property transformation is denoted:

$$z^0 = d^0(x) \tag{4}$$

This means that z^0 is the set of variables to be controlled whereas Δz is a vector of corrective variables which is introduced to achieve invertibility and stability. The design problem is to find a solution which yields a small Δz

Often the functions of (3) and (4) can be written:

$$z = D(x)x \tag{5}$$

where $D(x)$ is a matrix, which even in some cases, may be a constant matrix (D). For simplicity, in the following we assume that

$$\mathbf{D}(\mathbf{x}) = \mathbf{D} = \text{constant}$$

and

$$\mathbf{z}^0 = \mathbf{D}^0\mathbf{x} \tag{6}$$

and

$$\mathbf{z} = (\mathbf{D}^0 + \Delta\mathbf{D})\mathbf{x} = \mathbf{D}\mathbf{x} \tag{7}$$

The objective of END is to derive the control vector ($\mathbf{u}$) for a nonlinear dynamic process described by (1) such that the property vector defined by (7) follows a certain desired trajectory defined by its rate of change:

$$\dot{\mathbf{z}} = \mathbf{D}\dot{\mathbf{x}} = \dot{\mathbf{z}}_d \tag{8}$$

A solution to this problem is an iterative equation solver obeying the differential equation:

$$\dot{\mathbf{u}} = \mathbf{K}_\mathbf{u}(\cdot)(\dot{\mathbf{z}}_d - \mathbf{D}\dot{\mathbf{x}}) = \mathbf{K}_\mathbf{u}(\cdot)(\dot{\mathbf{z}}_d - \mathbf{D}\mathbf{f}(\mathbf{x}, \mathbf{u}, \mathbf{v})) \tag{9}$$

where the matrix $\mathbf{K}_u(\cdot)$ secures convergence of (9). A simple way of designing $\mathbf{K}_u(\cdot)$ is to require that the linearized loop described by (9) has prescribed and constant eigenvalues (Λ) such that:

$$\mathbf{K}_\mathbf{u}(\cdot) = -\Lambda\left(\mathbf{D}\frac{\partial \mathbf{f}(\cdot)}{\partial \mathbf{u}}\right)^{-1} = -\Lambda(\mathbf{DB})^{-1} \tag{10}$$

Other more sophisticated schemes can be suggested, but they turn out to be of little significance. We are only concerned that the solution is fast and stable.

The contents of (10) can be recognized in the lower left-hand corner of the block diagram in Figure 1. It also shows the function of the dynamic model of (1) together with the measurement model of (2) as part of the state estimator in which $\mathbf{K}_x(\cdot)$ represents the updating algorithm (e.g. Kalman Filter Gain Matrix).

Figure 1 also indicates that some elements in the process disturbance vector ($\mathbf{v}$) may be measurable and represented by $\bar{\mathbf{v}}$ and fed in a feed forward manner to the nonlinear function $\mathbf{f}(\cdot)$. Thereby the burden on the state estimator is reduced so that the signal energy in the output of the $\mathbf{K}_x(\cdot)$-matrix is reduced. This is significant in relation to a system's ability to react fast to high bandwidth disturbances.

Figure 1 also shows how the *actual property* vector ($\mathbf{z}^0$) and the *designed property* vector ($\mathbf{z}$) are generated from the estimated state vector ($\hat{\mathbf{x}}$).

In the feedback loop of the equation solver in Figure 1 there is a block labelled AW. This is an "anti windup facility" to prevent the integrators with the control variables as outputs being overcharged when the inputs ($\mathbf{u}$) go into saturation.

When (1) is linearized, around some working point, the following model will appear:

$$\dot{\mathbf{x}} = \mathbf{Ax} + \mathbf{Bu} + \mathbf{Cv} \tag{11}$$

$$\mathbf{y} = \mathbf{Hx} \tag{12}$$

which when applied to the structure of Figure 1, and assuming that the equation solver loop is made infinitely fast ($\Lambda \to \infty$), will give:

$$\dot{\mathbf{x}} = (\mathbf{I} - \mathbf{B(DB)}^{-1}\mathbf{D})(\mathbf{Ax} + \mathbf{Cv}) + \mathbf{B(DB)}^{-1}\dot{\mathbf{z}}_d = \tilde{\mathbf{A}}\mathbf{x} + \tilde{\mathbf{B}}\dot{\mathbf{z}}_d + \tilde{\mathbf{C}}\mathbf{v} \tag{13}$$

(13) describes the dynamics of the "inner system" which contains the n state variables. When multiplying (13) with the matrix D, it is observed that

$$\dot{\mathbf{z}} = \mathbf{D}\dot{\mathbf{x}} = \dot{\mathbf{z}}_d \tag{14}$$

which confirms that the structure of Figure 1 is converted into a set of r decoupled integrators.

Therefore, if we are able to design a property matrix $\mathbf{D}$, we have converted the system in Figure 1 to a new system which consists of r decoupled integrators i.e. (14) where: ($\dim \mathbf{z} = \dim \mathbf{u} = r$).

But while performing this transformation into a lower dimensional space ($r < n$) we still have the inner system which is of n'th order. And we must require that this inner system has acceptable behavior, such as stability, because it contains the actual state variables of the process.

In order to achieve stability of the inner system, we see from (13) that we must require that the eigenvalues of the matrix:

$$(\mathbf{I} - \mathbf{B}(\mathbf{DB})^{-1}\mathbf{D})\mathbf{A} = \tilde{\mathbf{A}}, \tag{15}$$

which we find in (13), must be located properly in the left half of the complex plane $(\text{Re}\tilde{\mathbf{A}} < 0)$.

Therefore there are strict limitations concerning which matrices ($\mathbf{D}$ or $\Delta\mathbf{D}$) we can choose.

In summary the END-algorithm performs the following functions:

- It generates a *new system* of order r with $\dot{\mathbf{z}}_d$ as inputs and $\mathbf{z}$ as outputs.

- The original system of order n still exists inside the *new system* and is referred to as the *inner system*.

- The *new system* is *linear* and *diagonal* and therefore easy to control as shown in Fig. 2.

2. DESIGN OF THE MATRIX D

It has been stated that the property transformation matrix $\mathbf{D}^0$ defining the properties ($\mathbf{z}^0$) which we want to control will most often be such that the product $\mathbf{D}^0\mathbf{B}$ is singular,

$$\text{i.e.} \quad \det\left(\frac{\partial \mathbf{d}^0(\cdot)}{\partial \mathbf{x}} \frac{\partial \mathbf{f}(\cdot)}{\partial \mathbf{u}}\right) = \det(\mathbf{D}^0\mathbf{B}) = 0$$

Therefore the matrix $(\mathbf{D}^0\mathbf{B})^{-1}$ does not exist and neither (10) nor (13) can be solved if $\mathbf{D} = \mathbf{D}^0$.

The correction matrix $\Delta\mathbf{D}$ in (7) is introduced in order to make the inversion possible. It has the equivalence of introducing zeros in the transfer functions of the feedback loop. This can easily be seen when studying a monovariable system (single-input single-output-system) as is done in Section 7 below. The problem is to find a matrix $\Delta\mathbf{D}$ which satisfies certain specifications regarding system performance, such as stability, robustness, bandwidth (agressiveness), constraint handling.

The first requirement in the design of $\mathbf{D}$ (or $\Delta\mathbf{D}$ since $\mathbf{D}^0$ is given) is derived from (10). We must require the product $(\mathbf{DB})$ is not close to singularity. By computing the largest and the smallest singular values ($\bar{\sigma}_{DB}$ and $\underline{\sigma}_{DB}$) we can determine the condition number

$$\kappa_{DB} = \frac{\bar{\sigma}_{DB}}{\underline{\sigma}_{DB}} \tag{16}$$

This number will directly represent the difference in gain in different directions of the matrix $\mathbf{K}_u(\cdot)$ of (10). If κ_{DB} is a large number, we will get a very unbalanced use of control actions in the different control variables. This is not acceptable. Therefore we wish to require $\kappa_{DB} < \kappa_0$ where $\kappa_0 = 10$ is reasonable.

Next, we recognize that the matrix $\mathbf{D}$ appears in (15) and we wish to design $\mathbf{D}$ (or $\Delta\mathbf{D}$) so that the stability of the "inner system" is adequate. This can be done in the following way:

- Assuming first that the "inner system" is stable, we can make a feedback system as shown in Figure 2 around the system of Figure 1 which has the END algorithm included. The matrix $\mathbf{G}$ can be chosen diagonal and it is not unreasonable in the end to select $\mathbf{G} = -\Lambda$ where Λ is the matrix of eigenvalues from (10). This is so because when the eigenvalues (Λ) of the equation solver loop are large but finite, the equivalent dynamic description of the relationship between $\dot{\mathbf{z}}_d$ and $\mathbf{z}$ will be as shown in Figure 3 indicating that due to limited bandwidth of the equation solver we will no longer have a pure integrations between $\dot{\mathbf{z}}_d$ and $\mathbf{z}$, but rather will have cascaded small first order delays.

- With the feedback system in Figure 2 operational, we can determine the difference $\Delta\mathbf{z}$ which will contain information about the behavior of the inner system. If, for example, the inner system is unstable, $\Delta\mathbf{z}$ will contain components which will increase indefinitely. Therefore we can design an optimization procedure based on an objective functional

$$\mathbf{J}_0 = \int_0^{t_1} (\Delta\mathbf{z}^T(\tau)\mathbf{Q}\Delta\mathbf{z}(\tau) + \alpha\Delta\mathbf{u}^T(\tau)\mathbf{P}\Delta\mathbf{u}(\tau))d\tau \tag{17}$$

282

where the matrices $\mathbf{Q}$ and $\mathbf{P}$ may be chosen as diagonal positive matrices and the factor α is used to control the relative influence of the state variables and the control variables.

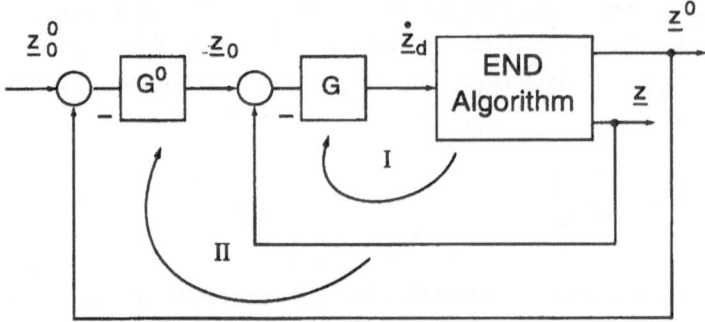

Figure 2 Feedback control based on the END-algorithm.

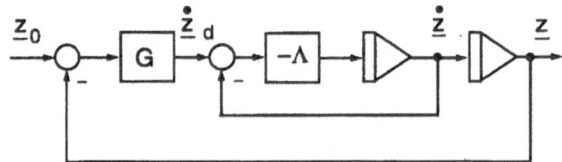

Figure 3 Equivalent dynamic system with finite $\wedge$.

An optimization of the elements of the matrix $\Delta\mathbf{D}$ can be performed based upon the objective functional J_0. The results of such an optimization will be dependent upon the excitation chosen $((\mathbf{z}_0)$ and/or $\mathbf{v})$ as seen in Figure 2. If an excitation of reasonable shape is chosen, the result will be rather independent of the shape.

- In order to succeed with the optimization described above, it is necessary to start with a $\Delta\mathbf{D}_{start}$ which yields a stable inner system. For most physical systems it is not difficult to find an acceptable $\Delta\mathbf{D}_{start}$ from a study of the structures of $\mathbf{D}^0$ and $\mathbf{B} = \dfrac{\partial\mathbf{f}(\cdot)}{\partial\mathbf{u}}$.

- Based on the optimal value $\Delta\mathbf{D}_{opt}$ found by the above procedure, it is now feasible to include other features in the optimization procedure such as:

- Imposing restrictions on the condition number $\kappa_{DB} < \kappa_0$

- Imposing restrictions on the size of the elements of $\Delta \mathbf{D}$ as measured for instance by the Frobenius-norm $F(\Delta \mathbf{D}) = \left(\sum_{i=1}^{r} \sum_{j=1}^{n} (\Delta d_{ij})^2 \right)^{1/2}$

- Imposing restrictions on some scalar measure $R > 0$ of the robustness of the complete system with respect to stability caused by specified variations in process parameters.

The new restrictions may be added to the original objective functional so as to produce a new functional

$$J_1 = J_0 + \beta_1 \kappa_{DB} + \beta_2 F(\Delta \mathbf{D}) + \beta_3 R \qquad (18)$$

in which β_1, β_2 and β_3 are weight factors controlling the influence of the new restrictions. The optimization procedure will now find $\min_{\Delta D} J_1$.

A block diagram illustrating the structure of the optimization procedure is shown in Figure 4. Here it is seen that a feedback system (loop I) is implemented from $\mathbf{z}$ through the matrix $\mathbf{G}$. This is important in order to move r eigenvalues in the origin into the left half plane. The integration period t_1 in (1) must be chosen to be long enough so that instabilities will be discovered, but not any longer because the total time needed for the optimization will be directly proportional to t_1. When starting the numerical optimization, the matrix $\mathbf{G}$ should be chosen diagonal and with rather small elements. After the optimization has been completed, the elements of $\mathbf{G}$ may be increased. The matrix Λ should be chosen diagonal and with elements

$$-\lambda_i \approx 10 |\lambda|_{Amax} - 100 |\lambda|_{Amax}$$

where λ_{Amax} is the largest eigenvalue of the process-matrix A.

The magnitude of the Frobenius-norm of $\Delta \mathbf{D}$, $(F(\Delta \mathbf{D}))$ must be compared to the equivalent Frobenius-norm of $(F(\mathbf{D}^0))$. An acceptable ratio between these two norms will be 0.1.

When computing the Frobenius-norm and the condition number, the matrices considered should be scaled as described in Section 6 below. But the optimization, as described in Figure 4, can preferably be done using the real model (not scaled).

284

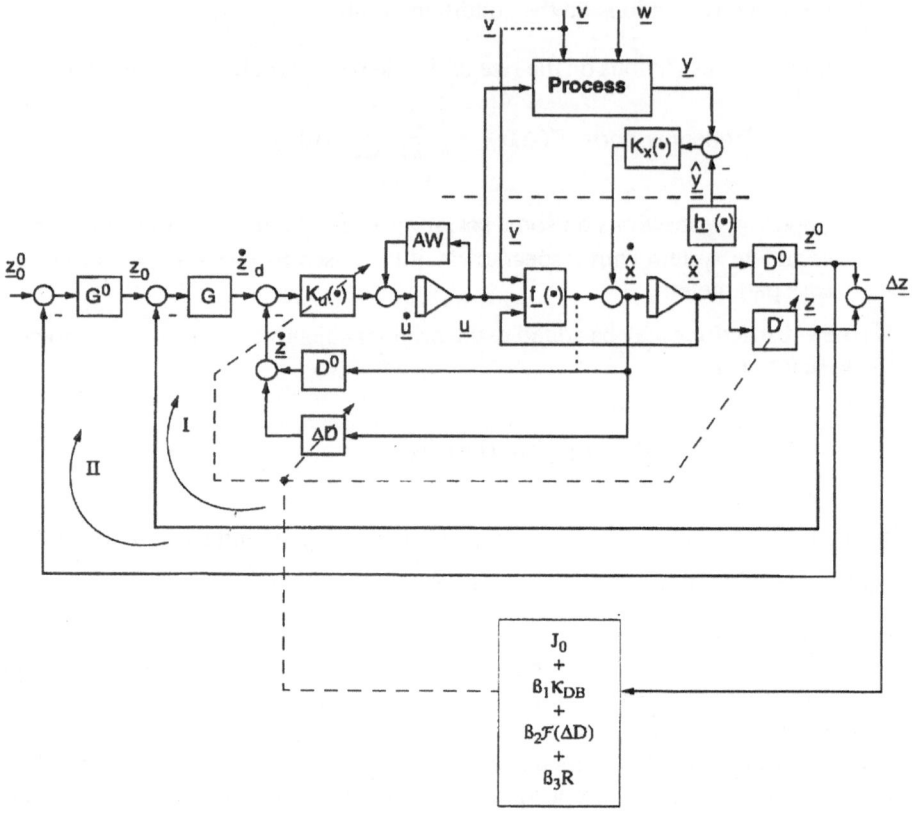

Figure 4 The optimization of the END-algorithm.

As is always the case when optimization is performed, the result will be entirely dependent upon the objective function chosen and the weighting matrices and parameters chosen (like $\mathbf{Q}$, $\mathbf{P}$, α, β_1, β_2, β_3, Λ). Changing any of these parameters will lead to a new result which must be critically analyzed to check whether it is acceptable or not. In the problem at hand, there will be an infinite number of acceptable solutions that all have the same structural basis.

Applications of the optimization technique described above have been demonstrated in relation to END-control of a ship steam boiler (Balchen and Larsen 1998), a binary distillation column (Balchen and Sandrib 1995a) and a catalytic cracking plant (Balchen and Sandrib 1995b).

It is highly desirable to develop simple rules for the choice of the weighting matrices and parameters. In doing that the process model should be scaled so that all variables become dimensionless (e.g. vary in the range 0-1) as described in section 6 below. Then $\mathbf{Q}$ and $\mathbf{P}$ matrices can be chosen $\mathbf{Q} = \mathbf{P} = \mathbf{I}$. Factor α determines to some extent the "aggressiveness" of the control system (large α, less "aggressive"). But since factor β_1 also influences the "aggressiveness" it has been found that rather small values of α (e.g. $\alpha = 0,05$ to $\alpha = 0,1$) can be used.

Factor β_1 which determines the final condition number κ_{DB} should be chosen so that the term $\beta_1 \kappa_{DB}$ contributes with about 50% to the total objective functional J_1 of (18). Chosing $\beta_1 = 0,1$ will result in $\kappa_{DB} < 5$ which is acceptable.

In Balchen and Larsen (1998) the choice of factor β_2 is discussed in some detail. The term $\beta_2 F(\Delta\mathbf{D})$ influences the resulting magnitude of the elements of $\Delta\mathbf{z}$ as does the first term in J_0 of (18). Therefore it may be concluded that as a first approximation $\beta_2 = 0$ could be selected.

3. THE END ALGORITHM WITH SATURATING CONTROL VARIABLES

Any physical control variable will be subject to limitations in its performance. In many cases control actions are implemented by means of "motors" which will have a limited speed of response and limited amplitude of response. It is not uncommon in industrial practice that both types of limitations become active during normal operation of the control system. The consequence of this is that such limitations have to be taken into account when analyzing the system performance. The limitations in the speed of response from the control devices are commonly underestimated in process control design leading to inferior control performance.

Figure 5 shows an elementary block diagram of a control device with both speed and amplitude limitation (saturation). As is seen, the amplitude limitation is achieved by a nonlinear function with infinite gain in the feedback loop. The input limitations will be properly taken care of if the system in Figure 5 is incorporated in the process model in the development of model based control.

The following will only consider amplitude saturation (Balchen and Sandrib 1995a).

When one or more of the control variables constituting $\mathbf{u}$ reaches a saturation level, the control system loses one or more degrees of freedom in control. This may or may not have serious consequences depending on the structure of the function $\mathbf{f}(\cdot)$ of (1) or the matrix

$$\mathbf{B} = \frac{\partial \mathbf{f}(\cdot)}{\partial \mathbf{u}}.$$

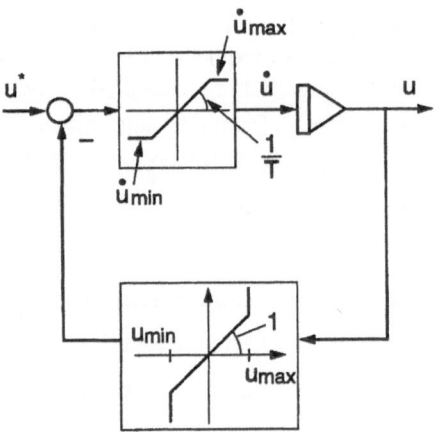

Figure 5 Elementary block diagram of control device with both speed and amplitude limitation.

A linearized version of (1) for small perturbations of the variables will be

$$\delta\dot{x} = A\delta x + B\delta u + C\delta v \tag{19}$$

The matrix B can be written in terms of parameter vectors b_i each representing the influence of the individual control variables (u_i) upon state equation (19).

$$B = [b_1, b_2 \ldots b_i \ldots b_r] \tag{20}$$

Therefore (19) can be written

$$\delta\dot{x} = A\delta x + B^{ij\cdots}\delta u^{ij\cdots} + b_i\delta u_{isat} + b_j\delta u_{jsat}\cdots + C\delta v \tag{21}$$

where $B^{ij\cdots}$ has been introduced for the matrix B where the columns i and j... are removed, $\delta u^{ij\cdots}$ is the control vector with elements i and j... removed and δu_{isat} and δu_{jsat} represent the saturating values of the control variables number i and j respectively.

Well known conditions for the linear state controllability show that the system with saturating control variables $u_i, u_j \ldots$ is controllable if

$$\text{rank}[\mathbf{B}^{ij\cdots} | \mathbf{AB}^{ij\cdots} | \ldots | \mathbf{A}^{n-1} \mathbf{B}^{ij\cdots}] = n \tag{22}$$

(22) should be tested for all possible combinations $\mathbf{B}^{ij\cdots}$.

If a non-controllable combination is discovered, the resulting system can be further analyzed to see if the noncontrollability is detrimental or not (unstable etc.).

If the noncontrollable modes are nondetrimental, a feedback control system can be implemented for each $\mathbf{B}^{ij\cdots}$ based upon the estimated states ($\hat{\mathbf{x}}$) and the system will eventually return to normal nonsaturating conditions.

If the noncontrollable modes are detrimental, there is usually no other way to solve the problem than by redesigning the process, for instance by adding more control variables.

The design of a control strategy can follow different ways:

- Strategy 1
  Use the "unsaturated" END-strategy, and *hope for the best*. It may well be that this solution will turn out to be satisfactory. This can be tested by simulation. On the other hand it may also be that this solution does not work, for a number of reasons. One of the most obvious reasons is that the "inner system" may become unstable when one or more control variables become saturated.

- Strategy 2
  Design a control algorithm for each possible saturation case which at least guarantees that the "inner system" is stable. A strategy belonging to this category is discussed below.

When one of the control variables u_i reaches saturation, the system loses one degree of freedom in control. This means that the number of property variables (z_i) which can be controlled by the END algorithm, must be reduced by one i.e. we must choose to sacrifice one of the property variables. An obvious choice in most cases is to sacrifice the property variable which has the least significance in terms of "survival" of the process rather than "economy" or "performance". This is so because saturation in one or more of the control variables is a transient phenomenon which will disappear after a short time. If the saturation phenomenon does not disappear, there is something rather wrong in the design of the process which calls for major modifications.

The property vector associated with a certain set ot saturating control variables symbolized by the associated control matrix $\mathbf{B}^{ij\cdots}$ will be designated as $\mathbf{z}^{ij\cdots}$ and the associated property transformation by the matrix $\mathbf{D}^{ij\cdots}$. The resulting system will have the same block diagram as that of Figure 1 and the convergence matrix $\mathbf{K}_u^{ij\cdots}$ will still be determined by the expression of (10) only with new matrices applied.

The contents of the new property transformation matrix may be determined in the same way as described for the unsaturated case above i.e. through an optimization procedure, but now with a reduced set of property variables. A logical way of doing this will in some cases be as follows:

Assume that the process has three control variables $(r = 3)$. In that case seven possible combinations of control matrices will exist, namely

$$\mathbf{B}, \mathbf{B}^1, \mathbf{B}^2, \mathbf{B}^3, \mathbf{B}^{12}, \mathbf{B}^{13}, \mathbf{B}^{23}, \mathbf{B}^{123}$$

where $\mathbf{B}$ represents the unsaturated and $\mathbf{B}^{123}$ the fully saturated case.

In the unsaturated case we have

$$\mathbf{z} = \begin{bmatrix} z_1 \\ z_2 \\ z_3 \end{bmatrix} = \mathbf{Dx} = \begin{bmatrix} \mathbf{d}_1^T \\ \mathbf{d}_2^T \\ \mathbf{d}_3^T \end{bmatrix} \mathbf{x} \tag{23}$$

where z_3 is the property with the lowest priority and z_1 has the highest priority. Thus it is proposed that

$$\mathbf{B}^1 \rightarrow \mathbf{D}^1 = \begin{bmatrix} \mathbf{d}_1^T \\ \mathbf{d}_2^T \end{bmatrix}, \ \mathbf{B}^2 \rightarrow \mathbf{D}^2 = \mathbf{D}^1, \ \mathbf{B}^3 \rightarrow \mathbf{D}^3 = \mathbf{D}^1 \tag{24}$$

$$\mathbf{B}^{12} \rightarrow \mathbf{D}^{12} = \mathbf{d}_1^T, \ \mathbf{B}^{13} \rightarrow \mathbf{D}^{13} = \mathbf{D}^{12}, \ \mathbf{B}^{23} \rightarrow \mathbf{D}^{23} = \mathbf{D}^{12} \tag{25}$$

This solution will work provided the requirements imposed on the unsaturated case, also apply to all combinations of the matrices $(\mathbf{D}^{ij}\mathbf{B}^{ij})$. All possible combinations, as described above, can be tested "off-line" and just implemented when the different conditions are detected in the estimator.

The technique of designing the elements of the $\Delta\mathbf{D}$ matrix by optimizing an objective functional as developed above, can also be applied to the case of saturating control variables.

First we have to decide about the priorities which shall be assigned to the different property variables when the system loses degrees of freedom due to saturating control.

Suppose that we have ordered the property variables

$$\mathbf{z} = \begin{bmatrix} z_1 \\ z_2 \\ \vdots \\ z_r \end{bmatrix} \tag{26}$$

such that z_1 is the property variable with the highest priority, that is, the variable which should be controlled when there is only *one* active control variable (irrespective of which) and all the others $(r-1)$ are saturated.

The property variables z_1 and z_2 are those variables that have priority when only *two* control variables are active. z_1, z_2 and z_3 are the property variables with priority when three control variables are active and so on.

A possible optimization procedure is as follows:

- Optimize the $\Delta\mathbf{D}$ matrix based on the objective functional

$$J = \int_0^{t_1} (\Delta\mathbf{z}^T\mathbf{Q}\Delta\mathbf{z} + \alpha\Delta\mathbf{u}^T\mathbf{P}\Delta\mathbf{u})dt + \beta_1\kappa_{DB} + \beta_2 F(\Delta\mathbf{D}) \tag{27}$$

when the system is excited in such a way that none of the control variables reach saturation. Thereby all the elements of the $\mathbf{D}$-matrix are determined so that in

$$D = \begin{bmatrix} \mathbf{d}_1^T \\ \mathbf{d}_2^T \\ \vdots \\ \mathbf{d}_r^T \end{bmatrix} \tag{28}$$

all the elements of $\mathbf{d}_i^T$ are determined.

- Then an excitation (disturbance) with such magnitude is applied so that all control variables except one, reach saturation for certain periods. The time intervals, in which only one control variable is unsaturated, is now used when applying (27) to optimize the elements of $\mathbf{d}_1^T$.

- With the same excitation as above (or another one), the time intervals in which all control variables, except *two* are saturated, are selected as basis for the optimization according to (27) to optimize the elements of $\begin{bmatrix} \mathbf{d}_1^T \\ \mathbf{d}_2^T \end{bmatrix}$

- The above procedure is continued until new elements of the matrix (28) are determined based on all possible combinations of saturating control variables. The matrix **D** determined in this way will be inferior to that determined originally, if the system never reaches input saturation. But it will be superior to the original **D** matrix when saturation occurs.

4. CONSTRAINTS IN STATE VARIABLES

A similar but more difficult situation arises when it is necessary (or desirable) to impose constraints on one or more of the state variables (most often physical process variables) for reasons related to safety or equipment protection. The main problem is that there is a dynamic relationship between control variables and state variables such that when a state variable reaches some limit, it may already be too late to apply a control action to prevent the state variable from exceeding the constraint value. However, this is a matter of the bandwidth of the controlled process. Applying the END algorithm to cope with constrained state variables, is quite simple and will be demonstrated below (Balchen and Sandrib 1995c).

One of the most convincing arguments for the application of "model-based *predictive* control" (MPC) is that this technique includes handling of constraints in both control and state variables. However, a comparison between the performance of an MPC algorithm and the END algorithm (which has not yet been done) may disclose that the END solution is the most advantageous. The reason for this is that the massive on-line computational demand to perform single/multiple shooting in MPC may in some cases lead to a low sampling rate and thus low control bandwidth. In END all demanding calculations are done off-line yielding high control bandwidth.

Handling constraints in state variables is done in much the same way as is described above for constraints (saturation) in control variables.

The proposed strategy is to replace the constraints in $\mathbf{x}$ by representative constraints in $\mathbf{z}$ and labelling these $\tilde{\mathbf{z}}$. In doing this it is observed that the number of state constraints that can be handled at any time is equal to or less than the number of unconstrained (unsaturated) control variables, i.e. degrees of freedom. In other words, if there are only two unconstrained control variables available, only two state variables can be controlled to stay on the constraints (maximal and minimal). Thus all the other state variables have to stay inside acceptable operating ranges. The way to achieve this is to design the $\mathbf{D}$ matrix with this in mind.

The property transformation to be applied when, as an example, the state variables x_i and x_j have reached their constraint values, is thus

$$\tilde{\mathbf{z}} = \tilde{\mathbf{D}}^{ij}\mathbf{x} \tag{29}$$

where the superscripts i and j refer to the constrained state variables x_i and x_j.

Thus, as an example, if the number of control variables is r = 3 and the number of states is n = 8, one could have a case with an actual unconstrained property transformation

$$\mathbf{D}^0 = \begin{bmatrix} 1 & 0 & 0 & 0 & 0 & 0 & 0 & 0 \\ 0 & 1 & 0 & 0 & 0 & 0 & 0 & 0 \\ 0 & 0 & 1 & 0 & 0 & 0 & 0 & 0 \end{bmatrix} \tag{30}$$

indicating that x_1, x_2, x_3 are to be controlled.

The designed property transformation for this case will be of the form

$$
\mathbf{D} = \begin{bmatrix} 1 & d_{12} & d_{13} & d_{14} & d_{15} & d_{16} & d_{17} & d_{18} \\ d_{21} & 1 & d_{23} & d_{24} & d_{25} & d_{26} & d_{27} & d_{28} \\ d_{31} & d_{32} & 1 & d_{34} & d_{35} & d_{36} & d_{37} & d_{38} \end{bmatrix} \tag{31}
$$

Assuming now that the states x_4 and x_6 are reaching their constraint values simultaneously, the appropriate $\tilde{\mathbf{D}}$ matrix could be

$$
\tilde{\mathbf{D}}^{46} = \begin{bmatrix} 1 & \tilde{d}_{12} & \tilde{d}_{13} & \tilde{d}_{14} & \tilde{d}_{15} & \tilde{d}_{16} & \tilde{d}_{17} & \tilde{d}_{18} \\ \tilde{d}_{21} & \tilde{d}_{22} & \tilde{d}_{23} & 1 & \tilde{d}_{25} & \tilde{d}_{26} & \tilde{d}_{27} & \tilde{d}_{28} \\ \tilde{d}_{31} & \tilde{d}_{32} & \tilde{d}_{33} & \tilde{d}_{34} & \tilde{d}_{35} & 1 & \tilde{d}_{37} & \tilde{d}_{38} \end{bmatrix} \tag{32}
$$

This $\tilde{\mathbf{D}}^{46}$ matrix indicates that the original emphasis on x_1 has been kept while x_2 and x_3 have been de-emphasized in order to be able to handle the constraints in x_4 and x_6.

The elements $\tilde{d}_{ij}$ introduced in $\tilde{\mathbf{D}}^{46}$ are mostly very small. Some of them will probably be so small that they can be neglected and made equal to zero. When none of the state variables reach their constraint values, the property transformation will return to that of (31) which has been optimized for the unconstrained case. The design of the matrix $\tilde{\mathbf{D}}^{ij\cdots}$ is done offline.

5. INCORPORATING MEASUREMENT WITH LONG DELAYS IN MODEL BASED MULTIVARIABLE CONTROL

Though it has not been properly observed by many process control system designers, the properties ($\mathbf{z}$) which one wants to control in an industrial process are not necessarily those that can be measured ($\mathbf{y}$). *This fact will often be the underlying motivation for using model based control.* Since the properties to be controlled ($\mathbf{z}$) are not available as measurements, they must be computed from model based estimates of internal states ($\mathbf{x}$).

Figure 6 shows the structure of such a model based state estimator which also contains models for measurements ($\mathbf{y}$) and properties ($\mathbf{z}$). A distinction has been made in Figure 6 between two kinds of measurements $\mathbf{y}_1$ and $\mathbf{y}_2$.

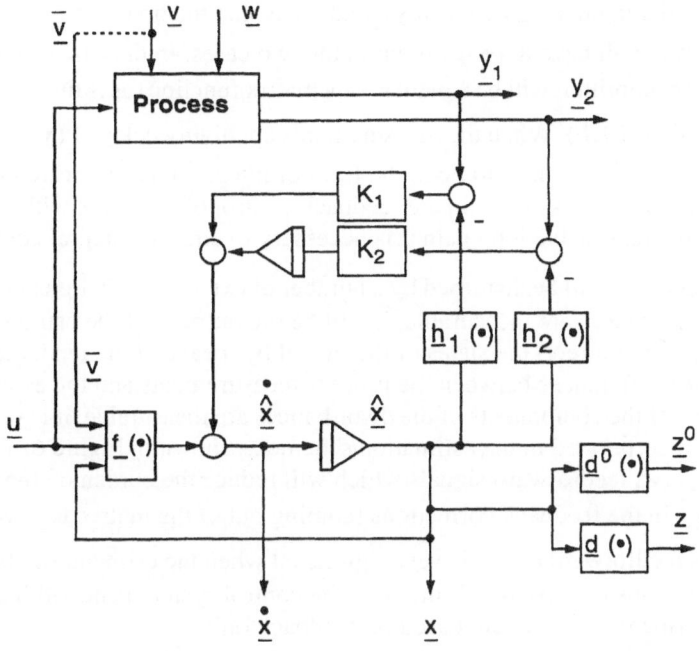

Figure 6 State estimator with instantaneous and delayed measurements.

$$\mathbf{y}_1 = \mathbf{h}_1(\mathbf{x}) \tag{33}$$

is characterized by the fact that they are related to the state vector ($\mathbf{x}$) through an algebraic (nondynamic) relationship $\mathbf{h}_1(\cdot)$.

The other type of measurement ($\mathbf{y}_2$) is characterized by a significant dynamic effect such as a delay, for example

$$\mathbf{y}_2 = \mathbf{h}_2(\mathbf{x}(t - \underline{\tau})) \tag{34}$$

in which the time parameter $\underline{\tau}$ indicates that delays of various lengths may be associated with the different measurements.

294

The reason for distinguishing between y_1 and y_2 is that the updating algorithm as indicated in Figure 6, will have to be different in the two cases. In the case of measurements y_1 the updating algorithm will be a matrix (or a matrix function) as is the case in an Extended Kalman Filter (EKF). When the measurements are highly delayed (y_2), the updating will, for stability reasons, have to be in the form of integrators as indicated in Figure 6. The reason for this is the same as in the feedback control of a process with pure delay. In order to obtain a reasonable loop gain it is necessary to employ integral control.

The physical process will be disturbed by a number of external excitations $(v(t))$. If none of of these disturbances are measurable, it will be the purpose of the estimator in Figure 6 to update (correct) the states of the model by means of the feedback actions shown from the differences between the process measurements and the estimates. If, however, some of the components of the disturbances are measurable, it is very important to have them incorporated in the estimation scheme as shown in Figure 6. These measurements will act as feedforward signals which will reduce the burden on the estimator so that the energy in the feedback corrections (coming out of the matrix K_x) will be highly reduced. This feedforward action is very significant when the estimated state (x) is used for control purposes because it will improve the control system bandwidth as compared with the performance of a system based on feedback only.

When measurements are made in the physical process, there will be uncertainties (w) which will influence the correction algorithm (K_x). This is an important aspect of the theory of Kalman filtering; the feedback gain (K_x) will be "proportional" to the inverse of the covariance of the uncertainty.

Measurements of higher precision will often be associated with some delay such as in analytical instrumentation which requires taking samples from the process stream, preparing samples for analysis, performing measurement etc. When using such high precision, delayed measurements, it is important that a dynamic model describing the delay, is incorporated in the updating scheme. But due to the delay the bandwidth of this type of updating will be low.

The combination of high bandwidth/low precision measurements and low bandwidth/high precision measurements is both feasible and desirable. Thereby requirements for both bandwidth and precision will be met.

As is seen in Figure 6, the property variables (z) are derived from the state variables (x) through the function

$$z = d(x) \tag{35}$$

In other words, there is not necessarily any relationship between the measurements ($\mathbf{y}$) and the properties ($\mathbf{z}$). If we have $\mathbf{z} = \mathbf{y}$, the END-solution will become somewhat simpler than in the general case, but not very much.

Considering the functional relationship given in (1), (2) and (3) we have an indication of how to choose the state space ($\mathbf{x}$) in modeling the system. The total state space may be divided into four subspaces defined by the vectors $\mathbf{x}_1, \mathbf{x}_2, \mathbf{x}_3$ and $\mathbf{x}_4$ such that the total state is

$$\mathbf{x} = [\mathbf{x}_1^T, \mathbf{x}_2^T, \mathbf{x}_3^T, \mathbf{x}_4^T]^T \tag{36}$$

The measurement function appearing in (33) could then be

$$\mathbf{y}_1 = \mathbf{h}_1(\mathbf{x}_2, \mathbf{x}_3) \tag{37}$$

and

$$\mathbf{z} = \mathbf{d}(\mathbf{x}_3, \mathbf{x}_4) \tag{38}$$

This means that the measurement vector and the property vector have one part of the total state space in common ($\mathbf{x}_3$), but are otherwise driven from two different parts of the total state space. Also as is seen in (36), the total state contains the component $\mathbf{x}_1$ which appears in neither (37) nor (38), but is significant in the dynamic description of the process. The total dimension of the state space will thus be the sum of the dimensions of the vectors appearing in (36).

6. THE SCALING OF VECTORS AND MATRICES IN MULTIVARIABLE SYSTEM MODELS

Many of the conclusions to be drawn from multivariable systems theory are dependent upon the scaling of the quantities involved. Therefore it is desirable to introduce a systematic procedure for scaling the different quantities involved such as state variables, control variables, measurements and property variables with reference to relevant ranges and biases. This is well known in conventional process control and instrumentation equipment where variables are represented in "percent" or "per unit".

The consequence of a systematic scaling will be that all scaled quantities will vary within the same range, say (0-1). Thus it is possible to compare responses directly.

In the following the physical quantities will be denoted by x_i, u_i... and the equivalent scaled quantities by $\tilde{x}_i$, $\tilde{u}_i$...

The maximal and the minimal values of a certain physical quantity will be denoted:

$$x_i^{max} \quad \text{and} \quad x_i^{min}$$

and the equivalent scaled quantities will be denoted

$$\tilde{x}_i^{max} \quad \text{and} \quad \tilde{x}_i^{min}$$

The relationship between the physical and the scaled variables will then be

$$\tilde{x}_i = \alpha_i^x x_i + \tilde{x}_i^b \tag{39}$$

where the "proportionality factor" is

$$\alpha_i^x = \frac{\tilde{x}_i^{max} - \tilde{x}_i^{min}}{x_i^{max} - x_i^{min}} \tag{40}$$

and the "bias" is

$$\tilde{x}_i^b = \frac{x_i^{max}\tilde{x}_i^{min} - x_i^{min}\tilde{x}_i^{max}}{x_i^{max} - x_i^{min}} \tag{41}$$

A vector of physical quantities will as usual be denoted

$$\mathbf{x} = \begin{bmatrix} x_1 \\ x_2 \\ \vdots \\ x_n \end{bmatrix} \tag{42}$$

Its equivalent scaled vector will be:

$$\tilde{x} = \begin{bmatrix} \alpha_1^x x_1 \\ \alpha_2^x x_2 \\ \vdots \\ \alpha_n^x x_n \end{bmatrix} + \begin{bmatrix} \tilde{x}_1^b \\ \tilde{x}_2^b \\ \vdots \\ \tilde{x}_n^b \end{bmatrix} = \widehat{X} x + \tilde{x}^b \tag{43}$$

where

$$\widehat{X} = \begin{bmatrix} \alpha_1^x & 0 & & 0 \\ 0 & \alpha_2^x & & \\ \vdots & & & \vdots \\ 0 & 0 & \ldots & \alpha_n^x \end{bmatrix} \tag{44}$$

$\tilde{x}^b$: scaled bias vector with elements defined in (41).

Similarly scaling transformations can be designed for all dynamic variables in a multivariable system leading to

$$\tilde{u} = \widehat{U} u + \tilde{u}^b \tag{45}$$

$$\tilde{y} = \widehat{Y} y + \tilde{y}^b \tag{46}$$

$$\tilde{z} = \widehat{Z} z + \tilde{z}^b \tag{47}$$

Based on (43) and (45) we find

$$x = \widehat{X}^{-1} (\tilde{x} - \tilde{x}^b) \tag{48}$$

298

$$\mathbf{u} = \widehat{\mathbf{U}}^{-1}(\tilde{\mathbf{u}} - \tilde{\mathbf{u}}^b) \tag{49}$$

When (48) and (49) are applied to

$$\dot{\mathbf{x}} = \mathbf{A}\mathbf{x} + \mathbf{B}\mathbf{u} \tag{50}$$

we get

$$\widehat{\mathbf{X}}^{-1}\dot{\tilde{\mathbf{x}}} = \mathbf{A}\widehat{\mathbf{X}}^{-1}\tilde{\mathbf{x}} + \mathbf{B}\widehat{\mathbf{U}}^{-1}\tilde{\mathbf{u}} - (\mathbf{A}\widehat{\mathbf{X}}^{-1}\tilde{\mathbf{x}}^b + \mathbf{B}\widehat{\mathbf{U}}^{-1}\tilde{\mathbf{u}}^b) \tag{51}$$

The left-hand side of (51) contains only the term shown because the time derivative of the constant bias term in (48) is zero. From (51) we get

$$\dot{\tilde{\mathbf{x}}} = \widehat{\mathbf{X}}\mathbf{A}\widehat{\mathbf{X}}^{-1}\tilde{\mathbf{x}} + \widehat{\mathbf{X}}\mathbf{B}\widehat{\mathbf{U}}^{-1}\tilde{\mathbf{u}} - (\widehat{\mathbf{X}}\mathbf{A}\widehat{\mathbf{X}}^{-1}\tilde{\mathbf{x}}^b + \widehat{\mathbf{X}}\mathbf{B}\widehat{\mathbf{U}}^{-1}\tilde{\mathbf{u}}^b) \tag{52}$$
$$= \widehat{\mathbf{A}}\tilde{\mathbf{x}} + \widehat{\mathbf{B}}\tilde{\mathbf{u}} + \dot{\tilde{\mathbf{x}}}^b$$

When the property transformation

$$\mathbf{z} = \mathbf{D}\mathbf{x} \tag{53}$$

together with (48) is applied to (47) the result will be

$$\tilde{\mathbf{z}} = \widehat{\mathbf{Z}}\mathbf{D}\widehat{\mathbf{X}}^{-1}\tilde{\mathbf{x}} + (\tilde{\mathbf{z}}^b - \widehat{\mathbf{Z}}\mathbf{D}\widehat{\mathbf{X}}^{-1}\tilde{\mathbf{x}}^b) = \widehat{\mathbf{D}}\tilde{\mathbf{x}} + \tilde{\mathbf{z}}^{b*} \tag{54}$$

In summary we have arrived at the new system matrices

$$\widehat{\mathbf{A}} = \widehat{\mathbf{X}}\mathbf{A}\widehat{\mathbf{X}}^{-1} \tag{55}$$

$$\widehat{\mathbf{B}} = \widehat{\mathbf{X}}\mathbf{B}\widehat{\mathbf{U}}^{-1} \tag{56}$$

$$\widehat{\mathbf{D}} = \widehat{\mathbf{Z}}\mathbf{D}\widehat{\mathbf{X}}^{-1} \tag{57}$$

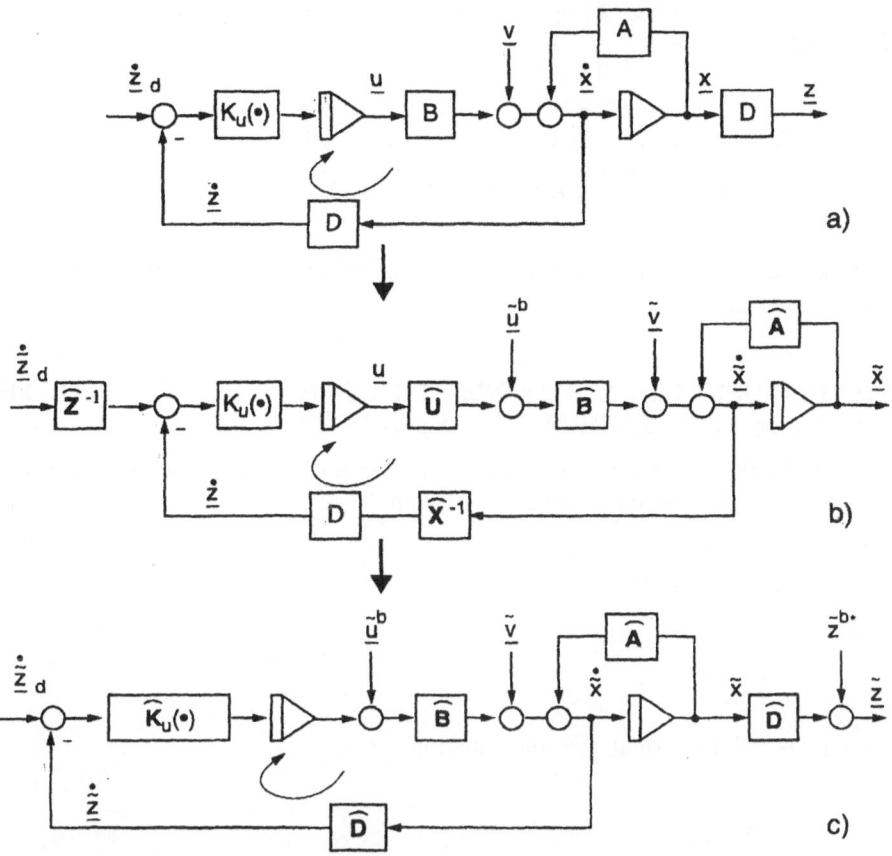

Figure 7 Process model and equation solver with
a) physical variables
b) mixed variables
c) scaled variables

To illustrate the consequence of the scaling transformations introduced above, we look at the functional block diagram of the END-algorithm for a linear multivariable system described by (11) and shown in Figure 7a. Employing the transformations of (48), (49), (9) and (14) we get the diagram of Figure 7b. And from there the transformation to Figure 7c is obvious. In addition to the transformations introduced above we now discover

$$\widehat{\mathbf{K}}_u(\cdot) = \widehat{\mathbf{U}} \, \mathbf{K}_u(\cdot) \, \widehat{\mathbf{Z}}^{-1} \tag{58}$$

We can test the validity of (58) by introducing the expressions for each of the three elements as follows

$$\widehat{\mathbf{K}}_u(\cdot) = -\widehat{\mathbf{U}}(\Lambda(\mathbf{DB})^{-1})\widehat{\mathbf{Z}}^{-1} = -\Lambda(\widehat{\mathbf{Z}}\,\mathbf{DB}\,\widehat{\mathbf{U}}^{-1})^{-1}$$
$$= -\Lambda(\widehat{\mathbf{Z}}\,\mathbf{D}\,\widehat{\mathbf{X}}^{-1}\,\widehat{\mathbf{X}}\,\mathbf{B}\,\widehat{\mathbf{U}}^{-1})^{-1} = -\Lambda(\widehat{\mathbf{D}}\,\widehat{\mathbf{B}})^{-1} \tag{59}$$

(59) shows that the system behavior is independent of the scaling of each of the variables as should be expected.

7. THE INFLUENCE OF THE NONMINIMUM-PHASE BEHAVIOR OF THE PROCESS ON END-ALGORITHM PERFORMANCE

The END-algorithm is essentially based on inversion of the dynamic model because it seeks to convert the original process description

$$\dot{\mathbf{x}} = \mathbf{f}(\mathbf{x}, \mathbf{u}, \mathbf{v}) \tag{60}$$

$$\mathbf{z} = \mathbf{d}(\mathbf{x}) \tag{61}$$

in which $\mathbf{x}$ is n-dimensional, into the equation

$$\dot{\mathbf{z}} = \dot{\mathbf{z}}_d \tag{62}$$

where $\mathbf{z}$ is r-dimensional $(r = \dim \mathbf{z} = \dim \mathbf{u} < n)$.

We assume for simplicity that the system is linear and

$$\mathbf{D} = \frac{\partial \mathbf{d}(\mathbf{x})}{\partial \mathbf{x}} \quad \text{and} \quad \mathbf{B} = \frac{\partial \mathbf{f}(\cdot)}{\partial \mathbf{u}}$$

We have seen that in order for the inverse solution to exist, the matrix $(\mathbf{DB})$ must be non-singular and the matrix

$$\widetilde{\mathbf{A}} = (\mathbf{I} - \mathbf{B}(\mathbf{DB})^{-1}\mathbf{D})\mathbf{A}$$

must have its eigenvalues in the left half plane.

If the process described by **A** and **B** displays nonminimum-phase behavior, this will influence the design of the **D**-matrix.

The property variables, which are actually the object of control, are defined by

$$\mathbf{z}^0 = \mathbf{D}^0 \mathbf{x} \tag{63}$$

But since the matrix $(\mathbf{D}^0\mathbf{B})$ is usually singular, we need to define

$$\mathbf{D} = \mathbf{D}^0 + \Delta\mathbf{D}$$

in which $\Delta\mathbf{D}$ is to be designed in order to achieve invertibility. The form of $\Delta\mathbf{D}$ will be different (the elements of $\Delta\mathbf{D}$ will be larger) for a system in which the process has non-minimum-phase behavior compared to one without. The consequence will be that the final feedback control system based upon the computed property variables $\mathbf{z}^0$ will have lower bandwidth when the process has nonminimum phase behavior.

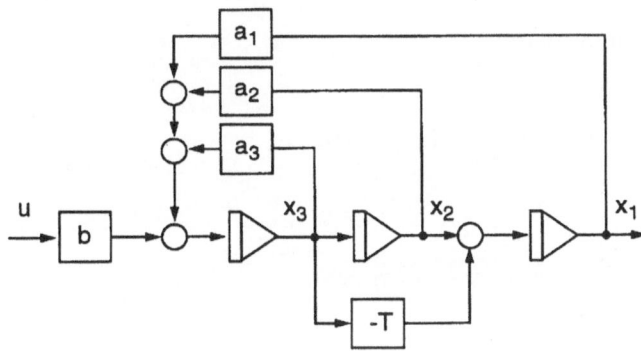

Figure 8 Simple third order process with non-minimum phase behavior.

This phenomenon will be illustrated by an example of a very simple linear single input-single output system shown in Figure 8. From the elementary block diagram of Figure 8 we see that the process is linear, of third order, and has the transfer function from the input (u) to the output, the last state (x_1)

$$\frac{x_1}{u}(s) = \frac{b(1 - Ts)}{(s - a_1)(s - a_2)(s - a_3)} \tag{64}$$

As is seen, the process contains a "negative feedforward" term leading to a zero in the right half plane in the transfer function, i.e. non-minimum phase behavior.

The state space model of this process will be

$$\dot{x}_1 = x_2 - T\dot{x}_2 = x_2 - Tx_3 \tag{65}$$

$$\dot{x}_2 = x_3$$

$$\dot{x}_3 = a_1 x_1 + a_2 x_2 + a_3 x_3 + bu$$

or

$$\dot{\mathbf{x}} = \begin{bmatrix} 0 & 1 & -T \\ 0 & 0 & 1 \\ a_1 & a_2 & a_3 \end{bmatrix} \mathbf{x} + \begin{bmatrix} 0 \\ 0 \\ b \end{bmatrix} u = \mathbf{A}\mathbf{x} + \mathbf{b}u \tag{66}$$

We actually want to control

$$z_1^0 = x_1 = \mathbf{d}^{0^T}\mathbf{x} = [1 \ 0 \ 0]\mathbf{x} \tag{67}$$

But since

$$\mathbf{d}^{0^T}\mathbf{b} = 0$$

we have to design the new property transformation

$$\mathbf{d}^T = \mathbf{d}^{0^T} + \Delta\mathbf{d}^T = [1 d_2 d_3] \tag{68}$$

The END-structure for this process is shown in Figure 9 where the "equation solver" is indicated on the left. Assuming that the "equation solver" is very fast ($\lambda \to \infty$), we will get the model for the inner system

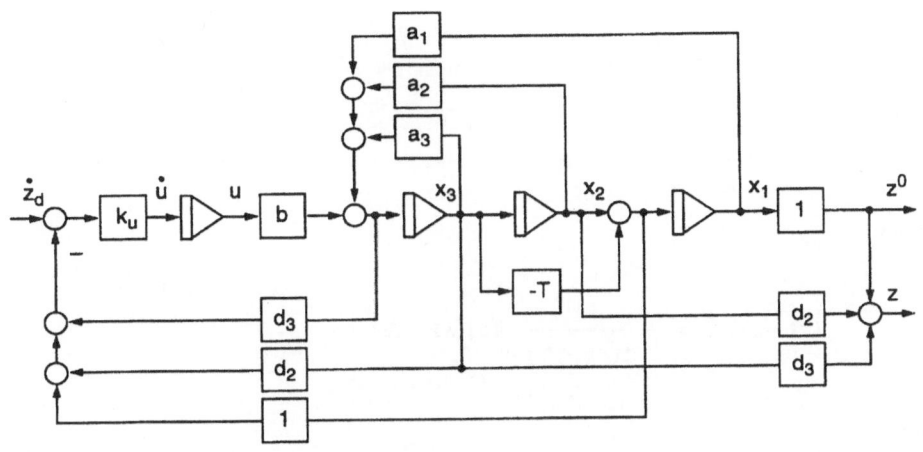

Figure 9 END structure for process of Figure 8.

$$\dot{\mathbf{x}} = (\mathbf{I} - \mathbf{b}(\mathbf{d}^T\mathbf{b})^{-1}\mathbf{d}^T)\mathbf{A}\mathbf{x} + \mathbf{b}(\mathbf{d}^T\mathbf{b})^{-1}\dot{z}_d = \tilde{\mathbf{A}}\mathbf{x} + \tilde{\mathbf{b}}\dot{z}_d \tag{70}$$

Applying (66) and (68) to (69) we get

$$\tilde{\mathbf{A}} = \begin{bmatrix} 0 & 1 & -T \\ 0 & 0 & 1 \\ 0 & -\dfrac{1}{d_3} & \dfrac{d_2-T}{d_3} \end{bmatrix} \tag{71}$$

$$\tilde{\mathbf{b}} = \begin{bmatrix} 0 \\ 0 \\ \dfrac{1}{d_3} \end{bmatrix} \tag{72}$$

By Laplace-transformation we get

$$\mathbf{x}(s) = (s\mathbf{I} - \tilde{\mathbf{A}})^{-1}\tilde{\mathbf{b}}\dot{z}_\mathrm{d} \qquad (73)$$

and

$$z(s) = \mathbf{d}^\mathrm{T}\mathbf{x}(s) \qquad (74)$$

We have

$$(s\mathbf{I} - \tilde{\mathbf{A}})^{-1} = \frac{1}{\det(s\mathbf{I} - \tilde{\mathbf{A}})} \cdot \mathrm{adj}(s\mathbf{I} - \tilde{\mathbf{A}}) \qquad (75)$$

where

$$\mathrm{adj}(s\mathbf{I} - \tilde{\mathbf{A}}) = \begin{bmatrix} s^2 + \dfrac{d_2 - T}{d_3}s + \dfrac{1}{d_3} & s + \dfrac{d_2}{d_3} & 1 - Ts \\ 0 & s\left(s + \dfrac{d_2 - T}{d_3}\right) & s \\ 0 & \dfrac{s}{d_3} & s^2 \end{bmatrix} \qquad (76)$$

Thereby we obtain

$$\mathbf{x}(s) = \begin{bmatrix} \dfrac{1 - Ts}{s} \\ 1 \\ s \end{bmatrix} \frac{1}{d_3\left(s^2 + \dfrac{d_2 - T}{d_3}s + \dfrac{1}{d_3}\right)}\dot{z}_\mathrm{d} \qquad (77)$$

Now applying (76) and (67) to (73) we get

$$\frac{z}{\tilde{z}_\mathrm{d}}(s) = \frac{[1\,d_2\,d_3]\mathbf{x}(s)}{\dot{z}_\mathrm{d}(s)} = \frac{1}{s} \qquad (78)$$

$$\frac{z^0}{\dot{z}_d}(s) = \frac{[1\ \ 0\ \ 0]\mathbf{x}(s)}{\dot{z}_d(s)} = \frac{1 - Ts}{s(1 + (d_2 - T)s + d_3 s^2)} \tag{79}$$

We can now see clearly the consequence of the choice of the two elements d_2 and d_3 in the END-algorithm. The second order term in the denominator of (78) can be written

$$(1 + (d_2 - T)s + d_3 s^2) = \left(1 + 2\zeta\frac{s}{\omega_0} + \left(\frac{s}{\omega_0}\right)^2\right) \tag{80}$$

where $\omega_0 = \dfrac{1}{\sqrt{d_3}}$ and $\zeta = \dfrac{d_2 - T}{2\sqrt{d_3}}$.

ω_0 must be chosen relative to the dynamics of the original process.

Assuming $-a_3 > -a_2 > -a_1$ it may be feasible to choose $\omega_0 = -a_3$. Thereby we have $d_3 = \dfrac{1}{a_3^2}$. Furthermore it makes sense to choose $\zeta = \dfrac{1}{2}$ in order to have reasonable damping of the inner system leading to

$$d_2 = T + \frac{1}{\omega_0} = T + \sqrt{d_3} = T - \frac{1}{a_3}$$

From (77) we see that we have achieved the original goal to transform the third order process into a system of first order (one single integrator). But we also see from (78) that the quantity which we actually want to control ($z^0 = x_1$) is still described by a third order transfer function which contains the integrator, two complex conjugate poles and a zero in the right half plane. *In other words, we have not removed the nonminimum phase behavior.* This is illustrated even clearer by considering the transfer function between $\dot{z}_d$ and $\Delta z = z - z^0$ which is

$$\frac{\Delta z}{\dot{z}_d}(s) = \frac{1}{s} - \frac{1 - Ts}{s(1 + (d_2 - T)s + d_3 s^2)} = d_2 \frac{1 + \dfrac{d_3}{d_2}s}{1 + (d_2 - T)s + d_3 s^2} \tag{81}$$

Introducing the above specification of   , (80) will yield

$$\frac{\Delta z}{\dot{z}_d}(s) = \frac{1 + \omega_0 T}{\omega_0} \cdot \frac{1 + \dfrac{s}{\omega_0(1 + \omega_0 T)}}{1 + \dfrac{s}{\omega_0} + \left(\dfrac{s}{\omega_0}\right)^2} \tag{82}$$

Figure 10 shows plots of $\left|\dfrac{\Delta z}{\dot{z}_d}(j\omega)\right|$ for $T = 0$ and for a case of $T > 0$. It is clearly seen that deviation (Δz) is larger for larger $T > 0$.

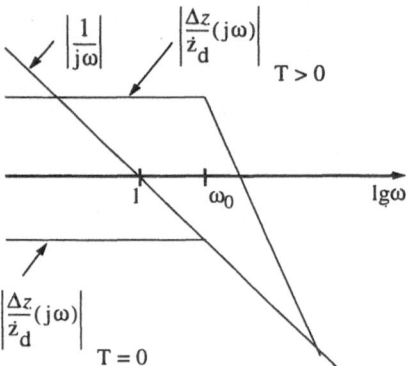

Figure 10 Asymptotic plot $\left|\dfrac{\Delta z}{\dot{z}_d}(j\omega)\right|$ for different T.

Similarly Figure 11 is an illustration of the frequency response of (78) with the assumption of $\zeta = \dfrac{1}{2}$. The phase diagram at the bottom particularly illustrates that ω_{180} (the frequency at which the phase angle is $-180°$) is lower with increasing values of T. This means that the bandwidth of the feedback control system from z^0 will be reduced with increasing nonminimum phase behavior.

The conclusions drawn above apply to monovariable systems (single input - single output). In multivariable systems the situation is different and may be less serious because of the multiplicity of channels of influence in such systems.

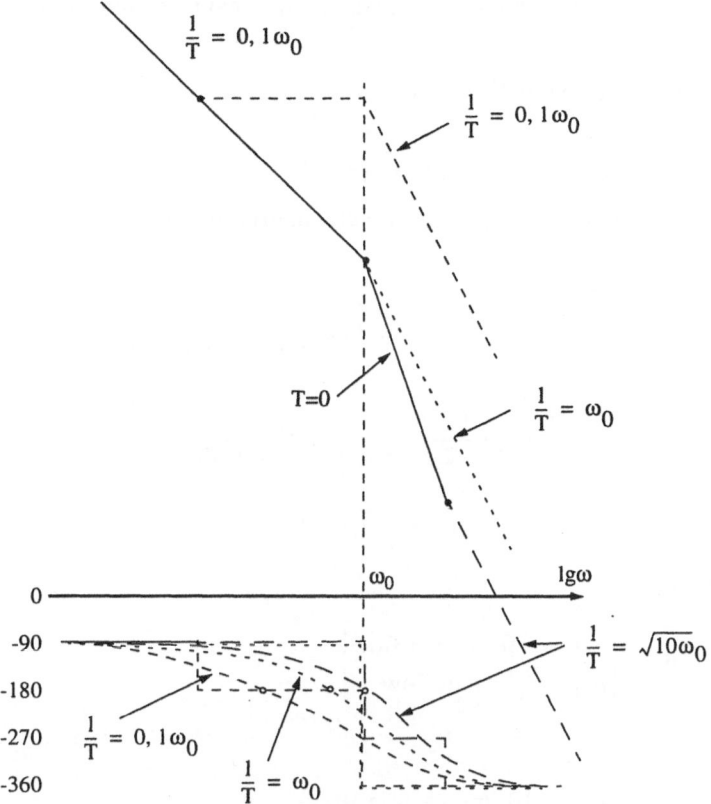

Figure 11 Asymptotic amplitude (upper) and phase plot (lower)

$$\text{of } \left| \frac{\Delta z}{\dot{z}_d}(j\omega) \right| \text{ for different T.}$$

In multivariable systems a control algorithm based upon the inverse of the process transfer matrix will not be realizable if the transfer matrix has *transmission zeros* in the right half plane. Transmission zeros of a transfer matrix are defined as the poles of the inverse matrix. This is derived as follows:

The process transfer matrix from **u** to **z** is denoted $\mathbf{H}_{zu}(s)$. Its inverse is

$$(\mathbf{H}_{zu}(s))^{-1} = \frac{\text{adj } \mathbf{H}_{zu}(s)}{\text{det } \mathbf{H}_{zu}(s)} \tag{83}$$

The poles of (82) are the zeros of the demoninator, that is the points in the s-plane giving

$$\det \mathbf{H}_{zu}(s) = 0 \tag{84}$$

These are the *transmission zeros* of $\mathbf{H}_{zu}(s)$.

A typical example of the above is the transfer matrix of a steam boiler (Chien et.al. 1958).

$$\mathbf{H}_{zu}(s) = \begin{bmatrix} k_1 \dfrac{1}{s} & , & -k_2 \dfrac{(1-122s)}{(1+3s)(1+40s)} \\[4mm] -k_3 \dfrac{1}{1+40s} & , & k_4 \dfrac{1}{1+40s} \end{bmatrix} \tag{85}$$

where

$$\mathbf{u} = \begin{bmatrix} u_1 \\ u_2 \end{bmatrix} = \begin{bmatrix} \text{feedwater flow} \\ \text{fuel flow} \end{bmatrix}$$

$$\mathbf{z} = \begin{bmatrix} z_1 \\ z_2 \end{bmatrix} = \begin{bmatrix} \text{drum water level} \\ \text{steam pressure} \end{bmatrix}$$

In (84) we see that one of the transfer functions has a dominant zero in the right half plane. This is caused by the so-called "shrink and swell effect" of the drum water level when the fuel flow is manipulated.

From (84) we establish

$$\det H_{zu}(s) = k_1 k_4 \frac{1}{s(1+40s)} - k_2 k_3 \cdot \frac{(1-122s)}{(1+3s)(1+40s)^2} \tag{86}$$

$$= \frac{k_2 k_3 \left(\dfrac{k_1 k_4}{k_2 k_3}(1+35)(1+40s) - s(1-122s) \right)}{s(1+3s)(1+40s)^2} = 0$$

k_1, k_2, k_3 and k_4 are positive numbers and for the boiler at hand we have

$$\frac{k_1 k_4}{k_2 k_3} = 1,11$$

We can now resolve the expression of (85) and get

$$\frac{k_2 k_3 \cdot 1,11(1 + 6,45s)(1 + 35,59s)}{s(1 + 3s)(1 + 40s)^2} = 0 \qquad (87)$$

yielding the zeros $s = -0,155$ and $s = -0,028$

Thus we see that even though the transfer matrix of (84) has a term with a dominant zero in the right half plane, the transmission zeros will both be located in the left half plane. Therefore the inverse (82) is realizable. In a monovariable system the zeros of the process transfer function are the transmission zeros of the feedback loop whereas in a multivariable system, they are not.

The conclusion to be drawn from the above discussion is as follows:

• In the design of the parameters of the END-algorithm, possible non-minimum phase behavior of the process is not considered explicitly.

• Non-minimum phase behavior of the process will not influence the "loop I" of the END control system, but will appear in the "loop II" and give this loop a lower bandwidth compared with a minimum phase system.

8. CONCLUSIONS

The END-algorithm offers a straightforward and logical procedure for the design of control strategies for a broad class of nonlinear multivariable processes. It is based on an off-line, computer-aided optimization of the critical control parameters $(\Delta \mathbf{D})$ and is very conveniently tuned via a small set of tuning parameters $\alpha, \beta_1, \beta_2 \dots$. The END-algorithm also provides a solution to handle constraints in control and state variables.

9. ACKNOWLEDGMENT

This research has been partly supported by STATOIL.

10. REFERENCES

1. Balchen, J.G. (1993). A modified LQG algorithm (MLQG) for robust control of nonlinear multivariable systems. *Modeling, Identification and Control (MIC)*, **14**, no. 3, pp 175-180.

2. Balchen, J.G. and Sandrib, B. (1995a). Elementary nonlinear decoupling control of composition in binary distillation columns. *J. Proc. Cont.*, 5, no. 4, pp 241-247

3. Balchen, J.G. and Sandrib, B. (1995b). Input saturation in nonlinear multivariable processes resolved by nonlinear decoupling. *3rd IEEE Conference on Control Applications*, Glasgow, Aug. 1994. Reprinted in *MIC*, **16**, no. 2, pp 95-106.

4. Balchen, J.G. and Sandrib, B. (1995c). Hard constraints in control and state variables of multivariable nonlinear processes resolved by Elementary Nonlinear Decoupling. *Proc. IFAC Symp. DYCORD+'95*, Helsingør, Denmark, June 1995.

5. Balchen, J.G. and Larsen, G. (1998). Control of a steam boiler by Elementary Nonlinear Decoupling (END). *Nonlinear Model Based Process Control*, editors: R. Berber and C. Kravaris, Kluwer Academic Publishers, Dordrecht

6. Chien, K.L., Ergin, E.I., Ling, C., and Lee, A. (1958). The noninteracting controller for a steam generating system. *Control Engineering*, **5** (Oct.), pp 95-101.

7. Henson, M.A., and Seborg, D.E. (1997). *Nonlinear Process Control*. Prentice-Hall, New Jersey

8. Isidori, A. (1989). *Nonlinear Control Systems* (2nd ed.). Springer Verlag, Berlin.

9. Qin, S.J., and Badgwell, T.A. (1996). An overview of industrial model predictive control technology. *Proc. Chemical Process Control (CPC-V)*, Tahoe Cal. Jan. 1996.

CONTROL OF NONLINEAR DIFFERENTIAL ALGEBRAIC EQUATION SYSTEMS : AN OVERVIEW

Aditya Kumar and Prodromos Daoutidis*

Department of Chemical Engineering and Materials Science
University of Minnesota, Minneapolis MN 55455 USA

1. Introduction

Chemical processes are inherently nonlinear and their dynamics are naturally described by systems of coupled differential and algebraic equations (DAEs); the differential equations arise from the standard dynamic balances of mass, energy and momentum, while the algebraic equations typically include thermodynamic relations, empirical correlations, quasi-steady-state relations etc. In many cases, the algebraic equations in the DAE model can be readily eliminated to obtain an equivalent ordinary differential equation (ODE) model, which can be used as the basis for controller design. On the other hand, there is a broad class of chemical processes for which the algebraic equations in the DAE models are "singular" in nature, and thus, inhibit a direct reduction of the DAE model into an ODE system. Such DAE systems with singular algebraic equations are said to have a high "index" and they are fundamentally different from ODE systems.

DAE systems have been a subject of extensive research over the past two decades, with particular emphasis on their numerical analysis and simulation (see the books [3, 14] and the references therein). It is now well-established that available numerical simulation methods for ODEs may exhibit poor convergence or fail altogether for DAEs [31]. For instance, unlike the case with ODEs, arbitrary initial conditions or non-smooth inputs in high-index DAEs may lead to impulsive solutions and cause failure of the simulation methods. The specification of "consistent" initial conditions that yield smooth solutions, is an important and non-trivial task in the simulation of DAEs [5, 24, 28]. An increased understanding of these differences between high-index DAEs and ODEs has led to the development of a variety of numerical simulation methods for specific classes of DAEs,

*Author to whom all correspondence should be addressed.
†Financial support for this work from the National Science Foundation, Grant CTS-9320402, is gratefully acknowledged.

R. Berber and C. Kravaris (eds.), Nonlinear Model Based Process Control, 311-344.
© 1998 *Kluwer Academic Publishers.*

ranging from the methods using a combination of index-reduction techniques with efficient ODE integration methods for the resulting index-one or index-zero DAE (e.g., [1, 6, 13]), to those employing nonlinear constrained optimization techniques [16, 33].

Research on the control of high-index DAE systems has focused primarily on linear systems, often referring to them as singular, semistate, or descriptor systems (see e.g., [4, 8, 25] for a survey of results). On the other hand, research on the analysis and control of nonlinear DAE systems had been quite limited, until recently. More specifically, basic system-theoretic issues like existence and uniqueness of solutions [10, 34] and stability analysis using Lyapunov techniques [2, 10] have been addressed for certain classes of nonlinear DAE systems. Initial work on control of nonlinear DAE systems focused on optimal control [7, 30], and the feedback stabilization [26] and tracking [17, 36] for some specific classes of DAE systems.

In this article, we review a control methodology for general high-index DAE systems developed in [20, 21], and illustrate through application examples, the effectiveness and advantages of such DAE model-based controllers. More specifically, in Section 2, we initially introduce the general description of DAE systems considered in this work and review the notion of index for such systems. While DAE systems with index one are essentially the same as ODE systems, high-index DAE systems are different. These differences are discussed in Section 3; Section 4 includes a few examples of various classes of chemical processes that are modeled by high-index DAEs. The design of feedback controllers for high-index DAE systems is addressed in Section 5, wherein two broad and fundamentally different classes of "regular" and "non-regular" DAE systems are identified. Finally, in Section 6, we illustrate the key advantages of using high-index DAE models of chemical processes for controller design, through simulations in two application examples.

2. DAE Systems – Preliminaries

The majority of chemical process applications are modeled by DAE systems in the so-called *semi-explicit* form, where the standard dynamic balances of mass, energy and momentum yield a set of explicit differential equations, whereas the thermodynamic relations, empirical correlations, quasi-steady-state relations etc., comprise the algebraic equations. Motivated by this, we consider nonlinear DAE systems with the following semi-explicit description:

$$\begin{aligned} \dot{x} &= f(x) + b(x)z + g(x)u \\ 0 &= k(x) + l(x)z + c(x)u \end{aligned} \tag{1}$$

where $x \in \mathcal{X} \subset \mathbb{R}^n$ is the vector of differential variables for which we have the explicit differential equations, e.g. holdup, composition and temperature, $z \in \mathcal{Z} \subset \mathbb{R}^p$ is the vector of algebraic variables which vary according to the algebraic equations ($\mathcal{X}$ and $\mathcal{Z}$ are open connected sets), $u \in \mathbb{R}^m$ is the vector of inputs to the system, $f(x)$ and $k(x)$ are analytic vector fields of dimensions n and p, respectively, and $b(x), g(x), l(x)$ and $c(x)$ are analytic matrices of appropriate dimensions. In this and the following section, we discuss the characteristics of the DAE system in Eq.1 with specified inputs $u(t)$. In sections 5 and 6, where we address the controller design, u denotes the vector of manipulated inputs which are not specified *a priori* as a function of time; rather they are determined by a feedback control law.

Note that in the above description, the inputs u and the algebraic variables z appear in the system equations in a linear (affine) fashion, which is typical of most practical applications. This linearity with respect to z and u, and the semi-explicit structure, facilitate the analysis and the derivation of explicit controller synthesis results. Systems that are nonlinear in u and/or z can also be easily recast in the above form through standard dynamic extension techniques (see e.g., [27], page 190). For instance, consider the following DAE system that is nonlinear in z:

$$
\begin{aligned}
\dot{x} &= f(x, z) + g(x, z)u \\
0 &= k(x, z) + c(x, z)u
\end{aligned}
\tag{2}
$$

Defining the extended vector of differential variables $\bar{x} = [x^T \ z^T]^T$ and the new vector of algebraic variables $\bar{z} = \dot{z}$, the following extended DAE system is obtained:

$$
\begin{bmatrix} \dot{x} \\ \dot{z} \end{bmatrix} = \begin{bmatrix} f(\bar{x}) \\ 0 \end{bmatrix} + \begin{bmatrix} 0 \\ I_p \end{bmatrix} \bar{z} + \begin{bmatrix} g(\bar{x}) \\ 0 \end{bmatrix} u
$$

$$
0 = k(\bar{x}) + c(\bar{x})u
\tag{3}
$$

which is clearly in the form of Eq.1 with $\bar{z}$ appearing in a linear fashion. Similarly, more general DAE systems with the following *fully-implicit* description:

$$
F(\dot{x}, x, u) = 0
\tag{4}
$$

where $F : \mathbb{R}^n \times \mathbb{R}^n \times \mathbb{R}^m \to \mathbb{R}^n$ is an analytic function, can also be transformed into a semi-explicit form by defining the algebraic variables $z = \dot{x}$ [13].

For the DAE system of Eq.1, we now recall the important notion of index [3].

Definition 1: The index ν_d of the DAE system in Eq.1 with specified smooth inputs $u(t)$, is the minimum number of times the algebraic equations or their subset have to be differentiated to obtain a set of differential equations for z, i.e. solve for $\dot{z} = \mathcal{F}(x, z, t)$.

The index ν_d of a DAE system provides a measure of the "singularity" of the algebraic equations, and the resulting differences from ODE systems. More specifically, consider the DAE system of Eq.1. Clearly, if the matrix $l(x)$ is non-singular, then the algebraic equations can be solved for z:

$$
z = -l(x)^{-1} [k(x) + c(x)u]
\tag{5}
$$

and one differentiation of the algebraic equations in Eq.1, or equivalently the solution for z, would yield the differential equations for z, i.e. the DAE system has an index $\nu_d = 1$. For such systems, a direct substitution of the solution for z in the differential equations for x, yields an equivalent ODE representation:

$$\dot{x} = \bar{f}(x) + \bar{g}(x)u \qquad (6)$$

where

$$\bar{f}(x) = f(x) - b(x)l(x)^{-1}k(x)$$
$$\bar{g}(x) = g(x) - b(x)l(x)^{-1}c(x)$$

Thus, index-one DAE systems of Eq.1 are essentially the same as ODE systems, and the simulation and control of such systems can be easily addressed on the basis of their ODE representation.

In contrast, DAE systems with singular algebraic equations, more specifically a singular matrix $l(x)$, cannot be readily reduced to an ODE system and they have a high index $\nu_d > 1$. In this article, we focus on such high-index DAE systems, which are fundamentally different from ODE systems. The above notion of index ν_d is also referred to as the differential index. Other notions of index that are related to the differential index, e.g. perturbation index [14] and structural index [35], are also used in the literature.

Remark 1: Consider the following linear time-invariant analogue of the DAE system in Eq.1:

$$\dot{x} = Ax + Bz + Gu$$
$$0 = Kx + Lz + Cu \qquad (7)$$

where A, B, G, K, L, C are constant matrices. The index of the above system is the same as the index of nilpotency [3] of the matrix pencil:

$$\mathcal{P} = \begin{bmatrix} \lambda I_n - A & -B \\ -K & -L \end{bmatrix}, \qquad (8)$$

which is regular for a solvable system, i.e. $\det \mathcal{P} \neq 0$ for some $\lambda \in \mathbb{C}$. For linear time-varying DAE systems, a notion of index was defined in [11].

3. Comparison of high-index DAE and ODE systems

An ODE system can be viewed as a special class of DAE systems with an index $\nu_d = 0$. While DAE systems with index one are also the same as ODE systems, DAE systems with a high index $\nu_d > 1$ are fundamentally different. More specifically, consider a high-index DAE system of Eq.1 with a singular matrix $l(x)$. For such systems, the singular algebraic equations imply the presence of underlying algebraic constraints in the differential variables x. Furthermore, if the index ν_d

exceeds two, then the algebraic equations have to be differentiated several times (depending on ν_d) to obtain a solution for z, and the differentiated equations impose additional constraints in x. These constraints restrict the solution for x in a state space of dimension less than n. Consequently, for arbitrary initial conditions $x(0)$, the DAE system exhibits an impulsive behavior at the initial time $t = 0$. This is a common cause of failure in the numerical simulation of DAE systems. Only the initial conditions $x(0)$ that satisfy the underlying algebraic constraints in x, allow smooth solutions $(x(t), z(t))$ and are referred to as "consistent" initial conditions. Clearly, this is in contrast with ODE systems, for which a well-defined solution exists for any initial condition $x(0)$. Furthermore, unlike ODE systems, the solution $(x(t), z(t))$ of high-index DAE systems in Eq.1 may also depend on the time-derivatives of the inputs $u(t)$, in which case these inputs must be sufficiently smooth. These differences between DAE and ODE systems are illustrated through the following simple example.

Example 1: Consider the DAE system:

$$\dot{x}_1 = x_2 \tag{9}$$
$$\dot{x}_2 = x_3 + x_4\, u(t) \tag{10}$$
$$\dot{x}_3 = x_1 \tag{11}$$
$$\dot{x}_4 = x_2 + z \tag{12}$$
$$0 = x_1 + x_2 \tag{13}$$

with the differential variables $x = [x_1\ x_2\ x_3\ x_4]^T$, an algebraic variable z and an input $u(t)$ specified as a function of time on an interval $[0, T]$. The algebraic equation $0 = x_1 + x_2$ is singular ($l(x) \equiv 0$) and specifies a constraint in x. Differentiating this constraint with respect to time, yields the following new equation:

$$0 = x_2 + x_3 + x_4\, u(t) \tag{14}$$

which again does not involve z and consequently, denotes another constraint in x. Another differentiation of the constraint in Eq.14 yields the following equation:

$$0 = x_1 + x_3 + (x_2 + x_4)u(t) + x_4\, \dot{u}(t) + u(t)\, z \tag{15}$$

which involves z.

Clearly, if $u(t) \neq 0$ on the time interval $[0, T]$, then Eq.15 can be solved for z:

$$z = -\frac{1}{u(t)}\left[x_1 + x_3 + (x_2 + x_4)u(t) + x_4\, \dot{u}(t)\right] \tag{16}$$

and the index is $\nu_d = 3$; another differentiation of Eq.15 would yield the solution for $\dot{z}$. The algebraic constraints in Eq.13,14 denote two constraints in x, one of which involves the input $u(t)$. These two constraints restrict the evolution of the four differential variables x in a subspace of dimension two, and a smooth solution for the DAE system exists only for "consistent" initial conditions $x(0)$

that satisfy these constraints. Furthermore, the solution for z (Eq.16) depends on the time-derivative of $u(t)$. Thus, the input $u(t)$ must be at least once continuously differentiable (as opposed to being just continuous for ODE systems).

On the other hand, if $u(t_1)=0$ at any time $t_1 \in [0, T]$, then at that instant Eq.15 becomes singular in z, i.e., it denotes an additional constraint in x, and $\nu_d \neq 3$. Thus, for nonlinear DAE systems with time-varying inputs, the index may depend on these inputs, or may even fail to be well-defined, leading to questions on the solvability of such systems. In contrast, the solvability and index ν_d of a linear DAE system of Eq.7 are independent of the inputs $u(t)$; they are completely determined by the matrix pencil in Eq.8.

4. Chemical processes modeled by high-index DAEs

In this section, we present a few examples of different classes of chemical processes that are modeled by DAE systems of the form in Eq.1 with a high index $\nu_d > 1$. While the presence of DAE models with a high index has been documented in the literature [12, 23, 29], often the source of high index is ascribed to "improper modeling" or "incorrect choice" of process specifications. In the DAE models of the processes described in this section, the high index arises naturally as a consequence of standard modeling assumptions like phase and reaction equilibrium, negligible pressure drop etc., corresponding to fast mass transfer/reactions and high pressure gaseous flow with small resistance.

4.1. TWO-PHASE REACTOR

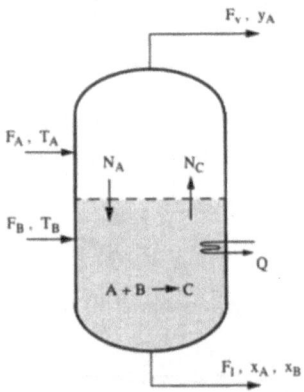

Figure 1. Two-phase reactor

Consider the two-phase (liquid/vapor) reactor in Figure 1, where reactants A and B are fed as pure vapor and liquid streams at molar flowrates F_A and F_B and temperatures T_A and T_B, respectively. An exothermic irreversible reaction

$$A + B \rightarrow C$$

occurs in the liquid phase, and the reactant B is non-volatile compared to reactant A and product C. The rate of production of C is given by the Arrhenius relation:

$$R_C = k_o \exp(-E_a/RT) M_l \rho x_A x_B \tag{17}$$

where M_l and ρ are the molar liquid holdup and density, T is the reactor temperature, and x_A and x_B are the liquid-phase mole fractions of A and B, respectively. For simplicity, it is assumed that all components have equal and constant molar density ρ, heat capacity c_p and heat of vaporization ΔH^v, and the liquid and vapor phases behave ideally. The product vapor and liquid streams are withdrawn from the reactor at molar flowrates F_v and F_l, respectively. Furthermore, the interphase mass transfer is much faster than the reaction, i.e. the vapor and liquid phases are close to equilibrium.

Under the assumption of well-mixed, ideal liquid and vapor phases that are at equilibrium, a dynamic model of the reactor is easily derived, which consists of the following set of differential and algebraic equations:

$$\dot{M}_v = F_A - N_A + N_C - F_v$$

$$\dot{y}_A = \frac{1}{M_v}[F_A(1 - y_A) - N_A(1 - y_A) - N_C y_A]$$

$$\dot{M}_l = F_B - F_l - R_C + N_A - N_C$$

$$\dot{x}_A = \frac{1}{M_l}[-F_B x_A - R_C(1 - x_A) + N_A(1 - x_A) + N_C x_A]$$

$$\dot{x}_B = \frac{1}{M_l}[F_B(1 - x_B) - R_C(1 - x_B) - N_A x_B + N_C x_B]$$

$$\dot{T} = \frac{1}{(M_l + M_v)c_p}[F_A c_p(T_A - T) + F_B c_p(T_B - T) - R_C(\Delta H_R^o - c_p(T - T_o))$$

$$+ (N_A - N_C)\Delta H^v - Q]$$

$$0 = P_A^s x_A - P y_A \tag{18}$$

$$0 = P_C^s(1 - x_A - x_B) - P(1 - y_A)$$

$$0 = M_v RT - P\left(V_T - \frac{M_l}{\rho}\right)$$

The differential equations in the above DAE model include the total and component mole balances in the liquid and the vapor phases, and the overall energy balance, while the algebraic equations include the phase equilibrium relations for the volatile components A and C and the ideal gas equation for the vapor holdup. In the above equations M_v and y_A denote the molar holdup and the mole fraction of A in the vapor phase, P is the reactor pressure, P_A^s and P_C^s denote the saturation vapor pressures of A and C at the reactor temperature T, given by the standard Antoine relations, and N_A, N_C are the inter-phase mole transfer rates for A and C, respectively. Clearly, the above model is a DAE system in the semi-explicit form of Eq.1, with the vector of differential variables $x = [M_v \ y_A \ M_l \ x_A \ x_B \ T]^T$ and the vector of algebraic variables $z = [N_A \ N_C \ P]^T$. Owing to the phase equilibrium assumption, the algebraic equations do not involve the inter-phase mole-transfer rates N_A or N_C. Thus, the algebraic equations are singular and the DAE model in Eq.18 has a high index. It can be verified that the index is two. Such high-index models arise similarly in the case of high-pressure absorption/ distillation columns and reactive distillation columns, where each tray is like a two-phase reactor discussed above.

4.2. REACTOR WITH FAST AND SLOW REACTIONS

Consider an isothermal CSTR, where reactant A is fed at a flowrate F_o and concentration C_{Ao}, the product is withdrawn at a flowrate F and the following first order reactions occur in series:

$$A \rightleftharpoons B \rightarrow C$$

with the net forward rates of reactions R_1 and R_2, respectively. The reversible reaction $A \rightleftharpoons B$ is much faster than the irreversible reaction $B \rightarrow C$. Thus, the fast reversible reaction is essentially at equilibrium. Under an assumption of reaction equilibrium, the following DAE model of the process is obtained:

$$\dot{V} = F_o - F$$
$$\dot{C}_A = \frac{F_o}{V}(C_{Ao} - C_A) - R_1$$
$$\dot{C}_B = -\frac{F_o}{V}C_B + R_1 - R_2$$
$$\dot{C}_C = -\frac{F_o}{V}C_C + R_2 \qquad (19)$$
$$0 = C_A - \frac{C_B}{K_{eq}}$$
$$0 = R_2 - k_2 C_B$$

The differential equations for the four differential variables $x = [V \ C_A \ C_B \ C_C]^T$ involve the two algebraic variables $z = [R_1 \ R_2]^T$. Note that while the second algebraic equation yields the explicit relation for the slow reaction rate R_2, the reaction rate R_1 for the fast reversible reaction does not appear in any of the algebraic equations, due to the quasi-steady-state assumption of reaction equilibrium. Thus, the algebraic equations are singular and the DAE model in Eq.19 has an index $\nu_d = 2$.

In many practical applications, e.g. chemical vapor deposition (CVD), catalytic crackers and combustion, several reactions occur simultaneously, some of which are often much faster than the others. The above example illustrates that the quasi-steady-state assumptions of reaction equilibrium for fast reversible reactions, and similarly complete conversion for fast irreversible reactions, lead to high-index DAE models of such processes.

4.3. REACTOR WITH FAST HEAT TRANSFER THROUGH A JACKET

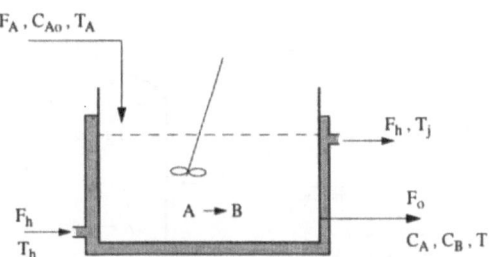

Figure 2. A CSTR with heating jacket

Consider the CSTR with heating jacket in Figure 2. Reactant A is fed at a flowrate F_A, molar concentration C_{Ao} and temperature T_A to the reactor, where the irreversible endothermic reaction $A \to B$ occurs. The rate of reaction is given by the Arrhenius relation:

$$r_A = k_o \exp(-E_a/RT)C_A$$

where C_A is the molar concentration of A in the reactor holdup, and T is the reactor temperature. The product stream is withdrawn at a flowrate F_o and heat is provided to the reactor through the heating jacket, where the heating fluid is fed at a flowrate F_h and temperature T_h. Consider the case when the heat transfer from the jacket to the reactor is much faster than the reaction, and correspondingly the reactor and the heating jacket are assumed to be in thermal equilibrium. Under this assumption, the following DAE model of the process is obtained:

$$\dot{V} = F_A - F_o$$

$$\dot{C}_A = \frac{F_A}{V}(C_{Ao} - C_A) - k_o \exp(-E_a/RT)C_A$$

$$\dot{C}_B = -\frac{F_A}{V}C_B + k_o \exp(-E_a/RT)C_A$$

$$\dot{T} = \frac{F_A}{V}(T_A - T) - k_o \exp(-E_a/RT)C_A \frac{\Delta H_r}{\rho c_p} + \frac{Q}{\rho V c_p}$$

$$\dot{T}_j = \frac{F_h}{V_h}(T_h - T_j) - \frac{Q}{\rho_h V_h c_{ph}} \tag{20}$$

$$0 = T - T_j$$

Clearly, due to the quasi-steady-state assumption of thermal equilibrium, the algebraic equation can not be solved for the algebraic variable $z = Q$, i.e. the heat transfer rate from the jacket to the reactor. It can be easily verified that the above DAE model of the process has an index $\nu_d = 2$.

4.4. CASCADE OF REACTORS WITH NEGLIGIBLE PRESSURE DROP

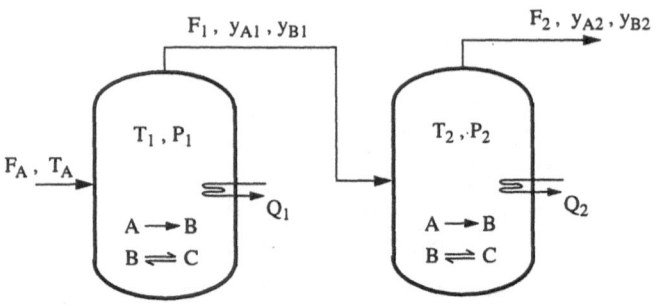

Figure 3. Cascade of two gas phase reactors

Consider the cascade of two gas phase CSTRs in Figure 3. Gaseous reactant A is fed to the first reactor at a molar flowrate F_A and temperature T_A, where the exothermic reactions $A \rightarrow B$, $B \rightleftharpoons C$ occur in series. The reactor operates at a high temperature T_1 to promote the conversion of A to B in the first irreversible reaction. However, the production of C is restricted by the thermodynamic equilibrium limitation for the reversible reaction. The gaseous mixture from the first reactor is fed to the second reactor, which operates at a lower temperature T_2 to shift the reaction equilibrium in the forward direction, favoring the production of C. The product vapor stream from the second reactor is withdrawn at a molar flowrate F_2. On the other hand, the molar flowrate F_1 of the vapor stream from the first reactor to the second reactor is governed by the pressure drop $\Delta P = P_1 - P_2$ between the two reactors:

$$F_1 = (\frac{P_1}{RT_1}) \frac{(\Delta P)^{4/7}}{\sigma}$$

where P_1/RT_1 is the molar density of the vapor.

At high operating pressures, this molar density is high, and consequently a small pressure drop may yield a high molar flowrate. Thus, at high operating pressures, the pressure drop between the two reactors is negligible and the following DAE model of the process is obtained:

$$\dot{M}_1 = F_A - F_1$$
$$\dot{y}_{A1} = \frac{1}{M_1}[F_A(1 - y_{A1}) - R_{B,1}]$$
$$\dot{y}_{B1} = \frac{1}{M_1}[-F_A y_{B1} + R_{B,1} - R_{C,1}]$$

$$\dot{T}_1 = \frac{1}{M_1 c_p}[F_A c_p(T_A - T_1) - R_{B,1}\Delta H_{R1} - R_{C,1}\Delta H_{R2} - Q_1]$$

$$\dot{M}_2 = F_1 - F_2$$

$$\dot{y}_{A2} = \frac{1}{M_2}[F_1(y_{A1} - y_{A2}) - R_{B,2}]$$

$$\dot{y}_{B2} = \frac{1}{M_2}[F_1(y_{B1} - y_{B2}) + R_{B,2} - R_{C,2}]$$

$$\dot{T}_2 = \frac{1}{M_2 c_p}[F_1 c_p(T_1 - T_2) - R_{B,2}\Delta H_{R1} - R_{C,2}\Delta H_{R2} - Q_2]$$

$$0 = M_1 R T_1 - P_1 V_{1T} \tag{21}$$

$$0 = M_2 R T_2 - P_2 V_{2T}$$

$$0 = P_1 - P_2$$

with the differential variables $x = [M_1\ y_{A1}\ y_{B1}\ T_1\ M_2\ y_{A2}\ y_{B2}\ T_2]^T$ and the algebraic variables $z = [P_1\ P_2\ F_1]^T$. In the above model, $R_{B,i}, R_{C,i}$ denote the rate of production of B and C through the irreversible and reversible reactions, respectively, in the first $(i = 1)$ and second $(i = 2)$ reactors, given by:

$$R_{B,i} = k_{1,o}\exp(-E_1/RT_i)M_i y_{Ai}$$

$$R_{C,i} = k_{2,o}\exp(-E_2/RT_i)M_i\left(y_{Bi} - \frac{y_{Ci}}{K_{eq}(T_i)}\right), \quad i = 1,2$$

Clearly, the vapor flowrate F_1 does not appear in any of the algebraic equations, due to the negligible pressure drop assumption. Thus, the algebraic equations are singular and the DAE model in Eq.21 has a high index. The ideal gas equations for the vapor holdups in the two reactors of fixed volumes V_{1T}, V_{2T}, respectively, are used for simplicity, and could be replaced by a non-ideal equation of state without affecting this conclusion. Such high-index DAE models arise in many staged processes with gaseous flow at high pressures, e.g. high-purity absorption, distillation, or reactive distillation columns, where the column pressure is high and the pressure drop between successive stages is small.

5. Feedback control of high-index DAE systems

In this section, we address the feedback control of multi-input multi-output (MIMO) DAE systems with the following general description:

$$\dot{x} = f(x) + b(x)z + g(x)u$$

$$0 = k(x) + l(x)z + c(x)u$$

$$y_i = h_i(x), \quad i = 1,\ldots,m \tag{22}$$

where $u \in \mathbb{R}^m$ is the vector of manipulated inputs and y_i is the i-th output to be controlled; $h_i(x)$ is an analytic scalar function. Throughout, we focus on high-index systems $(\nu_d > 1)$ for which the matrix $l(x)$ is singular.

Clearly, for a high-index DAE system of Eq.22, the analysis of basic system-theoretic and control-relevant properties like existence and uniqueness of solutions, stability, invertibility, minimum-phaseness etc., and the formulation of a controller synthesis problem, are obscure on the basis of this representation. A natural approach to this end, is to address the analysis and controller synthesis on the basis of a "state-space realization" of the DAE system, i.e. an equivalent ODE representation that describes the dynamics of the DAE system subject to the underlying algebraic constraints. This allows utilizing the available results for ODE systems, while systematically accounting for the differences between DAE and ODE systems.

The derivation of a state-space realization of the DAE system in Eq.22 involves:

(i) identifying the underlying algebraic constraints in the differential variables x, which specify the constrained state space region where these variables evolve, and

(ii) obtaining a solution for the algebraic variables z in terms of x and u that is consistent with the identified algebraic constraints in x.

However, a state-space realization of the DAE system in Eq.22 may or may not exist independently of the controller design. This motivates the following notion of regularity.

Definition 2: A DAE system of Eq.22 will be said to be regular, if

(i) it has a finite index ν_d, and

(ii) the state space region where the differential variables x are constrained to evolve is invariant under any control law for u.

Clearly, for a DAE system of Eq.22 to be regular, in particular for the constrained state space to be control-invariant, all the underlying algebraic constraints in x must be independent of the manipulated inputs u. For such regular systems, a state-space realization can be derived and then used as the basis for a controller synthesis [20]. On the other hand, if the constraints in x involve the inputs u (see Example 1), then the constrained state space *depends* on the control law for u and the DAE system is non-regular. For non-regular systems, the derivation of a state-space realization and the controller design are inherently coupled and can not be addressed sequentially. In light of this fact, a controller design methodology for non-regular systems was developed in [21], which involved as a key step, the design of a feedback compensator for the DAE system of Eq.22 such that the resulting modified system with a new vector of inputs is regular. Once the DAE system is regularized through feedback, a state-space realization of the feedback regularized system can be derived and used for the controller synthesis.

It should be mentioned that while the index ν_d of a regular DAE system of Eq.22 is independent of the inputs $u(t)$, the index of a non-regular system, specifically a system with constraints in x involving the inputs u, will, in general, depend on $u(t)$ (see Example 1). Throughout the article, it is assumed that for a DAE system of Eq.22, a finite and constant index ν_d exists, either independently of the inputs $u(t)$, or for some specified inputs $u(t)$, if the constraints in x involve u.

5.1. ALGORITHMIC PROCEDURE

Note that the classification of a DAE system of Eq.22 as regular or non-regular is not apparent on the basis of its original description. In this section, an algorithmic procedure (see also [20, 21]) is presented which

(i) yields a precise characterization of the class of high-index DAE systems of Eq.22 that are regular (or non-regular),

(ii) allows the derivation of state-space realizations for regular systems by identifying the constraints in x and yielding a solution for z, and

(iii) yields for a non-regular system, an equivalent DAE system where the algebraic equations explicitly include the constraints in x involving the inputs u, thereby identifying the cause of non-regularity.

The algorithmic procedure involves, in each iteration, (a) row operations on the algebraic equations to identify the underlying constraints in x, of which a *minimal* number, if any, involve the inputs u, and (b) selective differentiation of the constraints that do not involve u, to obtain the algebraic equations for the next iteration. The procedure yields an explicit algebraic condition that characterizes the class of regular systems, and for such systems, it converges with a final set of algebraic equations that can be solved for z in terms of x and u. On the other hand, for non-regular systems, the procedure converges with a final set of algebraic equations that is still singular with respect to z, i.e. it can not be solved for z, but is "non-singular" with respect to z and u together.

Iteration 1:

Consider the algebraic equations of the DAE system in Eq.22:

$$0 = k(x) + l(x)z + c(x)u \tag{23}$$

where rank $l(x) = p_1 < p$ and rank $[l(x) \ c(x)] = m_1 \leq p \ (m_1 \geq p_1)$. Then, there exists a nonsingular analytic $p \times p$ matrix $E^1(x)$ such that

$$E^1(x)[l(x) \ c(x)] = \begin{bmatrix} \bar{l}^1(x) & \bar{c}^1(x) \\ 0 & \hat{c}^1(x) \\ 0 & 0 \end{bmatrix} \tag{24}$$

where $\bar{c}^1(x), \hat{c}^1(x)$ are matrices of dimensions $p_1 \times m$, $(m_1 - p_1) \times m$, respectively, and the $p_1 \times p$, $m_1 \times (p+m)$ matrices

$$\bar{l}^1(x), \quad \begin{bmatrix} \bar{l}^1(x) & \bar{c}^1(x) \\ 0 & \hat{c}^1(x) \end{bmatrix}$$

have full row rank.

Step 1. Pre-multiply the algebraic equations (Eq.23) by the matrix $E^1(x)$ to obtain:

$$0 = \begin{bmatrix} \overline{k}^1(x) \\ \widehat{k}^1(x) \\ k^1(x) \end{bmatrix} + \begin{bmatrix} \overline{l}^1(x) \\ 0 \\ 0 \end{bmatrix} z + \begin{bmatrix} \overline{c}^1(x) \\ \widehat{c}^1(x) \\ 0 \end{bmatrix} u \qquad (25)$$

where $\overline{k}^1(x), \widehat{k}^1(x), k^1(x)$ are analytic vector fields of dimensions $p_1, (m_1 - p_1)$ and $(p - m_1)$, respectively. The last $p - p_1$ equations in Eq.25 denote underlying constraints in x; a minimal number of these constraints (the first $m_1 - p_1$) involve the inputs u in an irreducible fashion, i.e. $\widehat{c}^1(x)$ has full row rank.

Step 2. Differentiate the last $p - m_1$ constraints that do not involve u, to obtain the following new set of algebraic equations:

$$0 = \begin{bmatrix} \overline{k}^1(x) \\ \widehat{k}^1(x) \\ \widetilde{k}^2(x) \end{bmatrix} + \begin{bmatrix} \overline{l}^1(x) \\ 0 \\ \widetilde{l}^2(x) \end{bmatrix} z + \begin{bmatrix} \overline{c}^1(x) \\ \widehat{c}^1(x) \\ \widetilde{c}^2(x) \end{bmatrix} u \qquad (26)$$

In the above equation, $\widetilde{k}^2(x) = [L_f k_1^1(x) \cdots L_f k_{p-m_1}^1(\dot{x})]^T$ where $k_i^1(x)$ denotes the i-th component of the vector field $k^1(x)$, and $\widetilde{l}^2(x), \widetilde{c}^2(x)$ are matrices of dimensions $(p - m_1) \times p$ and $(p - m_1) \times m$, respectively, defined as:

$$\widetilde{l}^2(x) = \begin{bmatrix} L_{b_1} k_1^1(x) & \cdots & L_{b_p} k_1^1(x) \\ \vdots & & \vdots \\ L_{b_1} k_{p-m_1}^1(x) & \cdots & L_{b_p} k_{p-m_1}^1(x) \end{bmatrix}$$

$$\widetilde{c}^2(x) = \begin{bmatrix} L_{g_1} k_1^1(x) & \cdots & L_{g_m} k_1^1(x) \\ \vdots & & \vdots \\ L_{g_1} k_{p-m_1}^1(x) & \cdots & L_{g_m} k_{p-m_1}^1(x) \end{bmatrix}$$

where $b_j(x), g_j(x)$ denote the j-th columns of corresponding matrices, and $L_f k_i^1(x)$, $L_{b_j} k_i^1(x), L_{g_j} k_i^1(x)$ denote standard Lie derivatives defined as, e.g.

$$L_f k_i^1(x) = \sum_{\ell=1}^{n} \frac{\partial k_i^1(x)}{\partial x_\ell} f_\ell(x)$$

Step 3. Evaluate

$$\text{rank} \begin{bmatrix} \overline{l}^1(x) \\ \widetilde{l}^2(x) \end{bmatrix} = p_2, \qquad \text{rank} \begin{bmatrix} \overline{l}^1(x) & \overline{c}^1(x) \\ 0 & \widehat{c}^1(x) \\ \widetilde{l}^2(x) & \widetilde{c}^2(x) \end{bmatrix} = m_2, \quad m_2 \geq p_2$$

If $m_2 = p$ then stop, else proceed to the next iteration. For the case when $m_1 = p$, the procedure converges after Step 1, and all the $p - p_1$ constraints in x, i.e. $0 = \bar{k}^1(x) + \bar{c}^1(x)u$, involve u in an irreducible fashion.

Iteration q ($q \geq 2$):

Consider the algebraic equations from iteration $q - 1$:

$$0 = \begin{bmatrix} \bar{k}^{q-1}(x) \\ \hat{k}^{q-1}(x) \\ \tilde{k}^q(x) \end{bmatrix} + \begin{bmatrix} \bar{l}^{q-1}(x) \\ 0 \\ \tilde{l}^q(x) \end{bmatrix} z + \begin{bmatrix} \bar{c}^{q-1}(x) \\ \hat{c}^{q-1}(x) \\ \tilde{c}^q(x) \end{bmatrix} u \tag{27}$$

where the matrices:

$$L^q(x) = \begin{bmatrix} \bar{l}^{q-1}(x) \\ \tilde{l}^q(x) \end{bmatrix}, \qquad L^{q,e}(x) = \begin{bmatrix} \bar{l}^{q-1}(x) & \bar{c}^{q-1}(x) \\ 0 & \hat{c}^{q-1}(x) \\ \tilde{l}^q(x) & \tilde{c}^q(x) \end{bmatrix}$$

have ranks p_q and m_q, respectively ($p_q \leq m_q < p$). Then, there exists a nonsingular analytic $p \times p$ matrix $E^q(x)$ such that:

$$E^q(x)L^{q,e}(x) = \begin{bmatrix} \bar{l}^q(x) & \bar{c}^q(x) \\ 0 & \hat{c}^q(x) \\ 0 & 0 \end{bmatrix}$$

where the $p_q \times p$ matrix $\bar{l}^q(x)$ and the $m_q \times (p + m)$ matrix:

$$\begin{bmatrix} \bar{l}^q(x) & \bar{c}^q(x) \\ 0 & \hat{c}^q(x) \end{bmatrix}$$

have full row rank.

Step 1. Pre-multiply the algebraic equations (Eq.27) with the matrix $E^q(x)$ to obtain:

$$0 = \begin{bmatrix} \bar{k}^q(x) \\ \hat{k}^q(x) \\ k^q(x) \end{bmatrix} + \begin{bmatrix} \bar{l}^q(x) \\ 0 \\ 0 \end{bmatrix} z + \begin{bmatrix} \bar{c}^q(x) \\ \hat{c}^q(x) \\ 0 \end{bmatrix} u \tag{28}$$

where $\bar{k}^q(x), \hat{k}^q(x), k^q(x)$ are analytic vector fields of dimensions p_q, $(m_q - p_q)$, $(p - m_q)$, respectively, and the last $p - p_q$ equations denote underlying algebraic constraints in x.

Step 2. Differentiate the constraints in x that are independent of u, i.e. the last $p - m_q$ equations in Eq.28, to obtain the following new set of algebraic equations:

$$0 = \begin{bmatrix} \bar{k}^q(x) \\ \hat{k}^q(x) \\ \tilde{k}^{q+1}(x) \end{bmatrix} + \begin{bmatrix} \bar{l}^q(x) \\ 0 \\ \tilde{l}^{q+1}(x) \end{bmatrix} z + \begin{bmatrix} \bar{c}^q(x) \\ \hat{c}^q(x) \\ \tilde{c}^{q+1}(x) \end{bmatrix} u \tag{29}$$

326

where $\tilde{k}^{q+1}(x)$ is a vector field of dimension $p - m_q$ and $\tilde{l}^{q+1}(x)$, $\tilde{c}^{q+1}(x)$ are matrices of dimensions $(p - m_q) \times p$, $(p - m_q) \times m$, respectively, defined in a fashion analogous to that in Iteration 1.

Step 3. Evaluate the rank of the following matrices:

$$
\text{rank} \begin{bmatrix} \bar{l}^q(x) \\ \tilde{l}^{q+1}(x) \end{bmatrix} = p_{q+1}, \qquad
\text{rank} \begin{bmatrix} \bar{l}^q(x) & \bar{c}^q(x) \\ 0 & \hat{c}^q(x) \\ \tilde{l}^{q+1}(x) & \tilde{c}^{q+1}(x) \end{bmatrix} = m_{q+1}, \quad m_{q+1} \geq p_{q+1}
$$

If $m_{q+1} = p$ then stop, else repeat the above steps for the next iteration, starting with the algebraic equations in Eq.29.

For a DAE system of Eq.22 with a finite index ν_d, the above procedure will converge in a finite number of iterations s with $p_1 \leq p_2 \leq \cdots \leq p_{s+1} \leq p$ and $m_2 \leq m_3 \leq \cdots \leq m_{s+1} = p$, $(m_i \geq p_i, \forall i > 1)$. The algorithm identifies the following algebraic constraints among the differential variables x:

$$
\mathbf{k}(x) = \begin{bmatrix} \mathbf{k}^1(x) \\ \vdots \\ \mathbf{k}^s(x) \end{bmatrix} = 0 \tag{30}
$$

which do not involve the inputs u. The iterative procedure also yields the following set of algebraic equations:

$$
0 = \begin{bmatrix} \bar{k}^s(x) \\ \hat{k}^s(x) \\ \tilde{k}^{s+1}(x) \end{bmatrix} + \begin{bmatrix} \bar{l}^s(x) \\ 0 \\ \tilde{l}^{s+1}(x) \end{bmatrix} z + \begin{bmatrix} \bar{c}^s(x) \\ \hat{c}^s(x) \\ \tilde{c}^{s+1}(x) \end{bmatrix} u \tag{31}
$$

where the matrices:

$$
L^{s+1}(x) = \begin{bmatrix} \bar{l}^s(x) \\ \tilde{l}^{s+1}(x) \end{bmatrix}, \qquad
L^{s+1,e}(x) = \begin{bmatrix} \bar{l}^s(x) & \bar{c}^s(x) \\ 0 & \hat{c}^s(x) \\ \tilde{l}^{s+1}(x) & \tilde{c}^{s+1}(x) \end{bmatrix}
$$

have ranks p_{s+1} and $m_{s+1} = p$, respectively. Thus, there exists a nonsingular analytic $p \times p$ matrix $E^{s+1}(x)$ such that

$$
E^{s+1}(x)L^{s+1,e}(x) = \begin{bmatrix} \bar{l}(x) & \bar{c}(x) \\ 0 & \hat{c}(x) \end{bmatrix}
$$

where $\bar{c}(x)$ and $\hat{c}(x)$ are matrices of dimensions $p_{s+1} \times m$ and $(p - p_{s+1}) \times m$, respectively, and the $p_{s+1} \times p$, $p \times (p+m)$ matrices:

$$
\bar{l}(x), \qquad \begin{bmatrix} \bar{l}(x) & \bar{c}(x) \\ 0 & \hat{c}(x) \end{bmatrix}
$$

have full row rank. Pre-multiplying the algebraic equations in Eq.31 with the matrix $E^{s+1}(x)$, the following final set of algebraic equations are obtained:

$$0 = \begin{bmatrix} \overline{k}(x) \\ \widehat{k}(x) \end{bmatrix} + \begin{bmatrix} \overline{l}(x) \\ 0 \end{bmatrix} z + \begin{bmatrix} \overline{c}(x) \\ \widehat{c}(x) \end{bmatrix} u \tag{32}$$

where $\overline{k}(x)$ and $\widehat{k}(x)$ are analytic vector fields of dimensions p_{s+1} and $p - p_{s+1}$, respectively.

5.2. STATE-SPACE REALIZATION AND CONTROL OF REGULAR SYSTEMS

Note that the algorithmic procedure explicitly identifies the underlying constraints in x, which may or may not involve the manipulated inputs u, i.e. the DAE system may be regular or non-regular, depending on the matrix ranks p_i and m_i. The precise characterization of the class of DAE systems of Eq.22 that are regular is given in the following lemma.

Lemma 1: *Consider a DAE system of Eq.22. Then the DAE system is regular, if and only if (i) $m_i = p_i$ for every iteration $i \geq 1$ of the algorithmic procedure, and (ii) the algorithmic procedure converges in a finite number of iterations s with $p_{s+1} = p$.*

The fact that $m_i = p_i$ in every iteration implies that all the $\sum_{i=1}^{s}(p - p_i)$ algebraic constraints in x:

$$\mathbf{k}(x) = \begin{bmatrix} \mathbf{k}^1(x) \\ \vdots \\ \mathbf{k}^s(x) \end{bmatrix} = 0 \tag{33}$$

are independent of the manipulated inputs u. These constraints in x are linearly independent [20] and specify a control-invariant subspace:

$$\mathcal{M} = \{x \in \mathcal{X} \subset \mathbb{R}^n : \mathbf{k}(x) = 0\} \tag{34}$$

where the differential variables x are constrained to evolve. The problem of consistent initialization is essentially the one of specifying the initial conditions $x(0)$ such that $\mathbf{k}(x(0)) = 0$. Furthermore, the fact that the algorithmic procedure converges with the sequence of integers $p_1 \leq p_2 \leq \cdots \leq p_s < p_{s+1} = p$, implies that the procedure yields the following final set of algebraic equations:

$$0 = \overline{k}(x) + \overline{l}(x)z + \overline{c}(x)u \tag{35}$$

where the $p \times p$ matrix $\overline{l}(x)$ is non-singular. Thus, the algebraic equations in Eq.35 can be solved for z:

$$z = -\overline{l}(x)^{-1}[\overline{k}(x) + \overline{c}(x)u] \tag{36}$$

and another differentiation of Eq.35 would yield differential equations for z, i.e. the index is $\nu_d = s + 1$.

The specification of the constrained state space $\mathcal{M}$ (Eq.34), and the solution for the algebraic variables z (Eq.36) allow the derivation of a state-space realization of the DAE system in Eq.22. The resulting state-space realization is given in the following proposition [20].

Proposition 1: *Consider a regular DAE system of Eq.22 for which the algorithmic procedure converges after s iterations, with the constraints in Eq.33 and the solution for z in Eq.36. Then, the dynamic system:*

$$\dot{x} = \overline{f}(x) + \overline{g}(x)u$$
$$y_i = h_i(x), \quad i = 1, \ldots, m \tag{37}$$

where $x \in \mathcal{M} = \{x \in \mathcal{X} : \mathbf{k}(x) = 0\}$ and

$$\overline{f}(x) = f(x) - b(x)\,\overline{l}(x)^{-1}\,\overline{k}(x)$$
$$\overline{g}(x) = g(x) - b(x)\,\overline{l}(x)^{-1}\,\overline{c}(x) \tag{38}$$

is a state-space realization of the DAE system.

In view of the fact that the differential variables x are constrained to evolve on the manifold $\mathcal{M}$ of dimension $\kappa = n - \sum_{i=1}^{s}(p - p_i)$, the state-space realization in Eq.37 is not of minimal order. Such a realization can only be obtained in appropriate transformed coordinates. More specifically, given the vector fields $\mathbf{k}^i(x)$, $i = 1, \ldots, s$ with the components $\mathbf{k}_j^i(x)$, $j = 1, \ldots, (p - p_i)$, one can always find smooth functions $\phi_1(x), \ldots, \phi_\kappa(x)$ such that

$$\zeta = \begin{bmatrix} \zeta_1^{(0)} \\ \vdots \\ \zeta_\kappa^{(0)} \\ \zeta^{(1)} \\ \vdots \\ \zeta^{(s)} \end{bmatrix} = T(x) = \begin{bmatrix} \phi_1(x) \\ \vdots \\ \phi_\kappa(x) \\ \mathbf{k}^1(x) \\ \vdots \\ \mathbf{k}^s(x) \end{bmatrix} \tag{39}$$

is a local diffeomorphism. In these transformed coordinates ζ, the constraints $\mathbf{k}(x) = 0$ reduce to $\zeta^{(i)} = 0$, $i = 1, \ldots, s$, which allows obtaining the minimal-order state-space realization given in Proposition 2 [20].

Proposition 2: *Consider a regular DAE system of Eq.22 for which the proposed algorithmic procedure converges after s iterations, with the constraints in Eq.33 and the solution for z in Eq.36. Then, the dynamic system:*

$$\dot{\zeta}^{(0)} = f^{(0)}(\zeta^{(0)}) + g^{(0)}(\zeta^{(0)})u$$
$$y_i = h_i(x)|_{x=T^{-1}(\zeta^{(0)}, 0, \ldots, 0)}, \quad i = 1, \ldots, m \tag{40}$$

is a state-space realization of the DAE system, of dimension $\kappa = (n - \sum_{i=1}^{s}(p-p_i))$, where

$$f^{(0)}(\zeta^{(0)}) = \begin{bmatrix} L_{\overline{f}}\phi_1(x) \\ \vdots \\ L_{\overline{f}}\phi_\kappa(x) \end{bmatrix}_{x=T^{-1}(\zeta^{(0)},0,\ldots,0)}$$

$$g^{(0)}(\zeta^{(0)}) = \begin{bmatrix} L_{\overline{g}_1}\phi_1(x) & \cdots & L_{\overline{g}_m}\phi_1(x) \\ \vdots & & \vdots \\ L_{\overline{g}_1}\phi_\kappa(x) & \cdots & L_{\overline{g}_m}\phi_\kappa(x) \end{bmatrix}_{x=T^{-1}(\zeta^{(0)},0,\ldots,0)}$$

(41)

and $\overline{f}(x), \overline{g}(x)$ are defined in Eq.38.

Remark 2: In the coordinate change of Eq.39, the states $\zeta_i^{(0)}$, or equivalently the scalar functions $\phi_i(x)$, are chosen arbitrarily (they can be chosen simply as a proper subset $x_1, \ldots x_\kappa$ of the original differential variables x), and thus, the derivation of the state-space realization in Eq.40 requires an explicit solution for the algebraic variables z. However, in some cases (see Remark 3), it may be possible to choose the functions $\phi_i(x)$ in the coordinate change in a specific manner such that the differential equations for the corresponding state variables $\zeta_i^{(0)}$ do not involve the algebraic variables z, and thus, the minimal-order state-space realization can be derived without obtaining a solution for z. This is analogous to obtaining a representation of the zero dynamics of MIMO ODE systems independently of the solution for the manipulated inputs u, under appropriate involutivity conditions [15].

For a regular DAE system of Eq.22, various control-relevant issues (such as equilibrium points and their stability, zero dynamics and characterization of minimum phase behavior etc.) can be addressed directly on the basis of the equivalent state-space realizations in Eq.37 or Eq.40, using existing results for ODE systems. These state-space realizations can also be used as the basis for the formulation and solution of a variety of control problems for the DAE system, that are consistent with the underlying algebraic constraints in x. For further details on the controller design for a regular DAE system of Eq.22, the reader is referred to [20].

5.3. State-space realization of two-phase reactor example

Consider the two-phase reactor of section 4.1., for which it is desired to control the product vapor composition $y_1 = y_A$ and reactor temperature $y_2 = T$, using the product vapor flowrate $u_1 = F_v$ and heat output $u_2 = Q$ as the manipulated inputs. A dynamic model of this process is given by the DAE system in Eq.18, which is in the form of Eq.22 with the differential variables $x = [\, M_v \ y_A \ M_l \ x_A \ x_B \ T\,]^T$ and the algebraic variables $z = [\, N_A \ N_C \ P\,]^T$.

The algebraic equations are singular ($p_1 = 1$) and can not be solved for the inter-phase mole transfer rates N_A and N_C. Thus, they impose constraints in the differential variables x. More specifically, solving for the reactor pressure P from the ideal gas equation and substituting the solution in the phase equilibrium relations, or equivalently, pre-multiplying the algebraic equations with the matrix:

$$E^1(x) = \begin{bmatrix} 0 & 0 & 1 \\ 1 & 0 & -(\dfrac{x_2\rho}{V_T\rho - x_3}) \\ 0 & 1 & -(\dfrac{(1 - x_2)\rho}{V_T\rho - x_3}) \end{bmatrix}$$

the following two constraints are identified:

$$0 = \begin{bmatrix} \mathbf{k}_1^1(x) \\ \mathbf{k}_2^1(x) \end{bmatrix} = \begin{bmatrix} P_A^s(x_6)x_4 - (\dfrac{\rho R x_1 x_2 x_6}{V_T\rho - x_3}) \\ P_C^s(x_6)(1 - x_4 - x_5) - (\dfrac{\rho R x_1 (1 - x_2) x_6}{V_T\rho - x_3}) \end{bmatrix} \tag{42}$$

Differentiating the above constraints once, the resulting set of algebraic equations can be solved for z, i.e. the algorithmic procedure converges in one iteration with $p_2 = 3$, i.e., the index is two. Moreover, the two constraints in x are independent of the inputs u. Thus, the DAE system is regular, and substituting the solution for z, specifically N_A and N_C, in the differential equations of Eq.18, a state-space realization of the DAE model is obtained on the constrained state space $\mathcal{M}$ specified by the two constraints in Eq.42 (for further details, see [20]).

Given the fact that the constrained state space $\mathcal{M}$ is of dimension $\kappa = n-2 = 4$, the two constraints in Eq.42 can be used as a part of coordinate change to obtain a minimal-order state-space realization (of dimension $\kappa = 4$). In particular, since the Jacobian

$$\begin{bmatrix} \dfrac{\partial \mathbf{k}_1^1(x)}{\partial x_1} & \dfrac{\partial \mathbf{k}_1^1(x)}{\partial x_2} \\ \dfrac{\partial \mathbf{k}_2^1(x)}{\partial x_1} & \dfrac{\partial \mathbf{k}_2^1(x)}{\partial x_2} \end{bmatrix} = -\dfrac{\rho R x_6}{(V_T\rho - x_3)} \begin{bmatrix} x_2 & x_1 \\ 1 - x_2 & -x_1 \end{bmatrix} \tag{43}$$

is nonsingular, the choice of $\phi_1(x) = x_3$, $\phi_2(x) = x_4$, $\phi_3(x) = x_5$, $\phi_4(x) = x_6$ yields a valid coordinate change:

$$\zeta = \begin{bmatrix} \zeta_1^{(0)} \\ \zeta_2^{(0)} \\ \zeta_3^{(0)} \\ \zeta_4^{(0)} \\ \zeta_1^{(1)} \\ \zeta_2^{(1)} \end{bmatrix} = T(x) = \begin{bmatrix} x_3 \\ x_4 \\ x_5 \\ x_6 \\ \mathbf{k}_1^1(x) \\ \mathbf{k}_2^1(x) \end{bmatrix} \tag{44}$$

In these coordinates ζ, the states $\zeta_1^{(1)}, \zeta_2^{(1)}$ are identically zero on the state space

$\mathcal{M}$, and a state-space realization of dimension four is obtained by eliminating these zero states.

Remark 3: The differential equations for $x_3, \ldots, x_6$ in the DAE model of Eq.18 involve the algebraic variables $z_1 = N_A$ and $z_2 = N_C$, and thus the minimal-order state-space realization in coordinates ζ of Eq.44 requires the solution for these algebraic variables. It is possible to obtain a state-space realization of the DAE model without evaluating the solution for z_1 or z_2, in appropriate coordinates. More specifically, it can be verified that the distribution $B(x) = \text{span}\{b_1(x), b_2(x)\}$ is involutive ($b_i(x)$ denotes the i-th column of the matrix $b(x)$ in the DAE model). Thus, from Frobenius theorem (see, e.g. [15]), the states $\zeta_i^{(0)}$, or equivalently the functions $\phi_i(x)$, in the coordinate change of Eq.44 can be chosen such that $L_{b_j}\phi_i(x) \equiv 0$, $i = 1, \ldots 4$, $j = 1, 2$. One such choice of coordinates is

$$
\zeta = \begin{bmatrix} \zeta_1^{(0)} \\ \zeta_2^{(0)} \\ \zeta_3^{(0)} \\ \zeta_4^{(0)} \\ \zeta_1^{(1)} \\ \zeta_2^{(1)} \end{bmatrix} = T(x) = \begin{bmatrix} M \\ M_A \\ M_B \\ H \\ \mathbf{k}_1^1(x) \\ \mathbf{k}_2^1(x) \end{bmatrix}
\tag{45}
$$

where $M = x_1 + x_2$ is the total (in both phases) molar holdup, $M_A = x_1 x_2 + x_3 x_4$ is the total molar holdup of component A, $M_B = x_3 x_5$ is the total molar holdup of component B and $H = (x_1 + x_3)c_p(x_6 - T_o) + x_1 \Delta H^v$ is the total enthalpy. The corresponding state-space realization has the form:

$$
\begin{aligned}
\dot{\zeta}^{(0)} &= \widehat{f}(\zeta^{(0)}) + \widehat{g}(\zeta^{(0)})u \\
0 &= \widehat{h}(y, \zeta^{(0)})
\end{aligned}
\tag{46}
$$

where $y = [y_1 \cdots y_m]^T$ is the output vector, and the vector fields $\widehat{f}, \widehat{h}$ and the matrix $\widehat{g}$ depend only on $f(x), g(x), h_i(x)$ and $\phi_i(x)$. Note that the derivation of the above state-space realization essentially amounts to the "modeling" approach for phase-equilibrium processes proposed in [32], where the dynamic balances are written in terms of extensive variables like total holdup and enthalpy, instead of the intensive variables like composition and temperature. It should be mentioned however, that in the state-space realization of Eq.46, $\widehat{f}$ and $\widehat{g}$ are implicit functions of the extensive state variables $\zeta^{(0)}$, since the reaction rates, phase equilibrium relations etc. are functions of the intensive variables like T, x_A, x_B, which are related to $\zeta^{(0)}$ through highly nonlinear implicit functions. Furthermore, typically, the outputs y to be controlled are also intensive variables, which can not be expressed as explicit functions of the states $\zeta^{(0)}$. Thus, while the state-space realization in Eq.46 can be used for numerical simulation purposes, it is not suitable for feedback controller synthesis.

5.4. FEEDBACK REGULARIZATION AND CONTROL OF NON-REGULAR SYSTEMS

For a DAE system of Eq.22 that is non-regular, i.e. $m_j > p_j$, in some iteration $j \geq 1$ and all the subsequent iterations, the algorithmic procedure converges with the integer sequences $p_1 \leq p_2 \leq \cdots \leq p_{s+1} < p$ and $m_1 \leq m_2 \leq \cdots \leq m_s < m_{s+1} = p$, and the final set of algebraic equations in Eq.32. Thus, the algorithmic procedure yields the following new DAE system:

$$
\begin{aligned}
\dot{x} &= f(x) + b(x)z + g(x)u \\
0 &= \begin{bmatrix} \overline{k}(x) \\ \widehat{k}(x) \end{bmatrix} + \begin{bmatrix} \overline{l}(x) \\ 0 \end{bmatrix} z + \begin{bmatrix} \overline{c}(x) \\ \widehat{c}(x) \end{bmatrix} u \\
y_i &= h_i(x), \quad i = 1, \ldots, m
\end{aligned}
\tag{47}
$$

where $x \in \mathcal{X}$: $\mathbf{k}(x) = 0$. The above DAE system (Eq.47) is equivalent to the original DAE system in Eq.22, i.e., for consistent initial conditions $x(0)$ and smooth inputs $u(t)$, both systems have the same solution $x(t), z(t)$. Moreover, note that the algebraic equations in Eq.47 explicitly include the underlying constraints in x, $0 = \widehat{k}(x) + \widehat{c}(x)u$, that involve the inputs u in an irreducible fashion, i.e. $\widehat{c}(x)$ has a full row rank $p - p_{s+1}$.

For this equivalent DAE system in Eq.47, which is non-regular owing to the constraints $0 = \widehat{k}(x) + \widehat{c}(x)u$ that explicitly involve the inputs u, the aim is to design a feedback compensator such that the resulting feedback modified DAE system with a new vector of inputs is regular. Note however, that the DAE system in Eq.47 still has a high index $\overline{\nu}_d = \nu_d - s > 1$, and the constraints $0 = \widehat{k}(x) + \widehat{c}(x)u$ have to be differentiated at least once to obtain a set of algebraic equations solvable in z. Thus, the solution for the algebraic variables z is a function of the differential variables x, the manipulated inputs u and $at\ least$ one of their derivatives, i.e. it has the form:

$$
z(t) = \varphi(x(t), u(t), u^{(1)}(t), \ldots)
\tag{48}
$$

where $u^{(i)}$ denotes the i-th derivative of the manipulated input vector u. In the light of this fact, any causal feedback law for u must clearly be independent of z. Furthermore, any (regular) static feedback of the form:

$$
u = \mathcal{F}(x, v)
$$

where $v \in \mathbb{R}^m$ is the new input vector and $(\partial \mathcal{F}/\partial v)$ is nonsingular, will not regularize the DAE system of Eq.47, since the constraints $0 = \widehat{k}(x) + \widehat{c}(x)\mathcal{F}(x, v)$ in the resulting modified system would still involve the inputs v.

The above observations indicate the need for a $dynamic$ feedback compensator, to modify the DAE system in Eq.47, in particular the constraints:

$$
0 = \widehat{k}(x) + \widehat{c}(x)u
\tag{49}
$$

that cause the non-regularity, such that the resulting system with the inputs v is regular. A detailed discussion on the construction of such a compensator can be found in [21]. The following theorem states the result on the requisite feedback compensator and the resulting modified DAE system that is regular.

Theorem 1: *Consider a DAE system of Eq.22 for which the algorithmic procedure yields the equivalent DAE system of Eq.47. Then, for a choice of a constant matrix $S \in \mathbb{R}^{(p-p_s+1) \times n}$ such that:*

$$\text{rank} \begin{bmatrix} \bar{l}(x) \\ Sb(x) \end{bmatrix} = p \tag{50}$$

and an $m \times m$ matrix $M(x)$ such that:

$$\begin{bmatrix} \bar{c}(x) \\ \widehat{c}(x) \end{bmatrix} M(x) = \begin{bmatrix} \bar{c}_1(x) & \bar{c}_2(x) \\ \widehat{c}_1(x) & 0 \end{bmatrix} \tag{51}$$

where the $(p-p_{s+1}) \times (p-p_{s+1})$ matrix $\widehat{c}_1(x)$ is nonsingular, the following dynamic feedback compensator:

$$\dot{w} = v_1$$
$$u = M(x) \begin{bmatrix} (\widehat{c}_1(x))^{-1} \left(-\widehat{k}(x) + Sx + w \right) \\ 0 \end{bmatrix} + M(x) \begin{bmatrix} 0 \\ v_2 \end{bmatrix} \tag{52}$$

where $w, v_1 \in \mathbb{R}^{(p-p_{s+1})}$ and $v_2 \in \mathbb{R}^{m-(p-p_{s+1})}$, yields the modified DAE system:

$$\begin{bmatrix} \dot{x} \\ \dot{w} \end{bmatrix} = \begin{bmatrix} \tilde{f}(x,w) \\ 0 \end{bmatrix} + \begin{bmatrix} b(x) \\ 0 \end{bmatrix} z + \begin{bmatrix} 0 & \bar{g}_2(x) \\ I_{p-p_{s+1}} & 0 \end{bmatrix} \begin{bmatrix} v_1 \\ v_2 \end{bmatrix}$$

$$0 = \begin{bmatrix} \tilde{k}(x,w) \\ Sx + w \end{bmatrix} + \begin{bmatrix} \bar{l}(x) \\ 0 \end{bmatrix} z + \begin{bmatrix} 0 & \bar{c}_2(x) \\ 0 & 0 \end{bmatrix} \begin{bmatrix} v_1 \\ v_2 \end{bmatrix} \tag{53}$$

$$y_i = h_i(x), \quad i = 1, \dots, m$$

where

$$\tilde{f}(x,w) = f(x) + \bar{g}_1(x)\gamma(x) + \bar{g}_1(x)(\widehat{c}_1(x))^{-1} w$$
$$\tilde{k}(x,w) = \bar{k}(x) + \bar{c}_1(x)\gamma(x) + \bar{c}_1(x)(\widehat{c}_1(x))^{-1} w$$
$$\gamma(x) = (\widehat{c}_1(x))^{-1} \left\{ -\widehat{k}(x) + Sx \right\}$$
$$[\bar{g}_1(x) \quad \bar{g}_2(x)] = g(x)M(x) \tag{54}$$

and $x \in \mathcal{X} : k(x) = 0$. The DAE system of Eq.53 with the extended vector of differential variables $\tilde{x} = [x^T \ w^T]^T$ and a new vector of inputs $v = [v_1^T \ v_2^T]^T \in \mathbb{R}^m$ is regular.

The fact that the feedback modified DAE system in Eq.53 can be verified by observing that (*i*) the modified constraints in the differential variables $\tilde{x}$, $0 = Sx + w$ are independent of the new inputs v, and (*ii*) upon one differentiation of these constraints, the resulting algebraic equations are obtained:

$$0 = \begin{bmatrix} \tilde{k}(x,w) \\ S\tilde{f}(x,w) \end{bmatrix} + \begin{bmatrix} \bar{l}(x) \\ Sb(x) \end{bmatrix} z + \begin{bmatrix} 0 & \bar{c}_2(x) \\ I_{p-p_{s+1}} & S\bar{g}_2(x) \end{bmatrix} \begin{bmatrix} v_1 \\ v_2 \end{bmatrix} \tag{55}$$

which, owing to the condition in Eq.50, can be solved for z:

$$
\begin{aligned}
z &= - \begin{bmatrix} \bar{l}(x) \\ Sb(x) \end{bmatrix}^{-1} \left\{ \begin{bmatrix} \tilde{k}(x,w) \\ S\tilde{f}(x,w) \end{bmatrix} + \begin{bmatrix} 0 & \bar{c}_2(x) \\ I_{p-p_s+1} & S\bar{g}_2(x) \end{bmatrix} \begin{bmatrix} v_1 \\ v_2 \end{bmatrix} \right\} \\
&= R(x,w) + S_1(x)v_1 + S_2(x)v_2
\end{aligned}
\tag{56}
$$

i.e., there are no additional constraints in $\bar{x}$. For the feedback regularized DAE system of Eq.53, the constraints $\bar{x}$, $\mathbf{k}(x) = 0$, $Sx + w = 0$, which are independent of the inputs v, specify the $n -- \sum_{i=1}^{s}(p - m_i)$-dimensional subspace:

$$
\mathcal{M} = \left\{ (x,w) \in \mathcal{X} \times \mathbb{R}^{(p-p_s+1)} \; : \; \begin{matrix} \mathbf{k}(x) = 0 \\ Sx + w = 0 \end{matrix} \right\} \subset \mathbb{R}^{n+(p-p_s+1)}
\tag{57}
$$

which is invariant under any control law for the inputs v. Clearly, the result in Theorem 1 relies on the existence of the matrix S satisfying the condition in Eq.50. The existence of such a matrix is guaranteed for the solvable DAE system of Eq.47 [21].

For the feedback regularized DAE system in Eq.53 with the the constrained state space $\mathcal{M}$ (Eq.57) and the solution for z (Eq.56), a full- or minimal-order state-space realization can be derived (see section 5.2.). The derived state-space realizations of the feedback regularized system can be used as the basis for addressing the controller synthesis problem for the non-regular DAE system of Eq.22. For details on the controller design and its application to a chemical process example, the reader is referred to [21].

6.   Simulation studies

In section 4., high-index DAE models were derived for chemical processes with fast mass/heat transfer, fast reactions and fast gaseous flow, under the quasi-steady-state assumptions of phase, thermal, and reaction equilibrium and negligible pressure drop. It may be argued that the high index in these *equilibrium-based* DAE models can be avoided by replacing the equilibrium relations with explicit rate expressions for the fast reactions, mass transfer etc., to obtain index-one DAE or ODE models. However, standard inversion-based controllers designed on the basis of such *rate-based* models suffer from two key problems: (i) controller ill-conditioning, i.e. the controller is highly sensitive to small modeling/measurement errors due to the occurrence of large process parameters (e.g. large heat/mass transfer and reaction rate coefficients) that appear explicitly in the feedback law, and (ii) closed-loop instability in the case of slightly non-minimum phase systems, i.e. systems where the fast reactions, mass transfer etc. lead to a two-time-scale zero dynamics with a stable slow subsystem and an unstable fast subsystem. Both these problems are easily overcome by designing the controllers on the basis of the quasi-steady-state high-index DAE models; a rigorous justification of this is obtained through a singular perturbation analysis of the rate-based models [19]. In this section, we illustrate the above-mentioned problems and demonstrate the effectiveness of the controllers designed on the basis of the high-index DAE models, through simulations in two examples.

6.1. CSTR WITH HEATING JACKET

Consider the CSTR with heating jacket discussed in section 4.3., for which it is desired to control the product concentration $y_1 = C_B$ and reactor temperature $y_2 = T$ using the reactant flowrate $u_1 = F_A$ and the heating fluid flowrate $u_2 = F_h$ as the manipulated inputs. For this process, a detailed rate-based model is given by the following ODE system:

$$\dot{C}_A = \frac{F_A}{V}(C_{Ao} - C_A) - k_o e^{(-E/RT)} C_A$$

$$\dot{C}_B = -\frac{F_A}{V} C_B + k_o e^{(-E/RT)} C_A$$

$$\dot{T} = \frac{F_A}{V}(T_A - T) - k_o e^{(-E/RT)} C_A \frac{\Delta H_r}{\rho c_p} + \frac{UA}{\rho c_p}(\frac{T_j - T}{V})$$

$$\dot{T}_j = \frac{F_h}{V_h}(T_h - T_j) - \frac{UA}{\rho_h c_{ph}}(\frac{T_j - T}{V_h}) \tag{58}$$

which includes the explicit rate expression of the fast heat transfer between the jacket and the reactor $Q = UA(T_j - T)$, where $UA = (1/\epsilon)$ is a large parameter (see Table 1 for the nominal values of the parameters and process variables).

An input/output linearizing controller can be designed for the process, on the basis of the above ODE model [19]. More specifically, it can be verified that the relative orders of the two outputs y_1 and y_2 with respect to the manipulated input vector u are $r_1 = 1$ and $r_2 = 1$, respectively. However, both outputs are affected more directly by the same input u_1, and thus the characteristic matrix is singular. This implies the need for a dynamic feedback controller, which is designed by defining $v_1 = \dot{u}_1$ as a new manipulated input to obtain a system with an extended state vector $\bar{x} = [x^T \quad u_1]^T$, new inputs $v_1 = \dot{u}_1$, $v_2 = u_2$, relative orders $r_1 = 2$, $r_2 = 2$, and a characteristic matrix that is nonsingular (the reader may refer to [9] for details). An input/output linearizing dynamic state feedback controller was designed on the basis of the extended system to induce a well-characterized closed-loop response. However, owing to the fact that the relative orders of both outputs in the extended system are two, the resulting controller involves terms multiplied by the large factors $(1/\epsilon)$, $(1/\epsilon^2)$ and thus it is severely ill-conditioned as illustrated through simulations later in this section (see Figure 5).

The above-mentioned problem of controller ill-conditioning is overcome by designing the controller for the process on the basis of the index-two DAE model in Eq.20, where the fast (and stable) heat transfer is approximated by the quasi-steady-state thermal equilibrium condition $T_j = T$; this model does not involve the large parameter UA, and thus, the resulting controller does not suffer from ill-conditioning. Under this assumption of thermal equilibrium, the output $y_2 = T$ is approximated by the weighted average $y_{2s} = (VT + V_h T_h)/(V + V_h)$, whereas $y_{1s} = y_1$ (see [19]). It can be verified that the index-two DAE model in Eq.20 is regular, since the only constraint in the differential variables $0 = T - T_j$ does not involve the manipulated inputs. A state-space realization of the DAE model was derived and used for the controller design. More specifically, the relative orders of the two outputs with respect to the input vector u are $r_1 = 1$ and $r_2 = 1$, respectively, and the characteristic matrix is nonsingular. Thus, an input/output linearizing controller with an external linear controller with integral action was designed, to induce the following decoupled first-order responses:

$$y_{is} + \beta_{i1}^i \dot{y}_{is} = y_{isp}, \quad i = 1, 2 \tag{59}$$

where $\beta_{11}^1 = 10$ and $\beta_{21}^2 = 15$. The performance and robustness of this controller was compared with that of the controller synthesized on the basis of the rate-based model in Eq.58 to induce the following decoupled second-order responses:

$$y_i + \beta_{i1}^i \dot{y}_i + \beta_{i2}^i \ddot{y}_i = y_{isp}, \quad i = 1, 2 \tag{60}$$

The controller parameters were tuned for critically damped responses with the same time constants as in the response of Eq.59, i.e. $\beta_{11}^1 = 20$, $\beta_{12}^1 = 100$, $\beta_{21}^2 = 30$, $\beta_{22}^2 = 225$. In both cases, the detailed rate-based model of Eq.58 was used to simulate the process.

The first run was used to study the performance of the two controllers in the nominal process, for a 15% increase in the setpoint for y_1 at $t = 0$. The corresponding profiles for the controlled outputs and the manipulated inputs are shown in Figure 4. Clearly, the controller designed on the basis of the rate-based model in Eq.58, induces the decoupled second-order responses in the nominal process, as expected. The controller based on the index-two DAE model in Eq.20 also yields excellent performance in this case.

The second run was performed for the same setpoint change, in the presence of a 5% error in the process parameters ΔH_r, ρ and c_p. The corresponding profiles are shown in Figure 5. Clearly, the controller designed on the basis of the index-two DAE model successfully rejects the effects of these small parametric errors without much performance degradation, and the calculated control action is also close to the nominal case, establishing the fact that the controller is not sensitive to small errors. On the other hand, the controller designed on the basis of the rate-based model was very sensitive to these errors, since the effects of the small errors were magnified through the large factors $(1/\epsilon)$, $(1/\epsilon^2)$ in the control law. As a result, in the presence of these small modeling errors, the controller calculates a very large control action and leads to instability.

The problem of controller ill-conditioning illustrated in the above example also arises in the case of reactors with fast reactions, multi-phase systems with fast mass transfer etc., where the controller designed on the basis of the detailed rate-based

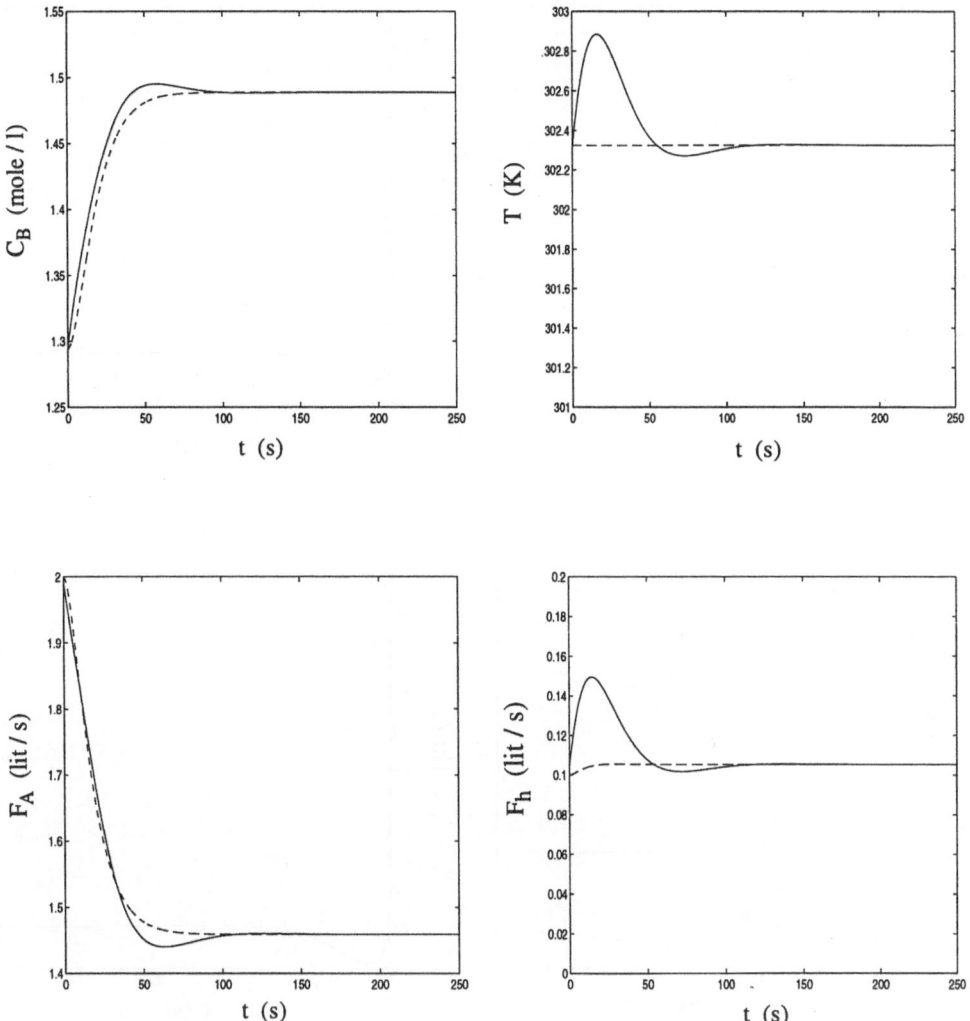

Figure 4. Comparison of closed-loop input and output profiles for setpoint tracking in the nominal system, under controllers designed on the basis of the rate-based ODE- (dashed) and the equilibrium-base high-index DAE- (solid) models

model explicitly involves the large process parameters, e.g. mass transfer/reaction rate coefficients. As shown through the simulations, this problem is easily over-come by designing the controller on the basis of the high-index DAE models (see section 4.), derived under the quasi-steady state assumption of phase/reaction equilibrium etc.

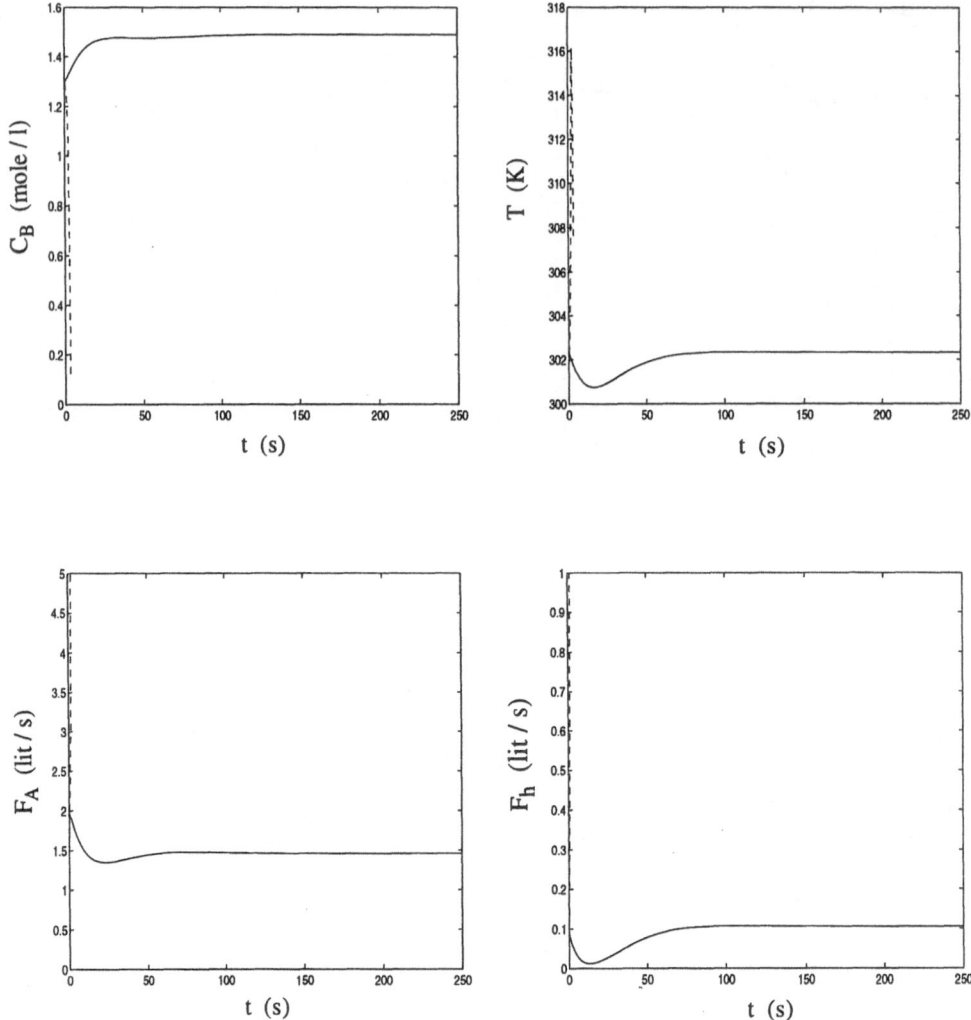

Figure 5. Comparison of closed-loop input and output profiles for setpoint tracking in the presence of modeling errors, under controllers designed on the basis of the rate-based ODE- (dashed) and the equilibrium-based high-index DAE- (solid) models.

6.2. CSTR WITH FAST AND SLOW REACTIONS

Consider the CSTR discussed in section 4.2. where it is desired to control the reactor holdup $y_1 = V$ and the product concentration $y_2 = C_B$ using the manipulated inputs $u_1 = F$ and $u_2 = F_o$. For this process, a detailed rate-based model is given by:

$$\dot{V} = F_o - F$$

$$\dot{C}_A = \frac{F_o}{V}(C_{Ao} - C_A) - R_1$$

$$\dot{C}_B = -\frac{F_o}{V}C_B + R_1 - R_2$$

$$\dot{C}_C = -\frac{F_o}{V}C_C + R_2 \tag{61}$$

$$0 = R_1 - k_1(C_A - \frac{C_B}{K_{eq}})$$

$$0 = R_2 - k_2 C_B$$

where the rate of the fast reaction, i.e. R_1, is explicitly included and it involves the large rate constant $k_1 = 1/\epsilon$ (see Table 2 for nominal values of parameters and variables). The above DAE model is of index one and is easily reduced to an ODE model by substituting the relations for R_1 and R_2 in the differential equations. Moreover, it can be verified that the relative orders of the two outputs with respect to the manipulated inputs are $r_1 = 1$ and $r_2 = 1$, and the characteristic matrix is nonsingular. Thus, an input/output linearizing controller can be designed on the basis of the above model, to induce the first-order responses:

$$y_i + \beta_{i1}^i \dot{y}_i = y_{isp}, \quad i = 1, 2$$

However, the resulting controller leads to closed-loop instability. This instability occurs due to the fact that due to the fast reaction in the model of Eq.61, the two-dimensional zero dynamics is slightly non-minimum phase; the zero dynamics has a slow stable mode and a fast unstable mode. This is illustrated in Figure 6, which shows an inverse response in the output C_B in the fast boundary layer, for a step increase in the input F_o, while keeping the holdup constant, i.e. $F = F_o$.

If the fast reversible reaction is approximated by the quasi-steady-state condition of reaction equilibrium, and correspondingly the output y_2 is approximated by $y_{2s} = K_{eq}(C_A + C_B)/K_{eq} + 1$, then the index-two DAE model of Eq.19 is obtained. Owing to the reaction equilibrium constraint, the zero dynamics for this model is one-dimensional and stable. Thus, a controller designed on the basis of this model does not suffer from instability. The controller was tuned with the parameters $\beta_{11}^1 = 10$ and $\beta_{21}^2 = 15$, and the closed-loop profiles for C_B and F_o for a 10% increase in the setpoint y_{2sp} at $t = 0$ are shown in Figure 7; the process was simulated with the detailed rate-based model of Eq.61. Clearly, the controller yields excellent tracking performance with a slight ($\mathcal{O}(\epsilon)$) discrepancy from the requested first-order response owing to the quasi-steady-state approximation, and the closed-loop system is stable.

7. Conclusions

A comprehensive feedback controller design methodology was presented for non-linear high-index DAE systems. The general approach involves the derivation of

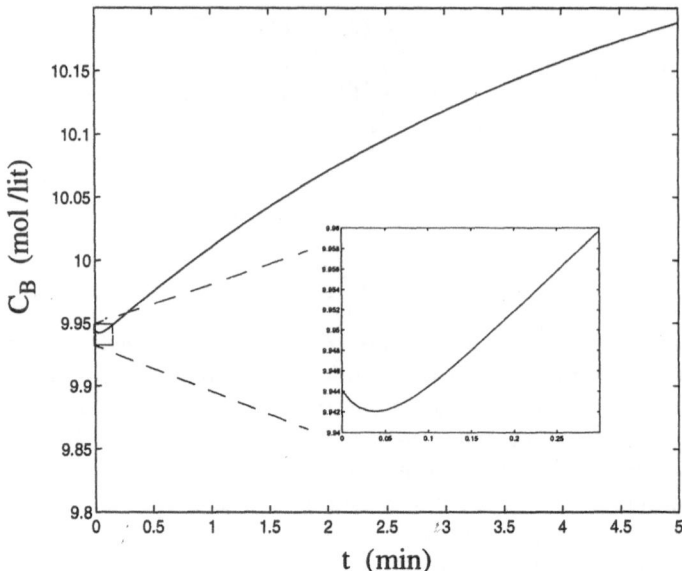

Figure 6. Inverse response in C_B in the fast boundary layer, for step increase in F_o.

state-space realizations that can be used as the basis for controller synthesis, using available results for nonlinear ODE systems. Within this approach, two broad classes of regular and non-regular systems were identified which are fundamentally different. While for regular systems, a state-space realization can be derived independently of the controller design, for non-regular systems, a state-space realization does not exist; a feedback compensator was proposed to regularize such systems so that state-space realizations can be derived.

DAE models, specifically high-index models with singular algebraic equations, arise naturally in a wide range of chemical processes with fast mass/heat transfer, fast reactions, fast gaseous flow etc., where the fast phenomena are approximated by the quasi-steady-state equilibrium conditions. If these equilibrium conditions in the DAE models are replaced by explicit rate expressions for the fast reactions, mass/heat transfer etc., then the resulting DAE model is typically of index one and can be easily reduced to an ODE model. However, owing to the presence of large parameters like large mass transfer or reaction rate coefficients, such rate-based models exhibit a stiffness/time-scale multiplicity. Thus, while these models can be used for numerical simulation using appropriate methods for stiff systems, controllers designed on the basis of these models often suffer from (i) controller ill-conditioning, and (ii) instability in the case of slightly non-minimum phase systems. On the other hand, controllers designed on the basis of the equilibrium-based high-index DAE models do not suffer from these problems and yield good performance with stability.

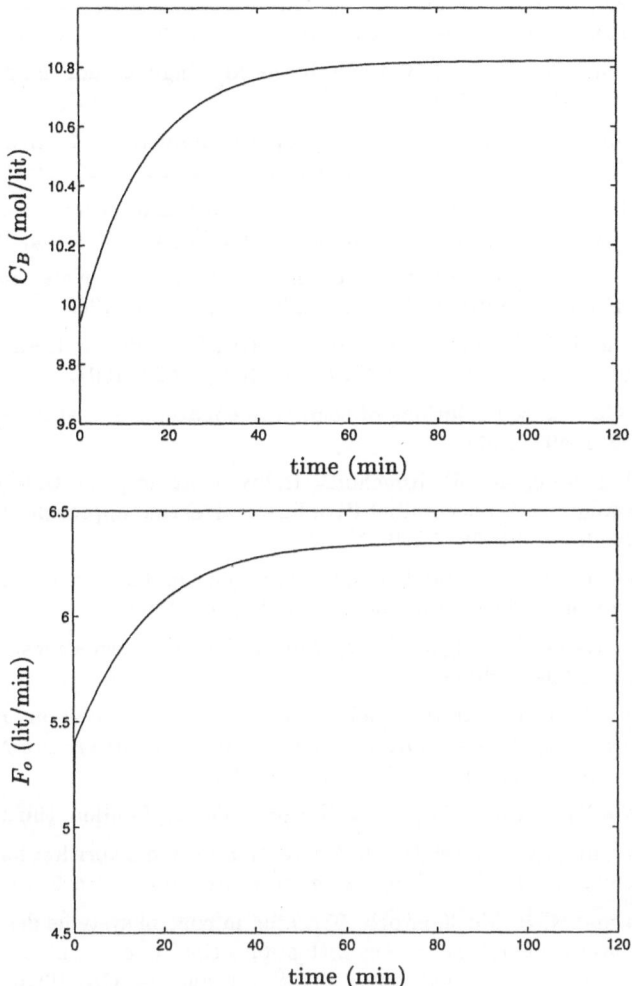

Figure 7. Closed-loop profiles for 10% increase in y_{2sp} under controller designed on the basis of equilibrium-based index-two DAE model.

8. References

[1] R. Bachmann, L. Brull, Th. Mrzigold, and U. Pallaske. On methods for reducing the index of differential algebraic equations. *Comput. chem. Engng.*, 14:1271–1273, 1990.

[2] V. B. Bajic. *Lyapunov's Direct Method in The Analysis of Singular Systems and Networks*. Shades Technical Publications, Hillcrest, Natal, 1992.

[3] K. E. Brenan, S. L. Campbell, and L. R. Petzold. *Numerical Solution of Initial-Value Problems in Differential-Algebraic Equations*. Classics in Applied Mathematics. Society for Industrial and Applied Mathematics, Philadelphia, 1996.

[4] S. L. Campbell. *Singular Systems of Differential Equations II*, volume 61 of *Research Notes in Mathematics*. Pitman Books Ltd., London, 1982.

[5] S. L. Campbell. Consistent initial conditions for singular and nonlinear systems. *Circ. Sys. Signal Proc.*, 2:45–55, 1983.

[6] Y. Chung and A. W. Westerberg. A proposed numerical algorithm for solving nonlinear index problems. *Ind. Eng. Chem. Res.*, 29:1234–1239, 1990.

[7] J.E. Cuthrell and L. T. Biegler. Simultaneous optimization and solution methods for batch reactor control profiles. *Comput. chem. Engng.*, 13:49–62, 1989.

[8] L. Dai. *Singular Control Systems*, volume 118 of *Lecture Notes in Control and Information Sciences*. Springer-Verlag, Berlin, Heidelberg, 1989.

[9] P. Daoutidis and A. Kumar. Structural analysis and output feedback control of nonlinear multivariable processes. *AIChE J.*, 40:647–669, 1994.

[10] V. Dolezal. Generalized solutions of semistate equations and stability. *Circ. Syst. Sig. Proc.*, 5:391–403, 1986.

[11] M. Fliess, J. Lévine, and P. Rouchon. Index of an implicit time-varying linear differential equation: A noncommutative linear algebraic approach. *Linear Algebra and its Applications*, 186:59–71, 1993.

[12] R. Gani and I. T. Cameron. Modelling for dynamic simulation of chemical processes: The index problem. *Chem. Eng. Sci.*, 47:1311–1315, 1992.

[13] C. W. Gear. Differential-algebraic equation index transformations. *SIAM J. Sci. Stat. Comput.*, 9:39–47, 1988.

[14] E. Hairer, C. Lubich, and M. Roche. *The Numerical Solution of Differential-Algebraic Systems by Runge-Kutta Methods*, volume 1409 of *Lecture Notes in Mathematics*. Springer-Verlag, Berlin, Heidelberg, 1989.

[15] A. Isidori. *Nonlinear Control Systems*. Springer-Verlag, London, third edition, 1995.

[16] R. B. Jarvis and C. C. Pantelides. A differentiation-free algorithm for solving high-index DAE systems. In *AIChE annual meeting 92*, Miami Beach, FL, 1992.

[17] H. Krishnan and N. H. McClamroch. Tracking in control systems described by nonlinear differential-algebraic equations with applications to constrained robot systems. In *Proc. of Amer. Contr. Conf.*, page 837, San Francisco, CA, 1993.

[18] A. Kumar. *Control of Nonlinear Differential Algebraic Equation Systems : Theory and Chemical Process Applications*. PhD thesis, Dept. of Chem. Eng. & Mat. Sci., University of Minnesota, Minneapolis MN, 1997.

[19] A. Kumar, P. D. Christofides, and P. Daoutidis. Singular perturbation modeling of nonlinear processes with non-explicit time-scale separation. *Chem. Eng. Sci.*, page in press, 1997.

[20] A. Kumar and P. Daoutidis. Feedback control of nonlinear differential-algebraic-equation systems. *AIChE J.*, 41(3):619–636, 1995.

[21] A. Kumar and P. Daoutidis. Dynamic feedback regularization and control of nonlinear differential-algebraic-equation systems. *AIChE J.*, 42:2175–2198, 1996.

[22] A. Kumar and P. Daoutidis. State-space realizations of linear differential-algebraic-equations systems with control-dependent state space. *IEEE Trans. Automat. Contr.*, 41:269–274, 1996.

[23] A. Lefkopoulos and M. A. Stadherr. Index analysis of unsteady-state chemical process systems – ii. strategies for determining the overall flowsheet index. *Comput. chem. Engng.*, 17:415–430, 1993.

[24] B. Leimkuhler, L. R. Petzold, and C. W. Gear. Approximation methods for the consistent initialization of differential-algebraic equations. *SIAM J. Numer. Anal.*, 28:205–226, 1991.

[25] F. L. Lewis. A survey of linear singular systems. *Circ. Syst. Sig. Proc.*, 5:3–36, 1986.

[26] N. H. McClamroch. Feedback stabilization of control systems described by a class of nonlinear differential-algebraic equations. *Syst. & Contr. Lett.*, 15:53–60, 1990.

[27] H. Nijmeijer and A. J. van der Schaft. *Nonlinear Dynamical Control Systems.* Springer-Verlag, New York, 1990.

[28] C. C. Pantelides. The consistent initialization of differential-algebraic systems. *SIAM J. Sci Stat. Comput.*, 9:213–231, 1988.

[29] C. C. Pantelides, D. Gritsis, K. R. Morison, and R. W. H. Sargent. The mathematical modelling of transient systems using differential-algebraic equations. *Comput. chem. Engng.*, 12:449–454, 1988.

[30] C. C. Pantelides, R. W. H. Sargent, and V. S. Vassiliadis. Optimal control of multistage systems described by differential-algebraic equations. In *AIChE annual meeting 92*, Miami Beach, FL, 1992.

[31] L. R. Petzold. Differential/algebraic equations are not ode's. *SIAM J. Sci. Stat. Comput.*, 3:367–384, 1982.

[32] J. W. Ponton and P. J. Gawthrop. Systematic construction of dynamic models for phase equilibrium processes. *Comput. chem. Engng.*, 15:803–808, 1991.

[33] J. G. Renfro, A. M. Morshedi, and O. A. Asbjornsen. Simultaneous optimization and solution of systems described by differential/algebraic equations. *Comput. chem. Engng.*, 11:503–517, 1987.

[34] W. C. Rheinboldt. Differential-algebraic systems as differential equation on manifolds. *Math. Comput.*, 43:473–482, 1984.

[35] J. Unger, A. Kröner, and W. Marquardt. Structural analysis of differential-algebraic equation systems – theory and applications. *Comput. chem. Engng.*, 19:867–882, 1995.

[36] W. Yim and S. N. Singh. Feedback linearization of differential-algebraic systems and force and position control of manipulators. In *Proc. of Amer. Contr. Conf.*, pages 2279–2283, San Francisco, CA, 1993.

Table 1. Parameters and variables for reactor with heating jacket.

Variable	Description	Nominal value
C_{Ao}	feed reactant concentration $(mole/lit)$	2.5
C_A	reactant concentration in reactor $(mole/lit)$	1.205
C_B	product concentration in reactor $(mole/lit)$	1.295
c_p	specific heat capacity $(J/g\ K)$	8.0
E	activation energy $(J/mol\ K)$	60000
F_A	outlet flowrate from reactor (lit/s)	2.0
F_h	heating fluid flowrate (lit/s)	0.1
k_o	pre-exponential factor in reaction rate $(lit/mole\ s)$	5×10^{10}
T_A	feed reactant temperature (K)	305
T	reactor temperature (K)	302.3
T_h	heating fluid temperature (K)	330
T_j	jacket temperature (K)	302.6
V	reactor holdup volume (lit)	1.0
V_h	jacket volume (lit)	0.2
ρ	liquid density (g/lit)	800
ΔH_r	heat of reaction $(J/mole)$	20000
UA	product of heat transfer coeff. and area $(J/K\ s)$	64000

Table 2. Parameters and variables for CSTR with fast and slow reactions.

Variable	Description	Nominal value
C_{Ao}	conc. of A in feed (mol/lit)	20.0
C_A	conc. of A in reactor (mol/lit)	2.1
C_B	conc. of B in reactor (mol/lit)	9.94
C_C	conc. of C in reactor (mol/lit)	7.96
F_o	feed flowrate (lit/min)	5.0
k_1	forward rate constant of fast reaction (min^{-1})	20.0
k_2	rate constant of slow reaction (min^{-1})	0.1
K_{eq}	equilibrium constant	5.0
V	reactor holdup volume (lit)	40.0

PROMISES AND LIMITATIONS OF FUNCTIONAL EXPANSIONS IN NONLINEAR MODEL-BASED CONTROL

KENNETH R. HARRIS AND AHMET PALAZOĞLU

Department of Chemical Engineering and Material Science
University of California
Davis, CA 95616, USA

Abstract. The application of functional expansion (FEx) models for the analysis and control of nonlinear processes is reviewed. Nonlinear analysis tools analogous to the linear pole/zero and frequency response concepts are presented, as well as the concept of a nonlinearity measure. FEx model-based controllers are developed based on the internal model control structure. Strengths and weaknesses of the methods are discussed, and the developed concepts are applied to a simulation example.

1. Introduction

Certainly all chemical processes have some form of nonlinearity inherent in their operation. Common practice has been to approximate these processes with a linear model around a particular equilibrium point. Although this greatly simplifies the analysis and control of the system, often this approximation has substantial performance and economic impacts on process operation. This is the common motivation for nonlinear control and system analysis that has recently been addressed by a number of different approaches [3, 6].

Among these many methods, here we focus on the application of functional expansions to address both the analysis and control problems. Several types of functional expansion models have already been applied for nonlinear system analysis and control, the most common being the Volterra model. Methods exist for identifying the Volterra kernels from either input/output data or first principle models [28]. However, in practice, the Volterra model typically requires a large number of terms to capture the

345

R. Berber and C. Kravaris (eds.), Nonlinear Model Based Process Control, 345-369.
© *1998 Kluwer Academic Publishers.*

rich behavior of many nonlinear systems. Recent efforts have utilized orthogonal functions to achieve parsimonious models [32].

The functional expansion models utilized in this work (FEx models) are inspired by the work of Fliess *et al.* [13]. Utilizing several tools of noncommutative algebra, the resulting models yield an algebraic expression for the input/output relationship for nonlinear systems. Although the Volterra kernels are related to the theory of the FEx models through a Taylor expansion [13], FEx models are straightforward to compute to high orders from first principle models. The developed FEx models have been used to solve nonlinear ODEs algebraically [5], as well as for analysis [15] and control [18, 17] of nonlinear systems.

Being tutorial in nature, the goal of this paper is to provide the reader with an outline of the applications of FEx models in nonlinear analysis and control. The majority of the concepts covered are topics of current research. The topics are organized into three sections. The first provides a brief overview of the FEx models and discusses the relevant properties. The second utilizes the FEx models for the analysis of nonlinear systems, while the third develops control laws based on FEx models. The latter two sections also include a case study.

2. Overview of FEx Models

Here, only a brief review of the basic properties and practical considerations of FEx models will be covered. The interested reader is referred to the references for a complete presentation [13, 5, 15].

FEx models can be computed using two different formulations. The first method derives algebraic FEx models using the tools of noncommutative algebra. This algebraic relationship describes the dynamic input/output behavior of the system. Alternately, numerical FEx models can be developed for use in feedback systems. Although the algebraic and numerical FEx models are equivalent, this latter type does not reveal the dependence of the model on process parameters.

We will focus on single input-single output (SISO) unbiased systems of the form

$$\begin{aligned} \dot{x} &= Ax + bu + \xi(x, u) \\ y &= h(x) \end{aligned} \tag{1}$$

where $x \in R^n$, $u \in R$, $y \in R$, $\xi(x, u)$ is vector-valued function of u and x, $h(x)$ is a scalar function of u and x, and the matrix A and vector b are of appropriate dimensions. We have limited the analysis to SISO systems in this paper, although the extension to multivariable (MIMO) systems is straightforward.

Additionally, we will assume that the system of Equation 1 has fading memory. This implies that the system does not depend on the infinite past. Boyd and Chua [9] provide a concrete definition and analysis on the approximation of fading memory systems by nonlinear operators. The fading memory assumption has several implications in this context, the most relevant being the guarantees of causality and the existence of a FEx model for all systems in the form of Equation 1.

2.1. ALGEBRAIC FEX MODELS

For the development of algebraic FEx models, a specific system form is required, such that $\xi(x, u)$ and $h(x)$ must contain only polynomial nonlinearities in x and u. If the system is not in this form, the Carleman linearization [22] or Taylor expansion methods can be used at the expense of an additional approximation.

Computation of the models relies on two tools, the Laplace-Borel ($\mathcal{LB}$) transform and the shuffle product. The $\mathcal{LB}$ transform is an integral transform defined as

$$F(x_0) = \mathcal{LB}\left[f(t)\right] = \frac{1}{x_0} \int_0^\infty f(t) e^{-t/x_0} dt \tag{2}$$

where x_0 is the transformation variable. Table 1 provides a few useful $\mathcal{LB}$ transform pairs. The shuffle product provides a means to represent polynomial nonlinearities in the $\mathcal{LB}$ domain

$$\mathcal{LB}\left[f(t)g(t)\right] = \mathcal{LB}\left[f(t)\right] \amalg \mathcal{LB}\left[g(t)\right] \tag{3}$$
$$= F(x_0) \amalg G(x_0) \tag{4}$$

In essence, multiplication in the time domain corresponds to the shuffle operation in the $\mathcal{LB}$ domain. Several methods can be used to compute the shuffle product, and the operation can be conveniently automated using a symbolic programming language [15].

Taking the $\mathcal{LB}$ transform of Equation 1 yields

$$Y = \amalg\left[X_0 + X_N, \amalg\right] \tag{5}$$
$$X_0 = \left(\frac{I}{x_0} - A\right)^{-1} bU$$
$$X_N = \left(\frac{I}{x_0} - A\right)^{-1} \Xi(X_0 + X_N, U, \amalg)$$

The state vector X has been partitioned into a linear portion, X_0, and a nonlinear contribution, X_N. X_0 is equivalent to the linearization of the state vector of the system about the equilibrium point, while X_N captures the

348

TABLE 1. $\mathcal{LB}$ transform pairs

$f(t)$	$F(x_0)$
$\delta(t)$	$1/x_0$
$1(t)$	1
$\exp(\alpha t)$	$\frac{1}{1-\alpha x_0}$
$\frac{df}{dt}$	$\frac{F(x_0)-f(0)}{x_0}$

nonlinearities in the system. The nonlinear functions $\xi(x,u)$ and $h(x)$ have been replaced by the corresponding functions Ξ and H in the $\mathcal{LB}$ domain with the shuffle operation representing the nonlinearities. Because it is not possible to explicitly solve for X in Equation 5, as there is no "inverse shuffle operation", the nonlinear term is expanded around the linear term resulting in a series solution with q terms

$$Y = Y_0 + Y_1 + \cdots + Y_q \qquad (6)$$

These expressions can then be converted back to the time domain yielding the system response, written as a series

$$y(t) = y_0(t) + y_1(t) + \cdots + y_q(t) \qquad (7)$$

In the sequel, a FEx model using q terms to capture the nonlinearities in the system will be denoted FEx^q.

2.2. NUMERICAL FEX MODELS

For the computation of feedback loops, the numerical form of the FEx models is required. Here we will allow non-polynomial nonlinearities in the state and output equations. For the numerical FEx models the use of an operator notation is convenient, so that the FEx^q operator mapping an input u to an output y can be written as

$$y = Y[u] = \sum_{m=0}^{q} Y_m[u] \qquad (8)$$

Examination of Equation 5 reveals that the higher order expansions can be interpreted in general as an output of a static nonlinearity $\xi(x,u)$ operated on by a linear filter. Figure 1 illustrates the block structure for the computation of the state vector operators for a general system. The linear

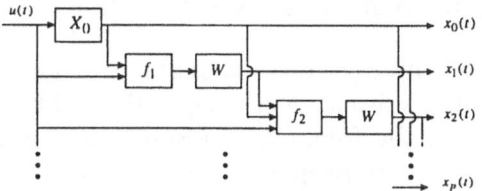

Figure 1. Block diagram for computation of the FEx model.

operators X_0 and W are defined in the $\mathcal{LB}$ domain as

$$W = (\frac{I}{x_0} - A)^{-1} \tag{9}$$

$$X_0 = (\frac{I}{x_0} - A)^{-1}b \tag{10}$$

while the functions f_i are defined in the time domain as

$$
\begin{aligned}
f_1\left[x_0(t), u(t)\right] &= \xi\left[x_0(t), u(t)\right] \\
f_2\left[x_0(t), x_1(t), u(t)\right] &= \xi\left[x_0(t) + x_1(t), u(t)\right] - \\
&\quad \xi\left[x_0(t), u(t)\right] \\
f_q\left[x_0(t), \ldots, x_{q-1}(t), u(t)\right] &= \xi\left[\sum_{i=0}^{q-1} x_i(t), u(t)\right] - \\
&\quad \xi\left[\sum_{i=0}^{q-2} x_i(t), u(t)\right]
\end{aligned}
\tag{11}
$$

Here $x_i(t)$ denotes the output of the state vector from the ith FEx model, and the zeroth contribution should not be confused with the $\mathcal{LB}$ transformation variable x_0. Using this scheme allows high order FEx models to be computed in a straightforward manner for use in feedback systems. The details of this method are left to the references [15].

FEx models can be used to capture nonlinearities in a wide range of systems. However, they do suffer from similar convergence limitations as the Volterra series. In particular, convergence is not guaranteed for all ranges of input magnitudes [28]. Determining this radius of convergence is a difficult problem, however for many processes the radius is adequate for control and analysis purposes. These convergence properties, among other motivations, have prompted research into orthogonal functional expansions [28, 32].

3. Nonlinear System Analysis via FEx Models

In nonlinear system analysis, one attempts to determine the behavior of the system within the critical operating regime. The behavior of interest could be the stability, the frequency response, or for nonlinear systems in particular, the bifurcation properties of the process. For linear systems, these properties are either nonexistent or are independent of the forcing signal into the process. These properties do depend on the forcing function of the process for nonlinear systems, making this analysis a non-trivial problem.

A variety of methods have been utilized in nonlinear system analysis. Early studies mapped the phase space through numerical methods [4], singularity and bifurcation theory has been used to delineate steady state solutions of nonlinear systems [27], and perturbation methods have been applied for bifurcation and limit cycle detection [26].

Additionally, the frequency response of nonlinear systems has been investigated through the use of the Volterra series [7, 8]. The method is a multidimensional extension of the amplitude and phase shifts common to linear system analysis, and can be used to determine nonlinear frequency phenomena such as intermodulation and amplitude compression/expansion. Although theoretically rigorous, the multidimensional nature makes intepretation of the results difficult. The dimensionality of the analysis problem increases as the number of expansion terms increases.

Here, algebraic FEx models are utilized to develop pole/zero and frequency response concepts, analogous to the linear methods. The pole/zero structure of the models can be viewed as the set of poles and zeros corresponding to the linearized model, augmented by a set of higher order poles and zeros to capture the nonlinearities in the system. The frequency response is developed in a similar fashion, with the nonlinearities incorporated as bounds on the amplitude ratio and phase shift. Consequently, for single-tone inputs, the frequency analysis remains one dimensional and intepretation is straightforward. For multi-tone inputs, the dimensionality increases with the number of input frequencies considered, but remains constant with respect to the number of terms in utilized in the FEx model. The algebraic nature of the expressions implictly captures the nonlinear frequency phenomena of the system as a function of process parameters.

While these methods yield theoretically attractive results, it is often unclear how to practically interpret or use the determined properties. To address this issue, the analysis problem can be restated to focus on the more fundamental question "Is nonlinear control necessary?" The answer is usually subjective, as nonlinear controller design involves an increased effort in analysis and design. Therefore, the problem can be simplified by considering the following set of questions:

- Given the physical properties of the process, can the nonlinearity of the process be quantitatively measured?
- How does the nonlinearity measure change as a function of the physical or design variables?
- Although the system may be nonlinear, is it possible to select a linear control law to achieve satisfactory performance?

This analysis relies on defining an appropriate nonlinearity measure. Several measures have been proposed, with applications ranging from determining control law nonlinearity [30] to defining objective functions for feedback linearization via optimization [2, 1].

This section will outline the possibilities of nonlinear systems analysis using algebraic FEx models. First, analogs to the well-know pole/zero and frequency response of linear systems are developed using the FEx models. Then, the problem of defining an appropriate nonlinearity measure is addressed. Finally, a nonlinear CSTR example is introduced and the developed concepts are applied.

3.1. $\mathcal{LB}$ POLES AND ZEROS

For linear systems, the location of the poles and zeros completely characterize the system response. Using the algebraic FEx models, the output of a nonlinear system of the form of Equation 1, in response to an impulse input, can be expressed as [10]

$$Y = K x_0^n \frac{\prod_{j=1}^{\nu}(\tau_{z,j} x_0 + 1)}{\prod_{i=1}^{\mu}(\tau_{p,i} x_0 + 1)} \tag{12}$$

Here n is the degree of the system and K is a constant gain. With the output written in this form, the following definitions of the $\mathcal{LB}$ pole and zero are presented.

Definition 1 *A $\mathcal{LB}$ pole is defined as the solutions for x_0 of*

$$\prod_{j=1}^{\nu}(\tau_{p,j} x_0 + 1) = 0 \tag{13}$$

and a $\mathcal{LB}$ zero is defined as the solutions for x_0 of

$$\prod_{i=1}^{\mu}(\tau_{z,i} x_0 + 1) = 0 \tag{14}$$

corresponding to the expression of Equation 12

Note that Equation 12 can be interpreted as a linear transfer function, augmented by a set of poles and zeros to capture the nonlinearities of the system. Consequently, by examining the location of the $\mathcal{LB}$ poles and zeros of the system, local stability of the nonlinear system can be deduced, and the speed of response and inherent modes determined. However, note that these properties will depend on the magnitude of the impulse used in the forcing. Consequently, care must be taken in choosing an impulse magnitude that spans the input domain of interest.

3.2. NONLINEAR FREQUENCY RESPONSE

To develop the frequency response framework from algebraic FEx models, the output to a system forced by a single sinusoidal input of amplitude A_{in} can be written in the time domain as

$$y = A\left(\omega, t\right) \sin\left[\omega t + \Phi(\omega, t)\right] + \Psi_0 \tag{15}$$

Calculation of the time-dependent amplitude, $A(\omega, t)$, and phase shift, $\Phi(\omega, t)$, can be determined in a straighforward manner using the algebraic FEx models [15]. Ψ_0 is an offset term that is a result of the mean value of the forced output not being equal to the nominal steady state value, as it is in linear systems. Note that Equation 15 is analogous to the linear case, except with the amplitude and phase being time-varying, and the addition of the offset term. The amplitude ratio, AR, and phase shift, Φ, can then be bounded over all time by

$$\frac{\min\left[A(\omega, t)\right]}{A_{in}} \leq AR(\omega) \leq \frac{\max\left[A(\omega, t)\right]}{A_{in}} \quad \forall t \tag{16}$$
$$\min\left[\Phi(\omega, t)\right] \leq \Phi(\omega) \leq \max\left[\Phi(\omega, t)\right] \quad \forall t \tag{17}$$

In this way, the time dependence of the nonlinear amplitude and phase functions has been removed. Instead they are represented within a bound of these functions.

This type of analysis can be used to provide an uncertainty to the common linear stability analysis when applied to nonlinear systems. Additionally, examination of the bounds of these functions as a function of the forcing frequency provides insight as to where the nonlinearities occur within the frequency response of the system. These concepts will be illustrated in the case study that follows.

Again, because these properties depend on the magnitude of the input, the correct magnitude must be chosen that represents the domain of interest. Here only single tone inputs have been considered. For multi-tone inputs the same development can be followed, and because superposition does not hold for nonlinear systems, the results will differ in general. This extension is discussed further in the references [15].

3.3. NONLINEARITY MEASURES

To formulate the concept of a nonlinearity measure using FEx models, we will adopt the nonlinearity measure as defined by Allgöwer [1]:

Definition 2 *The nonlinearity measure $\phi_{\mathcal{U}}$ of a BIBO-stable, causal I/O system $N : \mathcal{U} \to \mathcal{Y}$ for input signals $u \in \mathcal{U}$ is defined by the non-negative number*

$$\phi_{\mathcal{U}} = \inf_{G \in \mathcal{G}} \sup_{u \in \mathcal{U}} \frac{\|G[u] - N[u]\|_{p\mathcal{Y}}}{\|u\|_{p\mathcal{U}}} \tag{18}$$

with $G : \mathcal{U} \to \mathcal{Y}$ a linear operator in the set of linear operators $\mathcal{G}$, $\| \cdot \|_{p\mathcal{Y}}$ a p norm in $\mathcal{Y}$, $\| \cdot \|_{p\mathcal{U}}$ a p norm in $\mathcal{U}$ and $\mathcal{U}$ the set of inputs considered.

In the above definition we have implicitly assumed that the system is input/output stable, unbiased, and has finite power. This definition sets the size of a nonlinearity of an input/output system as the normalized largest difference between the output of a nonlinear system N and a linear system G, measured in a norm $\| \cdot \|_{p\mathcal{Y}}$ in $\mathcal{Y}$. Note that the difference is taken with respect to the worst case input in $\mathcal{U}$, and with respect to the best linear approximation $G^*[u]$ chosen from a parameterization of linear models $\mathcal{G}$.

For linear systems, $\phi_{\mathcal{U}}$ is zero. However, a zero nonlinearity measure does not imply that the system is linear because the set $\mathcal{U}$ may not contain all admissible inputs. The value of the nonlinearity measure can, in general, depend on the chosen set of input signals, $\mathcal{U}$.

While the results from Definition 2 are certainly intuitive and useful, their interpretation is difficult due to the lack of a reference point. The following normalized measure, which can be shown to be bounded by $[0, 1]$, is often easier to interpret. A measure of zero implies linearity in the input set $\mathcal{U}$. A measure of unity implies that the input set encompasses a switch in the sign of the system gain, which is identified as a highly nonlinear region.

Definition 3 *The normalized nonlinearity measure $\phi_{\mathcal{U}}^*$ of the system described by Equation 1 is given by*

$$\phi_{\mathcal{U}}^* = \inf_{G \in \mathcal{G}} \sup_{u \in \mathcal{U}} \frac{\|G[u] - N[u]\|_{p\mathcal{Y}}}{\|N[u]\|_{p\mathcal{Y}}} \tag{19}$$

where the signals and norms are as described above.

Typically the calculation of these measures requires extensive computational efforts [2]. Using this numerical approach, determining the dependence of the measure on system parameters can be formidable. Because the

FEx models provide an algebraic input/output relationship, they provide a natural framework for these computations. Several methods for calculation of the measure is possible through the use of FEx models [16], however here we will review the calculation of an upper and approximate lower bound on the measure.

An approximate lower bound can be determined by assuming the input set $\mathcal{U}_{LB}$ to be a set of sinusiods parameterized by $\mathcal{A}$ and Ω

$$
\begin{aligned}
\mathcal{U}_{LB} &= \{u \mid u(t) = A\sin(\omega t), A \in \mathcal{A}, \omega \in \Omega\} \\
\mathcal{A} &= \{A \in R^+ \mid A_{\min} \leq A \leq A_{\max}\} \\
\Omega &= \{\omega \in R^+ \mid \omega_{\min} \leq \omega \leq \omega_{\max}\}
\end{aligned}
\tag{20}
$$

The steady state output can then be written as

$$
y(t) = A_0 + \sum_{k=1}^{\infty} A_k \sin(k\omega t + \phi_k)
\tag{21}
$$

This expression can be derived from the time-varying amplitude and phase functions of the FEx models given by Equation 15 [15]. The approximate lower bounds on the nonnormalized and normalized measures, respectively, are calculated from [1]

$$
\phi_{\mathcal{U}_{LB}} \geq \sup_{A \in \mathcal{A}, \omega \in \Omega} \frac{1}{A} \sqrt{A_0^2(\omega, A) + \sum_{k=2}^{\infty} \frac{A_k^2(\omega, A)}{2}}
\tag{22}
$$

$$
\phi_{\mathcal{U}_{LB}}^* \geq \sup_{A \in \mathcal{A}, \omega \in \Omega} \sqrt{1 - \frac{A_1^2(\omega, A)}{2A_0^2(\omega, A) + \sum_{k=1}^{\infty} \frac{A_k^2(\omega, A)}{2}}}
\tag{23}
$$

These measures use the quasi-linearization for the best linear plant $G^*[u]$, which is derived using the principle of harmonic balance [31]. By measuring the norm of the terms that are not part of the quasi-linearization, the nonlinearity of the system is deduced.

An upper bound can be calculated directly from the definition of the measure. By choosing a suitably defined input set $\mathcal{U}_{UB}$ and the Jacobian linearization as the best linear model $G^*[u]$, an upper bound on the measure is given by

$$
\phi_{\mathcal{U}_{UB}} \leq \sup_{u \in \mathcal{U}_{UB}} \frac{\|\sum_{i=1}^{q} y_i[u]\|_{p\mathcal{Y}}}{\|u\|_{p\mathcal{U}}}
\tag{24}
$$

$$
\phi_{\mathcal{U}_{UB}}^* \leq \sup_{u \in \mathcal{U}_{UB}} \frac{\|\sum_{i=1}^{q} y_i[u]\|_{p\mathcal{Y}}}{\|\sum_{i=0}^{q} y_i[u]\|_{p\mathcal{Y}}}
\tag{25}
$$

The upper bound results from selecting a fixed linear model rather than the best linear model $G^*[u]$. Clearly there may be better linear approximations than the Jacobian linearization that will lead to a less conservative bound, however we have found that for most systems the Jacobian leads to satisfying results.

The input set on the lower bound $\mathcal{U}_{LB}$ has been restricted to sinusoidal parameterizations, however we have placed no restrictions on the input set for the upper bound, $\mathcal{U}_{UB}$. Ideally, these inputs sets should be identical for the computation of the bounds on the measure of nonlinearity in the input set $\mathcal{U}$. However, this then restricts both sets to be a set of sinusoids. Our experience, and the experience of others [2], have shown that the nonlinearity measure in general is rather invariant to the choice of the input set $\mathcal{U}$, provided the input set spans the input domain of interest adequately.

Implicitly, we have assumed that the FEx model will accurately represent the system, and that the set of inputs $\mathcal{U}$ lie inside the radius of convergence. Additionally, the Fourier coefficients must be truncated in practice, leading to a possible source of further approximation. It is straightfoward to check these conditions by using FEx models with more terms to capture the nonlinearities of the system, or through the use of numerical simulations.

3.4. CASE STUDY

Here, we introduce a simple continuous stirred tank reactor (CSTR) model. The reactor is an isothermal, perfectly mixed, constant volume CSTR, governed by the van de Vusse reactions

$$B_1 \xrightarrow{k_1} B_2 \xrightarrow{k_2} B_3 \qquad (26)$$
$$B_1 \xrightarrow{k_3} B_4$$

where the first and second reactions have first order kinetics, and the third reaction is second order. The following state equations then describe the component mass balances in the system

$$\dot{c}_{B_1} = -k_1 c_{B_1} - k_3 c_{B_1}^2 + \frac{F}{V}(c_{B_1,f} - c_{B_1}) \qquad (27)$$
$$\dot{c}_{B_2} = k_1 c_{B_1} - k_2 c_{B_2} - \frac{F}{V} c_{B_2}$$

The objective is to control the outlet concentration c_{B_2} (deviation variable y) by manipulating the inlet flowrate F (deviation variable u). Additionally, the inlet concentration of B_1 will be considered as an unmeasured disturbance (deviation variable d). Relevant physical parameters are outlined in Table 2. Two operating points were chosen as provided in Table 3, the first

TABLE 2. Parameter set
for CSTR

k_1	50	1/h
k_2	20	1/h
k_3	8	L/mol/h
$c_{B_1,f}$	10	mol/L
V	1	L

TABLE 3. Operating points for
CSTR

	NMP	MP	
F_0	15	55	L/h
$c_{B_1,0}$	1.88	4.00	mol/L
$c_{B_2,0}$	2.70	2.68	mol/L

displaying strong nonlinearities and nonminimum phase behavior (NMP), the second only mild nonlinearities and minimum phase behavior (MP).

Figure 2 illustrates the steady state locus for each operating point, as well as the predicted loci corresponding to the FEx and linear models. For the nonminimum phase case of Figure 2a, note that both the FEx[1] and the FEx[2] models accurately capture the steady-state behavior for a range of input moves, but outside this range the models diverge. This trend is not as apparent with the minimum phase case of Figure 2b, however it is always present if the input range is taken large enough. Because the FEx model is a temporal functional approximation, FEx models having an even number of terms (FEx[2], FEx[4], etc) display an odd type functional behavior, while FEx models with an odd number of terms capturing the nonlinearities have an even functional behavior. This can also be observed in Figure 2a.

Applying the above analysis to the developed CSTR model, the τ_p and τ_z locations as given by Definition 1 are shown in Figures 3 and 4. It is important to recall that because this analysis is nonlinear, the pole/zero locations will depend on the magnitude of the impulse used as the input. Here we have chosen an input magnitude of 10 L/h. Note that the $-\tau_p$ and $-\tau_z$ locations rather than the $\mathcal{LB}$ poles are zeros have been plotted so that they may be intuitively interpreted as Laplace poles and zeros.

The linearized model shown in Figures 3a and 4a correctly identifies the nonminimum and minimum phase behavior of the two operating points in

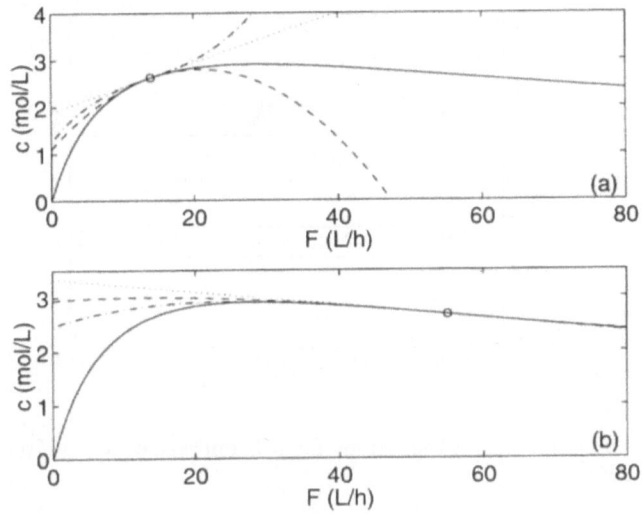

Figure 2. Steady state loci, (—) numerical, (···) linear, (- - -) FEx^1, (- · -) FEx^2. (a) NMP operation. (b) MP operation.

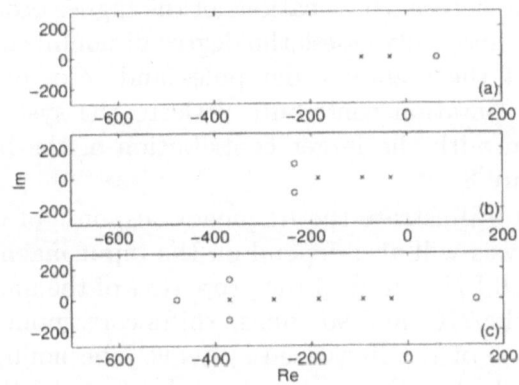

Figure 3. $-\tau_p$ (x) and τ_z (o) locations for NMP operation, $u = 10\delta(t)$ L/h. (a) Linear. (b) FEx^1. (c) FEx^2.

the pole/zero structure. The FEx^1 model has one added pole/zero pair to capture the nonlinearities of the system, while the FEx^2 model has three added pole/zero pairs. Note that the FEx models maintain the pole/zero excess of the system, however, for the FEx^1 model, the right half plane zero

358

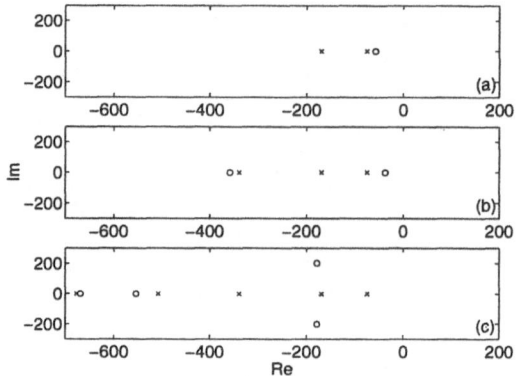

Figure 4. $-\tau_p$ (x) and $-\tau_z$ (o) locations for MP operation, $u = 10\delta(t)$ L/h. (a) Linear. (b) FEx[1]. (c) FEx[2].

is not present in Figure 3b. This is due to the magnitude of the forcing. Indeed, as the forcing or nonlinearity of the system decreases, the pole/zero structure transforms to the linear pole/zero structure, with the higher order poles and zeros cancelling each other. Consequently, in comparison with the linear pole/zero structure, the locations of the higher order poles and zeros can be used to qualitatively assess the degree of nonlinearity in the system. Here, we see that the highest order poles and zeros of Figure 4 for the minimum phase operation contribute little to the system response. This can be contrasted with the larger contribution of the higher order poles and zeros in Figure 3.

Figures 5 and 6 illustrate the frequency response of the two operating points. This analysis will also depend on the input magnitude of the sinusiod, here $A_{in} = 10$ L/h. Many of the properties of the linear Bode plots are present, such as the 270° and 90° phase shifts corresponding to the correct pole/zero structure of the linearized systems. The nonlinearities manifest themselves through the minimum and maximum bounds on both the amplitude and phase plots. In addition to providing an uncertainty element to the traditional linear frequency analysis, these plots can also be used to assess the degree of nonlinearity in the system. For both systems, the major impact of the nonlinearities appear in the lower frequency range. At the higher frequencies the bounds converge on the linearized system, indicating that the system behaves relatively linearly in these frequency ranges for single tone inputs. For the minimum phase operating point, the bounds are significantly closer than those of the nonminimum phase case, indicating a lesser degree of nonlinearity. Finally, because the analysis results in an

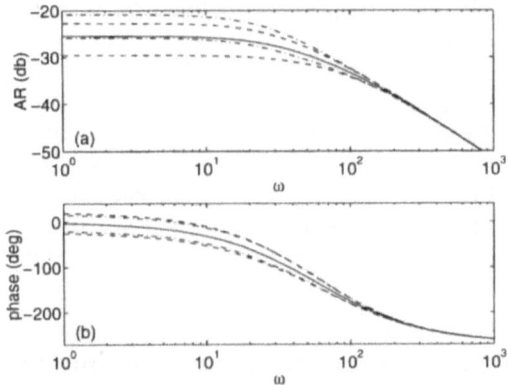

Figure 5. Nonlinear frequency response for NMP operation, $A_{in} = 10$ L/h. (—) Linearized model, (- - -) Maximum and minimum for FEx[1], (- · -) Maximum and minimum for FEx[2].

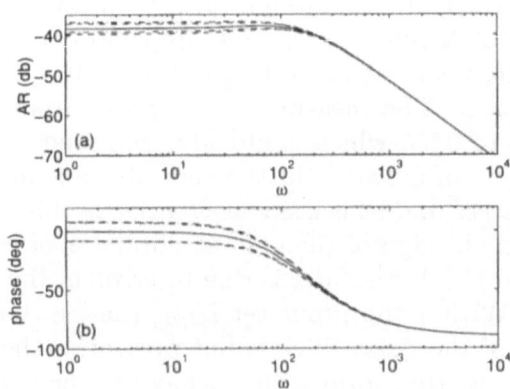

Figure 6. Nonlinear frequency response for MP operation, $A_{in} = 10$ L/h. (—) Linearized model, (- - -) Maximum and minimum for FEx[1], (- · -) Maximum and minimum for FEx[2].

algebraic expression for the frequency response functions, the dependence of key system parameters on the frequency response of the system can be investigated [15].

We have commented on the concept of determining the degree of nonlinearity in both the pole/zero and frequency response analysis, however, no quantitative analyses were made. The bounds on the nonlinearity measure can be calculated in a straightforward manner to provide this quantitative

TABLE 4. Bounds on nonlinearity measure for CSTR

	NMP	MP
FEx^1 ϕ	$[0.0213, 0.052]$	$[0.0003, 0.001]$
FEx^2 ϕ	$[0.0216, 0.075]$	$[0.0003, 0.002]$
FEx^1 ϕ^*	$[0.550, 1.00]$	$[0.030, 0.015]$
FEx^2 ϕ^*	$[0.471, 0.757]$	$[0.030, 0.032]$

measure. For the approximate lower bound, an input set $\mathcal{U}_{LB}$ of sinusoids is assumed, with a maximum amplitude of 10 L/h and a wide range of frequencies. The input set $\mathcal{U}_{UB}$ for computation of the upper bound is a set of step functions with maximum magnitude 10 L/h.

Table 4 provides the non-normalized and normalized measure bounds for both operating points. As inferred above, the measure confirms that the nonminimum phase operating point displays significant nonlinear behavior. However, although the nonnormalized measures appear small in magnitude, they alone provide little information as there is no point of reference. The normalized measure, being bounded by $[0, 1]$, indicates that the NMP system is quite nonlinear. The measure as computed from the FEx^1 model has a lower bound of 0.55, which is already high, and an upper bound of 1, which is indicative of a switch in the sign of the gain. The measure as provided from the FEx^2 model is less conservative in this case, and does not indicate a switch in the sign of the gain at normal operation. This switch, as predicted by the FEx^1 measure, is due to error in the model as can be seen in Figure 2. Within the input set $\mathcal{U}_{UB}$, the FEx^1 model predicts a switch in the sign of the gain which is not present in the original system. The FEx^2 model correctly captures this behavior. The normalized measure bounds of the FEx^1 model for the MP case reveals the slight dependence on the input set, as the approximate lower bound is actually higher than the upper bound.

Again the algebraic nature of the FEx models can be exploited to yield the functional dependence of system parameters. Figure 7 illustrates the normalized nonlinearity measure as a function of the operating point of the system. The upper bound of the normalized measure identifies the range of operation where the input set encompasses the switch in the sign of the steady-state gain as a highly nonlinear region. The lower bound verifies the operating point at which this critical phenomenon occurs at approximately 28 L/h, which can also be seen from the locus of Figure 2. The majority of linear and nonlinear control schemes would have difficulty operating in this

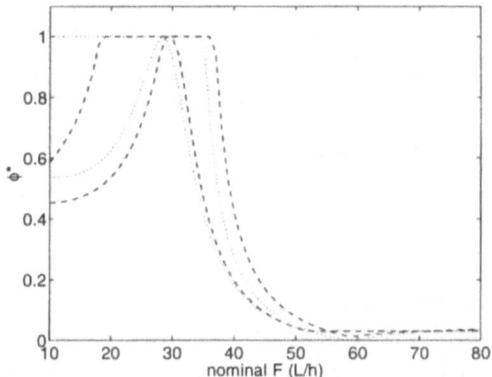

Figure 7. Normalized nonlinearity measure as a function of operating point, (· · ·) FEx[1], (- · -) FEx[2].

region, as any controller with integral action would prove to be unstable. Operating with nominal flowrates greater than 50 L/h yields quite linear behavior over the input set $\mathcal{U}$. In this region, the lower bound actually is higher than the computed upper bound.

4. Model-based Control via FEx Models

Development of the FEx control law is based on the internal model control (IMC) scheme [24]. Extension of the IMC structure to nonlinear systems relies on computation of the inverse to nonlinear dynamic systems. This inversion has been accomplished numerically, in the discrete case, by solving a set of nonlinear algebraic equations at each sampling instant [12]. Additionally for the continuous case, the inverses can be computed analytically [21], and the control law can be calculated using the tools of differential geometry [19]. Additionally, there has been a host of papers investigating the training of neural networks to learn the inverse process dynamics [25, 20].

Recently, a partitioned model inverse has been proposed to yield a flexible nonlinear model-based controller [11]. This controller consists of a standard linear controller augmented by a nonlinear correction term. The choice of the linear controller is flexible, and this element can be chosen as to address the control of nonminimum phase systems [11, 18], nonlinear decoupling schemes in MIMO systems [11, 17], the use of nonlinear model predictive control schemes [23], as well as the use of neural networks [29]. Furthermore, the nonlinear correction term can be "turned off" on-line, with the control law reverting to the chosen linear control scheme. It is this

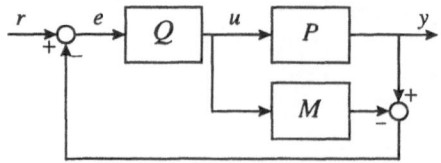

Figure 8. General IMC structure.

flexibility, that the numerical and analytical inversion techniques lack, that gives partitioned model inverses great promise in nonlinear control schemes.

The developed FEx controllers are applicable to a wide range of nonlinear systems. However, because the IMC structure implicitly includes integral action, changes in the steady-state gain of the process are forbidden. Additionally, one should be aware of the limitations of the FEx models themselves, as discussed above.

4.1. FEX CONTROL LAWS

The general IMC structure, either linear or nonlinear, can be described schematically as in Figure 8. Here P is the process and M is the model of this process. Q is the controller whose design is based upon the model M. The disturbance signal has been omitted owing to the fact that the effects of disturbances in general are indistinguishable from plant/model mismatch [14]. Note that by the parallel model/plant connection, plant/model mismatch is explicitly accounted for in the feedback loop. Indeed, in the absence of plant/model mismatch and disturbances, there is no need for feedback control.

If the control objective is to minimize the setpoint tracking error, then Q is selected as the inverse of M, such that

$$M\left[Q\left[u\right]\right] = u \tag{28}$$

The theoretical and practical issues associated with this choice of controller are well known [24], here we briefly discuss the implications. A controller determined from Equation 28 results in perfect control, requiring an infinite controller gain. Consequently the controller will lack the robustness to deal with plant/model mismatch. Furthermore, there is no guarantee that M^{-1} will be causal, stable, or because the model is nonlinear, will even exist.

The robustness and causality issues are addressed by augmenting the inverse controller by a low-pass filter, F. Additionally, this filter provides tuning parameters for shaping the desired closed-loop response. Typically this filter is chosen of the form

$$F(s) = \frac{1}{(\alpha s + 1)^r} \qquad (29)$$

where r is the relative degree of the system and α is the tuning parameter. For discrete systems an equivalent filter is utilized.

If the inverse of M is unstable, an indication of nonminimum phase dynamics, for the linear case the solution to yield a stable closed-loop is well known. In the transform domain, either continuous or discrete, the model is factored into an all-pass element M_+ containing all the nonminimum phase dynamics, and a minimum phase portion M_- which can be inverted

$$M = M_- M_+ \qquad (30)$$

Selection of Q as the inverse of the minimum phase elements is then a solution to the minimization of the setpoint tracking error. This inverse is commonly called the pseudo-inverse and is denoted as $M^\dagger$.

For nonlinear systems, the inversion of nonminimum phase systems is not so straightfoward. Clearly the numerical and analytical inversions previously used cannot discern between the minimum and nonminimum phase elements. However, recently a partitioned model inverse was shown to yield a stable inverse that also minimizes the setpoint tracking error [11].

Consider a model that can be partitioned into two portions, one consisting of solely linear elements M_L and the other of the remaining nonlinear dynamics M_N

$$M = M_L + M_N \qquad (31)$$

The inverse can then be shown to be

$$M^{-1} = \left(I + M_L^{-1}M_N\right)^{-1} M_L^{-1} \qquad (32)$$

Note that only the inverse of the linear portion of the model is required, and this computation in general is a straightfoward task. The attractive features of this inverse formulation lies in the fact that pseudo-inverses can be computed in the case when the full model inverse is unstable. Doyle *et al.* [11] demonstrated that the same factoring procedure described above for the linear case can be performed on the linear term M_L of the partitioned model, while the higher order terms in the feedback loop remain unchanged. Consequently, the pseudo-inverse for the nonlinear case can be calculated as

$$M^\dagger = \left(I + M_L^\dagger M_N\right)^{-1} M_L^\dagger \qquad (33)$$

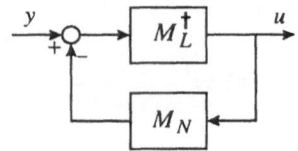

Figure 9. Partitioned model inverse.

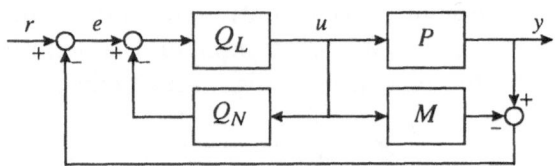

Figure 10. FEx models integrated with the IMC structure.

Additionally, this inverse can be computed on-line using the feedback loop illustrated in Figure 9. Finally, note that the numerical FEx models of Equation 8 explicitly fit the partitioned form of Equation 31, with

$$M_L = Y_0[u] \tag{34}$$

$$M_N = \sum_{i=1}^{p} Y_i[u]$$

Integration of the numerical FEx models into the IMC structure is shown in Figure 10, where

$$M = M_L + M_N \tag{35}$$

$$Q_L = M_L^{\dagger} F \tag{36}$$

$$Q_N = F^{-1} M_N \tag{37}$$

Several practical issues are involved with the controller design. In particular, note that the inverse of the robustness filter augments the nonlinear terms in the controller's feedback loop. This is required to achieve the equivalent robustness filtering to that of the linear scheme. For the common choice of F given by Equation 29, this inverse is non-causal. However, when integrated into the linear operator W of the FEx models, defined by Equation 9, the resulting transfer function is often realizable. If this is not the case, the inverse of the filter in the controller's feedback loop can be taken as a lead/lag element with fast lag dynamics to have a minimal impact on the control law.

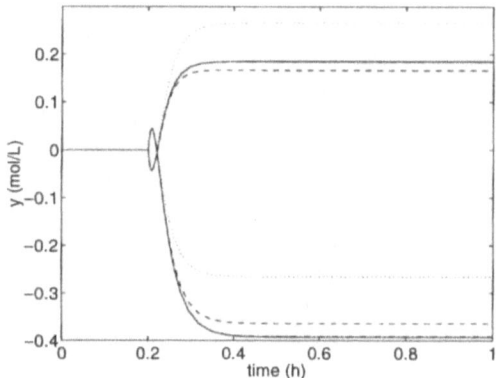

Figure 11. Open-loop responses for NMP operation with $u = \pm 5$ L/h. (—) Numerical, ($\cdots$) Linear, (- - -) FEx^2, (- $\cdot$ -) FEx^3.

4.2. CASE STUDY REVISITED

Using the same steady states of Table 3, numerical FEx models through FEx^3 were computed using the block diagram of Figure 1. The previous analysis identified that the first nonminimum steady state was significantly nonlinear, while the second was only mildly nonlinear.

Starting with the first steady state, Figure 11 illustrates the open-loop performance of the linear, FEx^2, and FEx^3 models for a step change of $u = \pm 5$ L/h. Note the inverse response characteristic of nonminimum phase systems, as well as the asymmetrical response. The FEx^2 model fits significantly better than the linear model, while the FEx^3 model essentially overlaps the numerical solution.

Using the FEx models, the nonlinear FEx controllers were developed as discussed above. Because the system is nonminimum phase, the pseudo-inverse of the linear term M_L was used, while the full nonlinear terms M_N remain unchanged in the controller's feedback loop as required. For all controllers, a first order robustness filter was sufficient to ensure causality of the nonlinear controller. A filter time constant of 0.03 h was used.

Figure 12 illustrates a reference move of $r = -0.85$ mol/L. The solid curve is the trajectory of the ideal controller, that is the product of the nonminimum phase elements and the robustness filter. The linear controller overshoots this trajectory, with indications of oscillations. The FEx^2 controller tracks significantly better, with the FEx^3 control law showing slightly better results. Also, offset is eliminated as expected since all the controllers implicitly include integral action. For larger reference moves, the linear

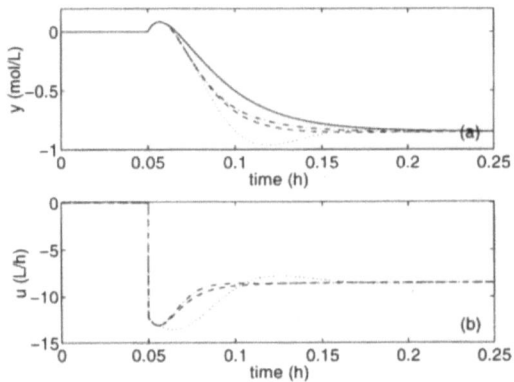

Figure 12. Closed-loop set-point responses for NMP operation with $r = -0.85$ mol/L. (—) Ideal trajectory, ($\cdots$) Linear IMC, (- - -) FEx^2, (- $\cdot$ -) FEx^3.

controller saturates at the no-flow constraint, while the FEx control laws explicitly avoid this situation.

The closed-loop response to a pulse disturbance of magnitude $d = -0.85$ mol/L and duration of 0.2 h is shown in Figure 13. Here we see that the linear controller actually performs better than the nonlinear FEx controllers. This is because the linear model under-estimates the system gain in the direction of decreasing inlet flowrate as shown in Figure 2a. This results in a higher controller gain than desired. Although the results are appealing for this particular simulation, for a more aggresively tuned controller this may result in excessive oscillations or even instability.

Disturbances are often omitted from the closed-loop analysis in nonlinear systems because they may be considered as plant/model mismatch. Figure 13 illustrates that the disturbance rejection properties of the FEx controllers and the linear controllers are similar. Indeed, further studies on the effect of plant/model mismatch have indicated that the FEx controllers are at least as robust as the corresponding linear controller [18].

The controller design for the minimum phase operating point using the FEx controllers is straighforward as there is no nonminimum phase behavior. The FEx controllers yielded almost identical closed-loop behavior as the linear controller for a wide range of reference moves and disturbance loads. This is to be expected, as the analysis of this operating point revealed nearly linear behavior. In this case, the nonlinear controller design is probably not beneficial, as the increased analysis and design effort yield little performance benefits.

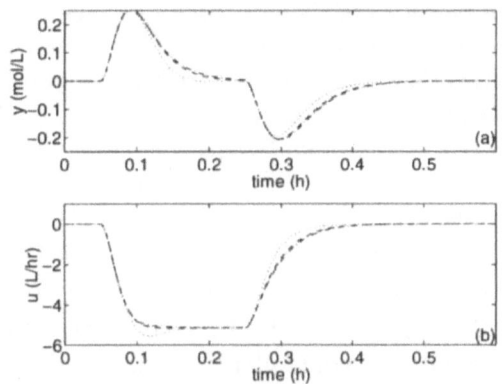

Figure 13. Closed-loop disturbance responses for NMP operation with $d = -0.85$ mol/L. (···) Linear IMC, (- - -) FEx^2, (- · -) FEx^3.

5. Summary

We have reviewed the analysis and control of nonlinear systems through the use of FEx models. The FEx models provide a framework to algebraically represent nonlinear input/output dynamic systems in a transform domain. The models are applicable to a wide range of both polynomial and non-polynomial nonlinear systems, and are straighforward to compute to high accuracy. They do suffer from having a limited input range, and are only applicable to fading memory systems. Although these limitations are important practical considerations, the models still remain applicable to a large class of systems.

Algebraic FEx models were used in the analysis to develop pole/zero and frequency response concepts for nonlinear systems that are analogous to their linear counterparts. Shortcomings of the analysis using FEx models lies mainly in the interpretation of the results. Although the results can be interpreted in a fashion similar to the linear concepts, the dependence on the selection of an appropriate input magnitude must be considered. The concept of a nonlinearity measure provides practical results that can be utilized in practice as illustrated in the case study.

Model-based controllers based on FEx models have been successful for a variety of systems. Here we utilized a partitioned model inverse to design nonlinear control laws that demonstrated superior performance to the equivalent linear controller. The flexibility of the partitioned inverse can be exploited to address the control of nonminimum phase systems as done in the case study, as well as decoupling schemes for nonlinear MIMO systems [17].

Stability of the closed-loop has not been discussed in this work, as this is an area of current research. Unfortunately, even the nominal stability problem is not trival for nonlinear controllers in general. This is in part because the model can have a limited output range. Consequently, when the model is inverted, the input domain of the controller is only defined over a subset of the entire possible domain. Selecting input moves into the controller outside of this domain leads to instabilities. Therefore, one formulation of the stability problem focuses constraining the closed-loop such that input moves outside this limited domain are not selected.

In addition to the closed-loop stability analysis, current work is focused on discrete FEx models and controller design. Discrete FEx models can be developed for both the algebraic and numerical FEx models. In the former case, a nonlinear continuous model can be discretized so that the resulting discrete FEx model is both stable and accurate for large sampling periods. The numerical FEx model and controller can be developed so that the discretization step is implicit in the method.

6. Acknowledgement

This work is supported by the National Science Foundation (CTS-9400304).

References

1. F. Allgöwer. Definition and computation of a nonlinear measure. *IFAC NOLCOS*, 1:279–284, 1995.
2. F. Allgöwer. Definition and Computation of a Nonlinearity Measure and Application to Appoximate I/O-Linearization. Technical Report Nr. 95-1, Universität Stuttgart, 1995.
3. F. Allgöwer and F. Doyle. Nonlinear process control – Which way to the promise land? In *Proceedings of CPC V*, 1995.
4. R. Aris and N. R. Amundson. An analysis of chemical reactor stability and control. *Chem. Engng. Sci.*, 7:121, 1958.
5. A. Batigün, K. R. Harris, and A. Palazoğlu. Studies on the dynamics of nonlinear processes via functional expansions: I. Solution to nonlinear ODEs. *Chem. Eng. Sci.*, 52:3183–3195, 1997.
6. B. Bequette. Nonlinear control of chemical processes: A review. *Ind. Eng. Chem. Res.*, 30:1391–1413, 1991.
7. S. A. Billings and K. M. Tsang. Spectral analysis for nonlinear systems, part i: Parametric nonlinear spectral analysis. *Mech. Sys. Sig. Proc.*, 3:319–339, 1989.
8. S. A. Billings and K. M. Tsang. Spectral analysis for nonlinear systems, part ii: Interpretation of nonlinear frequency response functions. *Mech. Sys. Sig. Proc.*, 3:341–359, 1989.
9. S. Boyd and L. Chua. Fading memory and the problem of approximating nonlinear operators with Volterra series. *IEEE Trans. Cir. Sys.*, CAS-32:1150–1161, 1985.
10. S. Can and A. Ünal. Transfer functions for nonlinear systems via Fourier-Borel transforms. Technical Report 100034, NASA, 1988.

11. F. Doyle, B. A. Ogunnaike, and R. K. Pearson. Nonlinear model-based control using second-order Volterra models. *Automatica*, 31:697, 1995.

12. C. G. Economou, M. Morari, and B. O. Palsson. Internal model control. 5. Extension to nonlinear systems. *Ind. Eng. Chem. Process Des. Dev.*, 25:403–411, 1986.

13. M. Fliess, M. Lamnabhi, and F. Lamnabhi-Lagarrigue. An algebraic approach to nonlinear functional expansions. *IEEE Trans. Cir. Sys.*, CAS-30:554–567, August 1983.

14. C. E. Garcia, D. M. Prett, and M. Morari. Model predictive control: Theory and practice - a survey. *Automatica*, 25(3):335–48, 1989.

15. K. R. Harris, A. Batigün, and A. Palazoğlu. Studies on the dynamics of nonlinear processes via functional expansions: II. Forced dynamic responses. *Chem. Eng. Sci.*, 52:3197–3207, 1997.

16. K. R. Harris, M. C. Colantonio, and A. Palazoğlu. Nonlinearity measures via functional expansions. Presented at the 1997 Annual AIChE Meeting, Los Angeles, CA, 1997.

17. K. R. Harris and A. Palazoğlu. Control of MIMO nonlinear systems via functional expansions. In *Proceedings of the 1997 ACC*, 1997.

18. K. R. Harris and A. Palazoğlu. Model-based control of nonlinear processes via functional expansions. In *Proceedings of the 1997 ECC*, 1997.

19. M. A. Henson and D. E. Seborg. An internal model control strategy for nonlinear systems. *AIChE Journal*, 37:1065, 1991.

20. E. Hernandez and Y. Arkun. Study of the control-relevant properties of backpropagation neural network models of nonlinear dynamical systems. *Computers Chem. Engng.*, 16:227–240, 1992.

21. R. A. Hirschorn. Invertibility of multivariable nonlinear control systems. *IEEE Trans. Auto. Control*, AC-24:855, 1979.

22. K. Kowalski and W Steeb. *Nonlinear Dynamical Systems and Carleman Linearization*. World Scientific, New Jersey, 1991.

23. B. R. Maner, F. J. Doyle, B. A. Ogunnaike, and R. K. Pearson. Nonlinear model predictive control of a simulated multivariable polymerization reactor using second-order Volterra models. *Automatica*, 32:1285–1301, 1996.

24. M. Morari and E. Zafiriou. *Robust Process Control*. Prentice Hall, New Jersey, 1989.

25. E. P. Nahas, M. A. Henson, and D. E. Seborg. Nonlinear internal model control strategy for neural network models. *Computers Chem. Engng.*, 16:1039–1057, 1992.

26. A. H. Nayfeh. *Introduction to Perturbation Techniques*. John Wiley and Sons, New York, 1981.

27. L. F. Razón and R. A. Schmitz. Multiplicities and instabilities in chemical reaction systems – a review. *Chem. Engng. Sci.*, 42:1005, 1987.

28. M. Schetzen. *The Volterra and Wiener Theories of Nonlinear Systems*. John Wiley and Sons, New York, 1980.

29. A. M. Shaw, F. J. Doyle, and J. S. Schwaber. A dynamic neural network approach to nonlinear process modeling. *Computers Chem. Engng.*, 21:371–385, 1997.

30. A. J. Stack and F. J. Doyle. Application of a control-law nonlinearity measure to the chemical reactor analysis. *AIChE J.*, 43:425–39, 1997.

31. M. Vidyasagar. *Nonlinear Systems Analysis*. Prentice-Hall, New Jersey, 1993.

32. Q. Zheng and E. Zafiriou. Identification of MIMO Volterra series and application to FCC unit. In *Proceedings of the 1996 IFAC, San Francisco, CA*, 1996.

NONLINEAR FEEDBACK CONTROL OF PARABOLIC PDE SYSTEMS

Panagiotis D. Christofides[1] and Prodromos Daoutidis[2]

[1]Department of Chemical Engineering
University of California, Los Angeles, CA 90095-1592

[2]Department of Chemical Engineering and Materials Science
University of Minnesota, Minneapolis, MN 55455

1. Introduction

Transport-reaction processes with significant diffusive and dispersive mechanisms (e.g. packed-bed reactors, rapid thermal processing systems, chemical vapor deposition reactors, etc.) are typically characterized by strong nonlinearities and spatial variations, and are naturally modeled by nonlinear parabolic PDE systems. The main feature of parabolic PDEs is that the eigenspectrum of the spatial differential operator can be partitioned into a finite-dimensional slow one and an infinite-dimensional stable fast complement [18, 1, 29]. This implies that the dynamic behavior of such systems can be approximately described by finite-dimensional systems. Motivated by this, the standard approach to control parabolic PDEs involves the application of Galerkin's method to the PDE system to derive ODE systems that describe the dynamics of the dominant (slow) modes of the PDE system, which are subsequently used as the basis for the synthesis of finite-dimensional controllers [20, 1, 26, 8]. However, there are two key controller implementation and closed-loop performance problems associated with this approach. First, the number of modes that should be retained to derive an ODE system that yields the desired degree of approximation may be very large, leading to high dimensionality of the resulting controllers [19]. Second, there is a lack of a systematic way to characterize the discrepancy between the solutions of the PDE system and the approximate ODE system in finite time, which is essential for characterizing the transient performance of the closed-loop PDE system.

A natural framework to address the problem of deriving low-dimensional ODE systems that accurately reproduce the solutions of parabolic PDEs is based on the concept of inertial manifold (IM) (see [29] and the references therein). An IM is

Financial support for this work in part by UCLA through the SEAS Dean's Fund, and National Science Foundation, CTS-9624725, is gratefully acknowledged.

R. Berber and C. Kravaris (eds.), Nonlinear Model Based Process Control, 371-399.
© 1998 *Kluwer Academic Publishers.*

a finite-dimensional manifold which attracts every trajectory exponentially. If an IM exists, the dynamics of the parabolic PDE system restricted on the inertial manifold is described by a set of ODEs called the inertial form. However, the explicit derivation of the inertial form requires the computation of the analytic form of the IM. Unfortunately, IMs have been proven to exist only for certain classes of PDEs (for example Kuramoto-Sivashinsky equation and some diffusion-reaction equations [29]), and even then it is almost impossible to derive their analytic form. In order to overcome the problems associated with the existence and construction of IMs, the concept of approximate inertial manifold (AIM) has been introduced [29, 17, 16] and used for the derivation of ODE systems whose dynamic behavior approximates the one of the inertial form.

In the area of control of nonlinear parabolic PDE systems, few papers have appeared in the literature dealing with the application of IM for the synthesis of finite-dimensional controllers. In particular, in [28] the problem of stabilization of a parabolic PDE with boundary finite-dimensional feedback was studied; a standard observer-based controller augmented with a residual mode filter [3] was used to induce an inertial manifold in the closed-loop system, and thus reduce the stabilization problem for the PDE system to a stabilization problem for the finite-dimensional inertial form. In [6], the theory of inertial manifolds was utilized to determine the extent to which linear boundary proportional control influences the dynamic and steady-state response of the closed-loop system.

In this paper, we review a methodology, developed in [14], for the synthesis of nonlinear low-dimensional output feedback controllers for nonlinear parabolic PDE systems. Singular perturbation methods are initially employed to establish that the discrepancy between the solutions of an ODE system of dimension equal to the number of slow modes, obtained through Galerkin's method, and the PDE system is proportional to the degree of separation of the fast and slow modes of the spatial operator. Then, a procedure, motivated by the theory of singular perturbations, is proposed for the construction of AIMs for the PDE system. The AIMs are used for the derivation of ODE systems of dimension equal to the number of slow modes, that yield solutions which are close, upto a desired accuracy, to the ones of the PDE system, for almost all times. These ODE systems are used as the basis for the synthesis of nonlinear output feedback controllers that guarantee stability and enforce the output of the closed-loop system to follow upto a desired accuracy, a prespecified response for almost all times. The methodology is successfully employed to stabilize the temperature profile of a catalytic rod, where an exothermic reaction takes place, around an unstable steady-state, and to control the temperature profile in a non-isothermal packed-bed reactor [12].

2. Preliminaries

2.1. Description of parabolic PDE systems

We consider parabolic PDE systems of the form:

$$\frac{\partial \bar{x}}{\partial t} = A\frac{\partial \bar{x}}{\partial z} + B\frac{\partial^2 \bar{x}}{\partial z^2} + wb(z)u + f(\bar{x})$$

$$y^i = \int_{z_i}^{z_{i+1}} c^i(z)k\bar{x}dz, \quad i = 1,\ldots,l \tag{1}$$

subject to the boundary conditions:

$$C_1\bar{x}(\alpha, t) + D_1\frac{\partial \bar{x}}{\partial z}(\alpha, t) = R_1$$

$$C_2\bar{x}(\beta, t) + D_2\frac{\partial \bar{x}}{\partial z}(\beta, t) = R_2 \tag{2}$$

and the initial condition:

$$\bar{x}(z, 0) = \bar{x}_0(z) \tag{3}$$

where $\bar{x}(z,t) = [\bar{x}_1(z,t) \cdots \bar{x}_n(z,t)]^T$ denotes the vector of state variables, $z \in [\alpha, \beta] \subset \mathbb{R}$ is the spatial coordinate, $t \in [0, \infty)$ is the time, $u = [u^1 \, u^2 \cdots u^l]^T \in \mathbb{R}^l$ denotes the vector of manipulated inputs, and $y^i \in \mathbb{R}$ denotes the i-th controlled output. $\frac{\partial \bar{x}}{\partial z}, \frac{\partial^2 \bar{x}}{\partial z^2}$ denote the first- and second-order spatial derivatives of $\bar{x}$, $f(\bar{x})$ is a nonlinear vector function, w, k are constant vectors, A, B, C_1, D_1, C_2, D_2 are constant matrices, R_1, R_2 are column vectors, and $\bar{x}_0(z)$ is the initial condition. $b(z)$ is a known smooth vector function of z of the form $b(z) = [b^1(z) \, b^2(z) \cdots b^l(z)]$, where $b^i(z)$ describes how the control action $u^i(t)$ is distributed in the interval $[z_i, z_{i+1}] \subset [\alpha, \beta]$, and $c^i(z)$ is a known smooth function of z which is determined by the desired performance specifications in the interval $[z_i, z_{i+1}]$. Whenever the control action enters the system at a single point z_0, with $z_0 \in [z_i, z_{i+1}]$ (i.e. point actuation), the function $b^i(z)$ is taken to be nonzero in a finite spatial interval of the form $[z_0 - \epsilon, z_0 + \epsilon]$, where ϵ is a small positive real number, and zero elsewhere in $[z_i, z_{i+1}]$. Figure 1 shows the location of the manipulated inputs and controlled outputs in the case of a prototype example. Throughout the paper, we will use the order of magnitude notation $O(\epsilon)$. In particular, $\delta(\epsilon) = O(\epsilon)$ if there exist positive real numbers k_1 and k_2 such that: $|\delta(\epsilon)| \leq k_1|\epsilon|$, $\forall |\epsilon| < k_2$.

Referring to the system of Eq.1, several remarks are in order: a) the spatial differential operator is linear; this assumption is valid for diffusion-convection-reaction processes where the diffusion coefficient and the conductivity can be taken independent of temperature and concentrations, b) the manipulated input enters the system in a linear and affine fashion; this is typically the case in many practical applications where, for example, the wall temperature is chosen as the manipulated input, and c) the nonlinearities appear in an additive fashion (e.g., complex

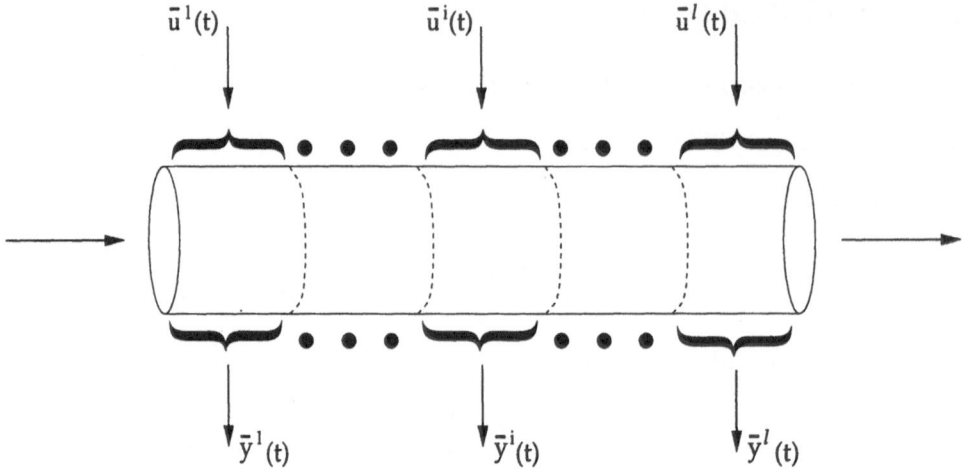

Figure 1. Specification of manipulated inputs and controlled outputs in the case of a prototype example.

reaction rates, Arrhenius dependence of reaction rates on temperature). Finally, we note that the present paper will not address the issue of compensation of model uncertainty; the reader may refer to [10, 31] for results on control of parabolic PDE systems with uncertainty.

2.2. Formulation of parabolic PDE system as infinite-dimensional system - Eigenvalue problem

In this subsection, we precisely characterize the class of parabolic PDE systems of the form of Eq.1 which we consider in the manuscript. To this end, we formulate the parabolic PDE system of Eq.1 as an infinite dimensional system in the Hilbert space $\mathcal{H}([\alpha, \beta], \mathbb{R}^n)$ (this will also simplify the notation of the paper, since the boundary conditions of Eq.2 will be directly included in the formulation; see Eq.8 below), with $\mathcal{H}$ being the space of n-dimensional vector functions defined on $[\alpha, \beta]$ that satisfy the boundary condition of Eq.2, with inner product and norm:

$$(\omega_1, \omega_2) = \int_\alpha^\beta (\omega_1(z), \omega_2(z))_{\mathbb{R}^n} \, dz$$

$$\|\omega_1\|_2 = (\omega_1, \omega_1)^{\frac{1}{2}}$$

$$(4)$$

where ω_1, ω_2 are two elements of $\mathcal{H}([\alpha, \beta]; \mathbb{R}^n)$ and the notation $(\cdot, \cdot)_{\mathbb{R}^n}$ denotes the standard inner product in $\mathbb{R}^n$. Defining the state function x on $\mathcal{H}([\alpha, \beta], \mathbb{R}^n)$ as:

$$x(t) = \bar{x}(z, t), \quad t > 0, \quad z \in [\alpha, \beta],$$

$$(5)$$

the operator $\mathcal{A}$ in $\mathcal{H}([\alpha, \beta], \mathbb{R}^n)$ as:

$$\mathcal{A}x \quad = A\frac{\partial \bar{x}}{\partial z} + B\frac{\partial^2 \bar{x}}{\partial z^2},$$

$$x \in D(\mathcal{A}) = \left\{ x \in \mathcal{H}([\alpha, \beta]; \mathbb{R}^n); \quad z = \alpha, \quad C_1 x + D_1 \frac{\partial x}{\partial z} = R_1; \right. \tag{6}$$

$$\left. z = \beta, \quad C_2 x + D_2 \frac{\partial x}{\partial z} = R_2 \right\}$$

and the input and output operators as:

$$\mathcal{B}u = wbu, \quad \mathcal{C}x = (c, kx) \tag{7}$$

where $c = [c^1 \; c^2 \; \cdots c^l]$, the system of Eqs.1-2-3 takes the form:

$$\dot{x} = \mathcal{A}x + \mathcal{B}u + f(x), \quad x(0) = x_0$$

$$y = \mathcal{C}x \tag{8}$$

where $f(x(t)) = f(\bar{x}(z,t))$ and $x_0 = \bar{x}_0(z)$. We assume that the nonlinear terms $f(x)$ are locally Lipschitz with respect to their arguments and satisfy $f(0) = 0$.

For $\mathcal{A}$, the eigenvalue problem is defined as:

$$\mathcal{A}\phi_j = \lambda_j \phi_j, \quad j = 1, \ldots, \infty \tag{9}$$

where λ_j denotes an eigenvalue and ϕ_j denotes an eigenfunction; the eigenspectrum of $\mathcal{A}$, $\sigma(\mathcal{A})$, is defined as the set of all eigenvalues of $\mathcal{A}$, i.e. $\sigma(\mathcal{A}) = \{\lambda_1, \lambda_2, \ldots, \}$. Assumption 1 that follows states that the eigenspectrum of $\mathcal{A}$ can be partitioned into a finite-dimensional part consisting of m slow eigenvalues and a stable infinite-dimensional complement containing the remaining fast eigenvalues, and that the separation between the slow and fast eigenvalues of $\mathcal{A}$ is large.

Assumption 1:

1. *$Re\{\lambda_1\} \geq Re\{\lambda_2\} \geq \cdots \geq Re\{\lambda_j\} \geq \cdots$, where $Re\{\lambda_j\}$ denotes the real part of λ_j.*

2. *$\sigma(\mathcal{A})$ can be partitioned as $\sigma(\mathcal{A}) = \sigma_1(\mathcal{A}) + \sigma_2(\mathcal{A})$, where $\sigma_1(\mathcal{A})$ consists of the first m (with m finite) eigenvalues, i.e. $\sigma_1(\mathcal{A}) = \{\lambda_1, \ldots, \lambda_m\}$, and $\frac{|Re\{\lambda_1\}|}{|Re\{\lambda_m\}|} = O(1)$.*

3. *$Re\,\lambda_{m+1} < 0$ and $\frac{|Re\{\lambda_m\}|}{|Re\{\lambda_{m+1}\}|} = O(\epsilon)$ where $\epsilon < 1$ is a small positive number.*

The assumption of finite number of unstable eigenvalues is always satisfied for parabolic PDE systems [18], while the assumption of discrete eigenspectrum and the assumption of existence of only a few dominant modes that describe the dynamics of the nonlinear parabolic PDE system are usually satisfied by the majority of diffusion-convection-reaction processes. We note that assumption 1 is not satisfied in the cases of: a) first-order hyperbolic PDE systems (i.e. convection-reaction processes) where the eigenvalues cluster along vertical, or nearly vertical, asymptotes in the complex plane (e.g. [11]), and b) parabolic PDE systems for which the spatial coordinate is defined in the infinite domain, where the eigenspectrum is continuous and wave-like behavior is usually exhibited (e.g. [24]).

3. Examples of processes modeled by nonlinear parabolic PDEs

In this section, we present two examples of processes modeled by nonlinear parabolic PDE systems of the form of Eq.1 and compute the eigenvalues and eigenfunctions of the corresponding spatial differential operators.

3.1. Catalytic rod

Consider a long, thin rod in a reactor (Figure 2). The reactor is fed with pure species A and a zeroth order exothermic catalytic reaction of the form $A \to B$ takes place on the rod. Since the reaction is exothermic, a cooling medium which

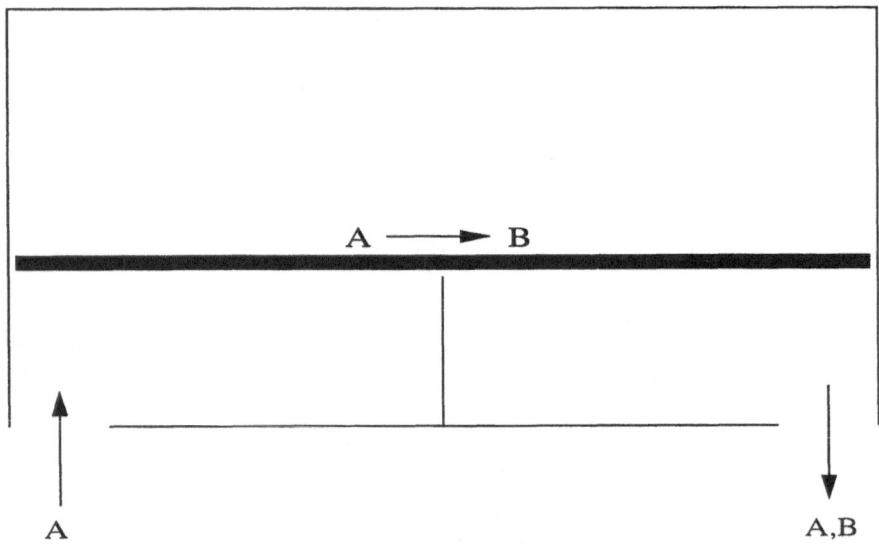

Figure 2. Catalytic rod.

is in contact with the rod is used for cooling. Under the assumptions of constant density and heat capacity of the rod, constant conductivity of the rod, and constant temperature at both ends of the rod, the mathematical model which describes the spatiotemporal evolution of the dimensionless rod temperature consists of the following parabolic PDE:

$$\frac{\partial \bar{x}}{\partial t} = \frac{\partial^2 \bar{x}}{\partial z^2} + \beta_T e^{-\frac{\gamma}{1+\bar{x}}} + \beta_U (b(z)u(t) - \bar{x}) - \beta_T e^{-\gamma} \tag{10}$$

subject to the Dirichlet boundary conditions:

$$\bar{x}(0,t) = 0, \quad \bar{x}(\pi, t) = 0 \tag{11}$$

and the initial condition:

$$\bar{x}(z, 0) = \bar{x}_0(z) \tag{12}$$

where $\bar{x}$ denotes the dimensionless temperature in the reactor, β_T denotes a dimensionless heat of reaction, γ denotes a dimensionless activation energy, β_U denotes a dimensionless heat transfer coefficient, and u denotes the manipulated input (temperature of the cooling medium). The following typical values were given to the process parameters:

$$\beta_T = 50.0, \quad \beta_U = 2.0, \quad \gamma = 4.0 \tag{13}$$

For the above values, the operating steady state $\bar{x}(z,t) = 0$ is an unstable one (Figure 3 shows the profile of evolution of open-loop rod temperature starting from initial conditions close to the steady state $\bar{x}(z,t) = 0$; the process moves to another stable steady state characterized by a maximum in the temperature profile, *hot-spot*, in the middle of the rod). Therefore, the control objective is to stabilize the rod temperature profile at the unstable steady state $\bar{x}(z,t) = 0$. Since the maximum open-loop temperature occurs in the middle of the rod (Figure 3), the controlled output was defined as:

$$y(t) = \int_0^{\pi} \sqrt{\frac{2}{\pi}} sin(z)\bar{x}(z,t)dz \tag{14}$$

and the actuator distribution function was taken to be $b(z) = \sqrt{\frac{2}{\pi}} sin(z)$, in order to apply maximum cooling to the middle of the rod. The eigenvalue problem for the spatial differential operator of the process:

$$\mathcal{A}x = \frac{\partial^2 \bar{x}}{\partial z^2}, \quad x \in D(\mathcal{A}) = \{x \in \mathcal{H}([0, \pi]; \mathbb{R}); \ z = 0, \ x = 0; \ z = \pi, \ x = 0\} \tag{15}$$

can be solved analytically and its solution is of the form:

$$\lambda_j = -j^2, \quad \phi_j(z) = \sqrt{\frac{2}{\pi}} sin(j z), \quad j = 1, \ldots, \infty \tag{16}$$

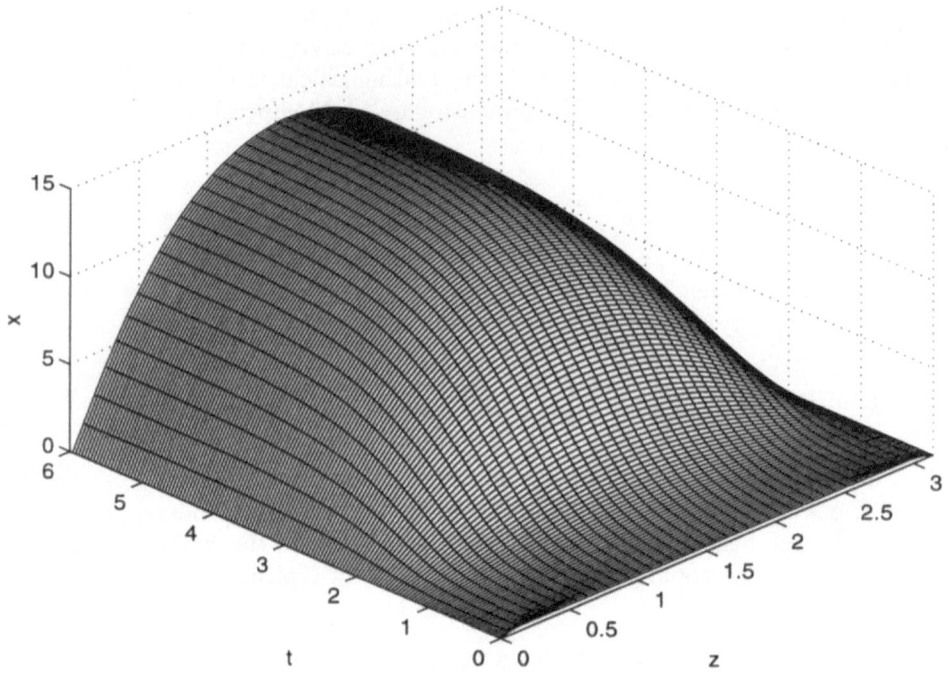

Figure 3. Profile of evolution of open-loop rod temperature.

3.2. Non-isothermal packed-bed reactor

Consider a non-isothermal packed-bed reactor, shown in Figure 4, where an ele-

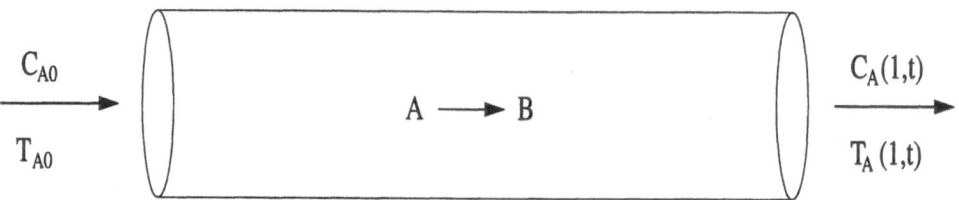

Figure 4. Non-isothermal packed-bed reactor.

mentary endothermic reaction of the form $A \to B$ takes place. Under standard modeling assumptions, the dynamic model of the process expressed in dimensionless variables takes the form [26]:

$$\frac{\partial \bar{x}_1}{\partial t} = -\frac{\partial \bar{x}_1}{\partial z} + \frac{1}{Pe_T}\frac{\partial^2 \bar{x}_1}{\partial z^2} - B_T B_C exp^{\gamma\frac{\bar{x}_1}{1+\bar{x}_1}}\bar{x}_2 + \beta_T(u - \bar{x}_1)$$

$$\frac{\partial \bar{x}_2}{\partial t} = -\frac{\partial \bar{x}_2}{\partial z} + \frac{1}{Pe_C}\frac{\partial^2 \bar{x}_2}{\partial z^2} - B_C exp^{\gamma\frac{\bar{x}_1}{1+\bar{x}_1}}\bar{x}_2$$

(17)

subject to the boundary conditions:

$$Pe_T\bar{x}_1(0,t) = \frac{\partial \bar{x}_1}{\partial z}(0,t), \quad Pe_C(\bar{x}_2(0,t) - 1) = \frac{\partial \bar{x}_2}{\partial z}(0,t);$$

$$\frac{\partial \bar{x}_1}{\partial z}(1,t) = \frac{\partial \bar{x}_2}{\partial z}(1,t) = 0$$

(18)

and the initial condition:

$$\bar{x}(z,0) = \bar{x}_0(z) \tag{19}$$

where $\bar{x}_1, \bar{x}_2$ denote dimensionless temperature and concentration in the reactor, Pe_T, Pe_C are the heat and mass Peclet numbers, B_T, B_C denote a dimensionless heat of reaction and a dimensionless pre-exponential factor, γ is a dimensionless activation energy, β_T is a dimensionless heat transfer coefficient, and u is the manipulated input (dimensionless jacket temperature) which is assumed to be spatially uniform. The control objective is to control the temperature profile along the length of the reactor, i.e. the controlled output is of the form:

$$y(t) = \int_0^1 \bar{x}_1 dz \tag{20}$$

Assuming that $Pe = Pe_T = Pe_C$, the solution to the eigenvalue problem of the spatial differential operator:

$$\mathcal{A}x = \begin{bmatrix} \dfrac{1}{Pe_T}\dfrac{\partial^2 \bar{x}_1}{\partial z^2} - \dfrac{\partial \bar{x}_1}{\partial z} & 0 \\ 0 & \dfrac{1}{Pe_C}\dfrac{\partial^2 \bar{x}_2}{\partial z^2} - \dfrac{\partial \bar{x}_2}{\partial z} \end{bmatrix},$$

$$x \in D(\mathcal{A}) = \left\{ x \in \mathcal{H}^2([0,1];\mathbb{R}^2); \quad z = 0, \quad Pe_T x_1 = \frac{\partial x_1}{\partial z}, Pe_C(x_2 - 1) = \frac{\partial x_2}{\partial z}; \right.$$

$$\left. z = 1, \quad \frac{\partial x_1}{\partial z} = \frac{\partial x_2}{\partial z} = 0 \right\}$$

(21)

can be obtained by utilizing standard techniques from linear operator theory (see for example [26]) and is of the form:

$$\lambda_j = -\frac{\bar{a}_j^2}{Pe} - \frac{Pe}{4}, \quad j = 1, \ldots, \infty$$

$$\phi_j(z) = B_j e^{Pe\frac{z}{2}}(cos(\bar{a}_j z) + \frac{Pe}{2\bar{a}_j}sin(\bar{a}_j z)), \quad j = 1, \ldots, \infty$$

(22)

where λ_j is an eigenvalue of multiplicity two and $\bar{a}_j, B_j$ can be calculated from the following formulas:

$$tan(\bar{a}_j) = \frac{Pe\bar{a}_j}{\bar{a}_j^2 - (\frac{Pe}{2})^2}, \quad j = 1,\ldots,\infty$$

$$B_j = \left\{ \int_0^1 \left(cos(\bar{a}_j z) + \frac{Pe}{2\bar{a}_j} sin(\bar{a}_j z) \right)^2 dz \right\}^{-\frac{1}{2}}, \quad j = 1,\ldots,\infty \tag{23}$$

Using the following typical values for the process parameters:

$$Pe = 5.0, \quad B_C = 0.00001, \quad B_T = 1.0, \quad \beta_T = 15.62, \quad \gamma = 22.14, \tag{24}$$

the first ten eigenvalues of $\mathcal{A}$ assume the following values: $\lambda_1 = -1.94$, $\lambda_2 = -4.80$, $\lambda_3 = -10.42$, $\lambda_4 = -20.93$, and $\lambda_5 = -34.78$.

4. Nonlinear Galerkin's method

We will now review the application of Galerkin's method to the system of Eq.8 to derive an approximate finite-dimensional system. Let $\mathcal{H}_s$, $\mathcal{H}_f$ be modal subspaces of $\mathcal{A}$, defined as $\mathcal{H}_s = span\{\phi_1, \phi_2, \ldots, \phi_m\}$ and $\mathcal{H}_f = span\{\phi_{m+1}, \phi_{m+2}, \ldots, \}$ (the existence of $\mathcal{H}_s$, $\mathcal{H}_f$ follows from assumption 1). Defining the orthogonal projection operators P_s and P_f such that $x_s = P_s x$, $x_f = P_f x$, the state x of the system of Eq.8 can be decomposed as:

$$x = x_s + x_f = P_s x + P_f x \tag{25}$$

Applying P_s and P_f to the system of Eq.8 and using the above decomposition for x, the system of Eq.8 can be equivalently written in the following form:

$$\frac{dx_s}{dt} = \mathcal{A}_s x_s + \mathcal{B}_s u + f_s(x_s, x_f)$$

$$\frac{\partial x_f}{\partial t} = \mathcal{A}_f x_f + \mathcal{B}_f u + f_f(x_s, x_f) \tag{26}$$

$$y = \mathcal{C}x_s + \mathcal{C}x_f$$

$$x_s(0) = P_s x(0) = P_s x_0, \quad x_f(0) = P_f x(0) = P_f x_0$$

where $\mathcal{A}_s = P_s \mathcal{A} P_s$, $\mathcal{B}_s = P_s \mathcal{B}$, $f_s = P_s f$, $\mathcal{A}_f = P_f \mathcal{A} P_f$, $\mathcal{B}_f = P_f \mathcal{B}$ and $f_f = P_f f$ and the notation $\frac{\partial x_f}{\partial t}$ is used to denote that the state x_f belongs in an infinite-dimensional space. In the above system, $\mathcal{A}_s$ is a diagonal matrix of dimension $m \times m$ of the form $\mathcal{A}_s = diag\{\lambda_j\}$, $f_s(x_s, x_f)$ and $f_f(x_s, x_f)$ are Lipschitz vector functions, and $\mathcal{A}_f$ is an unbounded differential operator which is exponentially stable (following from part 3 of assumption 1 and the selection of $\mathcal{H}_s, \mathcal{H}_f$). Neglecting the fast modes, the following finite-dimensional system is derived:

$$\frac{dx_s}{dt} = \mathcal{A}_s x_s + \mathcal{B}_s u + f_s(x_s, 0)$$

$$y_s = \mathcal{C} x_s$$

(27)

where the subscript s in y_s denotes that the output is associated with the slow system. The above system can be directly used for controller design employing standard control methods for ODEs [1, 26, 8].

Remark 1: We note that the large separation of slow and fast modes of the spatial operator in parabolic PDEs ensures that a controller which exponentially stabilizes the closed-loop ODE system, also stabilizes the closed-loop parabolic PDE system [1]. This result is not true for first-order hyperbolic PDE systems (e.g. convection-reaction processes) because assumption 1 is not satisfied for these systems (i.e. the eigenvalues cluster along vertical, or nearly vertical, asymptotes in the complex plane). The controller design problem for first-order hyperbolic PDE systems is addressed directly on the basis of the PDE model itself (see for example [21, 11, 15, 13]).

5. Accuracy of ODE system obtained from Galerkin's method

In this section, we use singular perturbation methods to establish that if the finite-dimensional system of Eq.27 is exponentially stable, then the system of Eq.26 is also exponentially stable and the discrepancy between the solution of the x_s-subsystem of the system of Eq.26 and the solution of the system of Eq.27, is proportional to the spectral separation of the slow and fast eigenvalues.

Defining $\epsilon = \dfrac{|Re\ \lambda_1|}{|Re\ \lambda_{m+1}|}$, the system of Eq.26 can be written in the following form:

$$\frac{dx_s}{dt} = \mathcal{A}_s x_s + \mathcal{B}_s u + f_s(x_s, x_f)$$

$$\epsilon \frac{\partial x_f}{\partial t} = \mathcal{A}_{f\epsilon} x_f + \epsilon \mathcal{B}_f u + \epsilon f_f(x_s, x_f)$$

(28)

where $\mathcal{A}_{f\epsilon}$ is an unbounded differential operator defined as $\mathcal{A}_{f\epsilon} = \epsilon \mathcal{A}_f$. Since ϵ is a small positive number less than unity (assumption 1, part 3), the system of Eq.28 is in the standard singularly perturbed form, with x_s being the slow states and x_f being the fast states.

Introducing the fast time-scale $\tau = \dfrac{t}{\epsilon}$ and setting $\epsilon = 0$, we obtain the following infinite-dimensional fast subsystem from the system of Eq.28:

$$\frac{\partial x_f}{\partial \tau} = \mathcal{A}_{f\epsilon} x_f \tag{29}$$

From the fact that $Re \ \lambda_{m+1} < 0$ and the definition of ϵ, we have that the above system is globally exponentially stable. Setting $\epsilon = 0$ in the system of Eq.28 and using that the operator $\mathcal{A}_{f\epsilon}$ is invertible, we have that:

$$x_f = 0 \tag{30}$$

and thus the finite-dimensional slow system takes the form:

$$\frac{dx_s}{dt} = \mathcal{A}_s x_s + \mathcal{B}_s u + f_s(x_s, 0) \tag{31}$$

We note that the above system is identical to the one obtained by applying the standard Galerkin's method to the system of Eq.8, keeping the first m ODEs and completely neglecting the x_f-subsystem. Assumption 2 that follows states a stability requirement on the system of Eq.31.

Assumption 2: *The finite-dimensional system of Eq.31 with $u(t) \equiv 0$ is exponentially stable i.e., there exist positive real numbers (K_2, a_2, a_4), with $a_4 > K_2 \geq 1$ such that for all $x_s \in \mathcal{H}_s$ that satisfy $|x_s| \leq a_4$, the following bound holds:*

$$|x_s(t)| \leq K_2 e^{-a_2 t} |x_s(0)|, \quad \forall \, t \geq 0 \tag{32}$$

Proposition 1 that follows establishes that the solutions of the open-loop systems of Eqs.31-29, after a short finite time interval required for the trajectories of the system of Eq.28 to approach the quasi steady-state of Eq.30, consist an $O(\epsilon)$ approximation of the solutions of the open-loop system of Eq.28. The proof of the proposition is given in [14].

Proposition 1: *Consider the system of Eq.28 with $u(t) \equiv 0$ and suppose that assumptions 1 and 2 hold. Then, there exist positive real numbers μ_1, μ_2, ϵ^* such that if $|x_s(0)| \leq \mu_1$, $\|x_f(0)\|_2 \leq \mu_2$ and $\epsilon \in (0, \epsilon^*]$, then the solution $x_s(t), x_f(t)$ of the system of Eq.28 satisfies for all $t \in [0, \infty)$:*

$$x_s(t) = \bar{x}_s(t) + O(\epsilon)$$
$$x_f(t) = \bar{x}_f(\frac{t}{\epsilon}) + O(\epsilon) \tag{33}$$

where $\bar{x}_s(t), \bar{x}_f(t)$ are the solutions of the slow and fast subsystems of Eqs.31-29 with $u(t) \equiv 0$, respectively.

Remark 2: An estimate of ϵ^* can be obtained, in principle, from the proof of the theorem. However, such an estimate is typically conservative, and thus, it is useful to check its appropriateness through computer simulations.

Remark 3: The counterpart of the result of Proposition 1 in finite-dimensional spaces is well-known (Tikhonov's theorem, [30]), while a similar result has also been established for linear infinite-dimensional systems [2]. The main technical difference in establishing this result between linear and quasi-linear infinite-dimensional systems is that, for quasi-linear systems the proof is based on Lyapunov arguments, while for linear systems the proof is obtained using combination of estimates of the states, obtained from the application of variations of constants formula [2]. This is a consequence of the fact that for quasi-linear systems it is not possible to derive a coordinate change that transforms the system of Eq.28 into a cascaded interconnection where the fast modes are decoupled from the slow modes, which allows to derive an exponentially decaying estimate, for sufficiently small ϵ, for the fast state, which is independent of the one of the slow state, and thus to prove the result through a direct combination of these estimates.

Remark 4: We note that it is possible, using standard results from center m anifold theory for infinite-dimensional systems of the form of Eq.8 [7] to show that if the system of Eq.31 is asymptotically stable, then the system of Eq.8 is also asymptotically stable and the discrepancy between the solution of the system of Eq.31 and the x_s-subsystem of the closed-loop full-order system is asymptotically (as $t \to \infty$) proportional to ϵ. Although this result is important because it allows establishing asymptotic stability of the closed-loop infinite-dimensional system by performing a stability analysis on a low-order finite-dimensional system, it does not provide any information about the discrepancy between the solutions of these two systems for finite t.

6. Construction of ODE systems of desired accuracy via AIMs

In this section, we propose an approach originating from the theory of inertial manifolds for the construction of ODE systems of dimension m which yield solutions that are arbitrarily close (closer than $O(\epsilon)$) to the ones of the infinite-dimensional system of Eq.8, for almost all times. An inertial manifold $\mathcal{M}$ for the system of Eq.8 is a subset of $\mathcal{H}$, which satisfies the following properties [29]: $i)$ $\mathcal{M}$ is a finite-dimensional Lipschitz manifold, $ii)$ $\mathcal{M}$ is a graph of a Lipschitz function $\Sigma(x_s, u, \epsilon)$ mapping $\mathcal{H}_s \times \mathbb{R}^l \times (0, \epsilon^*]$ into $\mathcal{H}_f$ and for every solution $x_s(t), x_f(t)$ of Eq.28 with $x_f(0) = \Sigma(x_s(0), u, \epsilon)$, then

$$x_f(t) = \Sigma(x_s(t), u, \epsilon), \quad \forall\, t \geq 0 \tag{34}$$

and $iii)$ $\mathcal{M}$ attracts every trajectory exponentially. The evolution of the state x_f on $\mathcal{M}$ is given by Eq.34, while the evolution of the state x_s is governed by the following finite-dimensional inertial form:

$$\frac{dx_s}{dt} = \mathcal{A}_s x_s + \mathcal{B}_s u + f_s(x_s, \Sigma(x_s, u, \epsilon)) \tag{35}$$

Assuming that $u(t)$ is smooth, differentiating Eq.34 and utilizing Eq.28, $\Sigma(x_s, u, \epsilon)$ can be computed as the solution of the following partial differential equation:

$$\epsilon \frac{\partial \Sigma}{\partial x_s}[\mathcal{A}_s x_s + \mathcal{B}_s u + f_s(x_s, x_f)] + \epsilon \frac{\partial \Sigma}{\partial u} \dot{u} = \mathcal{A}_{f\epsilon} x_f + \epsilon \mathcal{B}_f u + \epsilon f_f(x_s, x_f) \qquad (36)$$

which Σ has to satisfy for all $x_s \in \mathcal{H}_s$, $u \in \mathbb{R}^l$, $\epsilon \in (0, \epsilon^*]$. However, even for parabolic PDEs for which it is known that $\mathcal{M}$ exists, the derivation of an explicit analytic form of $\Sigma(x_s, u, \epsilon)$ is an extremely difficult (if not impossible) task.

Motivated by this, we will now propose a procedure, motivated by singular perturbations [23], to compute approximations of $\Sigma(x_s, u, \epsilon)$ (approximate inertial manifolds) and approximations of the inertial form, of desired accuracy. To this end, consider an expansion of $\Sigma(x_s, u, \epsilon)$ and u in a power series in ϵ:

$$u = u_0 + \epsilon u_1 + \epsilon^2 u_2 + \cdots + \epsilon^k u_k + O(\epsilon^{k+1})$$

$$\Sigma(x_s, u, \epsilon) = \Sigma^0(x_s, u) + \epsilon \Sigma^1(x_s, u) + \epsilon^2 \Sigma^2(x_s, u) + \cdots + \epsilon^k \Sigma^k(x_s, u) + O(\epsilon^{k+1})$$
$$(37)$$

where u_k, Σ^k are smooth functions. Substituting the expressions of Eq.37 into Eq.36, and equating terms of the same power in ϵ, one can obtain approximations of $\Sigma(x_s, u, \epsilon)$ up to a desired order. Substituting the expansion for $\Sigma(x_s, u, \epsilon)$ and u up to order k into Eq.35, the following approximation of the inertial form is obtained:

$$\frac{dx_s}{dt} = \mathcal{A}_s x_s + \mathcal{B}_s \left(u_0 + \epsilon u_1 + \epsilon^2 u_2 + \cdots + \epsilon^k u_k \right)$$
$$+ f_s(x_s, \Sigma^0(x_s, u) + \epsilon \Sigma^1(x_s, u) + \epsilon^2 \Sigma^2(x_s, u) + \cdots + \epsilon^k \Sigma^k(x_s, u)) \qquad (38)$$

In order to characterize the discrepancy between the solution of the open-loop finite-dimensional system of Eq.38 and the solution of the x_s-subsystem of the open-loop infinite-dimensional system of Eq.28, we will impose a stability requirement on the system of Eq.38.

Assumption 3: *The finite-dimensional system of Eq.38 with $u(t) \equiv 0$ is exponentially stable i.e., there exist positive real numbers $(\bar{K}_2, \bar{a}_2, \bar{a}_4)$, with $\bar{a}_4 > \bar{K}_2 \geq 1$ such that for all $x_s \in \mathcal{H}_s$ that satisfy $|x_s| \leq \bar{a}_4$, the following bound holds:*

$$|x_s(t)| \leq \bar{K}_2 e^{-\bar{a}_2 t}|x_s(0)|, \quad \forall\, t \geq 0 \qquad (39)$$

Proposition 2 that follows establishes that the discrepancy between the solutions obtained from the open-loop system of Eq.38 and the expansion for $\Sigma(x_s, u, \epsilon)$ of Eq.37, and the solutions of the infinite-dimensional open-loop system of Eq.28 is of $O(\epsilon^{k+1})$, for almost all times. The proof can be found in [14].

Proposition 2: *Consider the system of Eq.28 with $u(t) \equiv 0$ and suppose that assumptions 1 and 3 hold. Then, there exist positive real numbers $\bar{\mu}_1, \bar{\mu}_2, \bar{\epsilon}^*$ such*

that if $|x_s(0)| \leq \bar{\mu}_1$, $||x_f(0)||_2 \leq \bar{\mu}_2$ *and* $\epsilon \in (0, \bar{\epsilon}^*]$, *then the solution* $x_s(t), x_f(t)$
of the system of Eq.28 satisfies for all $t \in [t_b, \infty)$:

$$x_s(t) = \tilde{x}_s(t) + O(\epsilon^{k+1})$$
$$x_f(t) = \tilde{x}_f(t) + O(\epsilon^{k+1})$$
(40)

where t_b *is the time required for* $x_f(t)$ *to approach* $\tilde{x}_f(t)$, $\tilde{x}_s(t)$ *is the solution of
Eq.38 with* $u(t) \equiv 0$, *and* $\tilde{x}_f(t) = \epsilon\Sigma^1(\tilde{x}_s, 0) + \epsilon^2\Sigma^2(\tilde{x}_s, 0) + \cdots + \epsilon^k\Sigma^k(\tilde{x}_s, 0)$.

Remark 5: The result of proposition 2 provides the means for characterizing the discrepancy between the solution of the open-loop infinite-dimensional system of Eq.8, $x(t)$, (and thus, the solution of the parabolic PDE system of Eq.1 with $u(t) \equiv 0$) and the solution $\tilde{x}(t) = \tilde{x}_s(t) + \tilde{x}_f(t) = \tilde{x}_s(t) + \epsilon\Sigma^1(\tilde{x}_s(t), 0) + \cdots + \epsilon^k\Sigma^k(\tilde{x}_s(t), 0)$ which is obtained from the $O(\epsilon^k)$ approximation of the open-loop inertial form (i.e. Eq.38 with $u(t) \equiv 0$). In particular, substituting Eq.40 into the equation $x(t) = x_s(t) + x_f(t)$, we have that $x(t) = \tilde{x}_s(t) + \tilde{x}_f(t) + O(\epsilon^{k+1})$ for $t \geq t_b$. Utilizing the definition of order of magnitude, we finally obtain the following characterization for the discrepancy between $x(t)$ and $\tilde{x}(t)$: $||x(t) - \tilde{x}(t)||_2 \leq k_1\epsilon^{k+1}$ for $t \geq t_b$, where k_1 is a positive real number.

Remark 6: Following the proposed approximation procedure, it can be shown that the $O(\epsilon)$ approximation of $\Sigma(x_s, 0, \epsilon)$ is $\Sigma^0(x_s, 0) = 0$ and the corresponding approximate inertial form is identical to the system of Eq.31 (obtained via Galerkin's method) with $u(t) \equiv 0$. This system does not utilize any information about the structure of the fast subsystem, thus yielding solutions which are only $O(\epsilon)$ close to the solutions of the open-loop system of Eq.8 (proposition 1). On the other hand, the $O(\epsilon^2)$ approximation of $\Sigma(x_s, 0, \epsilon)$ can be shown to be of the form:

$$\Sigma(x_s, 0, \epsilon) = \Sigma^0(x_s, 0) + \epsilon\Sigma^1(x_s, 0) = \epsilon(\mathcal{A}_{f\epsilon})^{-1}[-f_f(x_s, 0)]$$
(41)

The corresponding open-loop approximate inertial form does utilize information about the structure of the fast subsystem, and thus allows to obtain solutions which are $O(\epsilon^2)$ close to the solutions of the open-loop system of Eq.8 (proposition 2).

Remark 7: The standard approach followed in the literature for the construction of AIMs for systems of the form of Eq.26 with $u(t) \equiv 0$ (see for example [5]) is to directly set $\dfrac{\partial x_f}{\partial t} \equiv 0$, solve the resulting algebraic equations for x_f and substitute the solution for x_f to the x_s-subsystem of Eq.26, to derive the following ODE system:

$$\frac{dx_s}{dt} = \mathcal{A}_s x_s + f_s(x_s, (\mathcal{A}_f)^{-1}[-f_f(x_s)])$$
(42)

It is straightforward to show that the slow system of Eq.42 is identical to the one obtained by using the $O(\epsilon^2)$ approximation for $\Sigma(x_s, 0, \epsilon)$ for the construction of the approximate inertial form.

Remark 8: The expansion of u in a power series in ϵ is motivated by our intention to modify the synthesis of the feedback controller appropriately such that the output of the $O(\epsilon^{k+1})$ approximation of the closed-loop inertial form will be arbitrarily close to the output of the closed-loop PDE system for almost all times (see also remark 9).

7. Finite-dimensional control

7.1. A general result

In this subsection, we use the result of proposition 2 to establish that a nonlinear finite-dimensional output feedback controller that guarantees stability and enforces output tracking in the ODE system of Eq.38, exponentially stabilizes the closed-loop PDE system and ensures that the discrepancy between the output of the closed-loop ODE system and the output of the closed-loop PDE system is of $O(\epsilon^{k+1})$, provided that ϵ is sufficiently small.

The finite-dimensional output feedback controller which achieves the desired objectives for the system of Eq.38 is constructed through a standard combination of a state feedback controller with a state observer. In particular, we consider a state feedback control law of the general form:

$$
\begin{aligned}
u &= u_0 + \epsilon u_1 + \cdots + \epsilon^k u_k \\
&= p_0(x_s) + Q_0(x_s)v + \epsilon[p_1(x_s) + Q_1(x_s)v] + \cdots + \epsilon^k[p_k(x_s) + Q_k(x_s)v]
\end{aligned}
\tag{43}
$$

where $p_0(x_s), \ldots, p_k(x_s)$ are smooth vector functions, $Q_0(x_s), \ldots, Q_k(x_s)$ are smooth matrices, and $v \in \mathbb{R}^l$ is the constant reference input vector (see remark 9 for a procedure for the synthesis of the control law i.e. the explicit computation of $[p_0(x_s), \ldots, p_k(x_s), Q_0(x_s), \ldots, Q_k(x_s)]$). The following m-dimensional state observer is also considered for the implementation of the state feedback law of Eq.43:

$$
\begin{aligned}
\frac{d\eta}{dt} &= A_s\eta + B_s\left(p_0(\eta) + Q_0(\eta)v + \epsilon[p_1(\eta) + Q_1(\eta)v] + \cdots + \epsilon^k[p_k(\eta) + Q_k(\eta)v]\right) \\
&\quad + f_s(\eta, \epsilon\Sigma^1(\eta, u) + \epsilon^2\Sigma^2(\eta, u) + \cdots + \epsilon^k\Sigma^k(\eta, u)) \\
&\quad + L(y - [\mathcal{C}\eta + \mathcal{C}\{\epsilon\Sigma^1(\eta, u) + \epsilon^2\Sigma^2(\eta, u) + \cdots + \epsilon^k\Sigma^k(\eta, u)\}])
\end{aligned}
\tag{44}
$$

where $\eta \in \mathcal{H}_s$ denotes the observer state vector, and L is a matrix chosen so that the eigenvalues of the matrix $C_L = A_s + \dfrac{\partial f_s}{\partial \eta}\Big|_{(\eta=\eta_s)}$

$$
-L\left[\mathcal{C}\eta_s + \mathcal{C}\{\frac{\partial}{\partial\eta}\left(\epsilon\Sigma^1(\eta, u(\eta)) + \epsilon^2\Sigma^2(\eta, u(\eta)) + \cdots + \epsilon^k\Sigma^k(\eta, u(\eta))\right)_{(\eta=\eta_s)}\}\right] \text{ lie in}
$$

the open left-half of the complex plane, where η_s denotes the steady-state for the system of Eq.44. The finite-dimensional output feedback controller resulting from the combination of the state feedback controller of Eq.43 with the state observer of Eq.44 takes the form:

$$\frac{d\eta}{dt} = A_s\eta + B_s \left(p_0(\eta) + Q_0(\eta)v + \epsilon[p_1(\eta) + Q_1(\eta)v] + \cdots + \epsilon^k[p_k(\eta) + Q_k(\eta)v] \right)$$

$$+ f_s(\eta, \epsilon\Sigma^1(\eta, u) + \epsilon^2\Sigma^2(\eta, u) + \cdots + \epsilon^k\Sigma^k(\eta, u))$$

$$+ L(y - [\mathcal{C}\eta + C\{\Sigma^0(\eta, u) + \epsilon\Sigma^1(\eta, u) + \epsilon^2\Sigma^2(\eta, u) + \cdots + \epsilon^k\Sigma^k(\eta, u)\}])$$

$$u = p_0(\eta) + Q_0(\eta)v + \epsilon[p_1(\eta) + Q_1(\eta)v] + \cdots + \epsilon^k[p_k(\eta) + Q_k(\eta)v]$$

$$(45)$$

We note that the static component of the above controller does not use feedback of the fast state vector x_f in order to avoid destabilization of the fast modes of the closed-loop system. Assumption 4 states the desired control objectives under the controller of Eq.45.

Assumption 4: *The finite-dimensional output feedback controller of the form of Eq.45 exponentially stabilizes the $O(\epsilon^{k+1})$ approximation of the closed-loop inertial form and ensures that its outputs $y_s^i(t)$, $i = 1, \ldots, l$, are the solutions of a known l − dimensional ODE system of the form $\phi(y_s^{i\,(r_i)}, y_s^{i\,(r_i-1)}, \cdots, y_s^i, v) = 0$, where ϕ is a vector function and r_i is an integer.*

Theorem 1 provides a precise characterization of the stability and closed-loop transient performance enforced by the controller of Eq.45 in the closed-loop PDE system (the proof can be found in [14].).

Theorem 1: *Consider the PDE system of Eq.8, for which assumptions 1 and 4 hold. Then, there exist positive real numbers $\bar{\mu}_1, \bar{\mu}_2, \bar{\epsilon}^*$ such that if $|x_s(0)| \leq \bar{\mu}_1$, $\|x_f(0)\|_2 \leq \bar{\mu}_2$ and $\epsilon \in (0, \bar{\epsilon}^*]$, then the controller of Eq.45:*
a) guarantees exponential stability of the closed-loop system, and
b) ensures that the outputs of the closed-loop system satisfy for all $t \in [t_b, \infty)$:

$$y^i(t) = y_s^i(t) + O(\epsilon^{k+1}) , \quad i = 1, \ldots, l \qquad (46)$$

where $y_s^i(t)$ is the i-th output of the $O(\epsilon^{k+1})$ approximation of the closed-loop inertial form.

Remark 9: The construction of the state feedback law of Eq.43, to ensure that the control objectives stated in assumption 4 are enforced in the $O(\epsilon^{k+1})$ approximation of the closed loop inertial form, can be performed following a sequential procedure. Specifically, the component $u_0 = p_0(x_s) + Q_0(x_s)v$ can be initially synthesized on the basis of the $O(\epsilon)$ approximation of the inertial form (Eq.31); then the component $u_1 = p_1(x_s) + Q_1(x_s)v$ can be synthesized on the basis of the $O(\epsilon^2)$ approximation of the inertial form. In general, at the k-th step, the component $u_k = p_k(x_s) + Q_k(x_s)v$ can be synthesized on the basis of the $O(\epsilon^k)$ approximation of the inertial form (Eq.38). The synthesis of $[p_\nu(x_s), Q_\nu(x_s)]$, $\nu = 0, \ldots, k$, can be performed, at each step, utilizing standard geometric control methods for nonlinear ODEs.

Remark 10: The implementation of the controller of Eq.45 requires to explicitly compute the vector function $\Sigma^k(\eta, u)$. However, $\Sigma^k(\eta, u)$ has an infinite-dimensional range and therefore cannot be implemented in practice. Instead a finite-dimensional approximation of $\Sigma^k(\eta, u)$, say $\Sigma_t^k(\eta, u)$, can be derived by keeping the first $\bar{m}$ elements of $\Sigma^k(\eta, u)$ and neglecting the remaining infinite ones. Clearly, as $\bar{m} \to \infty$, $\Sigma_t^k(\eta, u)$ approaches $\Sigma^k(\eta, u)$. This implies that by picking $\bar{m}$ to be sufficiently large, the controller of Eq.45 with $\Sigma_t^k(\eta, u)$ instead of $\Sigma^k(\eta, u)$ guarantees stability and enforces the requirement of Eq.46 in the closed-loop infinite-dimensional system.

7.2. Controller synthesis

7.2.1 Preliminaries

In this subsection, we will synthesize a finite-dimensional output feedback controller for the system of Eq.8 on the basis of the system of Eq.38, using geometric control methods. To this end, we will initially review the concepts of relative order and characteristic matrix that will be used in the subsequent subsection to synthesize the controller. Referring to the system of Eq.31, we set, in order to simplify the notation, $A_s x_s + f_s(x_s, 0) = f_0(x_s)$, $\mathcal{B}_s = g_0$, $\mathcal{C}x_s = h_0(x_s)$. The relative order of the output y_s^i with respect to the vector of manipulated inputs u is defined as the smallest integer r_i for which

$$\left[L_{g_0^1} L_{f_0}^{r_i-1} h_0^i(x_s) \quad \cdots \quad L_{g_0^l} L_{f_0}^{r_i-1} h_0^i(x_s) \right] \neq [0 \ \cdots \ 0] \tag{47}$$

or $r_i = \infty$ if such an integer does not exist. Furthermore, the matrix:

$$C(x_s) = \begin{bmatrix} L_{g_0^1} L_{f_0}^{r_1-1} h_0^1(x_s) & \cdots & L_{g_0^l} L_{f_0}^{r_1-1} h_0^1(x_s) \\ L_{g_0^1} L_{f_0}^{r_2-1} h_0^2(x_s) & \cdots & L_{g_0^l} L_{f_0}^{r_2-1} h_0^2(x_s) \\ \vdots & & \\ L_{g_0^1} L_{f_0}^{r_l-1} h_0^l(x_s) & \cdots & L_{g_0^l} L_{f_0}^{r_l-1} h_0^l(x_s) \end{bmatrix} \tag{48}$$

is the characteristic matrix of the system. For simplicity, we will assume that $det(C(x_s)) \neq 0$.

7.2.2 Controller formula

Theorem 2 provides the synthesis formula of the output feedback controller and conditions that guarantee closed-loop stability in the case of considering an $O(\epsilon^2)$ approximation of the exact slow system for the synthesis of the controller. The derivation of synthesis formulas for higher-order approximations of the output feedback controller is notationally complicated, although conceptually straightforward, and thus will be omitted for reasons of brevity.

Theorem 2 [9]: *Consider the parabolic PDE system of Eq.8, for which assumptions 1 and 2 hold. Consider also the $O(\epsilon^2)$ approximation of the inertial form, and assume that its characteristic matrix $\tilde{C}(x_s, \epsilon)$ is invertible $\forall\ x_s \in \mathcal{H}_s,\ \epsilon \in [0, \epsilon^*]$. Suppose also that the following conditions hold:*

1. *The roots of the equation:*

$$det(B(s)) = 0 \qquad (49)$$

   where $B(s)$ is a $l \times l$ matrix, whose (i,j)-th element is of the form $\sum_{k=0}^{r_i} \beta_{jk}^i s^k$, lie in the open left-half of the complex plane.

2. *The unforced ($v \equiv 0$) zero dynamics of the $O(\epsilon^2)$ approximation of the inertial form is locally exponentially stable.*

Then, there exist constants μ_1, μ_2, ϵ^* such that if $|x_s(0)| \le \mu_1$, $\|x_f(0)\|_2 \le \mu_2$ and $\epsilon \in (0, \epsilon^*]$, then if $\eta(0) = x_s(0)$, the dynamic output feedback controller:

$$\frac{d\eta}{dt} = A_s\eta + B_s u(\eta) + f_s(\eta, \epsilon(A_{f\epsilon})^{-1}[-B_f u_0(\eta) - f_f(\eta, 0))$$

$$+ L\left(y - [C\eta + \epsilon(A_{f\epsilon})^{-1}[-B_f u_0(\eta) - f_f(\eta, 0)]\right)$$

$$u = u_0(\eta) + \epsilon u_1(\eta) := \{[\beta_{1r_1} \cdots \beta_{lr_l}]C(\eta)\}^{-1}\left\{v - \sum_{i=1}^{l}\sum_{k=0}^{r_i} \beta_{ik} L_{f_0}^k h_0^i(\eta)\right\} \qquad (50)$$

$$+ \epsilon\left\{[\beta_{1r_1} \cdots \beta_{lr_l}]\tilde{C}(\eta, \epsilon)\right\}^{-1}\left\{v - \sum_{i=1}^{l}\sum_{k=0}^{r_i} \beta_{ik} L_{f_1}^k h_1^i(\eta, \epsilon)\right\}$$

a) guarantees exponential stability of the closed-loop system, and
b) ensures that the outputs of the closed-loop system satisfy for all $t \in [t_b, \infty)$:

$$y^i(t) = y_s^i(t) + O(\epsilon^2),\ i = 1, \ldots, l \qquad (51)$$

where t_b is the time required for the off-manifold fast transients to decay to zero exponentially, and $y_s^i(t)$ is the solution of:

$$\sum_{i=1}^{l}\sum_{k=0}^{r_i} \beta_{ik} \frac{d^k y_s^i}{dt^k} = v,\quad i = 1, \ldots, l \qquad (52)$$

Remark 11: In the case of open-loop stable systems, a more convenient way to reconstruct the state of the system is to consider the observer of Eq.44 with the matrix L set identically equal to zero. This is motivated by the fact that the open-loop stability of the system guarantees the convergence of the estimated values to the actual ones with transient behavior depending on the location of the spectrum of the matrix $\bar{C}_L = \frac{\partial f_1}{\partial \eta_s}(\eta_s, \epsilon)$.

Remark 12: Note that in the presence of small initialization errors of the observer states (i.e., $\eta(0) \neq x_s(0)$), uncertainty in the model parameters and external disturbances, although a slight deterioration of the performance may occur, (i.e., the requirement of Eq.51 will not be exactly imposed in the closed-loop system), the output feedback controller of theorem 2 will enforce exponential stability and asymptotic output tracking in the closed-loop system.

Remark 13: Whenever the eigenfunctions ϕ_j of the operator $\mathcal{A}$ can not be calculated analytically, one can still use Galerkin's method to perform model reduction by using the "empirical eigenfunctions" of the PDE system as basis functions in $\mathcal{H}_s$ and $\mathcal{H}_f$ (such "empirical eigenfunctions" can be extracted from numerical simulations or experimental data using Karhunen-Loeve expansion; see for example [8, 25, 4, 27]). Whenever empirical eigenfunctions are used as basis functions within Galerkin's model reduction framework, the interconnection of Eq.26 takes the form

$$\frac{dx_s}{dt} = \mathcal{A}_s x_s + \mathcal{A}_{sf} x_f + \mathcal{B}_s u + f_s(x_s, x_f)$$

$$\frac{\partial x_f}{\partial t} = \mathcal{A}_{fs} x_s + \mathcal{A}_f x_f + \mathcal{B}_f u + f_f(x_s, x_f) \tag{53}$$

$$y = \mathcal{C} x_s + \mathcal{C} x_f$$

where $\mathcal{A}_{sf} = P_s \mathcal{A} P_f, \mathcal{A}_{fs} = P_f \mathcal{A} P_s$ are bounded operators, and the terms $\mathcal{A}_{sf} x_f$, $\mathcal{A}_{fs} x_s$ represent the modeling errors resulting from the use of empirical eigenfunctions in the model reduction instead of the exact eigenfunctions of $\mathcal{A}$. The problem of synthesizing low-dimensional output feedback controllers for systems of Eq.53 can be addressed by employing the proposed method, provided that the modeling errors $\mathcal{A}_{fs}, \mathcal{A}_{sf}$ are sufficiently small.

8. Simulation studies

In this section, we illustrate, through computer simulations, the application of the developed method for control of nonlinear parabolic PDE systems to the catalytic rod and the non-isothermal packed-bed reactor.

8.1. Catalytic rod

In this simulation, we study the stabilization of the rod temperature at the unstable spatially uniform steady state, $\bar{x}(z,t) = 0$. For this system, we considered the first eigenvalue as the dominant one (and thus, $\epsilon = \dfrac{\lambda_1}{\lambda_2} = 0.25$) and used Galerkin's method to derive a scalar ODE that was used for controller design. The controller of Eq.50 was employed in the simulations with $\beta_0 = 2.0$, $\beta_1 = 1.0$ and $L = 4$.

Figures 5 and 6 show the closed-loop output and manipulated input profiles, respectively, and Figure 7 shows the evolution of the closed-loop rod temperature profile for the nonlinear controller starting from a non-zero initial condition. Clearly, the nonlinear controller regulates the temperature profile at $\bar{x}(z,t) = 0$, achieving the control objective. We also implemented on the process a linear controller obtained from the Taylor linearization of the nonlinear controller around the operating steady state $\bar{x}(z,t) = 0$. Figures 8 and 9 show the closed-loop output and manipulated input profiles, respectively, and Figure 10 shows the evolution of the closed-loop rod temperature profile for the linear controller starting from the same initial condition as in the previous simulation run. It is clear that the linear controller cannot regulate the temperature profile at the desired steady state, $\bar{x}(z,t) = 0$; the process moves to another stable steady state characterized by a maximum in the temperature profile ($hot - spot$) in the middle of the rod.

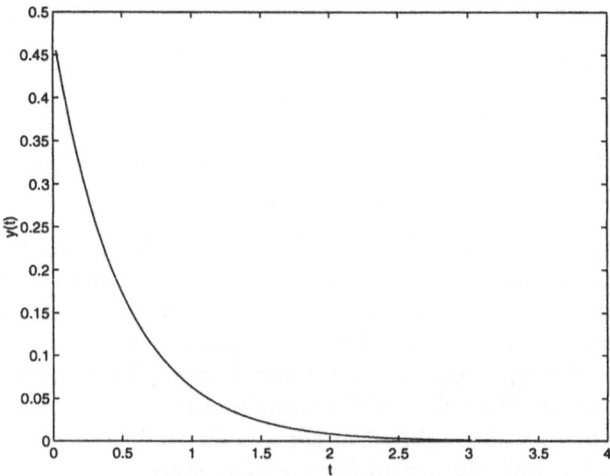

Figure 5. Closed-loop output profile for nonlinear controller (catalytic rod).

We finally note that the superiority of nonlinear control over linear control was also shown in [22], where the problem of attenuation of the effect of bifurcations on the output of the closed-loop system, in a typical diffusion-reaction system, was studied.

8.2. Non-isothermal packed-bed reactor

In this simulation, we evaluate the output tracking capabilities of nonlinear controllers synthesized on the basis of the $O(\epsilon)$ and $O(\epsilon^2)$ approximations of the exact slow system. The first four eigenvalues were considered as the dominant ones which implies that $\epsilon = \dfrac{\lambda_{11}}{\lambda_{13}} = \dfrac{\lambda_{21}}{\lambda_{23}} = 0.19$. The proposed method was used to derive $O(\epsilon)$ and $O(\epsilon^2)$ approximations of the exact slow system. The $O(\epsilon^2)$ approximation of $\Sigma(x_s, u, \epsilon)$ was constructed by retaining the first three of the fast modes for each PDE, and discarding the remaining infinite ones (this is because the use of more

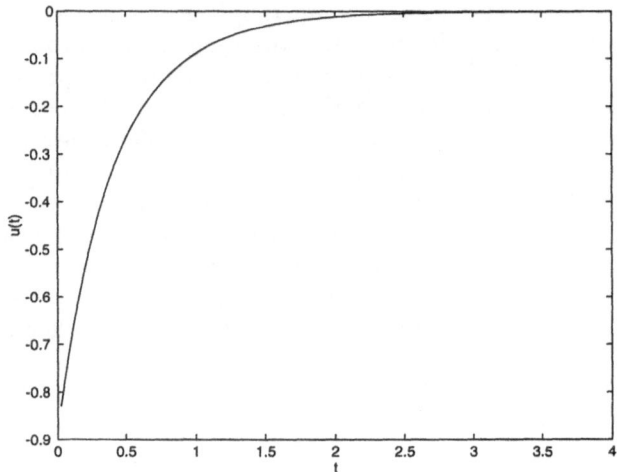

Figure 6. Manipulated input profile for nonlinear controller (catalytic rod).

than three fast modes provides negligible improvement in the accuracy of the $O(\epsilon^2)$ approximation of the fourth-order model). The zero dynamics of the systems corresponding to the $O(\epsilon)$ and $O(\epsilon^2)$ approximations of the exact slow system were found to be exponentially stable. Therefore, the nonlinear finite-dimensional controller of Eq.50 with $\beta_1 = 0.2$, $\beta_0 = 1.0$ and $L = 0$ (this is possible because the open-loop process and thus the $O(\epsilon^2)$ approximation of the exact slow system are exponentially stable) was employed in the simulations. The process was initially assumed to be at a spatially non-uniform steady state.

Figure 11 [12] shows the output and manipulated input profiles for an 8.0% decrease in the value of the set-point (the new set-point value is $v = 1.11$). It is clear that the controller synthesized on the basis of the system which uses an $O(\epsilon^2)$ approximation for $\Sigma(x_s, u, \epsilon)$ provides an excellent performance driving the output (solid line) very close to the new set-point (note that as expected, $lim_{t\to\infty}|y - v| = O(\epsilon^2)$). On the other hand, the controller of the form of Eq.50 with $\epsilon = 0$ drives the output (dotted line) to a neighborhood of the set-point (note that $lim_{t\to\infty}|y - v| = O(\epsilon)$) leading to significant offset (compare with set-point value). Figure 12 [12] displays the evolution of the dimensionless temperature of the reactor for the case of using an $O(\epsilon^2)$ approximation for $\Sigma(x_s, u, \epsilon)$. The controller achieves excellent performance, regulating the temperature at each point of the reactor to a new steady-state value, which is close to 8% lower than the one of the original steady-state.

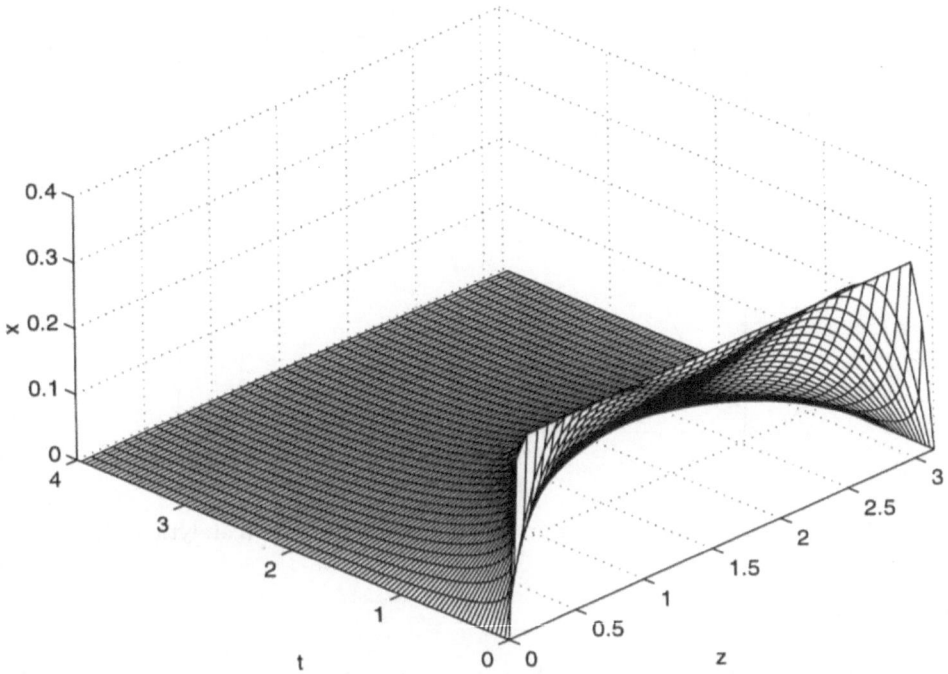

Figure 7. Profile of evolution of rod temperature for nonlinear controller.

9. Conclusions

In this article, we reviewed a method, proposed in [14], for the synthesis of non-linear low-dimensional output feedback controllers for nonlinear parabolic PDE systems, for which the eigenspectrum of the spatial differential operator can be partitioned into a finite-dimensional slow one and an infinite-dimensional stable fast one. Combination of Galerkin's method with a novel procedure for the construction of AIMs was used, for the derivation of ODE systems of dimension equal to the number of slow modes, that yield solutions which are close, upto a desired accuracy, to the ones of the PDE system, for almost all times. These ODE systems were used as the basis for the synthesis of output feedback controllers that guarantee stability and enforce the output of the closed-loop system to follow upto a desired accuracy, a prespecified response for almost all times. The methodology is successfully employed to stabilize the temperature profile of a catalytic rod around an unstable steady-state, and to control the temperature profile in a non-isothermal packed-bed reactor.

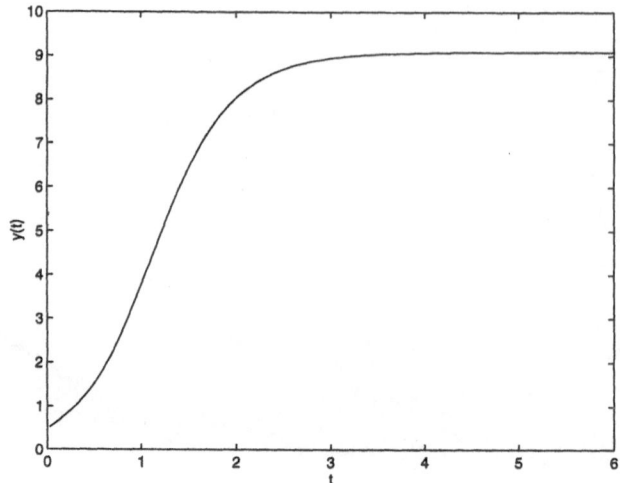

Figure 8. Closed-loop output profile for linear controller (catalytic rod).

10.   References

[1] M. J. Balas. Feedback control of linear diffusion processes. *Int. J. Contr.*, 29:523–533, 1979.

[2] M. J. Balas. Stability of distributed parameter systems with finite-dimensional controller-compensators using singular perturbations. *J. Math. Anal. Appl.*, 99:80–108, 1984.

[3] M. J. Balas. Nonlinear finite-dimensional control of a class of nonlinear distributed parameter systems using residual mode filters: A proof of local exponential stability. *J. Math. Anal. Appl.*, 162:63–70, 1991.

[4] A. K. Bangia, P. F. Batcho, I. G. Kevrekidis, and G. E. Karniadakis. Unsteady 2-D flows in complex geometries: Comparative bifurcation studies with global eigenfunction expansion. submitted, 1996.

[5] H. S. Brown, I. G. Kevrekidis, and M. S. Jolly. A minimal model for spatio-temporal patterns in thin film flow. In *Pattern and Dynamics in Reactive Media*, pages 11–31, R. Aris, D. G. Aronson, and H. L. Swinney, ed., Springer-Verlag, 1991.

[6] C. A. Byrnes, D. S. Gilliam, and V. I. Shubov. On the dynamics of boundary controlled nonlinear distributed parameter systems. In *Proc. of Symposium on Nonlinear Control Systems Design'95*, pages 913–918, Tahoe City, CA, 1995.

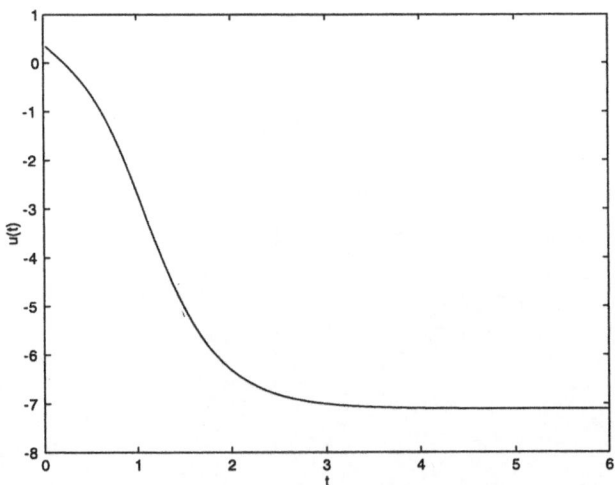

Figure 9. Manipulated input profile for linear controller (catalytic rod).

[7] J. Carr. *Applications of Center Manifold Theory.* Springer-Verlag, New York, 1981.

[8] C. C. Chen and H. C. Chang. Accelerated disturbance damping of an unknown distributed system by nonlinear feedback. *AIChE J.*, 38:1461–1476, 1992.

[9] P. D. Christofides. *Nonlinear Control of Two-Time-Scale and Distributed Parameter Systems.* PhD thesis, Dept. of Chem. Eng. & Mat. Sci., University of Minnesota, Minneapolis, MN, 1996.

[10] P. D. Christofides. Robust control of parabolic PDE systems. *Chem. Eng. Sci.* accepted; a version of this paper also appeared in the Proceedings of 36th Conference on Decision and Control, 1074-1081, San-Diego, CA, 1997.

[11] P. D. Christofides and P. Daoutidis. Feedback control of hyperbolic PDE systems. *AIChE J.*, 42:3063–3086, 1996.

[12] P. D. Christofides and P. Daoutidis. Nonlinear control of diffusion-convection-reaction processes. *Comp. Chem. Engng.*, 20(s):1071–1076, 1996.

[13] P. D. Christofides and P. Daoutidis. Distributed output feedback control of two-time-scale hyperbolic PDE systems. *J. of Appl. Math. & Comp. Sci.* in press, 1997.

[14] P. D. Christofides and P. Daoutidis. Finite-dimensional control of parabolic PDE systems using approximate inertial manifolds. *J. Math. Anal. Appl.*, in press; a version of this paper also appeared in the Proceedings of 36th Conference on Decision and Control, 1068-1073, San-Diego, California, 1997.

396

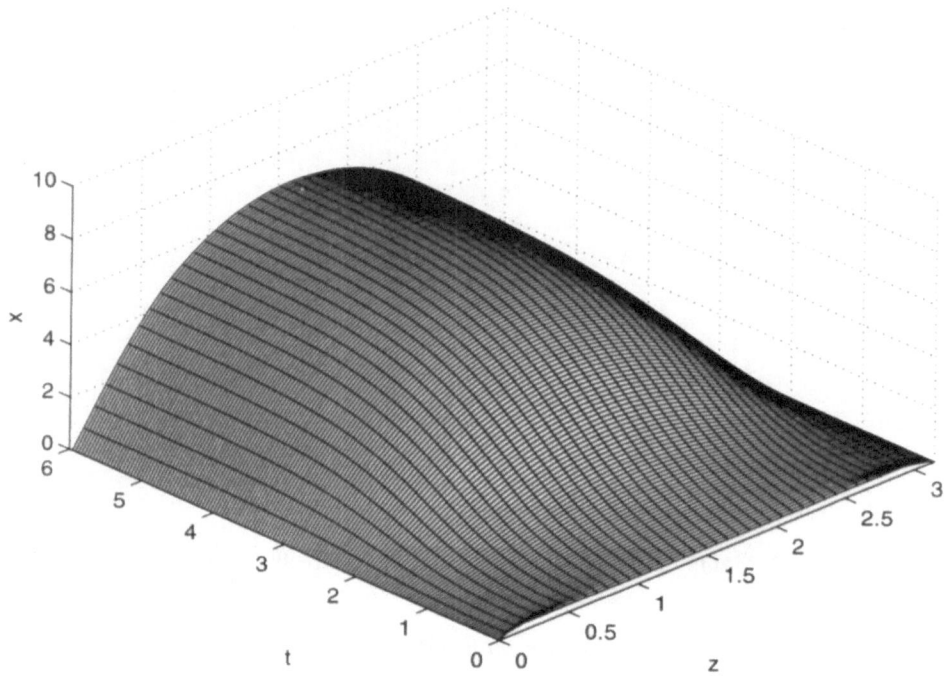

Figure 10. Profile of evolution of rod temperature for linear controller.

[15] P. D. Christofides and P. Daoutidis. Robust control of hyperbolic PDE systems. *Chem. Eng. Sci.,* in press, 1997.

[16] C. Foias, M.S. Jolly, I.G. Kevrekidis, G.R. Sell, and E.S. Titi. On the computation of inertial manifolds. *Phys. Lett. A*, 131:433–437, 1989.

[17] C. Foias, G.R. Sell, and E.S. Titi. Exponential tracking and approximation of inertial manifolds for dissipative equations. *J. Dynamics and Differential Equations*, 1:199–244, 1989.

[18] A. Friedman. *Partial Differential Equations.* Holt, Rinehart & Winston, New York, 1976.

[19] D. H. Gay and W. H. Ray. Identification and control of distributed parameter systems by means of the singular value decomposition. *Chem. Eng. Sci.,* 50:1519–1539, 1995.

[20] G. Georgakis, R. Aris, and N. R. Amundson. Studies in the control of tubular reactors: Part I & II & III. *Chem. Eng. Sci.,* 32:1359–1387, 1977.

[21] E. M. Hanczyc and A. Palazoglu. Sliding mode control of nonlinear distributed parameter chemical processes. *I & EC Res.,* 34:557–566, 1995.

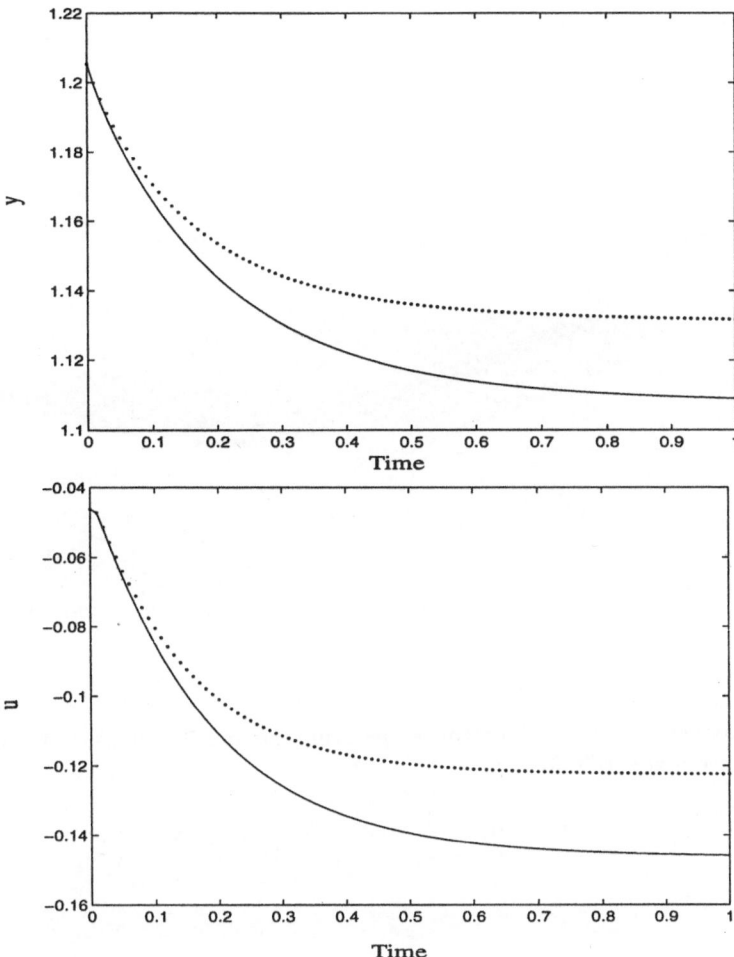

Figure 11. Comparison of output (top figure) and manipulated input (bottom figure) profiles of the closed-loop system, for an 8% decrease in the set point. The dotted lines correspond to a controller based on the slow ODE system with an $O(\epsilon)$ approximation for Σ, while the solid lines correspond to a controller based on the slow ODE system with an $O(\epsilon^2)$ approximation for Σ (packed-bed reactor).

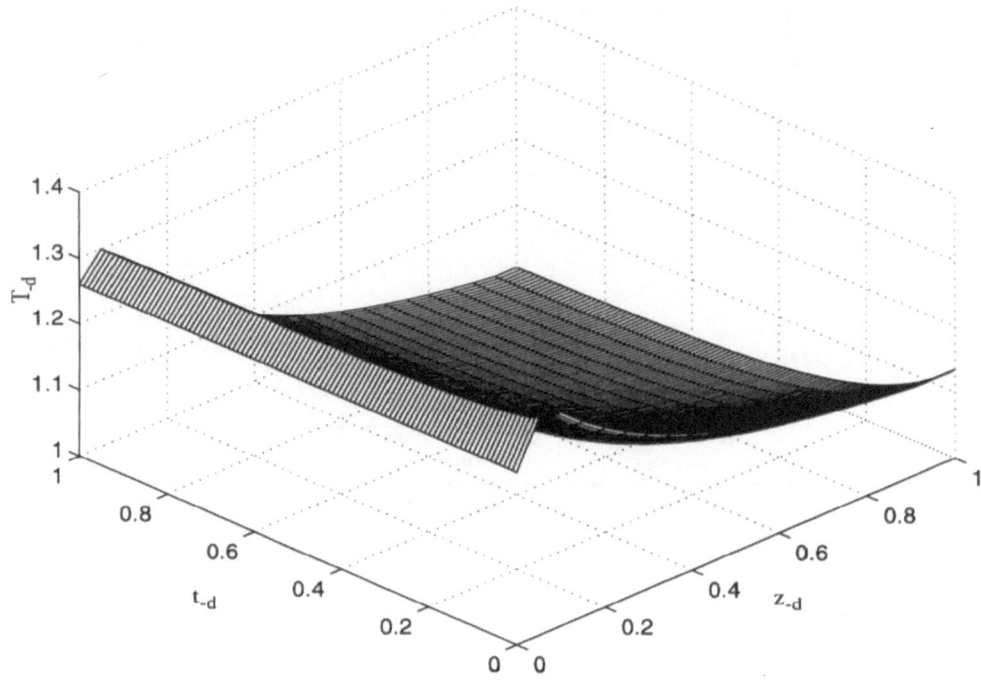

Figure 12. Profile of evolution of reactor temperature, for an 8% decrease in the reference input ($O(\epsilon^2)$ approximation for Σ).

[22] I. Karafyllis, P. D. Christofides, and P. Daoutidis. Dynamical analysis of a reaction diffusion system with brusselator kinetics under feedback control. In *Proceedings of American Control Conference*, pages 2213–2217, Albuquerque, NM, 1997.

[23] P. V. Kokotovic, H. K. Khalil, and J. O'Reilly. *Singular Perturbations in Control: Analysis and Design.* Academic Press, London, 1986.

[24] W. Marquardt. Traveling waves in chemical processes. *Int. Chem. Engng.*, 4:585–606, 1990.

[25] H. M. Park and D. H. Cho. The use of the karhunen-loeve decomposition for the modeling of distributed parameter systems. *Chem. Eng. Sci.*, 51:81–98, 1996.

[26] W. H. Ray. *Advanced Process Control.* McGraw-Hill, New York, 1981.

[27] A. Rigopoulos and Y. Arkun. Principal components analysis in estimation and control of paper machines. *Comp. Chem. Engng.*, 20(s):1059–1064, 1996.

[28] H. Sano and N. Kunimatsu. An application of inertial manifold theory to boundary stabilization of semilinear diffusion systems. *J. Math. Anal. Appl.*, 196:18–42, 1995.

[29] R. Temam. *Infinite-Dimensional Dynamical Systems in Mechanics and Physics.* Springer-Verlag, New York, 1988.

[30] A. N. Tikhonov. On the dependence of the solutions of differential equations on a small parameter. *Mat. Sb.*, 22:193–204, 1948.

[31] E. B. Ydstie and A. A. Alonso. Process systems and passivity via the Clausius-Planck inequality. *Syst. & Contr. Lett.*, 30:253–264, 1997.

Part III

**ON-LINE OPTIMIZATION APPROACHES
FOR NONLINEAR CONTROL**

CONTRACTIVE MODEL PREDICTIVE CONTROL WITH LOCAL LINEARIZATION FOR NONLINEAR SYSTEMS

SIMONE LOUREIRO DE OLIVEIRA
California Institute of Technology,
Pasadena, California 91125, USA.
Phone: 818-395-4576. Fax:818-568-8743.

AND

MANFRED MORARI
Automatic Control Laboratory,
ETH-Zentrum, 8092 Zürich, Switzerland.
Phone:41-1-632-7626. Fax: 41-1-632-1211.
E-mail: morari@aut.ee.ethz.ch

Abstract. Most model predictive control (MPC) schemes in the literature designed for constrained nonlinear systems involve the solution of nonlinear programming problems (NLPs). Due to the difficulties inherent to solving NLPs and since MPC requires the optimal (feasible) solution to be computed on-line and at each time step, it is important to simplify the controller from a computational point of view. Furthermore, many nonlinear MPC controllers do not provide any stability or performance guarantees for the closed-loop system. In this paper we propose a stabilizing MPC controller which can be implemented as a quadratic programming (QP) problem. Because we use local linear approximations of the nonlinear plant for prediction and because stability is guaranteed through the introduction of an additional state constraint in the optimization called *contractive constraint*, this scheme has been denoted *contractive MPC with local linearization*.

R. Berber and C. Kravaris (eds.), Nonlinear Model Based Process Control, 403-431.
© 1998 *Kluwer Academic Publishers.*

1. INTRODUCTION

Some of the features which have greatly contributed to the industrial success of MPC are its ability to handle process constraints, its simple extension from the single-input single-output (SISO) to the multi-input multi-output (MIMO) case and its variable structure in the event of faults.

MPC is an optimal-control based technique which computes the control law on-line. Regarding the formulation of the optimal control problem to be solved at each time step, MPC algorithms can be divided into the following main categories:

(1) Finite prediction horizon for:

  – Linear plants [20, 28, 15];
  – Nonlinear plants [27, 24, 10, 11];

(2) Infinite prediction horizon for:

  – Linear plants [48, 57, 32];
  – Nonlinear plants [43, 1, 3];

(3) Finite prediction horizon with end constraints[1] (also known as stability constraints) for:

  – Linear plants [34, 13, 29, 53, 54, 6, 18, 47, 57];
  – Nonlinear plants [16, 30, 37, 41, 38, 39, 40, 44, 45, 55, 1, 2, 31, 42].

In the first category a simple finite horizon objective function is employed which does not, per se, guarantee stability. [50] underline the poor stability properties of finite prediction horizon schemes.

In the second category, Lyapunov arguments are used to show asymptotic stability of the infinite prediction horizon scheme. Although these are simple and powerful results, one of the great restrictions of setting the prediction horizon to infinity in the control of nonlinear systems is naturally computational.

In the third category, asymptotic stability is added to finite prediction horizon schemes via the introduction of end constraints. The main difference amongst the various works in this area is the type of end constraint considered. Some of these constraints allow extension to the robust case. A comprehensive review of all these methods can be found in [33].

In this paper, we will address issues such as robust stability and implementability of MPC by proposing a different end constraint called *contractive constraint* and using a local linear approximation of the dynamics of

[1]By end constraint we mean any state constraint imposed at the end of the prediction horizon.

the system in the control computations. Both state and output feedback cases will be considered.

In [27] one can find for the first time an implementation of MPC to nonlinear systems where the model used for prediction is a linearization of the original nonlinear dynamics around the states and inputs at the current time step. Combining that with a quadratic objective function and linear constraints on the control variables, the MPC problem is formulated as a quadratic programming problem (QP). The resulting MPC strategy is known as quadratic dynamic matrix control (QDMC). This is a very simple and effective implementation of MPC from a computational point of view but the resulting closed-loop may be unstable. In this paper, we will look into how to combine the attractive computational features of Garcia's method in [27], and subsequently used in [35, 49], with the strong stability guarantees which can be obtained from the use of the contractive constraint [21, 22]. We will see that even though the contractive constraint is a quadratic constraint (and not a linear one as required for the problem to be posed as a QP), there are ways to incorporate this constraint into the optimization in combination with a quadratic objective function, a linear prediction model and other linear constraints, and still obtain a QP. Furthermore, we will show that if the QP problem at a given time step is feasible, the subsequent $P - 1$ (where P is known as the prediction horizon) are guaranteed to be feasible as well. Thus, we only need to be concerned with the feasibility of the optimization at the beginning of each prediction horizon.

Since the model used for prediction is a linear approximation of the original nonlinear system, this linearization procedure which highly simplifies the algorithm from a computational point of view, makes the stability analysis considerably more complex than the one presented in [21, 22] because of the model/plant mismatch caused by the linear approximation. It is no longer possible to guarantee convergence of the closed-loop states to the origin. We will show that the states can be steered to a neighborhood of the origin whose size is proportional to the mismatch between the nonlinear system and its linear approximation.

The paper is organized in the following manner: We will start with a description of the nonlinear systems treated here. Then we will present the contractive MPC algorithm with local linearization (which involves solving a convex programming problem at each time step) and show how the optimization can be reduced to a QP. Finally we will study the stability properties of the closed-loop system in both the state and output feedback cases. We will test the proposed controller on a well-known chemical process example. The proofs of all lemmas and theorems can be found in the appendix section.

2. CONTRACTIVE MPC ALGORITHM WITH LOCAL LINEARIZATION

2.1. DESCRIPTION OF THE SYSTEM

The nonlinear systems considered in this paper are described by the following equations:

$$\dot{x}^p(t) = f(x^p(t), u(t), d(t)) \tag{1}$$

where $x^p(t) \in \Re^n$ is the vector of state variables, $u(t) \in \Re^m$ are the manipulated variables, $d(t) \in \Re^d$ are unknown time-varying disturbances and/or parameters and $f : \Re^n \times \Re^m \times \Re^d \to \Re^n$ is a continuously differentiable function.

It is assumed that $d(t) \in \mathcal{D}$, $\forall t \in [0, \infty)$, with the set $\mathcal{D}$ defined by:

$$\mathcal{D} := \{ d(t) \in \Re^d \mid \; \| d(t) \| \leq \epsilon_d, \;\; \forall t \in [0, \infty) \}; \quad \epsilon_d \in [0, \infty) \tag{2}$$

The hard constraints on the manipulated variables, $u(t)$, will be expressed in the usual manner:

$$u(t) \in \mathcal{U} := \{ u \in \Re^m : u_{min} \leq u \leq u_{max} \}, \; \forall t \in [0, \infty); \;\; u_{min}, \; u_{max} \in \Re^m \tag{3}$$

Linear constraints on the rates of change of the manipulated variables are also commonly present and will be used here.

2.2. NOTATION

The solution of (1) at time t, corresponding to the initial time/state pair $\{t_0, x_0^p\}$ and the input $u(\tau)$, $\tau \in [t_0, t]$, is denoted by $x^p(t, t_0, x_0^p, u, d)$ or, in a simplified notation, $x_0^p(t)$.

Let T be the sampling time and P, M the prediction and control horizons, respectively.

Given any sampling time $t_k^0 := t_k := t_0 + kPT$, $k \in [0, \infty)$, and $t_k^j := t_k + jT$, $j = 0, \ldots, P$, with $t_k^P = t_{k+1}^0 = t_{k+1}, \forall k \geq 0$, let us adopt the following notation $x_k^p := x_k^{p,0} := x^p(t_k^0, t_0, x_0^p, u, d)$, $x_k^{p,j} := x^p(t_k^j, t_k, x_k^p, u, d)$, $x_k^{p,j}(t) := x^p(t, t_k^j, x_k^{p,j}, u, d)$ and $u_k^j(t)$ is the continuous control law for $t \in [t_k^j, t_k^j + PT]$. In order to conform to MPC's usual implementation scheme, let us consider a discontinuous control law of the kind $u_k^j(t) = u(kP + j + i|kP + j)$ for $t \in [t_k^j + iT, t_k^j + (i + 1)T]$, $i = 0, \ldots, P - 1$, i.e., $u_k^j(t)$ is constant during one sampling time. Moreover, $u(kP+j+i|kP+j) = u(kP + j + M - 1|kP + j)$, $\forall i = M, \ldots, P - 1$.

$\mathcal{P}(t_k^j, x_k^{p,j})$ denotes the optimal control problem at time t_k^j with initial condition $x_k^{p,j}$.

Throughout this paper, $|.|$ is the scalar norm, $\| . \|$ denotes the Euclidean norm of a vector and $\| x \|_{\tilde{P}} := \sqrt{x' \tilde{P} x}$, with $\tilde{P} \in \Re^{n \times n}$ positive definite, is the weighted Euclidean norm of $x \in \Re^n$.

The symbol := represents that the left-hand side of an equation is defined as the right-hand side. The converse applies to =:.

2.3. CONTROL ALGORITHM

Data: Initial Conditions: t_0, x_0^p; Controller Parameters: horizons P, M ($1 \le M \le P < \infty$), weights $Q, R, S, \hat{P} > 0$, contractive parameter $\alpha \in [0, 1)$, sampling time $T \in (0, \infty)$ and control constraints $u_{min}, u_{max}, \Delta u_{max} \in \Re^m$.

Step 0: Set $k = 0$, $j = 0$.

Step 1: Assuming that the optimal control problem $\mathcal{P}(t_k^j, x_k^{p,j})$ is feasible for the chosen set of parameters, then at $t = t_k^j$ solve $\mathcal{P}(t_k^j, x_k^{p,j})$ which is specified by:

$$\min_{u(.)} J_k^j := \sum_{i=1}^P \| x(kP + j + i|kP + j) \|_Q^2 +$$
$$+ \sum_{i=0}^{P-1} \| u(kP + j + i|kP + j) \|_R^2 +$$
$$+ \sum_{i=0}^{P-1} \| \Delta u(kP + j + i|kP + j) \|_S^2 \qquad (4)$$

subject to:

$$\left\{ \begin{array}{l} \dot{x}_k^j(t) = C_k^j + A_k^j x_k^j(t) + B_k^j u_k^j(t), \ x_k^j := x_k^{p,j} \\ u_{min} \le u(kP + j + i|kP + j) \le u_{max}, \ i = 0, \ldots, M - 1 \\ |\Delta u(kP + j + i|kP + j)| \le \Delta u_{max}, \ i = 0, \ldots, M - 1 \\ \Delta u(kP + j + i|kP + j) = 0, \ i = M, \ldots, P - 1 \\ \| \bar{x}_k^j(t_k^P) \|_{\hat{P}} \le \alpha \| x_k^p \|_{\hat{P}}, \ \alpha \in [0, 1) \end{array} \right. \qquad (5)$$

where $x(kP + j + i|kP + j)$ are the predicted states at time t_k^{j+i} computed with information up to time t_k^j, i.e., $x(kP + j + i|kP + j) := x_k^j(t_k^{j+i})$ and

$$\dot{\bar{x}}_k^j(t) = C_k^0 + A_k^0 \bar{x}_k^j(t) + B_k^0 u_k^j(t), \qquad (6)$$

with $\bar{x}_k^0 := x_k^p := x^p(t_k)$ and $\bar{x}_k^j = \bar{x}_k^{j-1}(t_k^j)$ for $j \ge 1$, is the linear model which is only updated with the states of the plant at $t = t_k^P$, i.e., at intervals of one prediction horizon. Moreover, while the matrices A, B, C are re-calculated at every t_k^j, $j = 0, \ldots, P - 1$, for computation of the predicted trajectories, they are only re-calculated at the beginning of prediction horizons for purpose of computation of the contractive constraint.

The result of this step is an optimal sequence of control moves $\{u(kP + j|kP + j), \ldots, u(kP + j + M - 1|kP + j)\}$.

Step 2: Apply the first control move, $u(kP + j | kP + j)$, to the plant (1) for $t \in [t_k^j, t_k^{j+1}]$ and measure the states at t_k^{j+1}. Set x_k^{j+1} equal to the measured states, $x_k^{p,j+1} := x^p(t_k^{j+1})$, and $\bar{x}_k^{j+1} = \bar{x}_k^j(t_k^{j+1})$, $j \geq 0$.

Step 3: If $j < P - 1$, set $j = j + 1$ and go back to **Step 1**. If $j = P - 1$ set $x_{k+1}^0 =: x_{k+1} = x_{k+1}^P$, $t_{k+1}^0 = t_{k+1} = t_k^P$, $k = k + 1$, $j = 0$, and go back to **Step 1**.

Remark 1 *The hard constraints on the rates of change of the inputs are linear and can be re-written component-wise as:*

$$|\Delta u_l(kP + j + i | kP + j)| \leq \Delta u_{max,l}, \ \forall j = 0, \dots, P - 1,$$
$$i = 0, \dots, M - 1, k \geq 0, \ with \ l = 1, \dots, m \tag{7}$$

In the vector representation, we have committed some abuse of notation since $\Delta u(kP + j + i | kP + j)$ is a vector and we have defined $|.|$ to be a scalar norm. The reason for this notation is to avoid confusion with the $2-norm$.

Remark 2 *The matrices A_k^j, B_k^j, C_k^j are given by:*

$$\begin{aligned}
A_k^j &:= \frac{\partial f}{\partial x}(x_k^{p,j}, u_k^j, 0) \\
B_k^j &:= \frac{\partial f}{\partial u}(x_k^{p,j}, u_k^j, 0) \\
C_k^j &:= f(x_k^{p,j}, u_k^j, 0) - A_k^j x_k^{p,j} - B_k^j u_k^j
\end{aligned} \tag{8}$$

i.e., the linearization is computed at nominal values of disturbances and parameters.

Remark 3 *The optimization step of the control algorithm is a convex programming problem in the control variables. The convexity of the optimization is due to the fact that the objective function is quadratic in the decision variables $\Delta u(kP + j | kP + j), \dots, \Delta u(kP + j + M - 1 | kP + j)$, the trajectory and input constraints are linear and the contractive constraint is quadratic and convex. It is a well-known result that every local solution to a convex programming problem is a global solution, and the set of global solutions is convex. Furthermore, if the function to be minimized is strictly convex on a given convex set, then any global solution is unique.*

Remark 4 *Because the contractive constraint does not change from t_k to t_{k+1} (see how the trajectory $\bar{x}(t)$ is computed in (6)), if $\mathcal{P}(t_k, x_k^p)$ is feasible, then $\mathcal{P}(t_k^j, x_k^{p,j})$ is also feasible for each $j = 1, \dots, P - 1$. Moreover, the properly restricted optimal solution of $\mathcal{P}(t_k, x_k^p)$ is a feasible solution to the following $P - 1$ control problems.*

3. CONTRACTIVE MPC POSED AS A QP

Even though, posed as it is, the contractive MPC algorithm is much simpler than a general NLP from a computational point of view (see **Remark 3**), further improvement can still be achieved as we can transform the convex optimization problem into a QP.

A QP is an optimization problem in which the objective function is quadratic and the constraint functions are linear. Thus the problem is to find a solution z^* to

$$
\begin{aligned}
\text{minimize} \quad & J(z) := \tfrac{1}{2} z' H z + h' z \\
\text{subject to} \quad & a_i' z = b_i, \quad i \in E, \\
& a_i' z \geq b_i, \quad i \in I
\end{aligned}
\tag{9}
$$

where z are the decision variables, $J(z)$ is the performance criterion, H is a symmetric positive (semi-)definite matrix (Hessian matrix), h' is known as the gradient vector, E and I are the sets of equality and inequality constraints, respectively, and the matrix A (with rows a_i) and vector b define the linear equality and inequality constraints on the optimization variable z.

Thus, since the prediction model and the constraints on the control variables used here are linear, the optimization would be a QP in the absence of the contractive constraint (which is quadratic in the control variables). There are, however, alternative ways to implement this constraint into the optimization step such that the optimization can still be posed as a QP.

3.1. PROCEDURES FOR TRANSFORMATION OF THE OPTIMIZATION INTO A QP

We have identified three different ways to incorporate the contractive constraint into the optimization step of our contractive MPC scheme in order to pose it as a QP:

Procedure 1: Approximate it by a more restrictive 1−norm constraint which can then be re-written as a set of $2n + 1$ linear constraints (where n is the state vector dimension).

Procedure 2: Add it to the objective function, pre-multiplied by a chosen scalar weight, and remove it from the list of constraints. This leads to an iterative procedure on the weight which is carried out until the contractive constraint is satisfied (*penalty function approach*). These computations must be repeated at each time step and therefore this procedure can become very expensive computationally.

Procedure 3: Re-write it as:

$$
\| \, \bar{x}_k^j(t_k^P) \, \|_1 \leq \alpha \, \delta_k^j \, \| \, x_k^p \, \|_{\hat{P}}
\tag{10}
$$

where $\delta_k^j := \frac{\|\bar{x}_k^j(t_k^P)\|_1}{\|\bar{x}_k^j(t_k^P)\|_{\hat{P}}}$, with $\bar{x}_k^j(t)$ being the trajectory used in the computation of the contractive constraint at $j = 0, \ldots, P-1$ for a given k.

The modified contractive constraint (10), can then be written as a set of $2n+1$ linear constraints in the control variables as in **Procedure 1**. The computation of the newly introduced parameter δ_k^j is done through a uni-directional search resulting from a first-order sensitivity analysis of the optimality conditions of the QP with respect to this parameter. Unlike the penalty function approach, the QP needs to be solved only once at each time step. Then, a linear system is solved repeatedly and δ_k^j is updated until the desired equality, $\delta_k^j = \frac{\|\bar{x}_k^j(t_k^P)\|_1}{\|\bar{x}_k^j(t_k^P)\|_{\hat{P}}}$, is satisfied [12]. For further reduction of the computational load, we do not need to re-compute δ_k^j at every time step. Since the contractive constraint needs to be satisfied only at $j = P-1$ for stability purposes, we can solve the optimization problems at $j = 1, \ldots, P-2$ with $\delta = \delta_k^0$, and only check for satisfaction of the contractive constraint at t_k^{P-1}. If it is not satisfied (which implies that $\delta_k^{P-1} := \frac{\|\bar{x}_k^{P-1}(t_k^P)\|_1}{\|\bar{x}_k^{P-1}(t_k^P)\|_{\hat{P}}} < \delta_k^0$), then we should calculate δ_k^{P-1} using the same iterative procedure described for computation of δ_k^0. In this case, the computation of δ using sensitivity analysis of the QP would be repeated only twice for a whole prediction horizon and stability would still be assured.

For a detailed description of the three previous approaches to transforming the optimization problem into a QP we refer the reader to the first author's Ph.D. thesis [21].

3.2. QP FORMAT

By using **Procedures 1, 2** or **3** in the implementation of the contractive constraint, we have a QP problem as in (9). With the rates of change of the control variables defined as the decision variables, i.e., $z = \{ \Delta u(kP + j|kP+j), \ldots, \Delta u(kP+j+M-1|kP+j) \}$, we need to compute the Hessian matrix H, the gradient vector h and the constraint matrix and vector, A and b, respectively. The detailed algebraic developments for computation of these controller parameters for **Procedures 1, 2** and **3** can be found in [21].

4. BASIC IDEA BEHIND THE CONTROLLER DESIGN

Figures 1 and *2* illustrate the closed-loop state trajectories resulting from implementation of the contractive MPC algorithm with local linearization

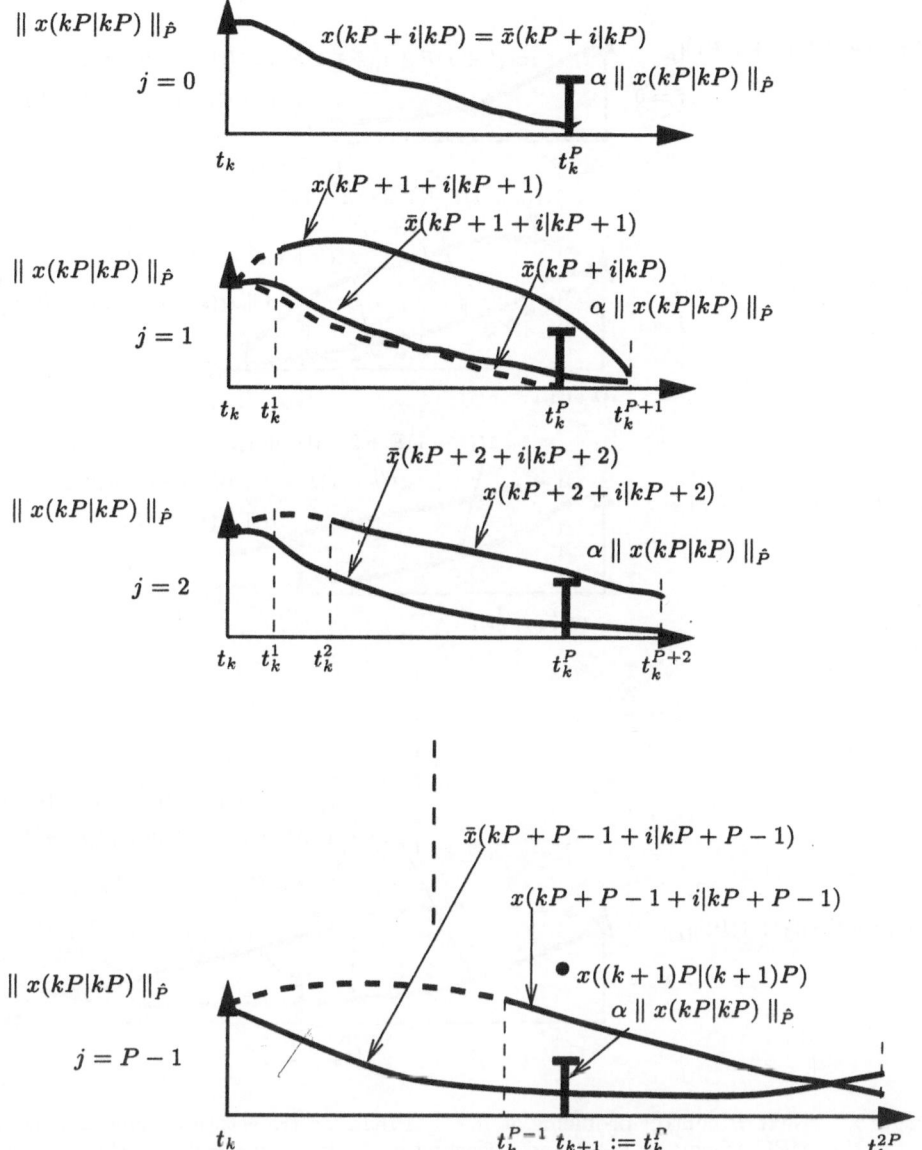

Figure 1. P control problems for a fixed k. Predicted trajectories generated by the contractive MPC scheme for a fixed k and j varying in the interval $j = 0, \ldots, P - 1$.

to system (1).

Thus, while the optimization problem remains solved over P time steps for different values of j and for a constant k, the number of steps between

412

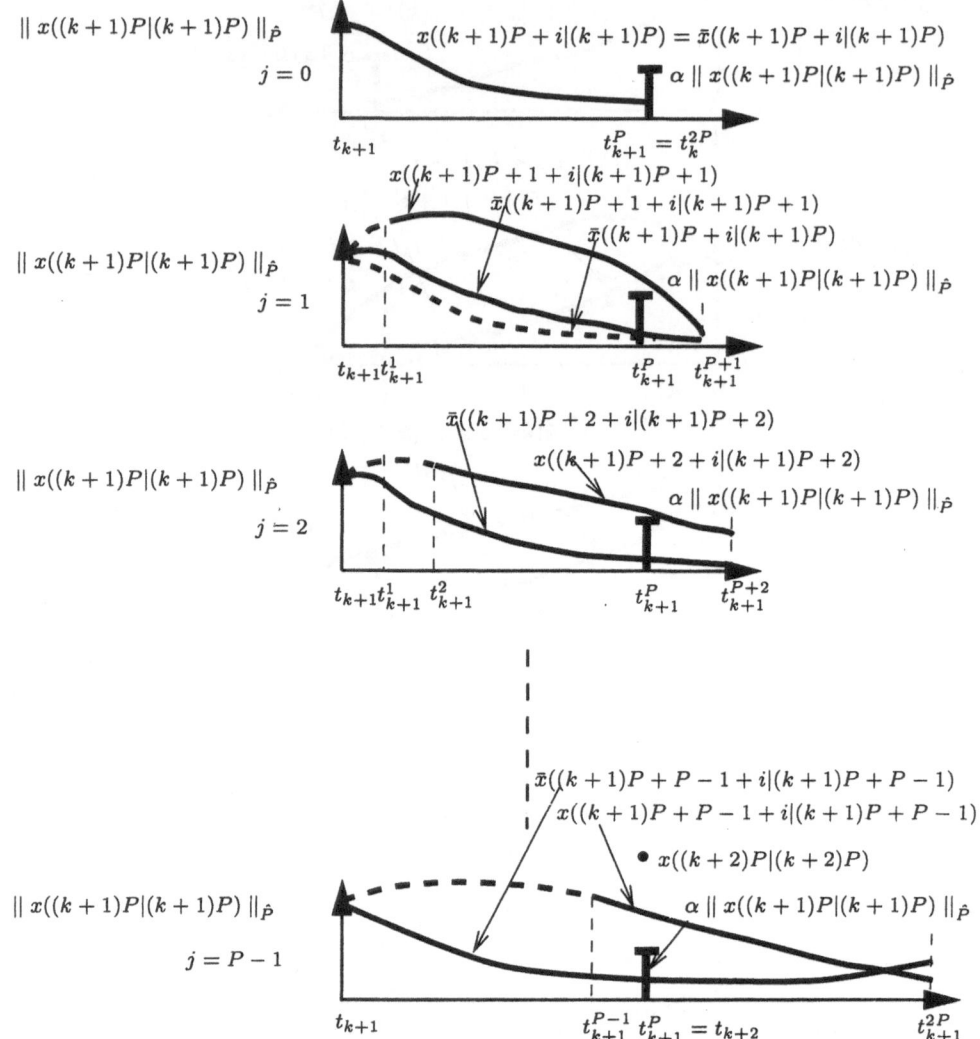

Figure 2. Next P control problems at $k+1$. Predicted trajectories generated by the contractive MPC scheme at $k+1$ and j varying in the interval $j = 0, \ldots, P-1$.

the beginning of the prediction and the location of the contractive constraint is equal to $P - j$ and therefore decreases as j increases, as we can clearly see from *Figures 1* and *2* (unlike what generally happens with end constraints, which are moved along with the prediction horizon).

5. STABILITY ANALYSIS

As previously discussed, in the attempt to simplify the controller from a computational point of view, we have introduced a model/plant (linear/nonlinear) mismatch which needs to be quantified and dealt with by the controller.

Finding appropriate uncertainty descriptions for nonlinear systems is an area only quite recently explored and much remains to be done. For example, in [7, 19, 26, 36], in order to achieve either stabilization or tracking, some restrictive assumptions often referred to as matching conditions were introduced regarding the structure of the uncertainties. Recently, [9] brings up the so-called generalized matching condition for a class of nonlinear systems and [8, 51, 17, 56, 52] conduct the robustness analysis of uncertain dynamical systems for mismatched uncertainties.

Here we will express the linear/nonlinear mismatch through a conic sector bound.

5.1. BASIC ASSUMPTIONS

Without loss of generality, let us consider the regulation problem where the desired operating point is the origin $(x^p, u, d) = (0, 0, 0)$. Then, the following assumptions are needed to ensure local stability:

Assumption 1 *It is assumed that there exists a constant $\rho \in (0, \infty)$ such that for all x_k^p, $\bar{x}_k \in B_\rho := \{x \in \Re^n \mid \| x \|_{\hat{P}} \leq \rho\}$, the optimization problem at time step k, $\mathcal{P}(t_k, x_k^p)$, is feasible for all $k \geq 0$.*

The basic assumptions on the nonlinear system are:

Assumption 2 *We assume that if x_k^p, $\bar{x}_k \in B_\rho$, $\forall k \geq 0$, then there exists a constant $\beta \in (0, \infty)$ so that the transient states of the model used in the computation of the contractive constraint remain inside the set $B_{\beta \| x_k \|_{\hat{P}}}$, i.e., $\| \bar{x}_k^j(t) \|_{\hat{P}} \leq \beta \| x_k \|_{\hat{P}} \leq \beta \rho$, $\forall j = 0, \ldots, P-1$, $k \geq 0$.*

Remark 5 *Since u is constrained, **Assumption 2** is always satisfied except for systems with finite escape time. So, systems with finite escape time are ruled out from our investigation.*

Assumption 3 *The linearization of the plant characterized by the pair $(A, B) := (\frac{\partial f}{\partial x}(x^*, u^*, 0), \frac{\partial f}{\partial u}(x^*, u^*), 0)$ is stabilizable for all points $(x^*, u^*, 0) \in \Re^n \times \Re^m \times \Re^d$ around which the linearization is performed.*

It is assumed that there exists a Lipschitz constant $L \in [0, \infty)$ and a so-called modeling bound $\gamma \in [0, \infty)$ such that for all x^p, $\bar{x} \in \Omega \subset \Re^n$ (where Ω is a compact set), $u \in \mathcal{U}$ and $d \in \mathcal{D}$, the following bounds hold:

Assumption 4

$$\| C + A\bar{x} + Bu \|_{\hat{P}} \leq L \left[\| \bar{x} \|_{\hat{P}} + \| u \| \right] \tag{11}$$

where $A := \frac{\partial f}{\partial x}(x^*, u^*, 0)$, $B := \frac{\partial f}{\partial u}(x^*, u^*, 0)$ and $C := f(x^*, u^*, 0) - Ax^* - Bu^*$.

Assumption 5

$$\| f(x^p, u, d) - f(\bar{x}, u, 0) \|_{\hat{P}} =: \tag{12}$$
$$=: \| f(x^p, u, d) - C - A\bar{x} - Bu - F(\bar{x}, u, 0) \|_{\hat{P}} \leq L[\| x^p - \bar{x} \|_{\hat{P}} + \| d \|]$$

where $F(\bar{x}, u, 0)$ represents the second and higher order terms of the Taylor series expansion of $f(\bar{x}, u, 0)$ around the point $(x^*, u^*, 0)$.

Assumption 6 *Growth condition on F:*

$$\| F(\bar{x}, u, 0) \|_{\hat{P}} \leq \gamma \left[\| \bar{x} \|_{\hat{P}} + \| u \| \right] \tag{13}$$

Remark 6 *Let the reachable set of states $\mathcal{X}$ be defined by:*

$$\mathcal{X} := \{ x_k^{p,j}(t), x_k^j(t), \bar{x}_k^j(t) \in \Re^n \mid x_k^{p,j}(t) := x_k^p(t, t_k^j, x_k^{p,j}, u_k^j(t), d_k^j(t)),$$
$$x_k^j(t) := x_k^j(t, t_k^j, x_k^{p,j}, u_k^j(t), 0) \text{ and } \bar{x}_k^j(t) := \bar{x}_k^j(t, t_k^j, \bar{x}_k^{j-1}(t_k^j), u_k^j(t), 0),$$
$$t \in [t_k^j, t_k^{j+1}], x_k^p, \bar{x}_k \in B_\rho, u_k^j(t) \in \mathcal{U}, d_k^j(t) \in \mathcal{D}; j = 0, \ldots, P-1, k \geq 0 \}$$

*Then, it is only necessary to satisfy **Assumptions 4, 5** and **6** along the trajectories generated by the contractive MPC algorithm, i.e., $x_k^{p,j}(t)$ and $\bar{x}_k^j(t) \in \mathcal{X}$ for all $j = 0, \ldots, P-1$ and $k \geq 0$. Because this is difficult to check beforehand (since we do not know a priori which control moves will be computed by the contractive MPC controller and, consequently, which trajectories will be generated), we have posed the assumptions in a more conservative way, as valid for all x^p, $\bar{x} \in \Omega$ with $\Omega \subset \Re^n$ large enough so that $\mathcal{X} \subset \Omega$.*

5.2. STABILITY RESULTS IN THE STATE FEEDBACK CASE

In the next theorem we will derive an upper bound on the difference between the states of the plant and those of the linear model used in the computation of the contractive constraint, at the end of prediction horizons. We will see that this difference is caused by the linear/nonlinear mismatch (quantified through γ) and the presence of disturbances and/or parameter uncertainty (quantified through ϵ_d).

Theorem 1 (Bound on the difference between plant and linear model states at t_k) *Let $\rho \in (0, \infty)$ and $L, \gamma \in [0, \infty)$ satisfy* **Assumptions 1, 4** *and* **5, 6,** *respectively. Then if x_k^p, $\bar{x}_k \in B_\rho, \forall k \geq 0$, there exist $\lambda_1, \lambda_2 \in [0, \infty)$ so that*

$$\| x_{k+1}^p - \bar{x}_{k+1} \|_{\hat{P}} \leq \lambda_1 \| x_k^p \|_{\hat{P}} + \lambda_2, \quad \forall k \geq 0 \tag{14}$$

with $\lambda_1 \to 0$ as $\gamma \to 0$ or $PT \to 0$ and $\lambda_2 \to 0$ as γ, $\epsilon_d \to 0$ or $PT \to 0$. Thus, the state difference disappears if the model is linear and there are no disturbances. $\square$

The result in **Theorem 1** was derived under the assumption that the optimization problem at each time step is well-posed, i.e., that x_k^p, $\bar{x}_k \in B_\rho$, $\forall k \geq 0$. In the next theorem we will derive conditions on the contractive parameter α (which can also be seen as conditions on γ and ϵ_d) so that this feasibility condition is satisfied and stability holds.

Theorem 2 (Stability and feasibility condition in the state feedback case) *If* **Assumptions 1-6** *hold and if the combined effect of the nonlinearities and disturbances is such that the following condition is satisfied*

$$\frac{\gamma}{L} e^{LPT} (e^{LPT} - 1) + \frac{[\gamma \tilde{u} e^{LPT} + L\epsilon_d] PT e^{LPT}}{\rho} < 1 \tag{15}$$

(with $\tilde{u} := \max\{\| u_{min} \|, \| u_{max} \|\}$), then there exists $\bar{\alpha} \in [0, 1)$ such that if $\alpha < \bar{\alpha}$ the control algorithm at each time step is feasible and the closed-loop system is stable with the states of the plant being steered to a control invariant set $B_{\hat{\rho}}$ where $\hat{\rho}$ is given by:

$$\hat{\rho} := \frac{LPT e^{LPT} [\gamma \tilde{u} e^{LPT} + L\epsilon_d]}{L(1 - \alpha) + \gamma e^{LPT} (e^{LPT} - 1)} \tag{16}$$

Thus, $\hat{\rho} \to 0$ as γ, $\epsilon_d \to 0$ or $PT \to 0$. $\square$

Remark 7 *From equation (16) notice that $\hat{\rho}$ decreases for smaller values of α. This is reasonable since we should be able to drive the states of the plant to a smaller control invariant set by requiring a stronger contraction on the model states. $\hat{\rho}$ increases as γ increases for $\alpha < 1 - \frac{\epsilon_d}{\tilde{u}} (e^{LPT} - 1)$ (we can see that by examining $\frac{\partial \hat{\rho}}{\partial \gamma}$). Thus, if $\epsilon_d = 0$, $\hat{\rho}$ increases as γ increases for any chosen α. Moreover, $\hat{\rho}$ always increases as ϵ_d increases.*

The results derived in **Theorems 1** and **2** were based on the assumption that there exists a non-empty set B_ρ of initial conditions for which feasibility of the optimization step of the control algorithm is guaranteed. Now we will take advantage of the fact that the model used in the computation of the contractive constraint is linear to derive a lower bound $\underline{\alpha}$ on $\alpha \in [0, 1)$

which establishes a sufficient condition for feasibility. Thus, we no longer need **Assumption 1** or, in other words, $\rho \to \infty$. This lower bound can only be derived under a more restrictive assumption than **Assumption 3**, namely that:

Assumption 7 $A := \frac{\partial f}{\partial x}(x^*, u^*, 0)$ *is stable (i.e., it has all eigenvalues located in the left half plane) for all points* $(x^*, u^*, 0) \in \Re^n \times \Re^m \times \Re^d$ *around which the linearization is performed.*

Theorem 3 (Stability and feasibility condition in the state feedback case for systems satisfying Assumption 7) *If* **Assumptions 1, 2, 4, 5, 6** *and* **7** *hold and if the nonlinearities are such that* γ *satisfies the following bound*

$$\gamma < \frac{L}{e^{LPT}(e^{LPT} - 1)} [1 - \sqrt{\lambda_{max}(\hat{P}^{\frac{1}{2}} e^{APT} \hat{P}^{-\frac{1}{2}})}] \qquad (17)$$

(with $\sqrt{\lambda_{max}(\hat{P}^{\frac{1}{2}} e^{APT} \hat{P}^{-\frac{1}{2}})} := \sup_{k \geq 0} \sqrt{\lambda_{max}(\hat{P}^{\frac{1}{2}} e^{A_k^0 PT} \hat{P}^{-\frac{1}{2}})}$*), then there exist* $\underline{\alpha}$, $\bar{\alpha} \in (0, 1)$ *such that if* $\underline{\alpha} \leq \alpha < \bar{\alpha}$ *then the control algorithm at each time step is feasible and the closed-loop system is stable with the states of the plant being steered to a control invariant set* $B_{\hat{\rho}}$ *(where* $\hat{\rho}$ *is defined as in equation (16)).* $\qquad\qquad\square$

5.3. STABILITY RESULTS IN THE OUTPUT FEEDBACK CASE

In the state feedback case, the states of the model used in the computation of the contractive constraint at step k after one sampling time are given by:

$$\bar{x}_k(t_k + T) = \Phi_k x_k^p + \Psi_k[B_k^0 u_k + C_k^0] \qquad (18)$$

where the matrices Φ_k, Ψ_k are defined as:

$$\Phi_k := e^{A_k^0 T} \qquad \text{and} \qquad \Psi_k := \int_0^T e^{A_k^0(T-t)} dt \qquad (19)$$

In the output feedback case, these model states are given by:

$$\bar{x}_k(t_k + T) = \Phi_k \hat{x}_k + \Psi_k[B_k^0 u_k + C_k^0] \qquad (20)$$

The difference between the two model dynamics can be represented by an additive disturbance, i.e., the state evolution of the model in the output feedback case is equivalent to the state feedback case modified to:

$$\dot{\bar{x}}_k(t) = A_k^0 \bar{x}_k(t) + B_k^0 u_k(t) + C_k^0 + \tilde{d}_k(t) \quad \text{with} \quad \bar{x}_k(t_k) = x_k^p \qquad (21)$$

If $\tilde{d}_k(t) = \tilde{d}_k$ = constant for $t \in [t_k, t_k + T]$, integration of (21) results in:

$$\bar{x}_k(t_k + T) = \Phi_k x_k^p + \Psi_k[B_k^0 u_k + C_k^0 + \tilde{d}_k] \tag{22}$$

Thus, we want to compute $\tilde{d}_k$ so that it represents the difference in the dynamic behavior of the model caused by the estimation, i.e., the states in equation (22) have to be equal to the states in (20). Thus, by subtracting equation (20) from (22), we have:

$$\tilde{d}_k = \Psi_k^{-1} \Phi_k e_k = -[e^{-A_k^0 T} - I_n]^{-1} A_k^0 e_k \tag{23}$$

where I_n is the identity matrix of dimension n and e_k is the estimation error defined as $e_k := \hat{x}_k - x_k^p$.

Applying the $\hat{P}$−norm to equation (23), we get:

$$\| \tilde{d}_k \|_{\hat{P}} \leq \| [e^{-A_k^0 T} - I_n]^{-1} A_k^0 \|_{\hat{P}} \| e_k \|_{\hat{P}} =: \phi_k \| e_k \|_{\hat{P}} =: \rho_k^d \tag{24}$$

Thus, the additive disturbance is proportional to the estimation error. In [21] the author proposes a nonlinear observer for continuous-time systems with discrete observations which produces asymptotically convergent estimates if the initial estimation error is not too large and the nonlinearities are reasonably weak (the exact sufficient conditions can be found in that reference). In this case, there exists $K \in [0, \infty)$ such that the estimation error at any sampling time t_k satisfies the following inequality:

$$\| e_k \|_{\hat{P}} \leq K \| e_0 \|_{\hat{P}}, \quad \forall k \geq 0 \tag{25}$$

From equations (24) and (25) we obviously have:

$$\| \tilde{d}_k \|_{\hat{P}} \leq \phi_k K \| e_0 \|_{\hat{P}} \leq \phi K \| e_0 \|_{\hat{P}} =: \rho^d \tag{26}$$

where $\phi := \max_{k \geq 0} \phi_k$.

Thus, if an asymptotically stable nonlinear observer is used (such as the one proposed in [21]) then its effect is to introduce an additive asymptotically decaying disturbance into the dynamics of the models used in the prediction and in the computation of the contractive constraint. In the stability analysis of the output feedback scheme we will consider $\epsilon_d = 0$ since the nonlinear observer can only be shown asymptotically stable in the absence of external disturbances and parameter uncertainty.

Assumption 8 *Let the asymptotically decaying properties of the discrete disturbance sequence $\{\tilde{d}_k\}_{k \geq 0}$ introduced by the observer be expressed as:*

For any $\tilde{\epsilon} > 0, \exists$ a finite $\bar{k} := \bar{k}(\tilde{\epsilon}) \in \mathcal{N}$ so that $\rho_k^d \leq \tilde{\epsilon}, \quad \forall k \in [\bar{k}, \infty)$, and $\bar{k}(\tilde{\epsilon}) \to \infty$ if $\tilde{\epsilon} \to 0$

418

Our stability results in the output feedback case will reveal that we can still drive the states of the plant to the same control invariant set $B_{\hat{\rho}}$ to which they could be driven in the state feedback case.

Theorem 4 (Stability and feasibility condition in the output feedback case) *Let ρ, L, γ,$\rho^d \in (0, \infty)$ be as defined in **Assumptions 1, 4** and **5, 6** and equation (26), respectively, and let the state estimator be asymptotically stable (such that **Assumption 8** holds). Then, if the norm of the additive disturbance caused by introduction of the observer into the closed-loop is bounded by,*

$$\| \tilde{d}_k \|_{\hat{P}} \leq \rho^d < \frac{[L(1-\alpha) - \gamma e^{LPT}(e^{LPT}-1)]}{LPT\, e^{LPT}}\, [\rho - K \| e_0 \|_{\hat{P}}] - \gamma \tilde{u} e^{LPT}$$
(27)

the output feedback control problem is well-posed (i.e., x_k^p, $\bar{x}_k$, $\hat{x}_k \in B_\rho$, $\forall k \geq 0$, and $x_0^p \in B_{\rho_0} \subset B_\rho$; with B_{ρ_0} a non-empty set) and the states of the resulting closed-loop system converge asymptotically to $B_{\hat{\rho}}$ (with $\hat{\rho}$ given by equation (16) with $\epsilon_d = 0$). □

6. EXAMPLE

This example has been widely used in the literature (see, e.g., [25, 23]) for testing linear and nonlinear control schemes due to its underlying dynamic characteristics which give rise to well-known control difficulties.

The objective is to exemplify the implementation of the contractive MPC scheme with local linearization (CNTMPC) and to show the stabilizing effect of the contractive constraint and how it influences the closed-loop performance.

We will show that the MPC controller with local linearization without the contractive constraint (let us denote this scheme as standard MPC with local linearization - STNMPC) can generate an unstable response.

We will also demonstrate how the performance improves as we decrease the contractive parameter α, the limitation being that the optimal control problem can become infeasible for α too small.

6.1. REACTOR DYNAMICS

The process consists of an ideal continuous stirred tank reactor (CSTR) where the reversible exothermic reaction $A \rightleftharpoons R$ is carried out. The reactor is modeled by the following differential equations:

$$\frac{dC_A}{dt} = \frac{C_{Ai} - C_A}{\tau} - k_1 C_A + k_{-1} C_R$$
(28)

$\tau = 60$ s	$\mathcal{R} = 1.987$ cal mol^{-1} K^{-1}
$K_1 = 5 \times 10^3$ s^{-1}	$K_{-1} = 1 \times 10^6$ s^{-1}
$Q_1 = 1 \times 10^4$ cal mol^{-1}	$Q_{-1} = 1.5 \times 10^4$ cal mol^{-1}
$\Delta H = -5 \times 10^3$ cal mol^{-1}	$\rho = 1$ kg/l
$C_p = 1 \times 10^3$ cal kg^{-1} K^{-1}	
$C_{Ai} = 1$ mol/l	$C_{Ri} = 0$ mol/l

TABLE 1. Parameters and operating conditions for the reactor.

$$\frac{dC_R}{dt} = \frac{C_{Ri} - C_R}{\tau} + k_1 C_A - k_{-1} C_R \tag{29}$$

$$\frac{dT}{dt} = \frac{-\Delta H}{\rho C_p}[k_1 C_A - k_{-1} C_R] + \frac{T_i - T}{\tau} \tag{30}$$

where $k_1 := K_1 e^{-\frac{Q_1}{\mathcal{R}T}}$ and $k_{-1} := K_{-1} e^{-\frac{Q_{-1}}{\mathcal{R}T}}$ and the specific parameters and operating conditions used in our simulations are shown in TABLE 1.

The reactor equilibrium conversion as a function of temperature has a well-defined maximum (see *Figure 3*) at which the steady state gain is zero.

Thus, because of the conversion maximum, the system gain changes sign from one side of the maximum to the other. Systems of this type are not "integral controllable" [46], i.e., they cannot be stable over the entire operating region when controllers with integral action are employed.

6.2. SIMULATION RESULTS

The simulations shown here represent a steady state change from an equilibrium point of low conversion and high temperature (located to the right side of the point of maximum conversion on the equilibrium curve) to a target equilibrium point of lower temperature and significantly higher conversion (located to the left of the point of maximum conversion), a much more desirable operating condition. The state and input coordinates of these two equilibria are shown in TABLE 2 and they are also marked on the equilibrium curve in *Figure 3*.

The simulation results for different values of the contractive constraint parameter α, are shown in *Figure 4*.

In these graphs, we are plotting deviation variables with respect to the target steady state coordinates. Thus, our goal is to control the plotted state and input variables to zero.

The controller parameters used in all the simulations are given in TABLE 3. No input constraints are enforced.

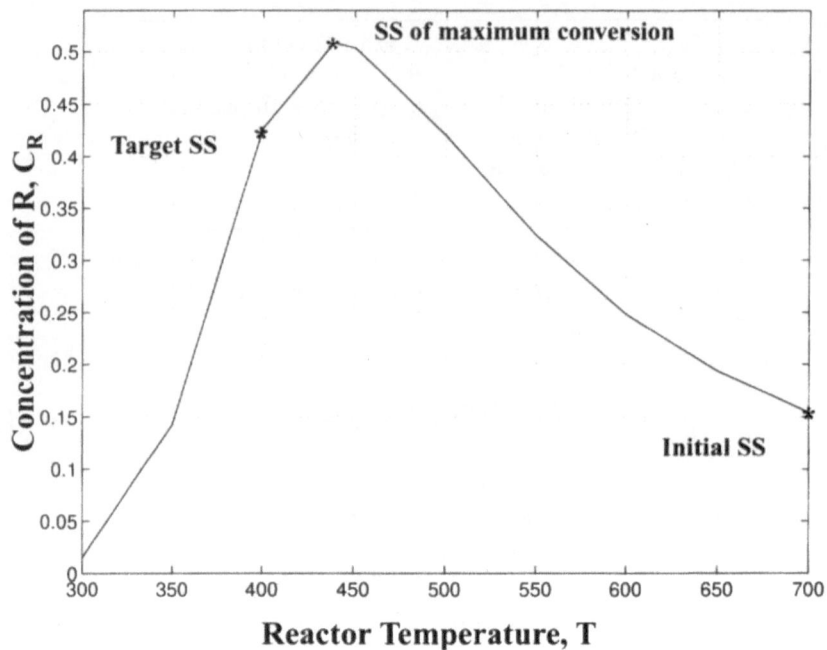

Figure 3. Equilibrium diagram for the reactor (SS = steady state).

State and input coordinates of the initial steady state			
$C_A = 0.8461$ mol/l	$C_R = 0.1539$ mol/l	$T = 700$ K	$T_i = 699.2305$ K
State and input coordinates of the target steady state			
$C_A = 0.5729$ mol/l	$C_R = 0.4271$ mol/l	$T = 400$ K	$T_i = 397.8644$ K

TABLE 2. Coordinates of the initial and target steady states.

For $\alpha = 0$ the optimal control problem is infeasible at the first time step

$Q = \text{diag}([0\ 1\ 0])$	$R = 0$	$S = 0$	$T = 5$
$P = 2$	$M = 2$		

TABLE 3. Controller parameters for simulations in *Figure 4*.

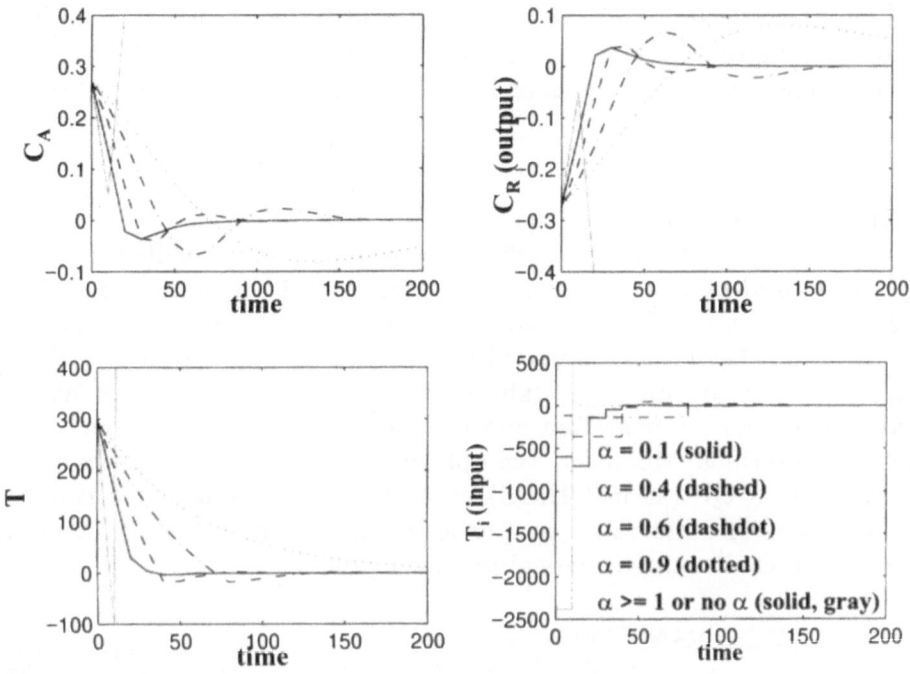

Figure 4. State and control responses for CNTMPC with different values of α (CNTMPC with $\alpha \to \infty$ corresponds to STNMPC).

(this means that the deviation state variables of the linear model used in the computation of the contractive constraint cannot be brought exactly to zero in only P time steps). For $\alpha \in (0, 1)$ the control problem is feasible at all time steps and the closed-loop system is stable. As shown in *Figure 4*, the performance is improved for smaller values of α. For this example, if $\alpha = 0.9$ the control effort is almost zero which means that the system is basically left open-loop for values of α in the range $\alpha \in [0.9, 1)$.

If $\alpha \geq 1$, the closed-loop system becomes unstable and the simulation stops after only two time sampling times because the control move computed is very large, pushing the output C_R to values below 0 (not physically meaningful).

The results obtained with $\alpha \geq 1$ are equivalent to the results obtained with STNMPC. Thus, we have demonstrated that for the given choice of control parameters ($P = M$, output weight only and small sampling time T) STNMPC goes unstable while the implementation of the contractive constraint introduces stability guarantees whenever the optimization is feasible. For this example and control parameter choices, the mismatch between the nonlinear system and the local linearizations used in the computation of the contractive constraint is small and CNTMPC can indeed drive the deviation states of the nonlinear system to the origin, without offset.

7. CONCLUSIONS

In this paper we addressed the problems of stability and implementability of MPC by introducing a stabilizing state constraint called contractive constraint and by using a linear approximation of the nonlinear system in the prediction step of the so-called contractive MPC algorithm. Extension to the output feedback case was also presented. We have seen that the contractive constraint preserves stability even in the presence of the mismatch between the nonlinear system and the local linearization used in the control computations. Moreover, the optimization problem is convex and can be posed as a QP (as in QDMC) even though the contractive constraint is quadratic. The stabilizing properties of the proposed MPC scheme were tested on a challenging chemical process example.

ACKNOWLEDGMENTS

Financial support from the Coordenadoria de Aperfeiçoamento de Pessoal de Nível Superior (CAPES), the US Department of Energy, the National Science Foundation and the Swiss Federal Institute of Technology (ETH) is gratefully acknowledged.

APPENDIX - PROOFS OF THEOREMS 1, 2, 3, 4

The following lemma will be very useful in proving **Theorems 2 and 4.**

Lemma 1 *Consider the discrete linear system:*

$$z_{k+1} \leq a_k z_k + b_k, \quad k \in \mathcal{N} \tag{31}$$

where $\mathcal{N}$ is the set of non-negative integers.

If $a_k \in [0,1)$ and $b_k \geq 0$, $\forall k \in \mathcal{N}$, then system (31) is stable in the practical sense, i.e.,

1. $z_k < z_0 + \frac{b_{max}}{1-a_{max}}, \quad \forall k \in \mathcal{N}$
2. $\lim_{k \to \infty} z_k \leq \frac{b_{max}}{1-a_{max}}$

where $a_{max} := \max_{k \in \mathcal{N}} a_k$ and $b_{max} := \max_{k \in \mathcal{N}} b_k$.

Proof: *The proof is straightforward. It can be found in [21].* □

Proof of Theorem 1

The optimal control problem $\mathcal{P}(t_k, x_k^p)$ has a solution for all $k \geq 0$ since we assume that $x_k^p, \bar{x}_k \in B_\rho$.

Then, since the states $\bar{x}(t)$ are set to x_k^p at $t = t_k$, we have that for all $t \in [t_k, t_{k+1}]$, the following inequality holds:

$$\| x_k^p(t) - \bar{x}_k(t) \|_{\hat{P}} \leq \int_{t_k}^t \| f(x_k^p(\tau), u_k(\tau), p_k(\tau), d_k(\tau)) - C_k - A_k \bar{x}_k(\tau) -$$
$$-B_k u_k(\tau) \|_{\hat{P}} \, d\tau \leq \int_{t_k}^t \| f(x_k^p(\tau), u_k(\tau), p_k(\tau), d_k(\tau)) - C_k - A_k \bar{x}_k(\tau) -$$
$$-B_k u_k(\tau) - F(\bar{x}_k(\tau), u_k(\tau), 0, 0) \|_{\hat{P}} \, d\tau +$$
$$+ \int_{t_k}^t \| F(\bar{x}_k(\tau), u_k(\tau), 0, 0) \|_{\hat{P}} \, d\tau \leq L \epsilon_d PT +$$
$$+ L \int_{t_k}^t \| x_k^p(\tau) - \bar{x}_k(\tau) \|_{\hat{P}} \, d\tau + \gamma \tilde{u} PT + \gamma \int_{t_k}^t \| \bar{x}_k(\tau) \|_{\hat{P}} \, d\tau$$

Besides, for all $t \in [t_k, t_{k+1}]$, we have:

$$\| \bar{x}_k(t) \|_{\hat{P}} \leq \| x_k^p \|_{\hat{P}} + \int_{t_k}^t \| C_k + A_k x_k(\tau) + B_k u_k(\tau) \|_{\hat{P}} \, d\tau \leq \| x_k^p \|_{\hat{P}} +$$
$$+ L \tilde{u} PT + L \int_{t_k}^t \| \bar{x}_k(\tau) \|_{\hat{P}} \, d\tau \tag{32}$$

Now, using the Bellman-Grownwall (BG) inequality, we get:

$$\| \bar{x}_k(t) \|_{\hat{P}} \leq [\| x_k^p \|_{\hat{P}} + L\tilde{u} PT] \, e^{L(t-t_k)} \tag{33}$$

Integration of both sides of inequality (33) leads to:

$$\int_{t_k}^t \| \bar{x}_k(\tau) \|_{\hat{P}} \, d\tau \leq \frac{[\| x_k^p \|_{\hat{P}} + L\tilde{u} PT]}{L} (e^{LPT} - 1) \tag{34}$$

By substituting (34) into (32), it results that:

$$\| x_k^p(t) - \bar{x}_k(t) \|_{\hat{P}} \leq L\epsilon_d PT + \gamma PT\tilde{u}e^{LPT} + \frac{\gamma}{L}(e^{LPT} - 1) \| x_k^p \|_{\hat{P}} +$$

$$+ L\int_{t_k}^t \| x_k^p(\tau) - \bar{x}_k(\tau) \|_{\hat{P}} \, d\tau \qquad (35)$$

Finally, using the BG inequality once more and setting $t = t_{k+1}$, we get:

$$\| x_{k+1}^p - \bar{x}_{k+1} \|_{\hat{P}} \leq \frac{\gamma}{L}e^{LPT}(e^{LPT} - 1) \| x_k^p \|_{\hat{P}} + L\epsilon_d PTe^{LPT} + \gamma PT\tilde{u}e^{2LPT} \qquad (36)$$

Thus, λ_1, λ_2 in the statement of the theorem are given by:

$$\lambda_1 := \frac{\gamma}{L} e^{LPT} (e^{LPT} - 1) \qquad \text{and} \qquad \lambda_2 := [\, \gamma\tilde{u}e^{LPT} + L\epsilon_d \,] PTe^{LPT} \qquad (37)$$

From these definitions we clearly see that $\lambda_1 \to 0$ as $\gamma \to 0$ or $PT \to 0$ and $\lambda_2 \to 0$ as γ, $\epsilon_d \to 0$ or $PT \to 0$. Moreover, both λ_1 and λ_2 increase as L and γ increase, which is natural since these constants "quantify" the strength of the nonlinearities in the system. $\qquad\qquad\square$

Proof of Theorem 2
The proof is constructive, i.e., we calculate $\bar{\alpha}$ so that the statement of the theorem holds.

From the triangle inequality it follows that:

$$\| x_{k+1}^p \|_{\hat{P}} \leq \| x_{k+1}^p - \bar{x}_{k+1} \|_{\hat{P}} + \| \bar{x}_{k+1} \|_{\hat{P}} \qquad (38)$$

Now, using the contractive constraint, i.e., $\| \bar{x}_{k+1} \|_{\hat{P}} \leq \alpha \| x_k^p \|_{\hat{P}}$, we get:

$$\| x_{k+1}^p \|_{\hat{P}} \leq \| x_{k+1}^p - \bar{x}_{k+1} \|_{\hat{P}} + \alpha \| x_k^p \|_{\hat{P}} \qquad (39)$$

Then by substituting (36) into (39), we obtain:

$$\| x_{k+1}^p \|_{\hat{P}} \leq (\lambda_1 + \alpha) \| x_k^p \|_{\hat{P}} + \lambda_2 =: \alpha^* \| x_k^p \|_{\hat{P}} + \lambda_2 \qquad (40)$$

Using the Contraction Mapping Principle we conclude that stability holds if $\alpha^* = \alpha + \lambda_1 < 1$, which implies that:

$$\alpha < 1 - \frac{\gamma e^{LPT} (e^{LPT} - 1)}{L} \; =: \; \bar{\alpha}^{(1)} \qquad (41)$$

Now, applying the results of **Lemma 1**, we get:

1.

$$\| x_k^p \|_{\hat{P}} < \| x_0^p \|_{\hat{P}} + \frac{\lambda_2}{1 - \alpha^*} \leq \rho_0 + \frac{\lambda_2}{1 - \alpha^*} \qquad (42)$$

for all initial conditions $x_0^p \in B_{\rho_0} \subset B_\rho$.

2.

$$\lim_{k \to \infty} \| x_k^p \|_{\hat{P}} \leq \frac{\lambda_2}{1 - \alpha^*} =: \hat{\rho} \qquad (43)$$

Thus, $\hat{\rho}$ is given by equation (16).

Using equations (16) and (43), we conclude that:

$$\lim_{\gamma, \epsilon_d \to 0} \hat{\rho} = \lim_{\gamma, \epsilon_d \to 0} [\lim_{k \to \infty} \| x_k^p \|_{\hat{P}}] = 0 \text{ or } \lim_{PT \to 0} \hat{\rho} = \lim_{PT \to 0} [\lim_{k \to \infty} \| x_k^p \|_{\hat{P}}] = 0$$

$$(44)$$

Our next step is to establish conditions which guarantee that x_k^p, $\bar{x}_k \in B_\rho$, $\forall k \geq 0$. Using inequality (42), a sufficient condition on the control and plant parameters so that x_k^p remains inside B_ρ for all $k \geq 0$ is given by:

$$0 < \rho_0 + \hat{\rho} < \rho \qquad (45)$$

Since $\| \bar{x}_{k+1} \|_{\hat{P}} \leq \alpha \| x_k^p \|_{\hat{P}}$, if inequality (45) is satisfied then the states $\bar{x}_k$ also remain inside B_ρ for all $k \geq 0$ (since $\bar{x}_0$ is set to $x_0^p \in B_{\rho_0}$).

Given the chosen controller parameters α, $\tilde{u}$, P and T, a necessary condition on the nonlinearities, disturbances and/or parameter mismatch so that $B_{\rho_0} \subset B_\rho$ (with ρ_0 satisfying inequality (45)) is a non-empty set is obviously that:

$$\frac{LPTe^{LPT}[\gamma \tilde{u}e^{LPT} + L\epsilon_d]}{L(1 - \alpha) + \gamma e^{LPT}(e^{LPT} - 1)} < \rho \qquad (46)$$

Or, expressed in terms of the contractive parameter α, we have:

$$\alpha < 1 - \frac{\gamma}{L}e^{LPT}(e^{LPT} - 1) - \frac{[\gamma \tilde{u}e^{LPT} + L\epsilon_d]PTe^{LPT}}{\rho} =: \bar{\alpha}^{(2)} \quad (47)$$

Since $\bar{\alpha}^{(2)} < \bar{\alpha}^{(1)}$ for all $\rho \in [0, \infty)$ and $\bar{\alpha}^{(2)} \to \bar{\alpha}^{(1)}$ if $\rho \to \infty$ (i.e., if the optimization problems $\mathcal{P}(t_k, x_k^p)$ are feasible for all initial conditions $x_k^p \in \Re^n$, then the bound on the contractive parameter α is dictated only by the stability requirement as in (41)) then we have that $\bar{\alpha}$ is defined by:

$$\alpha < \bar{\alpha} := 1 - \frac{\gamma}{L}e^{LPT}(e^{LPT} - 1) - \frac{[\gamma \tilde{u}e^{LPT} + L\epsilon_d]PTe^{LPT}}{\rho} \qquad (48)$$

Thus, if condition (15) is satisfied then there exists $\bar{\alpha} \in [0, 1)$ such that for $\alpha < \bar{\alpha}$ both feasibility and asymptotic stability to the control invariant set $B_{\hat{\rho}}$ are guaranteed. $\square$

Proof of Theorem 3

Discretization of the linear model used in the computation of the contractive constraint leads to:

$$\bar{x}(kP + j + i + 1 | kP + j) = \Phi_k^{i+1} \bar{x}(kP + j | kP + j) +$$
$$+ \sum_{l=0}^{i} \Phi_k^l \left[\Gamma_k u(kP + j + l | kP + j) + \eta_k \right] \tag{49}$$

for all i, $j = 0, \ldots, P - 1$. $\bar{x}(kP + j | kP + j) = \bar{x}(kP + j | kP + j - 1)$ for $j = 1, \ldots, P-1$ and $\bar{x}(kP|kP) = x_k^p$ with the matrices Φ_k, Γ_k and η_k given by:

$$\Phi_k := e^{A_k^0 T}, \quad \Gamma_k := \int_0^T e^{A_k^0 (T-t)} B_k^0 dt, \quad \eta_k := \int_0^T e^{A_k^0 (T-t)} C_k^0 dt \tag{50}$$

Because the local linearization is assumed stable, the worst case scenario in terms of trying to satisfy the contractive constraint is if the applied control action is such that $B_k u_k^j(t) = -C_k$ for all $t \in [t_k^j, t_k^{j+P}]$ and $j = 0, \ldots, P - 1$ (i.e., there are no driving terms in the system). Obviously, we are not considering the case when one may be trying to drive the states of the system away from the origin.

In this case, we have:

$$\bar{x}(kP + j + i + 1 | kP + j) = \Phi_k^{i+1} \bar{x}(kP + j | kP + j) \tag{51}$$

Since the trajectories $\bar{x}_k^j(t)$ do not differ for different values of j, if $B_k u_k^j(t)$ lies in the range of C_k, then we can drop the superscript j and by applying the $\hat{P}$−norm we have:

$$\| \bar{x}(kP+i+1|kP) \|_{\hat{P}} \leq \sqrt{\lambda_{max}(\hat{P}^{\frac{1}{2}} \Phi_k^{i+1} \hat{P}^{-\frac{1}{2}})} \| x_k^p \|_{\hat{P}}, \quad i = 0, \ldots, P-1 \tag{52}$$

Thus, if $i = P - 1$, we have:

$$\| \bar{x}((k + 1)P|kP) \|_{\hat{P}} \leq \sqrt{\lambda_{max}(\hat{P}^{\frac{1}{2}} \Phi_k^P \hat{P}^{-\frac{1}{2}})} \| x_k^p \|_{\hat{P}} =$$
$$= \sqrt{\lambda_{max}(\hat{P}^{\frac{1}{2}} e^{A_k^0 PT} \hat{P}^{-\frac{1}{2}})} \| x_k^p \|_{\hat{P}} =: \underline{\alpha}_k \| x_k^p \|_{\hat{P}} \tag{53}$$

Since our contractive constraint is given by:

$$\| \bar{x}((k + 1)P|kP) \|_{\hat{P}} \leq \alpha \| x_k^p \|_{\hat{P}} \tag{54}$$

Then, if

$$\alpha \geq \underline{\alpha} := \sqrt{\lambda_{max}(\hat{P}^{\frac{1}{2}} e^{APT} \hat{P}^{-\frac{1}{2}})} \tag{55}$$

the optimization is feasible at time step k, $\forall k \geq 0$. Since we have assumed that A_k^0 is stable for all $k \geq 0$, there always exists a finite prediction horizon P long enough such that $\underline{\alpha} \in [0,1)$.

The upper bound on α for stability is derived in the same way as in the proof of **Theorem 2** with $\rho \to \infty$. Therefore, $\bar{\alpha}$ is given as in equation (41).

Thus, if condition (17) holds then there exists $\bar{\alpha}$ which satisfies $1 > \bar{\alpha} > \underline{\alpha} \geq 0$ such that for $\underline{\alpha} \leq \alpha < \bar{\alpha}$, feasibility and asymptotic stability to the control invariant set $B_{\hat{\rho}}$ (with $\hat{\rho}$ given by equation (16)) are guaranteed.

$\square$

Proof of Theorem 4

Following the same procedure used to prove **Theorem 1**, if the model used in the computation of the contractive constraint is now given by (21) due to the state estimation error, we obtain:

$$\| x_{k+1}^p - \bar{x}_{k+1} \|_{\hat{P}} \leq \tfrac{\gamma}{L} e^{LPT}(e^{LPT} - 1) \| x_k^p \|_{\hat{P}} + [\gamma \tilde{u} e^{LPT} + \rho_k^d] PT e^{LPT} =:$$

$$=: \lambda_1 \| x_k^p \|_{\hat{P}} + \lambda_{2,k}$$

where (56) follows directly from (36) by making $\epsilon_d = 0$ and by adding the term resulting from integration of the additive discrete disturbance sequence $\tilde{d}_k, \ldots \tilde{d}_{k+P-1}$ (which satisfies equation (24)).

Using the contractive constraint and the triangle inequality in equation (56), we have:

$$\| x_{k+1}^p \|_{\hat{P}} \leq (\alpha + \lambda_1) \| x_k^p \|_{\hat{P}} + \lambda_{2,k} \tag{56}$$

Since the state estimation error is such that for any $\tilde{\epsilon} > 0$, $\exists \bar{k} := \bar{k}(\tilde{\epsilon}) \in \mathcal{N}$ large enough so that $\rho_k^d \leq \tilde{\epsilon}$ for $k \in [\bar{k}, \infty)$, then from (56) it follows that:

$$\| x_{k+1}^p \|_{\hat{P}} \leq (\alpha + \lambda_1) \| x_k^p \|_{\hat{P}} + [\gamma \tilde{u} \, e^{LPT} + \tilde{\epsilon}] PT \, e^{LPT}, \quad \forall k \in [\bar{k}, \infty) \tag{57}$$

Then, if $\alpha + \lambda_1 =: \alpha^* \in [0,1)$, we can use the results of **Lemma 1** to obtain:

$$\| x_{\bar{k}(\tilde{\epsilon})+l}^p \|_{\hat{P}} \leq (\alpha^*)^l \| x_{\bar{k}(\tilde{\epsilon})}^p \|_{\hat{P}} + [\sum_{i=0}^{l-1}(\alpha^*)^i][\gamma \tilde{u} e^{LPT} + \tilde{\epsilon}] PT e^{l,PT} <$$

$$< (\alpha^*)^l \| x_{\bar{k}(\tilde{\epsilon})}^p \|_{\hat{P}} + \frac{[\gamma \tilde{u} e^{LPT} + \tilde{\epsilon}] PT e^{LPT}}{1 - \alpha^*}, \quad \forall l > 0 \tag{58}$$

Thus, by taking the limit as $\tilde{\epsilon} \to 0$, we have:

$$\lim_{\tilde{\epsilon} \to 0} \| x_{\bar{k}(\tilde{\epsilon})+l}^p \|_{\hat{P}} < (\alpha^*)^l [\lim_{\tilde{\epsilon} \to 0} \| x_{\bar{k}(\tilde{\epsilon})}^p \|_{\hat{P}}] + \frac{\gamma \tilde{u} PT \, e^{2LPT}}{1 - \alpha^*} =$$

$$= (\alpha^*)^l [\lim_{\tilde{\epsilon} \to 0} \| x_{\bar{k}(\tilde{\epsilon})}^p \|_{\hat{P}}] + \frac{\gamma \tilde{u} LPT e^{2LPT}}{L(1-\alpha) + \gamma e^{LPT}(e^{LPT}-1)} =:$$

$$=: (\alpha^*)^l [\lim_{\tilde{\epsilon} \to 0} \| x_{\bar{k}(\tilde{\epsilon})}^p \|_{\hat{P}}] + \hat{\rho} \tag{59}$$

and if now we take the limit as $l \to \infty$ knowing that $\bar{k}(\tilde{\epsilon}) \to \infty$ for $\tilde{\epsilon} \to 0$ and that $\alpha^l \to 0$ exponentially fast as $l \to \infty$, we finally obtain:

$$\lim_{l \to \infty} [\lim_{\tilde{\epsilon} \to 0} \| x^p_{\bar{k}(\tilde{\epsilon})+l} \|_{\hat{P}}] < (\lim_{l \to \infty} \alpha^l)[\lim_{\tilde{\epsilon} \to 0} \| x^p_{\bar{k}(\tilde{\epsilon})} \|_{\hat{P}}] + \hat{\rho} = \hat{\rho}$$

$$\implies \quad \lim_{k \to \infty} \| x^p_k \|_{\hat{P}} < \hat{\rho}$$

which means asymptotic convergence of the plant states to the control invariant set $B_{\hat{\rho}}$.

Now, it remains to be shown that x^p_k, $\bar{x}_k$, $\hat{x}_k \in B_\rho$, $\forall k \geq 0$. From equation (56) and the definition of ρ^d, we have:

$$\| x^p_{k+1} - \bar{x}_{k+1} \|_{\hat{P}} \leq \tfrac{\gamma}{L} e^{LPT}(e^{LPT} - 1) \| x^p_k \|_{\hat{P}} + [\gamma \tilde{u} e^{LPT} + \rho^d] PT e^{LPT} =:$$
$$=: \lambda_1 \| x^p_k \|_{\hat{P}} + \lambda_2$$

Thus, using the contractive constraint and the results of **Lemma 1**, we obtain:

$$\| x^p_k \|_{\hat{P}} < \| x^p_0 \|_{\hat{P}} + \tfrac{\lambda_2}{1-\alpha^*} \leq \rho_0 + \tfrac{\lambda_2}{1-\alpha^*} = \rho_0 + \tfrac{[\gamma \tilde{u} e^{LPT} + \rho^d]LPT\,e^{LPT}}{L(1-\alpha)-\gamma e^{LPT}(e^{LPT}-1)},$$
$$\forall k > 0 \text{ and } x^p_0 \in B_{\rho_0} \tag{60}$$

We have then found an upper bound on the states of the plant at the end of horizons. Since the estimated states are given by $\hat{x}_k = x^p_k + e_k$ and the estimation error satisfies equation (25), we have the following bound on the estimated states:

$$\| \hat{x}_k \|_{\hat{P}} < \rho_0 + \frac{[\gamma \tilde{u} e^{LPT} + \rho^d]LPTe^{LPT}}{L(1-\alpha) - \gamma e^{LPT}(e^{LPT} - 1)} + K \| e_0 \|_{\hat{P}}, \forall k > 0, x^p_0 \in B_{\rho_0}$$
$$\tag{61}$$

Thus, if $\rho_0 := \rho - K \| e_0 \|_{\hat{P}} - \frac{[\gamma \tilde{u} e^{LPT} + \rho^d]LPT\,e^{LPT}}{L(1-\alpha)-\gamma e^{LPT}(e^{LPT}-1)}$ and the disturbance $\{\tilde{d}_k\}_{k \geq 0}$ satisfies (27), it follows that $\rho_0 > 0$, $\{x^p_k\}^\infty_{k=0}$, $\{\hat{x}_k\}^\infty_{k=0} \in B_\rho$ and, due to the contractive constraint, we also have $\{\bar{x}_k\}^\infty_{k=0} \in B_\rho$, which means that the control problem $\mathcal{P}(t_k, \hat{x}_k)$ is feasible and well-defined for all $k \geq 0$.

$\square$

References

1. M. Alamir and G. Bornard. New sufficient conditions for global stability of receding horizon control for discrete-time nonlinear systems. In David Clarke, editor, *Advances in Model-Based Predictive Control*, pages 173–181. Oxford University Press, 1994.

2. M. Alamir and G. Bornard. On the stability of receding horizon control of nonlinear discrete-time systems. *Syst. Control Lett.*, 23:291–296, 1994.

3. M. Alamir and G. Bornard. Stability of a truncated infinite constrained receding horizon scheme: The general discrete nonlinear case. *Automatica*, 31(9):1353–1356, September 1995.

4. A. Astolfi. *Asymptotic Stabilization of Nonholonomic Systems with Discontinuous Control.* PhD thesis, Swiss Federal Institute of Technology (ETH), 1995. Diss. ETH No. 10983.

5. A. Astolfi. Discontinuous control of nonholonomic systems. *Syst. Control Lett.*, 27:37–45, 1995.

6. V. Balakrishnan, Z. Zheng, and M. Morari. Constrained stabilization of discrete-time systems. In David Clarke, editor, *Advances in Model-Based Predictive Control*, pages 205–216. Oxford University Press, 1994.

7. B. R. Barmish, M. Corless, and G. Leitmann. A new class of stabilization controller for uncertain dynamical systems. *SIAM Journal of Control and Optimization*, 21:246–255, 1983.

8. B. R. Barmish and G. Leitmann. On ultimate boundedness control of uncertain systems in the absence of matching condition. *IEEE Trans. Aut. Control*, 27:153–158, 1982.

9. S. Behtash. Robust output tracking for nonlinear systems. *Int. J. Control*, 51:1381–1407, 1990.

10. B. W. Bequette. Nonlinear control of chemical processes: A review. *Ind. Eng. Chem. Res.*, 30:1391–1413, 1991.

11. L. Biegler and J. Rawlings. Optimization approaches to nonlinear model predictive control. In *Conf. Chemical Process Control (CPC-IV)*, pages 543–571, South Padre Island, Texas, 1991. CACHE-AIChE.

12. L. T. Biegler. Sensitivity analysis of QPs. Personal communication, 1994.

13. M. Bouslimani, M. M'Saad, and L. Dugard. Stabilizing receding horizon control: A unified continuous/discrete time formulation. In *Conf. on Decision and Control*, page 1298, 1993.

14. R.W. Brockett. Asymptotic stability and feedback stabilization. In R.W. Brockett, R.S. Millman, and H.J. Sussman, editors, *Differential Geometric Control Theory*, pages 181–193. Birkhäuser, 1983.

15. P. J. Campo and M. Morari. Robust model predictive control. In *American Control Conf.*, pages 1021–1026, 1987.

16. C. Chen and L. Shaw. On receding horizon feedback control. *Automatica*, pages 349–352, 1982.

17. Y. H. Chen and G. Leitmann. Robustness of uncertain systems in the absence of matching assumptions. *Int. J. Control*, 45:1527–1544, 1987.

18. D. W. Clarke. Advances in model-based predictive control. In D. W. Clarke, editor, *Advances in Model-Based Predictive Control*, pages 3–21. Oxford University Press, 1994.

19. M. Corless and G. Leitmann. Continuous state feedback guaranteeing uniform ultimate boundedness for uncertain dynamical systems. *IEEE Trans. Aut. Control*, 32:763–771, 1987.

20. C. R. Cutler and B. L. Ramaker. Dynamic matrix control– A computer control algorithm. In *Joint Automatic Control Conf.*, San Francisco, California, 1980.

21. S. L. De Oliveira. *Model Predictive Control (MPC) for Constrained Nonlinear Systems.* PhD thesis, California Institute of Technology, Pasadena, CA, March 1996.

430

22. S. L. De Oliveira and M. Morari. Contractive model predictive control for constrained nonlinear systems. *IEEE Trans. Aut. Control*, 1996. accepted.

23. F. Doyle III, F. Allgöwer, S. L. De Oliveira, E. Gilles, and M. Morari. On nonlinear systems with poorly behaved zero dynamics. In *American Control Conf.*, pages 2571–2575, 1992.

24. W. Eaton and J. Rawlings. Feedback control of nonlinear processes using on-line optimization techniques. *Comp. and Chem. Eng.*, 14:469–479, 1990.

25. C. Economou, M. Morari, and O. Palsson. Internal model control. 5. Extension to nonlinear systems. *Ind. Engng. Chem. Process Des. Dev.*, 25(1):403–411, 1986.

26. L. C. Fu and T. L. Liao. Globally stable tracking of nonlinear systems using variable structure control and with an application to a robotic manipulator. *IEEE Trans. Aut. Control*, 35:1345–1351, 1990.

27. C. E. García. Quadratic dynamic matrix control of nonlinear processes. An application to a batch reactor process. In *AIChE Annual Meeting*, San Francisco, 1984.

28. C. E. García and A. M. Morshedi. Quadratic programming solution of dynamic matrix control (QDMC). *Chem. Eng. Communications*, 46:73–87, 1986.

29. H. Genceli and M. Nikolaou. Robust stability analysis of constrained L_1−norm model predictive control. *AIChE Journal*, 39(12):1954–1965, 1993.

30. S. Keerthi and E. Gilbert. Optimal infinite-horizon feedback laws for a general class of constrained discrete-time systems: Stability and moving-horizon approximations. *Journal of Optimization Theory and Applications*, pages 265–293, 1988.

31. L. Kershenbaum, D. Q. Mayne, R. Pytlak, and R. B. Vinter. Receding horizon control. In D. W. Clarke, editor, *Advances in Model-Based Predictive Control*, pages 233–246. Oxford University Press, 1994.

32. M. V. Kothare, V. Balakrishnan, and M. Morari. Robust constrained model predictive control using linear matrix inequalities. *Automatica*, 32(10):1361–1379, October 1996.

33. W. H. Kwon. Advances in predictive control: Theory and application. In *1st Asian Control Conf.*, Tokyo, July 1994. (updated in October, 1995).

34. W.H. Kwon and A.E. Pearson. A modified quadratic cost problem and feedback stabilization of a linear system. *IEEE Trans. Aut. Control*, 22(5):838–842, 1977.

35. J. H. Lee and N. L. Ricker. Extended Kalman Filter Based Nonlinear Model Predictive Control. *Ind. Eng. Chem. Res.*, 33:1530–1541, 1994.

36. T. L. Liao, L. C. Fu, and C. F. Hsu. Adaptive robust tracking of nonlinear systems and with an application to a robotic manipulator. *Syst. Control Lett.*, 15:339–348, 1990.

37. D. Mayne and H. Michalska. Receding horizon control of nonlinear systems. *IEEE Trans. Aut. Control*, 35:814–824, 1990.

38. D. Mayne and H. Michalska. Approximate global linearization of nonlinear systems via on-line optimization. In *European Control Conf.*, pages 182–187, Grenoble,France, 1991.

39. D. Mayne and H. Michalska. Model predictive control of nonlinear systems. In *American Control Conf.*, pages 2343–2348, Boston, MA, 1991.

40. D. Mayne and H. Michalska. Receding horizon control of nonlinear systems without differentiability of the optimal value function. *Systems and Control Letters*, 16:123–130, 1991.

41. D.Q. Mayne and H. Michalska. An implementable receding horizon controller for the stabilization of nonlinear systems. In *Conf. on Decision and Control*, pages 3396–3397, Honolulu, HI, 1990.

42. E.S. Meadows, M.A. Henson, J.W. Eaton, and J.R. Rawlings. Receding horizon control and discontinuous state feedback stabilization. *Int. J. Control*, 62(5):1217–1229, 1995.

43. E.S. Meadows and J.B. Rawlings. Receding horizon control with an infinite horizon. In *American Control Conf.*, pages 2926–2930, San Francisco, California, 1993.

44. H. Michalska and D. Q. Mayne. Moving horizon observers. In M. Fliess, editor, *IFAC Nonlinear Control Systems Design Symposium (NOLCOS '92)*, pages 576–581, Bordeaux, France, 1992.

45. H. Michalska and D.Q. Mayne. Robust receding horizon control of constrained nonlinear systems. *IEEE Trans. Aut. Control*, 38(11):1623–1633, November 1993.

46. M. Morari. Robust stability of systems with integral control. In *Conf. on Decision and Control*, pages 865–869, San Antonio,TX, 1983.

47. G. De Nicolao and R. Scattolini. Stability and output terminal constraints in predictive control. In *Advances in Model-Based Predictive Control*, pages 105–121. Oxford University Press, 1994.

48. J. B. Rawlings and K. R. Muske. The stability of constrained receding horizon control. *IEEE Trans. Aut. Control*, 38(10):1512–1516, October 1993.

49. N. L. Ricker and J. H. Lee. Nonlinear model predictive control of the Tennesee Eastman challenge process. *Comp. and Chem. Eng.*, 19(9):961–981, 1995.

50. Michel Gevers Robert R. Bitmead and Vincent Wertz. *Adaptive Optimal Control. The Thinking Man's GPC*. International Series in Systems and Control Engineering. Prentice Hall, 1990.

51. S. N. Singh and A. R. Coelen. Nonlinear control of mismatched uncertain linear systems and application to control of aircrafts. *ASME J. Dynamic System and Control*, 106:203–210, 1984.

52. L. C. Fu T. L. Liao and C-F. Hsu. Output tracking control of nonlinear systems with mismatched uncertainties. *Syst. Control Lett.*, 18:39–47, 1992.

53. T. H. Yang and E. Polak. Moving horizon control of linear systems with input saturation and plant uncertainty. *Int. J. Control*, 58(3):613–638, September 1993.

54. T. H. Yang and E. Polak. Moving horizon control of linear systems with input saturation, disturbances and plant uncertainty. *Int. J. Control*, 58(3):639–663, September 1993.

55. T. H. Yang and E. Polak. Moving horizon control of nonlinear systems with input saturation, disturbances and plant uncertainty. *Int. J. Control*, 58(4):875–903, September 1993.

56. S. H. Zak. On the stabilization and observation of nonlinear uncertain dynamic systems. *IEEE Trans. Aut. Control*, 35:604–607, 1990.

57. Z. Q. Zheng. *Robust Control of Systems Subject to Constraints*. PhD thesis, California Institute of Technology, Pasadena, California, 1995.

FEEDBACK LINEARIZATION + CONTRACTIVE MPC: STABILITY ANALYSIS / APPLICATION TO A POLYMERIZATION PROCESS

SIMONE LOUREIRO DE OLIVEIRA
PEQ/COPPE, Federal University of Rio de Janeiro
Ilha do Fundão, Cx. Postal 68502
Rio de Janeiro, RJ, 21945-970, Brazil

DOUGLAS MARCELO MERQUIOR
Instituto Militar de Engenharia
Praça General Tibúrcio, 80, Urca
Rio de Janeiro, RJ, 22290-270, Brazil

AND

ENRIQUE LUIS LIMA
PEQ/COPPE, Federal University of Rio de Janeiro
Ilha do Fundão, Cx. Postal 68502
Rio de Janeiro, RJ, 21945-970, Brazil

Abstract.

In this work a control technique for nonlinear systems which deals with process constraints in a computationally efficient way is proposed. The technique is a fusion of two control methods: input/output feedback linearization (FL) and contractive model predictive control (CNTMPC). In this configuration, CNTMPC guarantees closed-loop stability and robustness but it is a hard problem to solve on-line because it is usually subject to nonlinear and state dependent constraints. In order to deal with such constraints, different approximate methods are proposed. As a result of some of these approximations, the CNTMPC problem can be reduced to a quadratic programming problem (QP). Although approximate, these methods are implemented in a way through which satisfaction of the original constraints is guaranteed, while generating an optimization problem with the implementation and computational levels of difficulty nearly equivalent to those of the case where the plant and the constraints are linear. The controller is successfully applied to a batch solution polymerization reactor in both the nominal and robust cases.

R. Berber and C. Kravaris (eds.), Nonlinear Model Based Process Control, 433-464.
© *1998 Kluwer Academic Publishers.*

1. Introduction

Still today the vast majority of chemical industries employ linear controllers in the control of their plants, in spite of the fact that chemical processes are inherently (and many times, strongly) nonlinear. However, these industries have experienced that the performance attained with linear controllers can be unacceptably poor in the presence of severe process nonlinearities. Moreover, it is well known that, in real situations, controllers should deal adequately with constraints on input and state variables in order to guarantee safe operation, product quality, satisfaction of environmental regulations and efficient energy utilization. For both linear and nonlinear systems, constraint handling requires the controller to be nonlinear [13]. Thus, the improvement on nonlinear control design methods is of extreme practical (and not only academic) importance nowadays.

Model predictive control (MPC) is a control technique which does not require solution of a computationally untractable Halmiton-Jacobi-Bellman partial differential equation to obtain an optimal feedback control law [13]. It is also well-known that one of the most special features of MPC is its capability of coping with various process constraints which can be incorporated directly into the on-line optimization problem solved at each time step [3]. Other special characteristics of MPC are the simple extension from the single-input single-output (SISO) to the multi-input multi-output (MIM0) case and its variable structure in the event of faults. These are some of the reasons why MPC has been widely used in the industry and largely studied and improved by researchers in the last decades.

MPC is an optimal control based technique that computes the control law on-line. With regards to the formulation of the optimal control problem to be solved at each time step, MPC algorithms can be divided into the following main categories (see [4] for a list of representative references on each category):

(a) Finite prediction horizon for linear/nonlinear plants
(b) Infinite prediction horizon for linear/nonlinear plants
(c) Finite prediction horizon with end constraints for linear/nonlinear plants

In the first category a simple finite horizon objective function is employed, but stability is not guaranteed. In the second one, Lyapunov arguments are used to show asymptotic stability of the infinite prediction horizon scheme. Although this is a powerful result, a serious restriction on using an infinite prediction horizon MPC scheme in the control of nonlinear systems is the prohibitive computational load. In the third one, asymptotic (or even exponential) stability is added to finite prediction horizon schemes via the introduction of end constraints.

A comprehensive discussion on the stability analysis of different MPC algorithms can be found in [3].

In the present work, an approximate control technique for constrained nonlinear processes implemented in a computationally efficient fashion is proposed. The technique is a fusion of two control methods: input/output feedback linearization (FL) [8] and contractive model predictive control (CNTMPC) [4, 5]. In [6, 16], the authors employ FL in combination with a standard nonlinear MPC (STNMPC) algorithm for which there are no stability guarantees. Here we combine FL with a contractive MPC scheme in order to add some degree of robust stability to the closed-loop system (see [3] for a complete analysis of the stability properties of CNTMPC algorithms). While FL in the internal loop linearizes the input/output connection and allows the reformulation of the prediction problem as a linear one (if only inputs and outputs are present in the objective function), the constrained contractive MPC in the external loop enforces the input/output and contractive constraints in an implicit and approximate way. In this configuration, contractive MPC is a hard problem to solve on-line because it is usually subject to nonlinear and state dependent constraints. In order to deal with such constraints, different approximate methods are proposed here. As a result of these approximations, the CNTMPC problem can sometimes be reduced to a quadratic programming problem (QP). Although approximate, these methods are implemented in a way through which satisfaction of the original constraints is guaranteed, while generating an optimization problem with the implementation and computational levels of difficulty nearly equivalent to those of the case where the plant and the constraints are linear.

Contractive MPC is a finite horizon nonlinear MPC algorithm which exponentially stabilizes the closed-loop system via the addition of an end constraint denoted contractive constraint [3, 4, 5]. There are other types of stabilizing end constraints and a rather comprehensive review of these methods can be found in [12]. The end constraint most commonly used is $x(k + P|k) = 0$, i.e., the predicted states of the model are set exactly to zero at the end of the prediction horizon P. In the stability analysis of the nonlinear MPC algorithm with this equality constraint, nominal asymptotic stability has been shown [9]. However, the end constraint $x(k + P|k) = 0$ only guarantees stability of a finite horizon nonlinear MPC algorithm, if this constraint is satisfied exactly. It is well known that in real time implementation of a nonlinear optimization, an equality constraint cannot be satisfied exactly in a finite number of algorithm iterations. Therefore, one has to accept a feasible solution within some tolerance and, in this case, all the nominal stability properties are lost. Moreover, no degree of robustness to model uncertainty can be assured with the use of this constraint for the same reason, i.e., since it is an equality constraint asymptotic stability of the model trajectories can be shown but nothing can be said with regards to the stability of the plant.

In [14] a relaxed version of this constraint, $x(k + P|k) \in W$ (where W is a neighborhood of the origin), is proposed. In this case, the MPC strategy loses its stabilizing properties inside W. Thus, a linear, locally stabilizing controller designed for the linearized system is used inside W. The resulting "hybrid" controller is shown to be globally stabilizing. The main difficulty in actually implementing this scheme is the computation of the region W.

The contractive constraint proposed in [3] is an inequality end constraint and, therefore, can be satisfied in a finite number of algorithm iterations. Moreover, contrary to the inequality end constraint in [14], no region of attraction W needs to be computed. This contractive constraint is itself a Lyapunov function for the closed-loop system and stability can be proven in a straightforward manner. In the nominal state feedback case the closed-loop system can be proven exponentially stable. In [21], a continuous-time formulation of the optimization problem was presented with an inequality end constraint which also had the idea of "contraction" of the states associated with it. In that work, the authors showed asymptotic stability and not exponential stability as done later in [5].

Thus, due to the strong stabilizing properties of CNTMPC, we have chosen to combine it with an FL scheme in order to create a more efficiently implementable MPC scheme (since FL linearizes the I/O map of the system without any approximations - unlike successive linearizations, as proposed in [4]) with a certain degree of robust stability/performance. The contractive constraint is imposed on the system states after the linearizing state/input transformations. In this way, the constraint can be posed as a set of $2n + 1$ (where n is the system dimension) approximate linear constraints. Since the FL coordinate transformation is diffeomorphic, the exponential stability of the transformed states implies asymptotic stability of the original system states.

Here, the proposed FL+CNTMPC technique was successfully applied to a batch polymerization process and its stability and robustness properties were analyzed. We also showed its higher computational efficiency and enhanced stability properties when compared to STNMPC algorithms in a heuristic way, via numerical simulations.

Polymerization reactors are known to exhibit strong nonlinearities and pose difficult control problems. Polymeric materials are employed in a vast number of applications. The end use of these materials depends on the molecular weight distribution of the polymer. This distribution can be controlled only during the polymer manufacturing process and it is a complicated function of the reactor conditions [1]. Thus, the control systems for these reactors should be able to give satisfactory performance over a wide range of operating conditions, which brings out the need for nonlinear control.

The polymerization of styrene by a free radical mechanism carried out in a solution isothermic batch reactor constitutes our system. This system is represented mathematically by a set of five ordinary differential equations. The internal states of the model are conversion of initiator, conversion of monomer, conversion of solvent, zero and second moments of the molecular distribution. The manipulated and controlled variables are the dimensionless initiator feed rate and the conversion of monomer, respectively. Our objectives are to control the conversion of monomer to a desired setpoint while keeping the polydispersity and average molecular weight of the final polymer within certain pre-specified limits (this is translated as final inequality state constraints in the formulation of the optimization problem) and maintain the initiator feed rate within hard control bounds.

Although we are dealing with kinetic parameters and physical properties of a specific monomer/solvent/initiator combination, the conclusions drawn here apply, at least qualitatively, to a relatively large class of systems (isothermic bulk/solution/suspension batch polymerization by a free radical mechanism).

2. Preliminaries

2.1. DESCRIPTION OF THE SYSTEM

The results in this paper apply to any SISO nonlinear time-invariant plant described by the following differential equations:

$$
\begin{aligned}
\dot{x}^p(t) &= f(x^p(t), p(t)) + g(x^p(t), p(t))\, u(t) \\
y^p(t) &= h(x^p(t))
\end{aligned}
\tag{1}
$$

where $x^p(t) \in \Re^n$ are the plant states, $p(t) \in \Re^p$ are unknown parameters, $u(t) \in \Re$ and $y^p(t) \in \Re$ are the manipulated and controlled variables, respectively. $f : \Re^n \times \Re^p \to \Re^n$, $g : \Re^n \times \Re^p \to \Re^n$, $h : \Re^n \to \Re \in C^1$.

The mathematical model used in the computation of the control law is given by:

$$
\begin{aligned}
\dot{x}(t) &= f(x(t), 0) + g(x(t), 0)\, u(t) \\
y(t) &= h(x(t))
\end{aligned}
\tag{2}
$$

i.e., the model is evaluated at the nominal values of the parameters (which are considered to be zero without loss of generality).

It is important to emphasize that this paper deals only with parameter uncertainty, i.e., structural uncertainty is not taken into account.

Throughout the paper we assume that the manipulated variable $u(t)$ is subject to the following hard constraints:

$$u(t) \in \mathcal{U} := \{u \in \Re : u_{min} \leq u \leq u_{max}\}, \quad \forall t \in [0, \infty) \qquad (3)$$

where u_{min}, $u_{max} \in \Re$ are known.

2.2. INPUT/OUTPUT FEEDBACK LINEARIZATION

In the control literature, there are excelent books and review papers on differential geometry applied in the control of nonlinear systems [2, 8, 11]. Thus, assuming that the reader is familiar with this theory, only a very brief review will be given here on how to construct an I/O feedback linearizing control law for a SISO system (an extension of these results to the MIMO case is presented in Chapter 5 of Isidori's book [8]). Unless otherwise indicated, standard notation will be used in this section and, therefore, it will not be defined.

Let $r \leq n$ be the well-defined relative degree of system (2) on a given set $\Omega \subset \Re^n$. Then, we can construct a diffeomorphic transformation from x- to z-coordinates in Ω, Φ_{xz} (and let Φ_{zx} be its inverse, i.e., $\Phi_{zx} = \Phi_{xz}^{-1}$), of the following form:

$$z = \Phi_{xz}(x) = \begin{bmatrix} h(x) & L_f h(x) & \dots & L_f^{r-1} h(x) & \phi_{r+1}(x) & \dots & \phi_n(x) \end{bmatrix}^T \qquad (4)$$

with the functions $\phi_{r+1}(x), \dots, \phi_n(x)$ being linearly independent solutions of the differential equations $\langle d\phi(x), g(x) \rangle = 0$ where $\langle d\phi(x), g(x) \rangle$ stands for the inner product between the vector fields $d\phi(x) = \begin{bmatrix} \frac{\partial \phi_{r+1}(x)^T}{\partial x} & \dots & \frac{\partial \phi_n(x)^T}{\partial x} \end{bmatrix}^T$ and $g(x)$.

Using the state coordinate transformation in (4), the system can be put into *normal form* in the new coordinates $z \in \Re^n$:

$$\dot{z}_1 = z_2$$
$$\vdots$$
$$\dot{z}_{r-1} = z_r$$
$$\dot{z}_r = b(z) + a(z)\, u \qquad (5)$$
$$\dot{z}_{r+1} = q_{r+1}(z)$$
$$\vdots$$
$$\dot{z}_n = q_n(z)$$
$$y = z_1$$

where $\{q_i(z)\}_{i=r+1}^n$, $a(z)$ and $b(z)$ are given by:

$$
\begin{aligned}
q_i(z) &= \left[\langle d\phi_i(x), f(x)\rangle\right]_{x=\Phi_{zx}(z)}, \quad i = r+1, \ldots, n \\
a(z) &= \left[L_g L_f^{r-1} h(x)\right]_{x=\Phi_{zx}(z)} \\
b(z) &= \left[L_f^r h(x)\right]_{x=\Phi_{zx}(z)}
\end{aligned}
\tag{6}
$$

The I/O feedback linearizing control law is then given by:

$$
u = \frac{v - b(z) - \beta_1 z_r - \beta_2 z_{r-1} - \ldots - \beta_{r-1} z_2 - \beta_r z_1}{a(z)}
\tag{7}
$$

where $v \in \Re$ is the newly defined input variable and $\beta_1, \ldots, \beta_r$ are freely chosen parameters.

Applying the input transformation (7) onto (6), we obtain the following closed-loop system:

$$
\begin{aligned}
\dot{z}_1 &= z_2 \\
&\vdots \\
\dot{z}_{r-1} &= z_r \\
\dot{z}_r &= v - \beta_1 z_r - \beta_2 z_{r-1} - \ldots - \beta_{r-1} z_2 - \beta_r z_1 \\
\dot{z}_{r+1} &= q_{r+1}(z) \\
&\vdots \\
\dot{z}_n &= q_n(z), \quad z_0 \text{ given} \\
y &= z_1
\end{aligned}
\tag{8}
$$

In system (9) we can observe that the first r equations are linear and they determine the input/output behavior of the system. Furthermore, they are completely independent of the last $n-r$ state equations which are nonlinear and constitute the internal dynamics of the system. These dynamics are unobservable and controlled only indirectly via the observable states $z_1, \ldots, z_r$ (i.e., the input u does not appear explicitly in these equations).

In order to adopt a shorter notation for system (9), let us define:

$$
\xi := \begin{pmatrix} z_1 \\ \vdots \\ z_r \end{pmatrix}, \quad \eta := \begin{pmatrix} z_{r+1} \\ \vdots \\ z_n \end{pmatrix}, \quad q(\xi, \eta) := \begin{pmatrix} q_{r+1}(z) \\ \vdots \\ q_n(z) \end{pmatrix}
\tag{9}
$$

which means that $z(t) = \left[\xi(t)^T \ \eta(t)^T\right]^T$.

Then, equation (9) can be written as:

$$
\begin{aligned}
\dot{\xi}(t) &= A\xi(t) + Bv(t) \\
\dot{\eta}(t) &= q(\xi(t), \eta(t)), \quad (\xi_0, \eta_0) \text{ given} \\
y(t) &= \xi_1(t)
\end{aligned}
\tag{10}
$$

where the matrices $A \in \Re^{r \times r}$ and $B \in \Re^r$ are defined as:

$$
A := \begin{pmatrix}
0 & 1 & 0 & 0 & \cdots & 0 & 0 \\
0 & 0 & 1 & 0 & \cdots & 0 & 0 \\
\vdots & \vdots & \vdots & \vdots & & \vdots & \vdots \\
0 & 0 & 0 & 0 & \cdots & 0 & 1 \\
-\beta_r & -\beta_{r-1} & -\beta_{r-2} & -\beta_{r-3} & \cdots & -\beta_2 & -\beta_1
\end{pmatrix}, \quad
B := \begin{pmatrix}
0 \\ 0 \\ \vdots \\ 0 \\ 1
\end{pmatrix}
$$

If the linear observable subsystem is discretized, the normal form can be written as:

$$
\begin{aligned}
\xi(i+1) &= \Phi\,\xi(i) + \Gamma\,v(i), \quad i = 0, 1, \ldots \\
\dot{\eta}(t) &= q(\xi(t), \eta(t)), \quad (\xi(0) := \xi_0, \eta_0) \text{ given} \\
y(t) &= \xi_1(t)
\end{aligned}
\tag{11}
$$

where $v(t)$ is assumed constant in a discretization interval T, i.e., $v(t) = v(i)$ for $t \in [iT, (i+1)T]$. The matrices $\Phi \in \Re^{r \times r}$ and $\Gamma \in \Re^r$ are defined as:

$$
\Phi := e^{AT} \qquad \Gamma := \left[\int_t^{t+T} e^{A(t+T-\tau)}\,d\tau \right] B
$$

The closed-loop system in (12) will be the one used in the optimization step of our FL+CNTMPC scheme with the discretization interval T equal to the sampling time of the control algorithm.

2.3. CONTRACTIVE MPC

Contractive MPC is implemented on-line according to the general structure of MPC schemes, in a *moving (or receding) horizon* fashion. The basic difference between CNTMPC and a standard finite-horizon nonlinear MPC (STNMPC) scheme is the introduction of the so-called *contractive constraint* which constitutes a robustly stabilizing quadratic constraint added to the optimization problem solved at each time step. We refer the reader to [3, 4, 5] for the details on the formulation and stability/robustness analysis of different implementations of the CNTMPC algorithm.

3. Notation

The following notation will be adopted throughout the paper.

$\| x \|_{\tilde{P}} := \sqrt{x^T \tilde{P} x}$, with $\tilde{P} \in \Re^{n \times n}$ positive definite, is the weighted Euclidean norm of $x \in \Re^n$.

$\mathcal{Z}_+$ is the set of non-negative integers.

T denotes the sampling time, t_0 the initial time for computations and P, M the prediction and control horizons, respectively.

Here we adopt the following notation for the sampling times: $t_k^j := t_0 + (j + kP)T$, with j varying in the interval $j = 0, \ldots, P-1$ while k is kept constant at $k = 0, 1, 2, \ldots$. Moreover, we adopt $t_k := t_k^0$, $\forall k \in \mathcal{Z}_+$. Thus, we have the following sequence of sampling times: $\{t_0, t_0^1, \ldots, t_0^{P-1}, t_1, t_1^1, \ldots, t_1^{P-1}, t_2, \ldots\}$. Although we realize that this is not a standard MPC notation, we have found it most convenient, because we are interested in emphasizing that the FL+CNTMPC controller is solved as sets of P problems where the contractive constraint remains constant. Thus, it is important that the time variable has two indexes, one which remains constant for each set of P problems (k) and the other which varies to indicate each sampling time within the set corresponding to a given $k(j)$ [5].

The following notation is used for the transformed state and input trajectories (since CNTMPC will be applied to the system in normal form, in z/v-coordinates): $z_k := z_k^0 := z(t_k^0 - t_0, z_0, v)$, $z_k^j := z(t_k^j - t_k, z_k, v)$, $z_k^j(t) := z(t - t_k^j, z_k^j, v)$ and $v_k^j(t)$ is the continuous control law for $t \in [t_k^j, t_k^{j+P}]$. In order to conform to MPC's usual implementation scheme, let us consider a discontinuous control law of the kind $v_k^j(t) = v(kP + j + i|kP + j)$ for $t \in [t_k^{j+i}, t_k^{j+i+1}]$, $i = 0, \ldots, P-1$, i.e., $v_k^j(t)$ is constant during one sampling time. Moreover, $v(kP + j + i|kP + j) = v(kP + j + M - 1|kP + j)$, $\forall i = M, \ldots, P-1$.

$\mathcal{P}(t_k^j, z_k^j)$ denotes the optimal control problem to be solved at time t_k^j with initial condition $z_k^j := \Phi_{xz}(x_k^{p,j})$.

4. FL+CNTMPC Controller

The proposed control structure, where FL is embedded in an external contractive MPC loop is shown in *Figure 1*.

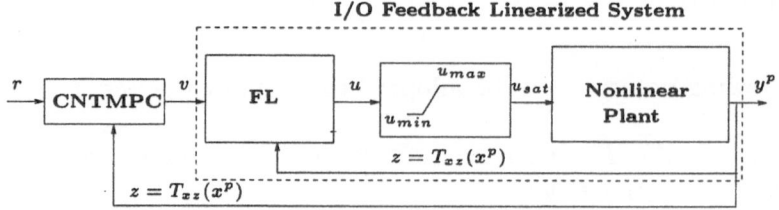

Figure 1. Proposed Controller Structure: FL+CNTMPC.

4.1. OPTIMIZATION STEP OF CNTMPC

The optimization problem at time t_k^j, $\mathcal{P}(t_k^j, z_k^j)$, $\forall j = 0, \ldots, P-1$, $k \in \mathcal{Z}_+$, with $z_k^j = \Phi_{xz}(x_k^{p,j})$ and $x_k^{p,j}$ being the measured states of the plant at t_k^j, is represented by:

$$\min_{v(.)} \sum_{i=0}^{P-1} \| z(kP+j+i+1|kP+j) \|_Q^2 + \| v(kP+j+i|kP+j) \|_R^2 \qquad (12)$$

subject to:

$$
\begin{aligned}
&\xi(kP+j+i+1|kP+j)=\Phi\xi(kP+j+i|kP+j)+\Gamma v(kP+j+i|kP+j) \\
&\dot{\eta}(t)=q(\xi(t),\eta(t)), \quad z_k^j=[\xi(kP+j|kP+j)^T \, (\eta_k^j)^T]^T=\Phi_{xz}(x_k^{p,j}) \\
&v_{min,k}^j(i) \leq v(kP+j+i|kP+j) \leq v_{max,k}^j(i), \quad i=0,\ldots,M-1 \\
&v(kP+j+i|kP+j)=v(kP+j+M-1|kP+j), \quad i=M,\ldots,P-1 \\
&F_l(z_k^j(t_k^{j+P})) \leq 0, \quad l=1,\ldots,n_{ineq} \\
&\| \bar{z}_k^j(t_{k+1}) \|_{\hat{P}} \leq \alpha \, \| z_k \|_{\hat{P}}, \quad z_k=\Phi_{xz}(x_k^p), \quad \alpha \in [0,1), \hat{P}>0
\end{aligned}
\qquad (13)
$$

where $v_{min,k}^j(i)$ is defined as

$$
\begin{aligned}
v_{min,k}^j(i) := \; & b(z(kP+j+i+1|kP+j)) + \beta^T z(kP+j+i+1|kP+j) \; + \\
& + a(z(kP+j+i+1|kP+j))u_{min}
\end{aligned}
\qquad (14)
$$

with $\beta := [\beta_r \beta_{r-1} \ldots \beta_1 0 \ldots 0]^T \in \Re^n$ (analogous definition applies to $v_{max,k}^j(i)$) and

$$
\begin{aligned}
\bar{\xi}(kP+j+i+1|kP+j) &= \Phi\bar{\xi}(kP+j+i|kP+j)+\Gamma v(kP+j+i|kP+j), \; i = 0,\ldots,P-1 \\
\dot{\bar{\eta}}(t) &= q(\bar{\xi}(t),\bar{\eta}(t)), \quad \bar{z}_k^j := \bar{z}_k^0 = \Phi_{xz}(x_k^p) \; \bar{z}_k^j = \bar{z}_k^{j-1}(t_k^j), \; j \geq 1 \quad (15)
\end{aligned}
$$

is the trajectory of the model used in the computation of the contractive constraint. The trajectory $\bar{z}_k^j(t)$ is obtained via numerical integration of the model in equation (15) with initial condition $\bar{z}_k^j$ from time t_k^j up to t_{k+1}

(notice that it is not necessary to integrate up to t_k^{j+P} because this model is used only for computation of $\bar{z}_k^j(t_{k+1})$, not in the prediction). The recursive definition in (15) shows that at t_k the initial condition for integration are the measured states $z_k = \Phi_{xz}(x_k^p)$. However, at t_k^j for $j = 1, \ldots, P-1$ the initial condition is given by the trajectory of the same model computed at the previous time step, $\bar{z}_k^{j-1}(t)$, evaluated at t_k^j. Thus, the model in (15) only takes feedback into account at every P steps (as opposed to the prediction model in the set of equations (13) which is updated with the plant states at every sampling time).

4.2. CNTMPC IMPLEMENTATION

The FL+CNTMPC controller is implemented according to the following scheme:

Control Algorithm 1

Data: *Initial Condition: $z_0 = \Phi_{xz}(x_0^p)$ at t_0; Controller Parameters: horizons P, M $(1 \leq M \leq P < \infty)$, weights $Q, R, \hat{P} > 0$, contractive parameter $\alpha \in [0, 1)$, sampling time $T \in (0, \infty)$ and hard control bounds $u_{min}, u_{max} \in \Re$.*

Step 0: *Set $k = 0$, $j = 0$.*

Step 1: *Assuming that the optimal control problem $\mathcal{P}(t_k^j, z_k^j)$ is feasible for the chosen set of parameters, then at $t = t_k^j$ solve $\mathcal{P}(t_k^j, z_k^j)$. Local optimal solutions or even feasible solutions are accepted. The result of this step is an optimal (or feasible) sequence of control moves $\{v(kP + j|kP + j), \ldots, v(kP + j + M - 1|kP + j)\}$.*

Step 2: *Transform $v(kP + j|kP + j)$ to the original control variable $u(kP + j|kP + j)$ via equation (7).*

Step 3: *Apply $u(kP + j|kP + j)$ to plant (1) for $t \in [t_k^j, t_k^{j+1}]$ and measure the states at t_k^{j+1}, $x_k^{p,j+1}$. Set $z_k^{j+1} = \Phi_{xz}(x_k^{p,j+1})$ and $z_k^{j+1} = \bar{z}_k^j(t_k^{j+1})$.*

Step 4: *If $j < P - 1$, set $j = j + 1$ and go back to **Step 1**. If $j = P - 1$ set $z_{k+1}^0 =: z_{k+1} = \Phi_{xz}(x_k^{p,P})$, $k = k + 1$, $j = 0$, and go back to **Step 1**.*

4.3. IMPLEMENTATION ASPECTS OF FL+CNTMPC

Even with the I/O feedback linearization of model (2), the optimal control problem $\mathcal{P}(t_k^j, z_k^j)$, specified by the set of equations (12), (13), (14) and (15), is a general nonlinear programming problem.

In this section we will discuss certain implementation aspects of the proposed FL+CNTMPC algorithm. In particular, we will investigate the scenarios in which the optimization problem solved at each time step can be simplified and implemented in an efficient manner.

Objective Function: If the objective function only takes into account the observable states ξ and the input v, if the internal dynamics is minimum-phase and if there are no constraints other than the model dynamics, then the optimal control problem has a closed-form solution. In other words, we have a linear unconstrained MPC problem. The internal dynamics, being minimum-phase, does not create internal stability problems and the system can be easily stabilized in the nominal case.

Model Dynamics: If the system can be fully feedback linearized (i.e., $r = n$) and if there are no constraints other than the linear model dynamics then the optimal control problem admits a closed-form solution.

Input Constraints: As pointed out in [15, 16], the FL puts the system into a simpler form in z-coordinates (it can even linearize the system completely if $r = n$) but adds the complication that simple hard bounds in u become nonlinear state dependent constraints in v. Thus, in order to justify the feedback linearization procedure we must deal with these constraints in v in a computationally efficient way. An obvious alternative is to approximate these constraints in some form (preferably, as linear constraints in v), solve the optimization subject to these simpler constraints and then check if the original nonlinear constraints were indeed satisfied. If they are satisfied at least for the first control move, $u(k|k)$, then we can move on to the next sampling time. If not, it is necessary to find an iterative procedure to update the optimal solution until the hard bounds in u are obeyed.

In [6] a computationally demanding iterative procedure and four relatively simple approximate procedures were suggested for implementation of the constraints in v. Here we opted for using the fourth suggestion, which consists in the linearization of the constraint functions, because we believe this option to have the potential to generate the best results. The linearization is performed around the measured states at the current time step and, therefore, it is expected that, at least for the first control move, the hard bounds in u should not be violated.

As we can see in the set of equations (13), the original nonlinear constraints in v for $i = 0, \ldots, M - 1$ are given by:

$$v_{min,k}^{j}(i) \leq v(kP+j+i|kP+j) \leq v_{max,k}^{j}(i) \qquad (16)$$

Then, if the functions $a(z)$ and $b(z)$ in $v_{min,k}^{j}(i)$ and $v_{max,k}^{j}(i)$ defined in equation (14) are linearized around z_k, the following linear constraints are obtained for $i = 0, \ldots, M - 1$:

$$v(kP+j+i|kP+j) \geq \bar{v}^k_{min} + \Delta\bar{v}^k_{min}z(kP+j+i+1|kP+j) \quad (17)$$
$$v(kP+j+i|kP+j) \leq \bar{v}^k_{max} + \Delta\bar{v}^k_{max}z(kP+j+i+1|kP+j)$$

where:

$$\bar{v}^k_{min} := b(z_k) + u_{min}a(z_k) - \left(\frac{\partial b(z)}{\partial z}|_{z=z_k} + u_{min}\frac{\partial a(z)}{\partial z}|_{z=z_k}\right) z_k$$

$$\Delta\bar{v}^k_{min} := \frac{\partial b(z)}{\partial z}|_{z=z_k} + u_{min}\frac{\partial a(z)}{\partial z}|_{z=z_k} + \beta^T \quad (18)$$

Analogous definitions apply to $\bar{v}^k_{max}$ and $\Delta\bar{v}^k_{max}$.

The approximate constraints in (18) are linear in v if the system is fully feedback linearizable. However, in the general case where $r < n$, then we either need to linearize the internal system dynamics as well or utilize the zero dynamics (i.e., the internal dynamics with $\xi = 0$) in the evaluation of the input constraints. The use of the linearization of the internal dynamics should be satisfactory at least for the first control moves. The use of the zero dynamics in the computation of approximate values of the non-observable states η for evaluation of the input constraints is only reasonable if the observable states are driven to the origin very rapidly. In fact, we can derive an upper bound on the norm of the error between the states of the internal dynamics, $\eta(t)$, and those of the corresponding zero dynamics, $\eta_z(t)$.

The difference between $\eta(t)$ and $\eta_z(t)$ for $t \in [t_k, t_{k+1}]$, i.e., during one prediction horizon, is given by:

$$\dot{\eta}_k(t) - \dot{\eta}_{z,k}(t) = q(\xi_k(t), \eta_k(t)) - q(0, \eta_{z,k}(t)) \quad (19)$$

Integrating both sides of equation (19), observing that $\eta_{z,k} = \eta_k$ due to the feedback at time t_k, and applying the $2-$norm, we have:

$$\| \eta_k(t) - \eta_{z,k}(t) \| \leq \int_{t_k}^t \| q(\xi_k(\tau), \eta_k(\tau)) - q(0, \eta_{z,k}(\tau)) \| \, d\tau \quad (20)$$

Thus, if we assume that the nonlinear internal dynamics in equation (11) is Lipschitz continuous, i.e., there exists a constant $L_q \geq 0$ such that

$$\| q(\xi_1, \eta_1) - q(\xi_2, \eta_2) \| \leq L_q[\| \xi_1 - \xi_2 \| + \| \eta_1 - \eta_2 \|] \quad (21)$$

then, from equation (20), it results that:

$$\| \eta_k(t) - \eta_{z,k}(t) \| \leq L_q \int_{t_k}^t [\| \xi_k(\tau) \| + \| \eta_k(\tau) - \eta_{z,k}(\tau) \|] \, d\tau \quad (22)$$

Since the control law $v(t)$ computed by CNTMPC is stabilizing (as we will prove later), there exist ν, $\sigma > 0$ such that the norm of the observable states $\xi(t)$ for $t \in [t_k, t_{k+1}]$, is bounded exponentially by:

$$\| \xi(t) \| \le \nu \, \| \xi_k \| \, e^{-\sigma(t-t_k)} \tag{23}$$

If we substitute equation (23) into (22) and apply the Bellman-Grownwall inequality, knowing that $t - t_k \le PT$, we finally have:

$$\| \eta_k(t) - \eta_{z,k}(t) \| \le L_q \, \nu \, \| \xi_k \| \, e^{L_q PT} \frac{(1 - e^{-\sigma PT})}{\sigma} \tag{24}$$

Thus, if the rate of decay of the observable states $\xi(t)$ to the origin, namely, σ, is very large, then the difference between the states of the internal and zero dynamics tends to zero, i.e.,

$$\lim_{\sigma \to \infty} \| \eta_k(t) - \eta_{z,k}(t) \| = 0, \quad t \in [t_k, t_{k+1}], \; \forall k \in \mathcal{Z}_+ \tag{25}$$

General Inequality End Constraints: The inequality end constraints in the set of constraints (13) are general constraints on the transformed state variables which are imposed at the end of the prediction horizon in order to be satisfied asymptotically. These constraints are useful in guaranteeing that some process variables remain inside a certain set as $t \to \infty$. This is indicated when it is not necessary to drive these variables to a chosen setpoint but it is important to guarantee that they satisfy certain asymptotic requirements.

These constraints, being general nonlinear state constraints, can greatly increase the complexity of the optimization problem. However, since we only need to satisfy them asymptotically, they may not be imposed at the initial time steps. Our suggestion is to, after the solution of the optimization problem at t_k without the inequality end constraints, compute the values of the functions $\{F_l(z_k^j(t_k^{j+P}))\}_{l=1}^{n_{ineq}}$. If some of these n_{ineq} constraints are satisfied (which means that they would have been inactive in the optimization at time step k) we can then explicitly impose them in the optimization at $k + 1$. If feasibility problems appear, we remove the constraints once again. If these state constraints are compatible with the desired steady state values, then there will exist a finite time after which they can be satisfied.

A further simplification in the implementation of these inequality end constraints is to submit them to the same approximate procedure as suggested for the input constraints. In other words, we would linearize the constraint functions $\{F_l\}_{l=1}^{n_{ineq}}$ and either linearize the internal dynamics or utilize the zero dynamics for evaluation of these linearized constraints.

Contractive Constraint: The contractive constraint imposed directly on the transformed variables z is quadratic in these variables. Thus, if full state feedback linearization is possible (i.e., $r = n$), the contractive constraint is quadratic in v and in [3, 4] the reader finds three different approaches to deal with the contractive constraint in an efficient way. In fact, if there are no constraints in u and no inequality end constraints, the optimization can be posed as a QP.

If $r < n$, once again we can use a linearized internal dynamics or the zero dynamics for evaluation of the contractive constraint. Since satisfaction of the exact contractive constraint is important in order to guarantee stability/performance, it is always necessary to verify if the original constraint is being satisfied after the solution of the optimization problem with the approximate constraint. Even if the constraint is not satisfied for the chosen α but a contraction by a factor α^* such that $\alpha < \alpha^* < 1$ is achieved, then the solution may be accepted. In this case, performance might suffer but nominal stability is still guaranteed.

4.4. STABILITY OF THE FL+CNTMPC TECHNIQUE

4.4.1. *Basic Assumptions*

Without loss of generality, let us consider the regulation problem for which the desired operating point is the origin. Then, the following assumptions are needed to ensure local stability:

Assumption 1 *The origin $(x^p, u, p) = (0, 0, 0)$ is an equilibrium point of the plant (1). Moreover, the origin in the $x/u/p$-space is mapped onto the origin in the $z/v/p$-space.*

Assumption 2 *The linearization of the original nonlinear system at the origin is stabilizable and all the system states are measured (state feedback case).*

Assumption 3 *There exists a constant $\rho \in (0, \infty)$ such that for all $z_k = \Phi_{xz}(x_k^p) \in B_\rho := \{z \in \Re^n \mid \| z \|_{\hat{P}} \leq \rho; \ \hat{P} > 0\}$, the optimization problem at t_k, $\mathcal{P}(t_k, z_k)$, is feasible for all $k \in \mathcal{Z}_+$.*

Assumption 4 *There exists $\gamma \in (0, \infty)$ such that $\| \bar{z}_k^j(t) \|_{\hat{P}} \leq \gamma \| z_k \|_{\hat{P}}$, $\forall j = 0, \ldots, P - 1, k \in \mathcal{Z}_+$.*

Assumption 5 *The relative degree of plant (1) is equal to that of model (2), $r \leq n$, and it is well-defined within the reachable set $\mathcal{X}$:*

$$\mathcal{X} := \{x^p(t), x(t) \in \Re^n \mid x^p(t) = x^p(t - t_0, x_0^p, u, p),$$
$$x(t) = x(t - t_0, x_0^p, u, 0), t \in [t_0, \infty); z_0 = \Phi_{xz}(x_0^p) \in B_\rho, u \in \mathcal{U}\} \quad (26)$$

Assumption 6 *The functions* $f : \Re^n \times \Re^p \to \Re^n$, $g : \Re^n \times \Re^p \to \Re^n$ *and the coordinate transformation* $\Phi_{xz} : \Re^n \times \Re^p \to \Re^n$ *are Lipschitz continuous on the reachable set* $\mathcal{X}$, *i.e., there exist* L_f, L_g, $L_T > 0$ *such that:*

$$\| f(x^p(t), p(t)) - f(x(t), 0) \|_{\hat{P}} \leq L_f [\| x^p(t) - x(t) \|_{\hat{P}} + \| p(t) \|_2]$$
$$\| g(x^p(t), p(t)) - g(x(t), 0) \|_{\hat{P}} \leq L_g [\| x^p(t) - x(t) \|_{\hat{P}} + \| p(t) \|_2] \quad (27)$$
$$\| \Phi_{xz}(x^p(t), p(t)) - \Phi_{xz}(x(t), 0) \|_{\hat{P}} \leq L_T [\| x^p(t) - x(t) \|_{\hat{P}} + \| p(t) \|_2]$$

4.4.2. *Nominal Stability*

Theorem 1 (Exponential stability) *Given* $\alpha \in [0, 1)$ *and* ρ, $\gamma \in (0, \infty)$ *satisfying* **Assumptions 3** *and* **4**, *respectively,* **Control Algorithm 1** *renders the closed-loop system exponentially stable in z-coordinates on the set* B_ρ *and uniformly asymptotically stable in x-coordinates, i.e., for any* $z_0 = \Phi_{xz}(x_0) \in B_\rho$, *the resulting trajectories* $z(t) := z(t - t_0, z_0, v)$ *and* $x := x(t - t_0, x_0, u)$ *satisfy the following conditions:*

$$\| z(t) \| \leq v \| z_0 \| e^{-(1-\alpha)\frac{(t-t_0)}{PT}}, \; with \; v \geq \gamma e^{(1-\alpha)}$$
$$\lim_{t \to \infty} x(t) = 0 \quad (28)$$

Proof: *For proof of exponential stability in the z-space the reader is referred to* PART 1 *of [5] and the asymptotic stability in the x-space follows from* **Assumption 5** *because, under this assumption, the transformation* Φ_{zx} *is diffeomorphic. Therefore, to a point in z-coordinates corresponds a unique point in x-coordinates and if* $\lim_{t \to \infty} z(t) = 0$ *then* $\lim_{t \to \infty} x(t) = 0$ *(using* **Assumption 1***).* □

4.4.3. *Robust Stability*

It is a well-known fact and it has been shown in [10] in a heuristic way, via numerical simulations of the Van der Pol oscillator, that a controller designed through an I/O feedback linearization procedure is very sensitive to modeling errors. In other words, I/O feedback linearizing controllers are known not to be robust to even small sources of model uncertainty. This occurs because the coordinate transformation from x/u- to z/v-coordinates on which the controller is based is performed over an uncertain model.

In [17] a framework for investigating the robustness properties of MPC algorithms (including an FL+MPC scheme) was presented.

Here we will show that the CNTMPC (instead of a standard finite horizon MPC algorithm) in the external control loop adds some degree of robustness to the closed-loop with respect to parametric uncertainty. Similar results can be derived if one considers unmodeled dynamics as well.

Asymptotically Convergent Parameters

Assumption 7 *The parameters $p(t)$ satisfy the following boundedness condition:*

$$p_k(t) \in B_{\rho_k^p} := \{p \in \Re^p : \|p\|_2 \le \rho_k^p\}, \rho_k^p \in [0, \infty), t \in [t_k, t_{k+1}], \forall k \in \mathcal{Z}_+ \quad (29)$$

and the asymptotic properties of $p(t)$ are described as:

For any $\epsilon > 0, \exists$ a finite $\bar{k} := \bar{k}(\epsilon) \in \mathcal{Z}_+$ so that $\rho_k^p \le \epsilon, \ \forall k \in [\bar{k}, \infty)$, and $\bar{k}(\epsilon) \to \infty$ if $\epsilon \to 0$

Under this assumption on the unknown parameters, the following result can be derived:

Theorem 2 (Uniform Asymptotic Stability) *Let Assumptions 1, 3, 4, 5, 6 and 7 be satisfied and let $z_k, \bar{z}_k \in B_\rho, \forall k \in \mathcal{Z}_+$. Then, the closed-loop system resulting from application of* **Control Algorithm 1** *to system (1) is uniformly asymptotically stable.*
Proof: *For a similar proof, the reader is referred to PART 2 of [5]. Here we will skip the proof for brevity but it is important to notice that some care must be taken with the fact that the contractive constraint is imposed on $\bar{z}_{k+1}$ and not directly on $\bar{x}_{k+1}$. However, because Φ_{xz} satisfies* **Assumption 6**, *the proof in [5] can be extended to this case in a simple manner.* □

Theorem 2 is valid under the assumption of feasibility. We shall now proceed to derive a sufficient condition on the parameters $p_k(t)$ under which $z_k, \bar{z}_k$ stay inside B_ρ for all $k \in \mathcal{Z}_+$.

Theorem 3 (Feasibility Condition) *Under* **Assumptions 6** *and* **7**, *if $\rho^p < \frac{\rho(1-\alpha)}{L_T(1+LPTe^{LPT})}$ (where $\rho^p := \max_{k \in \mathcal{Z}_+} \rho_k^p$ and $L := L_f + L_g \epsilon_u$ with $\epsilon_u := \max\{|u_{min}|, |u_{max}|\}$), then there exists $\rho_0 \in (0, \rho]$ such that for all $z_0 \in B_{\rho_0}$, the sequences $\{z_k\}_{k=0}^\infty$ and $\{\bar{z}_k\}_{k=0}^\infty$ resulting from implementation of* **Control Algorithm 1** *stay inside the set B_ρ.*
Proof: *A similar proof can be found in PART 2 of [5].* □

General Time-Varying Parameters

Assumption 8 *Here we assume that the unknown parameters $p(t)$ lie inside the following set $\Pi \subset \Re^p$:*

$$p(t) \in \Pi := \{p \in \Re^p : \| p(t) \|_2 \le \rho^p\}, \ \forall t \in [0, \infty); \ \rho^p \ge 0 \ known \quad (30)$$

Theorem 4 (Practical Stability) *Let* **Assumptions 1, 3, 4, 5, 6** *and* **8** *be satisfied and let* z_k, $\bar{z}_k \in B_\rho$, $\forall k \in \mathcal{Z}_+$. *Then, the trajectories of the closed-loop system in z-coordinates resulting from application of* **Control Algorithm 1** *to system (1) satisfy:*

$$\| z_k \|_{\hat{P}} \; < \; \| z_0 \|_{\hat{P}} + \frac{\rho^p L_T (1 + LPTe^{LPT})}{1 - \alpha} \tag{31}$$

$$\lim_{k \to \infty} \| z_k \|_{\hat{P}} \; < \; \frac{\rho^p L_T (1 + LPTe^{LPT})}{1 - \alpha} \tag{32}$$

From **Assumption 5** *we conclude that* x_k^p *is also bounded* $\forall k \in \mathcal{Z}_+$. *From (32) it follows that:*

$$\lim_{\rho^p \to 0} \left\{ \lim_{k \to \infty} z_k \right\} = 0 \tag{33}$$

And from **Assumption 1** *we finally have:*

$$\lim_{\rho^p \to 0} \left\{ \lim_{k \to \infty} x_k^p \right\} = 0 \tag{34}$$

Proof: *The proof is similar to that of* **Theorem 2**. $\qquad\qquad\square$

The feasibility condition is the same as that presented in **Theorem 3** for asymptotically convergent parameters.

5. Batch Polymerization Reactor Model

In this paper, we test the proposed FL+CNTMPC control technique on a batch polymerization reactor. The polymerization reactions under consideration may be described by the classical free radical mechanism. This mechanism includes initiation, propagation, chain tranfer and termination steps [7, 18]. The parameters and physical properties of the polymerization of styrene in toluene, carried out by chemical initialization with benzoil peroxide, are given in [7].

5.1. REACTOR MODEL

Using the kinetic mechanism mentioned above in an appropriate material balance for each of the chemical species involved, it was possible to derive a suitable mathematical model. The following assumptions were made in order to simplify the problem formulation: all the reactions are irreversible and elementary; the reaction rate constants are independent of the chain size; the reactor contents are perfectly mixed so that there are no temperature gradients; and the reactor operates isothermically.

The last assumption was considered acceptable in this work because our modeling was based on a pilot plant which we plan to use in a later comparison of the simulation results obtained here with real experimental data. The assumption of isothermic operation is realistic due to the relatively small dimensions of the reactor and the efficient cooling and mixing mechanisms.

The following dimensionless model was used in our simulations:

$$\frac{dx_I}{dt} = -u + k_d'(1 - x_I)$$

$$\frac{dx_M}{dt} = (k_p' + k_{fm}')\frac{(1 - x_M)}{(1 + \epsilon x_M)}\lambda_0' \tag{35}$$

$$\frac{dx_S}{dt} = k_{fs}'\frac{(1 - x_S)}{(1 + \epsilon x_M)}\lambda_0'$$

with the following dimensionless groups:

$$x_I := 1 - \frac{I}{I_0} \qquad x_M := 1 - \frac{M}{M_0} \qquad x_S := 1 - \frac{S}{S_0} \qquad t := \frac{\tau}{T}$$

$$u := \frac{FT}{I_0} \qquad k_d' := k_d T \qquad k_p' := \frac{k_p M_0 T}{V_0}$$

$$k_{fm}' := \frac{k_{fm} M_0 T}{V_0} \qquad \lambda_0' := \frac{\lambda_0}{M_0} \qquad k_{fs}' := \frac{k_{fs} M_0 T}{V_0}$$

where I, M and S (I_0, M_0 and S_0) are the total number of mols of initiator, monomer and solvent in the reactor at time t (at time t_0), respectively. F is the molar feed rate of initiator to the reactor, V_0 is the initial reactor volume and $V = V_0(1 + \epsilon x_M)$, where $\epsilon < 0$ is the volume contraction factor, is the total volume of the mixture in the reactor. τ is the real time of operation. k_d, k_i, k_{dm}, k_p, k_{fm} k_{fs}, k_{tc}, k_{td} are the kinetic constants for dissociation of initiator, chemical initiation, thermal initiation, propagation, chain tranfer to the monomer, chain transfer to the solvent, termination by combination and termination by disproportionation, respectively. All of these kinetic constants follow Arrhenius law in their dependence on the constant reactor temperature θ.

The variable λ_0 is defined as:

$$\lambda_0 := \sqrt{\frac{2(k_{dm}\frac{M^3}{V} + k_d f I V)}{k_{tc} + k_{td}}} \tag{36}$$

with f being the initiator efficiency.

The overall chain termination rate which appears in the computation of λ_0, defined as $k_t := k_{tc} + k_{td}$, has the special form $k_t := k_{t0} g$, where k_{t0} is the rate constant at very low monomer conversion and g is a function of the reactor composition which reflects the fact that k_t falls during the reaction course due to strong diffusion limitations of the "live" macroradicals as the viscosity of the polymeric mixture increases. This phenomenon, known as the *gel effect*, arises at high conversions in the reactor and is responsible for a strong nonlinear behavior of the system as the conversion increases.

The correlations which express g as a function of the monomer conversion x_M are in general empirical, obtained for a particular polymer/solvent combination in a certain range of conversion. In our case, the following correlation will be used [7]:

$$g(x_M) := \left(\alpha_g + \beta_g x_M + \gamma_g x_M^2\right)^2 \tag{37}$$

where the parameters α_g, β_g and γ_g are obatained via fitting of experimental data. In general, these parameter values are only known to be within certain bounds and they are the main source of uncertainty in polymerization reactor models.

In this model, u and $y := x_M$ are the manipulated and controlled variables, respectively. In a polymerization process, however, it is generally not enough to maximize the reactor productivity by controlling the monomer conversion to the chosen setpoint. There are other properties of the final polymer which should be controlled to stay within certain limits. Among these quantities we find the polidispersity (Pd) and the average molecular weight (Mw) of the final polymeric mixture. These variables do not need to be driven to setpoint values but it is desirable that the polidispersity is not too high and the average molecular weight is not too low.

5.2. FEEDBACK LINEARIZATION OF THE MODEL

The relative degree of the system with respect to the output $y = x_M$ is $r = 2$ because $L_g h(x) = 0$ and $L_g L_f h(x) \neq 0$ for $\ddot{x} \in U = \{x \in \Re^3 \mid x_I \geq 0 \cap x_M \neq 1\}$.

In the controller design, a full state feedback linearization of system (36) was achieved, because the solvent mass balance can be safely ignored at high solvent concentrations. Thus, under the assumption that $S(t) = S_0$ ($\forall t \geq 0$), the following linear system in z-coordinates is obtained:

$$\dot{z} = \begin{pmatrix} 0 & 1 \\ -\beta_2 & -\beta_1 \end{pmatrix} \begin{pmatrix} z_1 \\ z_2 \end{pmatrix} + \begin{pmatrix} 0 \\ 1 \end{pmatrix} v =: Az + Bv \tag{38}$$

As previously discussed, if a full scale industrial polymerization reactor were considered, the isothermic assumption would no longer be valid and it would be necessary to consider the overall energy balance in the reactor. This happens because polymerization reactions are, in general, very exothermic and due to the large industrial reactor dimensions, there are no practical cooling/mixing systems efficient enough to keep the temperature in the reactor constant. Thus, it is important to consider which extra difficulties would be introduced in the normal form of the reactor model which we are using in our simulations if a formal energy balance was considered. The temperature balance would create a one-dimensional internal dynamics. This internal dynamics would need to be stable (at least locally) by appropriate choice of the coolant temperature. Thus, the energy balance would increase the computational burden in the numerical simulations of the controller but would not introduce significant difficulties from a theoretical point of view if the coolant temperature was kept constant or manipulated within an appropriate range.

In fact, since we have three distinct goals: to lead the conversion to a given setpoint, to keep the polidispersity below a given pre-specified value and to keep the average polymer weight above a minimum value, it is helpful to have the temperature of the jacket as an additional manipulated variable. It is well-known that operation at high reactor temperatures leads to high conversions but at the expense of high polispersity and low average polymer weight. Low temperatures have the opposite effect. Thus, having the temperature of the cooling system as a second manipulated variable might, in fact, help establish a compromise between these two opposite scenarios. Thus, while we are dealing with a simpler problem from a computational point of view by having only one input variable, it is in fact a more challenging control problem because we dispose of less degrees of freedom to reach our multiple objectives.

6. Simulation Results

6.1. OPTIMIZATION PROBLEM FORMULATION FOR THE POLYMERIZATION REACTOR

The optimization problem of the FL+CNTMPC algorithm at time t_k^j, $\mathcal{P}(t_k^j, z_k^j = \Phi_{xz}(x_k^{p,j}))$, applied to the polymerization reactor is posed in the following form:

$$\min_{v(.)} \sum_{i=0}^{P-1} \| y(kP+j+i+1|kP+j) \|_Q^2 + \| v(kP+j+i|kP+j) \|_R^2 \quad (39)$$

454

subject to:

$$z(kP+j+i+1|kP+j)=\Phi z(kP+j+i|kP+j)+\Gamma v(kP+j+i|kP+j),$$
$$y(kP+j+i|kP+j)=z_1(kP+j+i|kP+j), \quad i=1,\ldots,P$$
$$v^j_{min,k}(i) \le v(kP+j+i|kP+j) \le v^j_{max,k}(i), \quad i=0,\ldots,M-1 \qquad (40)$$
$$v(kP+j+i|kP+j)=v(kP+j+M-1|kP+j), \quad i=M,\ldots,P-1$$
$$\| \bar{z}^j_k(t_{k+1}) \|_{\hat{P}} \le \alpha \| z_k \|_{\hat{P}}$$

and the linear model used for computation of the contractive constraint is given by:

$$\bar{z}(kP+j+i+1|kP+j)=\Phi\bar{z}(kP+j+i|kP+j)+\Gamma v(kP+j+i|kP+j) \qquad (41)$$

with $i=0,\ldots,P-1$, $\bar{z}(kP|kP) := \Phi_{xz}(x^p_k)$ and $\bar{z}(kP+j|kP+j) = \bar{z}(kP+j|kP+j-1)$, for $j \ge 1$.

Our simulation results were obtained using **Control Algorithm 1** with the optimization step represented by the set of equations (39), (40), (14) and (41). Naturally, in the nominal case, equation (41) was not used since $\bar{z}^j_k(t) = z^j_k(t)$. Furthermore, we will emphasize when approximate forms of the constraints in (40) were used or when some of these constraints were omitted.

Here we will compare simulation results obtained with a standard nonlinear MPC algorithm (STNMPC) to those obtained with the proposed FL+CNTMPC controller from the point of view of computational effort and stability/robustness properties.

The MPC-software used for implementation of the STNMPC and FL+ CNTMPC algorithms (in the latter case, only when solution of nonlinear programming problems was required), was developed at the Institut für Automatik, ETH-Zürich, in a project of Prof. Manfred Morari and under the supervision of the first author of this paper. The software documentation can be found in [20].

Now we proceed to present some of the most significant simulation results for both the nominal and robust cases.

The operating conditions and parameters of the reactor are shown in TABLE 1.

Unless otherwise indicated, the controller parameters in TABLE 2 and TABLE 3 were used in the simulations.

Naturally, the values for u_{min} and u_{max} are imposed only in the constrained case.

The appropriate choice of β_1 and β_2 in the FL+CNTMPC algorithm is very important. This was also observed in [11, 19]. Some parameter combinations lead to highly oscillatory behavior and physically meaningless results. In [16] the authors do not explore different choices of the parameters

TABLE 1. Operating conditons and parameters of the polymerization reactor.

θ	373.15 K	k_d	$7.12 \times 10^{13} e^{\left(\frac{-29589}{R\theta}\right)} \mathrm{s}^{-1}$
V_0	1 liter	k_{dm}	$2.190 \times 10^5 e^{\left(\frac{-27440}{R\theta}\right)} \mathrm{l}^2/\mathrm{mol}^2\mathrm{s}$
k_{td}	0	k_p	$1.051 \times 10^7 e^{\left(\frac{-7060}{R\theta}\right)} \mathrm{l}/\mathrm{mol~s}$
I_0	10^{-2} mol	k_{t0}	$1.255 \times 10^9 e^{\left(\frac{-1680}{R\theta}\right)} \mathrm{l}/\mathrm{mol~s}$
PM_{mon}	104.15 g/mol	k_{fm}	$2.31 \times 10^6 e^{\left(\frac{-12670}{R\theta}\right)} \mathrm{l}/\mathrm{mol~s}$
M_0	$\frac{3000 \rho_S \rho_M V_0}{PM_{mon}(\rho_M + 3\rho_S)}$ mol	k_{fs}	$5.92 \times 10^8 e^{\left(\frac{-17210}{R\theta}\right)} \mathrm{l}/\mathrm{mol~s}$
ρ_M	$(1.047 + 4.9 \times 10^{-4}\theta)^{-1}$ g/cm^3	ϵ	$-0.15 - 4.4 \times 10^{-4}(\theta - 273.15)$
ρ_S	$(0.8075 + 10^{-3}\theta)^{-1}$ g/cm^3	f	0.72
g	$\left(0.5093 + 2.4645 x_m - 3.7473 x_M^2\right)^2$		

TABLE 2. Controller parameters for simulations with FL+CNTMPC

Q	diag([1 0])	R	8
$\hat{P}$	I_n	α	0.7
P	5	M	5
u_{min}	0	u_{max}	0.1
T	1000	β_1	2
β_2	10^{-2}		

β_i, $i = 1, \ldots, \beta_r$, and we believe that their results could be further improved had different adjustments been tried out.

In all the figures displaying simulation results we plotted the variable $1 - x_I$ instead of x_I. The reason for this is that $1 - x_I = \frac{I}{I_0}$ and, therefore, it represents the amount of initiator in the reactor compared to how much was there in the beginning. This is more illustrative than to plot

TABLE 3. Controller parameters for simulations with STNMPC

Q	diag([0 1])	R	1
T	1000	P	5
M	5	u_{min}	0
u_{max}	0.1		

x_I which represents the initiator conversion but can be actually negative (since initiator may accumulate in the reactor if added in excess).

6.2. NOMINAL CASE

6.2.1. *Unconstrained*

In order to start comparing results, in *Figure 2* we used $R = 1$ for FL+ CNTMPC, i.e., the same controller parameters Q, R, P and M were adopted for both FL+CNTMPC and STNMPC.

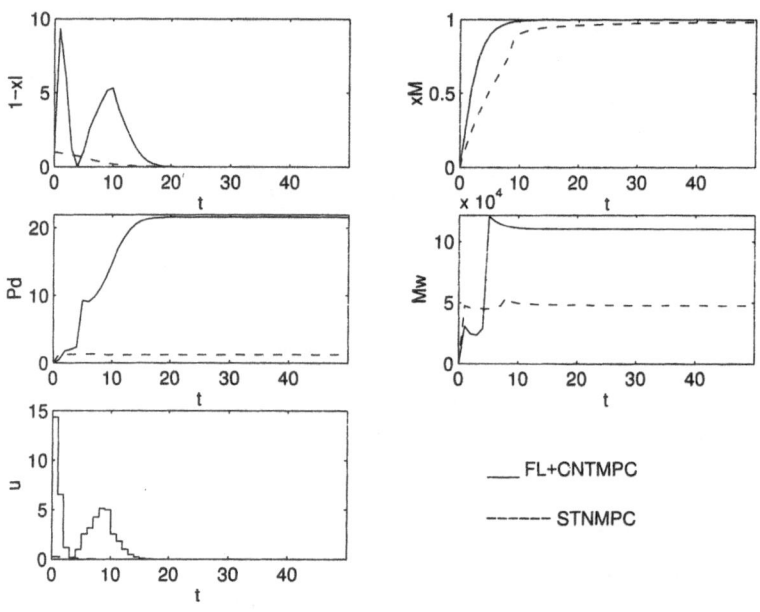

Figure 2. Unconstrained Nominal Case: Same controller parameters.

We notice from *Figure 2* that the two controllers behave dramatically differently. The FL+CNTMPC shows a much more aggressive response and the output x_M reaches the setpoint much faster than in the case of STNMPC. As a consequence, since too much initiator is being added to the reactor (which can be seen from the plots of $u \times t$ and $1 - x_I \times t$), the polidispersity (Pd) gets unacceptably high.

We also notice that the output response in the case of STNMPC displays the characteristic shape caused by the nonlinearities of the gel effect at high conversions. The same does not happen to the response obtained with FL+CNTMPC because of the linearization of the input/output map.

It is also worth noticing that the CPU time for running the FL+CNTMPC algorithm is about 375 times smaller than in the case of STNMPC.

Due to these results, we tried adjusting R in the case of FL+CNTMPC in order to make the control effort nearly equivalent to that of STNMPC. By doing this, we arrived at $R = 8$. The results of this simulation are shown in *Figure 3*.

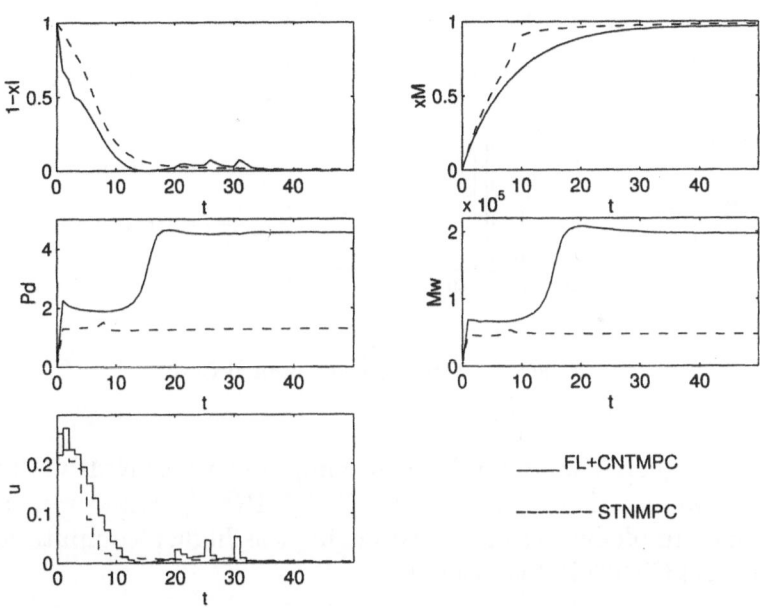

Figure 3.   Unconstrained Nominal Case: Same control effort.

We can see from *Figure 3* that the same control effort leads to a comparable speed of response for both x_I and x_M. However, the final polymer is quite different since the one obtained with FL+CNTMPC has high Pd and Mw and the one produced with STNMPC has low Pd and Mw. Thus, the polymer obtained with FL+CNTMPC is better with respect to Mw but worse with respect to Pd. This seems to be a trend in all the simulations performed with these controllers.

Here the CPU time difference is even larger: the FL+CNTMPC algorithm is approximately 750 times faster than STNMPC.

6.2.2. *Constrained*

Figure 4 shows the same trend as that observed in the unconstrained case displayed in *Figure 3*, i.e., FL+CNTMPC tends to generate a polymer

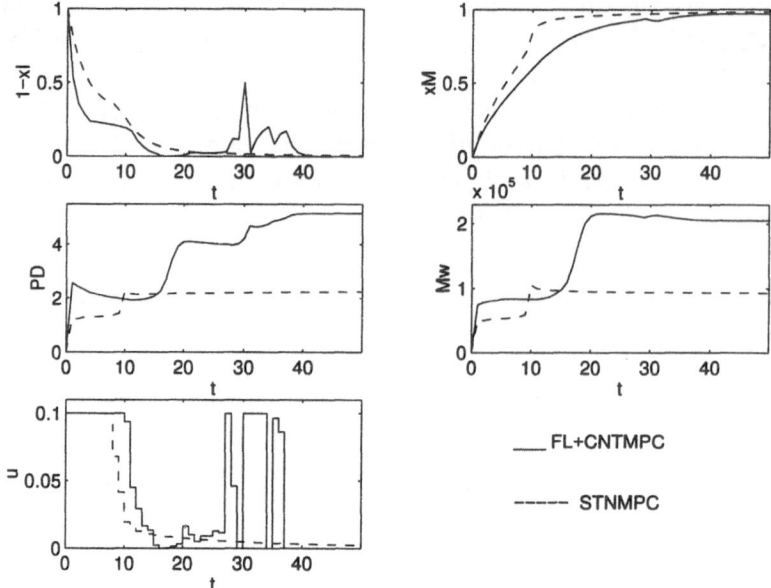

Figure 4. Constrained Nominal Case.

with much higher Mw and Pd when compared to STNMPC. The input constraints were satisfied for the FL+CNTMPC algorithm in spite of the approximate implementation. Once again, the higher computational efficiency of FL+CNTMPC is evident.

6.3. ROBUST CASE

6.3.1. *Robust Case 1: Exponentially Convergent Parameters*

In the nonlinear context, the influence of disturbances and parameters which appear nonlinearly in the system dynamics is not distinguishable. Therefore, since the temperature enters the system equations via the nonlinear expressions for the reaction rates, we will investigate here the system response to an exponentially decaying disturbance on the nominal value of the reactor temperature. Thus, the temperature used in the simulation of the model is $\theta = 373.15$ K and the real temperature of the plant $\theta^p(t)$ behaves according to the following expression:

$$\theta^p(t) = \theta + \Delta\theta_0\, e^{-\alpha_\theta\,(t-t_0)} \tag{42}$$

where $\Delta\theta_0 = \theta^p(t_0) - \theta$ represents the perturbation introduced on the reactor temperature at t_0. Notice that the temperature converges exponentially to its nominal value θ at a rate $\alpha_\theta > 0$.

In the legends for the next figures, FL+CNTMPC (model) (STNMPC (model)) represents the responses of the model used in the prediction of the FL+CNTMPC (STNMPC) controller algorithm, while FL+CNTMPC (plant) (STNMPC (plant)) represents the responses of the plant when the FL+CNTMPC (STNMPC) controller is employed.

In the following simulations we adopted $\Delta\theta_0 = 5$, $\alpha_\theta = 0.1$ (FL+CNTMPC) and $\alpha_\theta = 1$ (STNMPC).

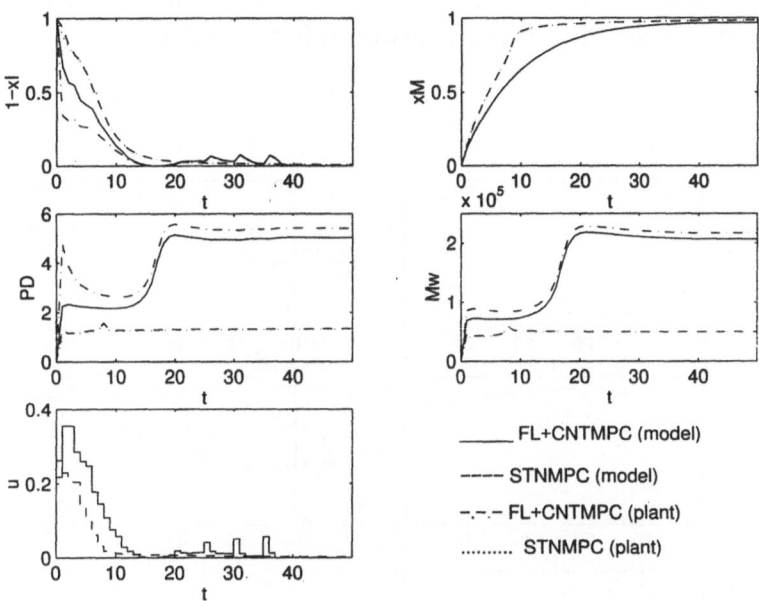

Figure 5. Unconstrained Robust Case 1.

Unconstrained In *Figure 5* we notice that both controllers repeat their characteristic behaviors. However, it is worth noticing that we used a rate of decay α_θ for STNMPC 10 times higher than that used for FL+CNTMPC, for the same initial perturbation $\Delta\theta_0 = 5$. Thus, the disturbance is much more persistent in the case of FL+CNTMPC. We tried to use the same α_θ for STNMPC but the simulation stopped after a few time steps (after several re-starts of NPSOL). We expect that a larger P would allow the STNMPC algorithm to deal with this more persistent disturbance. However, since the STNMPC schemes tend to be very computationally demanding, it was not practical to try to increase P any further.

In the case of FL+CNTMPC, it is interesting to notice that the plant output does not feel the disturbance effect because of the feedback lineariza-

460

tion (since the inverse transformation from z_1 to x_M is not affected by any parameters, $z_1 = x_M$). Thus, the plant and model output responses for FL+CNTMPC will always coincide in spite of any parameter uncertainty. The same does not apply for structural uncertainty.

Both controllers completely reject the disturbance. The plant responses of Pd and Mw show constant offset with respect to those of the model. This is due to the fact that the first three moments of the "dead" polymeric species in the reactor (used to calculate Pd and Mw) are computed based on an integration which accumulates effects of past disturbances on $x^p(t)$. As the disturbance vanishes and the states of the plant and those of the model tend to coincide, then a constant offset is observed.

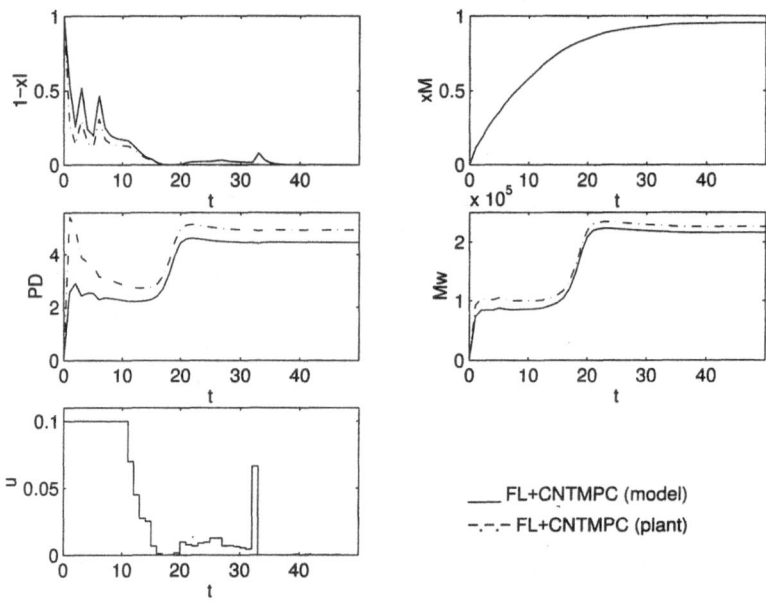

Figure 6. Constrained Robust Case 1.

Constrained From *Figure 6* we conclude that the FL+CNTMPC controller is still robust to the disturbance, in spite of the input constraints. As in the constrained nominal case, displayed in *Figure 4*, the input constraints are not violated in the robust case either even with the approximate implementation.

As with the unconstrained case (*Figure 5*), we were not able to run the constrained simulation for the STNMPC algorithm with $\alpha_\theta = 0.1$ because

NPSOL stops after several re-starts. Since the constraints do not alter much the response obtained with $\alpha_\theta = 1$, we skipped these results here.

6.3.2. *Robust Case 2: Constant Parameter Mismatch*

In the following simulations we chose the parameters α_g, β_g and γ_g (the parameters of the gel effect in equation (37)) to be 20% smaller than their nominal values displayed in Table 1.

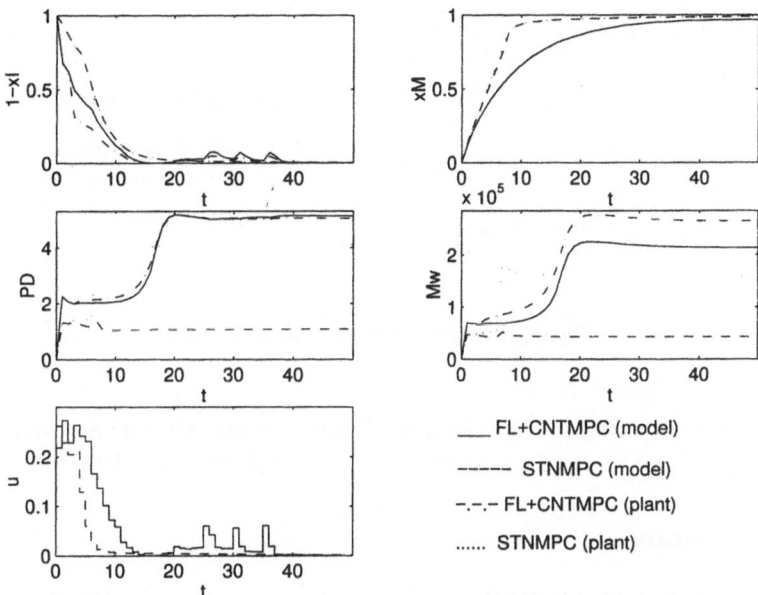

Figure 7. Unconstrained Robust Case 2.

Unconstrained In *Figure 7* we can really notice the effect of the contractive constraint on the robust stabilization of the system. The state responses show that both controllers are robustly stable. However, the Pd and Mw responses show that FL+CNTMPC is much more robust to the constant parameter mismatch. In the case of STNMPC, the model predicts that Pd and Mw will be much lower than their real plant values. For FL+CNTMPC this difference is much smaller and the prediction is more realistic.

Constrained Compared to *Figure 7*, *Figure 8* shows that the input constraints actually improve the behavior of STNMPC, decreasing significantly the difference between Pd and Mw of the plant and model. However, this naturally occurs at the expense of a slower output response.

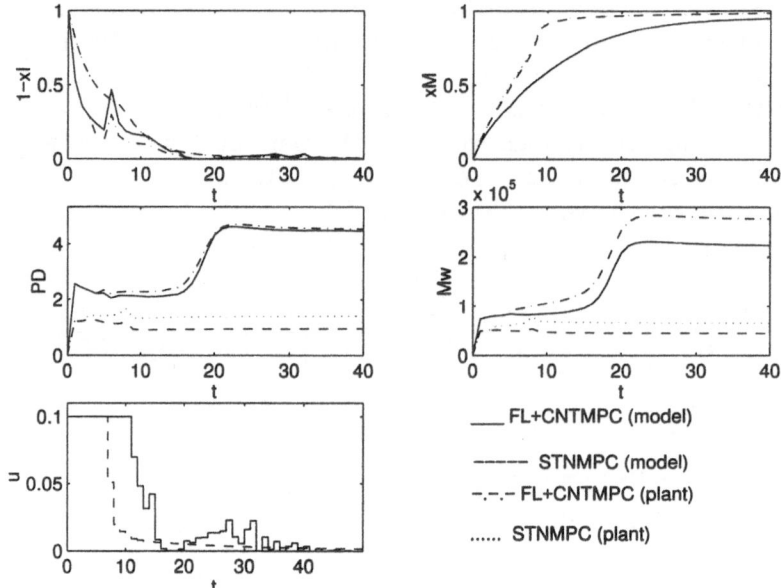

Figure 8. Constrained Robust Case 2.

Also by comparison with *Figure 7*, we can see that the constraints do not compromise the performance obtained with FL+CNTMPC.

7. Conclusions

In this paper a new controller scheme, namely, FL+CNTMPC, was proposed. The objectives of this new technique are twofold: first, use the I/O feedback linearization of the system to simplify the model used in the optimization step of the MPC algorithm without approximations (opposite to what happens when successive linearizations are employed); second, guarantee closed-loop stability and robustness via the introduction of the contractive constraint into the formulation of the MPC scheme. The contractive constraint applied onto the transformed state variables guarantees uniform asymptotic stability of the original nonlinear system in the nominal case and in the presence of uncertain, asymptotically convergent parameters. In the case of general time-varying, bounded, unknown parameters, state convergence to a bounded set whose size decreases as the parameter mismatch also does, can be assured.

The FL+CNTMPC scheme was applied to a batch polymerization reactor, an important chemical engineering process known to display highly nonlinear behavior. The reactor is only one example of a class of batch processes which can be significantly simplified via the use of I/O feedback linearization and then subsequently controlled by the CNTMPC algorithm which takes into account all the possible process constraints into its formulation. Since these constraints can dramatically increase the complexity of the optimization problem solved at each time step, we proposed here different approximate alternatives for efficient implementation of these constraints. In some special cases, the nonlinear programming problem can even be reduced to a QP.

Acknowledgements: The first author gratefully acknowledges the financial support provided by the Fundação de Amparo à Pesquisa do Estado do Rio de Janeiro (FAPERJ) under grants E-26/150.984/96 and E-26/170.943/96 and by the Fundação Coordenação de Aperfeiçoamento de Pessoal de Nível Superior (CAPES) under grant AEX0782/97-1. The second and third authors would like to thank the Instituto Militar the Engenharia (IME) and the Conselho Nacional de Desenvolvimento Científico (CNPq) for the financial aid provided, respectively.

References

1. D.K. Adebekun and F.J. Schork. Continuous Solution Polymerization Reactor Control. 1.Nonlinear Reference Control of Methyl Methacrylate Polymerization. 2.Estimation and Nonlinear Reference Control during Methyl Methacrylate Polymerization. *Ind. Eng. Chem. Res.*, 28:1308–1324,1846–1861, 1989.
2. W.M. Boothby. *An Introduction to Differentiable Manifolds and Riemannian Geometry.* Academic Press, 1975.
3. S. L. De Oliveira. *Model Predictive Control (MPC) for Constrained Nonlinear Systems.* PhD thesis, California Institute of Technology, Pasadena, CA, March 1996.
4. S. L. De Oliveira and M. Morari. Contractive Model Predictive Control with Local Linearization for Nonlinear Systems. *Automatica*, 1996. submitted.
5. S. L. De Oliveira and M. Morari. Contractive Model Predictive Control for Constrained Nonlinear Systems. *IEEE Transactions in Automatic Control*, 1997. (in press).
6. S. L. de Oliveira, V. Nevistic, and M. Morari. Control of Nonlinear Systems Subject to Input Constraints. In *Proceedings of the IFAC Nonlinear Control Design Symposium*, volume 1, pages 15–20, Lake Tahoe, CA, June 1995.
7. J. M. R. Fontoura. *Controle de um Reator de Polimerização Descontinuo.* PhD thesis, Federal University of Rio de Janeiro, April 1996.
8. A. Isidori. *Nonlinear Control Systems: An Introduction.* Lectures Notes in Control and Information Science. Springer-Verlag, 1985.
9. L. Kershenbaum, D.Q. Mayne, R. Pytlak, and R.B. Vinter. Receding Horizon Control. In *Advances in Model-Based Predictive Control*, volume 2, pages 233–246, September 1993.
10. M. V. Kothare, V. Nevistić, and M. Morari. Robust Control of Nonlinear Systems: A Comparative Study. In *1995 AICHE Annual Meeting (paper no. 180e)*, Miami, FL, November 1995.
11. C. Kravaris and J.C. Kantor. Geometric Methods for Nonlinear Process Control. 1. Background. 2. Controller Synthesis. *Ind. Eng. Chem. Res.*, 29:2295–2323, 1990.
12. W. H. Kwon. Advances in Predictive Control: Theory and Application. In *1st Asian*

464

Control Conf., Tokyo, July 1994. (updated in October, 1995).

13. D. Q. Mayne. Optimization in Model Based Control. In J. B. Rawlings, editor, 4th *IFAC Symposium on Dynamics and Control of Chemical Reactors, Distillation Columns, and Batch Processes (DYCORD+ '95)*, pages 229–242. Danish Automation Society, June 1995.

14. D.Q. Mayne and H. Michalska. Receding Horizon Control of Constrained Nonlinear Systems. *IEEE Trans. Aut. Control*, 38(11):1623–1633, 1993.

15. V. Nevistić and L. Del Re. Feasible Suboptimal Model Predictive Control for Linear Plants with State Dependent Constraints. In *American Control Conf.*, pages 2862–2866, Baltimore, MD, June 1994.

16. V. Nevistić and M. Morari. Constrained Control of Feedback-Linearizable Systems. In *Proceedings of the European Control Conference*, pages 1726–1731, Rome, Italy, 1995.

17. V. Nevistić and M. Morari. Robustness of MPC-based Schemes for Constrained Control of Nonlinear Systems. In *Proceedings of the IFAC 13th Triennial World Congress*, volume M, pages 25–30, San Francisco, CA, 1996.

18. H. Schuler and Z. Suzhen. Real-Time Estimation of the Chain Length Distribution in a Polymerization Reactor. *Chemical Engineering Science*, 40(10):1891–1904, 1985.

19. M. Soroush and C. Kravaris. Nonlinear Control of a Batch Polymerization Reactor: An Experimental Study. *AIChE Journal*, 38(9):1429–1448, 1992.

20. R. Weber. Prädiktive Regelung von nichtlinearen Systemen. Semester work (IfA 8781), ETH Zürich, Institut für Automatik, July 1995. supervised by S. De Oliveira, V. Nevistić and M. V. Kothare.

21. T. H. Yang and E. Polak. Moving horizon control of nonlinear systems with input saturation, disturbances and plant uncertainty. *Int. J. Control*, 58(4):875–903, September 1993.

NONLINEAR MODEL PREDICTIVE CONTROL SCHEMES WITH GUARANTEED STABILITY

H. CHEN*
Institut für Systemdynamik und Regelungstechnik
Universität Stuttgart, 70550 Stuttgart, Germany
email: chen@isr.uni-stuttgart.de

F. ALLGÖWER
Institut für Automatik, Eidg. Techn. Hochschule Zürich
CH-8092 Zürich, Switzerland
email: allgower@aut.ee.ethz.ch

1 An introduction to nonlinear model predictive control

Model predictive control (MPC), also referred to as moving horizon control or receding horizon control, has become an attractive feedback strategy, especially for linear or nonlinear systems subject to input and state constraints. In general, linear and nonlinear MPC are distinguished. Linear MPC refers to a family of MPC schemes in which linear models are used to predict the system dynamics, even though the dynamics of the closed-loop system is nonlinear due to the presence of constraints. Linear MPC approaches have found successful applications, especially in the process industries (Richalet, 1993). A complete overview on industrial MPC techniques with details and comparisons is given by Qin and Badgwell (1996), where more than 2200 applications in a very wide range from chemicals to aerospace industries are also summarized. By now, linear MPC theory is quite mature. Important issues such as stability are well addressed (see for example (Lee, 1996) for an overview).

Systems in the process industry are, however, in general inherently nonlinear, although linear models are widely used to solve control problems. In addition, higher product quality specifications and increasing productivity demands, tighter environmental regulations and demanding economical considerations in the process industry require systems to be increasingly complex and to be operated closer to the boundary of the admissible operating region. In these cases, linear models are

---

*currently with: Department of Electronic Engineering, Jilin University of Technology, 130025 Changchun, PR China. Email: chenh@jut.edu.cn

R. Berber and C. Kravaris (eds.), Nonlinear Model Based Process Control, 465-494.
© 1998 *Kluwer Academic Publishers.*

obviously not adequate to describe the process dynamics and nonlinear models have to be used.

In this paper, we want to review some of the available nonlinear MPC schemes with an emphasis on stability and robustness issues. We mainly concentrate on the continuous time case, but sometimes state the discrete version if this simplifies the exposition.

1.1 Principle and properties

In general, the model predictive control problem is formulated as solving on-line a finite horizon open-loop optimal control problem subject to system dynamics and constraints involving states and controls. Figure 1 shows the general principle of model predictive control. Based on the measurement obtained at time t and using

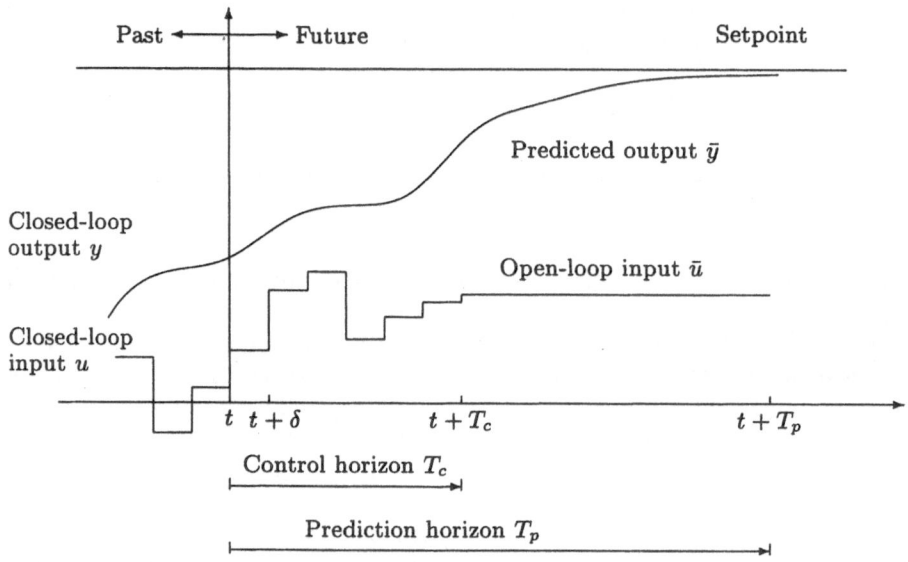

Figure 1: Principle of model predictive control.

a nominal model of the system being controlled, the controller predicts the future dynamic behavior of the system over a prediction horizon T_p and determines an open-loop manipulated input function (over a control horizon $T_c \leq T_p$) in such a way that some prespecified open-loop performance objective functional is optimized (for example, an integral square error between the predicted output and the setpoint). If the plant is stable and if there were no disturbances and model-plant mismatch, and if the optimization problem could be optimally solved for infinite prediction and control horizons, then we could apply the input function found at time $t = 0$ to the system for all times $t \geq 0$. However, this is not possible in general. Due to disturbances and/or model-plant mismatch, the true system behavior is different from the predicted behavior. In order to incorporate some feedback mechanism into the scheme, the open-loop manipulated input function found will be implemented only until the next measurement becomes available. We assume that this will be the case every δ time-units, where δ denotes the "sampling time". Updated with the new measurement, at time $t + \delta$, the whole procedure – prediction and optimization – is repeated to find a new input function, with the control and prediction horizons moving forward, (for this reason, MPC is also referred to as moving horizon control or receding horizon control). That results in a discrete feedback control with an implicit control law, because closed-loop inputs are calculated by solving on-line the optimization problem at each sampling time. Thus, MPC is characterized by the following features:

- **Prediction:** In contrast to other feedback controllers that calculate the control action based on the present or past state (or output) information, model predictive controllers determine the control action based on the predicted future dynamics of the system being controlled. The model used to complete the prediction can be linear or nonlinear, time-continuous or discrete, deterministic or stochastic, etc. Because the future response of the system is predicted, early control action can be taken so as to track, for example, a given reference trajectory with minimal tracking error.

- **Optimization:** An objective functional that mathematically specifies the desired control performance is minimized on-line at each sampling instance. A commonly used objective functional is an integral weighted square error between predicted controlled variables and their desired references.

- **Discrete manipulated inputs:** Manipulated inputs are presented in a finite horizon and discrete manner, because, in general, the involved optimization problem can only be solved with numerical methods, especially when nonlinear systems and/or constraints are considered. A time-continuous input parameterization and/or infinite prediction and control horizons lead to an infinite dimensional optimization problem that is numerically extremely demanding and often unsolvable. In order to get around that, the optimization problem is formulated with finite horizons and the inputs are discretized according to the sampling time and in the simplest case kept constant during the period between the sampling instances, as depicted in Figure 1.

- **Constraints:** In practice, most systems have to satisfy constraints on inputs and states. For example, an actuator reaches saturation and some states such as temperature and pressure are not allowed to exceed their limitations for the reason of safe operation, or some variables have to be held under certain threshold values to meet environmental regulations. Constraints automatically impose limitations on the achievable control performance, even if the system to be controlled is linear (Mayne, 1995). In extreme cases, the presence of constraints can destroy closed-loop stability, if they are not considered when a controller is designed (Chen, 1997). Even for cases where the constrained closed-loop systems are stable, the model predictive controller delivers better control performance, since it is able to take the current control action to minimize the errors caused by constraints that are predicted to become active in the future (Eaton and Rawlings, 1992). Thus, the feature of prediction provides in some sense an "active" handling of constraints, compared to other "passive" methods (e.g. (Gilbert and Tan, 1991; Mayne and Schroeder, 1994)).

- **Multivariable systems:** Multivariable controllers are often required to improve control performance. In the framework of MPC, multiple inputs and multiple outputs can be straightforwardly handled by using multivariable models for the prediction.

1.2   Model predictive control vs. optimal control

Theoretically, an optimal feedback control law can be found as a solution of the corresponding Hamilton-Jacobi-Bellman partial differential equation (Bryson and Ho, 1969). However, in practice, solving the Hamilton-Jacobi-Bellman partial differential equation is very difficult, if not impossible, unless linear systems, quadratic objective functionals and no constraints are considered. In MPC, instead of solving the Hamilton-Jacobi-Bellman partial differential equation, an *open-loop* optimal control problem is solved on-line repeatedly, initialized at the actual state. This is a considerably simpler task (Mayne, 1995), despite of the still expensive on-line computational demand. Especially for nonlinear systems subject to constraints, finite horizon nonlinear MPC provides a feasible approach to do optimal control in some sense.

1.3   State space models vs. I/O models

The history of MPC began with an attempt to use the powerful computer technology to improve the control of chemical processes that are constrained, multivariable and uncertain. In the late 70's and early 80's, Richalet and coworkers (Richalet *et al.*, 1976; Richalet *et al.*, 1978) developed Model Predictive Heuristic Control (MPHC) that is also referred to as Model Algorithmic Control (MAC) (Rouhani and Mehra, 1982), and Cutler and Ramaker (1980) developed Dynamic Matrix Control (DMC). Due to their obvious advantages and successful applications, MPC has attracted more and more researchers from both the industrial and academic sides. Many formulations of the MPC problem have been proposed and implemented such as Linear Dynamic Matrix Control (Morshedi *et al.*, 1985) (LDMC), Quadratic Dynamic Matrix Control (QDMC) (García and Morshedi, 1986), Receding Horizon Tracking Control (RHTC) (Kwon and Byun, 1989), Generalized Predictive Control (GPC) (Clarke *et al.*, 1987*a*; Clarke *et al.*, 1987*b*), Internal Model Control (IMC) (García and Morari, 1982; García and Morari, 1985*a*), Extended Horizon Adaptive Control (EHAC) (Ydstie, 1984), Extended Prediction Self-Adaptive Control (EPSAC) (De Keyser and Van Cauwenberghe, 1985). Some of these methods have different origins and different motivation. For example, GPC was meant as an extension to certain self-tuning control and adaptive control methods. For more complete surveys and comparisons between various MPC approaches we refer for example to (De Keyser *et al.*, 1988; García *et al.*, 1989; Lee, 1996; Qin and Badgwell, 1996).

The main difference between the early versions of MPC lies in the type of models used to predict the future dynamics of the system being controlled. Impulse response models are used in MPHC, DMC is based on step response models, while GPC includes ARIMAX (Auto-Regressive Integrated Moving-Average with eXogenous input) models as are standard in the theory of prediction of stochastic processes.

Convolution models (impulse or step response models), where only input-output behavior is considered, are used in almost all early versions of MPC, for the reason that they can be easily obtained from process tests and signal processing techniques without significant fundamental modeling effort, even for very complex processes. However, convolution models are in general nonminimal representations and merely suitable to describe strictly stable systems. A convolution model with an integrator is able to approximately model an unstable system (Eaton and Rawlings, 1992; Hovd *et al.*, 1993; Lee *et al.*, 1994), however, the structure error in the model imposes a limitation on the achievable control performance. In addition, it is difficult to extend linear MPC approaches with convolution models to nonlinear systems. Using the special case of second-order Volterra models, Genceli and Nikolaou (1995), Doyle III and coworkers (Doyle III *et al.*, 1995; Maner *et al.*, 1994) have extended the DMC scheme in that respect. However, in order to represent nonlinear systems with a certain accuracy, much higher order Volterra models are required, whose coefficients may not be easy to identify any more. Another shortcoming of using convolution models is that a theoretical analysis of the closed-loop properties such as stability is difficult.

The interpretation of MPC in the state space setting, presented for example by Morari and Lee (1991) and Ricker (1991), provides a unified framework to understand and generalize the existing MPC techniques, and more importantly, to systematically investigate the properties of the closed-loop system in the context of optimal control. In addition, state space model based MPC approaches can in principle be straightforwardly extended to nonlinear systems, just using nonlinear state space models.

1.4 Finite horizon predictive control

Consider a class of systems being described by the following general nonlinear set of ordinary differential equations

$$\dot{x}(t) = f\left(x(t), u(t)\right), \quad x(0) = x_0 \tag{1}$$

subject to input and state constraints

$$u(t) \in U, \ \forall t \geq 0 \tag{2a}$$

$$x(t) \in X, \ \forall t \geq 0, \tag{2b}$$

where $x(t) \in I\!\!R^n$ and $u(t) \in I\!\!R^m$ denote the vectors of states and manipulated inputs, respectively. In the simplest form, U and X are specified as follows:

$$U := \{u \in I\!\!R^m | u_{min} \leq u \leq u_{max}\} \tag{3a}$$

$$X := \{x \in I\!\!R^n | x_{min} \leq x \leq x_{max}\}, \tag{3b}$$

where $u_{min}, u_{max}, x_{min}$ and x_{max} are given constant vectors.

In order to distinguish clearly between the system and the system model used to predict the future "within" the controller, we denote the internal variables in the controller by a bar (for example $\bar{x}, \bar{u}$) to indicate that the predicted values need not and will not be the same as the actual values. This is also true for the undisturbed case with no model-plant mismatch, if only a finite horizon is used. Instead of the continuous time t a discrete time k will be used for discrete systems, i.e., $k, x(k), u(k)$ and so on will be used.

1.4.1 Controller setup

Usually, the finite horizon open-loop optimal control problem described in Section 1.1 is mathematically formulated as follows:

Problem 1 *Find*

$$\min_{\bar{u}(\cdot)} J\left(x(t), \bar{u}(\cdot), T_c, T_p\right) \tag{4}$$

with

$$J\left(x(t), \bar{u}(\cdot), T_c, T_p\right) := \int_t^{t+T_p} F\left(\bar{x}(\tau), \bar{u}(\tau)\right) d\tau \tag{5}$$

subject to

$$\dot{\bar{x}} = f(\bar{x}, \bar{u}), \ \bar{x}(t) = x(t) \tag{6a}$$

$$\bar{u}(\tau) \in U, \ \tau \in [t, t+T_c] \tag{6b}$$

$$\bar{u}(\tau) = \bar{u}(t+T_c), \ \tau \in [t+T_c, t+T_p] \tag{6c}$$

$$\bar{x}(\tau) \in X, \ \tau \in [t, t+T_p], \tag{6d}$$

where T_p and T_c are prediction and control horizons with $T_c \leq T_p$.

The function F specifies the desired control performance that can arise for example from economical and ecological considerations. The standard quadratic form is the simplest and most used one:

$$F(x, u) = (x - x_s)^T Q(x - x_s) + (u - u_s)^T R(u - u_s), \tag{7}$$

where x_s and u_s denote given reference trajectories, that can be constant or time-varying; Q and R denote positive definite, symmetric weighting matrices. More commonly, with output equation $y = g(x)$, function F can be for example chosen as

$$F(y, u) = (y - y_s)^T Q(y - y_s) + (u - u_s)^T R(u - u_s), \tag{8}$$

where y_s are the given output references.

In order for (x_s, u_s) to be feasible, the point (x_s, u_s) should be contained in the interior of $X \times U$. For the stabilization problem, it can be assumed, without loss of generality, that $(0,0)$ is the steady state point of interest of the system. Note the initial condition in (6a): The system model used to predict the future in the controller is initialized by the actual system states, that are in general assumed to be measured or must be estimated. Equation (6c) is not a constraint but implies that beyond the control horizon the predicted control takes a constant value equal to that at the last step of the control horizon. For a numerical implementation, the input function is in general parameterized in a step-shaped manner. As shown in Figure 1, $\bar{u}(\cdot)$ on $[t, t + T_c]$ satisfies $\bar{u}(\tau) = const$ for $\tau \in [t + j\delta, t + (j+1)\delta), j = 0, 1, \cdots, N_c$, where $N_c = \frac{T_c}{\delta}$ and δ denotes the sampling period.

An optimal solution to the optimization problem (existence assumed) is denoted by $\bar{u}^*(\cdot; x(t), T_c, T_p) : [t, t + T_c] \to U$. According to the principle of MPC, the open-loop optimal control problem will be solved repeatedly, when new measurements are available at the sampling instances $t = j\delta$, $j = 0, 1, \cdots$. The optimal *closed-loop* control is defined by

$$u^*(\tau) := \bar{u}^*(\cdot; x(t), T_c, T_p) , \tau \in [t, t + \delta]. \tag{9}$$

Model-plant mismatch and disturbances are generally not represented in the optimization problem. At most, they are indirectly considered in that their effect is assumed to be constant over the prediction horizon. For example, an additional unknown but constant state vector $d(t) \in \mathbb{R}^q$ that is used to represent disturbances and uncertainties is introduced into the system description:

$$\dot{x} = f(x, u, d) \tag{10a}$$
$$\dot{d} = 0 . \tag{10b}$$

1.4.2 Advantages and drawbacks

Since Problem 1 is formulated in a finite horizon manner and the input function can be finitely parameterized, a numerical solution is then possible in principle. Moreover, the F function in the cost functional can be chosen arbitrarily to meet any economical and ecological requirements. Trajectory (output) tracking problem can also be taken into account in this framework. However, because a finite horizon criterion is not designed to deliver an asymptotic property such as stability (Bitmead et al., 1990), a general finite horizon MPC scheme cannot guarantee closed-loop stability. Closed-loop stability can only be achieved by a suitable tuning of design parameters such as prediction horizon, control horizon and weighting matrices. This makes an independent specification of the desired control performance difficult. In addition, due to the presence of nonlinearities, a system behaves differently for different operating conditions and tuning for stability often cannot deliver satisfying results for various operating conditions. Thus, guaranteed stability, *i.e.*, stability independent of the choice of performance parameters, is of great interest not only in theory but also for practitioners.

In the following, we discuss nonlinear model predictive control schemes that deliver guaranteed stability.

2 Nominal stability of nonlinear MPC schemes

An intuitive way to achieve guaranteed stability is to use an infinite horizon cost functional (Bitmead *et al.*, 1990), that results in general in an infinite dimensional (or, approximated appropriately, a very high dimensional) optimization problem. For nonlinear systems, solving such an optimization problem is very difficult, if not impossible. For linear systems, using an additional terminal term in the cost functional the infinite horizon optimization problem can be transformed into a finite horizon one (Rawlings and Muske, 1993; Scokaert and Rawlings, 1996). In this approach, the prediction is exactly considered over an infinite horizon, but the input function to be determined on-line is only of finite horizon. This way, closed-loop stability can be guaranteed and the optimization problem is numerically solvable. In analogy to the linear case, nonlinear MPC schemes are developed, in which the open-loop optimal control problem is formulated in a finite horizon manner, but the prediction is considered over the infinite horizon in order to achieve closed-loop stability. Besides input and state constraints, so-called *stability constraints* (Mayne, 1995) have to be included into the finite horizon open-loop optimal control problem.

2.1 Predictive control with terminal equality constraints

The most widely suggested stability constraint is the *terminal equality constraint* (Keerthi and Gilbert, 1988; Kleinman, 1970; Mayne and Michalska, 1990)

$$\bar{x}(t + T_p) = \mathbf{0}, \tag{11}$$

that forces the state to be zero at the end of the finite horizon. In the context of LQ control, the terminal equality constraint is quite intuitive, since it is equivalent to the infinite initial condition of the backwards Riccati equation, whose solution constitutes a stabilizing state feedback controller for linear controllable systems (Kleinman, 1970). In fact, if the terminal equality constraint is feasible, *i.e.*, the system at the end of the finite horizon is at the origin and the input thereafter is set to be zero, then the unperturbed nominal system stays at the origin forever. This way, the prediction horizon can be thought of as expanding exactly to infinity.

For continuous systems, Mayne and Michalska (1990) show that under some rather strong assumptions, MPC with the terminal equality constraint (11) and a quadratic objective functional is able to stabilize a class of nonlinear constrained systems. The strong assumptions are needed to ensure that the optimal value function is continuously differentiable and can serve as Lyapunov function for the closed-loop system. Those assumptions are relaxed in (Michalska and Mayne, 1991) to ensure merely local Lipschitz continuity of the optimal value function. Analogous results for discrete systems are proposed by Keerthi and Gilbert (1988) and improved by Rawlings *et al.* (1994), where the optimization problem is formulated as follows:

Problem 2 *Find*

$$\min_{\pi_{N_p}} J\left(x(k), \pi_{N_p}\right) \tag{12}$$

with

$$J\left(x(k), \pi_{N_p}\right) := \sum_{j=0}^{N_p} F\left(\bar{x}(k+j), \bar{u}(k+j)\right) \tag{13}$$

subject to

$$\bar{x}(k+j+1) = f\left(\bar{x}(k+j), \bar{u}(k+j)\right), \ \bar{x}(k) = x(k) \tag{14a}$$
$$\bar{u}(k+j) \in U, \ j \in [0, N_p] \tag{14b}$$
$$\bar{x}(k+N_p) = 0, \tag{14c}$$

where $N_p = \frac{T_p}{\delta}$ *and* $\pi_{N_p} = \{\bar{u}(k), \bar{u}(k+1), \cdots, \bar{u}(k+N_p)\}$.

Under reasonable conditions, asymptotic stability for the discrete closed-loop system can be proven for both finite N_p and $N_p \to \infty$ (Meadows *et al.*, 1995).

Motivated by practical considerations, Genceli and Nikolaou (1995) derive an *end condition* for nonlinear SISO MPC with second-order Volterra models, where the system to be controlled is assumed to be stable. The optimization problem is formulated as follows:

Problem 3 *Find*

$$\min_{\pi_{N_c}} \sum_{j=1}^{N_p} |\bar{y}(k+j) - y_s|^l + \sum_{j=0}^{N_c} r_j |\Delta\bar{u}(k+j)|^l, \ l = 1, 2 \tag{15}$$

subject to

$$\bar{y}(k+j) = \bar{d}(k+j) + \sum_{i=1}^{N} g_i \bar{u}(k+j-i)$$
$$+ \sum_{i_1=1}^{M} \sum_{i_2-1}^{M} g_{i_1 i_2} \bar{u}(k+j-i_1)\bar{u}(k+j-i_2), \ j \in [1, N_p] \tag{16a}$$

$$\bar{d}(k+j) = \bar{d}(k) = y(k) - \sum_{i=1}^{N} g_i u(k-i)$$
$$- \sum_{i_1=1}^{M} \sum_{i_2=1}^{M} g_{i_1 i_2} \bar{u}(k-i_1)\bar{u}(k-i_2), \ j \in [1, N_p] \tag{16b}$$

$$\Delta u_{min} \le \Delta\bar{u}(k+j) \le \Delta u_{max}, \ j \in [0, N_c] \tag{16c}$$
$$\Delta\bar{u}(k+N_c+j) = 0, \ j \in [N_c+1, N_p] \tag{16d}$$
$$u_{min} \le \bar{u}(k+j) \le u_{max}, \bar{u}(k+j) = \bar{u}(k+j-1) + \Delta\bar{u}(k+j), j \in [0, N_c] \tag{16e}$$
$$\bar{d}(k) + Gu(k+N_c) + Gu(k+N_c)^2 = y_s \tag{16f}$$

In this formulation, g_i and $g_{i_1 i_2}$ are the coefficients of the second-oder Volterra model; N and M represent the memory length of the Volterra series; $\Delta \bar{u}$ denotes the increment of the control input $\bar{u}$, $i.e.$, $\Delta \bar{u}(k) = \bar{u}(k) - \bar{u}(k-1)$; $r_j, j = 0, 1, \cdots, N_c$ are control move suppressions. The additional constraint (16f) is referred to as end condition, where

$$G = \sum_{i=1}^{N} g_i, \quad G = \sum_{i_1=1}^{M} \sum_{i_2=1}^{M} g_{i_1 i_2}.$$

This end condition requires the input $u(k + N_c)$ at the end of the finite control horizon taking such a value that N (or M) steps thereafter the predicted output $y(k+N_c+N)$ (or $y(k+N_c+M)$) reaches the setpoint y_s, provided that the input and the disturbance are constant. Thus, the end condition is an extension of the terminal equality constraint to the output feedback case. With this end condition, closed-loop stability is achieved under some restrictions on control move suppressions in the objective functional. Since Volterra models describe in general the input-output behavior of the nonlinear system approximately, the achieved stability is in the input-output sense.

Guaranteeing stability by imposing a terminal equality constraint is by far the most widely used technique at present. For one, this is certainly due to the clear theoretical framework, but also due to the fact that *no* off-line computation or tuning is needed. On the other hand, a terminal equality constraint is an artificial additional burden that requires significant(!) extra on-line computation cost (see (Chen and Allgöwer, 1997a; Chen and Allgöwer, 1997b) for a comparison to other approaches) and even more importantly, leads to a mostly severely restricted region of operation due to feasibility problems. This approach can thus only be recommended with some reservations for practical applications.

2.2 Dual-mode variable horizon predictive control

From a computational point of view, an *exact* numerical satisfaction of terminal equality constraints used in the last section requires an infinite number of iterations in the nonlinear case. An approximate satisfaction means that the achieved stability is lost inside the region of approximation. In order to get around this, Michalska and Mayne (1993) propose a *terminal inequality constraint* such that the states are on the boundary of a *terminal region* at the end of a *variable* prediction horizon. The terminal region is a region of attraction for the nonlinear system controlled by a local linear state feedback law $u = Kx$ and may be small. They suggest a *dual-mode* receding horizon control scheme with the local linear state feedback controller being applied inside the terminal region and a receding horizon controller being applied outside the terminal region. Closed-loop control with this scheme is implemented by switching between the two controllers, depending on the states being inside or outside the terminal region, This is shown in Figure 2, where the solid line represents a finite (variable) horizon closed-loop trajectory generated by the predictive controller; the dotted line is the subsequent closed-loop trajectory generated by the local linear state feedback controller. Instead of terminal equality

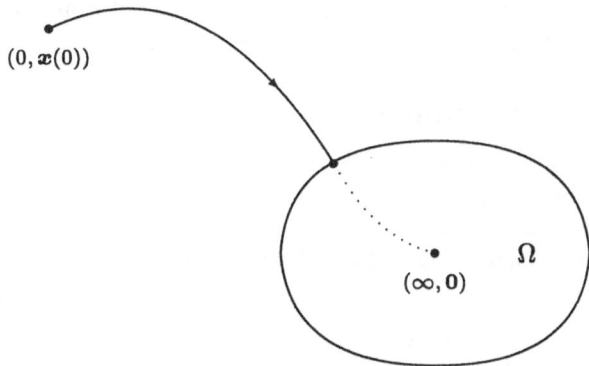

Figure 2: Dual-mode nonlinear MPC.

constraints (11), the terminal inequality constraint

$$\bar{x}(t + T_p) \in \Omega \tag{17}$$

is introduced into the optimization problem (Problem 1). The length of the control horizon (which is assumed to be equal to the length of the prediction horizon, $T_c = T_p$) is an additional minimizer, *i.e.*, (4) is replaced by

$$\min_{\bar{u}(\cdot),T_p} J\left(x(t), \bar{u}(\cdot), T_p\right) . \tag{18}$$

Thus, the feasibility of the terminal inequality constraint implies that the state at the end of the finite horizon is on the boundary of the terminal region. They have also pointed out that the feasibility of the optimization problem, and not necessarily the optimality, is needed for closed-loop stability. The controller algorithm can be summarized as follows:

Algorithm 1 *If $x(t) \notin \Omega$, solve the optimization problem (18) with (5) subject to (6) and (17) to find a feasible input function $\bar{u}^*(\cdot; x(t), T_p) : [t, t + T_p] \to U$ and a prediction horizon T_p^*. The closed-loop control is defined by (9).*
If $x(t) \in \Omega$, apply the linear feedback control law $u^* = Kx$.

Under fairly weak conditions, it can be shown that starting from outside the terminal region, the nonlinear system with the predictive controller will reach the terminal region in a finite time. Closed-loop stability follows from the use of the stabilizing local linear feedback law thereafter.

Computationally this approach is much more attractive than the one imposing a terminal equality constraint, as inequality constraints can be handled much more effectively during optimization than equality constraints. This holds despite the introduction of the additional scalar minimizer T_p. In addition, less feasibility problems have to be expected and hence the region of attraction will be larger. The most significant drawback of this approach is the involved implementation due to the required switching between control strategies and the need to determine a local stabilizing state feedback gain K and the terminal region Ω.

2.3 Contractive predictive control

Yang and Polak (1993) propose another way to achieve closed-loop stability of nonlinear MPC. They introduce a stability constraint imposing the state vector to contract by a prespecified factor. This constraint is referred to as *contraction constraint*. The optimization problem is then formulated as follows:

Problem 4 *Find*

$$\min_{\bar{u}(\cdot), T_c} J\left(x(t), \bar{u}(\cdot), T_c\right) \tag{19}$$

subject to

$$\dot{\bar{x}} = f(\bar{x}, \bar{u}), \ \bar{x}(t) = x(t) \tag{20a}$$

$$\bar{u}(\tau) \in U, \ \tau \in [t, t + T_c] \tag{20b}$$

$$T_c \in [t + T_1, t + T_2] \tag{20c}$$

$$\bar{x}(\tau) \in X, \ \tau \in [t, t + T_c] \tag{20d}$$

$$\|\bar{x}(t + T_c)\|^2 \leq \alpha^2 \|x(t)\|^2 \tag{20e}$$

$$\max_{\tau \in [t, t + T_c]} \|\bar{x}(\tau)\|^2 \leq \beta^2 \|x(t)\|^2, \tag{20f}$$

where T_1 is the least time needed to solve the optimization problem and T_2 is an a priori upper limit on the control horizon.

The constraints (20e) and (20f) with $\alpha \in (0, 1)$ and $\beta \in [1, \infty)$ are used to ensure closed-loop stability. Different from standard MPC, the *entire* input function over the interval $[t, t + T_c]$ found by solving the optimization problem (Problem 4) will be applied to the nonlinear system. Closed-loop exponential stability is proven by showing that $\|x(t + jT_c)\|^2 \leq \alpha^{2j} \|x(t)\|^2 \to 0$ as $j \to \infty$ holds, when the existence of a solution to the optimization problem at the time $t + jT_c$ with $j = 0, 1, 2, \cdots$ is assumed. However, this is a very strong assumption and cannot be guaranteed in general (Mayne, 1995). A complete algorithm and stability proof can be found in (De Oliveira and Morari, 1997).

The main advantage of this approach is its exponential stability property, because of which asymptotic stability for observer based contractive predictive controllers can be established (see Section 4). However, this approach is not very attractive for applications, since the feasibility of the optimization problem at each sampling instance cannot be guaranteed. Choosing large $\alpha \in (0, 1)$ might obtain the required feasibility, but large α implies a slow contraction of the state vector. Moreover, despite the free specification of the desired control performance by the objective functional, the achieved control performance will be influenced directly by the choice of design parameters α and β.

2.4 Linear MPC + feedback linearization

Nevistić and Morari (1995) combine state feedback linearization (FL) and the stability results of linear MPC with constraints to achieve stability for nonlinear, feedback linearizable systems. Figure 3 shows the schematic representation of this controller structure. A similar approach is taken independently by Kurtz and Henson (1996).

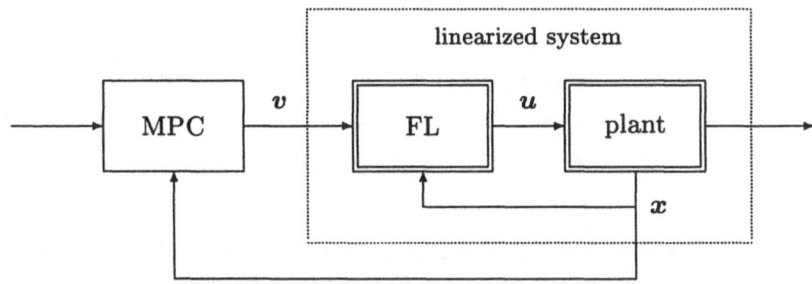

Figure 3: Controller structure: MPC + FL.

However, because the state linearization feedback law (or the input-output linearization feedback law)

$$u = \alpha(x) + \beta(x)v$$

is state-dependent and generally nonlinear, the originally linear input constraints are transformed into state-dependent and in general nonlinear constraints. In addition, the objective functional will not be quadratic any more and has to be approximated by a quadratic functional in order to apply quadratic programming techniques online. The theory of both feedback linearization and linear MPC being comparatively mature, the difficulty of using the "MPC+FL" technique remains in the effective approximation of input constraints and the objective functional in the new coordinates. Another drawback of this approach is that all nonlinearities are compensated independent of whether they might be helpful for achieving the control goal. Thus, this approach usually leads to a worse closed-loop performance, as a general nonlinear scheme is able to benefit from nonlinearities. From an implementational point of view, this scheme is also quite involved as it is known that feedback linearization control laws can become *very* complex.

2.5 Quasi-infinite horizon nonlinear model predictive control

In order to approximately extend the prediction horizon to infinity, a *terminal penalty term* $E(x(t + T_p))$ is introduced in (Chen and Allgöwer, 1996; Chen and Allgöwer, 1997a) into the finite horizon objective functional. This terminal penalty term is determined off-line such that it bounds from above the infinite horizon cost of the nonlinear system controlled by a local state feedback law in a terminal region Ω. The resulting optimization problem is formulated as follows:

Problem 5 *Find*

$$\min_{\bar{u}(\cdot)} J\left(x(t), \bar{u}(\cdot), T_p\right) \tag{21}$$

with

$$J\left(x(t), \bar{u}(\cdot), T_p\right) = \int_t^{t+T_p} F\left(\bar{x}(\tau), \bar{u}(\tau)\right) \, d\tau + E(\bar{x}(t+T_p)) \tag{22}$$

subject to

$$\dot{\bar{x}} = f(\bar{x}, \bar{u}), \quad \bar{x}(t) = x(t) \tag{23a}$$
$$\bar{u}(\tau) \in U, \quad \tau \in [t, t+T_p] \tag{23b}$$
$$\bar{x}(\tau) \in X, \quad \tau \in [t, t+T_p] \tag{23c}$$
$$\bar{x}(t+T_p) \in \Omega. \tag{23d}$$

The basic idea behind this setup is that it approximates an infinite horizon prediction for achieving closed-loop stability, while the input function to be determined on-line has to be of finite horizon only. This way, the resulting optimization problem can be solved numerically. Consider an infinite horizon cost functional defined by

$$J^\infty\left(x(t), \bar{u}(\cdot)\right) := \int_t^\infty F\left(\bar{x}(\tau), \bar{u}(\tau)\right) \, d\tau$$

with $\bar{u}(\cdot)$ on $[t, \infty)$. This cost functional can be split up into two parts

$$\min_{\bar{u}(\cdot)} J^\infty\left(x(t), \bar{u}(\cdot)\right) = \min_{\bar{u}(\cdot)} \left\{ \int_t^{t+T_p} F\left(\bar{x}(\tau), \bar{u}(\tau)\right) d\tau \right.$$
$$\left. + \int_{t+T_p}^\infty F\left(\bar{x}(\tau), \bar{u}(\tau)\right) d\tau \right\}. \tag{24}$$

The goal is to approximate the second term on the right hand side of (24) by a terminal penalty term. Without further restrictions, this is usually not possible for general nonlinear systems. However, if we can be sure that the trajectories of the closed-loop system remain within some neighborhood of the origin (terminal region) for the time interval $[t+T_p, \infty)$, then, an upper bound on this second term can be found. The terminal region Ω is constructed such that a local state feedback law $u = k(x)$ asymptotically stabilizes the nonlinear system in Ω and renders Ω invariant for the closed loop. Because of the invariance property of Ω, it suffices to impose an additional terminal (point) inequality constraint $x(t+T_p) \in \Omega$ (see (23d)) in order to ensure that the trajectories indeed remain in Ω beyond time $t+T_p$. With $u = k(x)$ for the time interval $[t+T_p, \infty)$, we now have

$$\min_{\bar{u}(\cdot)} J^\infty\left(x(t), \bar{u}(\cdot)\right) \leq \min_{\bar{u}(\cdot)} \left\{ \int_t^{t+T_p} F\left(\bar{x}(\tau), \bar{u}(\tau)\right) d\tau \right.$$
$$\left. + \int_{t+T_p}^\infty F\left(\bar{x}(\tau), k\left(\bar{x}(\tau)\right)\right) d\tau \right\}. \tag{25}$$

If the terminal region Ω and the terminal penalty term are chosen suitably (see below) such that

$$\int_{t+T_p}^{\infty} F\left(\bar{x}(\tau), k\left(\bar{x}(\tau)\right)\right) d\tau \leq E\left(\bar{x}(t+T_p)\right), \qquad (26)$$

then, substituting (26) into (25) leads to

$$\min_{\bar{u}(\cdot)} J^{\infty}\left(x(t), \bar{u}(\cdot)\right) \leq \min_{\bar{u}(\cdot)} J(x(t), \bar{u}(\cdot), T_p). \qquad (27)$$

This implies that the optimal value of the finite horizon optimization problem (Problem 5) bounds that of the corresponding infinite horizon optimization problem. In this sense, the prediction horizon in the discussed nonlinear MPC scheme can then be thought of as expanding quasi to infinity. Property (26) can be exploited to prove asymptotic stability for the closed-loop system independent of the specification of the desired control performance. See (Chen and Allgöwer, 1997a) for details.

The idea of using a local state feedback law and a terminal inequality constraint is inspired by the dual-mode nonlinear MPC scheme described in Section 2.2. However, in the dual-mode nonlinear MPC scheme, closed-loop control requires switching between the nonlinear predictive controller outside the terminal region and the linear feedback controller inside the terminal region. In the quasi-infinite horizon nonlinear MPC scheme, closed-loop control is determined by solving the optimization problem (Problem 5) on-line, no matter whether the system is inside or outside the terminal region.

Computation of a terminal region and a terminal penalty term

The local state feedback law can be linear or nonlinear and is only used to compute the terminal region and the terminal penalty term *off-line*. Because of the nonlinearity of the system, finding a terminal region and a terminal penalty term is in general not an easy task, except for some special cases. In the case of using a local *linear* feedback law $u = Kx$ and a *quadratic* objective functional with weighting matrices Q and R, the terminal penalty term can be also chosen to be quadratic with a terminal penalty matrix P that is the solution of a Lyapunov equation. For this quadratic case, a procedure to systematically compute a terminal region and a terminal penalty matrix *off-line* can be stated as follows: Assume that the Jacobian linearization (A, B) of (1) is stabilizable, where $A := \frac{\partial f}{\partial x}(0,0)$ and $B := \frac{\partial f}{\partial u}(0,0)$.

Step 1 : solve a control problem based on the Jacobian linearization (A, B) of (1) to get a locally stabilizing linear state feedback $u = Kx$,

Step 2 : choose a constant $\kappa \in [0, \infty)$ satisfying $\kappa < -\lambda_{max}(A_K)$ and solve the Lyapunov equation

$$(A_K + \kappa I)^T P + P (A_K + \kappa I) = -\left(Q + K^T R K\right) \qquad (28)$$

to get a positive definite and symmetric P, where $A_K := A + BK$,

Step 3 : find the largest possible α_1 defining a region by

$$\Omega_1 := \{x \in \mathbb{R}^n \mid x^T P x \le \alpha_1\}$$

such that $\Omega_1 \subseteq X$ and $Kx \in U$, for all $x \in \Omega_1$,

Step 4 : find the largest possible $\alpha \in (0, \alpha_1]$ specifying a terminal region by

$$\Omega := \{x \in \mathbb{R}^n \mid x^T P x \le \alpha\} \tag{29}$$

such that the optimal value of the following optimization problem is non-positive:

$$\max_{x}\{x^T P \phi(x) - \kappa \cdot x^T P x \mid x^T P x \le \alpha\}, \tag{30}$$

where $\phi(x) := f(x, Kx) - A_K x$.

If the nonlinear system is affine in u and feedback linearizable, then a terminal penalty term can be determined such that (26) is exactly satisfied with equality. A procedure to do that can be found in (Chen, 1997), where some case studies for the application of the discussed quasi-infinite horizon nonlinear MPC scheme can be found.

For discrete systems

Similar versions for discrete nonlinear systems can be found in (De Nicolao *et al.*, 1996*b*; De Nicolao *et al.*, 1997; Findeisen and Rawlings, 1997). In (De Nicolao *et al.*, 1996*b*; De Nicolao *et al.*, 1997), the terminal state penalty term is nonquadratic and defined by

$$E\left(\bar{x}(k + N_p)\right) := \sum_{i=k+N_p+1}^{\infty} F\left(\bar{x}(i), K\bar{x}(i)\right). \tag{31}$$

Clearly, this is in general not implementable. Thus, it is suggested to approximate (31) by replacing the infinite horizon by a *sufficiently long* finite horizon, *i.e.*, the terminal penalty term will be evaluated by integrating $\dot{\bar{x}} = f(\bar{x}, K\bar{x})$ on-line for a sufficiently long time horizon. An open question is how long the on-line integrating time horizon has to be in order to guarantee stability?

Like in the dual-mode approach, the use of the terminal inequality constraint gives the quasi-infinite horizon nonlinear MPC scheme computational advantages. The implementation is simpler than the dual-mode approach, because no switching between control strategies is needed. Moreover, the additional terminal penalty term is introduced to approximate an infinite horizon objective functional. Thus, its effect on the achieved control performance is not significant. The difficulty for a general implementation is to determine a local stabilizing feedback law $u = k(x)$, a terminal region Ω and a terminal penalty term $E(x(t+T_p))$. Although systematical procedures to determine these quantities are available for systems whose Jacobian linearization is stabilizable or for systems that are feedback linearizable, a semi-infinite optimization problem (see (Michalska and Mayne, 1993)) has to be solved off-line. For other cases, the problem might become very complicated.

3 Robustness of stability in nonlinear MPC

García and Morari (García and Morari, 1982; García and Morari, 1985a; García and Morari, 1985b) analyze the robustness of unconstrained MPC in the framework of internal model control (IMC) and develop tuning guidelines for the IMC filter to guarantee robust stability. Based on the contraction mapping principle, Zafiriou (1990) develops necessary and sufficient conditions for robust SISO QDMC with constraints. Specifying uncertainty as bounds on the coefficients of a finite impulse response model, Nikolaou and coworkers (Genceli and Nikolaou, 1993; Vuthandam *et al.*, 1995) present a robustness analysis of their DMC schemes with an end condition. Describing uncertainties as a modeling error bound $K_m \in [0, \infty)$ in the form of

$$\|f(x, u) - \bar{f}(x, u)\|_2 \leq K_m \left(\|x\|_2 + \|u\|_\infty\right), \ \forall x \in I\!\!R^n, \ \forall u \in U,$$

Polak and Yang (Polak and Yang, 1993a; Polak and Yang, 1993b; Yang and Polak, 1993) discuss robust stability of their moving horizon algorithm with a contraction constraint. The robustness of the "MPC+FL" scheme is discussed in a heuristic way in (Kothare *et al.*, 1995). In (De Nicolao *et al.*, 1996a), the robustness of nonlinear receding horizon controllers with equality terminal constraints is investigated with respect to gain and additive perturbations. Employing conservative terminal inequality constraints, Michalska and Mayne (1993) achieve robustness for their dual-mode receding horizon controller. A conservative terminal inequality constraint implies for example that Ω_α is replaced by $\Omega_{\frac{\alpha}{2}}$, with $\Omega_\alpha := \left\{ x \in I\!\!R^n \mid x^T P x \leq \alpha \right\}$. The basic idea is shown in Figure 4. In the presence of modeling errors, $x(t + \delta)$ does not coincide with $\bar{x}^*(t + \delta)$. The input function $\bar{u}^*(\cdot; x(t), T_p) : [t + \delta, t + T_p] \to U$ may not steer the model to a point on the boundary of $\Omega_{\frac{\alpha}{2}}$ but in Ω_α, if the plant-model mismatch is sufficiently small. The local linear feedback law $u = Kx$ can then be used to steer the model to the boundary of $\Omega_{\frac{\alpha}{2}}$ (dotted line in Figure 4). Thus, a feasible input function to the optimization problem at time $t + \delta$ can be constructed by concatenating $\bar{u}^*(\cdot; x(t), T_p) : [t + \delta, t + T_p] \to U$ with the local linear controller.

Even though the analysis of the robustness properties must still be considered as an unsolved problem for the general nonlinear case, some synthesis techniques to design robustly stabilizing nonlinear predictive controllers are known and will be discussed below. A philosophical difference of those techniques from what we have discussed above is the explicit description of uncertainties in the problem formulation.

482

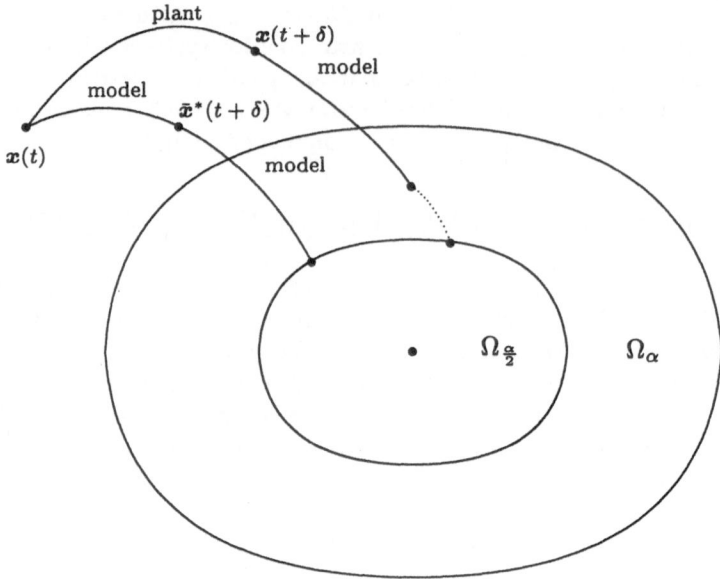

Figure 4: Dual-mode nonlinear MPC – Robust case

3.1 Synthesis of robust model predictive control

In order to synthesize a robust model predictive controller, an explicit description
of uncertainties in the MPC problem formulation is needed. A general concept is
to formulate the MPC problem as solving a minimax (instead of the minimization)
problem on-line (Lee, 1996; Mayne, 1996), where the maximization is taken over a
set of uncertainties. For linear systems, Campo and Morari (1987), Allwright and
Papavasiliou (1992) and Zheng and Morari (1993) propose robust MPC schemes
for SISO FIR plants. Uncertainties are described as bounds on the impulse or
step response coefficients. In the framework of linear matrix inequalities (LMIs),
Kothare *at el.* (Kothare *et al.*, 1996; Kothare *et al.*, 1994) present an approach for
synthesis of a robust linear model predictive controller. Instead of solving a minimax
problem, an upper bound of the standard quadratic objective functional for a set
of linear models is minimized over a finite horizon control input function. Lee and
Yu (1997) discuss minimax MPC formulations in both open-loop and closed-loop
manners for constrained linear systems, where uncertain parameters are assumed
to be bounded. Robust stability of *infinite horizon* minimax MPC formulations is
shown there.

Badgwell (Badgwell, 1997*a*; Badgwell, 1997*b*) proposes an alternative method
for achieving robust stability. The uncertain system is described by a set of stable
models

$$\{f_1, f_2, \cdots, f_p\} \tag{32}$$

which is assumed to contain the real plant and the nominal model used by the predictive controller. An additional *robustness constraint*

$$J_i\left(x(t), \bar{u}(\cdot), T_p\right) \leq J_i\left(x(t), \tilde{u}(\cdot), T_p\right) \quad \forall i = 1, \cdots, p \tag{33}$$

is introduced into the minimization problem so as to restrict the future behavior of the cost functions for all considered models. In the above, $\tilde{u}(\cdot)$ is a shifted version of the previous optimal input function. The feasibility of the robustness constraint leads to the plant cost function being nonincreasing, which in turn is used to show robust stability.

As a common deficiency of the nonlinear MPC schemes discussed in Section 2, disturbances and/or uncertainties either do not enter the problem formulation or they are only indirectly taken into account in that their effects are assumed to be constant over the whole prediction horizon. In the robust (linear) MPC schemes mentioned in this section, uncertainties are described either as a finite number of possible models or bounded uncertain parameters. This deficiency of MPC is exactly the strong point of $\mathcal{H}_\infty$ control. $\mathcal{H}_\infty$ theory (Ball *et al.*, 1993; Isidori, 1995; van der Schaft, 1992) provides an excellent theoretical framework for dealing with robust stability issues. Disturbance attenuation specifications can also be incorporated very naturally into the design setup. In standard nonlinear $\mathcal{H}_\infty$ control, however, it is not straightforward to satisfy constraints neither on control inputs nor on disturbances.

In the purpose of combing advantages of nonlinear MPC and nonlinear $\mathcal{H}_\infty$ control (guaranteed robust stability and constraint satisfaction), a robust nonlinear MPC scheme that can be conceptually viewed as a combination of both methods is proposed in (Chen *et al.*, 1997). In this robust MPC scheme, a finite horizon constrained minimax problem has to be solved on-line repeatedly and the uncertain system is described by the $M - \Delta$ structure as in $\mathcal{H}_\infty$ control. Thus, more general uncertainties can be considered. Moreover, uncertainties are explicitly represented in the objective functional as in $\mathcal{H}_\infty$ control again. In the next section, we discuss this robust nonlinear MPC scheme in a more detailed manner.

3.2   A game theoretical nonlinear model predictive control scheme

Assume that the nonlinear system under consideration is in standard affine form (that is standard in nonlinear $\mathcal{H}_\infty$ control)

$$\dot{x} = a(x) + B(x)u + G(x)w, \quad x(0) = x_0 \tag{34a}$$

$$z = \left[\begin{array}{c} h(x) \\ u \end{array} \right] \tag{34b}$$

with $a(0) = 0$, $h(0) = 0$, where $u(t) \in \mathbb{R}^m$ is the vector of control inputs, $w(t) \in \mathbb{R}^p$ is the vector of exogenous inputs and $z(t) \in \mathbb{R}^q$ is the vector of to-be-controlled outputs. Since uncertainties and/or disturbances are represented by the exogenous inputs w, they will not be distinguished in the following. The control inputs are constrained to vary in $U \subset I\!\!R^m$, where U is compact and contains $0 \in I\!\!R^m$ in its interior.

Description of uncertainties

For many practical systems, disturbances are bounded or uncertain parameters change in some region or only a set of uncertainties need to be considered. In these cases, arbitrarily time-varying uncertain parameters are usually not a good description of model-plant mismatch and may lead to overly conservative controller designs. Thus, we assume that the uncertainty enters in the nonlinear system as

$$w = \Delta(x)z, \tag{35}$$

where $\Delta(x)$ is an arbitrary (nonlinear) smooth matrix valued function satisfying

$$\|\Delta(x)\| \leq d_{max} \tag{36}$$

and $d_{max} < 1$ is a fixed bound on the size of the uncertainty.

Example 1 *As a simple example for the description of uncertainties, we consider the following nonlinear system*

$$\dot{x}_1 = x_2 + u(\mu + (1 - \mu)x_1) \tag{37a}$$

$$\dot{x}_2 = x_1 + u(\mu - 4(1 - \mu)x_2), \tag{37b}$$

where the model parameter μ is uncertain and its nominal value is 0.5. Let $\mu = 0.5 + \Delta\mu$, then the system (37) can be rewritten in form of (34a) with

$$a(x) = \begin{bmatrix} x_2 \\ x_1 \end{bmatrix}, \quad B(x) = \begin{bmatrix} 0.5(1 + x_1) \\ 0.5(1 - 4x_2) \end{bmatrix}$$

and

$$G(x) = \begin{bmatrix} a(1 - x_1) \\ a(1 + 4x_2) \end{bmatrix}, \quad w = \Delta(x)u,$$

where $\Delta(x) := \frac{1}{a}\Delta\mu$. Thus, $|\Delta\mu| < a$ implies $|\Delta(x)| < 1$.

The description of the uncertainty by (35) imposes constraints on the exogenous inputs that depend on the actual state of the system and on the actual control function. For a given initial condition x_0 and a given control input function $u(\cdot)$, an exogenous input function is admissible if, along the trajectory of (34), one has

$$w(t)^T w(t) \leq d_{max}^2 z(t)^T z(t).$$

The corresponding set of exogenous input functions is denoted as $\mathcal{W}(x_0, u(\cdot))$.

Precompensation

The basic idea of the game theoretical nonlinear MPC scheme is similar to that of the quasi-infinite horizon nonlinear MPC scheme. In the uncertain case, however, the terminal region Ω and the terminal penalty term E have to be chosen such that if $x(t + T_p) \in \Omega$ then

$$\int_{t+T_p}^{\infty} \frac{1}{2} \left(\|z(\tau)\|^2 - \|w(\tau)\|^2 \right) \, d\tau \leq E \left(x(t + T_p) \right) \tag{38}$$

and the invariance property of Ω has to hold *for all uncertainties considered*, *i.e.*, for all $w(\cdot)$ satisfying (35). As briefly outlined in the following, this can be achieved by a *precompensator* that solves locally in Ω the infinite horizon $\mathcal{H}_\infty$ suboptimal problem

$$\inf_{u(\cdot)} \sup_{w(\cdot)} \int_{t+T_p}^{\infty} \frac{1}{2} \left(\|z(t)\|^2 - \|w(t)\|^2 \right) \, d\tau$$

for the nonlinear system (34) with initial condition $x(t + T_p) \in \Omega$. The application of linear $\mathcal{H}_\infty$ techniques for the Jacobian linearization of (34) at the origin leads to a positive definite symmetric matrix P that satisfies the Riccati inequality

$$A^T P + PA + P \left(GG^T - BB^T \right) P + H^T H \leq 0, \tag{39}$$

where $A := \frac{\partial a}{\partial x}(0)$, $B := B(0)$ and $G := G(0)$. Then, the terminal penalty term can be defined by

$$E(x) := \frac{1}{2} x^T P x. \tag{40}$$

Moreover, a constant $\alpha > 0$ can be determined such that the following Hamilton-Jacobi inequality

$$E_x(x)a(x) + \frac{1}{2} E_x(x) \left(G(x)G(x)^T - B(x)B(x)^T \right) E_x(x)^T + \frac{1}{2} h(x)^T h(x) \leq 0 \tag{41}$$

holds for all

$$x \in \Omega := \{ x \in \mathbb{R}^n \, | \, E(x) \leq \alpha \} . \tag{42}$$

This implies that (40) also solves the nonlinear $\mathcal{H}_\infty$ problem locally in Ω. From (41), by simple completion of the squares, we can conclude that if the nonlinear system is controlled by

$$u = k(x) := -B(x)^T E_x(x)^T , \tag{43}$$

then, along the trajectories of

$$\dot{x} = a(x) + B(x)k(x) + G(x)w \tag{44}$$

with initial condition $x(t + T_p) \in \Omega$, the following differential dissipation inequality holds:

$$\frac{d}{d\tau} E(x(\tau)) + \frac{1}{2} \left(\|z(\tau)\|^2 - \|w(\tau)\|^2 \right) \leq 0, \ \tau \geq t + T_p .$$

486

Hence, for all $w\,(\cdot) \in \mathcal{W}(x(t+T_p), k(x))$, (44) is locally exponentially stable, Ω as defined in (42) is invariant and inequality (38) holds. Thus, (43) can serve as a precompensator.

Controller setup

The game theoretical nonlinear MPC scheme can be thought of as having two hierarchically structured layers, as shown in Figure 5. The inner layer consists of the precompensator that ensures the invariance property of Ω for the uncertain system and the outer layer controller is a predictive controller that differs from the standard MPC approach in that a constrained minimax instead of the constrained minimization problem has to be solved on-line.

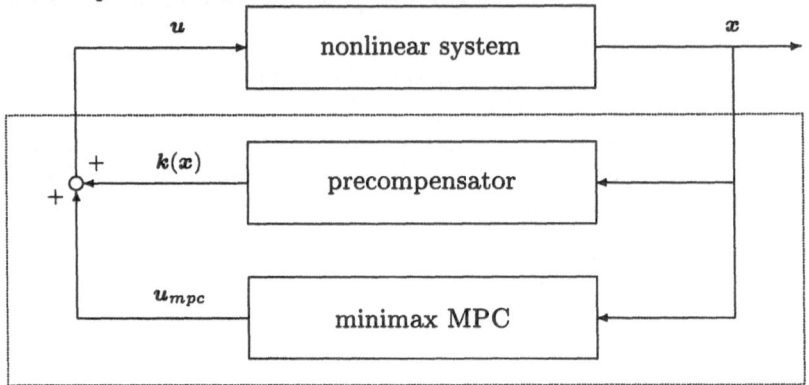

Figure 5: Game theoretical nonlinear MPC.

For any (playable) pairs $(\bar{u}, \bar{w})$ on $[t, t + T_p]$, the constrained minimax problem is then formulated as follows:

Problem 6 *Find*

$$\min_{\bar{u}_{mpc}(\cdot)} \max_{\bar{w}(\cdot)} J\left(\bar{u}_{mpc}(\cdot), \bar{w}(\cdot), x(t), T_p\right) \tag{45}$$

with

$$J(\bar{u}_{mpc}(\cdot), \bar{w}(\cdot), x(t), T_p) := \int_t^{t+T_p} \frac{1}{2}\left(\|\bar{z}(\tau)\|^2 - \|\bar{w}(\tau)\|^2\right)\,d\tau$$
$$+E\left(\bar{x}(t+T_p)\right), \tag{46}$$

subject to

$$\dot{\bar{x}} = a(\bar{x}) + B(\bar{x})k(\bar{x}) + B(\bar{x})\bar{u}_{mpc} + G(\bar{x})\bar{w}, \quad \bar{x}(t) = x(t) \tag{47a}$$

$$\bar{z} = \begin{bmatrix} h(\bar{x}) \\ \bar{u}_{mpc} + k(\bar{x}) \end{bmatrix} \tag{47b}$$

$$\bar{u}_{mpc}(\tau) + k\left(\bar{x}(\tau)\right) \in U, \ \tau \in [t, t + T_p] \tag{47c}$$

$$\bar{w}(\cdot) \in \mathcal{W}\left(x(t), \bar{u}_{mpc}(\cdot)\right) \tag{47d}$$

$$\bar{x}(t + T_p) \in \Omega. \tag{47e}$$

By the invariance property of Ω and (38), the minimax predictive controller achieves that the closed-loop system satisfies the following integral dissipation inequality

$$S\left(x(t_1), T_p\right) \geq \int_{t_1}^{t_2} \frac{1}{2} \left(\|z(t)\|^2 - \|w(t)\|^2\right) dt + S\left(x(t_2), T_p\right) \qquad (48)$$

for any $0 \leq t_1 < t_2 < \infty$ and for any admissible $w(\cdot)$ on $[t_1, t_2]$, where the function $S(x, T_p)$ is defined by

$$S(x, T_p) := \min_{\bar{u}_{mpc}(\cdot)} \max_{\bar{w}(\cdot)} J\left(\bar{u}_{mpc}(\cdot), \bar{w}(\cdot), x, T_p\right). \qquad (49)$$

Under some hypotheses, it can be shown that there exists a finite time T such that for the closed-loop system $x(T) \in \Omega$. Then, from the robust exponential stability in Ω, the closed-loop system is robustly stable. For more details see (Chen *et al.*, 1997).

Due to the conceptual combination of nonlinear MPC and nonlinear $\mathcal{H}_\infty$, this approach potentially combines the strong points of both methods and thus enables us to design controllers that have guaranteed robust stability and achieve good disturbance rejection properties in the face of input constraints. This scheme is however still far from a practical approach that can be recommended for and applied to industrial control problems. The major obstacle at present is the high on-line computational demand which results from the solution of the constrained minimax problem. Moreover, since the precompensation is realized by a state feedback law, originally linear input constraints will become state-dependent constraints (see (47c)), that are computationally very expensive.

4   Observer based nonlinear model predictive control

All the nonlinear MPC schemes discussed previously have in common that the optimization problem will be updated with the actual system state at each sampling instance, (except for the one proposed by Genceli and Nikolaou (1995)). That is, a state feedback strategy is used. However, it is in general impossible to measure the complete system state. Thus, observer based nonlinear MPC schemes are required, where the observer is used to reconstruct the states. The difficulty here lies not only in the design of stable nonlinear observers (see for example (Gauthier and Kupka, 1994; Misawa and Hedrick, 1989)) but also in a lack of a nonlinear separation principle. Stability of the closed-loop system has to be proven in each case even if the observer and the state feedback controller are by themselves stable/stabilizing. Thus, it is not surprising that only the surface of this field has been scratched so far.

For continuous nonlinear systems, Michalska and Mayne (1995) propose a moving horizon observer and discuss the stability property of the dual-mode controller with the moving horizon observer. The optimization problem of the nonlinear moving horizon observer is formulated for a given past input function $u(\cdot) : [t - T_e, t] \to U$ as follows:

488

Problem 7 *Find*

$$\min_{v(t-T_e)} J\left(v(t - T_e), T_e, y(\cdot)\right) \tag{50}$$

with

$$J\left(v(t - T_e), T_e, y(\cdot)\right) := \int_{t-T_e}^{t} \|\hat{y}(\tau) - y(\tau)\|^2 \, d\tau \tag{51}$$

subject to

$$\dot{\hat{x}} = f(\hat{x}, u), \ \hat{x}(t - T_e) = v(t - T_e) \tag{52a}$$

$$\hat{y} = g(\hat{x}), \tag{52b}$$

where T_e is referred to as observation horizon and $y(\cdot)$ on $[t - T_e, t]$ is the system measurement.

In analogy to MPC, the optimization problem (Problem 7) will be solved on-line repeatedly, updated by the new measurement. An optimal (or approximate) solution (existence assumed) is denoted as $v^*(t - T_e)$. If $v^*(t - T_e)$ satisfies $\|x(t - T_e) - v^*(t - T_e)\| \to 0$ as $t \to \infty$, then the corresponding $\hat{x}^*(t)$ is a useful estimate of $x(t)$. The nominal convergence property of the estimated state to the true state is presented in (Michalska and Mayne, 1995). The stability property of the observer based dual-mode predictive controller is also established, where the basic idea is similar to that for achieving robustness of the dual-mode controller, namely that a conservative terminal inequality constraint is applied.

A nonlinear moving horizon state observer for discrete systems can be found in (Muske and Rawlings, 1995), where additive disturbances are also estimated and state constraints are considered in the formulation of the optimization problem.

In (Yang and Polak, 1993), closed-loop stability of the contractive nonlinear MPC scheme with a least squares estimator is discussed and in (De Oliveira and Morari, 1997), closed-loop stability of the contractive nonlinear MPC scheme with an observer designed by the extended Kalman filter technique is discussed. The advantage of the contractive nonlinear MPC approach is its exponential nominal stability. Thus, the observers reconstruct the states for the closed-loop system and the observer based contractive predictive controller is stabilizing, because of the exponential stability of the contractive controller and the *locally* asymptotic convergence of the observer. Clearly, initial estimation errors have to be within the convergence region of the observer.

5 Conclusions and Perspectives

In this paper we have reviewed a number of nonlinear MPC schemes, that address issues related to nominal or robust closed-loop stability. All of these approaches have in common that so-called stability constraints (terminal equality or inequality constraints, contraction constraints, etc.) are introduced into the problem formulation in order to guarantee stability. Unfortunately, still neither one of these approaches is completely satisfying from an application point of view. First of all, all of these artificially introduced constraints certainly limit the achievable control performance and also require more or less additional on-line computational effort. Secondly, these constraints impose feasibility problems that together with the fact that all stability results are only sufficient (and not necessary), lead to a very conservative estimate for the region of attraction. Despite significant progress during the last decade there are still many open problems concerning the theoretical properties of nonlinear MPC:

- **Region of attraction:** As pointed out, the achieved stability is in general only local, *i.e.*, the closed-loop system is asymptotically stable with a region of attraction that is restricted by the feasibility of the optimization problem. The latter is strongly related to the (constrained) stabilizability of nonlinear systems. For practical applications, nonlinear MPC schemes that allow for significantly enlarged operating regions are needed.

- **Robustness:** Despite a few results on robustness analysis of nonlinear MPC, a general theory is still lacking. With respect to synthesis of robust nonlinear predictive controllers, merely the surface has been scratched. One of the key problems in this area (that also holds for non-predictive control schemes) is the formulation of suitable uncertainty descriptions.

- **Tracking:** In most existing nonlinear MPC schemes, the tracking problem can only be dealt with, when the required state trajectory can be calculated from the desired (output) trajectory. Trajectory generation is an old but still challenging problem and the presence of constraints imposes additional difficulties in this respect.

- **Output feedback nonlinear MPC:** In practice not the whole state vector can be measured on-line. Therefore, the problem of output feedback control is essential from an application point of view. As discussed, only few preliminary stability results are known to date, and all of these results incorporate a nonlinear observer, which is very difficult to design by itself. Future advances must directly aim at deriving robustly stable output feedback schemes.

- **Disturbance attenuation:** At present, disturbances are only considered in a very rudimentary way in most nonlinear MPC schemes. One main goal of control is however the suppression of the effect of unknown and unmeasured disturbances, and thus nonlinear MPC schemes are needed for which disturbance attenuation requirements can be included in the performance objective.

There are of course many more open problems related to nonlinear MPC than the ones pointed out above. Among the most important ones are the development of problem specific optimization formulations and solvers, the incorporation of identification and adaptation schemes, problems related to performance monitoring and diagnosis, and the link to higher automation levels.

References

Allwright, J.C. and G.C. Papavasiliou (1992). On linear programming and robust model predictive control using impulse responses. *Syst. Contr. Lett.* **18**, 159–164.

Badgwell, T.A. (1997*a*). A robust model predictive control algorithm for stable linear plants. In: *Proc. Amer. Contr. Conf.*. Albuquerque. pp. 1618–1622.

Badgwell, T.A. (1997*b*). A robust model predictive control algorithm for stable nonlinear plants. In: *Proc. IFAC ADCHEM*. Banff. pp. 477–481.

Ball, J.A., J.W. Helton and M.L. Walker (1993). H_∞ control for nonlinear systems with output feedback. *IEEE Trans. Automat. Contr.* **AC-38**(4), 546–559.

Bitmead, R.R., M. Gevers and V. Wertz (1990). *Adaptive Optimal Control – The Thinking Man's GPC*. Prentice Hall. New York.

Bryson, A.E. and Y.-C. Ho (1969). *Applied Optimal Control*. Ginn and Company. Waltham, Massachusetts.

Campo, P.J. and M. Morari (1987). Robust model predictive control. In: *Proc. Amer. Contr. Conf.*. pp. 1021–1026.

Chen, H. (1997). *Stability and Robustness Considerations in Nonlinear Model Predictive Control*. Fortschr.-Ber. VDI Reihe 8 Nr. 674, VDI Verlag. Düsseldorf.

Chen, H. and F. Allgöwer (1996). A quasi-infinite horizon predictive control scheme for constrained nonlinear systems. In: *Proc. 16th Chinese Control Conference*. Qindao. pp. 309–316.

Chen, H. and F. Allgöwer (1997*a*). A quasi-infinite horizon nonlinear model predictive control scheme with guaranteed stability. *Automatica*. (in press).

Chen, H. and F. Allgöwer (1997*b*). A quasi-infinite horizon nonlinear predictive control scheme for stable systems: Application to a CSTR. In: *Proc. IFAC ADCHEM*. Banff. pp. 471–476.

Chen, H., C.W. Scherer and F. Allgöwer (1997). A game theoretic approach to nonlinear robust receding horizon control of constrained systems. In: *Proc. Amer. Contr. Conf.*. Albuquerque. pp. 3073–3077.

Clarke, D. W., C. Mohtadi and P. S. Tuffs (1987*a*). Generalized predictive control – Part I. The basic algorithm. *Automatica* **23**(2), 137–148.

Clarke, D. W., C. Mohtadi and P. S. Tuffs (1987*b*). Generalized predictive control – Part II. Extensions and interpretations. *Automatica* **23**(2), 149–160.

Cutler, C.R. and B.L. Ramaker (1980). Dynamic Matrix Control – A computer control algorithm. In: *Proc. Joint Automatic Control Conference*. San Francisco, CA.

De Keyser, R.M.C. and A.R. Van Cauwenberghe (1985). Extended prediction self-adaptive control. In: *Proc. 7th IFAC Symposium on Identification and System*

Parameter Estimation (H.A. Barker and P.C. Young, Eds.). Pergamon Press. Oxford. pp. 1255–1260.

De Keyser, R.M.C., G.A. Van de Velde and F.A.G. Dumortier (1988). A comparative study of self-adaptive long-range predictive control methods. *Automatica* **24**(2), 149–163.

De Nicolao, G., L. Magni and R. Scattolini (1996*a*). On the robustness of receding horizon control with terminal constraints. *IEEE Trans. Automat. Contr.* **41**(3), 451–453.

De Nicolao, G., L. Magni and R. Scattolini (1996*b*). Stabilizing nonlinear receding horizon control via a nonquadratic terminal state penalty. In: *Symposium on Control, Optimization and Supervision, CESA'96 IMACS Multiconference.* Lille. pp. 185–187.

De Nicolao, G., L. Magni and R. Scattolini (1997). Stabilizing receding-horizon control of nonlinear time-varying systems. In: *Proc. 4rd European Control Conference ECC'97.* Brussels.

De Oliveira, S.L. and M. Morari (1997). Contractive model predictive control for constrained nonlinear systems. *IEEE Trans. Automat. Contr.* (in press).

Doyle III, F. J., B. A. Ogunnaike and R. K. Pearson (1995). Nonlinear model-based control using second-order Volterra models. *Automatica* **31**(5), 697–714.

Eaton, J. W. and J. B. Rawlings (1992). Model predictive control of chemical processes. *Chem. Eng. Sci.* **47**(4), 705–720.

Findeisen, R.H. and J.B. Rawlings (1997). Suboptimal infinite horizon nonlinear model predictive control for discrete time systems. *NATO Advanced Study Institute on Nonlinear Model Based Process Control.*

García, C.E. and A.M. Morshedi (1986). Quadratic programming solution of dynamic matrix control (QDMC). *Chem. Eng. Commun.* **46**, 73–87.

García, C.E. and M. Morari (1982). Internal Model Control. 1. A unifying review and some new results. *Ind. Eng. Chem. Process Des. Dev.* **21**, 308–323.

García, C.E. and M. Morari (1985*a*). Internal Model Control. 2. Design procedure for multivariable systems. *Ind. Eng. Chem. Process Des. Dev.* **24**, 472–484.

García, C.E. and M. Morari (1985*b*). Internal Model Control. 3. Multivariable control law computation and tuning guidelines. *Ind. Eng. Chem. Process Des. Dev.* **24**, 484–494.

García, C.E., D.M. Prett and M. Morari (1989). Model Predictive Control: Theory and practice – A survey. *Automatica* **25**, 335–347.

Gauthier, J.P. and I.A.K. Kupka (1994). Observability and observers for nonlinear systems. *SIAM J. Contr. Optim.* **32**(4), 975–994.

Genceli, H. and M. Nikolaou (1993). Robust stability analysis of constrained L_1-norm model predictive control. *AIChE J.* **39**(12), 1954–1965.

Genceli, H. and M. Nikolaou (1995). Design of robust constrained model-predictive controllers with Volterra series. *AIChE J.* **41**(9), 2098–2107.

Gilbert, E.C. and K.C. Tan (1991). Linear systems with state and control constraints: The theory and application of maximal output admissible sets. *IEEE Trans. Automat. Contr.* **AC-36**, 1008–1020.

Hovd, M., J.H. Lee and M. Morari (1993). Truncated step response models for model predictive control. *J. Proc. Contr.* **3**(2), 67–73.

Isidori, A. (1995). H_∞ control via measurement feedback for general nonlinear systems. *IEEE Trans. Automat. Contr.*

Keerthi, S.S. and E.G. Gilbert (1988). Optimal infinite-horizon feedback laws for a general class of constrained discrete-time systems: Stability and moving-horizon approximations. *J. Opt. Theory and Appl.* **57**(2), 265–293.

Kleinman, D.L. (1970). An easy way to stabilize a linear constant system. *IEEE Trans. Automat. Contr.* **AC-15**, 692.

Kothare, M. V., V. Balakrishnan and M. Morari (1996). Robust constrained model predictive control using linear matrix inequalities. *Automatica* **32**(10), 1361–1379.

Kothare, M.V., V. Balakrishnan and M. Morari (1994). Robust constrained model predictive control using linear matrix inequalities. In: *Proc. Amer. Contr. Conf.*. Baltimore, Maryland. pp. 440–444.

Kothare, M.V., V. Nevistić and M. Morari (1995). Robust constrained model predictive control for nonlinear systems: A comparative study. In: *Proc. 34th IEEE Conf. Decision Contr.*. New Orleans, LO. pp. 2884–2885.

Kurtz, M.J. and A. Henson (1996). Linear model predictive control of input-output linearized processes with constraints. In: *5th International Conference on Chemical Process Control – CPC V.*

Kwon, W.H. and D.G. Byun (1989). Receding horizon tracking control as a predictive control and its stability properties. *Int. J. Contr.* **50**(5), 1807–1824.

Lee, J. H., M. Morari and C.E. García (1994). State-space interpretation of model predictive control. *Automatica* **30**(4), 707–717.

Lee, J.H. (1996). Recent advances in model predictive control and other related areas. In: *5th International Conference on Chemical Process Control – CPC V.*

Lee, J.H. and Z.-H. Yu (1997). Worst-case formulations of model predictive control for systems with bounded parameters. *Automatica* **33**(5), 763–781.

Maner, B. R., F. J. Doyle III, B. A. Ogunnaike and R. K. Pearson (1994). A nonlinear model predictive control scheme using second order Volterra models. In: *Proc. Amer. Contr. Conf.*. Baltimore, Maryland. pp. 3253–3257.

Mayne, D. Q. (1995). Optimization in model based control. In: *Proc. IFAC Symposium Dynamics and Control of Chemical Reactors, Distillation Columns and Batch Processes*. Helsingor. pp. 229–242.

Mayne, D. Q. (1996). Nonlinear model predictive control: An assessment. In: *5th International Conference on Chemical Process Control – CPC V*.

Mayne, D. Q. and H. Michalska (1990). Receding horizon control of nonlinear systems. *IEEE Trans. Automat. Contr.* **AC-35**(7), 814–824.

Mayne, D. Q. and W. R. Schroeder (1994). Nonlinear control of constrained linear systems. *Int. J. Contr.* **60**(5), 1035–1043.

Meadows, E.S., M.A. Henson, J.W. Eaton and J.B. Rawlings (1995). Receding horizon control and discontinuous state feedback stabilization. *Int. J. Contr.* **62**(5), 1217–1229.

Michalska, H. and D.Q. Mayne (1991). Receding horizon control of nonlinear systems without differentiability of the optimal value function. *Syst. Contr. Lett.* **16**, 123–130.

Michalska, H. and D.Q. Mayne (1993). Robust receding horizon control of constrained nonlinear systems. *IEEE Trans. Automat. Contr.* **AC-38**(11), 1623–1633.

Michalska, H. and D.Q. Mayne (1995). Moving horizon observers and observer-based control. *IEEE Trans. Automat. Contr.* **AC-40**(6), 995–1006.

Misawa, E.A. and J.K. Hedrick (1989). Nonlinear observers – a state-of-the-art survey. *ASME J. Dyn. Syst. Meas. Cntrl.* **111**, 344–352.

Morari, M. and J.H. Lee (1991). Model predictive control: The good, the bad, and the ugly. In: *Proc. 4th International Conference on Chemical Process Control – CPC IV* (Y. Arkun and W. Ray, Eds.). AIChE, CACHE. TX. pp. 419–444.

Morshedi, A.M., C.R. Cutler and T.A. Skrovanek (1985). Optimal solution of dynamic matrix control with linear programming techniques (LDMC). In: *Proc. Amer. Contr. Conf.*. pp. 199–208.

Muske, K.R. and J.B. Rawlings (1995). Nonlinear moving horizon state estimation. In: *Methods of Model Based Process Control* (R. Berber, Ed.). pp. 349–365. Kluwer Academic Publishers. Netherlands.

Nevistić, V. and M. Morari (1995). Constrained control of feedback-linearizable systems. In: *Proc. 3rd European Control Conference ECC'95*. Rome. pp. 1726–1731.

Polak, E. and T.H. Yang (1993a). Moving horizon control of linear systems with input saturation and plant uncertainty. Part 1. Robustness. *Int. J. Contr.* **58**(3), 613–638.

Polak, E. and T.H. Yang (1993b). Moving horizon control of linear systems with input saturation and plant uncertainty. Part 2. Disturbance rejection and tracking. *Int. J. Contr.* **58**(3), 639–663.

494

Qin, S.J. and T.A. Badgwell (1996). An overview of industrial model predictive control technology. In: *5th International Conference on Chemical Process Control - CPC V*.

Rawlings, J.B. and K.R. Muske (1993). The stability of constrained receding horizon control. *IEEE Trans. Automat. Contr.* **AC-38**(10), 1512–1516.

Rawlings, J.B., E.S. Meadows and K.R. Muske (1994). Nonlinear model predictive control: A tutorial and survey. In: *Proc. ADCHEM'94*. Kyoto, Japan.

Richalet, J. (1993). Industrial applications of model based predictive control. *Automatica* **29**(5), 1251–1274.

Richalet, J., A. Rault, J.L. Testud and J. Papon (1976). Algorithmic control of industrial processes. In: *Proc. 4th IFAC Symposium on Identification and System Parameter Estimation*. Tbilisi. pp. 1119–1167.

Richalet, J., A. Rault, J.L. Testud and J. Papon (1978). Model predictive heuristic control: Application to industrial processes. *Automatica* **14**, 413–428.

Ricker, N.L. (1991). Model predictive control: State of the art. In: *Proc. 4th International Conference on Chemical Process Control - CPC IV* (Y. Arkun and W. Ray, Eds.). AIChE, CACHE. TX. pp. 271–296.

Rouhani, R. and R.K Mehra (1982). Model algorithmic control (MAC): Basic theoretical properties. *Automatica* **18**(4), 401–414.

Scokaert, P.O.M. and J.B. Rawlings (1996). Infinite horizon linear quadratic control with constraints. In: *Proc. 13th IFAC World Congress*. Vol. M. San Francisco. pp. 109–114.

van der Schaft, A.J. (1992). L_2-gain analysis of nonlinear systems and nonlinear state feedback H_∞ control. *IEEE Trans. Automat. Contr.* **AC-37**(6), 770–784.

Vuthandam, P., H. Genceli and M. Nikolaou (1995). Performance bounds for robust quadratic dynamic matrix control with end condition. *AIChE J.* **41**(9), 2083–2097.

Yang, T.H. and E. Polak (1993). Moving horizon control of nonlinear systems with input saturation, disturbances and plant uncertainty. *Int. J. Contr.* **58**(4), 875–903.

Ydstie, B.E. (1984). Extended horizon adaptive control. In: *Proc. 9th IFAC World Congress* (J. Gertler and L. Keviczky, Eds.). Pergamon Press. Oxford. pp. 911–915.

Zafiriou, E. (1990). Robust model predictive control of processes with hard constraints. *Comp. & Chem. Eng.* **14**(4/5), 359–371.

Zheng, A. and M. Morari (1993). Robust stability of constrained model predictive control. In: *Proc. Amer. Contr. Conf.*. pp. 379–383.

A COMPUTATIONALLY EFFICIENT NONLINEAR MODEL PREDICTIVE CONTROL ALGORITHM WITH GUARANTEED STABILITY

ALEX ZHENG, ZZHENG@ECS.UMASS.EDU
Department of Chemical Engineering
University of Massachusetts
Amherst, MA 01003-3110

Abstract. A novel Nonlinear Model Predictive Control algorithm is proposed for control of *large* nonlinear constrained systems. The basic idea is to calculate *exactly* the first control move, which is implemented, and approximate all other control moves which are never implemented. *Regardless* of the control horizon, the number of decision variables for the on-line optimization problem equals the number of manipulated variables, resulting in significant savings in on-line computational time. Asymptotic stability of the closed loop system is guaranteed if and only if the on-line optimization problem is feasible initially, under reasonable assumptions. The feasibility for a practical implementation of the proposed algorithm is demonstrated on two examples including the Tennessee-Eastman Challenge problem involving ten inputs and ten outputs.

1. Introduction

Following successful industrial application of linear model based control algorithms, various Nonlinear Model Predictive Control (NMPC) algorithms have been proposed for control of nonlinear systems with constraints [8, 11, 16, 6, 13, 21, 15, etc]. Several excellent reviews have been published (see, for example, [1, 14]). In general, NMPC solves an on-line optimization problem in real-time at each sampling time. In terms of how the on-line optimization problem is solved, the algorithms can be divided into two main groups—one attempts to solve the optimization problem exactly (exact optimization approach) and the other tries to solve the problem via (usually linear) approximations (approximation approach).

R. Berber and C. Kravaris (eds.), Nonlinear Model Based Process Control, 495-511.
© 1998 *Kluwer Academic Publishers.*

The main advantages for the exact optimization approach are that performance is "optimal" and that strong and non-conservative nominal stability results have been proven [7, 15, 2, etc]. Since the on-line optimization problem is generally nonconvex, the on-line computational demand is high for any reasonably nontrivial systems. This results in practical implementation of the exact optimization approach difficult, if not impossible. One may argue that the problem with high on-line computational demand would vanish as faster and cheaper computers and more powerful software become available. This may be true for small problems but unrealistic for large problems. Even under the most optimistic assumption that the Moore's Law (i.e. computer speed doubles about every 18 months) will continue to hold, we may have to wait about 50 years for a *typical* computer to be able to solve the on-line optimization problem for a general 5-by-5 nonlinear system with a control horizon of ten and a sampling time of two minutes (Figure 1).[1]

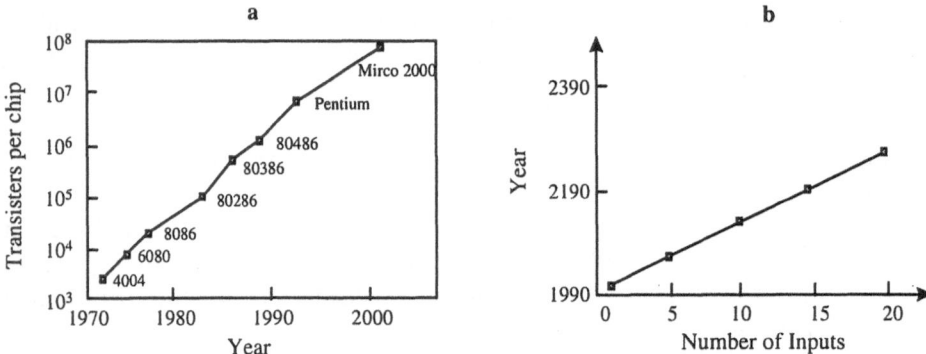

Figure 1. a. Moore's Law. b. Year in which computers will be powerful enough to solve the on-line optimization problem versus number of inputs.

Due to the prohibitively high on-line computational demand associated with the exact optimization approach, García [8] proposed to linearize the nonlinear model successively. Gattu and Zafiriou [9, 10], and Lee and Ricker [13] incorporated the state estimation techniques into this approach. By introducing a stability constraint and assuming the feasibility of the optimization problem, Olivera and Morari [3] have proposed a NMPC algorithm with guaranteed robust stability. Another approach [4, 12] is to linearize (either input/output or feedback) the nonlinear system. For a certain class

---

[1]We assume that a typical computer today can solve the on-line optimization problem for single-input single-output nonlinear systems with control horizon of ten and a sampling time of two minutes, and that the on-line computational time grows exponentially with the number of decision variables which equals the number of inputs times the control horizon.

of nonlinear systems, such transformation yields a linear dynamic system *but* with state-dependent (or nonlinear) constraints which need to be approximated in order to generate a computationally simpler optimization problem (otherwise, these methods would fall into the exact optimization approach and suffer the same disadvantage computationally). An attractive feature of this approach is that only a quadratic program needs to be solved on-line in real-time. However, one issue is that the design may be overly conservative in some situations: Linear approximation is only valid when both state and input do not deviate too much from where they are linearized. This implies that the control actions have to be close to their linearized values in order to preserve stability. Thus we may be sacrificing performance for computational simplicity.

In this paper, we propose an NMPC algorithm which combines the best features of the exact optimization approach (thus performance) and the approximation approach (thus low on-line computational demand). The key idea behind the algorithm is to calculate exactly the first control move, which is actually implemented, and to approximate the rest of control moves, which are never implemented. Thus the number of decision variables in the on-line optimization problem, *regardless* of the control horizon, equals the number of inputs, instead of the number of inputs *times* the control horizon for a conventional NMPC algorithm. This results in significant savings in on-line computational demand since the control horizon needs to be chosen reasonably large to ensure adequate performance and in some cases stability, and since the optimization problem is usually nonconvex. We show that asymptotic stability of the closed loop system is guaranteed.

2. Preliminaries

We consider a continuous-time nonlinear system described below.[2]

$$\begin{aligned} \dot{x}(t) &= f(x(t), u(t), d(t)) \\ y(t) &= g(x(t), d(t)) \end{aligned} \tag{1}$$

where x of dimension n_x is the state, u of dimension n_u is the input, y of dimension n_y is the output, and d of dimension n_d is the disturbance. We are interested in constructing a nonlinear feedback control law, $u(t) = h(x(t), e(t), d(t))$, either continuous or discrete,[3] with certain stability and performance properties. Here $e(t) \triangleq r(t) - y(t)$ is the tracking error and $r(t)$

---

[2]While we only consider a nonlinear system described by a set of ODE's, in principle the results are applicable to other types of models. The discrete version was presented in [23].

[3]For discrete controllers, $u(t)$ would be constant for $t \in [k, k+1)$ where k represents the kth sampling time.

is the desired trajectory. Throughout this paper, the following assumptions are made.

Assumption 1 *The origin is an equilibrium point, i.e.* $f(0,0,0) = 0$ *and* $g(0,0) = 0$.

Assumption 2 *The input, output, and disturbance are constrained, i.e.*

$$u(t) \in \mathcal{U} \triangleq \{u : 0 > u^{\min} \leq u \leq u^{\max} > 0\} \subset \Re^{n_u}$$
$$y(t) \in \mathcal{Y} \subset \Re^{n_y}$$
$$d(t) \in \mathcal{D} \subset \Re^{n_d}$$

The sets $\mathcal{U}, \mathcal{Y}$, *and* $\mathcal{D}$ *are described by sets of linear inequalities. In this paper, we treat the input constraints as hard constraints and the output constraints as soft constraints [20, 25] (i.e. they are allowed to be violated temporarily to satisfy other objectives).*

Assumption 3 *The vectors of real-valued functions* $f(x, u, d)$ *and* $g(x, d)$ *are Lipschitzian for* $u \in \mathcal{U}$ *and* $d \in \mathcal{D}$:

$$\|f(x', u', d') - f(x'', u'', d'')\| < L_1\|(x', u', d') - (x'', u'', d'')\|, L_1 < \infty$$

$$\|g(x', d') - g(x'', d'')\| < L_2\|(x', d') - (x'', d'')\|, L_2 < \infty$$

Assumption 4 *For any* $y^{ss} \in \mathcal{Y}$ *and* $d^{ss} \in \mathcal{D}$, *there exists a* finite *number of points* (x^{ss}, u^{ss}) *such that* $f(x^{ss}, u^{ss}, d^{ss}) = 0$ *and* $y^{ss} = g(x^{ss}, d^{ss})$. *Furthermore, there exist constants* $\alpha < \infty$ *and* $\beta < \infty$ *such that* $|(x^{ss}, u^{ss})| \leq \alpha|y^{ss}| + \beta$.

Remark 1 *The four assumptions are not restrictive:* **Assumption 1** *is made merely for convenience.* **Assumption 2** *is part of the problem specifications.* **Assumption 3** *rules out some pathological examples.* **Assumption 4** *allows a finite number of multiple steady-states and rules out the cases where* y *approaches some finite value in* $\mathcal{Y}$ *while either* u *or* x *approaches infinity.*

The objective function at sampling time k is defined as follows:

$$\Phi_k = \int_0^{H_p} \left\{ (r(t) - y(k+t|k))^T \Gamma_y(y(k+t|k))(r(t) - y(k+t|k)) + \right.$$
$$\left. \dot{x}^T(k+t|k)\Gamma_x \dot{x}(k+t|k) \right\} dt + \sum_{i=0}^{H_c-1} \Delta u(k+i|k)^T \Gamma_u \Delta u(k+i|k) \tag{2}$$

where H_p is the output horizon, H_c is the input (or control) horizon, k is the kth sampling time (the sampling period is assumed to be one for simplicity), r is the desired output trajectory, $\Delta u(k) \triangleq u(k) - u(k-1)$, and $\Gamma_y(e) \geq \Gamma_y(0) > 0, \Gamma_x > 0$ and $\Gamma_u > 0$ are diagonal weights. The output weight $\Gamma_y(y(k+t|k))$ depends on the predicted output $y(k+t|k)$. The reason behind it as well as its functional form will be discussed shortly. Notice that the inputs are discrete (i.e. $u(t)$ is constant for $t \in [k, k+1)$) while the outputs are continuous. Δu is penalized instead of u to guarantee integral control. The term penalizing $\dot{x}$ is there for purely technical reasons. While this objective is a little bit more complicated than the usual quadratic objective, its impact on computation is small as most of nonlinearity comes from the system itself. Now we define `Controller NMPC #1` as follows.

Definition 1 (Controller NMPC #1) *At each sampling time k, $\Delta u(k)$ equals the first control move $\Delta u(k|k)$ of the sequence $\{\Delta u(k|k), \cdots, \Delta u(k+ H_c - 1|k)\}$ which minimizes the following.*

$$\min_{\Delta u(k|k), \cdots, \Delta u(k+H_c-1|k)} \Phi_k \qquad \text{subject to}$$

$$\begin{cases} \dot{x}(k+t|k) = f(x(k+t|k), u(k+t|k), \hat{d}(k+t|k)), t \geq 0 \\ y(k+t|k) = g(x(k+t|k), \hat{d}(k+t|k)), t \geq 0 \\ u(k+i|k) \in \mathcal{U}, i = 0, \cdots, H_c - 1 \\ \Delta u(k+i|k) = 0, i \geq H_c \end{cases} \qquad (3)$$

where $\bullet(k+i|k)$ denotes the variable at time $k+i$ predicted at time k, $\hat{d}(k)$ denotes the estimated disturbance at time k, and the objective function Φ_k is defined by (2).

Notice that the output constraints are not included in the optimization problem. The reason is simply that the output is integrated and enforcing constraints on $\infty-$dimensional[4] variables would be expensive computationally. Thus, we choose to soften the output constraints through the error-dependent weight $\Gamma_y(e)$. As mentioned earlier, the weight is diagonal and each diagonal element has the following form (Figure 2):

$$\Gamma_y(y)_{\mathrm{ii}} = \begin{cases} \Gamma_y(0)_{\mathrm{ii}}[1 + \epsilon_i(y_i - y_i^{\min})^2] & \text{if } y_i \leq y_i^{\min} \\ \Gamma_y(0)_{\mathrm{ii}} & \text{if } y_i^{\min} \leq y_i \leq y_i^{\max} \\ \Gamma_y(0)_{\mathrm{ii}}[1 + \epsilon_i(y_i - y_i^{\max})^2] & \text{if } y_i \geq y_i^{\max} \end{cases} \qquad (4)$$

Here $\epsilon_i \geq 0$ can be chosen to achieve the desirable degree of softening: $\epsilon_i = 0$ indicates no constraints while $\epsilon_i = \infty$ indicates hard constraints. It can be easily shown that this is equivalent to the output softening discussed in [20, 25]. Clearly other functional forms can also be used.

---

[4]The $\infty-$dimensionality comes from the fact that $y(t)$ is a continuous variable, regardless of H_p.

500

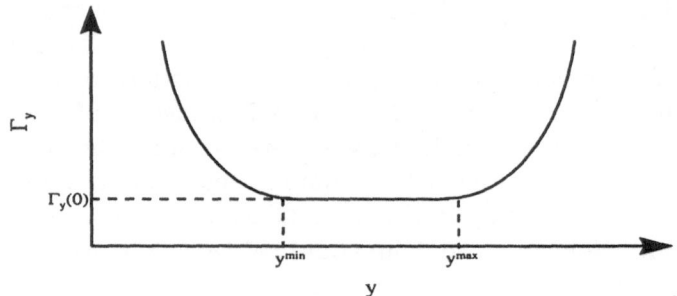

Figure 2. Output weight ($\Gamma_y(y)$) versus output (y).

Remark 2 *The objective function can be evaluated by simulating the nonlinear system for a given set of inputs.*

We can show, under some rather general conditions, that the closed loop system is asymptotically stable if and only if the optimization problem is feasible. It should be noted that similar results have been derived by several other researchers.

Theorem 1 *Assume*

- $H_p = \infty$
- *state measurement (and perfect disturbance estimate)*
- *constant disturbance (d) and setpoint (r)*[5]
- *perfect model*
- **Assumptions 1 - 4**

Then the close loop system of (1) with **Controller NMPC #1** *is asymptotically stable if and only if the optimization problem is feasible at the first sampling time.*

Proof. If the optimization problem is not feasible, then the feedback control law is undefined. Feasibility of the optimization problem at the first sampling time implies $J_0 = \min \Phi_0 < \infty$. Feasibility at time k implies feasibility at time $k+1$ since the sequence $\{\Delta u(k+1|k+1), \cdots, \Delta u(k+H_c - 1|k+1), \Delta u(k+H_c|k+1)\} = \{\Delta u(k+1|k), \cdots, \Delta u(k+H_c - 1|k), 0\}$ is feasible. Thus, feasibility of the optimization problem at all future times is guaranteed. Furthermore,

$$J_{k+1} + \int_k^{k+1} \left\{ (r - y(t))^T \Gamma_y(e)(r - y(t)) + \dot{x}^T(t)\Gamma_x \dot{x}(t) \right\} dt$$

$$+ \Delta u(k)^T \Gamma_u \Delta u(k) \leq J_k$$

[5]This assumption is necessary to show that feasibility at the first sampling time implies feasibility at all future sampling times.

From which we obtain,

$$J_{k+1} + \int_0^{k+1} \left\{ (r - y(t))^T \Gamma_y(e)(r - y(t)) + \dot{x}^T(t)\Gamma_x \dot{x}(t) \right\} dt$$

$$+ \sum_{i=0}^k \Delta u(i)^T \Gamma_u \Delta u(i) \leq J_0 < \infty$$

Since $\Gamma_y(e) \geq \Gamma_y(0)$, we have

$$J_{k+1} + \int_0^{k+1} \left\{ (r - y(t))^T \Gamma_y(0)(r - y(t)) + \dot{x}^T(t)\Gamma_x \dot{x}(t) \right\} dt$$

$$+ \sum_{i=0}^k \Delta u(i)^T \Gamma_u \Delta u(i) < \infty$$

Therefore, $\int_k^{k+1} (r - y(t))^T (r - y(t)) dt \to 0, \int_k^{k+1} \dot{x}(t)^T \dot{x}(t) dt \to 0$, and $\Delta u(k) \to 0$ asymptotically. By Assumption 3 (Lipschitzian conditions on f and g), we have $\dot{x} \to 0$ and $y(t) \to r$ asymptotically. Essentially, $f(x, u, d) \to 0, r - g(x, d) \to 0$, and $\Delta u \to 0$ asymptotically. Thus, the system approaches one of the steady-states, by Assumption 4. Furthermore, u^{ss} and x^{ss} are bounded (Assumption 4). Therefore, the closed loop system is asymptotically stable. □

Remark 3 Assumption 4 *implies that multiple steady-states are possible. The system is not guaranteed to approach the desired solution in the case of input multiplicity (i.e. there are at least two sets of inputs corresponding to a fixed output).*

Remark 4 *The global solution is not necessary for Theorem 1 to hold. A feasible local solution suffices.*

Remark 5 *If H_c is finite, then Theorem 1 holds if $\Gamma_u \geq 0$. By imposing additional technical assumptions on the system, Theorem 1 holds for $\Gamma_x \geq 0$.*

3. A Novel Algorithm

The optimization problem (3) has $n_u H_c$ decision variables. In general, for reasons of stability and performance, H_c should be chosen reasonably large: It is well known that, for a linear system with n unstable poles (including those on the $j\omega$ axis for a continuous-time system or on the unit circle for a discrete-time system), closed loop stability is guaranteed only if $H_c \geq n$ [18, 24]. Also as we will demonstrate in Example 1 of this paper, choosing small H_c may result in worst performance than that can be achieved by a linear controller. The question is: Can we significantly reduce the on-line computational time for *any* H_c with minimal loss in performance?[6]

3.1. BASIC IDEA

Since all future control moves $\Delta u(k+1|k), \cdots, \Delta u(k+H_c-1|k)$ are not implemented, the following question naturally arises: Do we have to determine them exactly? For linear systems, we showed that the answer is almost always "no" [22]. Efficient approximations of these control moves can result in significant savings in on-line computational demand. Controller NMPC #2, defined below, assumes that future control moves can be approximated by a controller C (i.e. $u(k+i|k) = \text{sat}(C(x(k+i|k), e(k+i|k))), i = 1, \cdots, H_c-1$).

Definition 2 (Controller NMPC #2) *At each sampling time k, $\Delta u(k)$ is the solution to the following optimization problem.*

$$\min_{\Delta u(k|k)} \quad \Phi_k \qquad \text{subject to}$$

$$\begin{cases} \dot{x}(k+t|k) = f(x(k+t|k), u(k+t|k), \hat{d}(k+t|k)), t \geq 0 \\ y(k+t|k) = g(x(k+t|k), u(k+t|k), \hat{d}(k+t|k)), t \geq 0 \\ u(k|k) \in \mathcal{U} \\ \mathbf{u(k+i|k)} = \text{sat}(\mathbf{C}(\mathbf{x(k+i|k)}, \mathbf{e(k+i|k)})), \mathbf{i} = \mathbf{1}, \cdots, \mathbf{H_c - 1} \\ \Delta u(k+i|k) = 0, i \geq H_c \end{cases} \tag{5}$$

where sat *denotes the saturation function, i.e.*

$$\text{sat}(u) = \left\{ \begin{array}{c} \text{sat}(u_1) \\ \vdots \\ \text{sat}(u_{n_u}) \end{array} \right. , \text{sat}(u_i) = \left\{ \begin{array}{ll} u_i^{\max} & u_i > u_i^{\max} \\ u_i & u_i^{\min} \leq u_i \leq u_i^{\max} \\ u_i^{\min} & u_i < u_i^{\min} \end{array} \right.$$

Notice that the on-line optimization problem for Controller NMPC #2 has *only* n_u decision variables, *regardless* of control horizon H_c. This results in significant savings in on-line computational time and Figure 3 compares on-line computational time for Controller NMPC #1 and #2. Asymptotic stability of the closed loop system with Controller NMPC #2 is stated in the following theorem. The proof follows directly from that for Theorem 1 and is thus omitted.

---

[6]Minimal loss in performance should be interpreted as the difference in performance between "global optimal" and "the best that can be achieved within *allowable* time for computation."

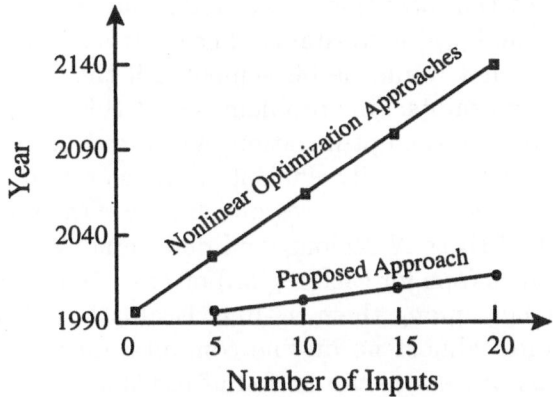

Figure 3. Comparison of on-line computational demand for `Controller NMPC #1` and #2: Year in which computers will be powerful enough to solve the on-line optimization problem versus number of inputs. We assume that $H_c = 5$ and that the on-line computational time grows *exponentially* with the number of decision variables.

Theorem 2 *Assume*

- $H_p = H_c = \infty$
- *state measurement (and perfect disturbance estimate)*
- *constant disturbance (d) and setpoint (r)*
- *perfect model*
- `Assumptions 1 - 4`

Then the close loop system of (1) with `Controller NMPC #2` *is asymptotically stable if and only if the optimization problem is feasible at the first sampling time.*

Remark 6 *While the infinite horizons may be impossible to implement in practice, the objective function can be approximated with finite horizon if both $r-y$ and $\dot{x}$ at the end of the finite horizon become sufficiently small. An alternative is to use results developed by Chen and Allgower [2] to guarantee stability by replacing infinite horizon with a finite horizon and an appropriate final weight.*

3.2. SYSTEMATIC CONSTRUCTION OF CONTROLLER C

Obviously, the overall performance of the closed-loop system depends on how the controller C is constructed. While in principle C can be *arbitrary*, we will discuss the simplest case where C is linear. In this case, we ask the following question: How should a *linear* controller be constructed such that $u(k + i|k)$ is "optimally" approximated? A simple answer is: Linearize the nonlinear system and analytically determine the controller corresponding to

Controller NMPC #1 that uses this linear model and assumes no input and output constraints (i.e. $\Gamma_y(y)$ is constant). The control action approximated by this controller is what would be implemented if the system were linear and there were no constraints. One problem is that this requires linearizing the system on-line during the optimization, which can be computationally expensive. A more computationally efficient approach is to divide the region of all possible operations into N sub-regions, linearize the nonlinear system at one point in each of these N regions, and determine N linear controllers for the N linear models (one for each region) off-line. Controller NMPC #2 automatically switches among them on-line. Evaluation of $u(k + i|k), i = 1, \cdots, H_c - 1$, requires almost no on-line computation (i.e. multiplication only). Details of this procedure can be found in [23].

The following chain of operations shows how Controller NMPC #2 approximates Controller NMPC #1 *within* allowable time for computation and why in general it should result in minimal loss in performance (in the sense defined earlier):

$$\text{NMPC \#1}: \min_{\Delta u(k|k), \cdots, \Delta u(k+H_c-1|k)} \Phi_k$$

$$\Downarrow \textit{Insufficient time for solution}$$

$$\min_{\Delta u(k|k)} \Phi_k, \textit{approximating } \Delta u(k + i|k), i = 1, \cdots, H_c - 1$$

$$\Downarrow \textit{Approximation by Controller } C$$

$$\text{NMPC \#2}: \min_{\Delta u(k|k)} \Phi_k, u(k + i|k) = \text{sat}(C(x(k + i|k), e(k + i|k)))$$

Obviously, that C is linear is the simplest case.

3.3. AN ALTERNATIVE INTERPRETATION

By writing $u(k|k) = \text{sat}(C(x(k), e(k))) + u(k|k)^{\text{NL}}$, the solution to (5) is equivalent to the solution to the following optimization problem.

$$\min_{u(k|k)^{\text{NL}}} \Phi_k \quad \text{subject to}$$

$$\begin{cases} \dot{x}(k + t|k) = f(x(k + t|k), u(k + t - 1|k), \hat{d}(k)), t \geq 0 \\ y(k + t|k) = g(x(k + t|k), u(k + t|k), \hat{d}(k)), t \geq 0 \\ u(k|k) \in \mathcal{U} \\ u(k + i|k) = \text{sat}(C(x(k + i|k), e(k + i|k))), i = 1, \cdots, H_c - 1 \\ \mathbf{u(k|k)} = \text{sat}(\mathbf{C}(\mathbf{x(k)}, \mathbf{e(k)})) + \mathbf{u(k|k)}^{\text{NL}} \\ \Delta u(k + i|k) = 0, i \geq H_c \end{cases} \quad (6)$$

With this transformation, we can see that the closed loop system for the nonlinear system (1) and `Controller NMPC #2` with $H_c = H_p$ is *equivalent* to the closed loop system for `Controller NMPC #1` with $H_c = 1$ and the following modified nonlinear system (Figure 4):

$$
\begin{aligned}
\dot{x}(t) &= f(x(t), u(t), d(t)) \\
y(t) &= g(x(t), d(t)) \\
u(k) &= \text{sat}(C(x(k), e(k))) + u(k)^{\text{NL}}
\end{aligned}
\tag{7}
$$

Thus, `Controller NMPC #2` essentially approximates `Controller NMPC #1` with two controllers—controller C and `Controller NMPC #1` with $H_c = 1$. Since C can be arbitrary, `Controller NMPC #2` can be implemented on top of an *existing* controller for a nonlinear system.

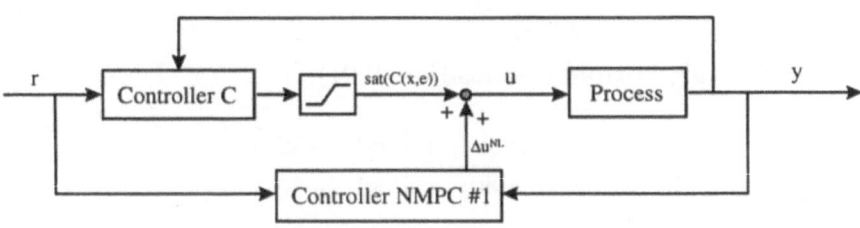

Figure 4. Series/Cascade Interpretation of `Controller NMPC #2`

Remark 7 *C can be a continuous-time controller. In fact, a continuous controller is used in all the examples in the next section.*

4. Examples

The proposed algorithm is illustrated on a distillation column dual composition control problem (LV-configuration) using a rigorous tray-by-tray model with 82 states[7], and the TE Challenge problem [5] involving 10 inputs and 10 outputs. We *emphasize* that the main purposes of these examples are to compare on-line computational time for `Controller NMPC #1` and #2 and to demonstrate the feasibility of a practical implementation of `Controller NMPC #2`, *not* to show that nonlinear controllers are needed to control these systems.

4.1. DISTILLATION DUAL COMPOSITION CONTROL

The sampling time is 5 minutes. The controller C is linear and is designed based on a linear model at the nominal operating condition (i.e. distillation

---

[7]See, http://www.kjemi.unit.no/skoge/, for details.

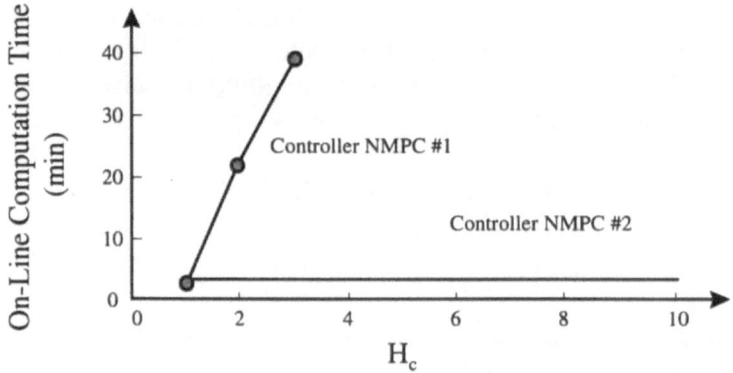

Figure 5. *Figure 5.* Comparison of on-line computational time for `Controller NMPC #1` and `#2` ($H_p = 400$). The on-line optimization problem is solved using a conjugate gradient method in MATLAB/SIMULINK on a Pentium 166 mHz PC.

composition: 0.99; bottom composition: 0.01):

$$C = \frac{190s + 1}{50s} \begin{bmatrix} 40 & -29 \\ 38 & -32 \end{bmatrix} \tag{8}$$

Figure 5 plots the on-line computational time versus H_c for the two algorithms for $H_p = 400$ based on *actual* simulations. The integration routine `Gear` is used.[8] Regardless of H_c, the computational time for `Controller NMPC #2` is about 3 minutes, which is larger than the computational time for `Controller NMPC #1` with $H_c = 1$ (about 2.5 minutes) because the inclusion of the controller C increases the time for evaluating the objective function which involves simulation. In general, since the nonlinear system is much more complex than the controller C, simulation time for the nonlinear system with or without C should be very similar. Therefore, including the controller C should have a very small impact on on-line computational time.

Figure 6 compares performance for `Controller NMPC #2` with $H_c = H_p$ and `Controller NMPC #1` with $H_c = 1$ for a setpoint change in distillate composition from 0.99 to 0.999. $H_p = 400$ in both cases. The reason for such big setpoint change is to illustrate the effect of nonlinearity, and the reason for this comparison is that in practice only implementing $H_c = 1$ is feasible for `Controller NMPC #1` while `Controller NMPC #2` can be implemented with any H_c. Performance for the linear controller C is also shown. Notice that `Controller NMPC #1` with $H_c = 1$ seems to give worse performance than the linear controller C.

[8]Other tuning parameters are $\Gamma_y = I, \Gamma_x = 0$, and $\Gamma_u = 0.1$.

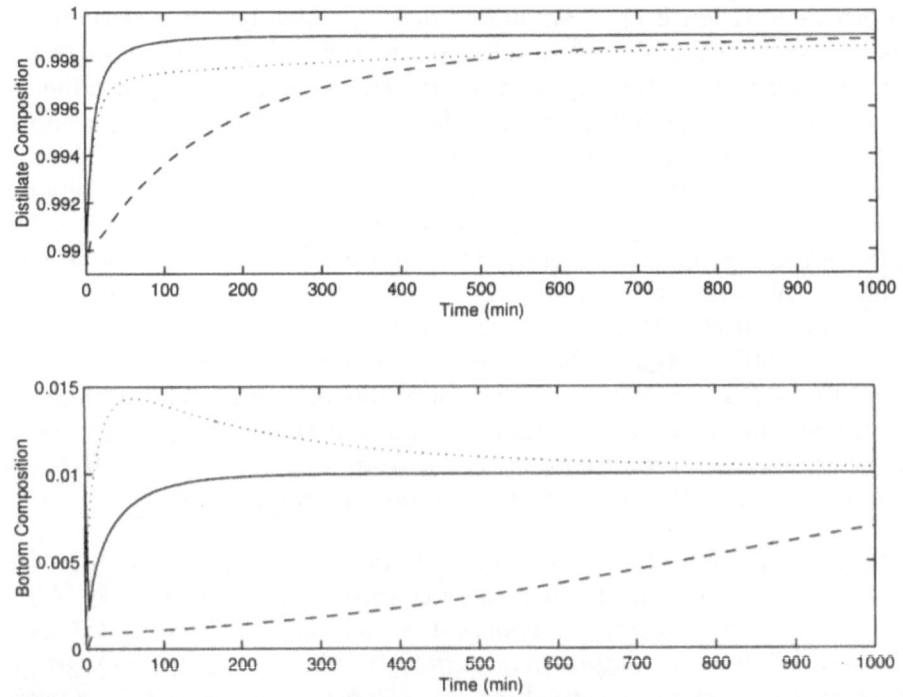

Figure 6. Performance for **Controller NMPC #2** (solid) with $H_c = H_p = 400$, **Controller NMPC #1** (dashed) with $H_c = 1$ and $H_p = 400$, and the linear controller C (dotted).

4.2. TENNESSEE-EASTMAN PROBLEM

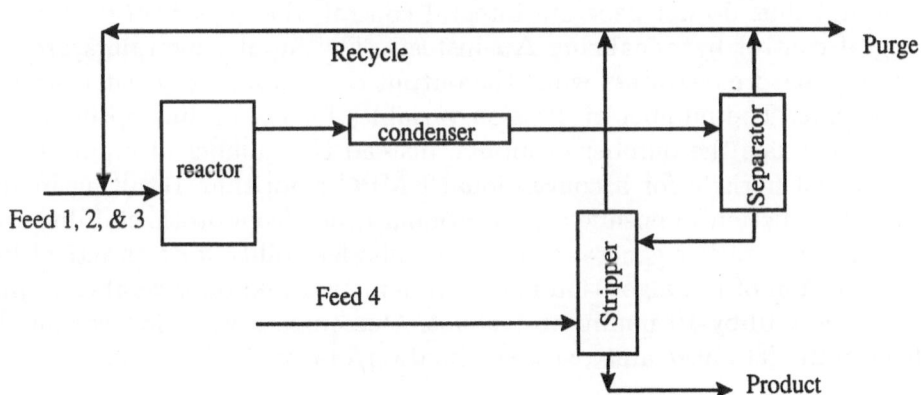

Figure 7. Simplified block diagram of the TE problem.

The TE Challenge problem was proposed by Downs and Vogel [5] for testing alternative control and optimization strategies for continuous chem-

508

ical processes (Figure 7). A simplified model, which has 26 states and consists of two PI controllers,[9] developed by Ricker [19] is used to test the proposed algorithm. Ten inputs are selected as manipulated variables and ten outputs as controlled variables. They are listed in Table 1. For simplicity, C consists of only three PI controllers to control reactor level, separator level, and stripper level by manipulating separator temperature, separator liquid flow, and product flow, respectively. It must be emphasized that these controllers are designed to merely stabilize the system and are *by no means* optimal. The interested readers are referred to the papers by Downs and Vogel [5] and Ricker [19] for details.[10] The sampling time is 5 minutes and $H_c = H_p = 600$.[11] Again, the optimization and simulation are carried out in MATLAB/SIMULINK on a Pentium 166 mHz PC. The on-line computational time is about 30 minutes. Figure 8 shows the performance for a step change in product concentration from 43.7%H to 49.0%H (molar) while maintaining the total production rate at 211.3 mole/hr.

Remark 8 *We found that the solution to the on-line optimization problem, determined via the Powell's method (a conjugate gradient method) [17], depends on the initial guess, implying that $\Delta u(k)$ may not be the global optimal solution. While performance may suffer, stability is still guaranteed since Theorem 1 only requires a local feasible solution, not the global solution.*

5. Conclusions

In this paper, we proposed a novel NMPC algorithm for control of large nonlinear constrained systems. Unlike many other methods which penalize u and thus do not generate integral control, the algorithm guarantees integral control by penalizing Δu instead. The input constraints are enforced as hard constraints while the output constraints are treated as soft constraints. The number of decision variables for the on-line optimization problem equals the number of inputs, instead the number of inputs times the control horizon for a conventional NMPC algorithm, resulting in significant reduction in on-line computational time. Asymptotic stability was proven under rather general conditions. The feasibility for a practical implementation of the algorithm has been demonstrated on several examples including a 10-by-10 nonlinear systems. Our current work focuses on the effect of model uncertainty, state estimation/output feedback, etc.

---

[9]The two PI controllers control reactor and separator temperatures by manipulating reactor and condenser coolant valves, respectively.

[10]A paper which discusses in details the application of the proposed algorithm to the TE problem is under preparation.

[11]Other tuning parameters are $\Gamma_y = I, \Gamma_x = 0$, and $\Gamma_u = 0$.

Number	Manipulated Variables	Controlled Variables
1	Feed 1 (pure A)	Reactor pressure
2	Feed 2 (pure D)	Reactor liquid Holdup
3	Feed 3 (pure E)	Separator pressure
4	Feed 4 (A & C)	Separator liquid holdup
5	Recycle flow	Stripper reboiler holdup
6	Purge	Feed zone pressure
7	Separator underflow	A in reactor feed
8	Product rate	C in reactor feed
9	Reactor temperature	G in product
10	Separator temperature	H in product

TABLE 1. Descriptions of manipulated and controlled variables.

References

1. L. T. Biegler and J. B. Rawlings. Optimization approaches to nonlinear model predictive control. In *Conf. Chemical Process Control (CPC-IV)*, pages 543–571, South Padre Island, Texas, 1991. CACHE-AIChE.
2. H. Chen and F. Allgöwer. A quasi-infinite horizon nonlinear model predictive control scheme with guaranteed stability. *Automatica*, 1997. in press.
3. S. L. De Oliveira and M. Morari. Robust model predictive control for nonlinear systems. In *Proceedings of the 33rd IEEE Conference on Decision and Control*, pages 3561–3567, Orlando, Florida, 1994.
4. S. L. De Oliveira, V. Nevistic, and M. Morari. Control of nonlinear systems subject to input constraints. In *Proceedings of IFAC Symposium on Nonlinear Control Systems Design*, pages 15–20, Tahoe City, CA, 1995.
5. J. J. Downs and E. F. Vogel. A plant-wide industrial-process control problem. *Computers and Chemical Engineering*, 17(3):245–255, 1993.
6. J. W. Eaton and J. B. Rawlings. Feedback control of nonlinear processes using on-line optimization techniques. *Comp. and Chem. Eng.*, 14:469–479, 1990.
7. J. W. Eaton and J. B. Rawlings. Model-predictive control of chemical processes. *Chem. Eng. Sci.*, 47:705–720, 1992.
8. C. E. García. Quadratic dynamic matrix control of nonlinear processes. An application to a batch reactor process. In *AIChE Annual Meeting*, San Francisco, CA, 1984.
9. G. Gattu and E. Zafiriou. Nonlinear quadratic dynamic matrix control with state estimation. *Ind. Eng. Chem. Res.*, 31(4):1096–1104, 1992.
10. G. Gattu and E. Zafiriou. Observer based nonlinear quadratic dynamic matrix

510

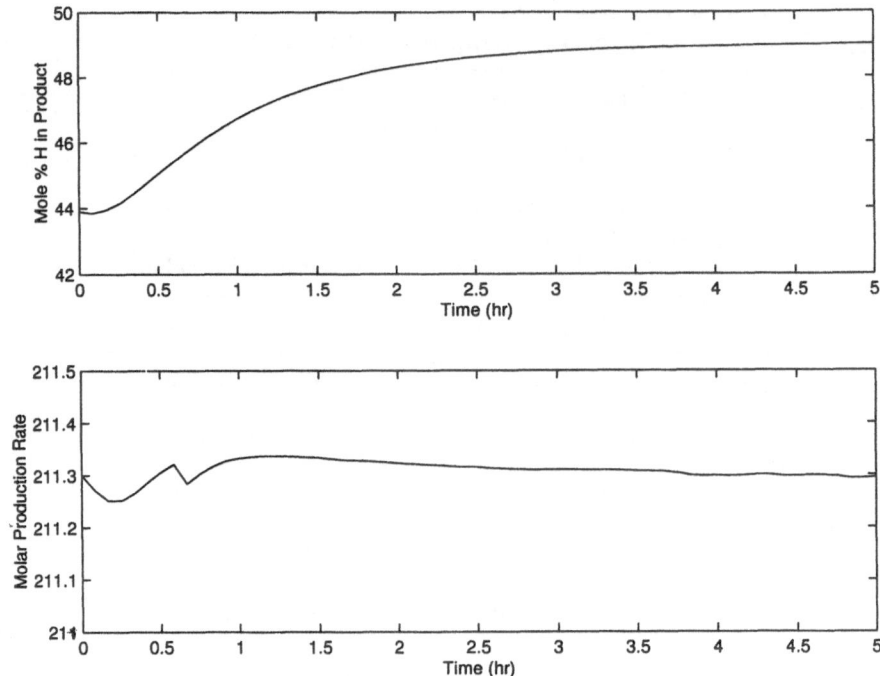

Figure 8. Performance for a production rate change $H_c = H_p = 240$.

control for state space and input/output models. *Canadian Journal of Chemical Engineering*, 73:883–895, Dec. 1995.

11. S. Jang, B. Joseph, and H. Mukai. On-line optimization of constrained multivariable chemical processes. *AIChE Journal*, 33:26–42, 1987.

12. M. J. Kurtz and M. A. Henson. Linear model predictive control of input-output linearized processes with constraints. In *CPC V*, Lake Tahoe, CA, 1996.

13. J. H. Lee and N. L. Ricker. Extended Kalman filter based on nonlinear model predictive control. *Ind. Eng. Chem. Res.*, 33:1530–1541, 1994.

14. D. Q. Mayne. Nonlinear model predictive control: An assessment. In *CPC V*, Lake Tahoe, CA, 1996.

15. H. Michalska and D. Mayne. Robust receding horizon control of constrained nonlinear systems. *IEEE Trans. Aut. Control*, 38(11):1623–1633, November 1993.

16. A. M. Morshedi. Universal dynamic matrix control. In *Proceedings of CPC III*, New York, 1986.

17. W. H. Press, B. P. Flannery, S. A. Teukolsky, and W. T. Vetterling. *Numerical Recipes—The Art of Scientific Computing*. Cambridge University Press, 1986.

18. J. B. Rawlings and K. R. Muske. The stability of constrained receding horizon control. *IEEE Trans. Aut. Control*, 38(10):1512–1516, October 1993.

19. N. L. Ricker. Decentralized control of the tennessee eastman challenge process. *Journal of Process Control*, 6:205–221, 1996.

20. N. L. Ricker, T. Subrahmanian, and T. Sim. Case studies of model-predictive control in pulp and paper production. In *Model Based Process Control - Proc. of the 1988 IFAC Workshop*. Pergamon Press, Oxford, 1989.

21. T. H. Yang and E. Polak. Moving horizon control of nonlinear systems with input saturation, disturbances and plant uncertainty. *Int. J. Control*, 58(4):875–903,

September 1993.

22. A. Zheng. Model predictive control: Is QP necessary? In *AIChE Annual Meeting*, Chicago, IL, 1996.

23. A. Zheng. A computationally efficient nonlinear model predictive control algorithm. In *Proceedings of American Control Conf.*, Albuquerque, NM, 1997.

24. A. Zheng, V. Balakrishnan, and M. Morari. Constrained stabilization of discrete-time systems. *International Journal of Robust and Nonlinear Control*, 5(5):461–485, Aug. 1995.

25. A. Zheng and M. Morari. Stability of model predictive control with mixed constraints. *IEEE Trans. Aut. Control*, 40(10):1818–1823, October 1995.

OPTIMIZATION APPROACHES TO CONTROL-INTEGRATED DESIGN OF INDUSTRIAL BATCH REACTORS

O. ABEL, A. HELBIG, W. MARQUARDT*
Lehrstuhl für Prozeßtechnik,
RWTH Aachen University of Technology,
D-52056 Aachen, Germany

Abstract. In this paper optimization based approaches to the design of batch reactors are presented. In addition to operational constraints on manipulated and quality related variables, safety requirements are explicitly incorporated into the design even in the case of possible failure scenarios. Furthermore, since many industrial design problems often have to cope with existing plants and low on-line computational resources, controller limitations should be considered already at the design stage. These aspects render the suggested design problem formulation a nonlinear mixed-integer stochastic optimal control problem for a hybrid discrete-continuous system. In order to make this problem computationally tractable several simplifications are discussed. A decomposition is proposed where inherently safe operation profiles for a number of given possible failure scenarios are calculated in an inner problem whereas all control related questions for the nominal mode of operation are dealt with in an outer problem. The ideas and concepts for control-integrated batch reactor design are illustrated by means of an industrial polymerization reactor.

1. Introduction

Batch and semibatch reactors are of growing importance in the chemical industry since they offer a quick response to changing market conditions [52]. They are widely used in the production of fine chemicals, specialties, and other high value products. Due to the transient nature their optimization involves the calculation of trajectories for input variables and/or controller

---

*Author to whom all correspondence should be addressed. Phone: +49-241-806712, Fax: +49-241-8888326, E-mail: marquardt@lfpt.rwth-aachen.de

R. Berber and C. Kravaris (eds.), Nonlinear Model Based Process Control, 513-551.
© 1998 *Kluwer Academic Publishers.*

514

setpoints rather than the determination of one single set of optimal operating conditions. This calculation requires methods from optimal control theory and thus strict separation of design and control is impossible for this class of problems. Throughout this paper the term "control" is therefore used in a broad sense covering trajectory design as well as closed loop control. The paper restricts itself to the treatment of exothermic reactions taking place in single batch processes. Scheduling problems or integration of design and scheduling (see e.g. [10, 12]) are not dealt with.

Over the past decades a vast amount of papers contributing to the field of batch reactor control has appeared which has been subject to intensive reviewing [9, 32, 71]. Literature is dominated by papers either on solution methods for and case studies of optimal control problems or on closed loop controller design for some specified trajectory or end-point problem, including observer design. Increasing interest in using optimization based methods also in the latter area can be observed in recent years. This includes optimization based control techniques like model predictive control for trajectory tracking [28, 31, 40, 44, 48], optimal feedback control [46, 49, 50], and on-line (re-)optimization strategies [54, 69].

In current industrial practice the number of industrial applications both of rigorous optimization strategies as well as advanced model and optimization based control techniques is still very limited. To the authors' understanding the main reasons are

- the high cost involved in derivation and validation of a process model and its adaptation to plant modifications and product switchover,
- the high implementation and maintenance cost associated with advanced control systems,
- the limited number of typically available on-line measurements,
- possible acceptance problems by the operating team, and
- the presumed low overall benefits of both rigorous optimization and advanced process control.

In addition to these rather process oriented arguments, the high computational effort necessary to solve large-scale dynamic optimization problems is still another important reason for the limited attraction optimization based techniques have gained in industrial batch processing. However, due to changing market conditions and tighter environmental constraints chemical industries are forced to permanently improve their batch processes. This necessitates the exploitation of all available process knowledge ranging from trivial process variable constraints to detailed physico-chemical models. In spite of the required process model little alternative approaches to optimization based technology exist to reach this goal. In order to maximize the overall benefits three issues can be identified:

– reduction of model development and maintenance cost,
– improvement of numerical methods, and
– derivation of appropriate problem formulations.

The first aspect requires an appropriate modeling methodology which is under investigation in our laboratory and treated in [13, 23, 39] whereas the second issue is currently tackled by a number of other research groups (see i.e. [36, 67, 74]). This paper tries to contribute to an increased applicability of optimization based control techniques by suggesting problem formulations for the design of batch reactors. These have to reflect fundamental industrial objectives going beyond purely economic optimization. In a recent review Bonvin [14] identified process safety to be of paramount importance. Furthermore, design and operation have to cope with simple controllers provided by an industrial control system [24]. Therefore, control and its possible limitations have to be included into the design problem formulations.

In this paper control-integrated design of batch processes is treated. Incorporation of safety and controllability aspects into the batch design problem leads to an extension of the concepts of dynamic resilience [42] and flexibility [41]. In section 2 a generic formulation for risk cautious control-integrated batch process and controller design is derived. Extending an approach introduced by Soroush and Kravaris [59, 60] the problem is decomposed into two steps. First, the reaction system including safety aspects is treated in Section 3 using the perfect control hypothesis. This is followed by the application of optimization based methods to the cooling system design and controllability analysis in Section 4.

2. Control-integrated design of flexible batch processes

Integration of design and control has been subject to intensive research. An excellent summary of the achievements up to 1994 can be found in [51]. The ultimate goal is to design processes which can operate safely and reliably under the effect of parameter variations and disturbances [47]. A variety of methods exist for closed loop performance assessment and the analysis of open loop performance limitations as well as for control structure selection and controller tuning. Most of the work treats continuous processes often analyzing a linear model valid in the vicinity of a specific operating point [56]. This is generally insufficient for transient processes like batch processes and thus integration of design and control is a far more difficult task for this class of processes. A first valuable approach has been presented by Soroush and Kravaris [59, 60]. Their formulation, however, does not allow the treatment of general safety constraints and controllability aspects are only assessed with respect to nonlinear systems theory. Therefore, in this

516

section an attempt is made to derive a more general problem formulation for control-integrated design of batch processes. This strongly necessitates the definition of several key expressions.

2.1. PRELIMINARIES AND DEFINITIONS

The design of a batch reaction process P for a given reaction system involves the determination of the following sets of process optimization variables:

- The set of equipment design parameters $\pi \in \mathbb{R}^{n_\pi}$ such as the reactor diameter, the cooling system dimensions or the corresponding cooling medium.
- Initial conditions $x_0 \in \mathbb{R}^{n_x}$, particularly the initial amount of the involved substances including the concentration of a possibly used initiator or catalyst.
- The trajectories for the operational design variables $u \in L_2 ([0, t_f])^{n_u}$.*

  u can be partitioned according to $u = \left[z^T, \ y_{set}^T \right]^T$. $z \in L_2^{n_z}$ includes elements which are directly accessed for open loop control whereas $y_{set} \in L_2^{n_y}$ refers to controller setpoints while the controller itself acts on manipulated variables $m \in L_2^{n_m}$. It should be noted that, depending on the chosen system boundary of the process, an operational variable may either be directly optimized as (open loop) manipulated variable z or as setpoint y_{set} of a corresponding controlled variable. A typical example for z is the feed flowrate whereas a setpoint y_{set} is usually calculated for the reactor temperature.
- Binary variables $X_c \in \mathbb{B}^{n_X}$ with $\mathbb{B} = \{0, 1\}$ describe the controller structure including pairing and type as well as related tuning parameters $K_c \in \mathbb{R}^{n_K}$.

Batch optimization has to account for both inevitable disturbances and uncertainties of stochastic nature as well as known failure situations. Assuming a structurally correct process model, three classes of variables can be distinguished:

- Uncertain but bounded time invariant model parameters $p \in \mathbb{P}^{n_p} \subset \mathbb{R}^{n_p}$ such as coefficients in the reaction kinetics.
- Bounded time varying disturbances $d \in \mathbb{D}^{n_d} \subset L_2^{n_d}$ comprising drifting model parameters, a changing environment, and measurement noise.
- An integer variable $\xi \in \Xi$ with $\Xi = \{1, 2, \ldots, n_\xi\}$ being the index set of failure scenarios to be considered, particularly mechanical failures, human errors and further abrupt deviations from nominal operation. The typical example in batch processing, which will be further treated

---

*For reasons of brevity the domain of the function space L_2 will be omitted in the sequel for all signals defined on $[0, t_f]$.

in Section 3, is a cooling system failure. In general the process dynamics after such a failure will have to be described by a model specific to the failure mode. The nominal and the failure modes together with the discrete transitions between them render the overall system hybrid. This adds a new dimension of complexity to the optimization of batch processes.

The determination of the values and trajectories of the process optimization variables has to account for several constraints on the input variables z and m as well as on other process variables related to safety, quality, and economics such as reactor temperature or product composition. These constraints hold either over the complete batch or only at the final time t_f. Constraints for the nominal operation and for the failure modes will usually be different, since it may not make sense to demand the enforcement of quality constraints in a failure situation for example.

In summary, a batch reaction process P is characterized by fixed π (e.g. equipment), x_0 and u (e.g. recipe). P may comprise the whole range from the isolated reactor content to the entire batch reactor with cooling system. Depending on the actual choice of the system boundary, P may include none, some, or all existing controllers. One single representation of a batch process, controlled or uncontrolled, with well defined exogenous input sequences, uncertain parameters, disturbances, and failures is called batch, B_P. Here nominal batches without safety related events can be distinguished from failure batches. The following important system properties of a batch process P or the corresponding batches B_P can now be defined:

Safety: A batch B_P is *safe* if all process variables considered to be safety relevant stay within certain predefined bounds during the whole operation cycle. Typical examples for such variables are reactor temperature and pressure.

Feasibility: A batch B_P is *feasible* if it is safe *and* all other operational constraints are enforced. As already stated the constraints may be failure mode dependent. In cases of severe failures they probably will reduce to safety constraints only.

Flexibility: A batch reaction process P is *flexible* if for *all* p, d, and any failure ξ being possible at any time $t \in [0, t_f]$ the corresponding batch B_P is feasible. This implies the operational constraints to be fulfilled under all possible variations in uncertain parameters, disturbances and failures.

Controllability: A batch process P, open loop or partly controlled, is (input-output) *controllable* if there exists a controller C which is able to meet some specified performance criteria for all nominal batches B_P. Thus controllability is a property independent of the controller C.

Optimal design: For the definition of optimal design two cases have to be distinguished.

- If the process P will not be connected to an external controller C, thus includes none or already some base controllers $\tilde{C}$, it is *designed optimally* if it is *flexible* and *design variables* are chosen to *minimize some objective function*.
- If the process P will be connected to an external controller C it is *designed optimally* if it is *flexible* and *controllable* and *design variables* are chosen to *minimize some objective function*.

If a partly controlled process already contains some controller $\tilde{C}$ different from C the objective function to be minimized will generally penalize structure and performance cost of the controller $\tilde{C}$ in addition to the primary (economic) objective.[†] Note that no separate definition of control-integrated optimal design is needed since the controller or at least controllability is already included in the definition of optimal design.

2.2. PROBLEM FORMULATION

Next, a general problem of designing a batch process including all necessary controllers and possible failures will be derived. Therefore, the optimal design involves the determination of the optimization variables π, x_0, z, y_{set}, X_c and K_c such that an objective function $\Phi \in \mathbb{R}$ is minimized in the presence of the uncertainties p, disturbances d, and failures ξ subject to the flexibility requirements, the model equations, and the control system equations. Mathematically this leads to a nonlinear mixed-integer stochastic optimal control problem and can be stated as[‡]

$$\min_{\substack{x_0,\, z(t),\, y_{set}(t), \\ \pi,\, X_c,\, K_c}} \quad \Phi(x(t), z(t), y_{set}(t), m(t), \pi, X_c, K_c, p, d(t), \xi, t_f) \quad (1)$$

$$\begin{aligned}
\text{s.t.} \quad 0 &= f(\dot{x}(t), x(t), z(t), m(t), \pi, p, d(t)), \quad x(t_0) = x_0, \\
0 &= m(t) - C(x(t), z(t), y_{set}(t), \pi, X_c, K_c, p, d(t)), \\
0 &\geq \mathcal{F}(x(t), z(t), y_{set}(t), m(t), \pi, X_c, K_c, p, d(t), \xi, t), \\
& \qquad \forall t \in [0, t_f].
\end{aligned}$$

[†]In principle these definitions permit a batch process $P_{\tilde{C}}$ with base controllers $\tilde{C}$ to be designed optimally such that the corresponding process P without $\tilde{C}$ is not controllable. In this case the economic objective function is minimized and the resulting design is flexible but violates the controllability criteria. This pathological case will not be considered and can always be excluded by choosing an appropriate objective function and appropriate controller performance criteria.

[‡]Since the optimization is carried out in a function space the use of *inf* and *sup* operators would be more appropriate here and subsequently. Since the use of these operators is rather uncommon in the optimal control literature, the *min* and *max* operators are used instead.

Here, $\Phi : L_2{}^{n_x} \times L_2{}^{n_z} \times L_2{}^{n_y} \times L_2{}^{n_m} \times \mathbb{R}^{n_\pi} \times \mathbb{B}^{n_X} \times \mathbb{R}^{n_K} \times \mathbb{P}^{n_p} \times \mathbb{D}^{n_d} \times \Xi \times \mathbb{R} \mapsto \mathbb{R}$ denotes the objective function, $f : L_2{}^{n_x} \times L_2{}^{n_z} \times L_2{}^{n_z} \times L_2{}^{n_m} \times \mathbb{R}^{n_\pi} \times \mathbb{P}^{n_p} \times \mathbb{D}^{n_d} \mapsto L_2{}^{n_x}$ describes the process model consisting of a set of differential and algebraic equations, $\mathcal{C} : L_2{}^{n_x} \times L_2{}^{n_z} \times L_2{}^{n_y} \times \mathbb{R}^{n_\pi} \times \mathbb{B}^{n_X} \times \mathbb{R}^{n_K} \times \mathbb{P}^{n_p} \times \mathbb{D}^{n_d} \mapsto L_2{}^{n_m}$ is an operator representing the control system, $\mathcal{F} : L_2{}^{n_x} \times L_2{}^{n_z} \times L_2{}^{n_y} \times L_2{}^{n_m} \times \mathbb{R}^{n_\pi} \times \mathbb{B}^{n_X} \times \mathbb{R}^{n_K} \times \mathbb{P}^{n_p} \times \mathbb{D}^{n_d} \times \Xi \times \mathbb{R} \mapsto L_2{}^{n_\mathcal{F}}$ is a flexibility test operator, t_f is the final time of the batch, and x_0 is chosen to be a consistent initial state for f. It has to be emphasized that the process dynamics after a failure can explicitly be incorporated into the flexibility test. Note that we do not assume a completely measurable state vector; for conciseness, the notation does not explicitly include a set of measurements and their defining equations.

Generally the calculation of the objective function $\Phi = \mathcal{S}(\phi)$ requires the evaluation of an operator $\mathcal{S}$ taking into account the stochastic nature of p and d and acting on the stochastic cost variable ϕ. As already stated above, ϕ reflects both economics as well as the controller structure and performance cost. The different possibilities for the operator $\mathcal{S}$ include the calculation of the expectation term of ϕ, the variance of ϕ, and the maximum of ϕ (worst case) as extensively discussed by Terwiesch et al. [68, 71]. Their treatment of batch trajectory optimization under uncertainty, however, only covers a subproblem of (1) since they do neither include controllability aspects nor controller design. Furthermore, in their formulation the operator $\mathcal{F}$ reduces to simple algebraic inequality constraints and particularly does not cover any failure situation.

With respect to the control system to be synthesized the formulation (1) is very generic and not restricted to a specific controller structure or type. For practical purposes however the restriction to controllers of limited complexity is unavoidable. In [41] the formulation of the general operator $\mathcal{C}$ for the case of multi-loop PI is given as an easily tractable example.

Finally the flexibility operator $\mathcal{F}$ has to be specified. For batch processes flexibility has first been introduced in [59] being defined as feasibility in the presence of parameter variations. Although this definition is formally equivalent to the one used here it should be noted that the underlying feasibility notation is totally different and therefore flexibility according to [59] also has a different meaning from our definition.

For continuous systems steady state flexibility analysis is a well developed area of research (see [26] for a review) and has recently been extended to dynamic systems [18]. It has been used as a subproblem in the control-integrated design problem of [41]. Adapting the flexibility analysis of [18] to the notation of problem (1) while neglecting failure modes ξ, the problem can mathematically be stated as

$$\max_{p,\, d(t)} \quad \min_{z(t),\, y_{set}(t)} \quad \max_{k} \quad g_k \tag{2}$$

$$
\begin{aligned}
\text{s.t.} \quad 0 \;&\geq\; g_k(x(t), z(t), y_{set}(t), m(t), \pi, X_c, K_c, p, d(t), t)\,, \\
0 \;&=\; f(\dot{x}(t), x(t), z(t), m(t), \pi, p, d(t)), \quad x(t_0) = x_0\,, \\
0 \;&=\; m(t) - C(x(t), z(t), y_{set}(t), \pi, X_c, K_c, p, d(t))\,, \\
& \quad\quad \forall t \in [0, t_f]\,.
\end{aligned}
$$

where $k \in K \subset \mathbb{N}$ is the index of inequality constraints $g_k : L_2^{n_x} \times L_2^{n_z} \times L_2^{n_y} \times L_2^{n_m} \times \mathbb{R}^{n_\pi} \times \mathbb{B}^{n_X} \times \mathbb{R}^{n_K} \times \mathbb{P}^{n_p} \times \mathbb{D}^{n_d} \times \mathbb{R} \mapsto L_2$ to be satisfied during the operation.

For the application to batch reactor design problems the general flexibility test can be simplified. Since the minimization of the stochastic objective function Φ in (1) requires the evaluation of (2) for all $p \in \mathbb{P}^{n_p}$ and $d \in \mathbb{D}^{n_d}$ the $max - min - max$ problem can be replaced by a simple enumeration of all inequalities. The flexibility test to be included in the optimal control problem thus reduces to

$$g_k(x(t), z(t), y_{set}(t), m(t), \pi, X_c, K_c, p, d(t), t) \;\leq\; 0\,, \tag{3}$$
$$\forall k \in K, \quad \forall p \in \mathbb{P}^{n_p}, \quad \forall d \in \mathbb{D}^{n_d}\,.$$

Similar simplifications from (2) to (3) have been used in [41] to derive a flexibility test which is computationally feasible as a subproblem of (1). Equation (3) is also in good agreement with the definition of flexibility given earlier in this section as long as failures ξ are not considered. Their incorporation into the flexibility test necessitates the extension of (3) covering nominal operation only to hybrid systems considering possible failure modes.

A system is called hybrid if it can be partitioned into continuous and discrete subsystems which mutually interact [22]. The continuous part can usually be described by a set of differential and algebraic equations while currently no unified framework seems to exist for the description of the discrete subsystem [5]. A large class of systems is covered by finite state machines. Often it is assumed that the transition from one discrete mode to another occurs instantaneously and that the time instant is not known a priori. The transitions are either triggered by switching conditions which are implicitly defined as functions of the continuous system states or by external inputs of the discrete subsystem.

The occurrence of a safety related failure can be viewed as a discontinuous state transformation. This transformation might involve simple parameter settings as well as changes in model structure and state space dimensionality. The flexibility test in the optimization of batch reactors

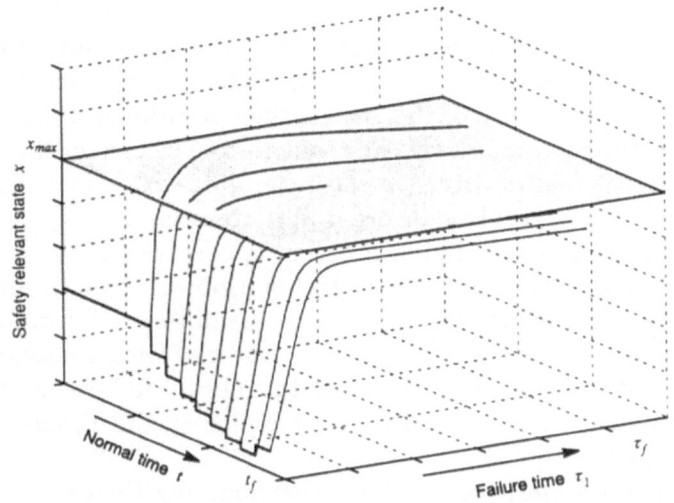

Figure 1. Flexibility test for one failure mode along two time axes.

has to consider the switching from nominal (i.e. faultless) operation to any possible failure mode ξ.[§]

Since the occurrence of failures is not known a priori, risk cautious optimization has to consider possible switches to any failure mode at any time instance $t \in [0, t_f]$. Therefore, n_ξ additional hypothetical failure batch processes P_ξ have to be analyzed at any t with regard to flexibility in a new time dimension $\tau_\xi \in [t, t + \tau_{\xi,f}]$. For the simple case of $n_\xi = 1$ the two time dimensions to be considered are illustrated in Figure 1 where a safety relevant state trajectory for nominal operation is shown (in the x-t-plane) together with trajectories of the same state extending into the second time dimension after failure. Mathematically, the flexibility requirements for the failure modes can be formulated as additional constraints to those in (3)

$$g_{k,\xi}(\boldsymbol{x}_\xi(\tau_\xi), \boldsymbol{z}_\xi(\tau_\xi), \boldsymbol{y}_{set,\xi}(\tau_\xi), \boldsymbol{m}_\xi(\tau_\xi), \boldsymbol{\pi}, \boldsymbol{X}_c, \boldsymbol{K}_{c,\xi}, \boldsymbol{p}_\xi, \boldsymbol{d}_\xi(\tau_\xi), \tau_\xi) \leq 0 \quad (4)$$

$$\text{with} \quad \begin{bmatrix} \boldsymbol{x}_\xi(\tau_\xi) \\ \boldsymbol{z}_\xi(\tau_\xi) \\ \boldsymbol{y}_{set,\xi}(\tau_\xi) \\ \boldsymbol{m}_\xi(\tau_\xi) \\ \boldsymbol{K}_{c,\xi} \end{bmatrix} = \mathcal{L}_\xi(\boldsymbol{x}(t), \boldsymbol{\pi}, \boldsymbol{X}_c, \boldsymbol{p}_\xi, \boldsymbol{d}_\xi(\tau_\xi), \tau_\xi) \,,$$

$$\forall k_\xi \in K_\xi, \quad \forall \boldsymbol{p}_\xi \in \mathbb{P}_\xi^{n_p}, \quad \forall \boldsymbol{d}_\xi \in \mathbb{D}_\xi^{n_d}, \quad \forall \xi \in \Xi,$$
$$\forall \tau_\xi \in [t, t + \tau_{\xi,f}], \quad \forall t \in [0, t_f] \,.$$

[§]Switching from one failure mode to another is not considered here, a combination of several failures may simply be introduced as an additional failure mode.

Here, $\mathcal{L}_\xi : {L_2}^{n_x} \times \mathbb{R}^{n_\pi} \times \mathbb{B}^{n_X} \times \mathbb{P}_\xi^{n_p} \times \mathbb{D}_\xi^{n_d} \times \mathbb{R} \mapsto {L_2}^{n_{x,\xi}} \times {L_2}^{n_{z,\xi}} \times {L_2}^{n_{m,\xi}} \times {L_2}^{n_{y,\xi}} \times \mathbb{R}^{n_{K,\xi}}$ denotes a generic operator mapping the initial conditions $x_\xi(\tau_\xi = 0)$ being a function of $x(t)$, the disturbances, and the fixed design variables to the trajectories of the states, the operational design variables, the controller tuning parameters and outputs of the hypothetical batch process after some failure. In the most general case $\mathcal{L}_\xi$ can involve an additional optimization problem in order to determine z_ξ, $y_{set,\xi}$ and $K_{c,\xi}$. Since the underlying optimization problem is constrained, it is not necessarily decoupled from the outer optimization problem for Φ. A lower degree of complexity arises if all operational design variables are fixed, for instance through existing emergency rules for the process. Then evaluating $\mathcal{L}_\xi$ reduces to the integration of a differential-algebraic failure model. In the simplest case dynamics are not considered explicitly and $\mathcal{L}_\xi$ corresponds to an algebraic equation.

Summarizing the derivations above and applying the expectation term formulation of Φ, the control-integrated design of flexible batch processes can be stated as stochastic nonlinear mixed-integer optimal control problem for a hybrid system:

$$\min_{\substack{x_0,\, z(t),\, y_{set}(t), \\ \pi,\, X_c,\, K_c}} \quad E_{\substack{p,\, d(t)}} \{\phi(x(t), z(t), y_{set}(t), \pi, X_c, K_c, p, d(t), \xi, t)\} \tag{5}$$

s.t.

$$0 = f(\dot{x}(t), x(t), z(t), m(t), \pi, p, d(t)), \quad x(t_0) = x_0,$$

$$0 = m(t) - C(x(t), z(t), y_{set}(t), \pi, X_c, K_c, p, d(t)),$$

$$0 \geq g_k(x(t), z(t), y_{set}(t), m(t), \pi, X_c, K_c, p, d(t), t),$$

$$0 \geq g_{k_\xi}(x_\xi(\tau_\xi), z_\xi(\tau_\xi), y_{set,\xi}(\tau_\xi), m_\xi(\tau_\xi), \pi, X_c, K_{c,\xi}, p_\xi, d_\xi(\tau_\xi), \tau_\xi),$$

$$\begin{bmatrix} x_\xi(\tau_\xi) \\ z_\xi(\tau_\xi) \\ y_{set,\xi}(\tau_\xi) \\ m_\xi(\tau_\xi) \\ K_{c,\xi} \end{bmatrix} = \mathcal{L}_\xi(x(t), \pi, X_c, p_\xi, d_\xi(\tau_\xi), \tau_\xi),$$

$$\forall k \in K, \quad \forall p \in \mathbb{P}^{n_p}, \quad \forall d \in \mathbb{D}^{n_d}, \quad \forall \xi \in \Xi,$$

$$\forall k_\xi \in K_\xi, \quad \forall p_\xi \in \mathbb{P}_\xi^{n_p}, \quad \forall d_\xi \in \mathbb{D}_\xi^{n_d},$$

$$\forall \tau_\xi \in [t, t + \tau_{\xi,f}], \quad \forall t \in [0, t_f].$$

Problem (5) offers a very holistic approach to control-integrated batch reactor design but its solution is practically impossible. Three directions to enhance tractability are recommended:

- By parameterization of the sets of the uncertainties p and disturbances d the expectation term can be replaced by a weighted summation term (multi-period approach) [71].
- The operators C and $\mathcal{L}_\xi$ should be chosen as simple as possible. Particularly, optimization problems, e.g. as those in model predictive control, should be avoided.
- Process knowledge and physical insight should be used to decompose the overall problem into subproblems.

This last aspect is treated in the following subsection in greater detail.

2.3. A DECOMPOSITION APPROACH TO BATCH REACTOR DESIGN

The ultimate goal of using a decomposition approach to the solution of (5) is to partition the overall problem in computationally feasible subproblems without loosing generality. This is easily achievable if the problems *flexible design* of a hybrid system and *control system design* can be decoupled. For simple batch reactors consisting of a reactor vessel with cooling system the most intuitive approach is to sequentially optimize the reaction system including all safety aspects first and then the cooling and control system. This is in line with the decomposition strategy proposed by Soroush and Kravaris [59, 60]. Triggered by the intention to reduce the model requirements and problem size for the optimization of the reaction system as much as possible they distinguish an inner and outer system as schematically shown in Figure 2.

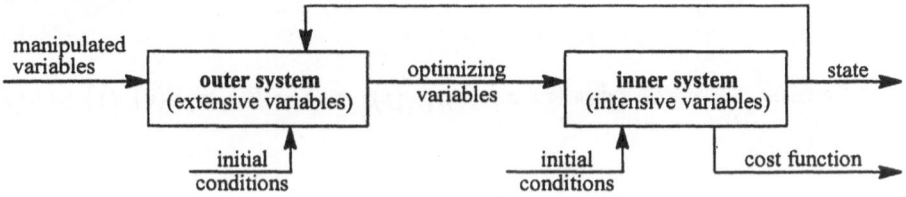

Figure 2. Structure of the design problem in [59].

The inner system describing the reaction system only contains intensive variables like temperature and concentrations whereas the outer system contains both intensive and extensive variables to describe the reactor vessel and the heating/cooling system. The inner and outer system are optimized sequentially obeying some feasibility aspects. The last step in their method is the synthesis of the control law and the test for regulatory and servo performance. The method of Soroush and Kravaris offers a powerful framework for the grass-root design of batch reactors although only heuristic suggestions are given for the case when their flexibility tests fail.

Furthermore, neither general safety constraints including failure situations are captured in the inner system nor is controllability analysis or integrated controller design regarded in the outer system.

Here an extended inner/outer decomposition strategy is proposed not restricting the inner system to intensive variables only. The optimization problem for the inner system can be stated as:

$$\min_{\substack{\bar{x}_0,\,\bar{z}(t),\,\bar{\pi} \\ \bar{p},\,\bar{d}(t)}} \quad E\left\{\bar{\phi}(\bar{x}(t),\bar{z}(t),\bar{\pi},\bar{p},\bar{d}(t),\xi,t)\right\} \tag{6}$$

s.t.
$$0 = \bar{f}(\dot{\bar{x}}(t),\bar{x}(t),\bar{z}(t),\bar{\pi},\bar{p},\bar{d}(t)), \quad \bar{x}(t_0) = \bar{x}_0,$$
$$0 \geq \bar{g}_{\bar{k}}(\bar{x}(t),\bar{z}(t),\bar{\pi},\bar{p},\bar{d}(t),t_f),$$
$$0 \geq \bar{g}_{\bar{k}_\xi}(\bar{x}_\xi(\tau_\xi),\bar{z}_\xi(\tau_\xi),\bar{\pi},\bar{p}_\xi,\bar{d}_\xi(\tau_\xi),\tau_\xi),$$

$$\begin{bmatrix} \bar{x}_\xi(\tau_\xi) \\ \bar{z}_\xi(\tau_\xi) \end{bmatrix} = \mathcal{L}_\xi(\bar{x}(t),\bar{\pi},\bar{p}_\xi,\bar{d}_\xi(\tau_\xi),\tau_\xi),$$

$$\forall \bar{k} \in \bar{K}, \quad \forall \bar{p} \in \bar{\mathbb{P}}^{n_{\bar{p}}}, \quad \forall \bar{d} \in \bar{\mathbb{D}}^{n_{\bar{d}}}, \quad \forall \xi \in \Xi,$$
$$\forall \bar{k}_\xi \in \bar{K}_\xi, \quad \forall \bar{p}_\xi \in \bar{\mathbb{P}}_\xi^{n_{\bar{p}}}, \quad \forall \bar{d}_\xi \in \bar{\mathbb{D}}_\xi^{n_{\bar{d}}},$$
$$\forall \tau_\xi \in [t, t+\tau_{\xi,f}], \quad \forall t \in [0,t_f].$$

The overbars in problem (6) indicate that the appearing variables only form a subset of the corresponding variables in (5) except for $\bar{z}$ which is a subset of $\{z, y_{set}\}$. The outer problem contains all variables of (5) except for those related to failures ξ:

$$\min_{\substack{x_0,\,z(t),\,y_{set}(t), \\ \pi,\,X_c,\,K_c}} \quad E\left\{\phi(x(t),z(t),y_{set}(t),\pi,X_c,K_c,p,d(t),t)\right\} \tag{7}$$

s.t.
$$0 = f(\dot{x}(t),x(t),z(t),m(t),\pi,p,d(t)), \quad x(t_0) = x_0,$$
$$0 = m(t) - \mathcal{C}(x(t),z(t),y_{set}(t),\pi,X_c,K_c,p,d(t)),$$
$$0 \geq g_k(x(t),z(t),y_{set}(t),m(t),\pi,X_c,K_c,p,d(t),t_f),$$

$$\begin{bmatrix} \bar{x}_0 \\ \bar{z}(t) \\ \bar{\pi} \end{bmatrix} = a(x_0,z(t),y_{set}(t),\pi,t)$$

$$\forall k \in K, \quad \forall p \in \mathbb{P}^{n_p}, \quad \forall d \in \mathbb{D}^{n_d}, \quad \forall t \in [0,t_f].$$

The connection between inner and outer problem is realized through the function a which fixes x_0, $z(t)$, $y_{set}(t)$, and π at least in part to $\bar{x}_0$, $\bar{z}(t)$, and $\bar{\pi}$. By appropriate discretization methods for the stochastic elements the resulting stochastic optimal control problems can be approximated by deterministic multi-period optimal control problems which can be solved using standard numerical methods [17, 75].

Problems (6) and (7) represent a generic formulation for a two step approach to flexible batch reactor design and control. It is especially suitable for exothermic batch reaction systems carried out in a reactor vessel with an adjoined cooling system. In contrast to [59] it explicitly covers grass-root design as well as retrofit. Furthermore the controller is directly included into the design problem. The boundary of the inner system is chosen in such a way that failure modes only have to be considered within the inner system. In consequence the reactor vessel and cooling system may at least in part belong to the inner system. In an extreme case, only the controller will not be included in the inner system.

The general concept of problem decomposition is based on the assumption that the outer problem contains appropriate free design variables which can be exploited to bring performance of the overall system close to the result of the general (one step) optimization problem (5). This will most probably be the case for grass-root design. Decomposition will also be appropriate for a realistic industrial scenario of redesigning the cooling system for a given reactor vessel after changes in the recipe. If the design problem of the outer system reduces to synthesizing a controller of a specified type, decomposition might be impossible at all necessitating the solution of (5). Though even problem (5) might not have a solution this is very unlikely for practical situations.

3. Design of optimal and inherently safe trajectories

In the chemical process industries batch and semibatch processes are much more involved in accidents than continuously operated plants [6, 73]. Therefore, the requirement for safe operation must be met well before other constraints are examined. Frequently, safety considerations motivate semibatch operation [14, 71]. In critical regimes rather conservative operation strategies are preferred in industry with typically low feed flowrates. This conservatism often reduces the economic performance significantly. As a result of a recent questionnaire among process control engineers in industry [71] it has been found that the impact of safety limitations is underestimated in many academic studies.

Therefore, this section shows possibilities to rigorously incorporate safety in the design and control of batch reactors going beyond coarse risk analysis methods used in industrial practice. This requires the treatment of failure scenarios as dynamic subproblems to the overall design problem. In the first subsection causes for safety problems are summarized. Available techniques for the analysis of and the design under safety aspects for batch reactors are briefly reviewed. In a second subsection an integrated optimization based solution method for the design of optimal and safe trajectories on the basis of problem (6) is proposed. Finally, two example problems are presented.

3.1. PREVIOUS WORK ON SAFE BATCH REACTOR OPERATION

Highly exothermic reactions are the dominant reason for safety problems encountered in batch reaction systems. These problems can be divided into at least three classes [27]. First, a low reactor temperature, a high feed flowrate, or bad mixing conditions may lead to high accumulation of un-reacted material in the reactor. If the reaction subsequently ignites a large amount of heat will be generated that can not be removed by the cooling system. Reactor runaway is the consequence. Second, impurities or high reactor temperatures trigger unwanted exothermic side or decomposition reactions which can not be controlled and hence also result in runaway. The third class comprises accidents and faults, particularly cooling system failures. In the latter case nearly adiabatic conversion will take place potentially driving the process again in dangerous regions.

During the past 15 years several researchers have worked in the area of safe batch reactor operation. In the treatment of exothermic reactions research has focused on three aspects:

- stability and sensitivity analysis of reaction systems,
- feasibility analysis with regard to nominal batch operation, and
- risk analysis in case of cooling system failures.

Regarding the first aspect Hugo and Steinbach [30, 63] have investigated thermal behavior of semibatch reactors in order to identify limits of safe operation using empirical stability criteria. In [55] a criterion based on first and second derivatives of the reactor temperature has been proposed to detect dangerous situations. This approach has been extended in [33] to on-line detection of hazardous process states by means of multiple Kalman filters. More recently Ljapunov exponents have been used to analyze the sensitivity of the process along its operating profiles [3, 65]. It is argued that high sensitivities correspond to possible safety or runaway problems.

The limitations of the cooling system capacity on nominal batch feasibility have been studied by several researchers. Bequette [7, 8] derives analytical expressions to determine appropriate jacket temperatures and feed flowrates under the assumption of perfect control and quasi steady state for the reaction system. In this work "operability" is ensured if the jacket temperature does not exceed certain limits. The computationally simple approach can be used as a screening method to enhance scale-up.

The closely related problem of allowable initial conditions for isothermal batch operation is studied in [35]. Additionally, viability theory [4] can in principle be used to enforce feasibility on-line through appropriate control actions. First results with respect to exact temperature control have been presented in [34]. All proposed methods share the restriction to simple process models. Furthermore, cooling system dynamics are simplified or completely neglected. Since the slow and nonlinear dynamics of the cooling system may even dominate the overall system dynamics [14], feasibility is examined for an idealized system only. An interesting possibility to circumvent problems through slow cooling system dynamics is presented in [73] where the pressure is used as a manipulated variable and the temperature results from thermodynamic equilibrium.

Comparatively few researchers have worked on the direct assessment of failure situations in batch processing. In [27] three basic questions are posed which usually affect trajectory design from the safety point of view:

- How large is the heat evolution over time?
- What temperature will be reached in case of a runaway after a cooling system failure?
- Which is the most critical point for a failure?

A similar statement can be found in [64] additionally addressing the problem of occurring side reactions. Both papers emphasize the key role of the adiabatic temperature rise ΔT_{ad} which will result from instantaneous adiabatic conversion of unreacted material in the reactor. ΔT_{ad} can be expressed as a function of mixture composition (mass fractions w), heat capacity (c_p), and heat of reaction (ΔH_R). The temperature T_{ad} to be reached after adiabatic conversion can therefore be calculated as

$$T_{ad} = T_r + \Delta T_{ad}(w, c_p, \Delta H_R) \tag{8}$$

with T_r being the reactor temperature [21]. Approximating system behavior after cooling system failure by adiabatic conversion the variables T_{ad} and ΔT_{ad} can directly be introduced as safety indicators bounded by upper limits. Both [27] and [64] recommend to monitor them on-line. An optimization based approach for the determination of the most critical points for system failures has only recently been studied by Dimitriadis et al. [19] in the framework of hybrid systems safety verification.

3.2. SAFETY AS CONSTRAINT IN OPTIMIZATION PROBLEMS

On the basis of the proposed two step approach problem (6) is the general framework to incorporate safety considerations into batch design. So far two possibilities have been presented in the literature which can both be interpreted as simplifications of problem (6). In the first case only simple algebraic safety constraints are considered in $\bar{g}_k$. A prominent example is to penalize the adiabatic temperature rise (8) leading to

$$0 \geq T_{ad,max} - T_{ad} . \tag{9}$$

Other examples like constraints on concentrations are given in [73]. In the second case the rigorous inclusion of parameter variations and disturbances in the optimization of batch reactors is proposed to avoid safety violations [70, 72]. Their impact is captured by an appropriate treatment of the stochastic elements in the objective function.

To the authors' knowledge process dynamics after possible failures have not been treated so far. As already discussed in Section 2.2 this requires the evaluation of a nontrivial operator $\mathcal{L}_\xi$ whose specific nature depends on the failure scenario and the plant equipment. Two cases can be distinguished. In the more general case several degrees of freedom $\bar{z}_\xi$ still exist to influence the state trajectories $\bar{x}_\xi$. The $\bar{z}_\xi$ may comprise some elements of $\bar{z}$ as well as possible additional manipulated variables like those of an emergency system. These degrees of freedom may be determined by the solution of an optimal control problem on $\tau_\xi \in [t, t + \tau_{\xi,f}]$. For any failure ξ and for any time instant t this can be formulated as

$$\min_{\bar{z}_\xi(\tau_\xi)} \; \max_{\bar{p}_\xi, \, \bar{d}_\xi(\tau_\xi)} \; \bar{\phi}_\xi(\bar{x}(\tau_\xi), \bar{z}(\tau_\xi), \bar{\pi}, \bar{p}_\xi, \bar{d}_\xi(\tau_\xi), \xi, \tau_{\xi,f}) \tag{10}$$

$$\begin{aligned}
\text{s.t.} \quad 0 \; &= \; \bar{f}_\xi(\dot{\bar{x}}_\xi(\tau_\xi), \bar{x}_\xi(\tau_\xi), \bar{z}_\xi(\tau_\xi), \bar{\pi}, \bar{p}_\xi, \bar{d}_\xi(\tau_\xi)) \;, \\
0 \; &\geq \; \bar{l}_{i_\xi}(\bar{x}_\xi(\tau_\xi), \bar{z}_\xi(\tau_\xi), \bar{\pi}, \bar{p}_\xi, \bar{d}_\xi(\tau_\xi), \tau_\xi) \;, \\
\bar{x}_\xi(\tau_\xi = t) \; &= \; \bar{h}_\xi(x(t), \xi, t) \;,
\end{aligned}$$

$$\forall i_\xi \in \bar{I}_\xi \subset \mathbb{N}, \quad \forall \bar{p} \in \bar{\mathbb{P}}^{n_{\bar{p}}}, \quad \forall \bar{d} \in \bar{\mathbb{D}}^{n_d}, \quad \forall \tau_\xi \in [t, t + \tau_{\xi,f}] \;.$$

Since a failure situation is examined it is appropriate to use the max operator to treat the stochastic elements in order to account for the worst case. According to the hybrid nature of the overall system different models $\bar{f}_\xi$ accompanied by the corresponding constraints $\bar{l}_{i_\xi}$ are considered allowing a difference in dimensionality between $\bar{f}_\xi$ and $\bar{f}$. The function $\bar{h}_\xi$ maps the state of the nominal model $\bar{f}$ to the initial state of the failure models $\bar{f}_\xi$.

In the second case all available operational variables are fixed either directly through the failure or indirectly through appropriate emergency counter actions. Therefore no optimization has to be performed and $\mathcal{L}_\xi$ reduces to the solution of a set of differential-algebraic equations given fixed inputs $\bar{z}_\xi$:

$$0 = \bar{f}_\xi(\dot{\bar{x}}_\xi(\tau_\xi), \bar{x}_\xi(\tau_\xi), \bar{z}_\xi(\tau_\xi), \bar{\pi}, \bar{p}_\xi, \bar{d}_\xi(\tau_\xi)) \,, \tag{11}$$

$$\bar{x}_\xi(\tau_\xi = t) = \bar{h}_\xi(x(t), \xi, t) \,,$$

$$\forall \bar{p}_\xi \in \bar{\mathbb{P}}, \quad \forall \bar{d}_\xi \in \bar{\mathbb{D}}, \quad \forall \tau_\xi \in [t, t + \tau_{\xi,f}] \,.$$

The solution of (10) as subproblem of (6) is rather tedious. Problem (11) however can be tackled in (6) by existing numerical solution methods if the probability distributions of the stochastic elements are discretized appropriately. Using a weighted sum to replace the expectation term [41, 71] results in

$$\min_{\bar{x}_0, \, \bar{z}(t), \, \bar{\pi}} \sum_{j=1}^{n_\omega} \omega_j \, \bar{\phi}_j(\bar{x}(t), \bar{z}(t), \bar{\pi}, \bar{p}_j, \bar{d}_j(t), \xi, t_f) \tag{12}$$

s.t.
$$0 = \bar{f}(\dot{\bar{x}}(t), \bar{x}(t), \bar{z}(t), \bar{\pi}, \bar{p}_j, \bar{d}_j(t)), \quad \bar{x}(t_0) = \bar{x}_0 \,,$$

$$0 \geq \bar{g}_{\bar{k}}(\bar{x}(t), \bar{z}(t), \bar{\pi}, \bar{p}_j, \bar{d}_j(t), t) \,,$$

$$0 \geq \bar{g}_{\bar{k}_\xi}(\bar{x}_\xi(\tau_\xi), \bar{z}_\xi(\tau_\xi), \bar{\pi}, \bar{p}_\xi, \bar{d}_\xi(\tau_\xi), \tau_\xi) \,,$$

$$0 = \bar{f}_\xi(\dot{\bar{x}}_\xi(\tau_\xi), \bar{x}_\xi(\tau_\xi), \bar{z}_\xi(\tau_\xi), \bar{\pi}, \bar{p}_\xi, \bar{d}_\xi(\tau_\xi)) \,,$$

$$\bar{x}_\xi(\tau_\xi = t) = \bar{h}_\xi(x(t), \xi, t) \,,$$

$$\forall \bar{k} \in \bar{K}, \quad \forall \bar{p} \in \bar{\mathbb{P}}^{n_{\bar{p}}}, \quad \forall \bar{d} \in \bar{\mathbb{D}}^{n_{\bar{d}}}, \quad \forall \xi \in \Xi,$$

$$\forall \bar{k}_\xi \in \bar{K}_\xi, \quad \forall \bar{p}_\xi \in \bar{\mathbb{P}}_\xi^{n_{\bar{p}}}, \quad \forall \bar{d}_\xi \in \bar{\mathbb{D}}_\xi^{n_{\bar{d}}},$$

$$\forall \tau_\xi \in [t, t + \tau_{\xi,f}], \quad \forall t \in [0, t_f] \,.$$

This is still an infinite dimensional optimization problem which can be transformed to a nonlinear programming problem (NLP) using standard discretization techniques. The inequality constraints in problem (12) will be evaluated at a finite number of mesh points t_i only, thus the number of necessary integrations of the failure models $\bar{f}_\xi$ will also be limited.

3.3. EXAMPLES

In order to demonstrate the capabilities of the above problem formulation two examples of different complexity shall be presented. For the sake of simplicity stochastics are omitted ($n_w = 1$) and only one failure scenario will be considered ($\Xi = \{1\}$). The arising dynamic optimization problems are solved using a sequential or control vector parameterization approach [74] where only the manipulated variables are discretized and the model equations are integrated in order to calculate the profiles of the state variables.

In our current implementation gradient information is obtained by simultaneously solving the adjoined sensitivity equations using the DDASAC code [16]. The inequalities are enforced at the discretization points of the manipulated variables only and NPSOL [25] is used for the solution of the nonlinear program. For free end-time problems the time is scaled by the final time t_f which then becomes a parameter of the model to be optimized simultaneously (see for instance [53] for details).

3.3.1. *Illustrative Example 1: Car Problem*

The first example is a simple test problem which has already been used in several studies on dynamic optimization (e.g. [66]). A car is supposed to cover a fixed distance of $300\,m$ in minimum time. It must start and stop at zero velocity, has a speed limit of $10\,\frac{m}{s}$, and is subject to limited acceleration. In order to make the problem nonlinear and thus a little bit more interesting a nonlinear acceleration function and a friction term proportional to the square of the velocity have been incorporated into the problem resulting in the following mathematical formulation

$$\min_{u(t)} t_f \tag{13}$$

$$\text{s.t.} \quad \dot{x}_1(t) = \frac{4}{\pi}\arctan(u(t)) - \alpha x_1^2(t)\,, \quad x_1(0) = 0\,m/s\,, \tag{14}$$

$$\dot{x}_2(t) = x_1(t)\,, \quad x_2(0) = 0\,m/s\,,$$

$$x_1(t_f) = 0\,m/s\,,$$

$$x_2(t_f) = 300\,m\,,$$

$$-2\,m/s^2 \leq u(t) \leq 2\,m/s^2\,.$$

with $\alpha = 0.0025\,\frac{1}{m}$. Using a piecewise constant profile of the manipulated variable and dividing the time horizon into 10 intervals for demonstration purpose, the nominal solution is calculated to be $37.31\,s$. The corresponding trajectories are shown in Figure 3.

In accordance with our physical insight the car is accelerated as fast as possible until the speed limit is reached. Then acceleration is reduced to a level which exactly compensates the speed loss due to the friction term in equation (14). This situation continues until the moment where maximum braking force has to be applied in order to stop the car at $300\,m$ with zero velocity. The intermediate values of the manipulated variables in the third and ninth interval are only caused by the discretization scheme and vanish as the grid is chosen to be finer.

In order to illustrate the effect of safety failures the car problem is now extended. Inspired by James Dean's "Rebel without a cause" a cliff is introduced at a distance of $350\,m$ from the starting point. In order to

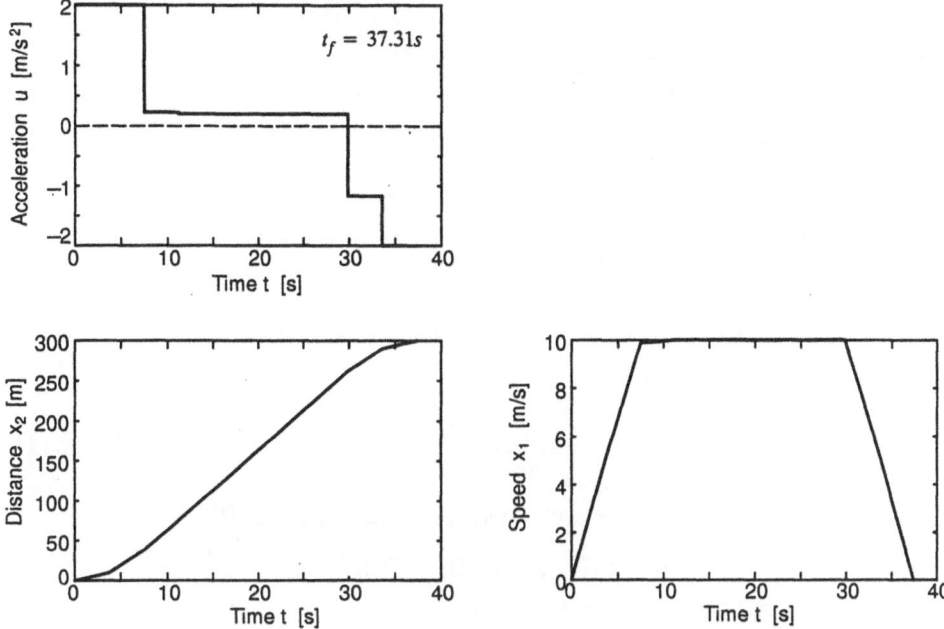

Figure 3. Car problem, nominal solution.

prevent the destruction of the car and severe injuries of the car passengers the vehicle has to be stopped in front of the cliff even in the case the brakes fail and only 10% of the brake force will be available ($u(\tau) = -0.2\,m/s^2$ as long as the velocity is greater than zero). We assume that the driver's insight will convince him to brake as hard as possible. Thus the profile of the manipulated variable in case of the safety related failure is fixed and the scenario corresponds to problem (12).

First, the inherent safety of the previously obtained optimal solution is tested. Departing from this solution the failure model is integrated over a sufficiently long horizon ($\tau_{1,f} = 50\,s$) at the end of each sampling interval. The results are presented in Figure 4 where the nominal trajectories are plotted in bold lines. As can be seen in the left part some trajectories, actually those starting after the seventh, the eighth and the ninth interval, enter the shaded area passing the cliff. Therefore the nominal trajectory can not be called inherently safe.

Now problem (12) is solved for the car. Although it would be possible in this case to solve the failure model analytically and to formulate an algebraic constraint for the distance, a second time dimension τ_1 is introduced here for demonstration purposes. This leads to the following mathematical formulation:

532

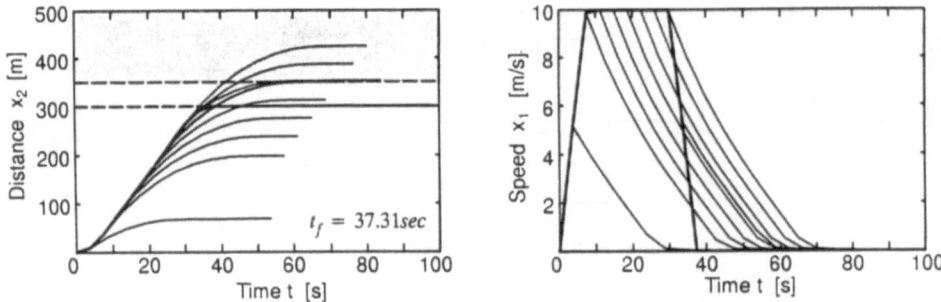

Figure 4. Car problem, consequences of brake failure.

$$\min_{u(t)} t_f \tag{15}$$

$$
\begin{aligned}
\text{s.t.} \quad \dot{x}_1(t) &= \frac{4}{\pi}\arctan(u(t)) - \alpha x_1^2(t), \quad x_1(0) = 0\,m/s, \\
\dot{x}_2(t) &= x_1(t), \quad x_2(0) = 0\,m/s, \\
x_1(t_f) &= 0\,m/s, \\
x_2(t_f) &= 300\,m, \\
-2\,m/s^2 &\le u(t) \le 2\,m/s^2, \\
\dot{x}_{1,1}(\tau_1) &=
\begin{cases}
4/\pi\arctan(-0.2) - \alpha x_{1,1}^2(\tau_1), \\
\quad \text{if } x_{1,1}(\tau_1) > 0, \\
\\
0, \text{ if } x_{1,1}(\tau_1) = 0,
\end{cases} \\
\dot{x}_{2,1}(\tau_1) &= x_{1,1}(\tau_1), \\
x_{1,1}(\tau_1 = t) &= x_1(t), \\
x_{2,1}(\tau_1 = t) &= x_2(t), \\
\tau_1 &\in [t, t+50s], \quad t \in [t, t_f].
\end{aligned}
$$

The resulting optimal trajectories are shown in Figure 5. As can clearly be seen the situation does not change within the first $20\,s$. Still the car reaches the speed limit as fast as possible and moves along this active constraint. But beginning at $t \approx 23\,s$ the speed is slowly reduced by taking into account the possible brake failure. The active constraint changes from the nominal speed limit to the endpoint distance in the safety scenario. This persists until the car stops at $300\,m$. The minimum time required to reach the desired endpoint has now increased to $38.89\,s$ which is approximately $4\,\%$ longer compared to the nominal case shown in Figure 3. However, the acceleration profile of Figure 5 is an inherently safe operation strategy and most car passengers would probably accept the slightly longer time being sure that the risk of injury due to brake failures is excluded.

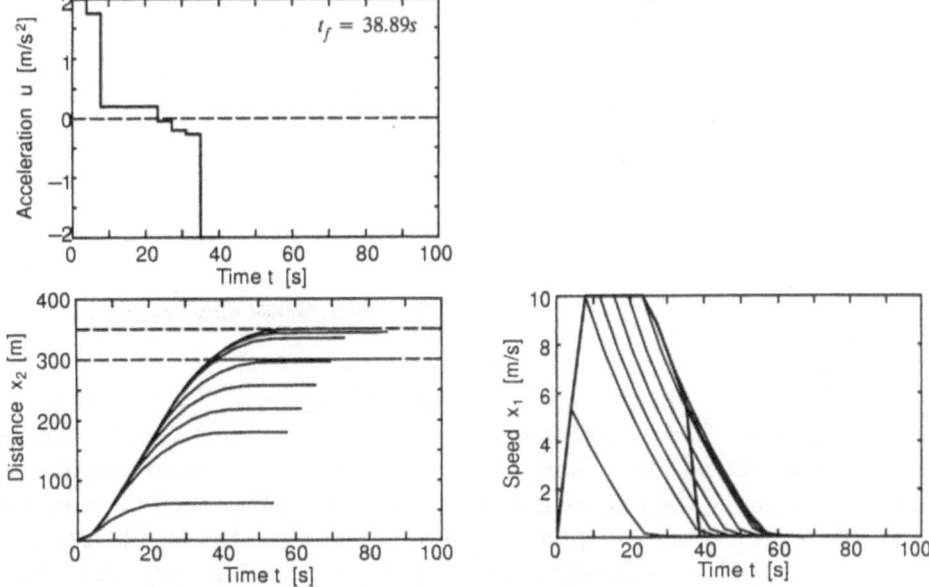

Figure 5. Car problem, inherently safe trajectories.

3.3.2. *Illustrative Example 2: Semibatch reactor*

The second example is a semibatch reactor schematically shown in Figure 6. Two consecutive exothermic reactions $A \mapsto B \mapsto C$ take place with A being fed into the reactor. The reactor is equipped with a simple cooling system. Temperature control is achieved through manipulation of the water inlet temperature $T_{w,in}$ of the cooling system. The process model is summarized in the Appendix. A first order lag has been included to approximate unmodeled dynamics of the water inlet temperature manipulation system.

For the recipe optimization we assume that the main goal is to reach a certain conversion, i.e. a fixed final concentration of component A in minimal batch time t_f. Additionally, a lower limit on the final concentration of component B is given. The molar feed flowrate F_a of A and the reactor temperature T_r are regarded as bounded free variables for the optimization, adopting the inner/outer approach described in Section 2. Furthermore, the entire amount of A to be fed to the reactor is set to be 20 $kmol$.

As a safety scenario ($\xi = 1$) a cooling system failure has to be considered. The temperature $T_{r,1}$ arising in this case is required to remain below a safety limit $T_{1,max} = 423\ K$ to avoid an (unmodeled) exothermic decomposition reaction leading to a runaway. In order to rigorously incorporate the dynamic process behavior after a possible cooling system failure an optimization problem of type (12) is stated. For the nominal case the reactor temperature is directly used as a manipulated variable and an energy bal-

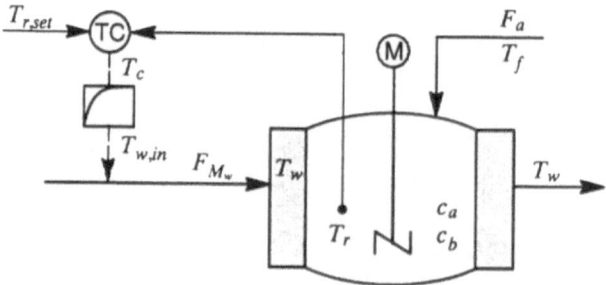

Figure 6. Semibatch reactor.

ance can be omitted in the model for the inner system [59]. On the other hand, the temperature evolution without exogenous influences is of particular interest for the failure scenarios making the energy balance unavoidable. A heat loss term to the environment has been included in the failure model whilst all terms related to F_a have been removed, since the feed is assumed to be switched off immediately after a cooling system failure occurs. Thus, in contrast to the car problem, the process model will change in the moment of failure rendering the overall system hybrid. The failure model and a summary of all constraints are included in the Appendix.

The results for a time horizon divided into 10 elements and piecewise constant profiles of the manipulated variables are presented in Figure 7. As can be seen rather complex trajectories of the feed flowrate and the reactor temperature evolve and the minimal batch time is calculated to be $t_f = 296.9 min$. The time horizon for the failure scenarios is fixed to be $\tau_{1,f} = 200 min$. It is assumed that this is sufficiently long for all reactions to have stopped and the temperature decrease due to heat losses to the environment will be the only phenomenon that can still be observed. Clearly, the process is operated along the safety constraint for the reactor temperature arising after a cooling system failure.

In the lower left part of the figure the profile of the adiabatic end temperature calculated according to equation (27) is shown. Due to the neglected heat loss to the environment it permanently violates the safety constraint. Applying this equation directly as constraint in the optimization would therefore lead to an operating strategy which is slightly more conservative than necessary. Another possibility would be to estimate the heat losses and to incorporate them into equation (27). Then, fairly similar results compared to those presented here can be expected.

It should be noted that the advantages of the proposed method can only partly be seen in this illustrative example since the maximum temperature which will be reached after a cooling system failure can nearly always be estimated by simple relations. This situation changes if further

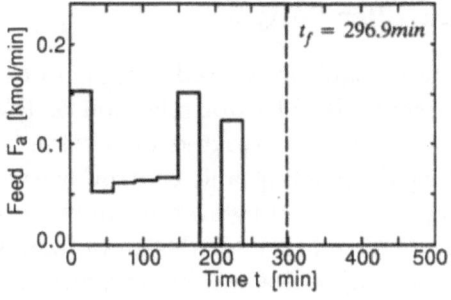
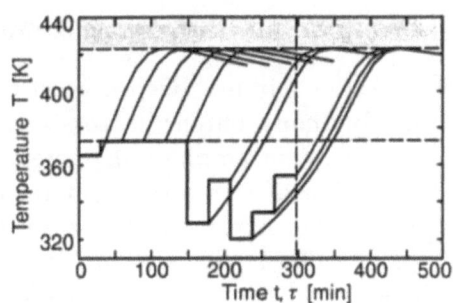

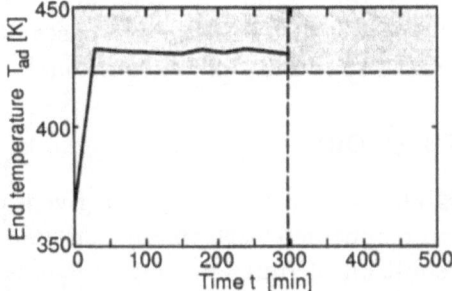

Figure 7. Semibatch reactor, inherently safe.

safety constraints exist on variables whose evolution can not be described in such an easy way. An example of this type is the pressure in a two phase system which can also rise to critical values in failure situations. This is the area where the real benefits of the proposed method can be found and appropriate examples are currently studied.

3.4. CONCLUDING REMARKS

In this section a generic problem formulation has been derived for the incorporation of safety aspects into the design of batch reactors. The proposed approach views the overall process as a hybrid system and treats failure situations as dynamic subproblems. Two simple examples have illustrated the capabilities of the method.

The computational requirements are comparatively high. Even without the treatment of stochastic elements and only considering one single safety scenario the calculations for the semibatch reactor example took a up to an hour on a Sun SPARCstation 10. Furthermore, the method requires models which do not only describe the nominal operation but also the dynamic behavior in usually unreached parts of the state space. However, these disadvantages have to be accepted if a rigorous treatment of safety aspects is desirable.

4. Controllability analysis and controller design

In Section 2 it has been argued that the control-integrated design procedure for many batch reactors can be decoupled. The optimization of the inner problem emphasizing safety aspects has been treated in Section 3. The design of the outer system including the cooling and control system has to be considered next. Since the design of the cooling system equipment does not offer a new problem dimension compared to those design problems already studied in the previous section, the emphasize will be on controllability analysis and optimization based controller design both depending on already specified trajectories of the inner system. First, previous work on controllability of batch reactors is briefly reviewed. Then the two cases of open and closed loop analysis will be treated in subsequent subsections.

4.1. CONTROLLABILITY OF BATCH REACTORS

As already defined in Section 2 controllability of a batch process is given if there exists a controller which meets some specified performance criteria. This is in good accordance with the definition of input-output controllability being the ability to achieve acceptable control performance by some controller; that is, to keep the outputs within specified bounds or displacements from their references in spite of unknown but bounded variations, such as disturbances and plant changes, using available inputs and available measurements [56]. This fundamentally differs from the concept of state controllability in nonlinear systems theory [45]. In the following it is assumed that state controllability is given at least in some vicinity of the considered process trajectories.

For continuous systems a variety of well established techniques for controllability analysis exists addressing the problem of control structure selection and identification of process inherent control performance limitations (see e.g. [56]). These techniques generally require a linear process model describing process dynamics around some chosen operating point; further, they are based on the assumption of perfect control. Controllability analysis for batch processes is much more difficult since

- batch reactors are not operated at steady state, making controllability analysis a dynamic and trajectory dependent problem, and
- disturbance rejection can not be studied independently of the tracking of time varying setpoints.

A straightforward extension from classical controllability analysis to batch processes can be achieved by linearizing process dynamics around the nominal trajectory at some points over batch time. Such a linearized process model has been used by Sørensen [57, 58] to investigate control structures for batch distillation. For a simple semi-batch reactor Bequette [7] uses a linearized model to analyze control performance limitations in temperature control. Non-minimum phase behavior is analyzed for the case of using the feed flowrate as manipulated variable.

In the literature on batch processes the term "controllability" is often used in a broader sense including aspects of feasibility. A number of papers address the problem of proper cooling system design in order to meet cooling requirements for nominal operation but also to ensure a high process gain and short response times. In [38] a hybrid multi-fluid cooling system with enhanced "controllability" is presented with experimental results being shown in [37].

In Section 2 only control-integrated design has been considered which requires the inclusion of a flexibility test for the *controlled system* in the optimization. A mathematical formulation of controllability has not been given yet. From the above controllability definition it is clear that this problem is closely related to an isolated flexibility test. Adapting (2), a controllability test for a completely designed batch process with given x_0, z and π can be stated as

$$\max_{p,\,d(t)} \quad \min_{X_c,\,K_c} \quad \max_k \quad b_k \tag{16}$$

$$
\begin{aligned}
\text{s.t.} \quad 0 &\geq b_k(x(t), z(t), y_{set}(t), m(t), \pi, X_c, K_c, p, d(t), t)\,, \\
0 &= f(\dot{x}(t), x(t), z(t), m(t), \pi, p, d(t)), \quad x(t_0) = x_0\,, \\
0 &= m(t) - C(x(t), z(t), y_{set}(t), \pi, X_c, K_c, p, d(t))\,, \\
&\forall t \in [0, t_f]
\end{aligned}
$$

with b_k including the controller performance criteria already mentioned in Section 2, constraints on manipulated variables, and any other additional constraints. It has to be noted that an isolated controllability test requires the search over all control structures and controller types X_c and is therefore practically impossible. Therefore the idea to examine the process inherent control performance limitations independently of the controller is appealing, but for nonlinear systems much more difficult than for their linear counterparts. This area of research is still at its infancy and an attempt to tackle this problem is presented in Subsection 4.2.

On the other hand the number of controlled and manipulated variables is typically very small for batch reactors and control structure selection is therefore often less important or even directly given. Furthermore, the set of eligible controller types may be rather limited through the industrial requirement of keeping controller implementation and maintenance cost small [24]. Therefore the general controllability test (16) is of little practical interest and it is often sufficient to restrict the controllability analysis to PI and PID controllers or simple extensions. This is referred to as closed loop controllability analysis and is actually based on designing a controller. This aspect is treated in Subsection 4.3.

4.2. OPEN LOOP ANALYSIS

Morari [42] identified manipulated variable constraints, non-minimum phase behavior and plant-model mismatch as those process characteristics limiting achievable control performance independent of the chosen controller. The classical approach to controllability analysis is based on analysis of these characteristics for a linearized plant model. The straightforward extension of linear analysis to batch processes has already been discussed [7, 57, 58]. Additionally other process characteristics can be analyzed on the basis of a linearized model. For instance it is well known that strong variations of the process gain over a batch is a major problem in controller design (see [20] for an industrial example).

Optimization offers the possibility to determine an upper bound of achievable performance directly taking into account the full nonlinear model. This involves the calculation of the optimal input trajectory which in case of invertibility of the process directly provides perfect control. Such an approach to controllability analysis with respect to manipulated variable constraints has recently been suggested by Cao et al. [15]. Their approach is based on calculating the optimal *inputs* trajectories for a set of *output* trajectories, e.g. step functions. The integral squared control error as cost function gives an indication on the best achievable controller performance with zero cost indicating "complete" input-output controllability. For the application of this approach to batch processes specific output trajectories like the nominal setpoint trajectory have to be chosen to cover the entire batch. An extension to include process variabilities and disturbances analogous to (7) is straightforward.

An approach different in nature is proposed in [29] and is based on determining a nonlinearity measure for batch reactors. This measure may indirectly be related to the difficulty of the control problem associated with a specified batch process trajectory. The definition, based on the work of Allgöwer [1, 2] for continuous systems, treats a batch process as a nonlinear dynamic input/output system N_B, mapping input signals u from the set of admissible input trajectories U_a into some outputs $y \in \mathcal{Y}$. In order to account for initial conditions, which might vary from batch to batch, a state space description is used for the nonlinear system:

$$\dot{x}(t) = f\left(x(t), u(t)\right), \quad x(0) = x_0, \tag{17}$$

$$y(t) = h\left(x(t), u(t)\right), \quad 0 \le t \le t_f. \tag{18}$$

Equations (17) and (18) define the nonlinear operator $N_B : y = N_B [u, x_0]$. Since nonlinearity of the process is to be analyzed around a nominal trajectory defined through an input trajectory u_{nom}, input sequences to be considered are those which keep the outputs within a specified vicinity of the nominal system outputs. The nonlinearity measure $\varphi_{N_B}^{U_B}$ is defined as a scaled norm of the output difference between N_B and a linear reference system $G_B [u, x_0] \in \mathcal{G}$. G_B is to be determined by minimizing the norm of the output difference for the "worst" input signals and initial conditions. Mathematically, this may be formulated as

$$\varphi_{N_B}^{U_B} (\lambda, u_{nom}, x_0, t_f) = \qquad\qquad (19)$$

$$\min_{G_B \in \mathcal{G}} \quad \max_{\substack{u \in \mathcal{U}_B \\ x_0 \in \mathcal{X}_0}} \quad \frac{\|G_B[u,x_0] - N_B[u,x_0]\|_{p_y}}{\|N_B[u,x_0]\|_{p_y}} \; .$$

The set of considered input signals is given by

$$\mathcal{U}_B = \left\{ u \in \mathcal{U}_a : \|N_B[u, x_0] - y_{nom}\|_{\infty_y} \leq \lambda \right\} \; .$$

with the not necessarily constant variable λ defining a tube around the nominal trajectory. A practicable mathematical solution approach leads to an approximation of the $min - max$ problem by a smooth minimization problem. For a discussion of the proposed definition and first results the reader is referred to [29]. Efficient numerical solution and interpretation of the measure with respect to controllability aspects are areas of ongoing research.

4.3. CLOSED LOOP ANALYSIS AND CONTROLLER DESIGN

Stack and Doyle [61] stress the fact that considering the open-loop system only may be misleading in control problem analysis. They present a novel approach to extend nonlinearity analysis to the closed loop system. Their approach is based on a structural analysis of the optimal controller derived from Lagrangian optimization, neglecting the boundary conditions of the arising two-point boundary-value problem. Through determination of an I/O-nonlinearity measure for the optimal control structure they present a first step to directly assess the required nonlinearity of the control law. The proposed method has already been applied to a small academic batch reactor example [62]. Although the overall approach is appealing the analysis method suffers from several drawbacks and the derivation of the optimal control structure is a nontrivial task for any process model of realistic size.

540

It has to be concluded that it is still unclear whether practicable controller independent closed-loop analysis methods can be derived giving insight into the controller requirements at a lower computational effort than designing a limited number of controllers. Therefore we will concentrate on the case where a number of controllers is specified a priori to the closed loop controllability analysis. Two approaches can then be distinguished depending on the remaining degrees of freedom in the outer system.

4.3.1. *Practical solution approaches*

In the first case *reactor and cooling system design is fixed* and the outer system reduces to the controller only. Specifying a controller type and hence X_c, (16) may be evaluated in an analysis step. This is referred to as a *controller suitability test*. If the suitability tests fail for all controllers considered, either more advanced and probably nonlinear controllers have to be chosen or the design has to be changed. If the test is passed, the result of (16), however, may not correspond to the best controller of this type, since no difference is made between the controller performance criteria and other constraints. Considering the case that all b_k are strictly less then zero ($b_k < 0$) it might be possible to improve the controller performance criteria at the expense of other constraints.

To synthesize a good control system for fixed reactor and cooling system design and with specified type and structure X_c, problem (7) has to be considered with K_c being the only free variables. This will be called optimization based controller tuning. The underlying idea is obviously not new. It is based on the definition of a scalar objective function ψ to represent controller performance. The cost function most frequently applied penalizes the integral squared error (ISE)

$$\psi = \int_0^{t_f} (y - y_{set})^T W (y - y_{set}) \, dt , \qquad (20)$$

with y denoting the controlled variables. Other control error norms like the ∞-norm can be applied, too, and might be much closer to practical needs [43]. Alternatively, a combination of the ISE and additional constraints on the maximum control error over the batch time may be used. Furthermore, a penalty on control action can be included in the cost function. Due to the general difficulty to describe meaningful objective functions and constraints reflecting controller performance, optimization based controller design has earned much criticism in literature (e.g. [42]). This also holds for control-integrated design problems. However, it has to be argued that this is a general bottleneck of any automated controller tuning procedure. On the other hand, optimization offers the possibility to rigorously integrate controllability issues into design although the computational effort might be high. Even if the actual controller parameters to be implemented may differ from those resulting from optimization at least the controller's structural capability to solve the given control problem should be guaranteed. For further discussions on the inclusion of simultaneous control structure and design parameter optimization for continuous processes the reader is referred to [47].

In the second case the outer system contains *free optimization variables in addition to those of the controller* and a (reasonably simplified) design problem (7) for the outer system has to be solved. For example, the control structure and single controller types are fixed to certain classes. Thus, the control system is synthesized directly within the given classes. If the flexibility requirements in (7) can be met no separate open loop controllability analysis is necessary since a suitable controller is already found. If a flexible control-integrated design of the outer system is impossible for the considered controller types different control structures and types have to be considered. Alternatively, decomposition has to be given up, necessitating the solution of the overall control-integrated design problem (5).

4.3.2. *Examples*

A simple example is treated to illustrate the concepts of optimization based controller tuning and control-integrated design. For this purpose the semi-batch reactor of Section 3 is revisited where optimal design of the inner system has been treated. Here, temperature control is studied. In a first step the cooling system is assumed to be fixed. Its dynamic model and all design parameters are given in the Appendix. The water supply rate to the cooling system is initially chosen to be $F_{M_w} = 4 \frac{kg}{s}$ and the manipulated variable (i.e. the water inlet temperature) is bounded by $283\,K \leq T_c \leq 423\,K$.[¶]

In this example it is assumed that only simple PID control can be realized on the plant. The only free optimization variables of the the outer optimization problem are the controller gain k_C, the reset time τ_R, and the derivative time constant τ_D. A weighted cost function ϕ penalizing the squared control error and the control action is used:

$$\phi = \int_0^{t_f} (T_r - T_{r,set})^2 + \beta \left(\frac{d\bar{T}_c}{dt}\right)^2 dt . \tag{21}$$

Here, $\frac{d\bar{T}_c}{dt}$ is an approximation of $\frac{dT_c}{dt}$. A first order lag is used to calculate $\bar{T}_c$ and $\frac{d\bar{T}_c}{dt}$:

$$\tau \frac{d\bar{T}_c}{dt} + \bar{T}_c = T_c . \tag{22}$$

[¶]The model parameters have been chosen for illustrative purpose and are of purely academic nature.

A time constant $\tau = 0.1\,s$ was found to be suitable. In Figures 8, 9 and 10 control results for the optimized recipe derived in Section 3 are given with the controlled variable in the left part and the manipulated variable in the right part of each figure. Three different values for the weighting factor β have been used. The obtained controller specifications are summarized in Table 1. Apparently, the control problem is rather difficult. Due to the large setpoint changes manipulated variable constraints are often active in all three cases. Whilst the results for $\beta = 0$ and $\beta = 1$ are nearly identical and lead to rather oscillatory control moves, $\beta = 100$ leads to a significant reduction in the controller gain and a more sluggish control behavior. In all three cases τ_R is at its upper bound of $100,000$, thus the optimal PID controller essentially is a PD controller. Hence, "steady state" offset can not be avoided.

Finally, the water supply rate F_{M_w} is introduced as additional degree of freedom of the optimization problem. The controller is designed for $\beta = 1$. The optimization is carried out in order to find the minimal water supply rate which leads to a 20 % reduction in ϕ compared to the case of $F_{M_w} = 4\,\frac{kg}{s}$. Control results are shown in Figure 11. As can be seen from Table 1, the water supply rate has to be doubled to reach the specification for ϕ.

The given example illustrates the principles of optimization based controller tuning and control-integrated design for a very simple control system only. It has to be stated that the use of piecewise constant temperature setpoints is an extreme case facilitating illustration of the technique. In practice it would be avoided, e.g. by choosing continuous and piecewise linear setpoint trajectories.

4.3.3. *Extensions to improve controller performance*
A key issue in the assessment of controllability and controller design is the treatment of uncertainties and disturbances, since their exclusion leads to inacceptable performance in practice. With regard to (7) this requires a suitable parameterization of p and d. Since disturbance rejection is a major task of the control system the proper consideration of the time varying

Design case	β	k_C	τ_R	τ_D	F_{M_w}	ϕ
Figure 8	0	22.59	100,000	7.48	4.0	2.634E6
Figure 9	1	22.81	100,000	4.01	4.0	2.652E6
Figure 10	100	10.20	100,000	8.52	4.0	2.968E6
Figure 11	1	16.96	100,000	4.04	8.1	2.122E6

TABLE 1. Controller design for the semibatch reactor.

543

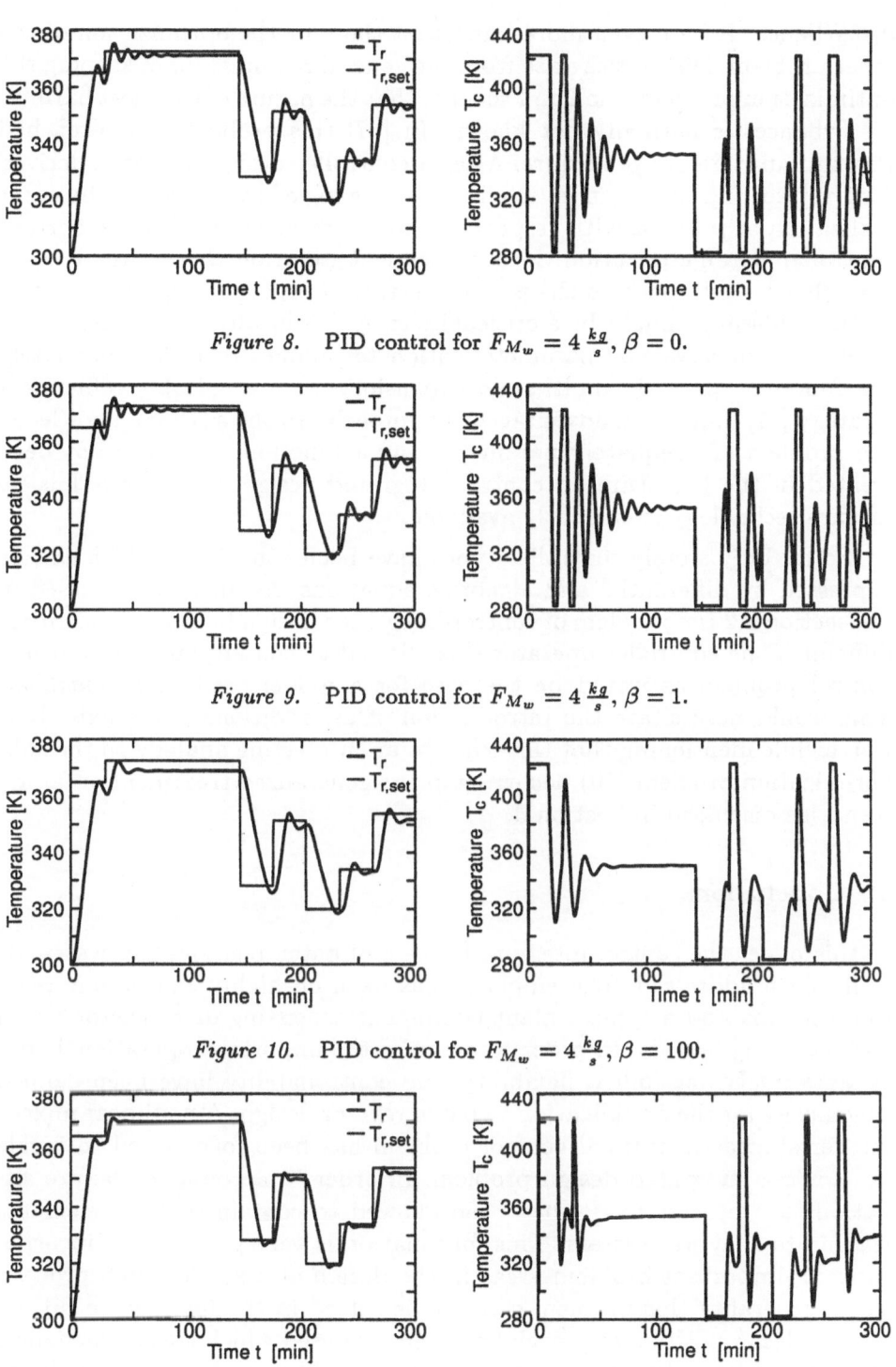

Figure 8. PID control for $F_{M_w} = 4 \frac{kg}{s}$, $\beta = 0$.

Figure 9. PID control for $F_{M_w} = 4 \frac{kg}{s}$, $\beta = 1$.

Figure 10. PID control for $F_{M_w} = 4 \frac{kg}{s}$, $\beta = 100$.

Figure 11. PID control for optimized F_{M_w}, $\beta = 1$.

disturbances is here even more important than for the optimization of the inner problem. Difficulties arise from the general requirement of keeping the optimization problem small and the fact that the nature of the most harmful disturbances is normally not known. In [47] this problem is approached through an iterative procedure. After a controller design has been derived for given set of p and d, an inverted problem is solved to determine the worst realization of p and d with respect to constraint violations for the derived controller parameterization. In a subsequent iteration these "worst case" disturbances are added to the previous set. Convergence properties for this method, however, might be a critical issue. A specifically tailored approach could rely on advanced parameterization techniques to cover adequately the function space $\mathbb{D}$; multi-scale expansions with a suitable refinement strategy [11] may be an advantageous approach. To the authors' knowledge the problem of adequately parameterizing a function space has not been treated in the literature on control-integrated design so far and thus no mature technology is currently available.

So far only simple controller types have been considered which can be expressed by differential and algebraic equations. As already discussed in Subsection 2.2 the problem of control-integrated design becomes much more difficult if the controller operator $\mathcal{C}$ is allowed to contain an inner optimal control problem as would be the case for a model predictive controller. This would necessitate the introduction of an additional time axis. It is worthwhile mentioning that this leads to an interesting analogy to the sub-optimization problem (10) occurring in the generalized treatment of failure scenarios discussed in Section 3.

5. Conclusions

In this paper the control-integrated design of batch reactors has systematically been addressed. The emphasis has been on highly exothermic reaction systems and a typical plant equipment consisting of a reactor vessel and a cooling system. In order to describe the important operational constraints safety, feasibility, flexibility, and controllability have been defined specifically for their application to batch reactor design. A nonlinear mixed-integer stochastic optimal control problem has been formulated to tackle the control-integrated design problem. In order to account for failure situations the process model has been allowed to contain discrete elements leading to a hybrid system. This formulation is very generic and incorporates all important problem areas in the design of a single batch process. In its generality this problem can not be solved in the foreseeable future. Thus a decomposition approach has been proposed which treats the hybrid elements of the safety scenarios in an inner problem and all controllability aspects in an outer problem. If enough degrees of freedoms are left to be determined in the outer problem, performance losses in comparison to the solution of the overall problem will usually be small.

Among the different possibilities for the treatment of safety failures in the inner problem the most general is to solve an optimization subproblem for each failure and time instant they can occur. Usually this will not be the preferred method of choice not only because it is computationally extremely expensive. In most failure cases it will be sufficient to exclude reactor runaway or other severe accidents. Any attempt to reach prescribed quality requirements might go too far after a failure has occurred. The other extreme is to approximate safety related variables after a failure through simple algebraic equations. This is industrial standard but may – like in the case of reactor pressure – not always be possible and sometimes lacks exactness. Therefore, it is reasonable to define emergency rules in order to fix the manipulated variables of every failure model and to solve these model equations as subproblems of the overall optimization problem. The presented results for this approach are encouraging and future research will concentrate on its capabilities and limitations. A major issue is obviously the availability of sufficiently validated batch reaction plant models not only for nominal but also for emergency operation.

Unfortunately, the development of methods for controllability analysis of nonlinear systems is still in its infancy. For batch reactors their inherent dynamic nature makes the problem even more difficult. The well known approach of linear systems theory to assume perfect control is difficult to apply here and a search over all possible controllers would be necessary. Alternatives may include methods based on the optimal controller or non-linearity measures but their capability is difficult to assess today. In short terms it may be more promising to analyze the performance of only a limited number of controllers. For batch reactors this approach may not be too restrictive and if the analysis satisfies the existing constraints, it directly yields a suitable initial guess for appropriate controller parameters.

ACKNOWLEDGEMENT

This research was supported by the German Bundesministerium für Bildung, Wissenschaft, Forschung und Technologie (FKZ 03D0020B/5).

References

1. F. Allgöwer. Definition and computation of a nonlinearity measure. In A.J. Krener and D.Q. Mayne, editors, *Proc. of Nonlinear Control System Design Symposium NOLCOS95*, pages 257–262, Oxford, UK, 1996. Pergamon.
2. F. Allgöwer. *Näherungsweise Ein-Ausgangs-Linearisierung nichtlinearer Systeme.* Reihe 8: Meß-, Steuerungs- und Regelungstechnik Nr. 582. VDI-Verlag, 1996.

3. M.A. Alós, F. Strozzi, and J.M. Zaldívar. A new method for assessing the thermal stability of semibatch processes based on Lyapunov exponents. *Chem. Eng. Sci.*, 51(11):3089–3094, 1996.

4. J.-P. Aubin. *Viability Theory*. Systems & Control: Foundations & Applications. Birkhäuser, Boston. 1991.

5. P.I. Barton and T. Park. Analysis and control of combined discrete/continuous systems: progress and challenges in the chemical processing industries. In J.C. Kantor, C.E. Garcia, and B. Carnahan, editors, *Proc. of Fifth International Conference on Chemical Process Control*, volume 93 of *AIChE Symposium Series No. 316*, pages 102–114. AIChE and CACHE, 1997.

6. A. Benuzzi and J.M. Zaldívar. *Safety of Chemical Batch Reactors and Storage Tanks*. Kluwer Academic Publishers, Dordrecht, The Netherlands, 1991.

7. B.W. Bequette. Operability analysis of an exothermic semi-batch reactor. *Comput. Chem. Eng.*, 20(Suppl.):S.1583–1588, 1996.

8. B.W. Bequette. Operability of batch reactors: Temperature profile feasibility. In J.C. Kantor, C.E. Garcia, and B. Carnahan, editors, *Proc. of Fifth International Conference on Chemical Process Control*, volume 93 of *AIChE Symposium Series No. 316*, pages 315–318, Tahoe City, USA, 1997. AIChE and CACHE.

9. R. Berber. Control of batch reactors: A review. In R. Berber, editor, *Methods of Model-Based Control*, NATO-ASI Series, pages 459–494. Kluwer Press, Dordrecht, The Netherlands, 1995.

10. T. Bhatia and L.T. Biegler. Dynamic optimization in the design and scheduling of multiproduct batch plants. *Ind. Eng. Chem. Res.*, 35(7):2234–2246, 1996.

11. T. Binder, L. Blank, W. Dahmen, and W. Marquardt. Towards multiscale dynamic data reconciliation. NATO ASI on Nonlinear Model Based Process Control. This volume.

12. D.B. Birewar and I.E. Grossmann. Incorporating scheduling in the optimal design of muli-product batch plants. *Comput. Chem. Eng.*, 13:141, 1989.

13. R. Bogusch, B. Lohmann, and W. Marquardt. Computer-aided process modeling with MODKIT. In *Proceedings of Chemputers Europe III*, Frankfurt, Germany, 29–30 October 1996.

14. D. Bonvin. Optimal operation of discontinuous reactors – A personal view. In *Preprints of the IFAC Symposium ADCHEM'97*, pages 155–169, Banff, Canada, 9–11 June 1997.

15. Y. Cao, D. Biss, and J.D. Perkins. Assessment of input-output controllability in the presence of control constraints. *Comput. Chem. Eng.*, 20(4):337–346, 1996.

16. M. Caracotsios and W.E. Stewart. Sensitivity analysis of initial value problems with mixed ODEs and algebraic equations. *Comput. Chem. Eng.*, 9(4):359–365, 1985.

17. J.E. Cuthrell and L.T. Biegler. On the optimization of differential-algebraic process systems. *AIChE J.*, 33(8):1257–1270, 1987.

18. V.D. Dimitriadis and E.N. Pistikopoulos. Flexibility analysis of dynamic systems. *Ind. Eng. Chem. Res.*, 34:4451–4462, 1995.

19. V.D. Dimitriadis, N. Shah, and C.C. Pantelides. Modelling, safety verification and design of discrete/continuous processing systems. Paper accepted for publication in *AIChE J.*, 1996.

20. P. Djavdan. Temperature control of a batch esterification reactor overcoming nonlinearities using adaptive gain. In J.B. Rawlings, editor, *Proc. of the IFAC Symposium DYCORD+'95*, pages 421–426, Helsingør, Denmark, 7–9 June 1995.

21. G. Eigenberger and H. Schuler. Reaktorstabilität und sichere Reaktionsführung. *Chem. Ing. Tech.*, 58(8):655–665, 1986.

22. S. Engell, S. Kowalewski, and B.H. Krogh. Discrete events and hybrid systems in process control. In J.C. Kantor, C.E. Garcia, and B. Carnahan, editors, *Proc. of Fifth International Conference on Chemical Process Control*, volume 93 of *AIChE Symposium Series No. 316*, pages 165–176. AIChE and CACHE, 1997.

23. B.A. Foss, B. Lohmann, and W. Marquardt. A field study of the industrial modeling

process. In *Preprints of the IFAC Symposium ADCHEM'97*, pages 570–585, Banff, Canada, 9–11 June 1997. To be published in *J. Proc. Cont.*, 1998.

24. M. Friedrich and R. Perne. Design and control of batch reactors - an industrial viewpoint. *Comput. Chem. Eng.*, 19(Suppl.):S357–S368, 1995.

25. P.E. Gill, W. Murray, M.A. Saunders, and M.H. Wright. *User's Guide for NPSOL, Version 4.0*. Systems Optimization Laboratory, Stanford University, Stanford, USA, 1996.

26. I.E. Grossmann and D.A. Straub. Recent developments in the evaluation and optimization of flexible chemical processes. In L. Puigjaner and A. Espuna, editors, *Proceedings Computer-Oriented Process Engineering*, page 49. Elsevier, Amsterdam, 1991.

27. R. Gygax. Chemical reaction engineering for safety. *Chem. Eng. Sci.*, 43(8):1759–1771, 1988.

28. A. Helbig, O. Abel, A. M'hamdi, and W. Marquardt. Analysis and nonlinear model predictive control of the Chylla-Haase benchmark problem. In *Proc. of CONTROL'96*, pages 1172–1177, Exeter, UK, 2–5 September 1996.

29. A. Helbig, W. Marquardt, and F. Allgöwer. Nonlinearity measures for batch reactors. Paper submitted to IFAC Symposium DYCOPS-5, 1997.

30. P. Hugo and J. Steinbach. A comparison of the limits of safe operation of a SBR and a CSTR. *Chem. Eng. Sci.*, 41(4):1081–1087, 1986.

31. P. Jarupintusophon, M.V. LeLann, M. Cabassud, and G. Casamatta. Realistic model-based predictive and adaptive control of batch reactor. *Comput. Chem. Eng.*, 18(Suppl.):S445–S449, 1995.

32. M.R. Juba and J.W. Hamer. Progress and challenges in batch process control. In M. Morari and T.J. McAvoy, editors, *Chemical Process Control - CPC III*, CACHE, pages 139–183. Elsevier, 1986.

33. R. King. *Modellgestützte Überwachung kritischer Reaktionssysteme*. Reihe 8: Meß-, Steuerungs- und Regelungstechnik Nr. 185. VDI-Verlag, Düsseldorf, 1989.

34. G. Labinaz, M.M. Bayoumi, and K. Rudie. A viable cascade controller design for a batch polymerization process. In *Preprints of the IFAC Symposium ADCHEM'97*, pages 195–200, Banff, Canada, 9–11 June 1997.

35. D.R. Lewin. Modelling and control of an industrial PVC suspension polymerization reactor. *Comput. Chem. Eng.*, 20(Suppl.):S865–S870, 1996.

36. J.S. Logsdon and L.T. Biegler. Decomposition strategies for large-scale dynamic optimization problems. *Chem. Eng. Sci.*, 47(4):851–864, 1992.

37. Z. Louleh, M. Cabassud, M.V. LeLann, and G. Casamatta. Experimental study of a new thermal control strategy for batch reactors. In *Preprints of the IFAC Symposium ADCHEM'97*, pages 119–124, Banff, Canada, 9–11 June 1997.

38. Z. Louleh, M.V. LeLann, A. Chamayou, and G. Casamatta. A new heating-cooling system to improve controllability of batch reactors. *Chem. Eng. Sci.*, 51(11):3163–3168, 1996.

39. W. Marquardt. Towards a process modeling methodology. In R. Berber, editor, *Methods of Model-Based Control*, volume 293 of *NATO-ASI Series*, pages 3–40, Dordrecht, The Netherlands, 1995. Kluwer Press.

40. A. M'hamdi, A. Helbig, O. Abel, and W. Marquardt. Newton-type receding horizon control and state estimation - a case study. In J.J. Gertler, J.B. Cruz, and M. Peshkin, editors, *Preprints of the 13th World Congress of IFAC*, volume M, pages 121–126, San Francisco, 30 June–5 July 1996.

41. M.J. Mohideen, J.D. Perkins, and E.N. Pistikopoulos. Optimal design of dynamic systems under uncertainty. *AIChE J.*, 42(8):2251–2272, 1996.

42. M. Morari. Design of resilient processing plants –III. A general framework for the assessment of dynamic resilience. *Chem. Eng. Sci.*, 38(11):1881–1891, 1983.

43. M. Morari and E. Zafiriou. *Robust Process Control*. Prentice-Hall, Englewood Cliffs, New Jersey, 1989.

548

44. Z. Nagy and Ş. Agachi. Model predictive control of a PVC batch reactor. *Comput. Chem. Eng.*, 21(6):571–591, 1997.
45. H. Nijmeijer and A.J. van der Schaft. *Nonlinear Dynamical Control Systems*. Springer, New York, 3. edition, 1996.
46. S. Palanki, C. Kravaris, and H.Y. Wang. Optimal feedback control of batch reactors with a state inequality constraint and free terminal time. *Chem. Eng. Sci.*, 49(1):86–97, 1994.
47. J.D. Perkins and S.P.K. Walsh. Optimization as a tool for design/control integration. *Comput. Chem. Eng.*, 20(4):315–323, 1996.
48. T. Peterson, E. Hernández, Y. Arkun, and F.J. Schork. A nonlinear DMC algorithm and its applications to a semibatch polymerization reactor. *Chem. Eng. Sci.*, 47(4):737–753, 1992.
49. A.K.M.S. Rahman and S. Palanki. On-line optimization of batch processes with nonlinear manipulated input. *Comput. Chem. Eng.*, 20(5):449–459, 1996.
50. S. Rahman and S. Palanki. On-line optimization of batch processes in the presence of measurable disturbances. *AIChE J.*, 42(10):2869–2882, 1996.
51. G.V. Reklaitis (editor) and E. Zafiriou (guest editor). *Selected Papers from the IFAC Workshop on Integration of Process Design & Control (IPDC'94)*. Comp. Chem. Eng., 20(4), 1996.
52. D.W.T. Rippin. Simulation of single- and multiproduct batch chemical plants for optimal design and operation. *Comput. Chem. Eng.*, 7(3):137–156, 1983.
53. O. Rosen and R. Luus. Evaluation of gradients for piecewise constant optimal control. *Comput. Chem. Eng.*, 15(4):273–281, 1991.
54. D. Ruppen. *A Contribution to the Implementation of Adaptive Optimal Operation for Discontinuous Chemical Reactors*. PhD thesis, ETH, No. 10634, Zürich, Switzerland, 1994.
55. H. Schuler. Frühzeitige Erkennung gefährlicher Reaktionszustände in chemischen Reaktoren. *Regelungstechnik*, 32(6):190–237, 1984.
56. S. Skogestad and I. Postlethwaite. *Multivariable Feedback Control*. John Wiley & Sons, Chicester, England, 1996.
57. E. Sørensen. *Studies on Optimal Operation and Control of Batch Distillation Columns*. PhD thesis, NTU, Trondheim, Norway, 1994.
58. E. Sørensen and S. Skogestad. Control strategies for reactive batch distillation. *J. Proc. Cont.*, 4(4):205–217, 1994.
59. M. Soroush and C. Kravaris. Optimal design and operation of batch reactors. 1. Theoretical framework. *Ind. Eng. Chem. Res.*, 32:866–881, 1993.
60. M. Soroush and C. Kravaris. Optimal design and operation of batch reactors. 2. A case study. *Ind. Eng. Chem. Res.*, 32:882–893, 1993.
61. A.J. Stack and F.J. Doyle III. Application of a control-law nonlinearity measure to the chemical reactor analysis. *AIChE J.*, 43(2):425–439, 1997.
62. A.J. Stack and F.J. Doyle III. The optimal control structure: an approach to measuring control-law nonlinearity. *Comput. Chem. Eng.*, 21(9):1009–1019, 1997.
63. J. Steinbach. Praxisorientierte Darstellung von Auslegungskriterien für Semibatch - Reaktoren. In *Dechema - Monographien Band 107*, pages 133–152. VCH Verlagsgesellschaft, 1987.
64. F. Stoessel. Design thermally safe semibatch reactors. *Chem. Eng. Progr.*, pages 46–53, September 1995.
65. F. Strozzi and J.M. Zaldívar. A general method for assessing the thermal stability of batch chemical reactors by sensitivity calculation based on Lyapunov exponents. *Chem. Eng. Sci.*, 49(16):2681–2688, 1994.
66. P. Tanartkit and L.T. Biegler. Stable decomposition for dynamic optimization. *Ind. Eng. Chem. Res.*, 34(4):1253–1266, 1995.
67. P. Tanartkit and L.T. Biegler. A nested, simultanous approach for dynamic optimization problems - I. *Comput. Chem. Eng.*, 20(6/7):735–741, 1996.

68. P. Terwiesch. *Dynamic Optimization of Batch Process Operations with Imperfect Modeling.* PhD thesis, ETH, No. 10857, Zürich, Switzerland, 1994.
69. P. Terwiesch. Cautious on-line correction of batch process operation. *AIChE J.*, 41(5):1337–1340, 1995.
70. P. Terwiesch and M. Agarwal. Robust input policies for batch reactors under parametric uncertainty. *Comput. Chem. Eng.*, 131:33–52, 1995.
71. P. Terwiesch, M. Agarwal, and D.W.T. Rippin. Batch unit optimization with imperfect modelling: A survey. *J. Proc. Cont.*, 4(4):238–258, 1994.
72. P. Terwiesch, D. Ravemark, and D.W.T. Rippin. Risk-cautious operation of batch processes. In J.B. Rawlings, editor, *Proc. of the IFAC Symposium DYCORD+'95*, pages 409–414, Helsingør, Denmark, 7–9 June 1995.
73. C. Toulouse, J. Cezerac, M. Cabassud, M.-V. LeLann, and G. Casamatta. Optimisation and scale-up of batch chemical reactors: Impact of safety constraints. *Chem. Eng. Sci.*, 51(10):2243–2252, 1996.
74. V.S. Vassiliadis, R.W.H. Sargent, and C.C. Pantelides. Solution of a class of multistage dynamic optimization problems. 1. Problems without path constraints. *Ind. Eng. Chem. Res.*, 33(9):2111–2122, 1994.
75. V.S. Vassiliadis, R.W.H. Sargent, and C.C. Pantelides. Solution of a class of multistage dynamic optimization problems. 2. Problems with path constraints. *Ind. Eng. Chem. Res.*, 33(9):2123–2133, 1994.

Appendix

In the sequel the model equations for the semibatch reactor treated in Section 3.3 and in Section 4.3.2 are given. The model parameters used are summarized in Table 2.

Reactor model for nominal operation

Inner system:

$$\frac{dc_a}{dt} = \frac{F_a}{V_r} - k_{01} \exp\left(-\frac{E_1}{RT_r}\right) c_a , \quad c_a(0) = 0 \, \frac{kmol}{m^3} , \qquad (23)$$

$$\frac{dc_b}{dt} = k_{01} \exp\left(-\frac{E_1}{RT_r}\right) c_a - k_{02} \exp\left(-\frac{E_2}{RT_r}\right) c_b , \qquad (24)$$

$$c_b(0) = 0 \, \frac{kmol}{m^3} ,$$

$$\frac{dV_r}{dt} = \frac{F_a M_{W_a}}{\rho_r} , \quad V_r(0) = V_{r0} , \qquad (25)$$

$$\frac{dn_{a,int}}{dt} = F_a , \quad n_{a,int}(0) = 0 \, kmol , \qquad (26)$$

$$T_{ad,max} = T_r + \frac{c_a(-\Delta H_1 - \Delta H_2) + c_b(-\Delta H_2)}{\rho_r c_{p_r}} . \qquad (27)$$

Additional equations for outer system:

$$(\rho_r V_r c_{p_r})\frac{dT_r}{dt} = V_r F_a M_{W_a} c_{p_r} (T_f - T_r) \qquad (28)$$

$$-V_r k_{01} \exp\left(-\frac{E_1}{RT_r}\right) c_a \Delta H_1$$

$$-V_r k_{02} \exp\left(-\frac{E_2}{RT_r}\right) c_b \Delta H_2$$

$$+\alpha_w A_{ro} \frac{V_r}{V_{r0}} (T_w - T_r), \quad T_r(0) = 300 \, K,$$

$$(\rho_w V_w c_{pw})\frac{dT_w}{dt} = F M_w c_{pw}(T_{w,in} - T_w) - \alpha_w A_{ro}\frac{V_r}{V_{r0}}(T_w - T_r), \quad (29)$$

$$T_w(0) = 300 \, K,$$

$$\tau_c \frac{dT_{w,in}}{dt} = T_c - T_{w,in}, \quad T_{w,in}(0) = 300 \, K. \quad (30)$$

Failure model of reactor content

$$\frac{dc_{a,1}}{d\tau_1} = -k_{01} \exp\left(-\frac{E_1}{RT_{r,1}}\right) c_{a,1}, \quad c_{a,1}(\tau_1 = t) = c_a(t), \quad (31)$$

$$\frac{dc_{b,1}}{d\tau_1} = k_{01} \exp\left(-\frac{E_1}{RT_{r,1}}\right) c_{a,1} - k_{02} \exp\left(-\frac{E_2}{RT_{r,1}}\right) c_{b,1}, \quad (32)$$

$$c_{b,1}(\tau_1 = t) = c_b(t),$$

$$(\rho_r c_{pr})\frac{dT_{r,1}}{d\tau_1} = -k_{01} \exp\left(-\frac{E_1}{RT_{r,1}}\right) c_{a,1}\Delta H_1 \quad (33)$$

$$-k_{02} \exp\left(-\frac{E_2}{RT_{r,1}}\right) c_{b,1}\Delta H_2$$

$$-U_{loss}(T_{r,1} - T_a), \quad T_{r,1}(\tau_1 = t) = T_r(t).$$

Constraints

$$c_a(t_f) = 0.5 \frac{kmol}{m^3}, \quad (34)$$

$$c_b(t_f) \geq 5.5 \frac{kmol}{m^3}, \quad (35)$$

$$n_{a,int}(t_f) = 20.0 \, kmol, \quad (36)$$

$$0\frac{kmol}{s} \leq F_a \leq 0.05 \frac{kmol}{s}, \quad (37)$$

$$300 \, K \leq T_r \leq 373 \, K, \quad (38)$$

$$T_{r,1} \leq 423 \, K, \quad (39)$$

$$283 \, K \leq T_c \leq 423 \, K. \quad (40)$$

Model parameters

Parameter	Value		Parameter	Value
Mw_a	$50\,\frac{kg}{kmol}$		ρ_r	$1000\,\frac{kg}{m^3}$
c_{pr}	$4.187\,\frac{kJ}{kgK}$		k_{01}	$15\,\frac{1}{s}$
T_f	$300\,K$		k_{02}	$85\,\frac{1}{s}$
T_a	$273\,K$		E_1	$30000\,\frac{kJ}{kmol}$
ΔH_1	$-40000\,\frac{kJ}{kmol}$		E_2	$40000\,\frac{kJ}{kmol}$
ΔH_2	$-50000\,\frac{kJ}{kmol}$		R	$8.314\,\frac{kJ}{kmolK}$
α_w	$0.7\,\frac{kJ}{sm^2K}$		ρ_w	$1000\,\frac{kg}{m^3}$
U_{loss}	$0.057\,\frac{kJ}{sm^3K}$		c_{pw}	$4.187\,\frac{kJ}{kgK}$
A_{ro}	$5.0\,m^2$		F_{Mw}	$4\,\frac{kg}{s}$
V_{r0}	$1.0\,m^3$		τ_c	$100\,s$

TABLE 2. Model parameters for the semibatch reactor.

Part IV

NONLINEAR STATE AND PARAMETER ESTIMATION

SOFTWARE SENSORS AND ADAPTIVE LINEARIZING CONTROL OF BIOREACTORS

D. DOCHAIN
Maître de Recherches FNRS
Cesame, Université Catholique de Louvain
Bâtiment Euler, 4-6 avenue G. Lemaître,
1348 Louvain-La-Neuve, Belgium

M. PERRIER
Département de Génie Chimique,
Ecole Polytechnique de Montréal
Succursale "Centre Ville", CP 6079,
Montréal H3C 3A7, Canada

Abstract

This paper is a short survey on methods which have been developed and applied in the field of dynamical modelling, analysis, monitoring and control design of bioprocesses over the past fifteen years. A key feature of the paper is to show how to incorporate the well-known knowledge about the dynamics of biochemical processes (basically, the reaction network and the material balances) in monitoring and control algorithms. These are moreover capable of dealing with the process uncertainty (in particular on the reaction kinetics) by introducing, for instance in the control algorithms, an adaptation scheme.

1 Introduction

Industrial-scale biotechnological processes have progressed vigorously over the last decades. Generally speaking, the problems arising from the implementation of these processes are similar to those of more classical industrial processes and the need for monitoring systems and automatic control in order to optimize production efficiency, improve products quality or detect disturbances in process operation is obvious. The application of automatic control to industrial biotechnological processes is facing two main difficulties :

1. The internal working and dynamics of these processes are as yet badly grasped and many problems of methodology in modelling remain to

R. Berber and C. Kravaris (eds.), Nonlinear Model Based Process Control, 555-597.
© 1998 *Kluwer Academic Publishers.*

556

be solved. It is difficult to develop models taking into account the numerous factors which can influence the specific bacterial growth rate. The modelling effort is often tedious and requires a great number of experiments before producing a reliable model. Reproductibility of experiments is often uncertain due to the difficulty in obtaining the same environmental conditions. Moreover, as these processes involve living organism, their dynamic behaviour is strongly non linear and non stationary. Model parameters cannot remain constant over a long period they will vary e.g. due to metabolic variations of biomass or to random and unobservable physiological or genetic modifications. It should also be noted that the lack of accuracy of the measurements often leads to identifiability problems.

2. Another essential difficulty lies in the absence, in most cases, of cheap and reliable instrumentation suited to real-time monitoring. To date, the market offers very few sensors capable of providing reliable on-line measurements of the biological and biochemical parameters required to implement high performance automatic control strategies. The main variables, i.e. biomass, substrate and synthesis product concentrations, generally need determining through laboratory analyses. The cost and duration of the analyses obviously limit the frequency of the measurements.

The classical monitoring and control methods do not prove very efficient to tide over these basic difficulties. Therefore, to reconstruct the state of the system from the only on-line available measurements and to control the biological variables (biomass, substrates or synthesis products), appropriate algorithms have to be developed. The efficiency of any monitoring or control system highly depends on the design of the control and monitoring techniques and the care taken in their design. Indeed, monitoring or control algorithms will prove to be efficient if they are able to incorporate the important well-known information on the process while being able to deal with the missing informations (lack of on-line measurements, uncertainty on the dynamics, ...) in a "robust" way, i.e. such that these missing informations will not significantly deteriorate the control performance of the process. The present generalized use of computers make possible and fairly easy to design and implement controllers which are more sophisticated (and susceptible of better performance) than the classical PID. These controllers may refer to quite complex theory (nonlinear control, adaptive control) but, as it will be shown, their structure and implementation may remain rather simple while including the key features of simple PID's.

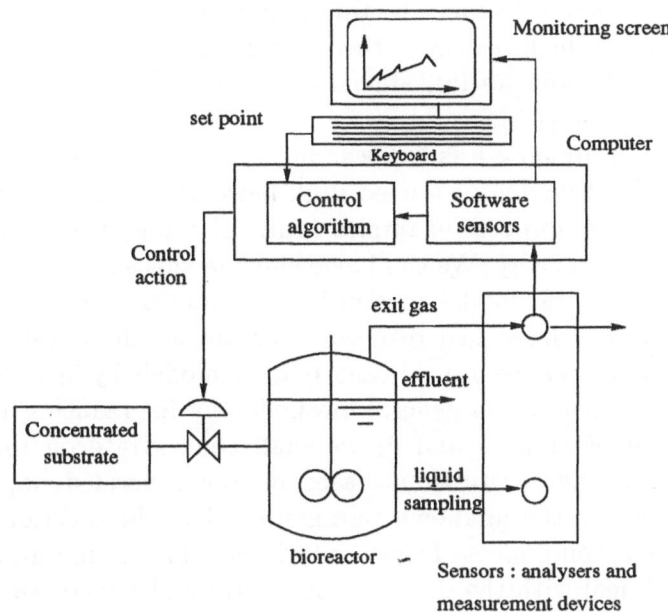

Figure 1: Schematic view of a computer controlled bioreactor

In this paper, we shall show how to incorporate the well-known knowledge about the dynamics of biochemical processes (basically, the reaction network and the material balances) in monitoring and control algorithms which are moreover capable of dealing with the process uncertainty (in particular on the reaction kinetics) by introducing, for instance in the control algorithms, an adaptation scheme). Figure 1 shows a schematic view of a computer controlled bioreactor.

A key feature of this paper is to emphasize the central role played by linear algebra which appears to be a very efficient tool in the design of monitoring and control algorithms for apparently complex nonlinear systems like bioprocesses, via the use of fairly simple and standard algebraic manipulations (based on the structural properties mentioned in Sections 3.1 ad 3.2). It is also important to draw the attention of the reader that these control systems are not just the object of academic research but are already used in several applications (see e.g. [1], [2], [3], [4], [5], [6], [7], [8], [9]). Adaptive as well as non-adaptive linearizing control of bioreactors has been a quite active research area over the last decade. In addition to the works of the authors and coworkers, let us also mention e.g. [10], [11], [12],

[13], [14], [15]. Key references to this paper are the book by Bastin and Dochain ([1]) in which a deeper theoretical analysis of many monitoring and control methodologies introduced here can be found, and more recent survey papers ([16], [17]).

The paper is organized as follows. In Section 2, we shall introduce the general dynamical model for stirred tank bioreactors and illustrate it with several practical bioprocess examples (anaerobic digestion, yeast process, activated sludge process). We shall also show in this section how to extend the general dynamical model to multi-tank processes or to non perfectly mixed reactors like fixed bed reactors. Section 3 will concentrate on the dynamical analysis of the stirred tank reactor models by introducing a key state transformation and a general methodology for reducing the order of the model. In Sections 4 and 5, we shall concentrate on the design of software sensors, i.e. algorithms based on the general dynamical model to estimate on-line the unknown parameters (like the reaction rates) and the unmeasured components from the few available on-line measurements. Section 4 will deal with the design of asymptotic observers for the process components, and Section 5 will concentrate on the on-line estimation of specific growth rates. In both cases, the proposed methodologies will be illustrated by real-life results on an anaerobic digestion process and a yeast fermentation, respectively. Finally, Section 6 will present the design of adaptive linearizing controllers for bioprocesses. The methodology will be illustrated with the activated sludge example.

2   General Dynamical Model

A biotechnological process can be defined as a set of M biochemical reactions involving N components. The reactions most often encountered in bioprocesses are microbial growth (in which the biomass plays the role of an autocatalyst) and enzyme catalysed reactions (in which the biomass can be viewed as a simple catalyst); but many other reactions can also take place, like microorganism death, maintenance, ... These reactions can be formalized into reaction schemes, as it is now illustrated. In the following we shall concentrate on three process examples which will be used in the rest of the text to illustrate the different themes of this paper:

1. yeast growth;

2. the two classical biological wastewater treatment processes : the anaerobic digestion and the activated sludge process.

The above choice is somewhat arbitrary (although the chosen processes correspond to typical processes which have been the object of research works over the past ten years and more, and for which estimation and/or control results are available). We could have considered indeed other process examples among which we would like to mention animal cell cultures and penicillin production. Let us mention to the interested reader some references concerning works related to the themes developed here : [18], [19], [20], [21] about the animal cell cultures, [22], [23] about penicillin G production, and [1] about PHB.

The dynamical model of bioprocesses in (batch, fedbatch and continuous) stirred rank reactors is obtained from mass balance considerations and can be written in the following matrix form :

$$\frac{d\xi}{dt} = -D\xi + Kr + F - Q \tag{1}$$

where ξ is the vector of the bioprocess component $(\dim(\xi) = N)$, D is the dilution rate (h^{-1}), K is the yield coefficient matrix $(\dim(K) = N \times M)$, r is the reaction rate vector $(\dim(r) = M)$, F is the feed rate vector and Q the gaseous outflow rate vector $(\dim(F) = \dim(Q) = N)$. The model (1) has been called the *General Dynamical Model* for stirred tank bioreactors (for further details on the notion scheme and the general dynamical model of bioreactors, see [1]). As it will be illustrated below, the derivation of the dynamical model from a reaction network is straightforward by noting that each component k_{ij} of the yield coefficient matrix:

$$K = [k_{ij}] \qquad i = 1 \text{ to } N, \ j = 1 \text{ to } M$$

is representative of the ith component : it is negative if the component is a reactant, it is positive if it is a product and it is equal to zero if the component does not intervene in the reaction.

2.1 Example #1 : Anaerobic Digestion

Anaerobic digestion is a biological wastewater treatment process which produces methane. Four metabolic paths ([24]) can be identified in this process: two for acidogenesis and two for methanization (Figure 2).

In the first acidogenic path (Path 1), glucose is decomposed into fatty volatile acids (acetate, propionate), hydrogen and inorganic carbon by acidogenic bacteria. In the second acidogenic path (Path 2), OHPA (Obligate Hydrogen Producing Acetogens) decompose propionate into acetate, hydrogen and inorganic carbon. In a first methanization path (Path 3),

560

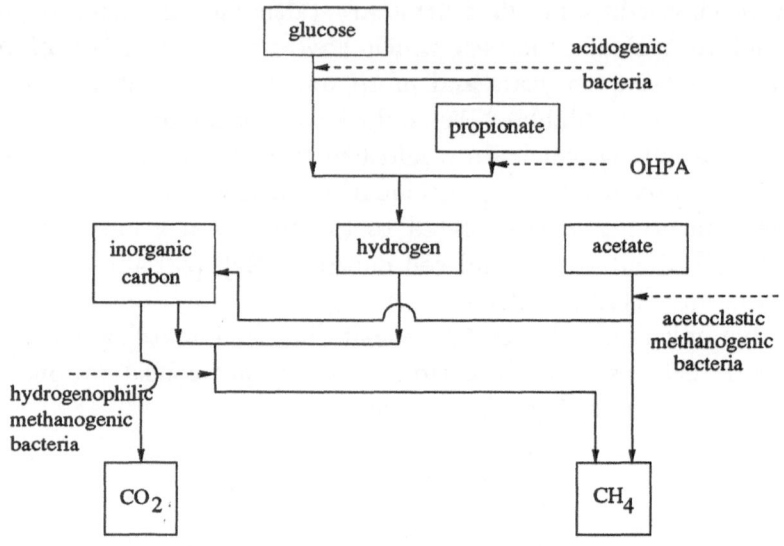

Figure 2: scheme of the anaerobic digestion

acetate is transformed into methane and inorganic carbon by acetoclastic methanogenic bacteria, while in the second methanization path (Path 4), hydrogen combines with inorganic carbon to produce methane under the action of hydrogenophilic methanogenic bacteria. The process can then be described by the following reaction network :

$$S_1 \longrightarrow X_1 + S_2 + S_3 + S_4 + S_5 \qquad (2)$$

$$S_2 \longrightarrow X_2 + S_3 + S_4 + S_5 \qquad (3)$$

$$S_3 \longrightarrow X_3 + S_5 + P_1 \qquad (4)$$

$$S_4 + S_5 \longrightarrow X_4 + P_1 \qquad (5)$$

where S_1, S_2, S_3, S_4, S_5, X_1, X_2, X_3, X_4 and P_1 are respectively glucose, propionate, acetate, hydrogen, inorganic carbon, acidogenic bacteria, OHPA (Obligate Hydrogen Producing Acetogens), acetoclastic methanogenic bacteria, hydrogenophilic methanogenic bacteria and methane. The dynamical model of the anaerobic digestion process ($N = 10, M = 4$) in a stirred tank reactor can be described within the above formalism (1) by using the following definitions :

$$
\xi = \begin{pmatrix} X_1 \\ S_1 \\ X_2 \\ S_2 \\ X_3 \\ S_3 \\ X_4 \\ S_4 \\ S_5 \\ P_1 \end{pmatrix}, K = \begin{pmatrix} 1 & 0 & 0 & 0 \\ -k_{21} & 0 & 0 & 0 \\ 0 & 1 & 0 & 0 \\ k_{41} & -k_{42} & 0 & 0 \\ 0 & 0 & 1 & 0 \\ k_{61} & k_{62} & 0 & -k_{63} \\ 0 & 0 & 0 & 1 \\ k_{81} & k_{82} & 0 & -k_{84} \\ k_{91} & k_{92} & k_{93} & -k_{94} \\ 0 & 0 & k_{03} & k_{04} \end{pmatrix}, F = \begin{pmatrix} 0 \\ DS_{in} \\ 0 \\ 0 \\ 0 \\ 0 \\ 0 \\ 0 \\ 0 \\ 0 \end{pmatrix} \tag{6}
$$

$$
Q = \begin{pmatrix} 0 \\ 0 \\ 0 \\ 0 \\ 0 \\ 0 \\ 0 \\ Q_1 \\ Q_2 \\ Q_3 \end{pmatrix}, r = \begin{pmatrix} r_1 \\ r_2 \\ r_3 \\ r_4 \end{pmatrix} = \begin{pmatrix} \mu_1 X_1 \\ \mu_2 X_2 \\ \mu_3 X_3 \\ \mu_4 X_4 \end{pmatrix} \tag{7}
$$

where μ_1, μ_2, μ_3, μ_4 are the specific growth rates (h^{-1}) of reactions (2), (3), (4), (5), respectively, and S_{in}, Q_1, Q_2 and Q_3 represent respectively the influent glucose concentration $(g\,l^{-1})$ and the gaseous outflow rates $(g\,l^{-1}h^{-1})$ of H_2, CO_2 and CH_4.

2.2 Example #2. Yeast Growth

Yeast (*Saccharomyces cerevisiae*) growth is usually characterized by the following three metabolic reactions (see e.g. [25]) :

1. respiratory growth on glucose;

2. reductive growth on glucose;

3. respiratory growth on ethanol.

These can be formalized by the following reaction schemes :

$$
S + C \longrightarrow X + P \tag{8}
$$

$$S \longrightarrow X + E + P \tag{9}$$

$$E + C \longrightarrow X + P \tag{10}$$

where S, C, X, P and E are, respectively, glucose, oxygen, yeast, carbon dioxide and ethanol. In absence of growth, substrate and oxygen may be consumed for maintenance, which may be formalized by the following reaction scheme :

$$S + C \longrightarrow P \tag{11}$$

The dynamical model of the yeast growth (N = 5, M = 4) deduced from material balances can be formalized within the general dynamical model framework (1) by considering the following definitions :

$$\xi = \begin{pmatrix} S \\ C \\ X \\ P \\ E \end{pmatrix}, \quad K = \begin{pmatrix} -k_1 & -k_2 & 0 & -k_{10} \\ -k_3 & 0 & -k_4 & -1 \\ 1 & 1 & 1 & 0 \\ k_5 & k_6 & k_7 & k_{11} \\ 0 & k_8 & -k_9 & 0 \end{pmatrix} \tag{12}$$

$$r = \begin{pmatrix} r_1 \\ r_2 \\ r_3 \\ r_4 \end{pmatrix} = \begin{pmatrix} \mu_o X \\ \mu_r X \\ \mu_e X \\ q_m X \end{pmatrix}, \quad F = \begin{pmatrix} DS_{in} \\ \Delta Q_{O_2} \\ 0 \\ 0 \\ 0 \end{pmatrix}, \quad Q = \begin{pmatrix} 0 \\ 0 \\ 0 \\ Q_1 \\ 0 \end{pmatrix} \tag{13}$$

where μ_o, μ_r and μ_e are the specific gowth rates of the respiratory growth on glucose (8), the reductive growth on glucose (9) and the respiratory growth on ethanol (10), respectively (h^{-1}), q_m is the specific maintenance rate (h^{-1}), k_i (i = 1 to 9) are yield coefficients, S_{in} is the influent glucose concentration $(g\,l^{-1})$, ΔQ_{O_2} the gaseous oxygen feed rate and Q_1 the gaseous outflow rate of carbon dioxide P $(g\,l^{-1}\,h^{-1})$.

2.3 Example #3. Activated Sludge Process

In the preceding examples, D is a scalar, but it may be a vector if there are more than one tank, as it is illustrated in the activated sludge process example.

The activated sludge process is one other classical biological (but aerobic) wastewater treatment process. It is usualy operated in two sequential tanks (see Figure 3) : an aerator (in which the degradation of the pollutants S takes place) and a settler (which is used to recycle part of the biomass

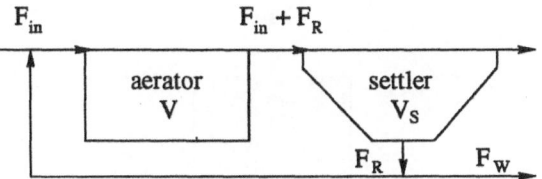

Figure 3: Schematic view of an activated sludge process

X to the aerator). The reaction in the aerator is usually described by a simple microbial growth (see e.g. [26], [27]).

$$S + C \longrightarrow X \qquad (14)$$

while the dynamics of the settler are described by the following mass balance equation :

$$\frac{dX_R}{dt} = \frac{F_{in} + F_R}{V_S} X - \frac{F_R + F_W}{V_S} X_R \qquad (15)$$

where X_R is the concentration of the recycled biomass $(g\, l^{-1})$, F_{in}, F_R and F_W are the influent, recycle and waste flow rates $(g\, l^{-1}\, h^{-1})$, respectively, and V_S the setttler volume (l). By considering the volume V (l) of the aerator and defining :

$$D_{in} = \frac{F_{in}}{V}, \ D_2 = \frac{F_R}{V}, \ D_1 = D_{in} + D_2, \ D_3 = \frac{F_{in} + F_R}{V_S}, \ D_4 = \frac{F_R + F_W}{V_S}$$
$$(16)$$

The dynamical equations of the process (N = 4, M = 1) can be rewritten in the formalism of the general dynamical model (1) with the following definitions :

$$\xi = \begin{pmatrix} S \\ C \\ X \\ X_R \end{pmatrix}, \ K = \begin{pmatrix} -k_1 \\ -k_2 \\ 1 \\ 0 \end{pmatrix}, \ F = \begin{pmatrix} D_{in}S_{in} \\ \Delta Q_{O_2} \\ 0 \\ 0 \end{pmatrix}, \ r = \mu X, \ Q = 0$$
$$(17)$$

$$D = \begin{pmatrix} D_1 & 0 & 0 & 0 \\ 0 & D_2 & 0 & 0 \\ 0 & 0 & D_1 & -D_2 \\ 0 & 0 & -D_3 & D_4 \end{pmatrix} \qquad (18)$$

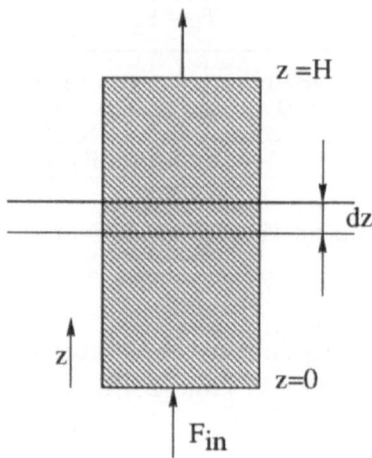

Figure 4: Schematic view of a fixed bed reactor

Note that D is now a matrix. In activated sludge processes, the oxygen feed rate ΔQ_{O_2} term in the dynamical equation of the dissolved oxygen is usually set equal to the liquid-gas oxygen transfer rate :

$$\Delta Q_{O_2} = k_L a (C_S - C) \tag{19}$$

where $k_L a$ is the oxygen mass transfer coefficient and C_S the saturation constant. In line with [28], we shall also consider in the following that $k_L a$ is a linear function of the air flow rate W :

$$k_L a = a_0 W, \ a_0 > 0 \tag{20}$$

2.4 Fixed Bed Reactors

Let us now see how to extend the General Dynamical Model (1) to non completely mixed reactors. As a matter of example, we shall concentrate on fixed bed reactors with axial diffusion.

Let us assume that among the N process components, N_{fi} are micro-organisms entrapped or fixed on some support, and N_{fl} other reactants (essentially substrates and products) flow through the reactor. For simplicity, we also consider the cross-section of the bioreactor to be constant and equal to A. From mass balance considerations on a section dz (see

Figure 4), we can deduce the following dynamical model :

$$\frac{\partial \xi_{fi}}{\partial t} = K_{fi} r(\xi_{fi}, \xi_{fl}) \tag{21}$$

$$\frac{\partial \xi_{fl}}{\partial t} = -\frac{F_{in}}{A}\frac{\partial \xi_{fl}}{\partial z} + D_{am}\frac{\partial^2 \xi_{fl}}{\partial z^2} + K_{fl} r(\xi_{fi}, \xi_{fl}) \tag{22}$$

with the following Danckwerts ([29]) boundary conditions :

$$D_{am}\frac{\partial \xi_{fl}}{\partial z} = -\frac{F_{in}}{A}(\xi_{fl,in} - \xi_{fl}) \qquad at \ z = 0 \tag{23}$$

$$\frac{\partial \xi_{fl}}{\partial z} = 0 \qquad at \ z = H \tag{24}$$

In the above equations, ξ_{fi} is the biomass concentration vector (dim $\xi_{fi} = N_{fi}$), ξ_{fl} is the other reactant concentration vector (dim $\xi_{fl} = N_{fl}$), $\xi_{fl,in}(t)$ is the influent concentration of ξ_{fl} (which is different from zero only for external substrates), $r(\xi_{fi}, \xi_{fl})$ is the reaction rate vector (dim $r = M$), K_{fi} and K_{fl} are the yield coefficient matrices (dim $K_{fi} = N_{fi} \times M$; dim $K_{fl} = N_{fl} \times M$), F_{in} the hydraulic flow rate ($m^3 h^{-1}$), D_{am} is the axial dispersion coefficient ($m^2 s^{-1}$) and z is the space variable (m) (z $\in$]0, H]).

The dynamics of the bioprocess are described by partial differential equations. Yet note the similarity between this model (21), (22) and the General Dynamical Model (1) of stirred tank reactors. An important difference lies in the hydrodynamics term (the partial derivative term $-\frac{F_{in}}{A}\frac{\partial \xi_{fl}}{\partial z} + D_{am}\frac{\partial^2 \xi_{fl}}{\partial z^2}$ instead of $-D\xi + F - Q$) and in the presence of boundary conditions (23), (24).

Remark : note that the dynamical model of the fixed bed reactor in absence of diffusion, i.e. of the plug flow reactor, is readily obtained from the equations (21), (22), (23) by simply setting the dispersion coefficient D_{am} to zero ($D_{am} = 0$). Since there is then only a first order derivative of the state variable ξ_{fl} with respect to the space variable z, only the boundary condition at the reactor input ($z - 0$) is kept (23) and the boundary condition at the reactor output (24) is dropped.

Comment : the rest of the paper will concentrate on stirred tank reactors. Let us suggest to the interested reader to look at the following references which deal with the dynamical modelling, analysis and control design for distributed parameter reactors [30], the asymptotic observer design for fixed bed reactors [31], the adaptive linearizing control for fixed bed reactors [32], [33] and fluidized bed reactors [34], and with the approximation of fixed bed reactor models [35] and the control analysis [36] via singular perturbation techniques.

3 Dynamical Analysis of Stirred Tank Bioreactor Models

3.1 A Key State Transformation

The key result of this section is the use of a state transformation by which part of the dynamical model (1) becomes independent of the reaction kinetics r (see [1], [37]). This transformation will play a very important role in the design of asymptotic observers (Section 4) and adaptive linearizing controllers (Section 6). The proposed transformation indeed readily derives from the notion of invariants in reaction systems (see e.g. [38], [39]).

The transformation is defined as follows. Let us denote the rank of the matrix K by R (rank(K) = R), and consider the following partition of the process component vector :

$$\xi = \left[\begin{array}{c} \xi_a \\ \xi_b \end{array} \right] \tag{25}$$

where ξ_a contains R (arbitrarily chosen) process variables and ξ_b the others, but such that the corresponding submatrix K_a is full rank (rank(K_a) = R). Let us define the state transformation ζ (dim(ζ) = N-R) :

$$\zeta = C_a \xi_a + C_b \xi_b \tag{26}$$

where C_a and C_b are solutions of the matrix equation :

$$C_a K_a + C_b K_b = 0 \tag{27}$$

In the particular (but quite general) situation of independent irreversible reactions, then R = M and C_b may be chosen to be a full rank square matrix. Let us derive the dynamics of ζ from (1) and the definition (26).

1) Single reactor : D = scalar

$$\frac{d\zeta}{dt} = -D\zeta + C_a(F_a - Q_a) + C_b(F_b - Q_b) \tag{28}$$

2) Multi-reactor : $D = matrix$

$$\frac{d\zeta}{dt} = -(C_b D_{bb} + C_a D_{ab})C_b^{-1}\zeta + C_a(F_a - Q_a) + C_b(F_b - Q_b)$$
$$+[(C_b D_{bb} + C_a D_{ab})C_b^{-1} - C_b D_{ba} - C_a D_{aa}]\xi_a \qquad (29)$$

with :

$$D = \begin{pmatrix} D_{aa} & D_{ab} \\ D_{ba} & D_{bb} \end{pmatrix} \qquad (30)$$

3) Fixed bed reactor

For simplicity reasons, let us consider here $C_b = I$ and let us put the vector ξ_{fi} of the fixed components in ξ_b :

$$\xi_b = \begin{pmatrix} \xi_{bf} \\ \xi_{fi} \end{pmatrix} \qquad (31)$$

Then we can rewrite the auxiliary variable ζ as follows:

$$\zeta = \begin{pmatrix} \zeta_{fl} \\ \zeta_{fi} \end{pmatrix} = \begin{pmatrix} \xi_{bf} \\ \xi_{fi} \end{pmatrix} + \begin{pmatrix} C_{af} \\ C_{ae} \end{pmatrix} \xi_a \qquad (32)$$

The dynamics of ζ can then be written as follows :

$$\frac{\partial \zeta_{fl}}{\partial t} = -\frac{F_{in}}{A}\frac{\partial \zeta_{fl}}{\partial z} + D_{am}\frac{\partial^2 \zeta_{fl}}{\partial z^2} \qquad (33)$$

$$\frac{\partial \zeta_{fi}}{\partial t} = -\frac{F_{in}}{A}C_{ae}\frac{\partial \xi_a}{\partial z} + D_{am}C_{ae}\frac{\partial^2 \xi_a}{\partial z^2} \qquad (34)$$

Note that the dynamical equations of ζ (28), (29) and (33), (34) are independent of the reaction kinetics $r(\xi)$.

3.2 Model Order Reduction

The examples of bioprocesses presented in the preceding sections have shown that a bioreactor dynamical model may be fairly complex in some instances and involve a large number of differential equations. But there are many practical applications where a simplified reduced order model is sufficient from an engineering viewpoint. Model simplification can be achieved by using the singular perturbation technique, which is a technique for transforming a set of $n + m$ differential equations into a set of n

568

differential equations and a set of m algebraic equations. This technique is suitable when neglecting the dynamics of substrates and of products with low solubility in the liquid phase. The method will be illustrated with one specific example (low solubility product) before stating the general rule for order reduction.

3.2.1 Singular perturbation technique for low solubility products

Let us consider a biochemical reaction described by the following reaction scheme :

$$S \longrightarrow P \qquad (35)$$

where P is a volatile product which can be given off in gaseous form and has low solubility in the liquid phase. The dynamical model is as follows :

$$\frac{dS}{dt} = -r - DS + DS_{in} \qquad (36)$$

$$\frac{dP}{dt} = kr - DP - Q \qquad (37)$$

The consistency of this model requires that the product concentration P be related to a saturation concentration representative of the product solubility, which is expressed as :

$$P = \Pi P_{sat} \qquad (38)$$

where P_{sat} is the saturation concentration which is constant in a stable physico-chemical environment. The model (36), (37) is rewritten in the standard singular perturbation form, with $\epsilon = P_{sat}$:

$$\frac{dS}{dt} = -r - DS + DS_{in} \qquad (39)$$

$$\epsilon \frac{d\Pi}{dt} = kr - \epsilon D\Pi - Q \qquad (40)$$

If the solubility is very low, we obtain a reduced order model by setting $\epsilon = 0$ and replacing the differential equation (40) by the algebraic one :

$$Q = kr \qquad (41)$$

3.2.2 A general rule for order reduction

The above example shows that the rule for model simplification is actually very simple and that an explicit singular perturbation analysis is not really needed. Consider that, for some i, the dynamics of the component ξ_i are to be neglected. The dynamics of ξ_i are described by equation (1) :

$$\frac{d\xi_i}{dt} = -D\xi_i + K_i r + F_i - Q_i \tag{42}$$

where K_i is the line of K corresponding to the component ξ_i. The simplification is then achieved by setting ξ_i and $d\xi_i/dt$ to zero i.e. by replacing the differential equation (42) by the following algebraic equation :

$$K_i r = -F_i + Q_i \tag{43}$$

It has been shown that the above model order reduction rule is not only valid for low solubility products but also for bioprocesses with fast and slow reactions. Then the above order reduction rule (43) applies to substrates of fast reactions (as long as they intervene only in fast reactions)(see [40] for further details).

Note the close connection between the singular pertubation reduction and the quasi-steady-state (QSS) approximation, which is largely used in (bio)chemical engineering. This suggests the following comment : singular perturbation can be viewed as an efficient mathematical tool to rigorously justify QSS approximations on a systematic basis via an appropriate analysis (including the choice of an appropriate *small* perturbation parameter).

3.2.3 Example : the anaerobic digestion

Let us see how to apply the above model order reduction rule (43) to a specific example, the anaerobic digestion. Let us first consider that the reactions (3) and (4) are fast. First of all, it is well-known that methane is a low solubility product. Therefore the above procedure applies. Futhermore, assume that the second methanization path (hydrogen consumption) is limiting, i.e. that the first three reactions (2), (3), (5) are fast, and then apply the model order reduction rule (43) to the propionate concentration S_2 and the hydrogen concentration S_4. By setting their values and their time derivatives to zero:

$$S_2 = S_4 = 0, \quad \frac{dS_2}{dt} = \frac{dS_4}{dt} = 0 \tag{44}$$

and by assuming that Q_1 is negligible, we reduce their differential equations to the following set of algebraic equations:

$$\begin{pmatrix} k_{41} & -k_{42} & 0 \\ k_{81} & k_{82} & -k_{84} \end{pmatrix} \begin{pmatrix} r_1 \\ r_2 \\ r_4 \end{pmatrix} = \begin{pmatrix} 0 \\ 0 \end{pmatrix} \tag{45}$$

By inverting the submatrix of the yield coefficients of the left hand side of (45), we can express the reaction rates r_2 and r_4 as functions of r_1 :

$$r_2 = \frac{k_{41}}{k_{42}} r_1, \quad r_4 = \frac{k_{41}k_{82} + k_{42}k_{81}}{k_{84}k_{42}} r_1 \tag{46}$$

Let us replace the reaction rates r_2 and r_4 by their above expressions (46) in the dynamical model (7), which is then rewritten as follows :

$$\xi = \begin{pmatrix} X_1 \\ S_1 \\ X_3 \\ S_3 \\ S_5 \\ P_1 \end{pmatrix} \quad K = \begin{pmatrix} 1 & 0 \\ -k_{21} & 0 \\ 0 & 1 \\ k_{61} + \frac{k_{41}}{k_{42}}k_{62} & -k_{63} \\ k_{91} + \frac{k_{41}}{k_{42}}k_{92} - \frac{k_{41}k_{82}+k_{42}k_{81}}{k_{84}k_{42}}k_{94} & k_{93} \\ k_{04}\frac{k_{41}k_{82}+k_{42}k_{81}}{k_{84}k_{42}} & k_{03} \end{pmatrix} \tag{47}$$

$$F = \begin{pmatrix} 0 \\ DS_{in} \\ 0 \\ 0 \\ 0 \\ 0 \end{pmatrix}, Q = \begin{pmatrix} 0 \\ 0 \\ 0 \\ 0 \\ Q_2 \\ Q_3 \end{pmatrix}, r = \begin{pmatrix} r_1 \\ r_3 \end{pmatrix} = \begin{pmatrix} \mu_1 X_1 \\ \mu_3 X_3 \end{pmatrix} \tag{48}$$

If k_{04} is negligible and if $k_{91} + \frac{k_{41}}{k_{42}}k_{92} - \frac{k_{41}k_{82}+k_{42}k_{81}}{k_{84}k_{42}}k_{94} > 0$ (this is the case when the methanization via the hydrogen is negligible), then we obtain the model of the two-step (acidogenesis and methanization) anaerobic digestion process. If we further consider that inorganic carbon is mainly CO_2, and that methane and CO_2 are low solubility product, the dynamical equations of S_5 and P_1 in the above dynamical model (47)(48) are replaced by the following algebraic equations (by setting S_5, P_1 and their time derivatives to zero) :

$$Q_2 = (k_{91} + \frac{k_{41}}{k_{42}}k_{92} - \frac{k_{41}k_{82} + k_{42}k_{81}}{k_{84}k_{42}}k_{94})\mu_1 X_1 + k_{93}\mu_3 X_3 \tag{49}$$

$$Q_3 = k_{03}\mu_3 X_3 \tag{50}$$

4 Monitoring of Bioprocesses. Part I : Asymptotic Observers

A key question in bioprocess control is how to monitor reactant and product concentrations in a reliable and cost effective manner. However, it appears that, in many practical applications, only some of the concentrations of the components involved and critical for quality control are available for on-line measurement. For instance, dissolved oxygen concentration and gaseous flowrates are available for on-line measurement while the values of the concentrations of biomass, substrates and/or synthesis products are often available via off-line analysis. An interesting alternative which circumvents and exploits the use of a model in conjunction with a limited set of measurements are the use of Luenberger or Kalman observers. In these techniques, a model, which includes states that are measured as well as states that are not measured, is used in parallel with the process and the model states may then be used for feedback. This configuration may be used to reduce the effect of noise on measurements as well as to reconstruct the states that are not measured. An introduction to these ideas can be found in e.g. [41]. These concepts were originally developed for linear problems. Because of the nonlinear characteristics of the bioprocess dynamics, it is of interest to extend these concepts and exploit particular structures for biochemical engineering application problems. Linearized versions (the linearized tangent model) of the process dynamics are computed from a Taylor's series expansion of a state space model around some equilibrium point and the observer theory referred to above can be applied. These modified observers, particularly the extended Kalman filter (EKF), has found applications in some biochemical processes (e.g. [42], [43], [44], [45], [46]).

One of the reasons for the popularity of the EKF is that it is easy to implement since the algorithm can be derived directly from the state space model. However, since (as the extended Luenberger observer) it is based on a linearized model of the process, the stability and convergence properties are essentially local and valid around some equilibrium point, and it is rather difficult to guarantee its stability over wide ranges of operation. [47] shows that the EKF for state and parameter estimation of linear systems may give biased estimates or even diverge if it is not carefully initialized. Let us also point out that the derivation of the EKF is based on some stochastic assumptions on the measurement and process noises, which might be questionable in practice.

One reason for the problem of convergence of EKF is that, in order to guarantee the (arbitrarily chosen) exponential convergence of the observer, the process must be locally observable, i.e. the linearized tangent model must be observable and fulfill the classical observability rank condition. This condition, as it turns out, is restrictive in many practical situations and may account for the failure of EKF to find widespread application (e.g. [1], [48], [49]).

Another problem is that the theory for the extended Luenberger and Kalman observers is developed using a perfect knowledge of the system parameters, in particular of the process kinetics : it is difficult to develop error bounds and there is often a large uncertainty on these parameters.

It appears from the above remarks that there is a clear incentive to develop new methodologies for the on-line estimation of the unmeasured concentration variables in biochemical reaction systems that do not rely on the explicit use of kinetic models. Indeed, the objective of this section is to propose an alternative to EKF and use process physics in a more direct manner to develop a nonlinear observer applicable to the estimation problem of stirred tank reactors (STR). The proposed observer is based on the well-known nonlinear model of the process without the knowledge of the process kinetics being necessary. In order to advance the application of this method, we discuss its stability and convergence properties. We would like to emphasize that the presented results are global (i.e. independent of the initial conditions) as opposed to the local properties for EKF. The approach (called *asymptotic observers*) proved to be very successful when applied to bioreactors (see e.g.[1]). The proposed asymptotic observer for STR's is an intermediate method between the "classical" observers (EKF or extended Luenberger observer) which require a full process model knowledge and the adaptive observers ([50], [51]) which include state and parameter estimation within the same estimation scheme. A review of the application of adaptive observers to biochemical processes is given by Dochain [52].

4.1 Asymptotic Observers for Single Tank Bioprocesses

The derivation of the asymptotic observer equations are based on the Key State Transformation (26), (27) introduced in Section 3.1 and on the following assumptions:

1. M components are measured on-line[1].

---

[1]the asymptotic observer when the number of measured component is larger than M is developed and discussed in [1]

2. The feedrates F, the gaseous outflow rates Q and the dilution rate D are known either by measurement or by choice of the user.

3. The yield coefficient matrix K is known.

4. The reaction rate vector r is unknown.

5. The M reactions are irreversible and independent, i.e. $\mathrm{rank}(K) = \mathrm{R} = \mathrm{M}$

From assumption 1, we can define the following state partition :

$$\xi = \left[\begin{array}{c} \xi_1 \\ \xi_2 \end{array} \right] \tag{51}$$

where ξ_1 and ξ_2 hold for the measured component concentrations and the unmeasured ones, respectively.

Let us consider one (arbitrarily chosen) transformation ζ defined by (26, 27). The variable ζ can be rewritten as a linear combination of the measured and unmeasured states ξ_1 and ξ_2, i.e. :

$$\zeta = C_a \xi_a + C_b \xi_b = A_1 \xi_1 + A_2 \xi_2 \tag{52}$$

Recall that the dynamics of ζ are independent of the reaction rate $r(\xi)$:

$$\frac{d\zeta}{dt} = -D\zeta + C_a(F_a - Q_a) + C_b(F_b - Q_b) \tag{53}$$

The equations (52), (53) are the basis for the derivation of the asymptotic observer. The dynamical equations of ζ are used to calculate an estimate of ζ on-line, which is used, via equation (52) and the on-line data of ξ_1, to derive an estimate the unmeasured component ξ_2. Let us further assume that the matrix A_2 is invertible. Then the asymptotic observer is written as follows :

$$\frac{d\hat{\zeta}}{dt} = -D\hat{\zeta} + C_a(F_a - Q_a) + C_b(F_b - Q_b) \tag{54}$$

$$\hat{\xi}_2 = A_2^{-1}[\hat{\zeta} - A_1 \xi_1] \tag{55}$$

Comment : if we consider the most simple and straightforward choice for the state transformation ζ, i.e. with $\xi_1 = \xi_a$ and $\xi_2 = \xi_b$, and with $C_b = I_{N-p}$ (i.e. the identity matrix of order N-p), then we have :

$$A_2 = I_{N-p}, \quad A_1 = -K_2 K_1^{-1} \tag{56}$$

And therefore the condition on the invertibility of A_2 is in fact a condition on the invertibility of the submatrix K_1 (i.e. K_1 is full rank or $\text{rank}(K_1) = M$).

The observer (54), (55) is completely independent of the process kinetics and can be implemented without the knowledge of the reaction rates $r(\xi)$ being required.

4.1.1 Theoretical convergence of the asymptotic observer

The convergence properties can summarized in the following theorem.

Theorem 4.
If the dilution rate D is a persistently exciting signal, i.e. if there exist positive constants δ and β such that:

$$\delta = \int_t^{t+\beta} D(\tau)d\tau \tag{57}$$

then :

$$\lim_{t \to \infty} (\xi_2 - \hat{\xi}_2) = 0 \tag{58}$$

Proof : the proof of the theorem is immediate if one observes that, from (28), (54), (55), the dynamics of the estimation error is equal to :

$$\frac{d(\xi_2 - \hat{\xi}_2)}{dt} = -D(\xi_2 - \hat{\xi}_2) \tag{59}$$

QED

Remark #1 : the persistence-of-excitation condition on D simply requires that D is not equal to zero for too long. This condition is clearly easily fulfilled in fedbatch and continuous reactors.

Remark #2 : Note that the stability of the asymptotic observer only depends on the dilution rate and not on the kinetics. In other words, the reactor may be unstable (due to the kinetics like in the Haldane inhibition model) while the asymptotic observer is asymptotically stable (because of stable hydrodynamics).

Remark #3 : the asymptotic observer is derived under the assumption that the yield coefficients are known and constant. This assumption is indeed very important, since an important preliminary step in the design of the observer is to consider a reliable reaction network of the process. This means that the chosen reaction network should be at the same time representative of the process and simple enough (in order to limit the number of required on-line measurements). "Representative" means that it reflects the major reaction mechanisms of the process (e.g. at the exclusion of side reactions that have a minor influence). In such a case, the yield coefficients should remain rather constant (ideally they would be constant by definition of the representativity of the reaction network). The yield coefficients considered for the asymptotic observer are then determined via preliminary (e.g. batch) experiments and/or stoichiometric arguments. If the yield coefficients are expected to change during the course of the process operation, our suggestion is to periodically re-evaluate them via the use of off-analyses combined to the on-line data.

Remark #4 (reversible reactions) : note that, in presence of reversible reactions, the matrix K will not be full column rank because it will contain two identical columns. However a simple way to treat the asymptotic observation problem of reversible reactions is to consider each reversible reaction as *one global reaction* (whose rate may then be positive or negative) and therefore characterized by *only one* column in the matrix . This means that if the "forward" and "backward" reactions are characterized by a reaction rate r_1 and r_2 respectively, we consider, for the observation, one global reaction characterized by the same stoichiometric coefficients but one global reaction $r_3 = r_1 - r_2$.

4.2 Application to an anaerobic digestion process

The asymptotic observer has been applied to a 60 liter pilot reactor by considering only two reactions (the first acidification reaction and the first methanization reaction)(i.e. model (47)(48) and measurements of acetate S_3 and methane P_1 once a day (Fig.5a-b).

The design of the asymptotic is indeed based on the model (47)(48) (i.e. before considering the low solubility of CH_4 and CO_2). The low solubility is considered in the implementation by setting P_1 and S_5 to zero in the asymptotic observer equations.

The results of the estimation of S_1, X_1 and X_3 are given in Fig.5e-g (see also [1]). In this application, the main objective was to validate the two step reaction network : the validation has been carried out by considering the difference between the total COD and the soluble COD (called COD_X), which is assumed to be representative of the total active biomass, i.e. here $X_1 + X_3$. More specifically the procedure consisted of finding values of the yield coefficients such that the estimation of $X_1 + X_3$ fits to the data of

576

COD_X (Fig.5h). Note that in the present instance the two step reaction scheme appears to be able to represent satisfactorily the process reaction network.

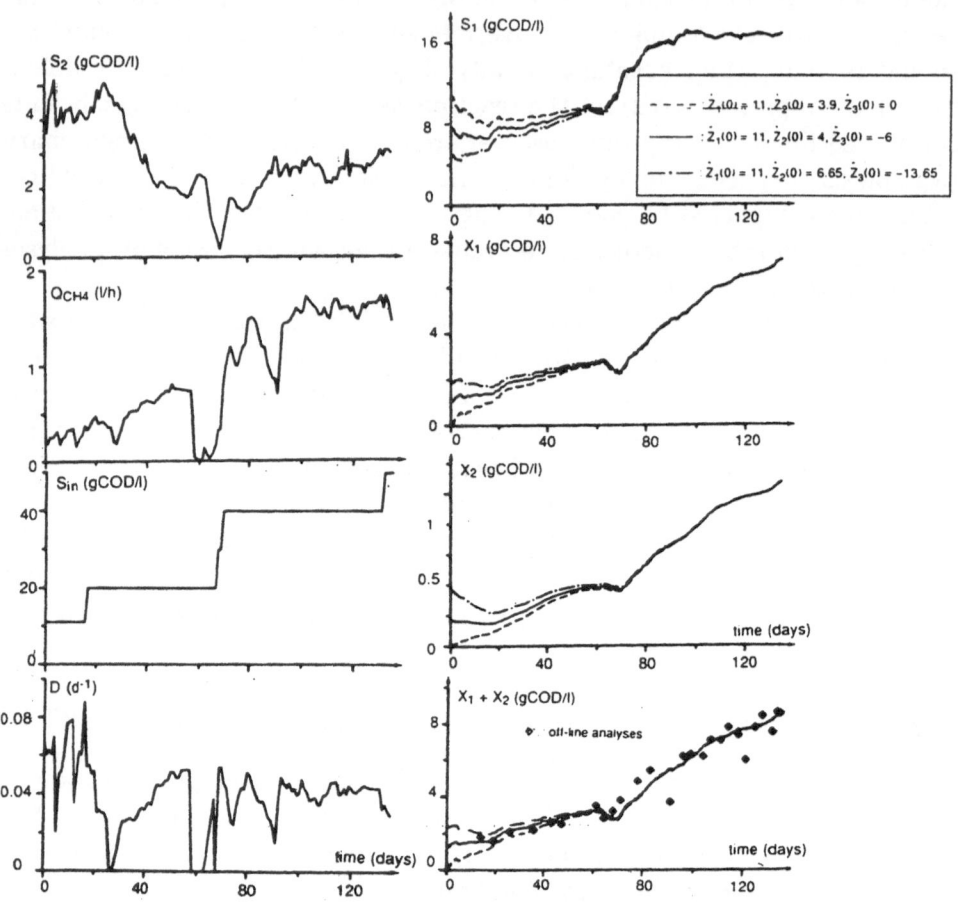

Figure 5. Experimental results of the asymptotic observer on an anaerobic digestion process

Remark : the design of the above asymptotic observer can be easily extended to multi-reactor processes, since the definition of the transformation ζ is the same. The main question is the stability of the dynamics of the auxiliary variables ζ (29). We shall not develop this point here, which is largely discussed in [37]. We shall simply concentrate on one example (activated sludge process) in Section 6, and discuss its stability.

5 Monitoring of Bioprocesses. Part II : On-line Estimation of Reaction Rates

In this section, we address the problem of estimating the reaction rates from on-line knowledge of the process components (knowledge available either from measurements or from state estimation). The statement of the estimation problem is presented first followed by the development of an observer-based estimator. Finally, an example on the estimation of specific growth rates for bakers' yeast is described in detail.

5.1 Statement of the estimation problem

We consider a biotechnological process described by the General Dynamical Model (1). In line with section 4, we assume that:

1. The feedrates F, the gaseous outflow rates Q and the dilution rate D are known either by measurement or by choice of the user.

2. The yield coefficient matrix K is known.

3. the process components ξ are known either by measurement or by estimation using an asymptotic observer (as described in Section 4.1).

4. the reaction rate vector $r(\xi)$ is partially unknown and written as follows:

$$r(\xi) = \begin{pmatrix} H(\xi)\rho(\xi) \\ h(\xi) \end{pmatrix} \qquad (60)$$

where $H(\xi)$ is a diagonal matrix of known functions of the state and $\rho(\xi)$ a vector of unknown functions of ξ with $\dim \rho(\xi) = n_u$. The known reaction rates are given by vector $h(\xi)$ with $\dim h(\xi) = M - n_u$.

Using equation (60), the General Dynamical Model is rewritten as :

$$\frac{d\xi}{dt} = K_u H(\xi)\rho(\xi) + K_k h(\xi) - D\xi - Q + F \qquad (61)$$

where K_u and K_k are matrices of yield coefficients associated with the unknown and known reaction rates respectively.

5.2 Observer-based estimator

A state observer form is used to provide on-line information for updating the estimate of $\rho(\xi)$. The estimation algorithm is written as follows:

$$\frac{d\hat{\xi}}{dt} = K_u H(\xi)\hat{\rho} + K_k h(\xi) - D\xi - Q + F - \Omega(\xi - \hat{\xi}) \qquad (62)$$

$$\frac{d\hat{\rho}}{dt} = [K_u H(\xi)]^T \Gamma(\xi - \hat{\xi}) \qquad (63)$$

The update of the parameter vector $\hat{\rho}$ is driven by the deviation $(\xi - \hat{\xi})$ which reflects the mismatch between $\hat{\rho}$ and ρ. The matrices Ω and Γ are tuning parameters for adjusting the rate of convergence of the algorithm. A common choice is :

$$\Omega = diag(-\omega_i), \ \Gamma = diag(-\gamma_j), \ \omega_i, \gamma_j > 0 \qquad (64)$$

With this choice, the stability of (62), (63) is satisfied (see [1] for further details). The tuning procedure may be simplified if the state equations are first decoupled using the following transformation:

$$\Psi = K_u^{-1}\xi \qquad (65)$$

Applying the estimation algorithm to the transformed state equations yields:

$$\frac{d\hat{\Psi}}{dt} = H\hat{\rho} + K_u^{-1}K_k h - D\Psi + K_u^{-1}(F - Q) - \Omega(\Psi - \hat{\Psi}) \qquad (66)$$

$$\frac{d\hat{\rho}}{dt} = H\Gamma(\Psi - \hat{\Psi}) \qquad (67)$$

5.3 Application to the bakers's yeast fed-batch process

A modified version of the process model given in Section 2.4 is considered. From a global point of view, the bakers' yeast fed-batch process can only be in an ethanol production regime or in an ethanol consumption regime. The process model is divided in two partial models to represent the two regimes. The first partial model corresponds to the ethanol production regime and is denoted as the oxido-reductive (RF) partial model:

$$\frac{d\xi}{dt} = K_{RF}.r_{RF} - D\xi - Q + F \qquad (68)$$

where

$$K_{RF} = \begin{pmatrix} -k_1 & -k_2 & -k_{10} \\ -k_3 & 0 & -1 \\ 1 & 1 & 0 \\ k_5 & k_6 & k_{11} \\ 0 & k_8 & 0 \end{pmatrix}, \quad r_{RF} = \begin{pmatrix} \mu_o X \\ \mu_r X \\ q_m X \end{pmatrix}$$

The second partial model, the respirative (R) partial model, corresponds to the ethanol consumption regime where oxidation of both glucose and ethanol may occur. The mass balance for this regime is written as:

$$\frac{d\xi}{dt} = K_R r_R - D\xi - Q + F \tag{69}$$

where

$$K_R = \begin{pmatrix} -k_1 & 0 & -k_{10} \\ -k_3 & -k_4 & -1 \\ 1 & 1 & 0 \\ k_5 & k_7 & k_{11} \\ 0 & -k_9 & 0 \end{pmatrix}, \quad r_R = \begin{pmatrix} \mu_o X \\ \mu_e X \\ q_m X \end{pmatrix}$$

The first step in the estimation procedure is to identify the measured variables. The concentration of ethanol, dissolved oxygen, and carbon dioxide are available for measurement. However, it was shown in [53] that those three measurements are not linearly independent. The two independent measured state variables selected are dissolved oxygen C and carbon dioxide P concentrations. Two specific growth rate estimation algorithms are needed, one for the estimation of $[\mu_o \ \mu_r]$ in the ethanol production regime and one for the estimation of $[\mu_o \ \mu_e]$ in the ethanol consumption regime. The maintenance coefficient q_m is assumed to be known [53]. Only the derivation for the ethanol production partial model is given below, the procedure being identical for the other partial model. The mass balance equations for the measured state variables are:

$$\frac{dC}{dt} = -DC - k_3 \mu_o X + q_m X + \Delta Q_{O_2} \tag{70}$$

$$\frac{dP}{dt} = -DP + k_5 \mu_o X + k_6 \mu_r X + k_{11} X - Q_1 \tag{71}$$

The biomass concentration is unknown and will be estimated by an asymptotic observer as described later. With reference to equation (61), we can define the following vector and matrices:

$$K_u = \begin{pmatrix} -k_3 & 0 \\ k_5 & k_6 \end{pmatrix}, \ H = \begin{pmatrix} \hat{X} & 0 \\ 0 & \hat{X} \end{pmatrix}, \ K_k = \begin{pmatrix} -1 \\ k_{11} \end{pmatrix}$$

$$\xi_1 = \begin{pmatrix} C \\ P \end{pmatrix}, \ \rho = \begin{pmatrix} \mu_o \\ \mu_r \end{pmatrix}, \ h = [q_m \hat{X}]$$

A linear transformation is applied to the measured state variables (ξ_1) set to decouple the equations in term of the specific growth rates:

$$\Psi = K_u^{-1} \xi_1 \tag{72}$$

The matrices of tuning parameters Ω and Γ in equations (66) and (67) are chosen as follows:

$$\Omega = -HC_1, \ \Gamma = C_2 \tag{73}$$

where C_1 and C_2 are diagonal matrices. The elements of these two matrices are chosen to ensure constant dynamics of the estimation error throughout the experiment. With C_1 and C_2 chosen as:

$$C_1 = \begin{pmatrix} C_{11} & 0 \\ 0 & C_{11} \end{pmatrix} = \begin{pmatrix} \frac{2\alpha}{\hat{X}} & 0 \\ 0 & \frac{2\alpha}{\hat{X}} \end{pmatrix}, \ C_2 = \begin{pmatrix} \frac{C_{11}^2}{4} & 0 \\ 0 & \frac{C_{11}^2}{4} \end{pmatrix} \tag{74}$$

the poles of the error dynamics are all located at $-\alpha$. A single tuning parameter (α) is thus needed.

The non-measured biomass concentration appears in the equations of the estimation algorithm. A biomass concentration observer is thus needed for the application of the estimation algorithm. First, the process model is partitioned in two subsets. The first subset includes the equations associated to the measured state variables ($\xi_1 = [C \ P]^T$), while the second subset is associated to the non-measured state variable ($\xi_2 = X$):

$$\frac{d\xi_1}{dt} = -D\xi_1 + K_u H \rho + K_k h + F_1 - Q_1 \tag{75}$$

$$\frac{d\xi_2}{dt} = -D\xi_2 + K_2 H \rho \tag{76}$$

Applying the following linear transformation between the unmeasured and measured state variables:

$$\zeta = \xi_2 - K_2 K_u^{-1} \xi_1 \tag{77}$$

we obtain:

$$\frac{d\zeta}{dt} = -(D + K_2 K_u^{-1} K_k)\zeta - K_2 K_u^{-1}(F_1 - Q_1 + K_k K_2 K_u^{-1} \xi_1) \quad (78)$$

By substituting $\hat{\zeta}$ for ζ, the biomass concentration observer is then:

$$\frac{d\hat{\zeta}}{dt} = -(D + K_2 K_u^{-1} K_k)\hat{\zeta} - K_2 K_u^{-1}(F_1 - Q_1 + K_k K_2 K_u^{-1} \xi_1) \quad (79)$$

$$\hat{X} = \hat{\zeta} + K_2 K_u^{-1} \xi_1 \quad (80)$$

The procedure proposed so far is based on the use of the proper partial model algorithm set according to the process state. The problems that remain to be solved are

- the detection of the proper regime,

- the transition between the two estimation algorithms

The first problem is resolved by looking at the values of the specific growth rate estimates. If the process is not in the corresponding regime, the specific growth rate estimate directly related to ethanol (μ_r in the oxido-reductive model and μ_e in the respirative model) will have negative value. The criterion for the transition between partial model algorithm sets is given by the transition between positive and negative values of these estimates.

The transition procedure between the two sets of algorithms is more complex. In each algorithm set, five variables have to be monitored in time: the transformed variable estimate in the biomass observer ($\hat{\zeta}$), the two transformed measured state estimates (Ψ) and the two specific growth rate estimates. It is important to monitor the time course of these variables also when the estimation algorithm does not correspond to the process regime to obtain appropriate values of these variables at transition time.

For the transformed variable in the biomass observer, the following technique is suggested. When the process regime changes, the transformed variable and the estimated biomass concentration issued from the other partial model algorithm set are used to provide a value of $\hat{\zeta}$ for the new partial model. Then the biomass observer algorithm of the new partial model is used. As an example, if the transition between respiro-fermentative model to respirative model occurs, then:

$$\hat{X}_{RF} = \hat{\zeta}_{RF} - K_2 K_u^{-1}{}_{RF} \xi_{1RF} \quad (81)$$

$$\hat{\zeta}_R = \hat{X}_{RF} + K_2 K_u^{-1}{}_R \xi_{1R} \quad (82)$$

For the specific growth rate estimates, distinct treatment is needed depending if the specific growth rate is related or not to ethanol. In the case of the estimation of μ_o, the time trajectory is still followed during the period when the partial model algorithm set is not used. Experiences have shown that the estimation error is small and that convergence following the transition is so rapid that a more sophisticated treatment is not required. For ethanol related specific growth rates, μ_e or μ_r, the estimated value is forced to zero when the partial model does not correspond to the process state.

For the transformed state variables (Ψ) of the specific growth rate estimator, the time trajectory have also to be followed during partial model and process mismatch, but the estimated value for biomass concentration used during this period is the one issued from the valid partial model observer. Also, the zero value of the ethanol related specific growth rate estimate is used in the prediction equations. This technique is required to avoid a too strong perturbation of the estimation algorithm at transition time.

5.3.1 Experimental verification

The experimental fermentations have been achieved in a 20 l BioEngineering reactor. The yeast strain, *Saccharomyces cerevisiae* and the carbon source (a mix of cane and beat molasses) were supplied by an industrial bakers' yeast producer. The operating conditions were chosen to reproduce the industrial processes. They were carried out under ethanol concentration regulation with different nonlinear adaptive control laws ([54]). The ethanol regulation keeps the process near the boundary between the ethanol production and ethanol consumption state. Set-point changes, agitation speed and aeration rate perturbations have been applied to test the estimation algorithm. These perturbations create a diversity of process conditions and some of these can be considered as extreme conditions which do not appear in normal operation. The dissolved oxygen concentration was not controlled and the agitation speed (700 RPM) and the aeration rate ($2\ l l^{-1} min^{-1}$) were kept constant except at the time of perturbation. The pH was kept constant at 5.0. At this low pH there is no influence on the equilibrium of the different forms of dissolved carbon dioxide.

The ethanol concentration was measured in the exit gas with a Figaro sensor (TGS822) and calibrated as a function of the liquid ethanol concentration in the reactor. Dissolved oxygen concentration was measured with an Ingold probe. Carbon dioxide concentration was presumed to be directly proportional to the carbon dioxide content of the exit gas. Oxygen transfer rate (ΔQ_{O2}) and carbon dioxide transfer rate (Q_1) were evaluated with off-gas analysis performed by a magnetic sector mass spectrometer (VG-MM8-80). The molasses feed rate was controlled by a variable speed peristaltic pump (Watson-Marlow 501U/R).

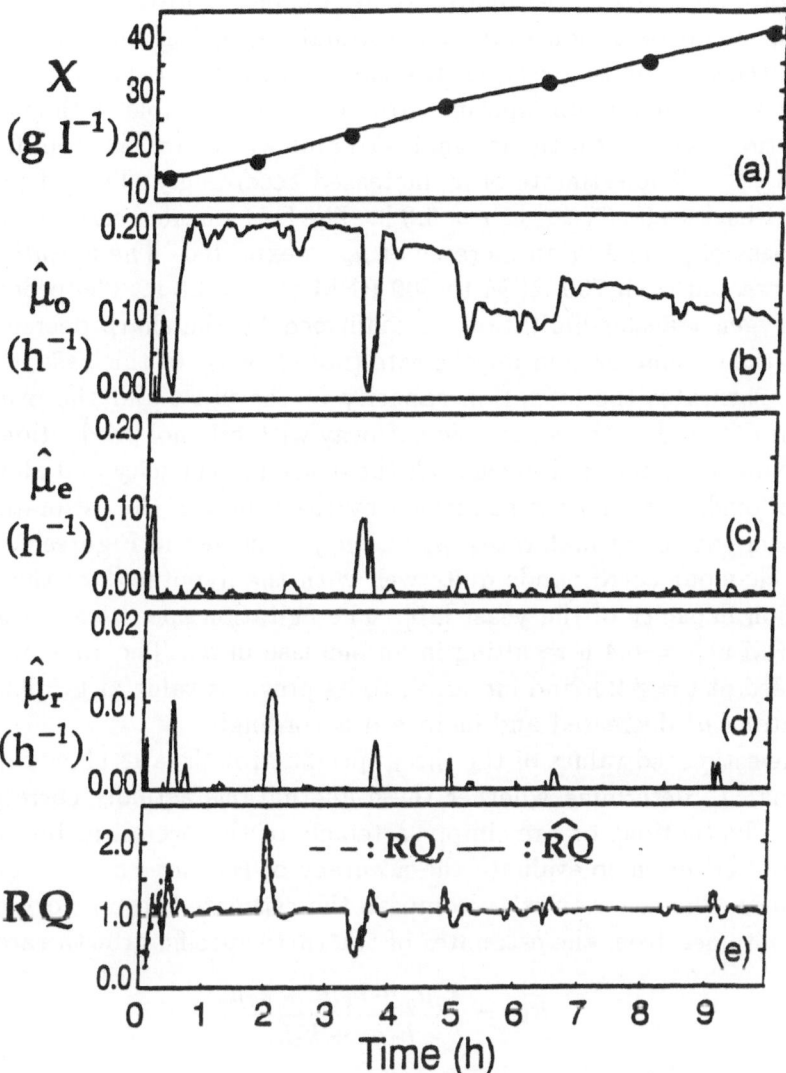

Figure 6. Estimation of biomass and specific growth rates for the bakers' yeast example

Figure 6a shows the biomass concentration estimates produced by the alternating use of the two sets of algorithms issued from the two partial models. The comparison of this time profile with the measured biomass concentration values from different samples is also shown. The precision of the biomass concentration estimate is within the precision of the measured values.

Figure 6b shows the estimated values of specific growth rate associated with the sugar oxidation (μ_o). The estimate of μ_o remains quite constant during the non-limiting oxygen transfer condition ($t \in [1,2]$ h)and corresponds to the maximum specific growth rate achievable without ethanol production. At $t = 2$ h the ethanol set-point was increased from 0.28 $g\,l^{-1}$ to 1.88 $g\,l^{-1}$. The estimate of μ_r increased accordingly. The set-point was brought back to 0.26 $g\,l^{-1}$ at $t = 3.3$ h. The feed was decreased resulting in a decrease of μ_o and in an increase of μ_e as expected. The agitation speed was decreased from 700 RPM to 600 RPM at t = 4.8 h. The process went into oxygen transfer limitation as confirmed by the sharp decrease in μ_o and in the specific oxygen uptake rate (not shown). In this case, the sugar uptake saturates the oxidation capacity of the yeast and the overflow of sugar is directed to the anaerobic pathway with ethanol production. When it is below, the sugar did not use all the oxidation capacity and allowed the ethanol oxidation. This is confirmed by the values of the estimates of the two other specific growth rates (μ_e and μ_r) presented in Figures 6c and 6d. This behaviour corresponds quite well with the hypothesis of the limiting oxidation capacity of the yeast [25]. The agitation speed was returned to 700 RPM at t = 6.4 h resulting in an increase in μ_o. The air flowrate was decreased at t = 8.2 h and increased to its previous value at t = 9.0 h. The estimate of μ_o decreased and increased accordingly.

The estimated values of the three specific growth rates show rapid fluctuations. To determine whether these fluctuations actually correspond to process fluctuations or are simply artefacts of the overall estimation procedure, a criterion to evaluate the accuracy of the three specific growth is proposed. The criterion is to compare the estimated respiratory quotient ($\widehat{RQ}$) obtained from the estimates of the three specific growth rates:

$$\widehat{RQ} = \frac{k_5\hat{\mu}_o + k_6\hat{\mu}_r + k_7\hat{\mu}_e}{k_3\hat{\mu}_o + k_4\hat{\mu}_e} \tag{83}$$

to the value evaluated from experimental data. The values of each of the three specific growth rate estimates have to be accurate in order to produced a good estimate of the respiratory quotient. Figure 6e shows the comparison between the experimental and the estimated respiratory quotient. A respiratory quotient over 1.06 indicates an ethanol production state and below 1.06 an ethanol consumption state. The agreement between the two curves is very good in spite of the inherent lag of the estimation procedure. This comparison allows us to be confident in the proposed estimation procedure.

6 Adaptive Linearizing Control of Bioprocesses

6.1 Design of the Adaptive Linearizing Controller

We shall now concentrate on the design of model-based controllers for bioreactors. The key idea of the control design is here again to take advantage of what is well known about the dynamics of bioprocesses (reaction network and mass balances) which are summarized in the General Dynamical Model (1) while taking into account the model uncertainty (mainly the kinetics). Since the model is generally nonlinear, the model-based control design will result in a *linearizing* control structure, in which the on-line estimation of the unknown variables (component concentrations) and parameters (reaction rates and yield coefficients) are incorporated. the resulting controller will be an *adaptive linearizing controller*. The design of the control algorithm is based on the General Dynamical Model (1) or on a reduced-order form of (1)(design examples based on a reduced-order model where the (unknown) kinetics terms are eliminated by incorporating the gas measurements are presented in [1], [16], [17], [55]).

The objective is to control the concentration of some reactant components by acting on flow rates under the following practical constraints:

1. the components to be controlled are assumed to be measured on-line;

2. the concentrations of the other components (particularly of the biomass) are not available for on-line measurement;

3. the reaction rate vector r is unknown;

4. most of the yield coefficients are unknown (only the ratio k_1/k_2 will be assumed to be known, in the example below);

5. the mass transfer coefficients are known;

6. the feedrates F, the dilution rate D and the gaseous outflow rates Q are known (either by user's choice or by measurement).

By defining y the controlled component(s), the dynamics of y are simply the equation(s) of y in model (1) and can be rewritten as follows :

$$\frac{dy}{dt} = -Dy + K_y r + F_y - Q_y \tag{84}$$

where the index y holds for the rows of K, F and Q corresponding to the controlled output y. By considering the above control problem and defining u the control input, the output equation (84) can be rewritten as follows :

$$\frac{dy}{dt} = f(F_y, Q_y, K_y r) + g(F_y, y)u \tag{85}$$

Assume now that we wish to have a linear stable closed-loop (process + controller) dynamical behaviour, i.e. :

$$\frac{dy}{dt} = C_1(y^* - y), \ C_1 > 0 \tag{86}$$

with y^* the desired value of y. By combining equations (85), (86), we readily obtain the control law :

$$u = g(F_y, y)^{-1}[C_1(y^* - y) - f(F_y, Q_y, K_y r)] \tag{87}$$

Since the kinetics and most of the yield coefficients are assumed to be unknown, they are replaced by on-line estimates of selected parameters :

$$u = g(F_y, y)^{-1}[C_1(y^* - y) - f(F_y, Q_y, \hat{K}_y \hat{r})] \tag{88}$$

The above controller (88) is also known as the model reference adaptive linearizing control law (see e.g. [1]). As we shall see in the examples, the unknown parameters appear linearly in the equations and will therefore be estimated by using linear regression techniques or via a Lyapunov design estimation approach. Moreover, the unknown components that may appear explicitly in the output equation (84) via the reaction rate r will be replaced either by an auxiliary variable easy to compute (see the example below) or by gaseous outflow rates (see e.g. [1], [16], [17], [55]); (in the latter instance, the effect will also to make disappear the kinetics terms).

6.2   Example #2 : Activated Sludge Process

The following example is treated in details in [56]. The control problem we shall consider now can be formulated as follows. Like in any aerobic fermentation processes, proper aeration is crucial to process efficiency, and adequate control of the dissolved oxygen concentration in the aerator is very important. But it is also important to limit load variations and substrate concentration variations by acting on them. As suggested by many authors (e.g. [27], [57]), load variations can be expected to be smoothed by using the recycle flow rate as a control input. We shall consider the above control algorithm under the following conditions :

1. the controlled outputs are the effluent BOD (Biological Oxygen Demand) and the dissolved oxygen concentrations, which are assumed to be measured on-line (see [58] for an on-line measuring device of the BOD in activated sludge processes).

2. the control inputs are the recycle flow rate and the air flow rate.

Therefore the input u and the output y are defined as follows :

$$u = \begin{pmatrix} F_R \\ W \end{pmatrix}, y = \begin{pmatrix} S \\ C \end{pmatrix} \tag{89}$$

and the control problem is multivariable (2 inputs, 2 outputs). The expressions of f and g in (85) are readily obtained from model (18) :

$$f = \begin{pmatrix} -k_1 \mu X + D_{in}(S_{in} - S) \\ -k_2 \mu X - D_{in}C \end{pmatrix}, g = \begin{pmatrix} -\frac{S}{V} & 0 \\ -\frac{C}{V} & a_0(C_S - C) \end{pmatrix} \tag{90}$$

g is invertible (i.e. g^{-1} exists) as long as S and $C_S - C$ are different from zero. Under these physically realistic conditions, the linearizing controller (88) can be applied. The remaining problem is how to deal with the "unknown" parameters k_1, k_2 and μ, and variable X. The solution proceeds as follows. Let us first define the auxiliary variables ζ (as introduced in Section 3.1).

$$z = \begin{pmatrix} \zeta_1 \\ \zeta_2 \\ \zeta_3 \end{pmatrix} = \begin{pmatrix} k_1 \\ k_2 \\ 0 \end{pmatrix} X + \begin{pmatrix} 1 & 0 & 0 \\ 0 & 1 & 0 \\ 0 & 0 & k_2 \end{pmatrix} \begin{pmatrix} S \\ C \\ X_R \end{pmatrix} \tag{91}$$

The dynamical equation of ζ are derived from model (18)) and definitions (91):

$$\frac{d}{dt} \begin{pmatrix} \zeta_1 \\ \zeta_2 \\ \zeta_3 \end{pmatrix} = \begin{pmatrix} -D_1 & 0 & D_2 k_1/k_2 \\ 0 & -D_1 & D_2 \\ 0 & D_3 & -D_4 \end{pmatrix} \begin{pmatrix} \zeta_1 \\ \zeta_2 \\ \zeta_3 \end{pmatrix} + \begin{pmatrix} D_{in}S_{in} \\ \Delta Q_{O_2} \\ -D_3 C \end{pmatrix} \tag{92}$$

with D_i (i = 1 to 4) as defined in Section 2.5. One important feature of the above dynamics (92) is that these are independent of the (unknown) kinetics. Equation (92) is a dynamical system linear-in-the-state-ζ. We can check that its state matrix is asymptotically stable if $F_{in} - F_R > 0$ and $F_W > 0$. Therefore equation (92) can be used to compute the value of ζ on-line from the knowledge of the flow rates F_{in}, F_R and F_W, the feedrates $F_{in}D_{in}$ and ΔQ_{O_2}, the dissolved oxygen concentration C and the ratio k_1/k_2. Let now rewrite the specific growth rate μ as follows:

$$\mu = \alpha SC \tag{93}$$

where α is an (unknown) positive function of the process components : the equation (93) simply implies that there is no growth in absence of one of the limiting substrate. By introducing (91), (93), the function f in (90) becomes :

$$f = \left(\begin{array}{c} -\alpha SC(\zeta_1 - S) + D_{in}(S_{in} - S) \\ -\alpha SC(z_2 - C) - D_{in}C \end{array} \right) \tag{94}$$

This expression of f is used in the computation of the control law (88) :

$$F_R = -\frac{V}{S}[C_{11}(S^* - S) - D_{in}(S_{in} - S) + \hat{\alpha}_1 SC(\zeta_1 - S)] \tag{95}$$

$$\begin{aligned} W = {} & \frac{1}{a_0(C_S - C)}[C_{12}(C^* - C) - \frac{C_{11}(S^* - S)C}{S} + \frac{D_{in}S_{in}C}{S} \\ & + \hat{\alpha}_2 C(S\zeta_2 - C\zeta_1)] \end{aligned} \tag{96}$$

C_{11} and C_{12} (> 0) are the control design parameters (we have considered here a decoupling diagonal matrix C_1). The variables ζ_1 and ζ_2 are computed via the use of equations (92) (which only requires the knowledge of the ratio k_1/k_2) and the parameter α is estimated on-line by using e.g. a recursive least square (RLS) algorithm. In discrete-time, if *alpha* is estimated via the substrate concentration equation S (α_1, to be used in equation (95)) or via the dissolved oxygen concentration C (α_2, to be used in equation (96)), the RLS algorithm is written as follows:

$$\hat{\theta}_{i,t+1} = \hat{\theta}_{i,t} + g_{i,t}\phi_{i,t}e_{i,t+1} \qquad i = 1, 2 \tag{97}$$

$$g_{i,t} = \frac{g_{i,t-1}}{\gamma_i + g_{i,t-1}\phi_{i,t}^2} \qquad 0 < \gamma_i \leq 1 \tag{98}$$

with :

$$\hat{\theta}_{1,t} = \hat{\alpha}_{1,t}, \quad \phi_{1,t} = TS_tC_t(S_t - z_{1,t}) \tag{99}$$

$$e_{1,t+1} = S_{t+1} - S_t + TD_{1,t}S_t - TD_{in,t}S_{in,t} - \hat{\alpha}_{1,t}\phi_{1,t} \qquad (100)$$

$$\hat{\theta}_{2,t} = \hat{\alpha}_{2,t}, \quad \phi_{2,t} = TS_tC_t(C_t - z_{2,t}) \qquad (101)$$

$$e_{2,t+1} = C_{t+1} - C_t + TD_{1,t}C_t - Ta_0W_t(C_s - C_t) - \hat{\alpha}_{2,t}\phi_{2,t} \qquad (102)$$

and t the time index, and γ_i (i = 1, 2) a forgetting factor. Note that now because of the variables ζ, the biomass X does not appear explicitly anymore in the controller (95), (96).

A typical simulation result is shown in Figure 7 (see also [56]). The activated sludge process has been simulated by numerical integration of the basic dynamical model equations (18) (using a 4th order Runge-Kutta method) with the following (Monod-type) model for the specific growth rate:

$$\mu = \mu_{max} \frac{S}{K_S + S} \frac{C}{K_C + C} \qquad (103)$$

and a decay rate $(-k_dX)$ which has been added to the biomass equation (in order to simulate biomass mortality). The specific growth rate model and the decay rate term are (obviously) completely ignored by the control algorithm. The decay rate term can be viewed as an unmeasured disturbance. The model parameters have been set to the following values (inspired from literature data, in particular from [57], [59]) :

$$k_1 = 1.2, \ k_2 = 0.565, \ \mu_{max} = 0.2\,h^{-1}, \ K_S = 75\,mg\,l^{-1}, \ k_d = 0.001\,h^{-1}$$

$$a_0 = 0.018\,m^{-3}, \ C_S = 10\,mg\,l^{-1}, \ V = 100\,m^3, \ V_S = 50\,m^3, \ K_C = 2\,mg\,l^{-1}$$

In Figure 7, a square wave signal of the influent BOD concentration (disturbance input) S_{in} (from $150\,mg\,l^{-1}$ to $200\,mg\,l^{-1}$) (in order to simulate the periodical variation of the pollutant load) has been applied over a period of 20 d (480 h). The sampling period has been set to 3 minutes following the constraints of the commercially available BOD measuring device proposed by [58]. The following initial process conditions have been considered in the simulation :

$$S = 5\,mg\,l^{-1}, \ C = 6\,mg\,l^{-1}, \ X = 1225\,mg\,l^{-1}, \ X_R = 2333mg\,l^{-1} \qquad (104)$$

The controller parameters have been set to the following values :

$$S^* = 5\,mg\,l^{-1}, \ C^* = 6mg\,l^{-1}, \ C_{11} = 1\,h^{-1}, \ C_{12} = 10\,h^{-1} \qquad (105)$$

The auxiliary variables ζ, the unknown parameters and tuning estimation variables have been initialized as follows :

$$\zeta_1 = 1400\,mg\,l^{-1}, \ \zeta_2 = 750\,mg\,l^{-1}, \ \zeta_3 = 1400\,mg\,l^{-1}, \ g_{1,0} = g_{2,0} = 10^{-3}$$

$$\gamma_1 = \gamma_2 = 0.9, \ \hat{\alpha}_{1,0} = \hat{\alpha}_{2,0} = 0.00025\,l^2\,mg^{-2}\,h^{-1}$$

590

Note the ability of the controller to maintain the controlled outputs S and C close to their desired values in spite of the unknown disturbance.

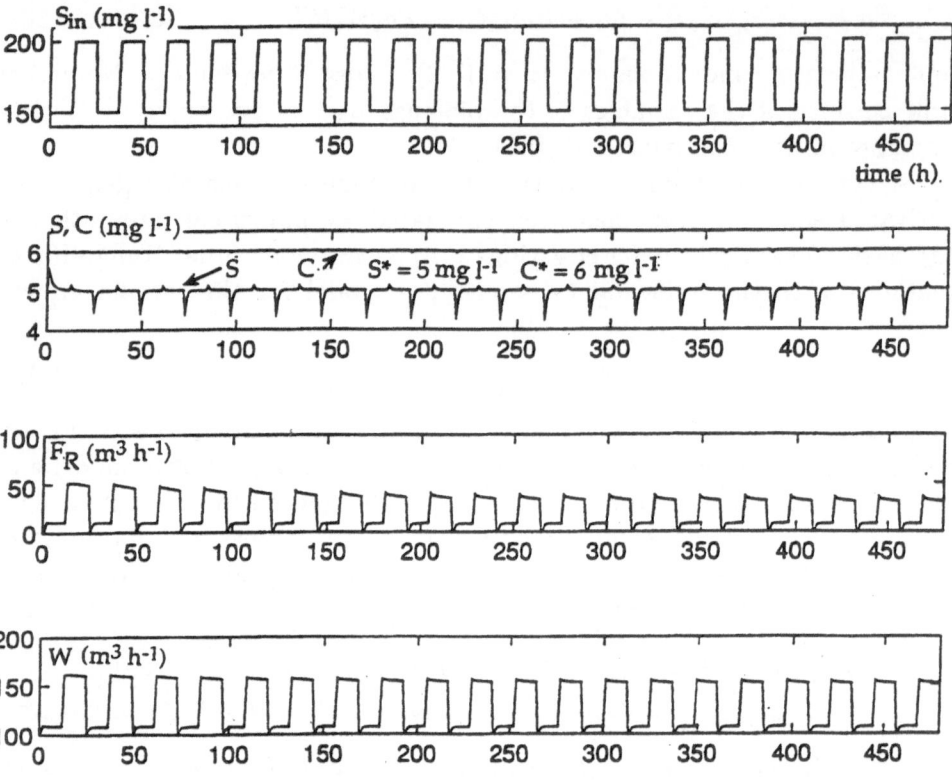

Figure 7. Control of an activated sludge process in presence of an unknown biomass mortality

Discussion

One of the key feature of the adaptive linearizing can be illustrated on the basis of the activated sludge example : it allows to incorporate the well-known characteristics of the process dynamics, while keeping the usual features of classical controllers :

- proportional action : via the terms $C_{12}(C^* - C)$ and $C_{11}(S^* - S)$;

- integral action : via the adaptation mechanism (97)(98);

- feedforward action : via the presence of S_{in}. Indeed the controller is capable of anticipating the effect of a variation of the influent substrate concentration : for instance, an increase of S_{in} will result in an increase of the control input F_R, proportional to this increase (see also Figures 7a and 7c).

Besides the controller contains a state "estimate" via the terms ζ_1 and ζ_2. If the process is in good working conditions, in particular if the biomass is in high concentrations, this results in high values of ζ_1. Then there is no need to recycle a lot of biomass into the aerator.

Finally note that beside the integral action, the estimation of "physical" parameters (here α_1 and α_2) has the further advantage of giving useful information, which can be used for monitoring the process, and possibly also for analyzing the internal working of the process.

7 Conclusions

The objective of this paper was to present a survey about recent approaches of model-based monitoring and control of bioreactors. The proposed results covers the whole range from theory (dynamical modelling, dynamical analysis, monitoring and control design) to practice (experimental results). They apply to different types of reactors, mainly stirred tank reactors and fixed bed reactors, but could also be extended to e.g. fluidised bed reactors. The dynamical model of the bioprocesses is based on material balances and is formalized into a General Dynamical Model framework which serves as a basis for dynamical analysis and for monitoring and control algorithm design. Monitoring was found to be a key question in bioprocess applications. The design of asymptotic observers for the concentrations of the process components has been developed, analyzed and illustrated. Due to the usually large uncertainty of some process parameters, namely e.g. the process kinetics, on-line estimation of the uncertain parameters has been considered either for monitoring purposes or to be included in an adaptive model-based control scheme. In the latter case, the incorporation of "physical" parameters presents the double advantage of introducing an integral action in the controller while giving extra information about the process behaviour and performance.

Acknowledgements : This paper presents research results of the Belgian Programme on Interuniversity Poles of Attraction initiated by the Belgian State, Prime Minister's Office, Science, Technology and Culture. The scientific responsibility rests with its authors.

References

[1] G. Bastin and D. Dochain. *"On-line Estimation and Adaptive Control of Bioreactors"*. Elsevier, Amsterdam, 1990.

[2] P. Renard, D. Dochain, G. Bastin, H. Naveau, and E.J. Nyns. Adaptive control of anaerobic digestion processes. a pilot-scale application. *Biotechnol. Bioeng.*, 31:287–294, 1988.

[3] P. Renard, V. Van Breusegem, N. Nguyen, H. Naveau, and E.J. Nyns. Implementation of an adaptive controller for the start-up and steady-state running of a biomethanation process operated in the cstr mode. *Biotechnol. Bioeng.*, 38:805–812, 1991.

[4] Y. Pomerleau and G. Viel. Industrial application of adaptive non-linear control for baker's yeast production. In N.M. Karim and G. Stephanopoulos, editors, *Proc. 5th Int. Conf. Computer Appl. in Biotechnol.*, pages 315–318. Pergamon Press, Oxford, 1992.

[5] B. Dahhou, G. Roux, and A. Chéruy. Linear and nonlinear adaptive control of alcoholic fermentation processes : experimental results. *Int. J. Adaptive Control and Signal Processing*, 7(3):213–233, 1993.

[6] L. Chen, G. Bastin, and V. Van Breusegem. A case study of adaptive nonlinear regulation of fed-batch biological reactors. *Automatica*, 31(1):55–65, 1995.

[7] I. Cornet, D. Dochain, B. Ramsay, and M. Perrier. Application of adaptive linearizing inferential control to a phb producing process. *Biotechnology and Bio E*, 1:96–102, 1995.

[8] Y. Pomerleau, M. Perrier, and D. Bourque. Dynamics and control of the fed-batch production of poly-β-hydroxybutyrique by *methylobacterium extorquens*. *Proc. 6th Int. Conf. Computer Appl. in Biotechnol.*, pages 107–112, 1995.

[9] S. Bourrel. *"Estimation et Commande d'un Procédé à Paramètres Répartis Utilisé pour le Traitement Biologique de l'Eau à Potabiliser"*. PhD thesis, Université Paul Sabatier, Toulouse, France, 1996.

[10] M.A. Henson and D. Seborg. Nonlinear control strategies for continuous fermenters. *Chem. Eng. Sci.*, 47(4):821–835, 1992.

[11] J.M. Flaus, A. Cheruy, and J.M. Engasser. An adaptive controller for batch feed bioprocess. applicaton to lysine production. *J. Proc. Cont.*, 1:271–281, 1991.

[12] B. Dahhou, J. Bordeneuve, and J.P. Babary. Multivariable long-range predictive control algorithm applied to a continuous flow fermentation process. *Proc. World IFAC Congress*, pages 393–397, 1991.

[13] M.P. Golden, B.J. Pangrie, and B.E. Ydstie. Nonlinear adaptive optimization of a continuous bioreactor. *Proc. AIChE 1986 National Meeting*, Pap.125b, 1986.

[14] J.G. Alvarez and J.G. Alvarez. Analysis and control of fermentation processes by optimal and geometric methods. *Proc. ACC*, 2:1112–1117, 1988.

[15] K.A. Hoo and J.C. Kantor. Linear feedback equivalence and control of an unstable biological reactor. *Chem. Eng. Comm.*, 46:385–399, 1986.

[16] D. Dochain and M. Perrier. Recent approaches to the dynamical modelling, analysis, monitoring and control design for nonlinear bioprocesses. *Recent Advances in Chemical Engineering*, 2:215–251, 1995.

[17] D. Dochain and M. Perrier. Dynamical modelling, analysis, monitoring and control design for nonlinear bioprocesses. *Advances in Biochemical Engineering/Biotechnology*, 56:147–197, 1997.

[18] M. DeTremblay and M. Perrier. Optimisation of fed-batch culture of hybridoma cells using dynamic programming : single and multi-feed cases. *Bioprocess Eng.*, 7:229–234, 1992.

594

[19] M. DeTremblay, M. Perrier, C. Chavarie, and J. Archambault. Fed-batch culture of hybridoma cells : comparison of optimal control approach and closed-loop strategies. *Bioprocess Eng. (in press)*, 1993.

[20] V. Chotteau and G. Bastin. Identification of a reaction mechanism for a class of animal cell cultures. *Proc. ICCAFT 5/IFAC-BIO 2*, pages 215–218, 1992.

[21] V. Chotteau. *"A General Modeling Methodology for Animal Cell Cultures"*. PhD thesis, Université Catholique de Louvain, Belgium, 1995.

[22] J. VanImpe. *"Modelling and Optimal Adaptive Control of Biotechnological Processes"*. PhD thesis, Katholieke Universiteit Leuven, Belgium, 1993.

[23] J. VanImpe, B. Nicolai, P. Van Rolleghem, J. Spriet, B. De Moor, and J. Van De Walle. Optimal control of the penicillin g fed-batch fermentation : an analysis of a modified unstructured model. *Chem. Eng. Comm.*, 117:337–353, 1992.

[24] F.E. Mosey. Mathematical modelling of the anaerobic digestion process : regulatory mechanisms for the formation of short-chain volatile acids from glucose. *Water Sci. Technol.*, 15:209–232, 1983.

[25] B. Sonnleitner and O. Kappeli. Growth of *saccharomyces cerevisiae* is controlled by its limited respiratory capacity : formulation and verification of an hypothesis. *Biotechnol. Bioeng.*, 28:927–937, 1986.

[26] R.P. Hamalainen, A. Halme, and A. Gyllenberg. A control model for activated sludge wastewater treatment process. *Proc. 6th IFAC World Congress, Boston, Paper 61:6*, 1975.

[27] S. Marsili-Libelli. Optimal control of the activated sludge process. *Trans. Inst. Meas. Control*, 6:146–152, 1984.

[28] A. Holmberg and J. Ranta. Procedures for parameter and state estimation of microbial growth process models. *Automatica*, 18:181–193, 1982.

[29] P.V. Danckwerts. Continuous flow systems. distribution of residence times. *Chem. Eng. Sci.*, 2 (1):1–13, 1953.

[30] D. Dochain. *Contribution to the Analysis and Control of Distributed Parameter Systems with Application to (Bio)chemical Processes and Robotics*. Technical report, Thèse d'Aggrégation de l'Enseignement Supérieur, UCL, Belgium, 1994.

[31] D. Dochain and M. Perrier. Asymptotic observers for fixed bed reactors. *Proc. ACC'93*, pages 1179–1183, 1993.

[32] D. Dochain, J.P. Babary, and M.N. Tali-Maamar. Modelling and adaptive control of nonlinear distributed parameter bioreactors via orthogonal collocation. *Automatica*, 68:873–883, 1992.

[33] D. Dochain, N. Tali-Maamar, and J.P. Babary. Design of adaptive linearizing controllers for fixed bed reactors. *Proc. ACC*, pages 335–339, 1994.

[34] H. Aoufoussi, M. Perrier, J. Chaouki, C. Chavarie, and D. Dochain. Feedback linearizing control of a fluidized-bed reactor. *Can. J. Chem. Eng.*, 70:356–367, 1992.

[35] D. Dochain and B. Bouaziz. Approximation of the dynamical model of fixed bed reactors via a singular perturbation approach. *Proc. IMACS Int. Symp. MIM-S2'93*, pages 34–39, 1993.

[36] B. Bouaziz and D. Dochain. Control analysis of fixed bed reactors : a singular perturbation approach. *Proc. ECC'93*, pages 1741–1745, 1993.

[37] L. Chen. *"Modelling, Identifiability and Control of Complex Biotechnological Processes"*. PhD thesis, Université Catholique de Louvain, Belgium, 1992.

[38] G.R. Gavalas. *"Nonlinear Differential Equations of Chemically Reacting Systems"*. Springer Verlag, Berlin, 1968.

[39] M. Fjeld, O.A. Asbjornsen, and K.J. Astrom. Reaction invariants and their importance in the analysis of eigenvectors, state observability and controllability of the continuous stirred tank reactor. *Chem. Eng. Sci.*, 29:1917–1926, 1974.

[40] V. VanBreusegem and G. Bastin. A singular perturbation approach to the reduced order dynamical modelling of reaction systems. *Submitted for publication*, 1995.

[41] H. Kwakernaak and R. Sivan. *"Linear Optimal Control Systems"*. John Wiley, New York, 1972.

[42] G. Stephanopoulos and K.-Y. San. Studies on on-line bioreactor identification. *Biotechnol. Bioeng.*, 26:1176–1188, 1984.

[43] S.H. Lee, P. Tsobanakis, J.A. Phillips, and C. Georgakis. Issues in the optimization, estimation and control of fed-batch bioreactors using tendency models. *Proc. ICCAFT 5/IFAC-BIO 2, Keystone, Colorado*, 1992.

[44] Y.J. Yoo, J. Hong, and R.T. Hatch. Sequential estimation of states and kinetic parameters and optimization of fermentation processes. *Proc. ACC*, 2:866–871, 1985.

[45] G. Caminal, F.J. Lafuente, J. Lopez-Santin, M. Poch, and C. Sola. Application of the extended kalman filter to identification of enzymatic deactivation. *Biotechnol. Bioeng.*, 24:366–369, 1987.

[46] G. Acuna, E. Latrille, and G. Corrieu. Biomass estimation using neural networks and the extended kalman filter. *Proc. 6th Int. Conf. Computer Appl. in Biotechnol.*, pages 209–212, 1995.

[47] L. Ljung. Asymptotic behavior of the extended kalman filter as a parameter estimator for linear systems. *IEEE Trans. Aut. Cont.*, 24:36–50, 1979.

[48] G. Bastin and J. Levine. On state reachability of reaction systems. *Proc. 29th CDC*, pages 2819–2824, 1990.

[49] D. Dochain and L. Chen. Local observability and controllability of stirred tank reactors. *J. Process Control*, 2 (3):139–144, 1992.

[50] K.S. Narendra and A.M. Annaswamy. *"Stable Adaptive Systems"*. Prentice-Hall, Englewood Cliffs, NJ, 1989.

[51] R. Marino. Adaptive observers for single output nonlinear systems. *IEEE Trans. Aut. Cont.*, 35:1054–1058, 1990.

[52] D. Dochain. *"On-line Parameter Estimation, Adaptive State Estimation and Adaptive Control of Fermentation Processes"*. PhD thesis, Université Catholique de Louvain, Belgium, 1986.

[53] Y. Pomerleau and M. Perrier. Estimation of multiple specific growth rates in bioprocesses. *AIChE J.*, 36(2):207–215, 1990.

[54] Y. Pomerleau. *"Modélisation et Contrôle d'un Procédé Fed-batch de Culture des Levures à Pain Saccharomyces cerevisiae"*. PhD thesis, Ecole Polytechnique de Montréal, Canada, 1990.

[55] M. Perrier and D. Dochain. Evaluation of control strategies for anaerobic digestion processes. *Int. J. Adaptive Cont. Signal Proc.*, 7 (4):309–321, 1993.

[56] D. Dochain and M. Perrier. Adaptive linearizing control of activated sludge processes. *Proc. Control Systems'92*, pages 211–215, 1992.

[57] A. Holmberg. A microprocessor-based estimation and control system for the activated sludge process. *In A. Halme (Ed.), Modelling and Control of Biotechnical Processes*, Pergamon:111–120, 1983.

[58] M. Khone. Practical experiences with a new on-line bod measuring device. *Env. Technol. Letters*, 6:546–555, 1985.

[59] S. Marsili-Libelli. Modelling, identification and control of the activated sludge process. *Advances in Biochemical Enginering/Biotechnology*, 38:90–148, 1989.

INPUT SEQUENCES
FOR NONLINEAR MODELING

R.K. PEARSON*
Institut für Automatik, Eidg. Techn. Hochschule Zürich
CH-8092 Zürich, Switzerland
email: pearson@aut.ee.ethz.ch

1   Introduction

Empirical modeling represents a practical, popular approach to the development
of the dynamic models required for nonlinear model-based process control. This
approach requires the selection of a nonlinear model structure, the determination of
dynamic order parameters, and the estimation of unknown model coefficients. The
results we obtain at each of these steps can depend strongly on the input/output
data on which they are based. This paper presents a brief but detailed discussion
of input sequences for nonlinear empirical model development, beginning with the
question of what constitutes a "good" input sequence. In particular, the following
three criteria are proposed:

(a) effectiveness in model structure discrimination;

(b) effectiveness in model parameter determination;

(c) conformance to practical constraints.

Examples are presented to illustrate these criteria and the differences between them;
one of the points demonstrated by these examples is that compromises must generally be made between input sequence performance with respect to these three
criteria. For simplicity, this paper restricts consideration to the problem of input sequence design for single-input, single-output nonlinear model identification;
multivariable problems are certainly of practical interest, but the SISO problem is
challenging enough to be worthy of a separate treatment first.

---

*on leave from the DuPont Company

R. Berber and C. Kravaris (eds.), Nonlinear Model Based Process Control, 599-621.
© 1998 *Kluwer Academic Publishers.*

600

2   Fundamental issues

The first of the "goodness criteria" listed above — structure discrimination effectiveness — refers to the ability of a given input sequence or class of input sequences
to distinguish between different models or model classes. A simple example will
be presented in Section 4 to illustrate this criterion in practical terms. The second
criterion — parameter estimation effectiveness — is similarly defined as the ability
of a given input sequence or class of input sequences to provide sufficiently infor-
mative responses for accurate estimation of model parameters. A brief summary
of a detailed case study [13, 14] will be presented in Section 5 that illustrates the
difference between this criterion and structure discrimination effectiveness. The
third effectiveness criterion — conformance to practical constraints — is frequently
in conflict with the other two. In particular, the examples discussed in Sections
4 and 5 both illustrate that input sequence effectiveness with respect to the first
two criteria *generally* (though not always) degrades as input sequences become "less
lively" in character. This observation is important because it is often argued that
"lively" input sequences (e.g., "white noise" sequences) are undesirable in practice
because of the unrelenting demands they make on control actuators. To quantify
this practical consideration, Heemstra *et al.* [6] define the *friendliness index f* for
an input sequence $\{u(k)\}$ of fixed length N as:

$$ f \;=\; 1 - \frac{n_T}{N-1}. $$

Here, n_T is the number of transitions (i.e., changes in value) that $\{u(k)\}$ exhibits,
a quantity that varies between 0 (for constant sequences) and $N-1$ (for typical
"white noise" sequences that change value at every sampling time). Thus, f varies
between 0 and 1, with large values of f corresponding to sequences that are "less
demanding of control actuators" than those with small values of f. This measure of
"plant friendliness" is particularly appropriate for the random step input sequences
introduced in Section 3.

To manage the "trade-offs" between these three effectiveness criteria, we have
the following four basic input sequence design parameters:

(a) the sequence length N;

(b) the sequence range $[u_-, u_+]$;

(c) the sequence "shape";

(d) the sequence distribution on $[u_-, u_+]$.

The first two of these criteria are self-explanatory, while the discussions presented
in Section 3 will attempt to clearly define and distinguish the last two. As a general
rule, we usually attempt to maximize the sequence length N, subject to practical
constraints. Typically, we sample uniformly at a constant sampling rate T, so the
maximum sample size N reflects the "longest tolerable identification experiment,"
usually defined on the basis of the expected "degree of interference with normal

operation" to be caused by our efforts. On the other hand, note that the longer our identification experiment lasts, the higher the probability that some form of "unmodelled disturbance" will occur (e.g., changes in feed composition, upstream or downstream process equipment failures, demand or product grade changes, etc.). The key point is that, while it is generally desirable to "acquire as much data as possible," practical constraints will usually limit the quantity of (good) data we can expect.

The second design criterion — input sequence range — is also generally dictated by practical concerns. That is, to maximize discrimination and parameter estimation effectiveness, it is generally desirable to maximize the input sequence range $[u_-, u_+]$. This strategy is particularly valid for nonlinear model identification, since "small amplitude" inputs generally do not excite process nonlinearities strongly, making the nonlinear character of the process dynamics difficult to identify. In fact, this idea has been used explicitly by Wigren [15] in the development of a two-stage identification procedure for Wiener models. There, the model of interest consists of the cascade connection of a linear, time-invariant dynamic model followed by a static (i.e., memoryless) nonlinearity. In the case of "smooth" nonlinearities, the Wiener model's response to input sequences of sufficiently small amplitude is not strongly influenced by the nonlinearity. Thus, Wigren's two-step approach is as follows:

(a) excite the process with a "low amplitude" sequence, and use the responses to identify the linear subsystem of the Wiener model;

(b) excite the process with a "high amplitude" sequence, use the previous identification results to predict the input to the static nonlinearity, and estimate the static nonlinearity from this predicted input and the observed process response.

In practice, the range of our input sequence is usually limited by physical actuator limits, concerns over the production of poor quality product during the identification experiment, and inherent operational safety limitations.

3   Shape *vs.* distribution

Given the practical scenario just described, it appears that the two input sequence design parameters over which we have the greatest control are likely to be the last two: the "shape" of the input sequence, and its distribution over the admissible range $[u_-, u_+]$. Because these design parameters sound somewhat alike, it is important to distinguish clearly between them. This distinction is perhaps clearest for stochastic input sequences, where the "shape" of a sequence represents its *dependence structure*, partially characterized by the autocorrelation function, while the distribution describes the probability that $u(k)$ assumes any value in a specified sub-interval of $[u_-, u_+]$. Alternatively, the "shape" of the sequence $\{u(k)\}$ may be viewed as its "frequency content," while the distribution may be viewed as the uniformity (or non-uniformity) with which the range $[a, b]$ of admissible inputs is

602

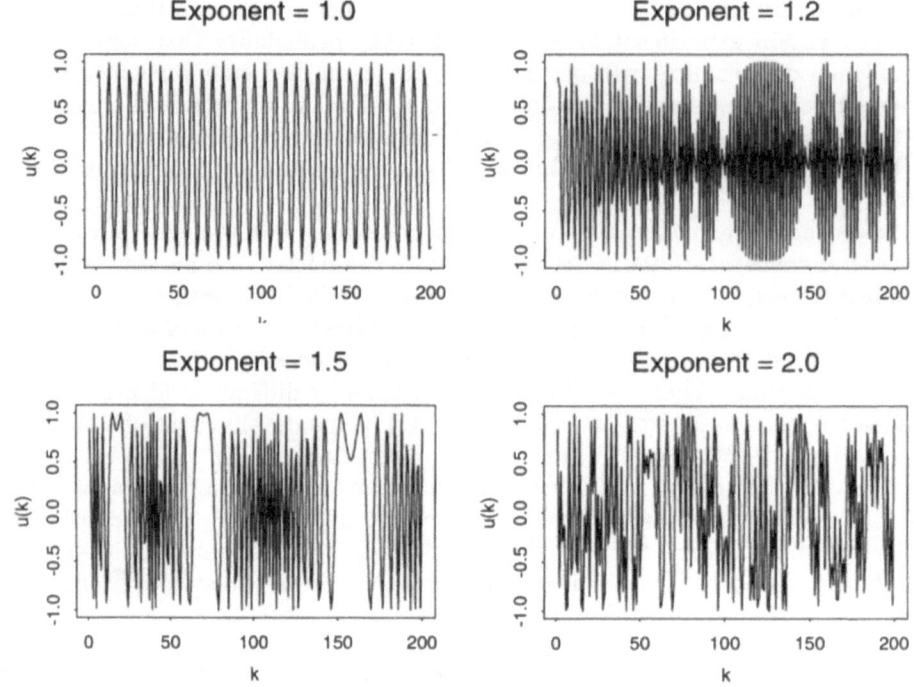

Figure 1: Four sine-power sequence examples

covered. To clarify this distinction, the following two subsections describe two input sequence examples — one deterministic and one stochastic — for which both criteria can be quantified. These sequences will then be used in subsequent discussions of structure discrimination and parameter estimation.

3.1 SINE-POWER SEQUENCES

The *sine-power sequences* constitute an interesting class of input sequences that clearly illustrate the difference between the distribution of values — which is essentially the same for all members of this class — and the "shape" or "frequency content" of the sequence. Specifically, the sine-power sequences $\{u(k)\}$ are defined by:

$$u(k) \;=\; A \sin\left[(\omega k)^{\lambda}\right]. \tag{1}$$

Here, A is a positive number defining the magnitude of the input sequence (specifically, note that $|u(k)| \leq A$ for all k), ω is a "frequency-like" parameter that determines the time scale (in discrete sample times) on which $u(k)$ varies, and λ is a "shape parameter" that determines the overall character of the sequence. Four

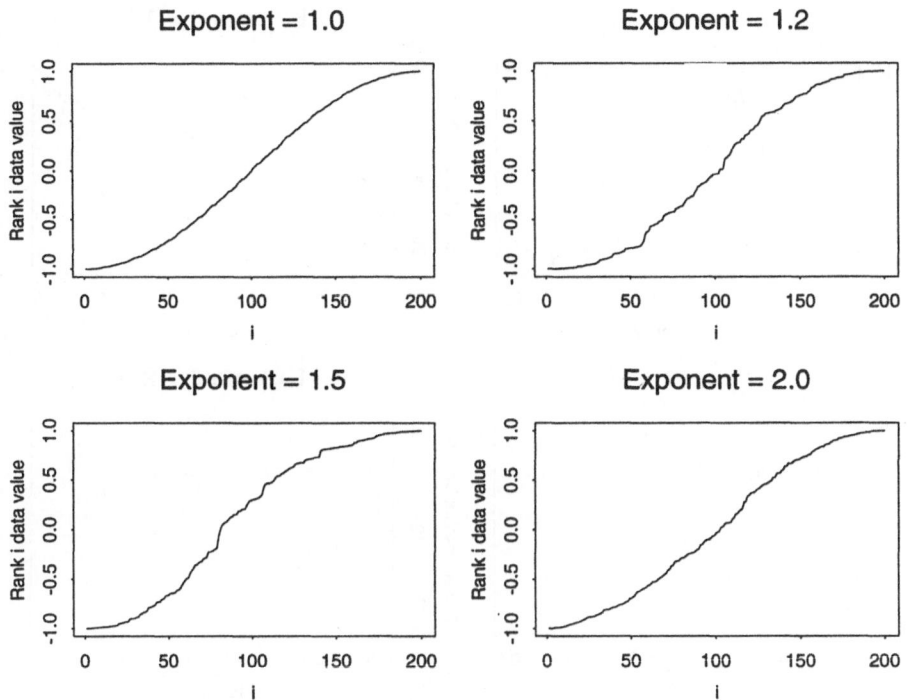

Figure 2: Distributions of sine-power sequence values

members of this family of sequences are shown in Fig. 1 for $\lambda = 1.0$, 1.2, 1.5, and 2.0; in all cases, $A = 1$ and $\omega = 1$ to facillitate comparison.

The *distribution* of the input sequence $\{u(k)\}$ refers to its "coverage" of the interval $[u_-, u_+]$. A useful graphical representation of the distribution of a sequence $\{u(k)\}$ — either deterministic or stochastic — is to plot the *rank-ordered* sequence $\{u_{(i)}\}$ against the rank i. Specifically, $u_{(i)}$ denotes the "i^{th} smallest" element of the input sequence when it is re-ordered as:

$$u_{(1)} \leq u_{(2)} \leq \cdots \leq u_{(N-1)} \leq u_{(N)}.$$

As a specific example, note that if $\{u(k)\}$ were uniformly distributed over the admissible range $u_- \leq u(k) \leq u_+$, the rank-ordered samples $\{u_{(i)}\}$ would be approximately evenly spaced, so the plot of $u_{(i)}$ *versus* i would be approximately linear. In fact, this construction corresponds very closely to the *quantile-quantile* plot for uniformly distributed random variables [3]. Plots of $u_{(i)}$ *versus* i are shown in Fig. 2 for the four sine-power sequences shown in Fig. 1; note that the shape of these plots exhibit only minor variations, in contrast to the plots of the data sequences themselves. This observation illustrates that for sine-power sequences, the distribution of values over the admissible range is essentially independent of the parameter γ.

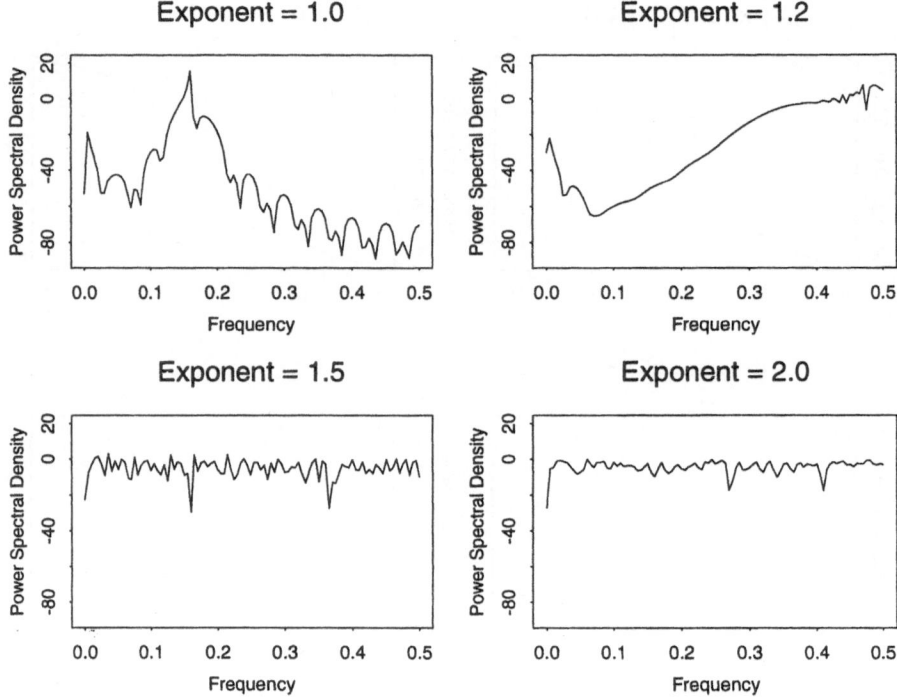

Figure 3: Estimated sine-power spectra

In contrast, the "frequency content" of the sine-power sequences varies quite strongly as a function of the exponent γ. This variation may be seen by considering the power spectra of these sequences, which are plotted in Fig. 3. Specifically, the four plots in this figure show the periodogram estimates of the power spectral density [10] computed for these sequences. For stochastic input sequences, it is important to note that the periodogram is a poor estimate of the power spectrum (in particular, it is an *inconsistent* estimator [12], implying that its variance does *not* approach zero as $N \to \infty$); here, however, because we are dealing with "noise-free" deterministic input sequences, comparing periodograms does yield useful insight into the differences in frequency content of these sequences. For example, when $\gamma = 1.0$, the input sequence is sinusoidal with frequency $f = 1/\pi \simeq 0.318$, and the corresponding periodogram exhibits a broad peak centered at this frequency, as seen in the upper left plot in Fig. 3 (the breadth of this peak reflects the finite sequence length $N = 200$ considered here). In contrast, as γ increases, the high-frequency content of the sine-power sequence increases and the power spectrum becomes essentially uniform (i.e., like that of "white noise") for sufficiently large γ. The key point is that these differences are pronounced, in contrast to the slight differences in distribution between the four sine-power sequences compared above.

3.2 RANDOM STEP INPUTS

A "standard" input sequence for the development and characterization of identification algorithms is "white noise," corresponding to an *independent, identically distributed* (i.i.d.) sequence $\{u(k)\}$. If the distribution of possible values for $u(k)$ is continuous (e.g., uniform, Gaussian, etc.), then the probability that $u(k) = u(j)$ for $j \neq k$ is precisely zero, implying the sequence changes value at every time step, so the expected value of the "friendliness index" f introduced in Section 1 is zero. To obtain sequences with nonzero expected values for f, consider the following class. First, suppose $z(k)$ is an i.i.d. sequence defined for $k = 1, 2, ..., N$, with any valid probability distribution, and let p be any real number in the range $0 \leq p \leq 1$. Define the input sequence $\{u(k)\}$ by $u(1) = z(1)$ and, for $k = 2, ..., N$:

$$u(k) = \begin{cases} z(k) & \text{with probability } p \\ u(k-1) & \text{with probability } 1 - p. \end{cases}$$

Here, p represents the *switching probability* for this sequence, since $u(k) \neq u(k-1)$ with probability p. Note that as $p \to 1$, the sequence $\{u(k)\}$ approaches the underlying i.i.d. sequence $\{z(k)\}$, while as $p \to 0$, the sequence $\{u(k)\}$ approaches the constant limit $u(k) = z(1)$ for all k.

This class of input sequences has been considered previously for uniformly distributed i.i.d. sequences $\{z(k)\}$ [7]; here, we generalize this input class by considering sequences $\{z(k)\}$ drawn from the beta distribution with parameters a and b [9, ch. 25]. On the interval $[0, 1]$, the probability density function for this distribution is given by:

$$p(x) = Cx^{a-1}(1-x)^{b-1},$$

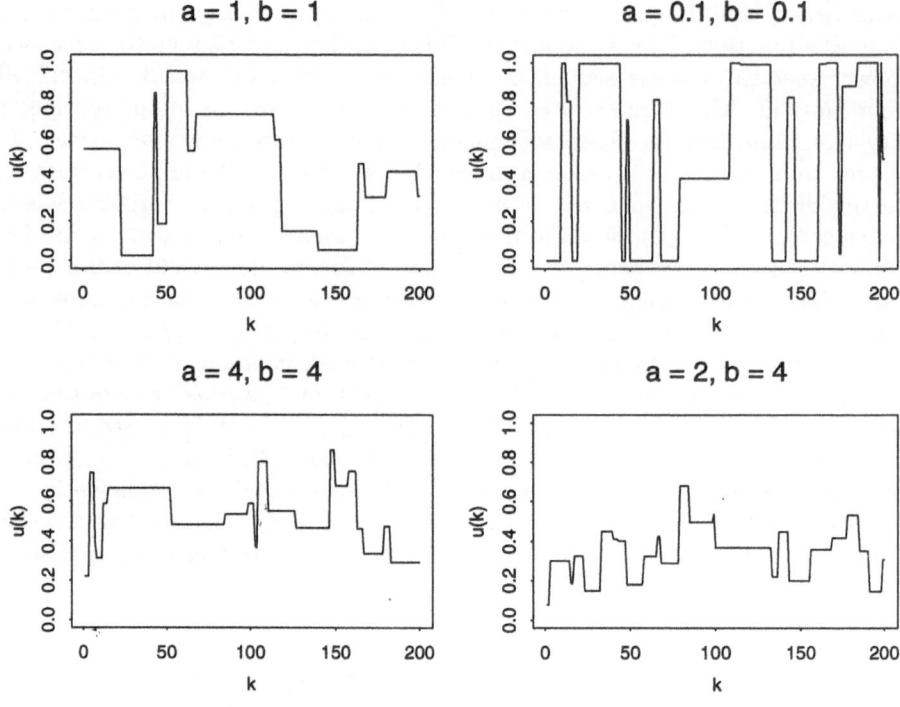

Figure 4: The random step input sequences, $p = 0.10$

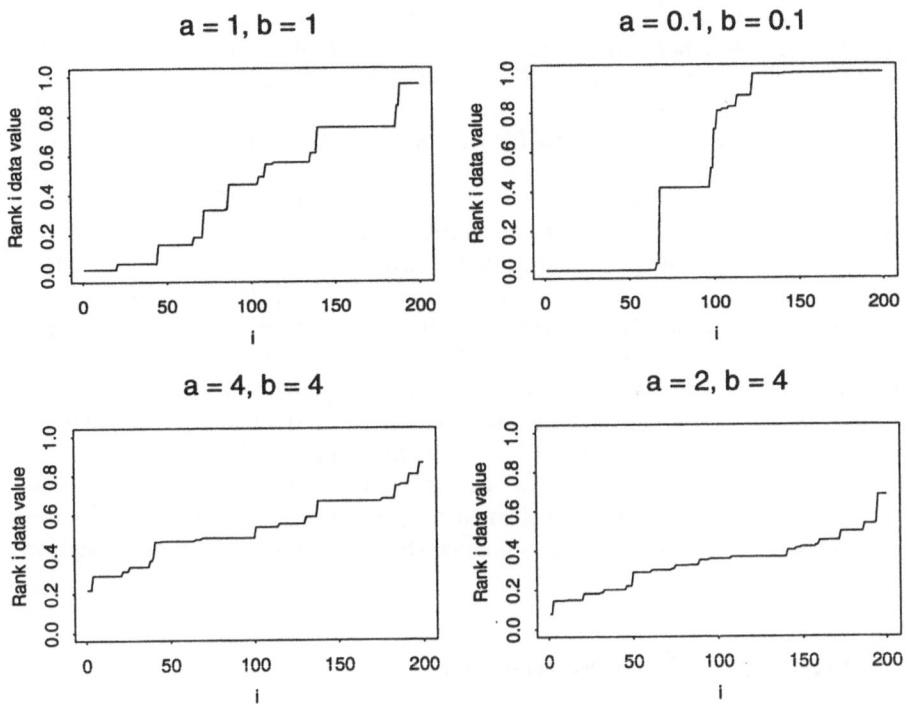

Figure 5: Distributions of the random-step inputs

where C is a normalization constant that depends on a and b. For $a = b = 1$, we recover the uniform distribution, while for $a = b \to 0$ the beta distribution approaches a degenerate limit concentrated at the two points $x = 0$ and $x = 1$. Similarly, for $a = b \to \infty$, the beta distribution approaches a Gaussian limit, while if $a \neq b$, the distribution is asymmetric. Here, four specific members of this family are considered, designated as follows: "Dxyz" indicates a sequence defined by distribution D and switching probablity $p = x.yz$ (e.g., U010 is uniformly distributed with a switching probability $p = 0.10$). The four distributions considered here are the uniform distribution U (corresponding to $a = b = 1$), a "nearly binary" distribution B (corresponding to $a = b = 0.1$), a "nearly Gaussian" distribution G (corresponding to $a = b = 4$), and an asymmetric distribution A (corresponding to $a = 2$, $b = 4$). Examples of these four sequences for $p = 0.10$ are shown in Fig. 4.

As with the deterministic input sequences considered previously, it is possible to gain some useful insights into the distribution of input sequence values over the admissible input range by plotting the rank-ordered data sequence $\{u_{(i)}\}$. These rank-ordered sequences are shown in Fig. 5 for the data sequences shown in Fig. 4. In contrast to the sine-power sequences considered above, these plots provide an indication of the pronounced distributional differences between the four random step input sequences considered here. Conversely, these four sequences all have *the same* dependence structure or "frequency content," again in marked contrast to the sine-power input sequences considered above. In particular, it can be shown fairly simply [13] that the autocorrelation function for these sequences is given by:

$$R_{uu}(m) \quad = \quad E\{u(k)u(k+m)\} \quad = \quad (1-p)^m \sigma^2,$$

where σ^2 is the variance of the i.i.d. sequence $\{z(k)\}$ on which the random step sequence is based. Thus, if we scale the input sequences to have the same variance (e.g., if we adjust the range $[u_-, u_+]$ to accomplish this objective), the autocorrelation function $R_{uu}(m)$ will be *independent of the underlying distribution*. The significance of this observation is that, for stochastic input sequences, the power spectral density is defined as the discrete Fourier transform of $R_{uu}(m)$, so this observation implies the "frequency content" of these sequences is identical, regardless of the distribution. Thus, for this example, the beta distribution parameters a and b specify the distribution of input values over the range $[u_-, u_+]$, while the switching probability p specifies the "shape" or "frequency content" of these sequences. Also, note that since p specifies the transition probability, the "friendliness index" for these sequences is simply $f = 1 - p$ since the expected number of transitions is $n_T = (N - 1)p$.

4 Structure discrimination

One of the critical issues in empirical modeling is that of model structure selection, particularly in the case of nonlinear models. This point is discussed at length in the reference [13], but the key corollary here is that different input sequences can vary greatly in their ability to discriminate between different models. In particular, if we are to decide whether "Model A" provides a better approximation to the dynamic behavior of a process than "Model B," it is important that the input sequences on which this determination is based yield substantially different responses when applied to these models. This point is illustrated in Figs. 6 and 7, which show the responses of four different models to the sine-power input sequence with $A = 1$, $\omega = 1$, and either $\lambda = 1.0$ or $\lambda = 1.6$. In particular, the four models compared here are special cases of the following "AR-Volterra" model:

$$\begin{aligned} y(k) \quad = \quad & a_1 y(k-1) + b_0 u(k) + b_1 u(k-1) \\ & + d_{01} u(k)u(k-1), \end{aligned} \tag{2}$$

with $a_1 = 0.8$, $b_0 = 1.0$, and the following values for the coefficients b_1 and d_{01}:

Model 1: $b_1 = 0$, $d_{01} = 0$;

Model 2: $b_1 = 1$, $d_{01} = 1$;

Model 3: $b_1 = 1$, $d_{01} = -1$;

Model 4: $b_1 = -1$, $d_{01} = 1$.

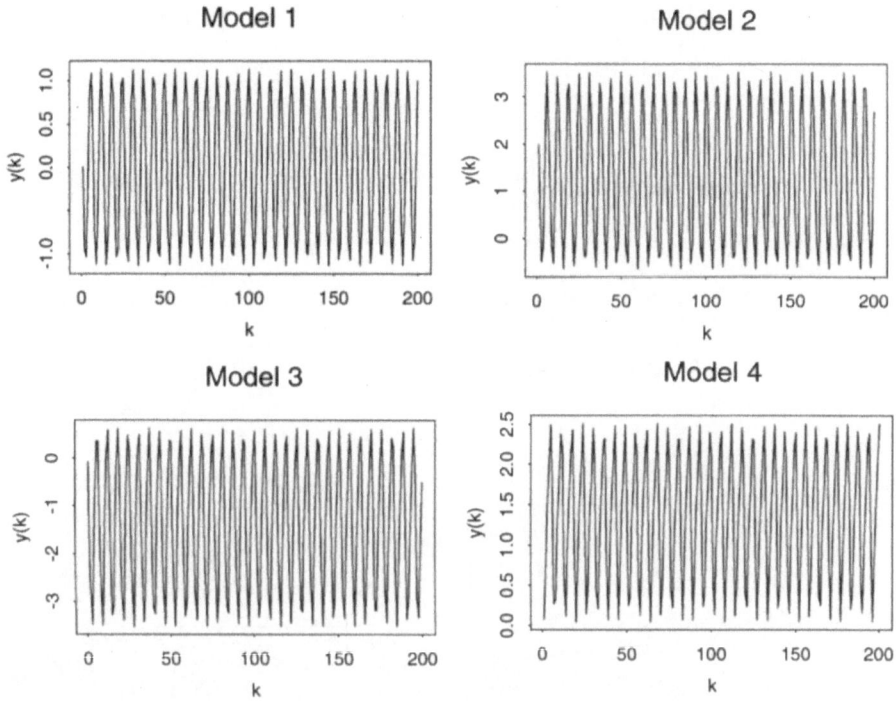

Figure 6: Four sine-power responses, $\lambda = 1.0$

Fig. 6 shows the responses of these models for $\lambda = 1$, corresponding to simple sinusoidal excitation. Aside from the the different ranges of these responses, there is essentially nothing to distinguish them; in contrast, this fact is immediately obvious from even a cursory glance at the responses for $\lambda = 1.6$ shown in Fig. 7. Since the lengths, ranges, and distributions of these two different sequences is identical, it is clear that the pronounced differences in their ability to discriminate between the four models considered here lies in their "shape" or "frequency content." The example presented in the next section illustrates that input sequence distribution can also profoundly influence our ability to discriminate between candidate models.

5   Parameter estimation

The following example is a synopsis of a short case study, described in more detail in the references [13, 14]. The basis of this case study is a simple CSTR model described by Eaton and Rawlings [4], for which an exact discretization is possible. The following subsections briefly describe this model, its discretization, and the results of empirical model identification experiments using the random step input sequences described above.

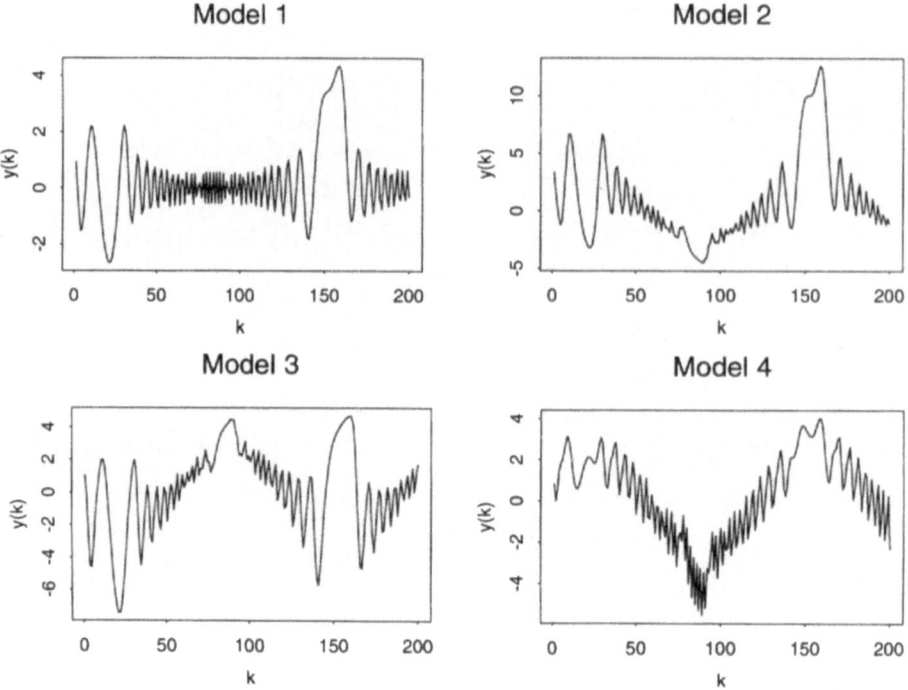

Figure 7: Four sine-power responses, $\lambda = 1.6$

5.1 THE REACTOR MODEL

The basis of this case study is the simple CSTR model considered by Eaton and Rawlings [4]:

$$\frac{dy}{dt} = -h[y^2 + 2\mu y - 2d\mu].$$ (3)

Here, $y(t)$ represents the reactant concentration in the reactor at time t, and $\mu(t) = u(t)/2hV$ is the input flow rate ($u(t)$), re-scaled to simplify subsequent results. If we assume that the manipulated variable $\mu(t)$ changes only at discrete sampling instants $t_k = kT$, for some fixed sampling interval T, Eq. (3) can be solved analytically for the reactor concentration $y(k)$ at time t_k, given the previous concentration $y(k-1)$ and the constant input $\mu(k-1)$ maintained during the interval $t_{k-1} \le t < t_k$. Details are given in [13], but the resulting recursion relation is:

$$
\begin{aligned}
y(k) &= \frac{[1 - \tau(k-1)\mu(k-1)]y(k-1) + 2d\tau(k-1)\mu(k-1)}{1 + \tau(k-1)[y(k-1) + \mu(k-1)]} \\
\tau(k-1) &= \frac{\tanh[hT\zeta(k-1)]}{\zeta(k-1)}, \\
\zeta(k-1) &= \sqrt{\mu^2(k-1) + 2d\mu(k-1)}.
\end{aligned}
$$ (4)

In the specific example considered here, $h = 1.5$ liters per mole-hour, $V = 10.5$ liters, and $d = 3.5$ moles per liter. The sampling rate was $T = 0.1$ hour, and the input range considered was 0.5 to 5.0 liters per hour; while this range is wider than that considered by Eaton and Rawlings, this change was made to enhance the nonlinearity of this model without losing its simplicity.

Two points are worth emphasizing here. First, the exact discretization considered here is *not* equivalent to the Euler discretization obtained by replacing derivatives with first differences, a procedure which does preserve the basic structure of the continuous-time model but only *approximates* the response to piecewise constant inputs. The second point is that the question of how different discretizations relate to the original continuous-time model is somewhat complex even in the linear case, where it is known that if the continuous-time model has m zeros and $n > m$ poles, the discretized model will exhibit non-minimum phase behavior for sufficiently small sampling intervals T, even if the original system is minimum phase [1]. Analogous results have been developed for the discretization of nonlinear continuous-time systems, with the relative degree of the original nonlinear system playing the role of the pole excess $n - m$ for linear systems [11]. Here, this particular difficulty does not arise because the Eaton-Rawlings CSTR model exhibits first-order dynamics (i.e., it is of relative degree 1).

No.	$\nu(k-1)$	Type
0	0	Affine (reference case)
1	$\mu^2(k-1)$	Hammerstein
2	$\mu^2(k-2)$	Modified Hammerstein
3	$\mu(k-1)\mu(k-2)$	AR-Volterra
4	$\mu(k-1)y(k-1)$	bilinear, diagonal
5	$\mu(k-1)y(k-2)$	bilinear, superdiagonal
6	$\mu(k-2)y(k-1)$	bilinear, subdiagonal
7	$\mu(k-2)y(k-2)$	bilinear, diagonal
8	$y^2(k-1)$	controlled logistic eq.
9	$y^2(k-2)$	additive NARX
10	$y(k-1)y(k-2)$	non-additive NARX

Table 1: The nonlinear model term $\nu(k-1)$

5.2 NONLINEAR EMPIRICAL MODELS

To illustrate some of the key practical issues that arise in nonlinear empirical modeling, eleven different special cases of the following general model form:

$$\begin{aligned} y(k) &= y_0 + \alpha y(k-1) + \beta u(k-1) + \\ &\quad \gamma \nu(k-1), \end{aligned} \tag{5}$$

were considered as approximations to the exact discrete-time dynamics defined in Eq. (4). Here, y_0, α, β, and γ are model parameters to be estimated from input-output data, while the term $\nu(k-1)$ is either identically zero, giving an affine reference model, or one of the ten nonlinear terms listed in Table 1. For a more detailed discussion of the different model types listed there, see [13]; the main point here is that both the qualitative behavior of these models and their suitability to different model-based control system design procedures varies greatly. The primary motivation for choosing this particular set of models is that these choices for $\nu(k-1)$ then represent all possible nonlinear terms in a NARMAX model of the form:

$$y(k) = F(y(k-1), y(k-2), \mu(k-1), \mu(k-2)),$$

if $F : R^4 \to R$ is restricted to a second-order polynomial in its arguments. This particular model form is a natural choice for a "moderate complexity" nonlinear model that could be identified and simplified by stepwise regression techniques [2].

5.3 MODEL STRUCTURE SELECTION

Fig. 8 shows box-plots [8] of the relative RMS prediction errors obtained for 100 simulation runs, for each of four different input sequences and three different model

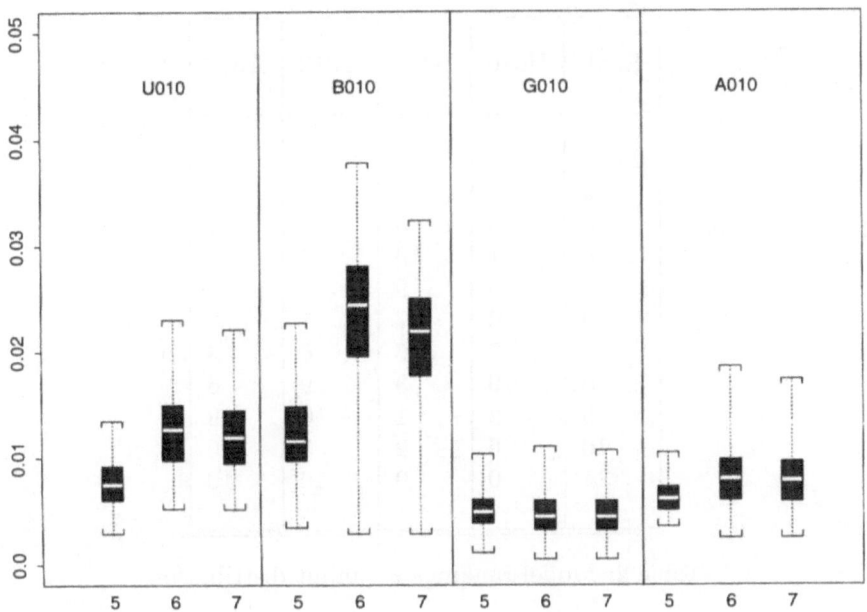

Figure 8: Prediction errors, Models 5, 6, and 7

Rank	U010	B010	G010	A010
1	8	8	8	8
2	4	4	4	4
3	5	5	1	1
4	1	10	3	5
5	3	9	2	3
6	9	7	7	7
7	7	6	6	2
8	10	3	9	6
9	2	1	0	9
10	6	2	5	0
11	0	0	10	10

Table 2: Model rankings *vs.* input distribution

structures. Specifically, the numerical quantities summarized in these plots are the square root of the estimated prediction error variance divided by the mean response to each input sequence under consideration. The white horizontal line at the center of each box in this figure represents the median relative prediction error over the 100 simulations, while the upper and lower limits of the dark boxes correspond to the upper and lower quartiles (thus, 50% of the simulation runs resulted in normalized prediction errors lying in the dark boxes in Fig. 8). The horizontal lines at the top and bottom of each box-plot (connected by vertical dashed lines) represent, respectively, the maximum and minimum relative prediction errors observed. The horzontal axis is labeled by the model number given in Table 1 (5, 6, or 7 here), and the vertical lines divide the plot into four sections, each corresponding to a specific input sequence (U010, B010, G010, and A010, respectively). The main point of this plot is that some of these input sequences have essentially no ability to distinguish between model structures 5, 6, and 7 (e.g., the G010 input sequences), while others are much more effective (e.g., the U010 and B010 input sequences).

Despite these differences in model discrimination effectiveness, if we compare the results obtained for all eleven of the models listed in Table 1, certain general conclusions emerge. In particular, Table 2 gives the relative model rankings obtained when all eleven models are ordered by median RMS prediction error. These results illustrate the influence of input sequence distribution on model structure selection. Note that the "best" two models obtained from these identifications were consistently Model 8 (best) and Model 4 (second best), independent of the distribution of the input sequence. The relative rankings of the other nine model structures do vary with the choice of input sequence, dramatically so in some cases. For example, Model 5 is ranked "third best" on the basis of the U010 input sequence results, but

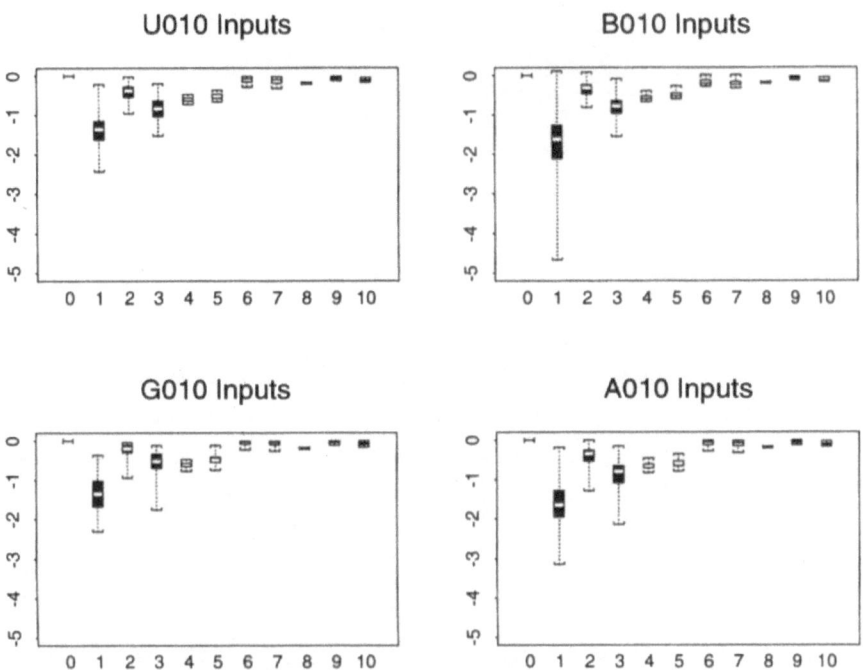

Figure 9: Estimated γ *vs.* input distribution

is "second worst" on the basis of the G010 input sequence results. Also, note that while Model 0 (the affine reference model) is consistently among the poorest of the models considered here, it is not uniformly *the* poorest. If we were identifying these models by directly minimizing the RMS prediction error, this observation would represent a contradiction, since the affine model may be viewed as a constrained special case of all of the other models considered here. The difference is that the models considered here are identified by the standard practice of minimizing the *one step-ahead* prediction error variance, and *not* the prediction error variance associated with the entire input sequence $\{u(k)\}$.

5.4   PARAMETER ESTIMATION

The estimated parameter values for a given model structure can also depend strongly on both the distribution and "shape" of the input sequence $\{u(k)\}$. This point is illustrated in Figs. 9 and 10, which show box-plots of the estimated model parameter γ *versus* the model number from Table 1, for several different input sequences. In particular, Fig. 9 compares the range of parameter estimates obtained for four different sequences of "constant shape" (i.e., constant switching probability $p = 10\%$), but four different distributions. The strongest dependence on input sequence

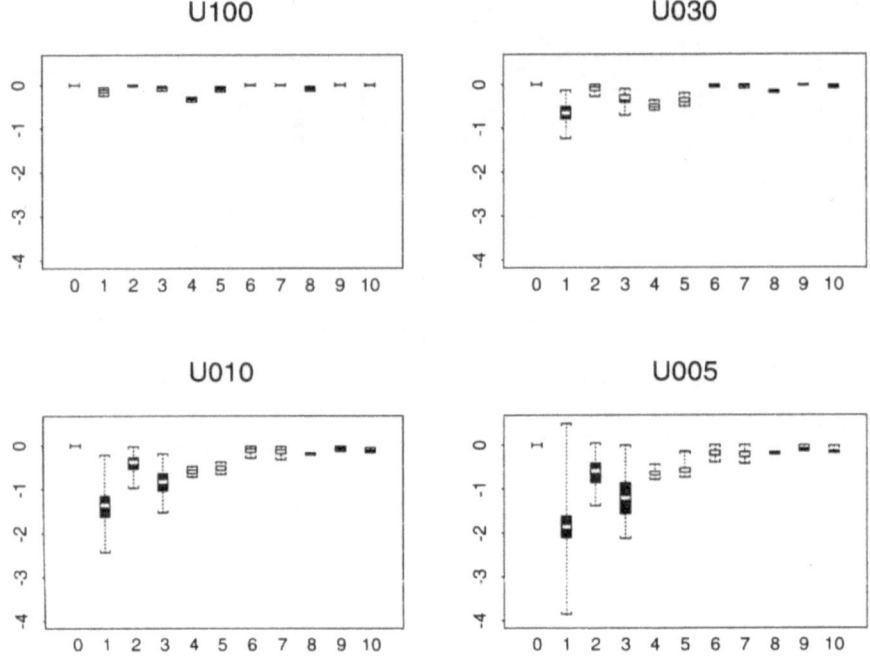

Figure 10: Estimated γ *vs.* switching probability p

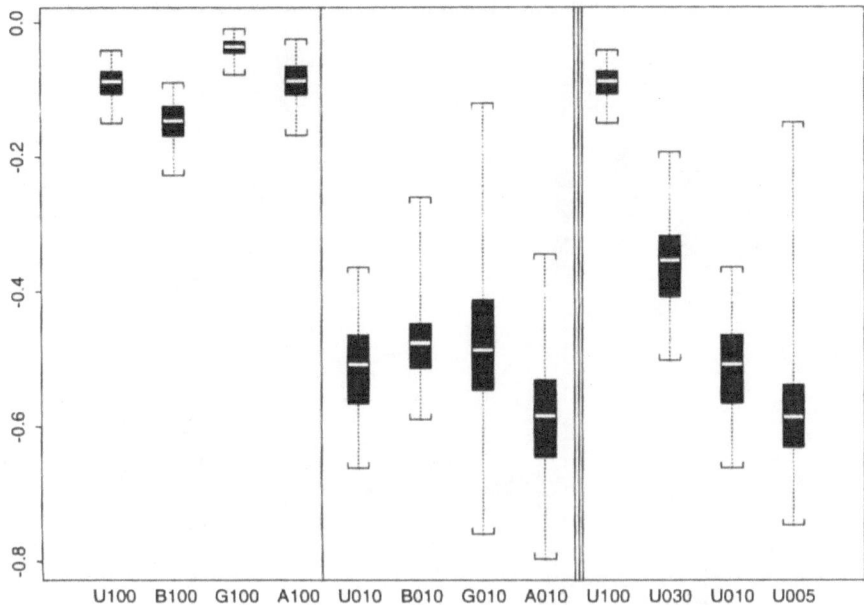

Figure 11: Estimated γ parameter, Model 5

appears for Model 1, which corresponds to a Hammerstein model consisting of a quadratic steady-state nonlinearity, followed by a first-order linear dynamic model. While the median estimate of γ is relatively consistent, the variability is large and depends strongly on the input sequence distribution (e.g., compare the results for the B010 and G010 input sequences). An even more pronounced dependence on input sequence is seen in Fig. 10, which presents box-plot summaries of the estimated γ parameters obtained for uniformly distributed random step inputs with different switching probabilities p. Here again, the greatest variability is exhibited by the Hammerstein model (Model 1), but here, almost all of the other models also exhibit parameter estimates that depend strongly on the switching probability p.

Both of these types of input sequence dependence may be seen somewhat more clearly by considering the dependence of a single model parameter on different input sequence choices. A typical result is shown in Fig. 11, which presents box-plot summaries of the estimated parameter γ for Model 5. Several observations are apparent from this figure. First, the pronounced dependence on switching probability p noted above is seen more clearly here. In particular, the four right-most plots summarize the parameter estimates obtained using the popular uniformly distributed random step input sequences, plotted against the switching probabilities $p = 100\%$, 30%, 10%, and 5%. It is clear from these results that both the median parameter esti-

618

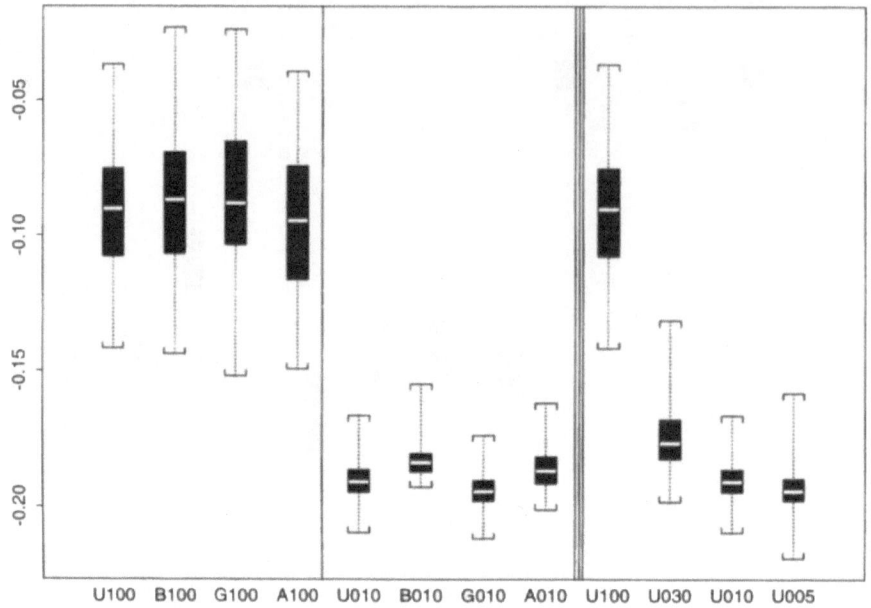

Figure 12: Estimated γ parameter, Model 8

mate and its variability increase significantly as $p \to 0$. In addition, distributional differences are also seen by comparing the first four plots and the middle four plots, each corresponding to four different distributions at the same switching probability. As a specific example, note the substantial differences between the estimated γ values obtained from the B100 and G100 input sequences.

Similar comparisons are presented in Fig. 12 for Model 8, consistently the "best" in terms of its relative RMS prediction error. As with Model 5, the dependence on switching probability p is much stronger than the distributional dependence, but it is particularly interesting to note that in this case, the nature of this dependence is opposite to that observed previously: parameter estimates obtained from input sequences with $p = 10\%$ are *more consistent* than those obtained from input sequences with $p = 100\%$. In fact, the same type of dependence is also observed for the y_0 and α parameters for this model, but not for any of the other model parameters considered in this case study. The reason for this unusual behavior appears closely related to the structural similarity of Model 8 to the *approximate discretization* obtained by applying Euler's method to the original Eaton-Rawlings reactor equations. The relation between these models is discussed further elsewhere [13, 14].

Finally, it is interesting to compare the estimation results obtained for the γ

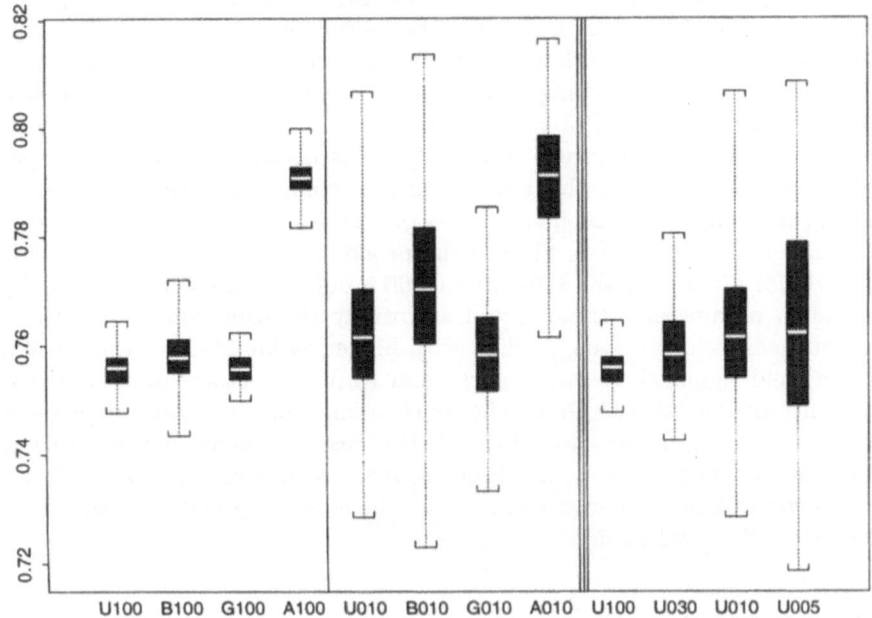

Figure 13: Estimated β parameter, Model 8

parameter in Model 8 with those obtained for the β parameter. These results are summarized in Fig. 13 and they illustrate two forms of pronounced input sequence dependence. The first and perhaps most obvious is the increase in parameter variability with decreasing switching probability observed for *all* parameters in all of the other models considered in this study. This point is seen most clearly in the right-most four plots, which show essentially no dependence of the median parameter estimate obtained from the 100 simulations compared here, but a consistent increase in the range of variability as $p \to 0$. Perhaps more interestingly, the second point to note in these box plots is the pronounced distributional dependence, particularly for $p = 100\%$: the U100 and G100 input sequences yield essentially the same median parameter estimates and essentially the same variability, the B100 input sequences yield the same median with higher variability, but the A100 input sequences yield significantly larger parameter estimates, clearly outside the range of variability of the others. Given the much weaker distributional dependence of the y_0, α, and γ parameters for Model 8, this result is somewhat surprising, but it emphasizes an important point: input sequences that are "good" or "bad" for *one parameter* of a particular model need not be equally "good" or "bad" for *other parameters of the same model.*

6 Summary

This paper has attempted to illustrate some of the key issues that arise in the selection of effective input sequences for nonlinear dynamic model identification. It was noted that "effectiveness" may be defined in at least three different ways, and that these "effectiveness measures" can be substantially in conflict. In particular, practical considerations often argue for using input sequences that "do not cause excessive control valve motion" — one reason for the great popularity of pseudo-random binary sequences (PRBS) in linear model identification [5, 10]. On the other hand, it is also well-known that "persistently exciting" input sequences are required to obtain consistent parameter estimates [10]. For nonlinear model identification, Section 1 of this paper suggested four "design parameters" to be considered in the design of input sequences satisfying both practical and theoretical constraints. It was argued that two of these parameters — the sequence length and the input range — are often constrained by practical factors much more than the other two: the distribution of values over the admissible range, and the "shape" or "frequency content" of the input sequence. The primary focus of this paper was on the influence of these last two design parameters on the three "effectiveness" criteria introduced at the beginning of the paper. Particular attention was given to the class of random step inputs because there it is possible to quantify the "plant friendliness" of the sequence in terms of the switching probability p. While the dependence of model discrimination effectiveness and parameter estimation effectiveness on p appears complicated and situation-dependent in general, it does appear that parameter estimation effectiveness generally decreases as p decreases, implying an inherent trade-off between "plant friendliness" and parameter estimation effectiveness.

References

[1] K.J. Astrom, P. Hagander, and J. Sternby, "Zeros of Sampled Systems," *Automatica*, v. 20, 1984, ppl 31 - 38.

[2] S.A. Billings and W.S.F. Voon, "A prediction-error and stepwise regression algorithm for nonlinear systems," *Int. J. Control*, v. 44, 1986, pp. 235 - 244.

[3] R.B. D'Agostino and M.A. Stephens, *Goodness-of-fit Techniques*, Marcel-Dekker, New York, 1986.

[4] J.W. Eaton and J.B. Rawlings, "Feedback control of chemical processes using on-line optimization techniques," *Comput. Chem. Eng.*, v. 14, 1990, pp. 469 - 479.

[5] K. Godfrey, *Perturbation Signals for System Identification*, Prentice-Hall, Englewood-Cliffs, NJ, 1993.

[6] D.G. Heemstra, F.J. Doyle, III, B.A. Ogunnaike, and R.K. Pearson, "The Identification of Nonlinear Plant Models for Process Control Using Tailored 'Plant-Friendly' Input Sequences," in preparation.

[7] E. Hernandez and Y. Arkun, "Control of Nonlinear Systems Using Polynomial ARMA Models," *AIChE J.*, v. 39, 1993, pp. 446 - 460.

[8] D.C. Hoaglin, F. Mosteller, and J.W. Tukey, *Fundamentals of Exploratory Analysis of Variance*, John Wiley and Sons, New York, 1991.

[9] N.L. Johnson, S. Kotz, and N. Balakrishnan, *Continuous Univariate Distributions*, vol. 2, John Wiley & Sons, New York, 1995.

[10] L. Ljung, *System Identification: Theory for the User*, Prentice-Hall, Englewood-Cliffs, NJ, 1987.

[11] S. Monaco and D. Normand-Cyrot, "Zero dynamics of sampled nonlinear systems," *Systems Control Lett.*, v. 11, 1988, pp. 229 - 234.

[12] A.V. Oppenheim and R.W. Schafer, *Digital Signal Processing*, Prentice-Hall, Englewood-Cliffs, NJ, 1975.

[13] R.K. Pearson, *Discrete-time Dynamic Models*, in preparation.

[14] R.K. Pearson and B.A. Ogunnaike, "A Case Study in Nonlinear Dynamic Model Identification," to appear in *Proc. DYCOPS-5*, Corfu, Greece, June, 1998.

[15] T. Wigren, "Recursive Prediction Error Identification Using the Nonlinear Wiener Model," *Automatica*, v. 29, 1993, pp. 1011 - 1025.

TOWARDS MULTISCALE DYNAMIC DATA RECONCILIATION

T. BINDER[1], L. BLANK[2],
W. DAHMEN[2], W. MARQUARDT[1]*
[1]*Lehrstuhl für Prozeßtechnik,*
[2]*Institut für Geometrie und Praktische Mathematik*
RWTH Aachen University of Technology,
D-52056 Aachen, Germany

Abstract. Although reconciliation of steady-state process data is routinely applied in industrial practice, the theoretical understanding of the problem and its adequate formulation in a dynamic setting is still not mature. Existing formulation approaches are based on stochastic filters, deterministic observers or mathematical programming techniques. In this contribution, we suggest a general problem formulation of dynamic data reconciliation based on the theory of ill-posed problems and their regularizations. It results in a large-scale dynamic optimization problem which requires efficient numerical solution methods in real-time under strict limitations of computational resources. We explore a novel mathematical framework for the discretization of the dynamic optimization problem and the solution of the discretized nonlinear programming problem based on multiscale approximation. The framework attempts to integrate signal processing and optimization finally leading to a fully adaptive and highly efficient numerical treatment which always provides the best possible estimate which is attainable in the allotted time period.

1. Introduction

Process measurements never reveal adequate information about the process state. This is due to the always limited number of process quantities accessible by measurement instruments as well as to unavoidable measurement

*Correspondence author. Phone: +49-241-806712, Fax: +49-241-8888326, E-mail: marquardt@lfpt.rwth-aachen.de

R. Berber and C. Kravaris (eds.), Nonlinear Model Based Process Control, 623-665.
© 1998 *Kluwer Academic Publishers.*

errors caused by imperfect instrumentation and signal processing. Hence, there are quantitative as well as a qualitative deficiencies which cannot be removed in an industrial environment by increased instrumentation or improved maintenance procedures due to cost and technical constraints. However, imperfect and incomplete measurement information will lead to inadequately performing or even unreliable processes. Therefore, there is a significant incentive to upgrade the information content in process measurements, since any increased understanding of the process state will enable improvement of process operations, monitoring and control. The upgraded measurement information can be directly used by operators to better manually control the process, it can be fed to some operator decision support system or to some automatic control system to lead to safer process operations of increased performance.

The measurement information can be upgraded, in principle, by means of process knowledge incorporated in some mathematical model which ideally predicts process behavior in some range of operating conditions. Such a model may either result from first principles or from a large set of previously recorded and reasonably correlated process data. Obviously, the model will never perfectly predict process behavior due to unmodeled dynamics, uncertain (and potentially time-varying) parameters, or unaccounted external disturbances. Model and measurements are incorporated in some algorithm to render an estimate of all measured process quantities as well as of all those unmeasured process quantities which are additionally required to fully describe the process behavior. The objective of these algorithms is therefore to remove any error from available measurements and to yield optimal estimates of the full process state as well as of unknown process parameters.

The problem stated has been treated extensively in various branches of mathematical and engineering science focusing on different problem formulations and objectives. The first notion of the problem in the context of chemical process operations has been given in 1961 by Kuehn and Davidson [26]. These authors formulated the problem of adjusting flow and temperature measurements on a crude oil distillation tower to satisfy steady-state material and energy balances. The problem of reconciling process measurements in order to obey steady-state material and energy balances as well as any other constraint has been subsequently called *data reconciliation*. The various problem formulations studied in recent years as well as the solution algorithms taking advantage of a specific problem structure are comprehensively summarized in a recent survey by Crowe [13]. Data reconciliation is nowadays routinely applied in industrial practice (see [70] for an industrial case study) relying on general-purpose commercial or in-house software packages.

A common assumption underlying all the established data reconciliation methods and software implementations is a stationary process. Therefore, the applicability of these methods is restricted to continuously operated processes during periods when the process is reasonably close to some steady-state. Since this assumption turns out to be very restrictive in many cases, the treatment of more general transient situations is an important extension of established technology as recently affirmed in an industrial assessment of the status of real-time process optimization [42].

So far, the reconciliation of dynamic process data by means of a dynamic process model has only received limited attention. Gertler and Almasy [34] seem to have been the first addressing the distribution of measurement error among process variables to satisfy dynamic mass and energy balances in a chemical engineering context. Explicit reference to *dynamic data reconciliation* (or the like) as an extension of the steady-state problem has been made only by a few researchers including Almasy [32, 34], Bagajewicz [5], Bequette [54], Biegler [2, 3], Darouach [20], Edgar [40, 41, 44, 48], as well as Himmelblau [35, 39] and coworkers. Due to the close relationship between dynamic data reconciliation and simultaneous parameter and state estimation in dynamic systems, there is a large body of related work in the chemical as well as, in particular, in the control engineering literature. One may distinguish deterministicly motivated observers (e.g. [52] for a review), stochastic filtering (e.g. [33, 37], or [52] for a review) and mathematical programming based estimators (e.g. [2, 12, 36, 55], and [52] for a review). Still, the theoretical understanding of dynamic data reconciliation, adequate formulations of the estimation problem, and the development of solution algorithms for large-scale industrial problems in process operations are still in their infancy.

This contribution does not attempt to give a comprehensive survey on the state of the art of dynamic data reconciliation. Such a survey would be quite difficult to focus due to a largely missing or at least not generally accepted problem statement on the one hand and due to the large body of literature on the closely related state and parameter estimation of dynamic systems on the other hand. The latter area has been summarized recently in an excellent survey paper by Muske and Edgar [52]. Instead, we suggest first a general problem formulation of dynamic data reconciliation in Section 2 which should reflect the requirements of realistic (industrial) scenarios not being biased by available solution approaches and their underlying assumptions. Relevant research issues will be pointed out. Section 3 suggests a novel mathematical approach to dynamic data reconciliation based on multiscale approximation. Some details on the numerical solution techniques in the multiscale framework will be given in Section 4. An illustrating example is provided in Section 5 to demonstrate the potential as well as current limitations of the novel approach.

2. Problem formulation

This section attempts to provide a concise problem statement of dynamic data reconciliation. It is based on previous work (e.g. see Section 1) but does not introduce restricting assumptions potentially needed for the derivation of a mathematical framework for problem solution. This general problem formulation should serve as a reference for future work where several simplifying assumptions need to be introduced to either permit theoretical analysis or to derive efficient solution algorithms taking advantage of a particular structure apparent in a restricted problem class.

Dynamic data reconciliation is *defined* as a procedure to optimally rectify the measurements taken from a process under transient conditions as well as to optimally adjust potential uncertainties in a mathematical process model (such as parameters or unmodeled disturbances) to render cleaned measurements which are consistent with predicted process dynamics. Typically, besides removing or at least smoothing unavoidable measurement error, an estimate of the process state, the unknown parameters and some additional quantities representing model uncertainty is achieved simultaneously. The measurement errors include (i) small random fluctuations caused by the irreproducibility of the measurement device, (ii) potentially time-varying offsets or biases due to miscalibration, incorrect installation or drift of the sensor, and (iii) gross errors resulting from discrete events such as malfunctioning measurement or signal processing devices on the one hand or from process upsets on the other hand [44].

The problem complexity ranges from situations, where the full process state is measured and the model is perfect, to those situations usually occurring in realistic applications, where only a small number of process quantities is accessible by instrumentation, and where there are model uncertainties of various kinds. In the first case, only rectification of erroneous measurements must be achieved, whereas combined state, parameter and other uncertainty estimates must be provided by the reconciliation algorithm in addition in the latter case. Obviously, measurements and model must reveal a certain degree of redundancy in the latter case in particular, in order to render improvement in the information content of the available measurements feasible.

In the sequel, we first discuss a priori knowledge to be coded in a process model. Then, the estimation problem will be set up and discussed.

2.1. PROCESS MODEL

The process model under consideration can be associated with different types of process quantities. They include

- the measured input variables $u \in (L_2)^{n_u}$, becoming dynamic forcing functions in the process model,
- the unmeasured inputs or time-varying parameters $\delta \in (L_2)^{n_\delta}$, leading to unknown forcing functions in the process model,
- the time-invariant parameters $\theta = [p^T, \pi^T]^T$ with $p \in \mathbb{R}^{n_p}$ and $\pi \in \mathbb{R}^{n_\pi}$ being the unknown and the known parameters respectively,
- the state variables $z \in (L_2)^{n_z}$, uniquely representing the process state at future times for given independent initial conditions, inputs and parameters, as well as
- the measurements $y \in (L_2)^{n_y}$, including all measured inputs u and some of the states z.

All the process quantities involved in the mathematical model are formally viewed as being real functions of continuous time t. Obviously, this is often not true for the really measured quantities which are subsequently denoted by $\check{\cdot}$. In most practical situations the measurements are sampled at discrete times. Usually (but not necessarily) there is a constant sampling rate for every measurement. In order to allow for the most general situation we assume that l_i measurements at given instants $t_i \in [\tau - T, \tau]$ are taken for every measured process quantity in some time interval $[\tau - T, \tau]$ of length T with τ being the current point in time. The values of the measured quantities $\check{Y}_{i,j}, j = 1, \ldots l_i, i = 1, \ldots n_y$ are catenated to the vectors $\check{Y}_i \in \mathbb{R}^{l_i}, i = 1, \ldots n_y$. The discrete measurements on the time interval $[\tau - T, \tau]$ form the vector $\check{Y} = [\check{Y}_1^T, \check{Y}_2^T, \ldots, \check{Y}_{n_y}^T]^T$. In contrast, $\breve{y}_i \in L_2$ denotes a real function of continuous time which suitably interpolates (or extrapolates at the boundaries) the discrete measurements $\check{Y}_i$ on the interval $[\tau - T, \tau]$. The interpolated (real) measurements form the vector $\breve{y} \in (L_2)^{n_y}$ which corresponds to the measurements y as predicted by the process model. This set-up allows the consideration of multi-rate sampling or even asynchronous sampling in a simple but formally sound manner. Furthermore, as we will see later, this continuous representation permits a more efficient numerical treatment of the estimation problem.

The mathematical model of the process is assumed to be given by the set of differential-algebraic equations

$$0 = f(\dot{x}, x, u, d, p) \tag{1}$$

and initial conditions

$$0 = b(\dot{x}_0, x_0, u_0, d_0, p) . \tag{2}$$

Here, $f : (L_2(\Omega))^{n_x} \times (L_2(\Omega))^{n_x} \times (L_2(\Omega))^{n_u} \times (L_2(\Omega))^{n_d} \times \mathbb{R}^{n_p} \to (L_2(\Omega))^{n_x}$, $\Omega = [\tau - T, \tau]$ represents the n_x differential-algebraic model equations whereas $b : \mathbb{R}^{n_x} \times \mathbb{R}^{n_x} \times \mathbb{R}^{n_u} \times \mathbb{R}^{n_d} \times \mathbb{R}^{n_p} \to \mathbb{R}^r, r \leq n_x$ refers to the set of initial conditions required to guarantee existence of the solution. In this notation the subscript 0 denotes the quantities at the left boundary of Ω, $\tau - T$.

The differential-algebraic system often has a more special structure [47]; in particular, it is of index one and of semi-explicit form in many cases; then r initial values are assigned to a subset of the states x at $t = \tau - T$ (see [65] for a more detailed treatment). The system of equations includes n_z equations for the state variables z as well as $n_x - n_z$ equations for some of the unmeasured time-varying inputs and parameters δ. Often, simple signal models, i.e. a sequence of one or more integrators, are assumed to model these quantities (i.e. [33] or [67] for an example in the context of batch reactor monitoring) when no physical explanation can be employed as for example in catalyst decay modeling [31]. The unknowns $x \in (L_2)^{n_x}$ denote an extended state vector containing the states z and the modeled quantities in the vector δ. The remaining unmodeled quantities in the vector δ are collected in the vector of unmodeled disturbances $d \in (L_2)^{n_d}$. For brevity, the known parameters π are not shown in the equations. For convenience, we assume that suitable reference values for the unmodeled disturbances and for the parameters are included in the model equations. Hence, the quantities d and p denote the difference between the reference values and the actual values; d and p are zero if the reference coincides with the actual value.

The number of equations is typically quite large; the Jacobian is always sparse reflecting the flowsheet topology on a coarse block matrix level and the structure of the unit model in every block matrix. More general models either including discontinuities or comprising also partial differential equations are possible [47] but will not be considered here in more detail.

The equations (1), (2) denote the nominal model of the process. Since the model is never perfect, there is always a mismatch between the real process and the model which needs to be taken into account explicitly. It should be noted that part of the model uncertainty has been already incorporated in the nominal model to the extent that it is transparent to the modeler by means of the unmodeled disturbances d as well as by the unknown constant parameters p. However, usually unstructured uncertainty is present in addition. The simplest model of unstructured uncertainty is an additive time-varying forcing function $w \in (L_2)^{n_w}$, $0 \leq n_w \leq n_x$, in the model equations and an additive constant $w_0 \in \mathbb{R}^{n_w}$ in the initial conditions:

$$0 = f(\dot{x}, x, u, d, p) + W w, \tag{3}$$
$$0 = b(\dot{x}_0, x_0, u_0, d_0, p) + W_0 w_0. \tag{4}$$

The rectangular matrices W and W_0 are indicator matrices containing a unit element in each column where a forcing function should be considered in a model equation and are otherwise zero. Note, that there is some arbitrariness in the uncertainty description regarding the quantities d and w. However, the disturbances d may enter nonlinearly in the model due to their physical interpretation whereas the error functions w always enter linearly by assumption. Neither of the two types of forcing functions are modeled. It is the decision of the modeler, which of these uncertainty representations should be included in the model. Obviously, the number of variables in d and w should be kept at a minimum for redundancy and computational efficiency reasons.

If a dynamic model for all or some of the error terms w is preferred, it can be covered by the model structure suggested above. In this case, the set of model equations (3), (4) and state vector x is appropriately extended and the indicator matrices W and W_0 are adjusted accordingly. Often, such dynamic models describing unstructured uncertainty are difficult to justify.

This model formulation generalizes previous work in dynamic data reconciliation such as that by Almasy [32], Edgar [44], or Biegler [2] and coworkers. The major difference is the way in which we are explicitly dealing with different types of uncertainties as well as with the measurement in general not covering the full process state in a general differential-algebraic setting. The measurements $y = [y_x^T, y_u^T]^T$, $y_x \in (L_2)^{n_{y_x}}$, $y_u \in (L_2)^{n_{y_u}}$, $n_{y_u} = n_u$ and $n_y = n_{y_u} + n_{y_x}$, as predicted by the model are

$$y_x = Cx + V_x v_x, \tag{5}$$

$$y_u = u + V_u v_u. \tag{6}$$

The measurement errors $v = [v_x^T, v_u^T]^T$, $v_x \in (L_2)^{n_{v_x}}$, $v_u \in (L_2)^{n_{v_u}}$, $0 \leq n_{v_x} \leq n_{y_x}$ and $0 \leq n_{v_u} \leq n_{y_u}$, are selected by rectangular indicator matrices v_x and V_u. In the differential-algebraic setting considered here, the rectangular matrix C is an indicator matrix selecting some of the state variables x without loss of generality. Nonlinear measurement functions as usually considered in nonlinear systems theory are here included as algebraic model equations in the vector f introducing auxiliary state variables.

If the model (3) is simplified to a system of pure ordinary differential equations, the measurement equation should be generalized to

$$y_x = g(x, u) + V_x v_x, \tag{7}$$

$$y_u = u + V_u v_u. \tag{8}$$

with $g : (L_2(\Omega))^{n_x} \times (L_2(\Omega))^{n_u} \to (L_2(\Omega))^{n_{y_x}}$. This measurement model explicitly considers uncertainties in the measured inputs as well as in the measured outputs. Both types of uncertainties are present in any practical situation and are therefore accounted for. This approach has been referred to as *error–in–variables* in the context of dynamic systems by Kim et al. [40]. Our measurement model coincides in part with the literature (i.e. [3] or [55]) but differs from the formulation used for example by Edgar and coworkers in [44, 48]. These authors define the measurement errors as the *difference between the true value of the observed process quantity and the measurement* whereas here the errors v are contaminating the *outputs predicted by the model*. We prefer this formulation since the true values of the observed variables are not accessible in practice, as they always differ from the values predicted by the model.

Besides the model equations, there is additional a-priori knowledge which should be stated and used in the estimation procedure. Following Kim et al. [40] or Robertson et al. [55], lower and upper bounds can be formulated for any of the process quantities such as the states x, the inputs u, the disturbances d, the parameters p, the measurements y, and the model and measurement errors v and w. These additional inequality constraints are of the type

$$s_{i,min} \leq s_i \leq s_{i,max}, \tag{9}$$

where s_i refers to some scalar process quantity [55]. These inequality constraints must be formulated to obtain a feasible set [52]. It should be noted that the constraints also refer to a bound on the function norm for the process quantities.

In addition to the simple bounds modeled by (9) statements on the stochastic properties of the process quantities may be included if a probabilistic interpretation of the estimation problem is pursued as done in classical filtering theory. In this case, stochastic processes with certain properties are associated with the model errors w and the measurement errors v. Pragmatically, uncorrelated zero mean random variables with Gaussian probability distribution and given variance-covariance information are assumed. This assumption is crucial for reducing the analytical and computational complexity of the estimation problem; at the same time it permits an interpretation of the estimation quality and supports the choice of the estimator's tuning parameters in the linear case. In the nonlinear case, which prevails in most realistic scenarios, the advantages of using such a probabilistic interpretation largely break down. The more general situation is extremely difficult to tackle even if perfect information on the probabilistic properties of the process quantities were available. For an in-depth discussion we refer to the monographs [33] and [37] or to Robertson et al. [55] who compare stochastic filtering with mathematical programming techniques.

The general problem formulation, we are interested in, aims at avoiding any assumptions on statistical properties of the process quantities and error functions. Therefore, a deterministic rather than a stochastic setting is envisaged. This decision is based on the fact that there is no justification for simple assumptions on the probabilistic properties of the model and measurement errors in almost any real situation. Since the model errors w account for unstructured uncertainty in the process model, the signal is typically correlated in time and reveals a non-zero mean on some time interval. The signals cannot be expected to be Gaussian. Similar arguments hold for the measurement errors v. In these situations the filtering approaches have shown poor performance and are extremely difficult to tune (see [52] for a summarizing discussion) aside from the loss of the theoretical foundation. We accept (at least for the moment) the loss of potentially useful information on the process statistics.

2.2. ESTIMATION PROBLEM

For reconciliation of the measured data with the process model we may require that the degrees of freedom in the model (3)-(6) are chosen in such a way that the predicted measurements y exactly coincide with the measurements $\breve{y}$, i.e.

$$r(t) := y(t) - \breve{y}(t) = 0, \quad \forall t \in [0, \tau] \tag{10}$$

from the startup of the process at $t_0 = 0$ to the current time τ. These degrees of freedom are w, v, d, p and those r initial values $\dot{x}_0, x_0$ collected in x_0' for later reference to which arbitrary values can be assigned. In the sequel, these degrees of freedom will be collected in the vector Δ for conciseness of notation.

2.2.1. *Least-squares formulation*
The requirement (10) would not be appropriate since the real measurements are not reflecting the measured process quantities precisely due to contamination with measurement error. A reasonable relaxed requirement is therefore a least-squares solution to equation (10), which is given by those $\hat{\Delta}$ minimizing a cost functional Υ according to

$$\hat{\Delta} = argmin \ \Upsilon(\Delta) = \| Q(y - \breve{y}) \| \tag{P1}$$

subject to the equality constraints provided by the model (3)-(6) and a feasible set of inequality constraints of type (9) for those process quantities and error functions which can be bounded by a-priori knowledge. The weighted norm of the residual r may be defined by

$$\| Qr \|_{L_2([0,\tau])}^2 = \int_0^\tau r(t)^T Q^T Q \, r(t) dt. \tag{11}$$

Then, (P1) is equivalent to a weighted least squares solution of (10) with the symmetric weighting matrix $Q^T Q$. A more general function norm could be used instead of the L_2-norm. An advantageous choice is the Sobolev norm for integer β in L_2 as given by

$$\| \, r \, \|^2_{H^\beta([0,\tau])} = \| \, r^{(0)} \, \|^2_{L_2([0,\tau])} + \cdots + \| \, r^{(\beta)} \, \|^2_{L_2([0,\tau])} \, . \tag{12}$$

Here $r^{(\beta)}$ denotes the β-th weak derivative of the function r with respect to time and is required to be square integrable. Note that the Sobolev norm with the associated Sobolev regularity β coincides with the usual L_2-norm for the choice $\beta = 0$; employing the Sobolev norm instead of the L_2-norm in (P1) therefore generalizes the least-squares formulation.

The estimation problem (P1) is a potentially large-scale dynamic optimization problem to be solved by an appropriate technique at current time τ to provide a reconciled set of measurements $\hat{y}$ as well as estimated $\hat{\Delta}$ after the computational delay time h at $t = \tau + h$. For on-line dynamic data reconciliation the problem must be solved repeatedly with a (not necessarily) constant sampling interval of $H \geq h$. Obviously, we would like to choose H as small as possible to get the estimate with minimal delay. The choice of H will be constrained by the algorithmic complexity and by computing resources.

This optimization problem may not be computationally tractable since the computational effort may grow unboundedly with time. On the other hand, it is also not necessary to include the measurements for $t < \tau - T_{max}$ in the reconciliation. This is due to the uncertainty in the measurements $\breve{y}$ which dominates for a sufficiently long horizon T_{max} (much larger than the dominating process time constant) the true process measurement rendering an unfavorable signal-to-noise ratio. Hence a modified least-squares problem is formulated which reads

$$\hat{\Delta} = argmin \, \Upsilon(\Delta) = \| \, Q(y - \breve{y}) \, \|^2_{H^\beta([\tau-T,\tau])} \tag{P2}$$

subject to the same equality and inequality constraints as in problem (P1) above. In order to capture all relevant measurement information, the length of the time interval T, the horizon, should be chosen as large as computationally tractable for good estimation quality as pointed out by various authors (i.e. [44] or [55]). As above, the problem is solved repeatedly with sampling interval $H > h$. This solution approach has been previously called *moving* or *receding horizon estimation* (i.e. [53, 55]).

2.2.2. *Ill-posedness of the least-squares problem*
Both estimation problems belong to the class of so-called *inverse problems* since we want to *determine causes for an observed effect* [27]. Most

often these inverse problems do not fulfill Hadamard's postulates of well-posedness [28]. Any *well-posed problem in the sense of Hadamard* must possess the following properties:

1. A solution exists for all admissible data.
2. The solution is unique for all admissible data.
3. The solution depends continuously on the data.

If one of these properties is not fulfilled the problem is termed *ill-posed*. In the context of dynamic data reconciliation *data* refers to the measured inputs and states. These requirements are minimalistic. Obviously, there are stronger requirements, such as the convergence of the nonmeasured states to the true states (at least in the nominal case where there is neither measurement error nor model uncertainty) which are related to observability (see e.g. [50, 51, 53, 64] for a discussion in the context of state estimation).

Obviously, the estimation problems (P1) or (P2) are not set up in a reasonable way, if they are not at least well-posed in the Hadamard sense. A rigorous analysis of the nonlinear problem is quite difficult and, to our knowledge, not available yet. However, some qualitative considerations easily reveal the ill-posedness of (P1) and (P2).

We are not concerned with existence, since we can expect that there is a minimum of the cost functional Υ in both, (P1) and (P2), if the constraints set is feasible. Uniqueness is more of a concern. Let us look at the very simple case given by

$$\dot{x} = -ax + w, \quad x(0) = x_0, \tag{13}$$
$$y = x + v \tag{14}$$

with known parameter a, unknown initial condition x_0 and error terms v and w. The predicted output y is given by the linear operator equation

$$y(t) = e^{-at}x_0 + \int_0^t e^{-a(t-s)}w(s)ds + v(t). \tag{15}$$

The cost functional (of problem (P1)) reads

$$\Upsilon(x_0, v, w) = \| (y - \breve{y}) \|^2_{H^\beta([0,\tau])} \tag{16}$$

with y taken from (15). We cannot expect to get a unique solution for the degrees of freedom since the same best $\hat{y}$ in the least-squares sense can be established by an infinite number of solutions. Remember, that there are no restrictions on the sign of the additive terms in (15). Hence, given one solution $\{\hat{y}, \hat{x}_0, \hat{v}, \hat{w}\}$, we can easily find many other solutions by arbitrarily changing x_0 or w and compensate for this influence by an adjusted v in

(15). Note that there are many more subtle possibilities to get solutions with the same $\hat{y}$.

This non-uniqueness of the least-squares solution is characteristic for *many* inverse problems. An established approach to enforce uniqueness is to define the *best-approximate solution* as a *least-squares solution* of (10) where a *suitable norm* of the degrees of freedom at the solution *takes its minimum* [27]. The best-approximate solution is therefore given by the least-squares solution with the smallest possible norm of all values of degrees of freedom which include the initial conditions as well as the model and measurement uncertainties. For linear inverse problems, the best-approximate solution is unique and closely related to the Moore-Penrose (generalized) inverse of the linear operator mapping the degrees of freedom to the measurements. An alternative related approach modifies the least-squares cost functional by a regularization term penalizing the norm of the degrees of freedom, or, more generally the norm of some differential operator applied to the degrees of freedom in order to enforce uniqueness.

So far, we did not address the third requirement for a well-posed problem which asks for a continuous dependence of the solution on the data. This property guarantees that small changes in the measured data will not lead to completely different reconciled measurements and estimated degrees of freedom. Again, we rely for the moment on the results obtained for general linear inverse problems as presented in [27] and briefly summarized below.

It can be shown that the solution to the regularized problem tends to the best approximation for error free measurements if the regularization penalty approaches zero. However, for a certain fixed level of the measurement errors the difference between the regularized solution for exact data and the regularized solution for perturbed data may assume very large values if the regularization penalty approaches zero. This difference is proportional to the measurement error level and may grow without bound depending on the spectral properties of the linear operator in the equality constraints. Hence, only with non-vanishing regularization, continuous dependence of the solution on the data can be achieved.

For given regularization parameters and fixed level of the measurement error, the estimation error can be bounded by a function of the regularization penalty and the order of magnitude of the measurement error. The estimation error approaches zero for a suitable regularization adjusted to the level of measurement error as the measurement error is reduced to zero. The estimation error depends therefore only on the level of the measurement error provided the regularization penalty can be chosen in some optimal manner.

Hence, in the linear case, regularization of the least-squares problems (P1),(P2) not only renders a unique solution but also reduces the sensitivity of the solution to the data and provides a theoretical basis to bound the estimation error if the order of magnitude of the measurement error is available. Therefore, a generalization of such an approach to nonlinear problems seems to be appropriate [28] and has already been applied in system identification problems [38].

2.2.3. *Regularized estimation problem*

Using the regularization approach as suggested in the theory of inverse problems [28] we could consider to (heuristically) modify the cost functional in (P1) and (P2) as follows

$$
\Upsilon(\mathbf{\Delta}) = \parallel \mathbf{Q}(\mathbf{y} - \breve{\mathbf{y}}) \parallel^2_{H^{\beta_y}([t_0,\tau])} + \alpha_v \parallel \mathbf{R}_v \mathbf{v} \parallel^2_{H^{\beta_v}([t_0,\tau])}
$$
$$
+ \alpha_w \parallel \mathbf{R}_w \mathbf{w} \parallel^2_{H^{\beta_w}([t_0,\tau])} + \alpha_d \parallel \mathbf{R}_d \mathbf{d} \parallel^2_{H^{\beta_d}([t_0,\tau])}
$$
$$
+ \alpha_p \parallel \mathbf{R}_p \mathbf{p} \parallel^2_{L_2} + \alpha_0 \parallel \mathbf{R}_0(\mathbf{x}'_0 - \mathbf{x}'_{0,ref}) \parallel^2_{L_2} . \tag{17}
$$

Here, t_0 is either 0 for (P1) or $\tau - T$ for (P2). $\mathbf{R}_v, \mathbf{R}_w, \mathbf{R}_d, \mathbf{R}_p$ and $\mathbf{R}_0$ are nonsingular weighting matrices whereas $\alpha_v, \alpha_w, \alpha_d, \alpha_p$, and α_0 are nonnegative constants called regularization parameters.

Let us look at this objective function in the nominal case where there are no uncertainties either in the model or in the measurements. Then, the minimization of the cost functional should lead to $\hat{\mathbf{v}} = \hat{\mathbf{w}} = \hat{\mathbf{d}} = 0, \forall t, \hat{\mathbf{p}} = 0$ and $\breve{\mathbf{y}} = \hat{\mathbf{y}}, \forall t$, if the reference values for $\mathbf{d}$ and $\mathbf{p}$ are chosen as the actual values (which is always possible in the nominal case).

We would expect the cost functional to be zero in the nominal case. This could only be achieved if we knew the correct initial values which lead to $\breve{\mathbf{y}}$ which is impossible. Therefore, the cost functional will take at the minimum some positive value $\hat{\Upsilon} \geq \alpha_0 \parallel \mathbf{R}_0(\hat{\mathbf{x}}'_0 - \mathbf{x}'_{0,ref}) \parallel^2_{L_2}$. In general we cannot expect to get correct estimates by minimizing the cost functional (17), since the minimum $\hat{\Upsilon}$ might be obtained with different $\hat{\mathbf{x}}'_0$ and some non-vanishing and hence incorrect solution $\hat{\mathbf{v}}, \hat{\mathbf{w}}, \hat{\mathbf{d}}$, or $\hat{\mathbf{p}}$. Based on this argument, we drop the regularization of the initial values in the cost functional and formulate the estimation problem as

$$
\hat{\mathbf{\Delta}} = argmin \ \Upsilon(\mathbf{\Delta}) \tag{18}
$$

with the cost functional

$$
\Upsilon(\mathbf{\Delta}) = \parallel \mathbf{Q}(\mathbf{y} - \breve{\mathbf{y}}) \parallel^2_{H^{\beta_y}([t_0,\tau])}
$$
$$
+ \alpha_v \parallel \mathbf{R}_v \mathbf{v} \parallel^2_{H^{\beta_v}([t_0,\tau])} + \alpha_w \parallel \mathbf{R}_w \mathbf{w} \parallel^2_{H^{s\beta}([t_0,\tau])}
$$
$$
+ \alpha_d \parallel \mathbf{R}_d \mathbf{d} \parallel^2_{H^{\beta_d}([t_0,\tau])} + \alpha_p \parallel \mathbf{R}_p \mathbf{p} \parallel^2_{L_2} \tag{P3}
$$

subject to the equality constraints (3)-(6) and inequality constraints (9).

Note, that Rawlings and coworkers [53, 55] found in the context of nonlinear state estimation that penalizing the initial conditions does not guarantee convergence in the nominal case whereas convergence can be shown if there is no such penalty. The penalty in the initial conditions is used in [55] to optimally reflect the past measurements in a receding horizon estimator by adjusting the weight matrix in the norm according to *linear* filtering theory. This strategy is useful for short estimation horizons whereas it is not really required if the estimation horizon is taken sufficiently large [52].

For later reference in the context of the discussion of the multiscale framework in Section 5 we formulate here a simplified version of (P3) where the model (3) is substituted by a system of ordinary differential equations and the measurement model is restricted to (7). Further, we assume that the indicator matrices W and V_x are identity matrices thus allowing the error functions to contaminate every state x and measurement y. V_u is assumed to be zero which corresponds to error free measured inputs. We are neither considering unknown parameters p nor disturbances d. The variables with specified bounds are catenated in the vector s with upper bounds s_{max} and lower bounds s_{min}. For simplicity of notation, the regularization parameters are included in the matrices Q, R_v, R_w. In summary, these modifications yield the simplified dynamic data reconciliation problem

$$\min \| Q(y - \breve{y}) \|^2_{H^{\beta_y}([t_0,\tau])} + \| R_v v \|^2_{H^{\beta_v}([t_0,\tau])} + \| R_w w \|^2_{H^{\beta_w}([t_0,\tau])} \quad (P4)$$

subject to the constraints

$$\dot{x} - f(x, u) - w = 0,$$
$$y - g(x, u) - v = 0,$$
$$s - s_{max} \leq 0,$$
$$s_{min} - s \leq 0.$$

2.2.4. *Relation to previous work*

Some authors (e.g. [44] or [54]) use the formulation of (P1) which without any regularization is ill-posed. Their solution does not obviously reveal amplification of measurement errors since the relatively coarse discretization of continuous time acts itself as some regularization of the estimation scheme [27]. In fact, Liebmann et al. [44] observe improved estimation quality in the context of bias estimation if they penalize the bias (corresponding to our v) in the least-squares objective function, i.e. if they regularize the ill-posed problem.

The formulation suggested by Rawlings and coworkers (see [55] for example) coincides with a specialization of (P3) (e.g. their formulation does not consider our p and d variables) *if we add the constraints* (10) to (P3). Inserting this equality of real and predicted measurements in (P3) eliminates the first term in the cost functional and hence leads to the objective function used by Robertson et al. [55]. These authors therefore determine the modeled measurement errors such that the real measurements coincide exactly with the predicted measurements whereas our formulation is less restrictive at the expense of a larger number of degrees of freedom.

Finally, we want to compare our approach with the work of Michalska and Mayne [50, 51]. These authors suggest a moving horizon observer for error free measurements and a perfect model using a formulation similar to (P2). They repeatedly solve an optimization problem but do not require to reach the minimum of the cost functional. Rather, they only attempt to reduce the cost functional at each step by a given factor which assures convergence of the estimated state to the real state at a given rate as time goes to infinity under an observability condition by adjusting either the initial conditions or in addition an observer input. They motivate their approach with the limited computational resources available in real-time applications. The approach suggested here is very similar to the moving horizon observer in the nominal case if the true minimum is computed in every sampling interval.

3. Research questions

In conclusion of the preceding discussion, we want to explicitly state that there is no guarantee that the formulation of the dynamic data reconciliation problem suggested here satisfies the requirements of well-posedness nor that there is convergence of all the estimated states to the true states for the nominal model. The formulation of the dynamic data reconciliation problem (P3) is still heuristic at this point although it is a reasonable extension of (P1) or (P2) which have been studied in the literature though they are definitely ill-posed. This leads us to open research questions to be tackled before the suggested problem formulation can be safely applied to dynamic data reconciliation in practice:

1. Definition of reconcilability: reconcilability should be a property of the set of measurements and the process model alone. The discussion above shows, that we cannot avoid to include the way the estimation problem is formulated. A first suggestion is as follows:

   > A set of measurements and a given process model are called reconcilable, if there exists an algorithm which renders a unique solution for the degrees of freedom in the model for all admissible data. The solution must depend continuously on the data. In the limit of van-

ishing measurement error it must converge to the best-approximate solution.

2. Relationship between reconcilability and observability: Obviously, observability is a much stricter requirement than reconcilability. This system property may, however, not be of very much use in situations where there is significant error (random, systematic or gross) in the measurements.

3. Analysis of (P3) and its specializations regarding estimation quality. Here, we may distinguish convergence of the estimate to the least-squares solution for vanishing measurement error as well as convergence of the estimated degrees of freedom to the true values at least in limiting situations. This analysis may reveal necessary modifications to the suggested problem formulation. One objective of the analysis is the minimization of the number of tuning parameters and constructive guide lines for their selection.

4. Criteria and strategies to guarantee sufficient redundancy in the data and the process model to result in useful reconciled measurements and useful estimates for the degrees of freedom (see [3] for a first treatment). The selection of a set of degrees of freedom Δ of minimal cardinality must also be addressed, trading off the requirements on the measurements needed for proper identification on the one and the model quality on the other hand.

5. Methods for preprocessing the measured data in order to reduce some norm of the measurement error and to provide a reasonable interpolation of the discrete measurements. The estimation quality achievable will depend on the level of measurement error as indicated by the results on linear inverse problems. Outlier removal and denoising are important requirements in this context.

6. Highly efficient numerical solution of (P3) with $t_0 = 0$ for a given set of measurements in $[0, \tau]$. This problem would be typically solved off-line to support mechanistic modeling of some (bench-scale, laboratory-scale or industrial-scale) process or to diagnose the performance of some industrial process. There are no strict limitations on computational resources. More recent contributions on numerical solution algorithms are [4, 14, 29, 61, 62, 63, 68, 69].

7. Highly efficient numerical solution of (P3) with $t_0 = \tau - T$ in a repetitive manner at a sampling interval of H for a long horizon T which takes into account the most recent measurements becoming available in $[\tau, \tau + H]$. This problem must be solved on-line in real-time obeying strict limitations of the computational resources. The numerical solution approach must take advantage of the repetitive nature and other structural properties of the problem in order to become feasi-

ble in practical implementations. A reasonable approximation to the solution of the problem must be provided within a real-time interval of H. The quality of the approximate solution should scale with H, e.g. the solution accuracy must increase with the time available for carrying out the numerical computations. Some recent contributions addressing the solution of dynamic optimization problems on a moving horizon are [2, 4, 44, 58, 71]. It should be noted that any optimization based algorithm has to somehow compete with computationally efficient observers or filters [52].

This is a demanding list. Some of these research issues have already been addressed in previous work. Nevertheless, there are no fully satisfactory solutions available and much work remains to be done. In our work to date, we have concentrated largely on the numerical aspects (items 5, 6 and 7 of the list) as presented in the next sections.

4. Multiscale dynamic data reconciliation - the framework

The numerical solution of dynamic data reconciliation problems (P3) is typically quite challenging, in particular, if on-line applications are envisaged where a reconciled set of measurements and an estimate of the degrees of freedom must be provided in real–time ideally with minimum delay to the last available measurement. Typically, the process model restricting the minimization of the cost function is large-scale. The length of the time interval T employed in moving horizon estimation should be chosen as long as possible to improve the estimation results. Obviously, due to the increasing number of discrete measurements, the computational complexity of the estimation problem increases with T. For a given problem, algorithm and hardware, the required solution time is governed by some h which may strongly vary over time depending on the availability of the process computer and on the frequencies included in the measured (and hence) estimated quantities. Despite these variations in h, the computations must *always* be accomplished in $h \leq H$. The sampling interval H for reconciled measurements and estimated values is usually much longer than most of the sampling intervals of the measurement devices but always fixed. H is chosen such that the information demand of the operating personal or of the control system (potentially including economic set-point optimization) can be satisfied.

The majority of the available algorithms for the solution of (P3) or (P4) can be subdivided into two classes with differing solution strategies. In the so-called *sequential* (or *control-vector parameterization*) *approach* (see [1, 49, 68, 69] for an example) the degrees of freedom are approximated by a polynomial or often piecewise constant function on a fixed time mesh to result in a set of algebraic decision variables. In every iteration, the differential-algebraic equations are integrated together with the sensitivity equations in order to evaluate the objective function and to provide the required gradient information with respect to the decision variables. In contrast, in the *simultaneous approach*, the state variables as well as the degrees of freedom are discretized on a fixed time mesh approximating the dynamic optimization problem by a large-scale nonlinear programming problem (NLP) (see [2, 62] for an example). Both approaches have been used to implement moving horizon estimation algorithms with varying performance. However, due to the inherently expensive integration of the model and sensitivity equations on the one hand and the significant advances in large-scale nonlinear programming on the other, the simultaneous approach may be considered more attractive in the longer run if the (typically superior) robustness of the sequential solution approach can be met by appropriate means.

All the suggested algorithms discretize all or some process quantities on a fixed time mesh. There are only few attempts to incorporate adaptivity in the discretization scheme [46, 61] which have not been used in moving horizon algorithms to our knowledge. Always, the solution is computed in a single step. Since the computation time varies with time and can only be estimated in a very rough sense, there is by no means a guarantee that a valid solution is found within $h \leq H$. To increase the chances that this restriction is met, the user can only relax the tolerances, coarsen the grid, or simplify the process model since the available upper bound on the computing time H can typically not be increased as argued above. This approach would lead to a solution of possibly inadequate accuracy in many cases; it might be over-restrictive because the algorithm most probably will not use the full amount of available computation time to further improve the solution quality.

Multiscale theory [11, 21] seems to offer attractive alternative techniques which might push the limit of large–scale moving horizon computation forward. Such techniques have just recently been applied to moving horizon estimation and control problems [43, 60, 59]. Inspired by the adaptivity inherent to multiscale theory [15] we focus in our work on a concept which steadily refines the optimization problem starting from a coarse approximation up to arbitrary degree of detail. This concept differs from the well established single step solution approach such that the optimization problem is solved repeatedly on various discretization meshes of increasing number of decision variables and approximation quality. This simultaneous approach is based on discretizing all time-varying functions using wavelets which support refinement on various scales to resolve the global trend of the functions as well as the local detail in an appropriate manner. The

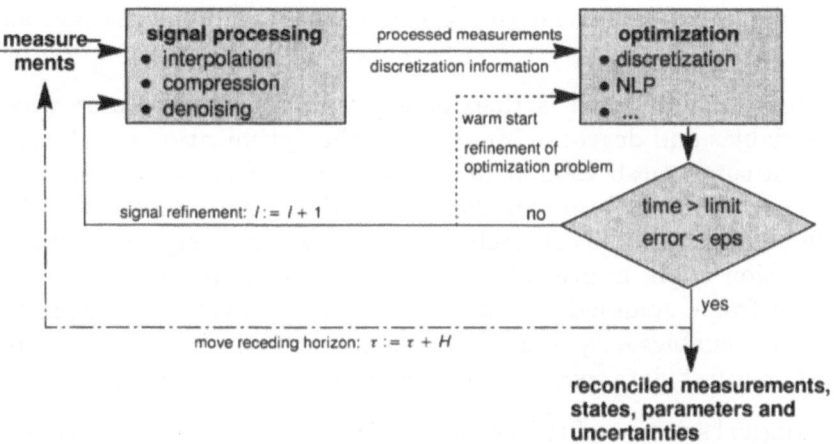

Figure 1. Schematical outline of the refinement strategy

suggested fully adaptive method will prompt during one cycle in the moving horizon approach a valid solution of some accuracy associated with the level of refinement at any time during the course of refinement. Moreover, the method always utilizes the available computation time H, since a solution of a coarse approximation quality on level $\ell - 1$ is always available if the time limit is reached while the computations on the current level ℓ have not been completed. In addition, preprocessing of the measured data is envisaged to be treated in an integrated manner with the solution of the optimization.

The general outline of the refinement strategy is schematically given in Figure 1. Since the measurements are sampled in discrete time an appropriate transformation into continuous time representation is necessary. As it will be seen later, this functional representation is the key for a substantially compressed representation of the measurements. Compression of the measurements is used here to avoid the substantial growth of computational complexity with an increasing number of discrete measurement samples in a different manner than explored recently in [2].

The measurements, once in continuous representation, can as well be discretized in various approximation levels with wavelet based multiscale techniques. Since the measurements are corrupted with noise, outliers and the like, they are subject to pretreatment. The objective is in general to reduce the norm of the measurement error or more precisely to eliminate noise or outliers or to interpolate missing data due to a malfunctioning measurement device. According to the theoretical considerations on the solution of inverse problems in the presence of erroneous data, the reduction of some norm of the measurement error not only eases the numerical burden but also leads to smaller estimation errors caused by regularization and measurement error.

The approximated signal is then passed to the optimizer unit where the state variables and degrees of freedom of the optimization problem are discretized on some mesh related to the resolution used to represent the measurements. Thus we obtain an algebraic optimization problem which is solved with an appropriate NLP technique. The two units signal processing and optimization might appear at first glance to be separated. However, a discretization frame spanned by wavelet functions provides an integrated approach which enables easy and fast compression of the problem formulation as well as a fast data exchange between signal and optimization unit.

The numerical solution of the former refinement cycle $\ell - 1$ is reused for a warm start of the optimization in the current cycle ℓ (see [44] for the computational savings with warm start with conventional discretization) avoiding significant computing loads as compared to a start from scratch. Ideally the computational complexity in each refinement step has to be closely related to the number of the additional decision variables included in the actual refinement step. An acceptable complexity is obtained, if the solution of the last approximation can be fully utilized. This includes the evaluation of the discretized equation systems including the Jacobian and Hessian matrices as well as the numerical solution of the updated equation system. For a certain problem class the wavelet discretized equation system can be completely reused while updating. In contrast, applying Runge-Kutta or orthogonal collocation on finite elements discretizations, popular for the discretization of dynamic optimization problems on moving horizons (i.e. [2]), the discrete system of equations have to be completely reformulated and solved in every refinement step.

The adaptive mesh refinement, which includes refinement of the compressed measurement functions as well as refinement of the discretization of the states and degrees of freedom leading to a discretized system of equations of steadily increasing dimension, has to ensure a small additional computational effort compared to the overall time span H available. That way the estimation accuracy is steadily improved with small increments. The procedure is repeated until either the overall approximation error is lower than a specified tolerance or the time restrictions are violated ($h = H$). Even in the latter case where the computation of the actual approximation level is aborted, a solution of the last approximation level is still available. This way, the best valid solution achievable with given computational resources can be always prompted. Finally, the horizon window is moved on, and the multiscale dynamic data reconciliation is repeated on the next time horizon.

5. Multiscale discretization and numerical solution techniques

5.1. INTRODUCTION TO WAVELETS

The conceptional ideas of refinement in the multiscale approach have been presented in the last section. Now the ideas have to be put on a solid mathematical basis in order to make the concept concrete. A rigorous treatment of the wavelet methodology is far beyond the scope of this paper. For a more thorough treatment, the reader is referred to [11, 15, 21]. Here we introduce only the notation and the basic concepts, which are relevant for the understanding of our approach. For simplicity, the case of scalar functions is treated, which straightforwardly extends to vector functions. Let the inner product in L_2 be defined by $\langle f, g \rangle := \langle f, g \rangle_{L_2(\Omega)} = \int_0^1 f(t)g(t)dt$ with $\Omega = [0, 1]$ and $f, g \in L_2$. Further, we define finite dimensional function spaces spanned by a finite number of basis functions $\varphi_{j,k}$

$$S_j := span\{\varphi_{j,k} : k \in \Delta_j\}$$

with the scale dependent index set Δ_j. The basis functions $\varphi_{j,k}$ are derived from a single so-called scaling function $\varphi^{(m)}$. At this point we employ pragmatically a B-spline of order m [22]. In the interior of $[0, 1]$ the $\varphi_{j,k}$ are obtained by dyadic dilation with powers of two and integer translation while certain special modifications are necessary near the end points of the interval. The spaces S_j are to form a multiresolution sequence

$$\ldots \subset S_j \subset S_{j+1} \subset \ldots$$

whose union $clos(\cup_j S_j) = L_2(\Omega)$ is dense. Due to the nestedness of the spaces S_j the basis $\Phi_j = \{\varphi_{j,k} : k \in \Delta_j\}$ and Φ_{j+1} (the set of functions will in the sequel also be interpreted as ordered array) form a two-scale relation which in compact form reads as

$$\Phi_j^T = \Phi_{j+1}^T M_{j,0}.$$

In other words, each basis function on a coarse scale can be written as a linear combination of basis functions on the next finer scale. Since only finitely many fine scale basis functions appear independently of j in these linear combinations, $M_{j,0}$ is a sparse matrix.

Let us assume that an arbitrary function $f \in S_{j+1}$ is approximated in some sense in S_j then some detail has to be added to get the more accurate approximation in S_{j+1}. This detail is embedded in the detail space

$$W_j := span\{\psi_{j,k} : k \in \nabla_j\}$$

such that W_j forms a complement space of S_j to S_{j+1}

$$S_{j+1} = S_j \oplus W_j.$$

The basis functions in $\mathbf{\Psi}_j = \{\psi_{j,k} : k \in \nabla_j\}$ of the complement space W_j are called wavelets. Since $W_j \subset S_{j+1}$ the wavelet basis on scale j can also be expressed as a linear combination by scaling functions in $\mathbf{\Phi}_{j+1}$ and hence

$$\mathbf{\Psi}_j^T = \mathbf{\Phi}_{j+1}^T M_{j,1}$$

with $M_{j,1}$ a sparse matrix. Repetition of the decomposition splits the space S_J into a sum of complement spaces

$$S_J = S_{j_0} \oplus W_{j_0} \oplus W_{j_0+1} \oplus \ldots \oplus W_{J-1}$$

where j_0 is some fixed coarsest scale and J an arbitrary but fixed finest scale. A thorough analysis [15] reveals that not all splittings are equally well suited. In fact, the complement spaces should be chosen in such a way that there exists a dual multiresolution of spaces

$$\tilde{S}_j := span\{\tilde{\varphi}_{j,k} : k \in \tilde{\Delta}_j\}$$

with corresponding complements splittings

$$\tilde{W}_j := span\{\tilde{\psi}_{j,k} : k \in \tilde{\nabla}_j\}.$$

They are constructed in a similar way based on a scaling function $\tilde{\varphi}^{(m,\tilde{m})}$ satisfying

$$\langle \varphi(\cdot)^{(m)}, \tilde{\varphi}^{(m,\tilde{m})}(\cdot - k)\rangle = \delta_{0,k}$$

for integer $k \in \mathbb{Z}$. Here and in the sequel, the notation $f(\cdot)$ defines a function with generalized argument $\cdot$. Moreover, the dual function spaces should possess also two-scale relations of the above form [16, 17]. For the sake of brevity we do not explicitly give the definitions for the dual function relations but will always employ a $\tilde{\ }$ to denote corresponding relations in the dual setting. The dual and primal single- and multiscale spaces are related through the biorthogonality property

$$\langle \mathbf{\Psi}_j, \tilde{\mathbf{\Psi}}_j\rangle = I, \quad \langle \mathbf{\Phi}_j, \tilde{\mathbf{\Phi}}_j\rangle = I, \quad j \in \mathbb{N} \tag{19}$$

and

$$\tilde{W}_j \perp S_j, \quad W_j \perp \tilde{S}_j \tag{20}$$

where I denotes the identity matrix. The inner products of the functions collected in vectors are defined here and in the sequel by their inner products associated with the corresponding index set and hence form a matrix, e.g. $\langle \mathbf{\Psi}, \tilde{\mathbf{\Psi}}\rangle := (\langle \psi_{j,k}, \tilde{\psi}_{l,i}\rangle)_{(j,k)\in\nabla, (l,i)\in\tilde{\nabla}}$. One can show that these biorthogonality properties form essential preconditions for important stability properties of the basis, sometimes referred to a Riesz basis, explained below.

Every function $f \in L_2$ has a unique expansion

$$f = \sum_{k \in \Delta_{j_0}} \langle f, \tilde{\varphi}_{j_0,k} \rangle \varphi_{j_0,k} + \sum_{j=j_0}^{\infty} \sum_{k=\nabla_j} \langle f, \tilde{\psi}_{j,k} \rangle \psi_{j,k} . \tag{21}$$

Upon subsuming all basis functions $\varphi_{j_0,k}, \psi_{j,k}$ in an infinite array $\mathbf{\Psi}$ and likewise, all expansion coefficients $\langle f, \tilde{\varphi}_{j_0,k} \rangle, \langle f, \tilde{\psi}_{j,k} \rangle$ in a corresponding array $\mathbf{d} := \langle f, \tilde{\mathbf{\Psi}} \rangle$, (21) can be written in a compact form as $f = \mathbf{d}^T \mathbf{\Psi}$. Moreover, in addition to the uniqueness of such expansions a Riesz basis has the remarkable property that the (Euclidean) norm of the discrete array $\mathbf{d}$ that characterize the continuous objects $f \in L_2$ behaves essentially as the function norm, i.e.

$$\| f \|_{L_2}^2 \sim \| \mathbf{d} \|_{l_2}^2 = \sum_{j,k} |\langle f, \tilde{\psi}_{j,k} \rangle|^2 . \tag{22}$$

Thus small variations in the coefficients result in small variations of the function and vice versa – a crucial property for numerical calculations. The basis constructed in [16] enjoys all these properties for any desired choices of m and $\tilde{m}$, $\tilde{m} \geq m$, $m + \tilde{m}$ even. These properties remain trivially valid for finite truncations $\mathbf{\Psi}^J = \Phi_{j_0} \cup_{j=j_0}^{J-1} \mathbf{\Psi}_j$ of $\mathbf{\Psi}$. Thus every function $f \in S_J$ has two equally stable representations

$$\begin{aligned} f &= \sum_{k \in \Delta_J} \langle f, \tilde{\varphi}_{J,k} \rangle \varphi_{J,k} =: \sum_{k \in \Delta_J} c_{J,k} \varphi_{J,k} \\ &= \sum_{j=j_0-1}^{J-1} \sum_{k=\nabla_j} \langle f, \tilde{\psi}_{j,k} \rangle \psi_{j,k} =: \sum_{j=j_0-1}^{J-1} \sum_{k=\nabla_j} d_{j,k} \psi_{j,k} \end{aligned}$$

with $\psi_{j_0-1,1} := \varphi_{j_0,k}$ and $\nabla_{j_0-1} := \Delta_{j_0}$. To keep the notation simple we express the above as the single- and multiscale representation of f in vector form according to

$$f(t) = \mathbf{c}^T \mathbf{\Phi}_J = \mathbf{d}^T \mathbf{\Psi}^J . \tag{23}$$

Both representations are related by the fast wavelet transformation (FWT) $\mathbf{c} = \mathbf{T}_J \mathbf{d}$ where $\mathbf{T}_J : \mathbf{d} \to \mathbf{c}$ is typically executed in cascadic fashion by subsequent applications of the two-scale matrices $\mathbf{M}_{j,0}, \mathbf{M}_{j,1}$ (see [16, 21]). When, as in the case referred to here, these matrices are sparse, it is not hard to verify that the execution of $\mathbf{T}_J$ requires $O(N_J)$ operations where N_J is the cardinality of the index set Δ_J and therefore denotes the dimension of the space S_J.

Let us now comment briefly on the role of the integers $m, \tilde{m}$ appearing in the above description of the basis. The m says that the spaces S_J are

646

exact of order m, i.e. all polynomials up to degree $m-1$ are represented exactly as

$$x^r = \sum_{k \in \Delta_j} \langle (\cdot)^r, \tilde{\varphi}_{j,k} \rangle \varphi_{j,k}, \quad r = 0, \ldots, m-1. \tag{24}$$

It is well known that polynomial exactness combined with the local supports of the $\varphi_{j,k}$ imply that smooth functions can be approximated with the order of 2^{-mj} by functions in S_j. The $\tilde{m}$ means that the dual multiresolution spaces S_j are exact of order $\tilde{m}$. On account of the biorthogonality relation (19) this in turn is equivalent to saying that

$$\int x^r \psi_{j,k}(x)dx = 0, \qquad r = 0, \ldots, \tilde{m}-1, \tag{25}$$

$$\int x^r \tilde{\psi}_{j,k}(x)dx = 0, \qquad r = 0, \ldots, m-1,$$

which are often referred to as moment conditions. An immediate consequence of (25) is that one can subtract any polynomial from a function without changing its wavelet coefficients, i.e. $\langle f, \tilde{\psi}_{j,k} \rangle = \langle f - p, \tilde{\psi}_{j,k} \rangle$ where p is any polynomial of order m. Specifically, choosing p as a Taylor polynomial of order m in the support of $\tilde{\psi}_{j,k}$ shows that $\|\langle f, \tilde{\psi}_{j,k} \rangle\|$ will be quite small when f is smooth on the support of $\tilde{\psi}_{j,k}$. Consequently by far most of the wavelet coefficients in the expansion (20) are rather small. Moreover, on account of the norm equivalence (22), discarding small coefficients causes only a small error in the function. Therefore such functions may be approximated very accurately by linear combinations of only relatively very few wavelet basis functions corresponding to the large coefficients. In fact, this number may be much smaller than the dimension of the full space S_J corresponding to the highest refinement scale among the significant coefficients. Note that working with the single scale representation in S_J corresponds to a discretization based on a uniform *mesh* refinement down to width 2^{-J}. The possibly much sparser wavelet representation determined by the significant coefficients of the approximated object means to employ an individually adapted trial space which could be much smaller than S_J and thus leads to a much more economic discretization concept. Such a possibly sparse basis is denoted by

$$\Psi_\Lambda := \{ \psi_\lambda : \lambda \in \Lambda \} \tag{26}$$

where $\Lambda \subseteq \Lambda^J := \Delta_{j_0} \cup_{j=j_0}^{J-1} \nabla_j$ is an arbitrary subset of the complete wavelet basis Λ^J. It is only assumed that Λ contains the basis functions in Δ_{j_0}, i. e. the basis functions of the coarsest single scale space S_{j_0}.

By now we have established the most important ingredients in our multiscale approach and continue with the signal processing.

5.2. SIGNAL PROCESSING

As mentioned earlier the tasks to be performed in the signal processing part are

- transformation of the discrete measurement samples into a continuous representation,
- denoising of the measurement function,
- outlier removal,
- extrapolation of missing signal parts,
- data compression, which refers to the representation of the measurement signal on various levels of resolution.

Recent developments show that wavelets are very well suited for dealing with these tasks (e.g. [6, 7, 9, 24, 45]). We restrict the following brief presentation to interpolation, denoising and compression only. For the sake of brevity we do not attempt to give a survey on the rich literature on wavelet based signal processing. Instead we illuminate only the basic principles needed to gain a more thorough understanding of the proposed approach.

The same function spaces for signal processing as well as for the discretization of the dynamic optimization problem are used. As it turns out later, this leads to an integrated approach where the processed measurement data can be passed on fast and easily to the optimization problem on various scales of resolution. To ease notation we project the time interval $[\tau - T, \tau]$ of the moving horizon onto the unity interval $\Omega = [0, 1]$ still assuming that the measurement samples are given by $\breve{Y} \in \Omega$. Further, we consider only one measured quantity $\breve{Y}_i$, since the discrete measurements are processed for each single quantity separately.

5.2.1. *Interpolation using a multiscale basis*

The discrete, noisy, possibly non-equidistant discrete measurements $\breve{Y}_i$ can be transformed into a continuous function by a basis spanned by scaling functions. Here, an approximation based on B-splines $\varphi^{(m)}$ is used to fit the basis used later for the discretization of the dynamic optimization problem. The continuous measurement function $\breve{y}_i$ is expanded in single scale representation in S_J where J is chosen sufficiently large to consider all l_i measurement samples $\breve{Y}_i$ such that $\breve{Y}_{i,j} = \breve{y}_i^J(t_j) = c_{\breve{y}_i}^T \Phi_J(t_j), \forall t_j, \ j = 1, \ldots, l_i$. This function can then be transformed into the equivalent multiscale representation $\breve{y}_i^J = d_{\breve{y}_i}^T \Psi^J$ applying the fast wavelet transformation.

Of course, the wavelet interpolation of the measurement function does not limit further signal cleaning to wavelet based techniques. Rather, any other classical approach may be applied. However, we expect that the use of wavelets offers a theoretically sound and unifying way of getting a reasonably preprocessed measurement signal to feed it into the optimization algorithm. It should be noted, that any preprocessing will modify the (unknown) statistics of the measurement signal.

5.2.2. *Denoising*

Now $\breve{y}_i^J$ is transformed into the signal $\bar{y}_i^J$ where the over bar denotes the processed interpolated signal after denoising. To accomplish this task various methods are suggested by several authors (e.g. [10, 24, 25]). Without going into detail we only mention one common method. There, the expansion coefficients $d_{\breve{y}_i^J}$ of the interpolated signal $\breve{y}_i^J$ are replaced by modified coefficients $d_{\bar{y}_i^J}$ according to the shrinkage algorithm

$$
d_{\bar{y}_i^J} := \begin{cases} d_{\breve{y}_i^J} - (\text{sign } d_{\breve{y}_i^J})\varepsilon & |d_{\breve{y}_i^J}| \geq \varepsilon, \\ 0 & |d_{\breve{y}_i^J}| < \varepsilon. \end{cases}
$$

This strategy interprets small wavelet coefficients as noise and reduces their magnitude by a nonlinear algorithm which leads to a smoothed (and compressed) signal. Of course, the question how to choose the threshold ε is subtle and has to be addressed with care. The more it is known about the noise and the smoothness of the signal the more rigorously ε can be determined. For example under the assumption of uncorrelated normal distributed white noise a level independent threshold ε can be derived only knowing the coefficients $d_{\breve{y}_i^J}$. Otherwise, if little or nothing is known about the noise, ε has to be tuned in a rather heuristic manner, usually depending on location, magnitude or scale.

5.2.3. *Representation of the signal in various resolutions*

Keeping the refinement strategy in mind, the signal $\bar{y}_i^J$ is represented in various levels of resolutions indicated by the refinement index ℓ. Because of the norm equivalences (22), approximations of $\bar{y}_i^J$ can easily be obtained in multiscale representation retaining only those coefficients $d_{j,k}$, $j, k \in \Lambda_\ell \subset \Lambda^J$ which exceed a certain threshold ε_ℓ in magnitude:

$$
\bar{y}_{i,\ell} = \sum_{|d_{j,k}| \geq \varepsilon_\ell} d_{j,k}\psi_{j,k} = d_{\bar{y}_{i,\ell}}^T \Psi_{\Lambda_{\bar{y}_{i,\ell}}}.
$$

The threshold ε_ℓ is chosen such that the approximation is compressed in a sense that a specified accuracy of the approximation is preserved [9, 23, 24, 45]. The signal $\bar{y}_i^J$ is first represented by a coarse approximation determined by a comparatively large threshold ε_0. The approximation accuracy is increased taking an extended set of coefficients into account which corresponds to $\varepsilon_{\ell+1} < \varepsilon_\ell$. Since the number of significant wavelet coefficients is often far less than the number of discrete measurement samples we do have a fairly compressed representation of the signal $\bar{y}_i^J$ which contributes to a reduction of the computational effort in the subsequent optimization. Since we choose the same function spaces for signal processing and optimization the information in $\Lambda_{\bar{y}_{i,\ell}}$ can be used as an information source to select an appropriate discretization of the states and degrees of freedom in the optimization problem in a truncated multiscale basis. In addition, only the expansion coefficients $d_{\bar{y}_{i,\ell}}$ instead of the whole sampled signal have to be exchanged and therefore fast and easy data exchange is facilitated.

5.3. OPTIMIZATION PROBLEM

5.3.1. *Weak problem formulation*
The optimization problem (P4) should be discretized in a way that the concept of adaptivity is supported. A multiscale formulation of the process model, which permits model representation of various resolutions is obtained by a Petrov–Galerkin discretization. Following this approach the weak formulation turns out to be appropriate:

$$\min \| \boldsymbol{Q}(\boldsymbol{y} - \bar{\boldsymbol{y}}) \|^2_{H^{\beta_y}(\Omega)} + \| \boldsymbol{R}_v \boldsymbol{v} \|^2_{H^{\beta_v}(\Omega)} + \| \boldsymbol{R}_w \boldsymbol{w} \|^2_{H^{\beta_w}(\Omega)} \qquad (27)$$

subject to the constraints

$$\langle \dot{\boldsymbol{x}} - \boldsymbol{f}(\boldsymbol{x}, \boldsymbol{u}) - \boldsymbol{w}, \boldsymbol{\nu}_x \rangle = 0 \quad \forall \, \boldsymbol{\nu}_x \in (L_2(\Omega))^{n_x}, \qquad (28)$$

$$\langle \boldsymbol{y} - \boldsymbol{g}(\boldsymbol{x}, \boldsymbol{u}) - \boldsymbol{v}, \boldsymbol{\nu}_y \rangle = 0 \quad \forall \, \boldsymbol{\nu}_y \in (L_2(\Omega))^{n_y}, \qquad (29)$$

$$\langle s_j - s_{max,j}, \delta_{t_{i,j}} \rangle \leq 0 \quad \forall \, j = 1 \ldots n_s, \ i = 1 \ldots n_{s_j}, \qquad (30)$$

$$\langle s_{min,j} - s_j, \delta_{t_{i,j}} \rangle \leq 0 \quad \forall \, j = 1 \ldots n_s, \ i = 1 \ldots n_{s_j}. \qquad (31)$$

Here $\delta_{t_{i,j}} = \delta(\cdot - t_{i,j})$ are Dirac distributions. $\boldsymbol{\delta}_{max}$ and $\boldsymbol{\delta}_{min}$ collect the Dirac functions in (30) and (31) respectively. They span together with $\boldsymbol{\nu}_x, \boldsymbol{\nu}_y$ the spaces of test functions. In this setting the j-th inequality constraint of the catenated vector $\boldsymbol{s}$ is forced to hold at the n_{s_j} mesh points $t_{i,j} \in \Omega, i = 1 \ldots n_{s_j}$. For finite dimensional approximations of s_j, this mesh can always be chosen fine enough to guarantee the inequality constraints with sufficient accuracy. Note, that the weak formulation holds on every refinement level ℓ. Hence, the processed measurements as well as the states and degrees of freedom should be indicated by the refinement index ℓ which is omitted here and subsequently to ease notation.

To further simplify the notation we define $\boldsymbol{\mu} := (\boldsymbol{x}^T, \boldsymbol{y}^T, \boldsymbol{v}^T, \boldsymbol{w}^T)^T \in U$ and $\boldsymbol{\nu} := (\boldsymbol{\nu}_x^T, \boldsymbol{\nu}_y^T, \boldsymbol{\delta}_{max}^T \boldsymbol{\delta}_{min}^T)^T \in M$, where U and M are the product spaces $U := U^x \times U^y \times U^v \times U^w$ and $M := M^x \times M^y \times M^{s_{max}} \times M^{s_{min}}$.

Before a discrete problem can be formulated the trial spaces U and test spaces M have to be chosen properly. This choice can take advantage of the richness of biorthogonal multiscale function spaces. The best choice of a suitable wavelet basis and its dual would lead to discretized equations with very sparse or ideally (almost) diagonal Jacobian matrices to ease the subsequent numerical solution step and at the same time showing good approximation properties in signal processing as well as in optimization. At this point we choose basis functions from the B-spline family and its dual as already employed for signal processing for both U and M and explicitly state that our current choice is one of many other (probably more advantageous) options. Further work has to be done to find the best suitable basis.

The hat function $\varphi^{(2)} \in C^1$ and the corresponding wavelet functions $\psi^{(2,2)}$ are selected for the discretization of x, v, and y taking advantage of their smoothness and the interpolation property in the mesh points 2^{-j}. Further, in order to obtain a simple discretization structure w is discretized with Haar wavelets $\psi^{(1,1)}$ which are constructed from $\varphi^{(1)} \in C^0$. With this (first and therefore preliminary) choice the trial spaces have the form

$$
\begin{aligned}
U_\Lambda &:= U_{\Lambda_x}^x \times U_{\Lambda_y}^y \times U_{\Lambda_v}^v \times U_{\Lambda_w}^w \\
&= span\{\psi_{j,k}^{(2,2)} : j, k \in \Lambda_x\} \times span\{\psi_{j,k}^{(2,2)} : j, k \in \Lambda_y\} \times \\
&\quad span\{\psi_{j,k}^{(2,2)} : j, k \in \Lambda_v\} \times span\{\psi_{j,k}^{(1,1)} : j, k \in \Lambda_w\}
\end{aligned}
$$

where the finite index sets $\Lambda_x, \Lambda_y, \Lambda_v$ and Λ_w can be chosen adaptively in a specific manner for every refinement level ℓ. As mentioned already before, we only assume that the index set includes all scaling functions $\psi_{j_0-1,k}$ and selected functions on higher scales of the multiscale expansion. At this point, it is open how to select the basis functions in the index sets in order to achieve a good approximation.

5.3.2. *Projection*

As it will be seen during the course of this section the flexibility offered by a biorthogonal wavelet basis can be fully utilized to obtain a suitable discretization structure. To ease the notation and to avoid an explosion of the indices, we only consider a scalar component of the vector functions x, y etc.. The extension to the vector case is straightforward.

First, we start with an inequality constraint (30). The trial spaces, defined in the last section, together with the test space $M_\Lambda^{s_{max,j}} = span\{\delta_{t_{i,j}}, i = 1, \ldots, n_{s_j}\}$ yield

$$
d_{s_j}^T \Psi_{\Lambda_{s_j}}^{(2,2)}(t_{i,j}) - s_{min,j}(t_{i,j}) \leq 0, \quad i = 1, \ldots, n_{s_j}. \tag{32}
$$

Note, that the choice of the mesh $t_{i,j}, i = 1, \ldots, n_{s_j}$ is independent of the measurement samples and reflects the locations where the inequality is exactly satisfied. The inequality constraints (31) are projected in a similar manner.

In order to minimize computational work the test space M_Λ^y of the algebraic model equation (29) is chosen such that the discretization matrix for y will be the identity matrix I. At this point biorthogonality of the wavelet spaces (19) comes into play. Recall that the approximation space U_Λ^y is spanned by the wavelet basis $\{\psi_{j,k}^{(2,2)} : j,k \in \Lambda_y\}$ and that there exist biorthogonal functions $\tilde{\psi}_{j,k}^{(2,2)}$ such that $\langle \psi_{j,k}^{(2,2)}, \tilde{\psi}_{i,l}^{(2,2)}\rangle = \delta_{i,j}\delta_{k,l}$. Consequently, if $M_\Lambda^y = span\{\tilde{\psi}_{i,l}^{(2,2)} : i,l \in \Lambda_y\}$ the algebraic model equation (29) is equivalent to

$$d_y - \langle g(x_\Lambda, u_\Lambda), \tilde{\Psi}_{\Lambda_y}\rangle - V_\Lambda d_v = 0, \qquad (33)$$

where $V_\Lambda = I$ for $\Lambda_y = \Lambda_v$.

The determination of the test space M_Λ^x for the differential equations (28) requires some care. For stability reasons the test space M_Λ^x has to capture the space spanned by $\dot{\Psi}_{\Lambda_x}$, i.e. the trial space of the differential operator in (28). Utilizing a Petrov-Galerkin approach we choose $M_\Lambda^x :=$ $span\{\psi_{j,k}^{(1,1)} : j,k \in \Lambda_x'\}$ as trial space such that $span\{\psi_{j,k}^{(1,1)} : j,k \in \Lambda_x'\} = span\{\dot{\psi}_{j,k}^{(2,2)} j,k \in \Lambda_x\}$ where Λ_x' denotes an adapted index set of the trial functions. This choice yields

$$X_\Lambda d_x - W_\Lambda d_w - \langle f(x_\Lambda, u_\Lambda), \Psi_{\Lambda_x'}^{(1,1)}\rangle = 0 \qquad (34)$$

with the matrices $X_\Lambda = \langle \dot{\Psi}_{\Lambda_x}^{(2,2)}, \Psi_{\Lambda_x'}^{(1,1)}\rangle$ and $W_\Lambda = \langle \Psi_{\Lambda_w}^{(1,1)}, \Psi_{\Lambda_x'}^{(1,1)}\rangle$. In summary, the test spaces are

$$
\begin{aligned}
M_\Lambda &= M_\Lambda^x \times M_\Lambda^y \times M_\Lambda^{s_{max}} \times M_\Lambda^{s_{min}} \\
&= span\{\psi_{j,k}^{(1,1)} : j,k \in \Lambda_x'\} \times span\{\tilde{\psi}_{j,k}^{(2,2)} : j,k \in \Lambda_y\} \times \\
&\quad span\{\delta_{t_{i,j}} : j = 1,\ldots,n_s, \ i = 1,\ldots,n_{s_j}\} \times \\
&\quad span\{\delta_{t_{i,j}} : j = 1,\ldots,n_s, \ i = 1,\ldots,n_{s_j}\}
\end{aligned}
$$

supporting a stable and adaptive discretization scheme.

5.3.3. *Discretization of the cost functional*
For the actual computations, the cost functional in (P4) has to be substituted by a discrete norm which can be computed utilizing only the wavelet coefficients. Straightforward substitution of the trial functions into (27) yields the discretized cost functional

$$\acute{e} = (d_y - d_{\bar{y}})^T \acute{Q}_\Lambda (d_y - d_{\bar{y}}) + d_v^T \acute{R}_{\Lambda_v} d_v + d_w^T \acute{R}_{\Lambda_w} d_w \qquad (35)$$

652

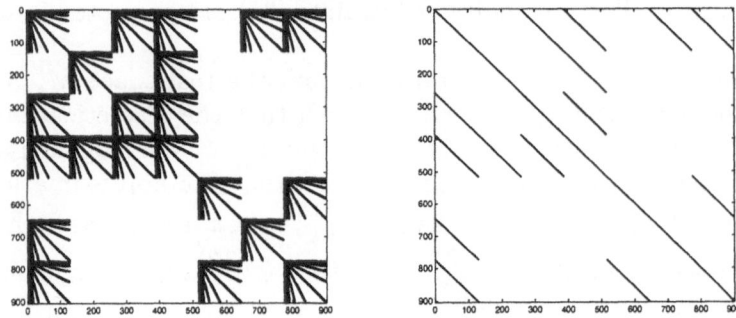

Figure 2. Structure of the discretization matrix $\acute{\boldsymbol{Q}}_{\Lambda_y}$ (left) and $\boldsymbol{Q}_{\Lambda_y}$ (right).

with $\acute{\boldsymbol{Q}}_{\Lambda_y} = \boldsymbol{Q}^T \boldsymbol{Q} \otimes \langle \boldsymbol{\Psi}_{\Lambda_y}, \boldsymbol{\Psi}_{\Lambda_y} \rangle$ and analogously for $\acute{\boldsymbol{R}}_{\Lambda_v}, \acute{\boldsymbol{R}}_{\Lambda_v}$. For the moment the Sobolev regularity is specialized to $\beta = 0$; i.e. differentiation of the function is avoided. The matrices in (35) exhibit sparse block finger structure which is typical for wavelet discretized systems. Applying the norm equivalences (22), the block finger structure can be simplified to a block diagonal structure due to biorthogonality of the wavelets and their duals. To allow for a more general cost functional involving Sobolev regularities $\beta \geq 0$ and thus weighting smoothness properties of the functions, we use another important property of wavelet bases which extends the Riesz basis property (22) to the Sobolev space H^β. In fact, for a certain range of regularity parameters β, depending only on the regularity of $\boldsymbol{\Psi}$ and $\tilde{\boldsymbol{\Psi}}$, one has the norm equivalence

$$\| f \|_{H^\beta(\Omega)}^2 \sim \sum_{j,k} 2^{2\beta j} |\langle f, \tilde{\psi}_{j,k} \rangle|^2, \quad f \in C^\beta(\Omega).$$

Thus the weighting of the smoothness of the functions in the cost functional can be easily obtained through a simple level dependent weighting of the wavelet coefficients. Therefore (35) can be replaced by

$$e = (d_y - d_{\bar{y}})^T \boldsymbol{Q}_\Lambda (d_y - d_{\bar{y}}) + d_v^T \boldsymbol{R}_{\Lambda_v} d_v + d_w^T \boldsymbol{R}_{\Lambda_w} d_w \tag{36}$$

where $\boldsymbol{Q}_{\Lambda_y} = \boldsymbol{Q} \otimes \boldsymbol{D}_{\Lambda_y}^\beta$ with $\left(\boldsymbol{D}_{\Lambda_y}^\beta \right)_{(j,k),(i,l)} = 2^{2\beta j} \delta_{j,i} \delta_{k,l}$ and analogously for $\boldsymbol{R}_{\Lambda_v}, \boldsymbol{R}_{\Lambda_w}$. Note, that $\boldsymbol{Q}_\Lambda, \boldsymbol{R}_{v_\Lambda}, \boldsymbol{R}_{w_\Lambda}$ in (36) are matrices with block diagonal structure and do not reveal the block finger structure of the direct discretization approach (35). For illustration the discretization matrices $\acute{\boldsymbol{Q}}_{\Lambda_y}$ and $\boldsymbol{Q}_{\Lambda_y}$ are shown in Figure 2 for a non-diagonal $\boldsymbol{Q}$ where the nonzero entries are represented by dots.

5.4. SOLUTION OF DISCRETIZED OPTIMIZATION PROBLEM

The discretization employing biorthogonal multiscale approximation and test spaces results in a potentially large number of simple inequalities and a large-scale sparse nonlinear equation system (32)–(34) constraining a quadratic cost functional (36). The resulting NLP can be written in the following compact form

$$\min e(\boldsymbol{\mu}_\Lambda) \tag{37}$$

subject to the equality and inequality constraints

$$\begin{aligned} g(\boldsymbol{\mu}_\Lambda) &= \mathbf{0}, \\ h(\boldsymbol{\mu}_\Lambda) &\leq \mathbf{0}. \end{aligned}$$

This large-scale NLP reveals a lot of structure which must be exploited to take full advantage of the multiscale approach. At this point, only preliminary results and explorative codes have been developed to assess some of the design issues. In the following we present a straightforward application of a general purpose NLP solver and give some remarks on the research questions to be tackled when tailoring an NLP solver to the particular problem structure.

5.4.1. *Solution of the NLP with a general purpose solver*
In principle, (37) can be solved with any available nonlinear programming technique. However, since the function evaluations are quite expensive we are interested in solution techniques which are known to take as few iterations as possible. Here, sequential quadratic programming (SQP) or interior point SQP methods seem to be more advantageous than other alternatives like gradient or generalized Gauss-Newton methods.

Hence, we have decided to employ a general purpose SQP solver which is suitable for large-scale problems. Our current implementation employs the SQP solver RND developed by Biegler and coworkers [46, 56, 57]. Only first order information needs to be provided in addition to the residuals since the Hessian is approximated by means of the BFGS update formula. The SQP method solves the QP subproblem

$$\min[\boldsymbol{b}_\Lambda(\boldsymbol{\mu}_\Lambda^i)^T + \frac{1}{2}\Delta\boldsymbol{\mu}_\Lambda^{i\,T}\boldsymbol{\mathcal{A}}_\Lambda(\boldsymbol{\mu}_\Lambda^i)\,\Delta\boldsymbol{\mu}_\Lambda^i] \tag{38}$$

subject to

$$\begin{aligned} g_\Lambda(\boldsymbol{\mu}_\Lambda^i) + \boldsymbol{\mathcal{G}}_\Lambda(\boldsymbol{\mu}_\Lambda^i)\Delta\boldsymbol{\mu}_\Lambda^i &= 0 \\ h_\Lambda(\boldsymbol{\mu}_\Lambda^i) + \boldsymbol{\mathcal{H}}_\Lambda(\boldsymbol{\mu}_\Lambda^i)\Delta\boldsymbol{\mu}_\Lambda^i &\leq 0 \end{aligned}$$

in every iteration i to generate the search direction $\Delta\mu_\Lambda^i$. Here, $\mathcal{A}_\Lambda$ is the Hessian of the Lagrangian and $b_\Lambda = \nabla e_\Lambda, \mathcal{G}_\Lambda = \nabla g_\Lambda^T, \mathcal{H}_\Lambda = \nabla h_\Lambda^T$ are defined as the local derivatives of the cost functional and constraints respectively. In an application of a general purpose code, the user has to provide the function evaluations $e(\mu_\Lambda^i), g(\mu_\Lambda^i), h(\mu_\Lambda^i)$ as well as the first order information $\mathcal{G}(\mu_\Lambda^i), \mathcal{H}(\mu_\Lambda^i), b(\mu_\Lambda^i)$ at the current iterate μ_Λ^i.

In order to make this information available, inner products of wavelet functions have to be computed. For example, consider equation (34) which is included in the set of equality constraints of the NLP (37):

$$X_\Lambda d_x^i - W_\Lambda d_w^i - \langle f(x_\Lambda^i, u_\Lambda), \Psi_{\Lambda_x'}^{(1,1)} \rangle = 0.$$

Note that $x_\Lambda^i = d_x^{i\,T}\Psi_\Lambda$, $w_\Lambda^i = d_w^{i\,T}\Psi_\Lambda$ are part of μ_Λ^i. Due to the linearity, X_Λ and W_Λ can be computed prior to the optimization run with known numerical techniques [18]. However, the evaluation of the nonlinearity is much more difficult. To date, there are no methods available which allow an efficient evaluation of this expression. The objective of current research is to find methods for the evaluation of nonlinear functions of arguments given as a wavelet expansion which avoid the complexity of the finest mesh (determined by the highest resolution in Λ, briefly referred to as full single scale dimension) but require only an effort proportional to the dimension of the adapted wavelet basis.

Pragmatically, the nonlinearity is evaluated by a method which is of the complexity of the single scale dimension at this stage of our investigations. For this purpose, the function f is evaluated pointwise first leading to a representation in terms of the single scale basis functions which are dual to the generators of $\tilde{\Psi}_{\Lambda_x'}^{(1,1)}$. Then a fast wavelet transformation is carried out to result in

$$f(d_x^{i\,T}\Psi_{\Lambda_x}, d_u^T\Psi_{\Lambda_u}) \simeq d_f^{i\,T}\tilde{\Psi}_{\Lambda_x'}^{(1,1)} \tag{39}$$

which is a multiscale approximation of f. Hence, the inner product simplifies to

$$\langle f(d_x^{i\,T}\Psi_{\Lambda_x}, d_u^T\Psi_{\Lambda_u}), \Psi_{\Lambda_x'}^{(1,1)} \rangle \simeq \langle \tilde{\Psi}_{\Lambda_x'}^{(1,1)}, \Psi_{\Lambda_x'}^{(1,1)} \rangle d_f^i = d_f^i.$$

5.4.2. Refinement of discretization

The exactness of the solution to problem (P4) is governed by the choice of the trial space U_Λ. The key element of the proposed strategy is the successive refinement. Let Λ_ℓ denote the "old" index set which gives rise to the trial space U_{Λ_ℓ}, while the "new" trial functions, added in the refinement step, are given by the set Γ_ℓ. The optimization problem has been solved already in resolution U_{Λ_ℓ} using the SQP approach. The solution is then refined taking the additional trial functions selected by Γ_ℓ into account. Hence the resulting trial space after one refinement step is $U_{\Lambda_{\ell+1}}$ which is determined by the index set $\Lambda_{\ell+1} = \Lambda_\ell \cup \Gamma_\ell$.

The update procedure ensures that at any time the solution to the last approximated problem is available. Furthermore, this solution can also be used as an initial guess for the refined problem. Since the moment condition (25) implies decreasing wavelet coefficients with increasing scales for smooth functions, the zero augmented solution vector of resolution U_{Λ_ℓ} is a good initial guess for the refined problem (37) in resolution $U_{\Lambda_{\ell+1}}$. Hence, $\mu_{\Lambda_{\ell+1}}^0 = [\mu_{\Lambda_\ell}^T, 0^T]^T$ is used as an initial guess and the changes in every entry of the old vector are expected to become smaller and smaller.

In principle, the trial functions spanning U_{Λ_ℓ} can be chosen arbitrarily. The index set Λ_ℓ might therefore contain *all* wavelet functions up to a scale J, which corresponds to an equidistant mesh. Alternatively, in an adaptive approach Λ_ℓ would consist a strict subset of all wavelet indices up to scale J. Thus, when progressing up to higher discretization levels, the trial spaces are augmented possibly only by those subsets of the complements bases $\Psi_j, j < J$, that within some tolerance recapture most of the searched object. This provides a higher resolution for all measurements, state variables and degrees of freedom only at places where needed due to the individual behavior of these functions. The number of trial functions needed to get a certain accuracy in a nonuniform adaptive scheme is often significantly lower compared to a uniform refinement depending on the properties of the function to be approximated.

The question immediately arises how to determine Γ_ℓ in an adaptive refinement approach. Ultimately, an error estimator will be constructed on the basis of the theoretical multiscale framework such that those trial functions are selected which guarantee the best possible reduction of the approximation error. At this point, no such error estimator is available instead we employ a heuristic adaptation strategy introduced in [30]. Given the index set Λ_ℓ and the solution μ_{Λ_ℓ} in refinement level ℓ we search for the expansion coefficients with the largest absolute values of μ_{Λ_ℓ}. Because of the norm equivalence property (22) these coefficients have the biggest impact on the quality of the approximation. The multiscale property together with smoothness assumptions suggest that neighbors of trial functions with large coefficients are excellent candidates for refinement. Here the wavelet functions $\psi_{j,k-1}$ and $\psi_{j,k+1}$ as well as the functions of the higher scale $\psi_{j+1,2k}$ are referred to as neighbors to the of $\psi_{j,k}$ and therefore correspond to the local neighborhood involving scale and translation. Typically not all the neighbors of the large coefficients not yet contained in Λ_ℓ are selected and added into Γ_ℓ. Rather, the additional basis functions are chosen to ensure reduction of the approximation error on the one side and a sufficiently small computation time h which is balanced against the cycle time H such that several update steps can be executed.

5.4.3. *Tailored NLP algorithm*

In order to establish efficiency in the algorithmical treatment the strategy for the solution of the NLP (37) presented so far is not sufficient. Rather, a tailored NLP algorithm need to be developed to take full advantage of our framework. Preliminary theoretical as well as experimental studies indicate, that there may be significant gains in efficiency if the following issues are considered during the development of a tailored NLP solver:

- exploitation of the sparse discretization structure,
- application of wavelet matrix compression techniques,
- tailored linear algebra solvers,
- efficient preconditioning of the linear systems in particular if iterative solvers are employed,
- updating of the Hessian and Jacobian, and
- the treatment of the inequality constraints.

In the following some more detailed remarks are given in order to render our current ideas a little bit more transparent.

The discretization matrices in the QP subproblem (38) exhibit a block sparse structure with sparse block matrices. The block structure reflects the structure of the cost functional and the constraints. Each block matrix reveals the particular wavelet discretization structure depending on the choice of the test and trial spaces. A proper combination may lead to more favorable patterns than those we have in our current approach. At the moment, some of the blocks of the matrices in (38) possess the wavelet typical finger structure while others are diagonal. The finger structure has $O(N \log N)$ non zero entries, where N denotes the dimension of each block. With known matrix compression techniques [19] the block finger structure can be represented by only $O(N)$ entries without significant loss of accuracy.

The underlying linear system of the QP subproblem can be solved either by factorization or by iterative techniques. Only the latter may allow for a reasonable order of complexity roughly proportional to the number of variables. Fast convergence of the iterative algorithm is a key issue. Convergence acceleration can be established through efficient preconditioning. For $J \to \infty$ diagonal preconditioners exist [8, 19] in a sense that the condition number will be independent from the number of trial functions used to discretize the system. However, more work remains to be done to develop efficient iterative solution approaches for the problem considered here.

Since we do not assume that the second order information of the model is available the Hessian is approximated using established techniques. During the course of updating the approximation to the Hessian $\mathcal{A}_{\Lambda_{\ell+1}}$ can also be warm started augmenting the Hessian $\mathcal{A}_{\Lambda_\ell}$ obtained from the old approximation such that positive definiteness is guaranteed. Although the elements of the Jacobian $\mathcal{G}_{\Lambda_{\ell+1}}$ and Hessian $\mathcal{A}_{\Lambda_{\ell+1}}$ change in each QP iteration, the parts which refer to the old trial functions given through the set Λ_ℓ stay almost constant. Therefore they only have to be updated rather than computed in each iteration step. However, this observation is based on the assumption that the mesh is already chosen fine enough such that the coarse scale components of $\mu_{\Lambda_\ell}^i$ remain nearly unchanged when augmenting Λ_ℓ.

6. Case-studies

In this section we illustrate the principles of the multiscale approach to dynamic reconciliation using a toy example. We consider a nonisothermal continuous stirred tank reactor (CSTR) with an exothermic first order reaction. The model equations can be written in dimensionless form [66]

$$
\begin{aligned}
\dot{x}_1 &= -x_1 + Da(1 - x_1)\exp(\frac{x_2}{1 + x_2/\gamma}) + u_1, \quad (40) \\
\dot{x}_2 &= -(1 + \beta)x_2 + BDa(1 - x_1)\exp(\frac{x_2}{1 + x_2/\gamma}) + \beta u_2.
\end{aligned}
$$

Here x_1, x_2 are defined as the dimensionless concentration and temperature in the reactor whereas u_1, u_2 are the associated feed concentration and cooling temperature respectively. Da is the Damkohler number, β is the dimensionless heat transfer coefficient, and B is the adiabatic temperature rise. Note, that the time in the above formulation is as well normalized. The process parameters used in this example are ($Da = 0.162, B = 14.0, \beta = 3.0, \gamma = \infty$). In our numerical experiments, we consider only one fixed horizon of the length $[0, T]$ with T = 10 and project this horizon on the normalized interval $\Omega = [0, 1]$. We do not yet consider a moving time window. The "true" (or nominal) process state variables are simulated by integration of the model equations with $u_1 = u_2 = 0$ starting with the initial conditions $x_0 = [0.35, 0.9]^T$. In this setting the dimensionless temperature x_2 is considered to be the measured variable. The measurements $\check{Y}$ which are corrupted by normally distributed uncorrelated noise are sampled on an equidistant mesh with step size 2^{-J}, J = 7. The goal of the optimization problem is to obtain reconciled estimates $\hat{x}_1, \hat{x}_2$ of concentration and temperature from the temperature measurements $\check{Y}$ for some Q, R_v, R_w utilizing the model (40).

The test and trial spaces are chosen as outlined in Section 5.3.1 and 5.3.2. In the first step, we interpolate the measurement samples $\check{Y}$ by means

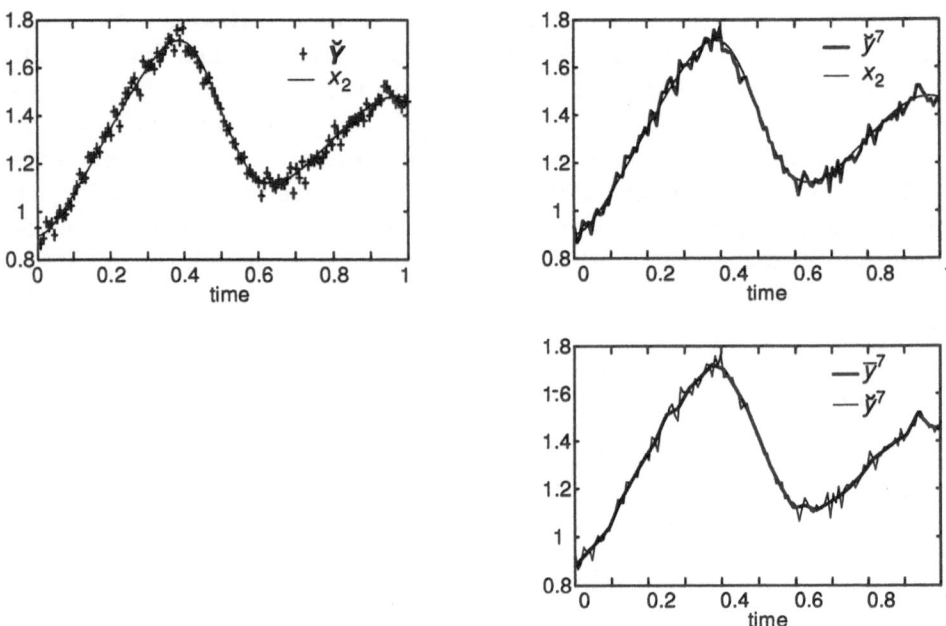

Figure 3. Measurements, original and processed measurements.

of B–splines $\varphi^{(2)}$ to obtain the continuous representation $\breve{y}^7$. Since this B–spline is a piecewise linear function the samples are interpolated also by a piecewise linear function. The measurement function $\breve{y}^7$ then is transformed into the multiscale representation and denoised with an appropriate chosen threshold parameter ε. The nominal temperature x_2, the measurement samples $\breve{Y}$, the interpolated measurements $\breve{y}^7$ and the processed measurements $\bar{y}^7$ are shown in Figure 3. Occasionally sharp spikes can be seen in the denoised signal. With the use of higher order trial functions, i.e. quadratic ($m = 3$) or cubic B–splines ($m = 4$), the representation will be smoother.

In the example considered here, the optimization problem is formulated and solved starting from a coarse approximation $\ell = 0$ with three subsequent refinement steps ($\ell = 0, \ldots, 3$) employing uniform refinement: Each quantity $\bar{y}, x, y, v, w$ in the optimization problem (P4) is discretized by wavelet functions of the index set $\Lambda_\ell^{J_\ell}$ with $J_\ell = 2, \ldots, 5$. This choice represents all trial functions of $S_1 \oplus W_1 \oplus \ldots \oplus W_{J_\ell - 1}$ and is equivalent to an equidistant mesh with step size 2^{-J_ℓ}. To ease notation we ommit the index J_ℓ in $\Lambda_\ell^{J_\ell}$. Figure 4 reveals the processed measurement function $\bar{y}_{\Lambda_\ell} = d_{\bar{y}_\ell}^T \Psi_{\Lambda_{\bar{y}_\ell}}$ which are used as input to the optimization problem. The reconciled states $\hat{x}_{i,\Lambda_\ell} = d_{\hat{x}_{i,\ell}}^T \Psi_{\Lambda_{\hat{x}_{i,\ell}}}$ of the optimization based on the choice of $Q = R_v = R_w = I$ are shown in the Figures 5-6. Note that starting

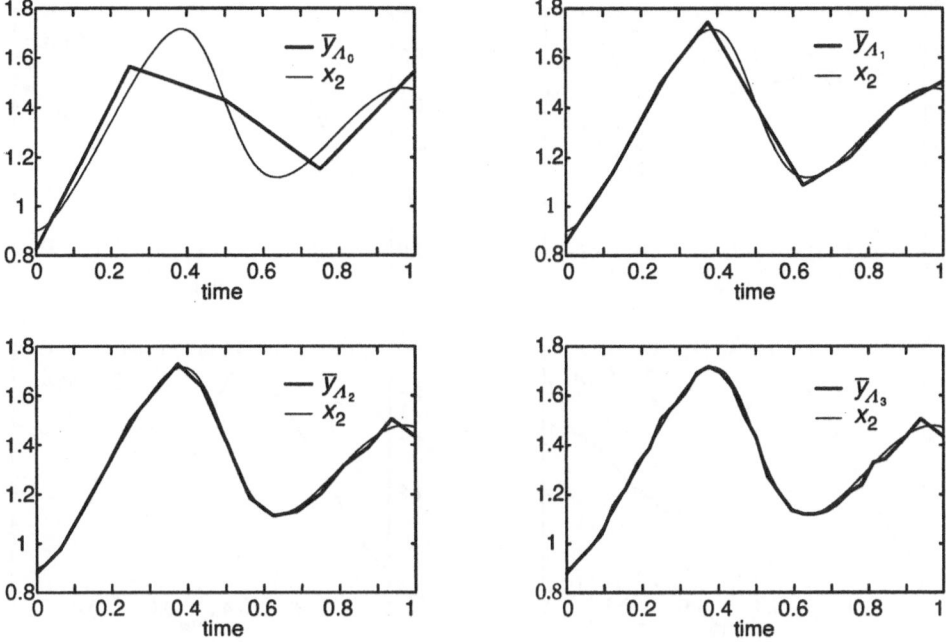

Figure 4. Processed measurement function on various scales.

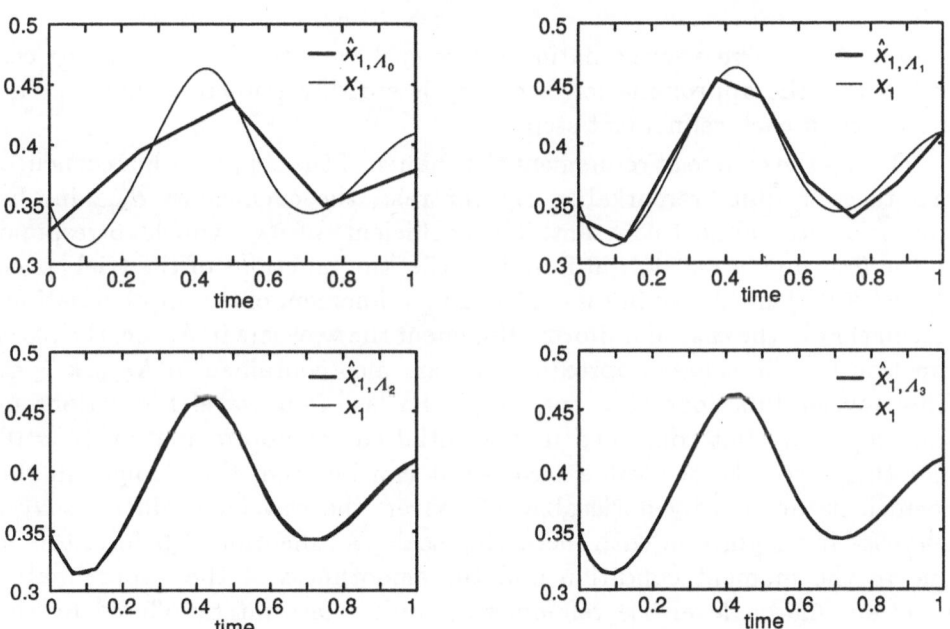

Figure 5. Refinements and original of state x_1.

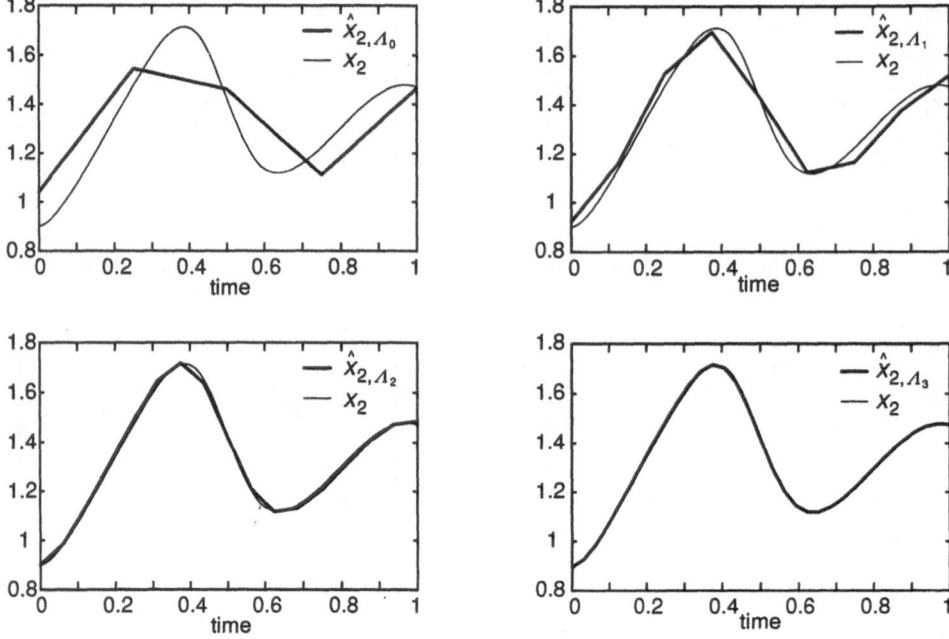

Figure 6. Refinements and original of state x_2.

from a very coarse approximation where each quantity has only 5 degrees of freedom the approximation accuracy is steadily refined by halving the mesh size in each refinement step.

During the course of refinement the change of the shape of the reconciled quantities is quite remarkable. By contrast, the coefficients $\boldsymbol{d}_{\hat{x}_{i,\ell}}$ in the expansion stay almost constant. The coefficients of $\boldsymbol{d}_{\hat{x}_{i,\ell}}$ which correspond to the first nine wavelet trial functions (i.e. the functions of the trial space $S_1 \oplus W_1 \oplus W_2$) are shown in Figure 7 during refinement of the approximation. Recall that in the case of uniform refinement the wavelets in Λ_ℓ, i.e. the basis function for the coarse approximation, are also contained in $\Lambda_{\ell+k}, k \geq 0$. The wavelet functions that span W_2 (the last four wavelet functions in the graph) are not contained in the initial coarse approximation ($\ell = 0$) and therefore take the value zero. As it can be seen, the changes in the coefficients are hardly noticeable. Moreover, the wavelet coefficients $\boldsymbol{d}_{\hat{x}_{i,\ell}}$ decrease in magnitude with increasing scale. As mentioned before, this is due to the moment condition and the smoothness of the approximated function. Again, levelwise refinement is only one strategy offered by the discretization scheme presented. It leads to a large number of degrees of freedom after few refinement steps. On the other hand, there is no need to increase the resolution uniformly. In this example, however, the levelwise

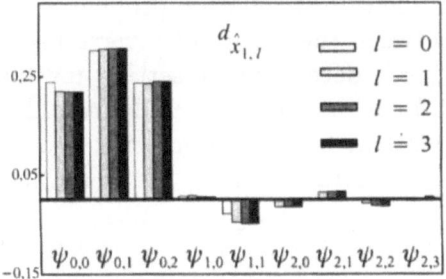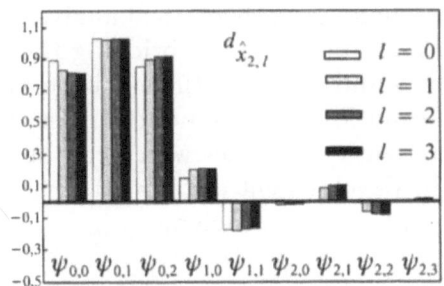

Figure 7. Wavelet coefficients during refinements of $\hat{x}_1$ and $\hat{x}_2$.

refinement strategy is very favorable due to the nature of the approximated process quantities and a nonuniform adaptive strategy does not result in significant reduction of the effort. In other examples we have studied, a significant reduction can be achieved by nonuniform refinement as outlined in Section 5.4.2.

7. Conclusions

We have sketched a novel treatment of dynamic data reconciliation problems. In the first part, it suggests a purely deterministic formulation of the problem based on the theory of inverse problems. At this point, the formulation is still partly heuristic and based on theoretical evidence gained from the understanding of general linear inverse problems. A theoretical analysis of the estimation problem for the linear as well as the nonlinear data reconciliation problem is subject to future work in order to show the appropriateness of the suggested formulation. In particular, an assessment of the estimation quality is essential.

The second part suggests a new framework to the numerical treatment of dynamic data reconciliation. It is based on multiscale techniques for the discretization of the dynamic optimization problem to result in a nonlinear programming problem as well as for the solution of the underlying system of equations. The presentation is limited to an outline of the general ideas underlying the multiscale framework. The technical realization of the conceptional ideas is still at the very early stage of research. Hence, all the techniques presented at this point must be viewed as being of preliminary nature. These first results demonstrate, however, convincingly the potential of the approach to finally result in a highly efficient numerical technique which is able to provide the best possible estimate for given computational resources.

Obviously, much work remains to be done. We will concentrate on the one hand on a theoretical analysis of the linear reconciliation problem first. We are interested in bridging the gap between the theory of inverse problems and observer theory, both employing a deterministic setting for the estimation. The numerical research will focus on improved multiscale discretization schemes which can be integrated with signal processing and on the development of tailored methods for the solution of the mathematical programming problem.

ACKNOWLEDGEMENT

The authors acknowledge the contribution by A. M'hamdi. The work presented here was supported by the "Deutsche Forschungsgemeinschaft" under grant no. MA1188/6 which is also gratefully acknowledged.

References

1. O. Abel, A. Helbig, and W. Marquardt. Optimization approaches to control-integrated design of industrial batch reactors. NATO ASI on Nonlinear Model Based Process Control. This volume. pp. 513-551.
2. J.S. Albuquerque and L.T. Biegler. Decomposition algorithmus for on–line estimation with nonlinear DAE models. *Comput. Chem. Eng.*, 19:1031–1039, 1995.
3. J.S. Albuquerque and L.T. Biegler. Data reconcilliation and gross–error detection for dynamic systems. *AIChE J.*, 42(10):2841–2856, 1996.
4. J.S. Albuquerque, V. Gopal, G.H. Stauss, L.T. Biegler, and B.E. Ydstie. Interior point SQP strategies for structured process optimization problems. *Comput. Chem. Eng.*, 21(Suppl.):853–859, 1997.
5. M.J. Bagajewicz and Q. Jiang. Integral approach to plant linear dynamic reconciliation. *AIChE J.*, 43(10):2546–2558, 1997.
6. B. Bakshi and G. Stephanopoulos. Representation of process trends: III multi–scale extraction of trends from process data. *Comput. Chem. Eng.*, 18(4), 1994.
7. B. Bakshi and G. Stephanopoulos. Compression of chemical process data by functional approximation and feature extraction. *AIChE J.*, 42(2):477–492, 1996.
8. J.H. Bramble and J.E. Pasciak. A preconditioning technique for indefinite systems resulting from mixed approximations for elliptic problems. *Math. Comp.*, 50:1–17, 1988.
9. A.G. Bruce, D.L. Donoho, H.Y. Gao, and R.D. Martin. Smoothing and robust wavelet analysis. In *Proceedings on Computational Statistics. 11th Symposium, Berlin*, volume 4, pages 531–547. COMPSTAT, 1994.
10. A. Chambolle, R. DeVore, N.Y. Lee, and J. Lucier. Nonlinear wavelet image processing: Varational problems, compression, and noise removal through wavelet shrinkage. 1996.
11. C.K. Chui. *An Introduction to Wavelets*. Academic Press, Boston, 1992.
12. H. Cox. On the estimation of state variables and parameters for noisy dynamic systems. *IEEE Trans. Auto. Cont.*, pages 5–12, January 1964.
13. C.M. Crowe. Data reconciliation – progress and challage. *J. Proc. Cont.*, 6:89–98, 1996.
14. S.A. Dadebo and K.B. McAuley. Dynamic optimization of constrained chemical engineering problems using dynamic programming. *Comput. Chem. Eng.*, 19(5):513–525, 1995.
15. W. Dahmen. Wavelet and Multiscale Methods for Operator Equations. *Acta Numerica*, pages 55–228, 1997.

16. W. Dahmen, A. Kunoth, and K. Urban. Biorthogonal spline-wavelets on the interval – stability and moment conditions. IGPM–Report 129, RWTH Aachen, 1996.

17. W. Dahmen, A. Kunoth, and K. Urban. Wavelets in Numerical Analysis and their Quantitative Properties. In A.Le Méhauté, C. Rabut, and L.L. Schumaker, editors, *Surface Fitting and Multiresolution Methods*, pages 93–130. Vanderbilt University Press, 1997.

18. W. Dahmen and C.A. Micchelli. Using the refinement equation for evaluating integrals of wavelets. *SIAM J. Numer. Anal.*, 30(2):507–537, 1993.

19. W. Dahmen, S. Prössdorf, and R. Schneider. Multiscale methods for pseudo-differential equations on smooth manifolds. In C.K. Chui, L. Montefusco, and L. Puccio, editors, *Proceedings of the International Conference on Wavelets: Theory, Algorithms, and Applications*, pages 385–424. Academic Press, San Diego, 1994.

20. M. Darouach and M. Zasadzinski. Data reconciliation in generalized linear dynamic systems. *AIChE J.*, 37:193–201, 1991.

21. I. Daubechies. *Ten Lectures on Wavelets*. Society for Industrial and Applied Math., Philadelphia, 1992.

22. C. de Boor. *A Practical Guide to Splines*. Springer-Verlag, New York, 1978.

23. R.A. DeVore, B. Jawerth, and B. Lucier. Image compression through wavelet transform coding. *IEEE Trans. Information Theory*, 38(2):719–746, 1992.

24. D. Donoho and I. Johnstone. Ideal spatial adaption by wavelet shrinkage. *Biometrika*, 81:425–455, 1994.

25. D. Donoho, I. Johnstone, G. Kerkyacharian, and D. Picard. Wavelet shrinkage: Asymptopia? *J. Roy. Statist. Assoc.*, 90:301–369, 1995.

26. D.R.Kuehn and H.Davidson. Computer control. *Chemical Engineering Progress*, 57(6):44–47, 1961.

27. H.W. Engl. Regularization methods for the stable solution of inverse problems. *Surv. Math. Ind.*, 3:71–143, 1993.

28. H.W. Engl, M.H. Hanke, and A. Neubauer. *Regularization of Inverse Problems*. Kluwer Academic Publishers, 1996.

29. W.F. Feehery and P.I. Barton. Dynamic simulation and optimization with inequality path constraints. *Comput. Chem. Eng.*, 20(Suppl.):S707–S712, 1996.

30. J. Fröhlich and K. Schneider. An adaptive wavelet–vaguelette algorithm for the solution of nonlinear PDEs. *Preprint SC. ZIB*, (95-28), 1997.

31. G.F. Froment and K.B. Bischoff. *Chemical Reactor Analysis and Design*. John Wiley & Sons, 1990.

32. G.A.Almasy. Principles of dynamic balancing. *AIChE J.*, 36:1321–1330, 1990.

33. A. Gelb. *Applied Optimal Estimation*. MIT Press, 1974.

34. J. Gertler and G.A. Almasy. Balance calculation through dynamic system modeling. *Automatica*, 9:79–85, 1973.

35. D.M. Himmelblau and T.W. Karjala. Rectification of data in a dynamic process using artificial neural networks. *Comput. Chem. Eng.*, 20:805–812, 1996.

36. S.-S. Jang, B. Joseph, and H. Mukal. Comparison of two approaches to on-line parameter and state estimation of nonlinear systems. *Ind. Eng. Chem. Proc. Des. Dev.*, 25:809–814, 1986.

37. A.H. Jazwinski. *Stochastic Processes and Filtering Theory*. Academic Press Inc., 1970.

38. T. Johansen. On Tikhonov regularization, bias and variance in nonlinear system identification. *Automatica*, 33(3):441–446, 1997.

39. T.W. Karjala and D.M. Himmelblau. Dynamic data rectification by recurrent neural networks vs. traditional methods. *AIChE J.*, 40(11):1865–1875, 1994.

40. I.-W. Kim, M.J. Liebmann, and T.F.Edgar. A sequential error-in variables method for non-linear dynamic systems. *Comput. Chem. Eng.*, 15:663–670, 1991.

41. I.-W. Kim, S.W. Park, and T.F.Edgar. Data reconciliation for input-output models in nonlinear dynamic systems. *Korean J. Chem. Engng*, 13:211–215, 1996.

42. J.L.A. van Koolen. Plant operation in the future. *Comput. Chem. Eng.*, 18:S477–S481, 1994.

43. J. Lee, Y. Chikkula, Z. Yu, and J. Kantor. Improving computational efficiency of model predictive control algorithm using wavelet transformation. *Int. J. Control*, 61(4):859–883, 1995.

44. M.J. Liebmann, T.F. Edgar, and L.S. Lasdon. Efficient data reconciliation and estimation for dynamic processes using nonlinear programming techniques. *Comput. Chem. Eng.*, 16:963–986, 1992.

45. L.C. Lin and C.-C. Kuo. Signal extrapolation in noisy data with wavelet representation. In *Proceedings of the SPIE*, volume 2028, pages 156–167. The International Society for Optical Engineering, 1993.

46. J.S. Logsdon and L.T. Biegler. Decomposition strategies for large-scale dynamic optimization problems. *Chem. Eng. Sci.*, 47(4):851–864, 1992.

47. W. Marquardt. Numerical methods for the simulation of differential-algebraic process models. In R. Berber, editor, *Methods of Model-Based Control*, NATO-ASI Series, pages 42–79. Kluwer Press, 1995.

48. K.F. McBrayer, T.F Edgar, and L.S. Lasdon. Bias detection and estimation in dynamic data reconciliation. *J. Proc. Cont.*, 5, 1995.

49. A. M'hamdi, A. Helbig, O. Abel, and W. Marquardt. Newton-type receding horizon control and state estimation - a case study. In J.J. Gertler, J.B. Cruz, and M. Peshkin, editors, *Preprints of the 13th World Congress of IFAC*, volume M, pages 121–126, San Francisco, 30 June–5 July 1996.

50. H. Michalska and D.Q. Mayne. Moving horizon observers. In *IFAC Symposium NOLCOS'92*, Bordeaux, France, 1992.

51. H. Michalska and D.Q. Mayne. Moving horizon observers and observer–based control. *IEEE Trans. Aut. Contr. AC-40*, pages 995–1006, 1995.

52. K.R. Muske and T.F. Edgar. Nonlinear state estimation. In M.A. Henson and D.E. Seborg, editors, *Nonlinear Process Control*. Prentice Hall, 1997.

53. K.R. Muske and J.B. Rawlings. Nonlinear moving horizon state estimation. In R. Berber, editor, *Methods of Model-Based Control*, NATO-ASI Series, pages 349–365. Kluwer Press, 1995.

54. Y. Ramamurthi, P.B. Sistu, and B.W. Bequette. Control-relevant dynamic data reconciliation and parameter estimation. *Comput. Chem. Eng.*, 17(1):41–59, 1993.

55. D. Robertson, J.H. Lee, and J.B. Rawlings. A moving horizon-based approach for least-squares estimation. *AIChE J.*, 42(8):2209–2223, 1996.

56. C. Schmid and L.T. Biegler. Quadratic programming methods for reduced hessian SQP. *Comput. Chem. Eng.*, 18(9):817–832, 1994.

57. C. Schmid, J. Nocedal, and L.T. Biegler. A reduced hessian method for large–scale constrained optimization. *SIAM J. Optimization*, 5(2):314, 1995.

58. P.B. Sistu, R.S. Gopinath, and J.B. Rawlings. Computional issues in nonlinear predictive control. *Comput. Chem. Eng.*, 17:361, 1993.

59. G. Stephanopoulos, M. Dyer, and O. Karsligil. Multi–scale aspects in linear and nonlinear estimation and control. NATO ASI on Nonlinear Model Based Process Control. This volume.

60. G. Stephanopoulos, M. Dyer, and O. Karsligil. Multi–scale modeling, estimation and control of processing systems. *Comput. Chem. Eng.*, 21(Suppl.):S797–S803, 1997.

61. O. von Stryk. *Numerische Lösung optimaler Steuerungsprobleme: Diskretisierung, Parameteroptimierung und Berechnung der adjungierten Variablen*. Reihe 8: Meß-, Steuerungs- und Regelungstechnik Nr. 441. VDI-Verlag, Düsseldorf, 1995.

62. P. Tanartkit and L.T. Biegler. Stable decomposition for dynamic optimization. *Ind. Eng. Chem. Res.*, 34(4):1253–1266, 1995.

63. D. Tieu, W.R. Cluett, and A. Penlidis. A comparison of collocation methods for solving dynamic optimization problems. *Comput. Chem. Eng.*, 19(4):375–381, 1995.

64. M.L. Tyler and M. Morari. Stability of constrained moving horizon estimation

schemes. In R. Berber, editor, *Nonlinear Model Based Process Control*, NATO-ASI Series. Kluwer Press, 1998.

65. J. Unger, A. Kröner, and W. Marquardt. Structural analysis of differential-algebraic equation systems - theory and applications. *Comput. Chem. Eng.*, 19(8):867–882, 1995.

66. A. Uppal, W. H. Ray, and A. B. Poore. On the dynamic behaviour of continuous stirred tank reactors. *Chem. Eng. Sci.*, 29:967–985, 1974.

67. P. de Vallière and D. Bonvin. Application of estimation techniques to batch reactors - III: Modelling refinements which improve the quality of state and parameter estimation. *Comput. Chem. Eng.*, 14(7):799–808, 1990.

68. V.S. Vassiliadis, R.W.H. Sargent, and C.C. Pantelides. Solution of a class of multistage dynamic optimization problems. 1. Problems without path constraints. *Ind. Eng. Chem. Res.*, 33(9):2111–2122, 1994.

69. V.S. Vassiliadis, R.W.H. Sargent, and C.C. Pantelides. Solution of a class of multistage dynamic optimization problems. 2. Problems with path constraints. *Ind. Eng. Chem. Res.*, 33(9):2123–2133, 1994.

70. G.H. Weiss, J.A. Romagnoli, and K.A. Islam. Data reconciliation - an industrial case study. *Comput. Chem. Eng.*, 20:1441–1449, 1996.

71. S.J. Wright. *Primal–Dual Interior–Point Methods*. SIAM, 1996.

MULTI-SCALE ASPECTS IN LINEAR AND NONLINEAR ESTIMATION AND CONTROL

G. STEPHANOPOULOS, O. KARSLIGIL AND M.S. DYER
Department of Chemical Engineering
Massachusetts Institute of Technology
Cambridge, MA 02139

Multi-scale models of processing systems offer an attractive alternative to models defined in the time- or frequency-domain. They are the outgrowth of a series of developments which came about with the advent of the wavelet decomposition for the analysis of discrete signals. Multi-scale models are defined on dyadic or higher-order trees. The nodes of such trees are used to index the values states, inputs and outputs, modeling errors, and measurement errors. These values are localized in both time and scale (range of frequencies), and thus they offer a hybrid domain that is particularly conducive for estimation and control problems. In this paper we introduce a formal framework for the formulation of multi-scale models on trees, which are consistent with their time-domain counterparts. Such models lead to a multi-scale control theory and the definition of the corresponding transfer functions, stability, controllability, and observability concepts for systems on trees. Fusion of control actions and measurements at different rates as well as their implications on the controllability and observability of dynamic systems are also examined. Of particular significance is the issue of closure between the models in the time- and the time-scale domains, which constrains the type of physico-chemical processes that can be modeled on a tree.

Based on these developments, we then proceed to address the solution of certain basic tasks in systems engineering, such as; simulation of linear systems, optimal control, and state estimation with optimal fusion of measurements. It is shown that the multi-scale models lend themselves nicely to parallelizable computations, which can produce algorithms of substantially lower computational load. Multi-scale Model Predictive Control is subsequently formulated and the ensuing estimation and optimal control sub-problems on trees are examined. The time-scale localization of the states, inputs, outputs and errors, in a multi-scale MPC allows a more explicit selection of the design specifications for the MPC formulation. Furthermore, it is shown that the reduced computational load offered by the parallelizability of many computational tasks, leads to a re-examination of the ways that classical MPC formulations are addressed in the time-domain, and thus a re-examination of how closed-loop stability, constraint satisfaction, horizon determination, and others, can be resolved. Finally, the paper discusses several tentative ideas and suggestions on how multi-scale aspects can be extended to handle nonlinear systems.

R. Berber and C. Kravaris (eds.), Nonlinear Model Based Process Control, 667-734.
© 1998 *Kluwer Academic Publishers.*

1. INTRODUCTION

Control of nonlinear processes through Model-Predictive Control (MPC) strategies, raises a series of formidable problems, the source for most of which can be traced back to the following two distinctive features [37].

(a)   The open-loop optimal control policy, computed at each step, must be globally optimum if it is to be admissible.

(b)   The closed-loop stability requirement imposes a "stability constraint", i.e. an equality constraint, $x(t=T) = 0$, on the terminal state, which can be achieved only asymptotically in time.

In addition, inequality constraints on inputs impose constraints on what states can be reached by the allowed control actions, while inequality constraints on outputs have an impact on the closed-loop stabilization of a process. Progress has been slow and only in fairly specialized classes of models (not Physico-chemical systems), such as the "finite response" systems, feasibility of control policies and terminal conditions for closed-loop stability can be readily established.

However, the raison-d'être for nonlinear process control research and technology must be founded on the answer of the following simple question: "What is to be gained by nonlinear control over what is accomplished by linear control"? The answers have not been very convincing. Nevertheless, substantive progress in nonlinear process control and development of credible technology may change the "cultural" inhibitions of process designers and allow the deployment of processes, which "must" posses nonlinear control systems.

Given the formidable difficulties, offered by the general nonlinear control problem, and its "cultural" affinity with the linear control tradition (e.g. MPC was developed as a control strategy for linear constrained systems, why should it be the correct paradigm to tackle nonlinear control systems?), in this paper we will attempt to explore an alternative framework to the classical linear systems theory, in an attempt to establish a formalism that can solve a wide class of control problems. Thus, instead of tackling nonlinear control problems directly, we will examine whether the multi-scale systems theory, the foundation of the methods in this paper, can solve a broader class of problems through linear methods, reducing in the process the need for nonlinear control.

The Need for Scale and Time Localization of Processes and Measurements

It is broadly accepted that physical phenomena occur at different time scales. However, it is not clear how to systematically incorporate this knowledge in the generation of adequate process models, or how to use it for the solution of some basic process engineering problems, e.g. control, estimation, diagnosis. The conventional models, causal and explicit, provide a convoluted representation of physics at various scales and hamper engineering analysis and interpretation. Furthermore, process models used in various engineering tools involve different time scales, e.g. closed-loop control, adaptive control, fault diagnosis, scheduling of control strategies, and planing of operating procedures, involved process models with dynamics of progressively increasing time constants. Current modeling practices are not very instructive on how to create consistent models for a sequence of interrelated engineering tasks, as the above, deployed at different time scales [50].

In addition, sensors may provide measurements of process behavior at different sampling rates, invoking control actions at correspondingly different rates. The optimal fusion of measurement information or control actions relies on the availability of process models at time scales, which are commensurable with the sampling rate of measurements and the application intervals of control actions.

The above requirements indicate that we need representations, which capture scale-based characteristics of process models, measurements and control actions. However, the classical Fourier analysis, which has been used to provide such representations, is not adequate. It provides frequency-based information on the behavior of processes and measurements, by integrating the time behavior of such entities over an infinite time horizon, while processes and their characteristics change over finite segments of time. Therefore, we need a framework, which can provide explicit representation of process dynamics, measurements, and control actions, localized in both time and frequency (scale).

Wavelets and the Transformation to a Time-Scale Domain

The advent of wavelets has introduced the basic capability in constructing models and signal representations, which are localized in both time and frequency (scale). Although the bulk of the work on wavelets has focused on the representation of signals (variables) in the time-scale domain [1], [2], [17], [31], and [36], recent work [10], has extended the original ideas to the formation of models, whose variables are localized in both time and scale. Such models, to be called "multi-scale" models, constitute the basis in the formation of an alternative framework for the solution of process control and estimation problems. This framework has several advantages:

(1)   Allows a direct correspondence between process behavior in time and scale, and the associated models describing such process.
(2)   Enables the deployment of naturally parallelizable algorithms for the solution of several basic tasks, e.g. simulation of process dynamics, optimal control, optimal state estimation.
(3)   Permits the optimal fusion of measurements or/and control actions which occur at different rates.

Outline of the Paper

In Section 2 we will discuss the wavelet-based multi-scale representation of signals and show that homogeneous trees is the natural basis for indexing the coefficients resulting from the wavelet decomposition of a signal. In Section 2.3. we discuss how one can formalize shifts on trees, and define the requisite operators for shifting signal values on the nodes of a tree. Section 3. introduces the concept of a multi-scale model and shows how such models in the time-scale domain can be generated from linear, time-invariant models, defined in the time-domain. Section 3.4 discusses the essentials of a system theory in the time-scale domain, e.g. transfer functions, stability, steady state, controllability, observability, etc. Of particular practical value is the creation of models with fusion of measurements or/and control actions at multiple rates, and of theoretical value the notion of "closure". The requirement of maintaining closure between the models in the time domain and those in the time-scale domain has some interesting implications, which affect the efficiency of basic tasks.

In Section 4 we develop a series of procedures for the solution of basic tasks, such as, simulation, optimal control, and estimation, using multi-scale models. It is shown that the multi-scale models lend themselves to easily and naturally parallelizable algorithms. Of particular value here is also the ease in fusing measurements and control actions at several rates. In Section 5 we discuss the application of multi-scale models in the deployment of multi-scale MPC. It is shown that the multi-scale representation of variables is naturally compatible with the design specification needs in the formulation of MPC, and that the reduced computational load allows the accomplishment of several theoretical targets, e.g. a practically meaningful infinitive horizon.

Section 6 is speculative in nature. It discusses several ideas on how multi-scale (or, multi-resolution) modeling and analysis could be used to address basic problems in nonlinear system theory.

2. MULTI-SCALE REPRESENTATION OF SIGNALS

In the last fifteen years, the wavelet transform has emerged as an important tool for the analysis of signals. Although the concept was not new to mathematicians and physicists, the implications to signal analysis were introduced by Morlet in his pioneering work on the analysis of seismic data [25]. The pivotal feature of the wavelet transform stems from its ability to analyze a signal by extracting its features in such a way that they are localized both in frequency and time. A number of good introductions exist to the topic [16], [26], [28], [49].

2.1. THE WAVELET DECOMPOSITION OF SIGNALS

Consider the continuous signal, $f(t)$, and generate the following sequence of approximations

$$f^{(m)}(t) = \sum_{n=-\infty}^{+\infty} f(m,n)\phi(2^m t - n) \qquad m = 0, 1, 2, \ldots \qquad (1)$$

Each approximation is expressed as the weighted sum of the shifted versions of the same function, $\phi(\tau)$, which is called the *scaling function*. The choice of the scaling function is far from arbitrary [36], [17]. In particular, if we require that the *(m+1)*th approximation is a refinement of the *m*th approximation, then the function $\phi(2^m t)$, should be a linear combination of the basis functions spanning the space of the (m+1)st approximation [7], i.e.

$$\phi(2^m t) = \sum_{k} h(k)\phi(2^{m+1} t - k) \qquad (2)$$

In addition, several other properties are required of the scaling function, such as: (a) have local support or decay very fast to zero; (b) its integer translates form an orthogonal set. These properties impose certain conditions that the coefficients, $h(k)$ must satisfy. In particular, $h(k)$ must be the impulse response of a *quadrature mirror filter* [36], [17]. Although local support (or fast decay to zero) is satisfied by all types of commonly used scaling functions, the orthogonality is not.

As we move from the coarser approximation, $f^{(m)}(t)$, to the finer approximation, $f^{(m+1)}(t)$, we add new information. If $V^{(m+1)}$ represents the space of all functions spanned by the orthogonal set, $\{\phi(2^{m+1}t-k) ; k \in Z$, the set of integers$\}$, and $V^{(m)}$ the space of the coarser functions spanned by the orthogonal set, $\{\phi(2^m t-\ell) ; \ell \in Z\}$, then $V^{(m)} \subset V^{(m+1)}$. Let

$$V^{(m+1)} = V^{(m)} \oplus W^{(m)}, \qquad (3)$$

then, $W^{(m)}$, is the space that contains the added information as we go from the coarser, $f^{(m)}(t)$, to the finer, $f^{(m+1)}(t)$, representation of the original function, $f(t)$. It can be shown [36] that the space, $W^{(m)}$ is spanned by the orthogonal translates of a single function, $\psi(2^m t)$, thus leading to the following equation

$$f^{(m+1)}(t) = f^{(m)}(t) + \sum_{n=-\infty}^{+\infty} \delta f(m,n) \psi(2^m t - n) \qquad (4)$$

The function, $\psi(2^m t)$, is called a *wavelet* and is related to the scaling function $\phi(2^{m+1} t)$, through the following relationship [36], [17],

$$\psi(2^m t) = \sum_k g(k)\phi(2^{m+1} t - k) \tag{5}$$

$h(k)$ and $g(k)$ from a *conjugate mirror filter* pair [49], [16].

Equation (4) constitutes the basis for the multitude of signal analysis techniques that have emerged over the last 8-10 years. It suggests the following:

(i)   A continuous signal can be filtered to a desired approximation, $f^{(m+1)}(t)$, using the scaling function, $\phi(2^{m+1} t)$, as a low-pass filter. The dilation parameter, 2^{m+1}, signifies the temporal *scale* of sampling the continuous function.

(ii)   The difference in information content between the two approximations, $f^{(m)}(t)$ and $f^{(m+1)}(t)$, is characterized by the coefficients, $\delta f(m,n)$. This implies that the wavelet function can be seen as a band-pass filter that extracts information at a given scale (or equivalently, a given range of frequencies).

(iii)   By allowing $m \rightarrow \infty$, equation (4) implies that the continuous signal, $f(t)$, can be expressed as,

$$f(t) = \sum_{m+-\infty}^{+\infty} \sum_{n=-\infty}^{+\infty} \delta f(m,n) \psi(2^m t - n) \tag{6}$$

i.e. a linear combination of dilated and translated versions of the same wavelet function. The coefficient, $\delta f(m,n)$, signifies the contribution of the corresponding wavelet, $\psi(2^m t - n)$, to the value of the signal, (a) over a localized segment of time, and (b) containing information in a localized range of frequencies. Equation (6) yields a non redundant representation of $f(t)$ for any orthogonal set of wavelets.

2.2. SIGNALS ON TREES AND LATTICES

Combining (1), (2) and (5) with equation (4) produces

$$\sum_{n=-\infty}^{+\infty} f(m+1,n)\phi(2^{m+1} t - n) = \sum_{n=-\infty}^{+\infty} f(m,n) \sum_k h(k)\phi(2^{m+1} t - n - k) + \sum_{n=-\infty}^{+\infty} \delta f(m,n) \sum_k g(k)\phi(2^{m+1} t - n - k) \tag{7}$$

Equation (7) implies that we have a dynamic relationship between the coefficients, $f(m,n)$, at one scale and those, i.e. $f(m+1,n)$, at the next. Indeed, as Benveniste et al. [7] have pointed out, this relationship defines a lattice on the points (m,n), where $(m+1, n)$

is connected to *(m,n)* if, *f(m,n)*, influences, *f(m+1,n)*. This observation has provided the motivation for the development of models on trees and lattices [12], [21], [27], [34], which define the time-scale domain.

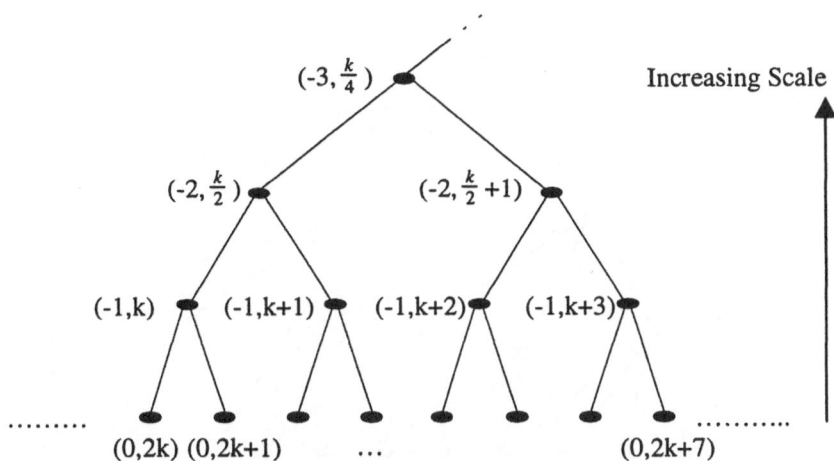

Figure 1. Indexing the coefficients of Haar-based decomposition on the nodes of binary tree.

The simplest examples of a scaling function and wavelet are the Haar function and corresponding wavelet defined as follows:

$$\phi(t) = \begin{cases} 1 & 0 \le t < 1 \\ 0 & \text{otherwise} \end{cases} \quad \text{and} \quad \psi(t) = \begin{cases} 1 & 0 \le t < \frac{1}{2} \\ -1 & \frac{1}{2} \le t < 1 \end{cases}$$

Let, $\{f^{(a)}(t)\} = \{f(0,0) (= f^{(0)}(t_0)), f(0,1) (= f^{(0)}(t_1)), ..., f(0,2k), f(0,2k+1), ...\}$ be a sequence of discrete-time values of *f(t)* for a given sampling interval. Filtering $f^0(t)$ with $h = \frac{1}{\sqrt{2}}$ [1 1] and $g = \frac{1}{\sqrt{2}}$ [1 −1] we generate the coefficients

$$f(-1,k) = \frac{1}{\sqrt{2}}\{f(0,2k) + f(0,2k+1)\}$$

$$\delta f(-1,k) = \frac{1}{\sqrt{2}}\{(f(0,2k) - f(0,2k+1)\}$$

Subsequent filtering of, *f(-1,k)* , leads to coarser depictions of *f(t)*. The Haar decomposition of $\{f^{(0)}(t)\}$ leads to a set of coefficients, which can be indexed to correspond to the nodes of a binary tree, as shown in Figure 1.

674

Consequently, in describing $\{f^{(0)}(t)\}$, we can replace the sequence $\{f(0,0)\,,\,f(0,1)\,,\,...\,f(0,2k)\,,\,f(0,2k+1)\,,\,...\}$ with the equivalent two-dimensional array of coefficients $\{f(m,n)\}$; m = -1, -2 , ... and n = $-\infty$, ..., $+\infty$.

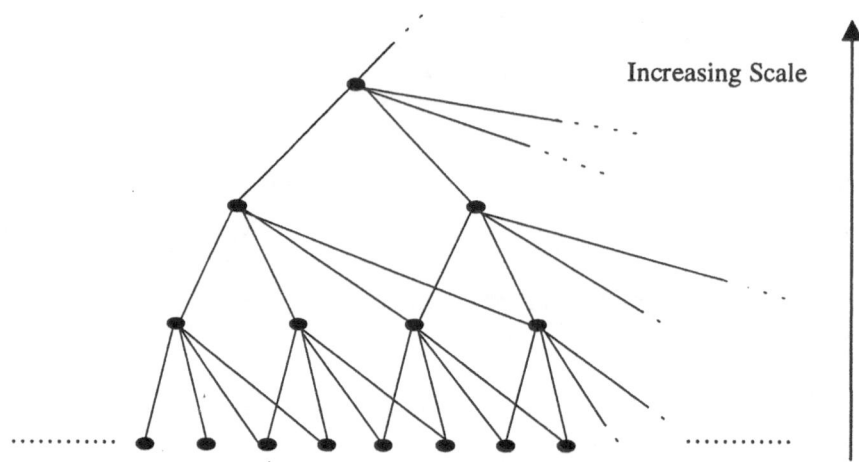

Figure 2. Lattice for the indexing of D_4-based decomposition of discrete-time functions

The Daubechies' D_4 scaling function and wavelet correspond to the filters

$$h = \left[\frac{1}{4}(1+\sqrt{3}),\frac{1}{4}(3+\sqrt{3}),\frac{1}{4}(3-\sqrt{3}),\frac{1}{4}(1-\sqrt{3})\right]$$

and

$$g = \left[+\frac{1}{4}(1+\sqrt{3}),-\frac{1}{4}(3+\sqrt{3}),+\frac{1}{4}(3-\sqrt{3}),-\frac{1}{4}(1-\sqrt{3})\right]$$

Decomposition of the sequence $\{f^{(0)}(t)\}$ produces coefficients which can indexed to correspond to the nodes of the lattice shown in Figure 2.

Basseville et al. [3], [4] and [5] were the first to realize the significance of homogeneous trees in (a), indexing the wavelet coefficients of a signal, and (b), creating a systems theory for processes defined directly on trees. It is important to realize that the representational framework for signals (and models), offered by homogeneous trees, is far more flexible than Figures 1 and 2 might suggest. For example, it is applicable to higher dimensional trees (for signals defined in multidimensional spaces, e.g. time, spatial coordinates), and to trees with asymmetry, or structures with unusual shapes. Such flexibility can be exploited to match the particular multi-scale structure of a signal's components, or to capture the localized behavior of physico-chemical phenomena with multi-scale features. Given the significant role that homogeneous trees and shifts on them will play, in the development of multi-scale process models for estimation and control, in the next section we will try to summarize the main definitions and properties to be used in subsequent sections. The material has been primarily drawn from the works of Basseville et al. [3], [4], Chou [12], Luettgen [34], Irving [27], and, Fieguth [21].

2.3. HOMOGENEOUS TREES AND SHIFT OPERATORS ON TREES AND SIGNAL VALUES

Homogeneous Trees
A homogeneous tree, T, is an infinite, acyclic, undirected, connected graph. If every node of T has exactly $(q + 1)$ branches to other nodes, the tree is of order, q. The tree of Figure 1 with $q = 2$ is called a *dyadic* tree and will be the primary focus in subsequent sections of this paper.

A homogeneous tree, T, possesses a natural notion of *distance*. If s_1 and s_2 are two nodes on a tree, T, the distance between them, $d(s_1, s_2)$, is equal to the number of branches along the shortest path between the nodes s_1 and s_2. For a given tree we need to define a *boundary point*, before we can specify a partial ordering of the nodes on the tree and thus a convention for indexing the nodes. When $q = 1$, the corresponding tree represents the set of integers and possesses two boundary points, $+\infty$ and $-\infty$. Indeed, any two sequences of integers increasing towards $+\infty$ or decreasing towards $-\infty$ become equivalent, i.e. differ by a finite number of nodes. For a dyadic tree with $q = 2$ the set of boundary points in uncountable. Choose one of them and denote it as $-\infty$.

Once a boundary point has been specified, draw the tree as shown in Figure 3(a). Nodes s_1 and s_2 in Figure 3(a) are at the same distance from the boundary point of $-\infty$, since

$$d(s_1, -\infty) = d(s_1, w) + d(w, -\infty) = 2 + d(w, -\infty)$$

$$d(s_2, -\infty) = d(s_2, w) + d(w, -\infty) = 2 + d(w, -\infty)$$

Nodes at the same distance from the boundary point are at the same *horocycle*, e.g. nodes s_1, s_2, s_3, s_4 in Figure 3(b). In the multi-scale representation of a signal, nodes at the same horocycle correspond to wavelet (or scaling function) coefficients at the same scale. An upward shift along the branches of the tree in Figure 3(a) represents a shift from a wavelet coefficient describing a signal at a finer resolution, to a wavelet coefficient describing the signal at a coarser one.

Having specified the notions of a boundary point and distance between two nodes on a homogeneous tree, it is a rather simple task to devise an indexing system that allows the unique and concise characterization of all nodes. We will discuss a specific indexing system after we have introduced the shift operators on trees.

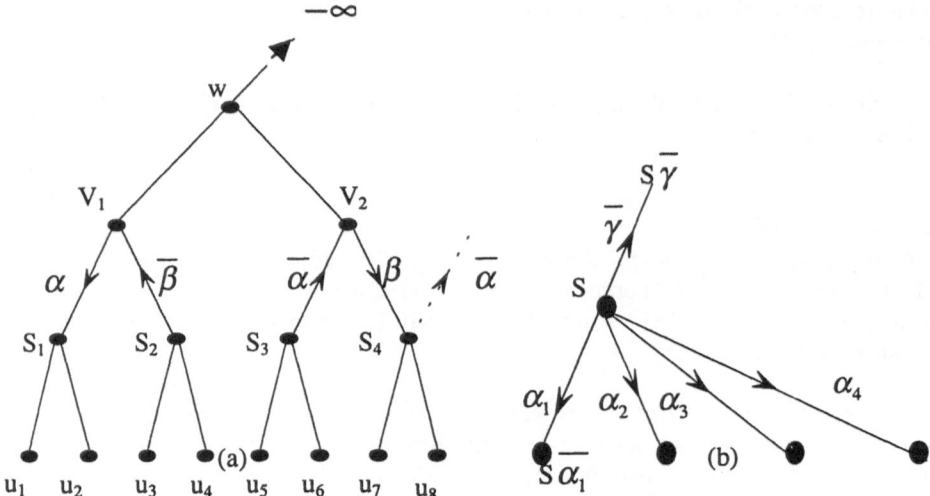

Figure 3. (a) Boundary point, horocycles, and shifts on a dyadic tree
(b) Shift operators on a tree of order 4.

Shift Operators on Trees

For a tree of order q we need two types of shift operators to traverse the tree as shown in Figure 3(b):

(a)   upward shift operator ; $\quad \bar{\gamma} : s \rightarrow s\bar{\gamma}$
       where node $s\bar{\gamma}$ is the parent node of s,

(b)   downward shift operator; $\alpha_i : s \rightarrow s\alpha_i$
       where node $s\alpha_i$ is the child of node s, $\ 1 \le i \le q$.

For the dyadic tree in Figure 3(a), we observe the following.

(i)   There exist two downward shift operators, which we will denote as follows:

       α : $s \rightarrow s\alpha \equiv$ left-offspring of s
                     e.g. $s_1 = V_1\alpha$ and $s_3 = V_2\alpha$

       β : $s \rightarrow s\beta \ \equiv$ right-offspring of s
                     e.g. $s_4 = V_2\beta$ and $s_2 = V_1\beta$

(ii)   There exist two upward shift operators, defined as follows:

       $\bar{\alpha}$: $s \rightarrow s\bar{\alpha} \equiv$ parent of s through a left up-shift
                           e.g. $V_2 = s_3\bar{\alpha}$ and $V_1 = s_1\bar{\alpha}$

$$\overline{\beta}: \quad s \rightarrow s\,\overline{\beta} \equiv \text{parent of s through a right up-shift}$$

$$\text{e.g. } V_1 = s_2\,\overline{\beta} \text{ and } V_2 = s_4\,\overline{\beta}$$

Note that any node on the tree can only have one parent through either a left- or a right-branch; e.g. there is no node, $s_4\,\overline{\alpha}$, since s_4 has a parent through a right-branch.

It should be noted that the shift operators, α and β, are the counterpart of the shift operator, z, in discrete Fourier transforms. Both shift the index of a value away from -∞. However, in discrete Fourier transform the corresponding tree is of first-order $(q = 1)$, requiring a single shift, while on a dyadic tree, with two offsprings from every node, we need two distinct shift operators to move away from the boundary point of -∞. Clearly, the upward-shift operators, $\overline{\alpha}$ and $\overline{\beta}$, correspond to z^{-1}.

Shift Operators on Signals
Consider two signals, $x(t)$ and $u(t)$, both of which are decomposed through Haar wavelet, with their wavelet (and scaling function) coefficients indexed by the nodes of the dyadic tree shown in Figure 3(a). Thus, by "signal" we will mean the array of scaling function (or, wavelet) coefficients indexed by the nodes of the dyadic tree. Let us define shift operators on the values of a signal which are consistent with the shift operators on trees, introduced in the previous paragraph. For any node, s, on the dyadic tree, we have the following:

1. If $\quad x(s) = u(s\alpha)$, then let $x(s) = \alpha u(s)$
    α is the left branch down-shift operator

2. If $\quad x(s) = u(s\beta)$, then let $x(s) = \beta u(s)$
    β is the right branch down-shift operator

3. If $\quad x(s) = u(s\,\overline{\alpha}\,)$ with $u(s\,\overline{\beta})= 0$,
    then let, $x(s) = \overline{\alpha}\,u(s)$
    $\overline{\alpha}$ is the left-branch up-shift operator

4. If $\quad x(s) = u(s\,\overline{\beta}\,)$ with $u(s\,\overline{\alpha}\,) = 0$,
    then let, $x(s) = \overline{\beta}\,u(s)$
    $\overline{\beta}$ is the right-branch up-shift operator.

It is rather straightforward to show that these shift operators on the signal values indexed by a dyadic tree satisfy the following equations:
$$\alpha\overline{\alpha} = \beta\overline{\beta} = 1$$
$$\beta\overline{\alpha} = \alpha\overline{\beta} = 0$$
$$\overline{\alpha}\alpha + \overline{\beta}\beta = 1$$

Finally, it should be noted that α and β are one-to-one but not onto operators. In the next section we will see how these shift operators on the values of signals provide the essential elements for the definition of multi-scale transfer functions of dynamic systems on trees.

3. MULTI-SCALE MODELS OF LINEAR SYSTEMS

The state, input, and measured variables of a dynamic process in a state-space or input-output model, are defined on the set of real numbers for continuous models, or the set of integers for discrete-time models. The wavelet transform of states, inputs and measurements allows us to define these variables in the time-scale domain, which is characterized by the nodes of a homogeneous tree. However, the character of the resulting dynamic model is not obvious, nor is the presence of any attractive characteristics that might facilitate or enhance the engineering tasks of simulating, estimating, controlling, and/or optimizing the dynamic behavior of linear processes. In this section we will attempt to elucidate these issues. Specifically, we will examine

(a) the conversion of state-space models to multi-scale models defined on dyadic trees,
(b) the direct generation of multi-scale models on dyadic trees, and
(c) the characteristics of such models, e.g. stability, controllability, observability.

3.1. FROM THE TIME DOMAIN TO THE TIME-SCALE DOMAIN

Consider the input-output model given by the convolution integral

$$y(t) = \int_o^t g(t-\tau)u(\tau)d\tau \tag{8}$$

It is well known that the Fourier transform deconvolves $g(t\text{-}\tau)$ and $u(\tau)$ and leads to the following transfer function-based model in the frequency domain,

$$\bar{y}(\omega) = \bar{g}(\omega)\bar{u}(\omega) \tag{9}$$

It is possible to establish a similar deconvolution of $g(t\text{-}\tau)$ and $u(\tau)$ under a wavelet transform and generate a transfer function-based model in the time-scale domain of the form

$$\tilde{y}(s) = \tilde{g}(s)\tilde{u}(s) \tag{10}$$

where, $\tilde{y}(s)$ and $\tilde{u}(s)$ are the wavelet coefficients of the corresponding variables at the node, s, of the associated homogeneous tree? The answer is no, as the following theorem states [35]

Theorem 1. There is no wavelet function, which can transfoim the dynamic system of equation (8) into one of the form given by equation (10), i.e. into one where the transfer function and the input are completely deconvolved and separable.

The result of Theorem 1 is not surprising. The transfer function model of equation (9) is possible because the $\exp(j\omega t)$ basis functions of the Fourier transform are also eigenfunctions of the differential operator. On the contrary, no wavelet function is an eigenfunction of the differential operator.

Simply, the transfer function we are looking for will not be related to the classical linear system theory, as is the Fourier-based transfer function in the frequency domain.

3.2. HOMOGENEOUS LINEAR SYSTEMS ON TREES

Consider the following multivariable, linear, discrete-time homogeneous model,

$$x(k+1) = Ax(k) \qquad (11)$$

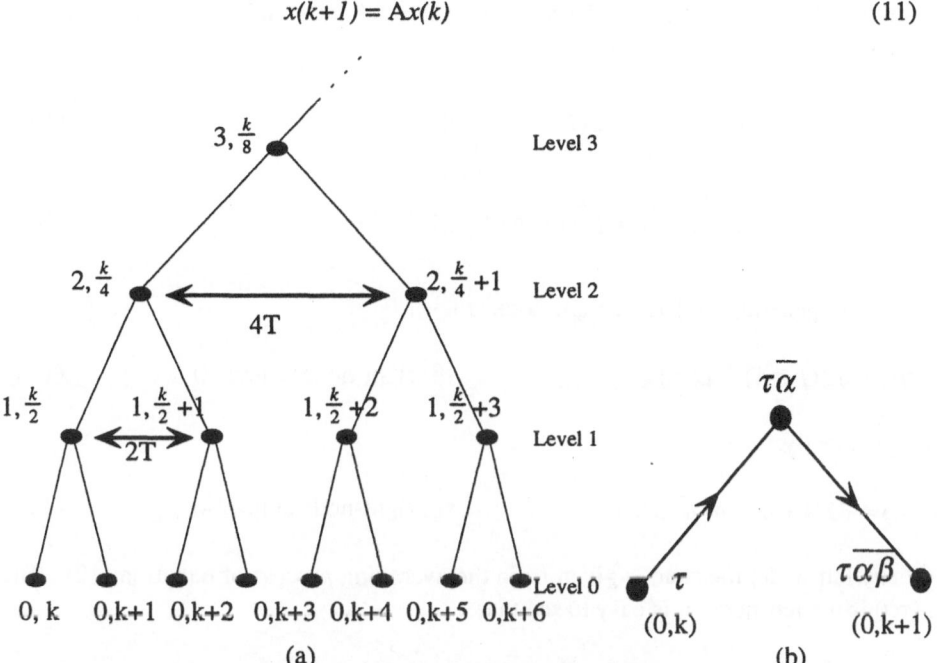

Figure 4. (a) Dyadic tree on which the multi-scale, linear, homogeneous models are defined. (b) Shifts on the dyadic tree leading to the formation of a transfer function.

Let us index the discrete-time values of the state, $x(k)$, by the nodes at level-0 (or equivalently, horocycle-0) of the dyadic tree in Figure 4(a), where the discretization interval is equal to the sampling period under which the model of equation (11) was generated. Then, equation (11) takes the following equivalent form,

$$x(0,k+1) = Ax(0,k) \qquad (11a)$$

Using Haar decomposition on the sequence $\{x(0,k)\}$ we can generate the scaling and wavelet coefficients at level-1. For any left-node, $\tau\alpha$, at level-0, it is easy to show that (Figure 4(b))

$$x(\tau\alpha) = \sqrt{2}\,x(\tau) - x(\tau\beta) \qquad (12)$$

e.g. (see Figure 4 (a))

$$x(0,k) = \sqrt{2}x(1,\frac{k}{2}) - x(0,k+1); \qquad (12a)$$

$$x(0,k+2) = \sqrt{2}x(1,\frac{k}{2}+1) - x(0,k+3)\ ,...\text{etc}$$

Substituting the value of a left-note state, given by equation (12), into the modeling equation (11a), we obtain a modeling relationship for the right-nodes at level-0, where the state is given by

$$x(0,k+1) = \sqrt{2}(I+A)^{-1}A \cdot x(1,\frac{k}{2}) \qquad (13a)$$

$$x(0,k+3) = \sqrt{2}(I+A)^{-1}Ax(1,\frac{k}{2}+1) \qquad (13b)$$

In general, if $(\tau\beta)$ is a right-node at level-0,

$$x(\tau\beta) = \sqrt{2}(I+A)^{-1}Ax(\tau) \qquad \qquad \tau\beta:\text{ right-node at level-0} \qquad (14a)$$

or equivalently,

$$x(\tau\beta) = \sqrt{2}(I+A)^{-1}A \cdot \overline{\beta} \cdot x(\tau\beta) \qquad \qquad \tau\beta:\text{ right-node at level-0} \qquad (15a)$$

For any left-node, the state is given from the averaging process of equation (12). Thus, if $(\tau\alpha)$ is the left-node, it is easy to see that

$$x(\tau\alpha) = \sqrt{2}(I+A)^{-1}x(\tau) \qquad \qquad \tau\alpha:\text{ left-node at level-0} \qquad (14b)$$

or equivalently,

$$x(\tau\alpha) = \sqrt{2}(I+A)^{-1} \cdot \overline{\alpha} \cdot x(\tau\alpha) \qquad \qquad \tau\alpha:\text{ left-node at level-0} \qquad (15b)$$

Given that at any node, τ, only one of the upward-shifts leads to a node on the tree, equations (14a) and (14b) lead to

$$x(\tau) = A_\alpha^{(0)} x(\tau\overline{\alpha}) + A_\beta^{(0)} x(\tau\overline{\beta}) \qquad\qquad \tau: \text{any node at level-0} \qquad (14)$$

and equations (15a) and 15(b) lead to

$$x(\tau) = (\overline{\alpha}A_\alpha^{(0)} + \overline{\beta}A_\beta^{(0)})x(\tau) \qquad\qquad \tau: \text{any node at level-0} \qquad (15)$$

where,

$$A_\alpha^{(0)} = \sqrt{2}(I + A)^{-1} \qquad \text{and} \qquad A_\beta^{(0)} = \sqrt{2}(I + A)^{-1} A \qquad (16)$$

It should be noted that, in generating the models of equation (14) or (15), the discrete-time modeling relationship (11a) was applied to the *right-nodes only*. Applying the modeling relationship (11a) to the left nodes at level-0, e.g. $\tau \equiv (0, k + 2)$, we take

$$x(0, k + 2) = Ax(0, k + 1)$$

or

$$\sqrt{2}(I + A)^{-1} x(1, \frac{k}{2} + 1) = A\left\{ \sqrt{2}(I + A)^{-1} A \cdot x(1, \frac{k}{2}) \right\}$$

or,

$$x(1, \frac{k}{2} + 1) = A^2 x(1, \frac{k}{2})$$

The last equation can now be applied to all right-nodes at level-1, yielding as before the multi-scale model

$$x(1, \frac{k}{2} + 1) = \sqrt{2}(I + A^2)^{-1} A^2 x(2, \frac{k}{4})$$

or, in general, for any node, τ, at level-1,

$$x(\tau) = A_\alpha^{(1)} x(\tau\overline{\alpha}) + A_\beta^{(1)} x(\tau\overline{\beta})$$

Equivalently, the last equation yields

$$x(\tau) = (\overline{\alpha}A_\alpha^{(1)} + \overline{\beta}A_\beta^{(1)})x(\tau)$$

where,

$$A_\alpha^{(1)} = \sqrt{2}(I + A^2)^{-1} \quad \text{and} \quad A_\beta^{(1)} = \sqrt{2}(I + A^2)^{-1} A^2$$

682

Recursive application of the above mechanism generates the complete set of multi-scale models at any node of the dyadic tree, i.e. for any node, τ, at level-m,

$$x(\tau) = A_\alpha^{(m)} x(\tau\overline{\alpha}) + A_\beta^{(m)} x(\tau\overline{\beta}) \tag{17}$$

or

$$x(\tau) = \left(\overline{\alpha} A_\alpha^{(m)} + \overline{\beta} A_\beta^{(m)}\right) x(\tau) \tag{18}$$

where,

$$A_\alpha^{(m)} = \sqrt{2}\left(I + A^{2^m}\right)^{-1} \qquad \text{and} \qquad A_\beta^{(m)} = \sqrt{2}\left(I + A^{2^m}\right)^{-1} A^{2^m} \tag{19}$$

Remarks

1. Equation (17) provides a *two-scale model* of the linear system, since it relates the states at two distinct scales.

2. The state of the system at the higher scale, e.g. $x(\tau\overline{\alpha})$ or $x(\tau\overline{\beta})$ captures all "past" history and plays a role analogous to that of $x(k)$ in the discrete-time model of equation (11).

3. The physical causality manifested in the form of model (11), i.e. $x(k)$ at time k affects the value of the state of the next time point, k+1, is lost in the models (17) or (18). These models should be seen as relating "averaged" process dynamics over different time-horizons. The multi-scale models preserve a notion of "computational causality", as will be seen in subsequent sections.

4. The shift operators on the state values have led to a form of a transfer function (see equation 18) in the time-scale domain. This transfer function cannot be related to the linear system theory since the shift operators are one-to-one but not onto. Basseville et al. [4] have indicated that these transfer functions are related to the realization theory for automata.

5. The models of equation (17) or (18) have been expressed in terms of the scaling function coefficients. Analogous steps can lead to models involving wavelet coefficients of the state variables. It is easy to show that equation (17) can take the following form for any node, τ

$$\delta x(\tau) = \left(I + A^{2^m}\right)^{-1}\left(I - A^{2^m}\right) x(\tau) \tag{20}$$

6. For an n-th order, discrete-time, linear homogeneous system, it can be easily shown that the state at any node, τ, at level-m of the dyadic tree is given by the following multi-scale model

$$x(\tau) = A_1 x(\tau\overline{\gamma}) + A_2 x(\tau\overline{\gamma}\overline{\gamma}) + \cdots + A_n x(\underbrace{\tau\overline{\gamma}\overline{\gamma}\cdots\overline{\gamma}}_{n})$$

or equivalently,

$$x(\tau) = \left(\bar{\gamma}A_1 + \bar{\gamma}\bar{\gamma}A_2 + \cdots + \underbrace{\bar{\gamma}\bar{\gamma}\cdots\bar{\gamma}}_{n}A_n \right)x(\tau)$$

where, $\bar{\gamma}$ denotes an upward-shift operator, which, depending on the branch (left, or right), could be equal to $\bar{\alpha}$ or $\bar{\beta}$. Figure 5 shows the support on a dyadic tree of 1*st*, 2*nd*, and 3*rd* order linear models applied at specific nodes.

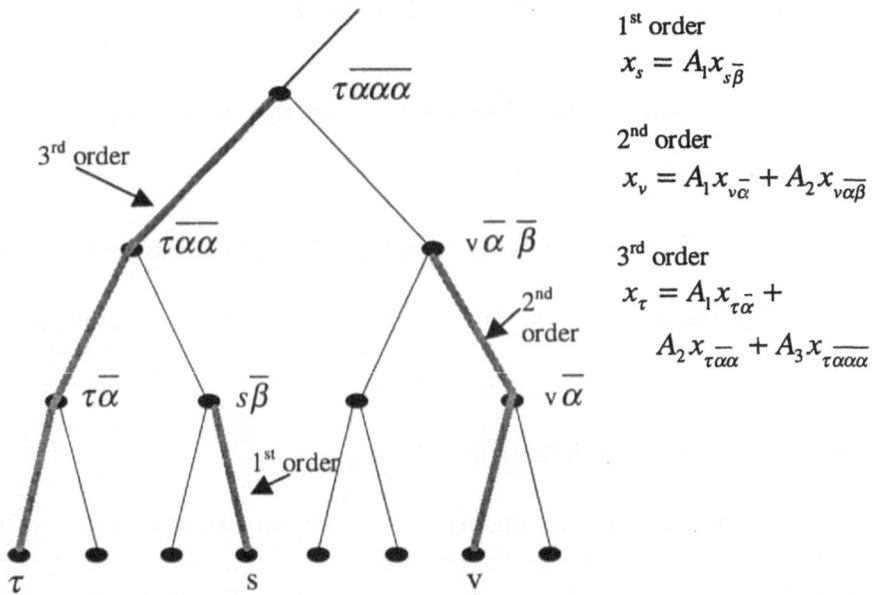

1st order
$$x_s = A_1 x_{s\bar{\beta}}$$

2nd order
$$x_v = A_1 x_{v\bar{\alpha}} + A_2 x_{v\bar{\alpha}\bar{\beta}}$$

3rd order
$$x_\tau = A_1 x_{\tau\bar{\alpha}} +$$
$$A_2 x_{\tau\bar{\alpha}\bar{\alpha}} + A_3 x_{\tau\bar{\alpha}\bar{\alpha}\bar{\alpha}}$$

Figure 5. The support on a dyadic tree of 1st, 2nd, and 3rd order linear homogeneous systems at specific nodes.

3.3. FORCED LINEAR SYSTEMS ON TREES

In this section we will extend the previous developments to multivariable, linear, discrete-time models of the form

$$x(k+1) = Ax(k) + Bu(k) \tag{21}$$

where $u(k)$ is an external input. Indexing the sequences $\{x(k)\}$ and $\{u(k)\}$ by the nodes at level-0 of the dyadic tree in Figure 4(a), model (21) becomes

$$x(0,k+1) = Ax(0,k) + Bu(0,k) \tag{21a}$$

In a manner similar to that used for the development of equation (13a), we can show that for the right-nodes at level-0, the state is given by,

$$x(0, k+1) = \sqrt{2}(I + A)^{-1} Ax(1, \frac{k}{2}) + (I + A)^{-1} Bu(0, k)$$

$$x(0, k+3) = \sqrt{2}(I + A)^{-1} Ax(1, \frac{k}{2}+1) + (I + A)^{-1} Bu(0, k+2)$$

In general, if $(\tau\beta)$ is a right node at level-0, then

$$x(\tau\beta) = \sqrt{2}(I + A)^{-1} Ax(\tau) + (I + A)^{-1} Bu(\tau\alpha)$$

or,

$$x(\tau\beta) = \sqrt{2}(I + A)^{-1} Ax(\tau) + (I + A^{-1}) Bu(\tau\alpha) \qquad \tau\beta: \text{right-node at Level} \qquad (22a)$$

Equivalently,

$$x(\tau\beta) = \sqrt{2}(I + A)^{-1} A\overline{\beta}x(\tau\beta) + (I + A)^{-1} Bu(\tau\alpha) \quad \tau\beta: \text{right-node at Level-0} \quad (23a)$$

The value of the state at any left-node of level-0, is given from the averaging process, i.e.

$$x(\tau\alpha) = \sqrt{2}\, x(\tau) - x(\tau\beta)$$

while, $x(\tau\beta)$ is given by equation (22a), thus yielding

$$x(\tau\alpha) = \sqrt{2}(I + A)^{-1} x(\tau) - (I + A)^{-1} Bu(\tau\alpha) \qquad \tau\alpha: \text{left node at level-0} \qquad (22b)$$

or equivalently

$$x(\tau\alpha) = \sqrt{2}(I + A)^{-1} \overline{\alpha}x(\tau\alpha) - (I + A)^{-1} Bu(\tau\alpha) \qquad \tau\alpha: \text{left node at level-0} \qquad (23b)$$

Equations (23a) and (23b) can be condensed for any node, τ, left or right into

$$x(\tau) = \left(\overline{\alpha}A_{\alpha}^{(0)} + \overline{\beta}A_{\beta}^{(0)}\right)x(\tau) - \alpha\left(\overline{\alpha}B^{(0)} - \overline{\beta}B^{(0)}\right)u(\tau) \qquad (24)$$

where, $A_{\alpha}^{(0)}$ and $A_{\beta}^{(0)}$ are given by equation (16) and

$$B^{(0)} = (I+A)^{-1}B$$

From equation (24) we take

$$\left(I - \overline{\alpha}A_{\alpha}^{(0)} - \overline{\beta}A_{\beta}^{(0)}\right)x(\tau) = -\alpha\left(\overline{\alpha}B^{(0)} - \overline{\beta}B^{(0)}\right)u(\tau)$$

thus leading to the following transfer function model at Level-0,

$$x(\tau) = -\left(I - \overline{\alpha}A_\alpha^{(0)} - \overline{\beta}A_\beta^{(0)}\right)^{-1}\alpha\left(\overline{\alpha}B^{(0)} - \overline{\beta}B^{(0)}\right)u(\tau) \qquad (25)$$

It should be noted again that model (24) was generated by applying the discrete-time model of equation (21) to the *right-nodes only*. Application of (21) to the left-nodes at level-0 leads to discrete-time models at the resolution of level-1. For example, applying (21) on the left-node, $x(0,k+2)$, we take:

$$x(0,k+2) = Ax(0,k+1) + Bu(0,k+1)$$

or

$$\sqrt{2}(I + A)^{-1}x(1,\frac{k}{2}+1) - (I+A)^{-1}Bu(0,k+2) =$$
$$A\left\{\sqrt{2}(I+A)^{-1}Ax(0,\frac{k}{2}) + (I+A)^{-1}Bu(0,k)\right\} + Bu(0,k+1)$$

or,

$$x(1,\frac{k}{2}+1) = A^2x(1,\frac{k}{2}) + \frac{1}{\sqrt{2}}\left\{ABu(0,k) + (I+A)Bu(0,k+1) + Bu(0,k+2)\right\}$$

Define

$$Bu(1,\frac{k}{2}) = \frac{1}{\sqrt{2}}(I+A)^{-1}\left\{ABu(0,u) + (I+A)Bu(0,k+1) + Bu(0,k+2)\right\} \qquad (26)$$

and take

$$x(1,\tfrac{k}{2}+1) = A^2x(1,\tfrac{k}{2}) + (I+A)Bu(1,\tfrac{k}{2}) \qquad (27)$$

which represents the discrete-time model at level-1, i.e. resolution which is half of that at level-0. Following a similar construction we can start from equation (27) and recursively generate the following generalized discrete-time model at level-m

$$x(m,\frac{k}{2^m}+1) = A^{2^m}x(m,\frac{k}{2^m}) + (I+A^{2^{m-1}})(I+A^{2^{m-2}})\cdots(I+A)Bu(m,\frac{k}{2^m}) \qquad (27a)$$

Starting with equation (27a) we can generate the following two-scale models for any node, τ, at level-m,

$$x(\tau\alpha) = A_\alpha^{(m)}x(\tau) - B^{(m)}u(\tau\alpha) \qquad (28a)$$
$$x(\tau\beta) = A_\beta^{(m)}x(\tau) + B^{(m)}u(\tau\alpha) \qquad (28b)$$

or equivalently

$$x(\tau) = -(I - \overline{\alpha}A_{\alpha}^{(m)} - \overline{\beta}A_{\beta}^{(m)})^{-1}\alpha(\overline{\alpha}B^{(m)} - \overline{\beta}B^{(m)})u(\tau) \qquad (28c)$$

where, $A_{\alpha}^{(m)}$ and $A_{\beta}^{(m)}$ are given by equations (19) and

$$B^{(m)} = (I + A^{2^{m-1}})(I + A^{2^{m-2}})\ldots(I + A)B \qquad (29)$$

The counterpart of model (20), employing wavelet coefficients, for a forced multi-scale system is given by:

$$\delta x(\tau) = (I + A^{2^m})^{-1}(I - A^{2^m})x(\tau) - \sqrt{2}(I + A^{2^m})^{-1}B^{(m)}u(\tau\alpha) \qquad (30)$$

Remarks:

1. From equation (26) we notice that the inputs are decomposed by a scheme different to Haar. The scaling and wavelet coefficients of the corresponding filters have the following properties

$$\frac{1}{\sqrt{2}}\{A + (I + A) + I = \sqrt{2}(I + A) \qquad (31)$$

$$-A + (I + A) - I = 0 \qquad (32)$$

$$\phi(t) = \tfrac{1}{2}\,\phi(2t-2) + \phi(2t-1) + \tfrac{1}{2}\,\phi(2t)$$

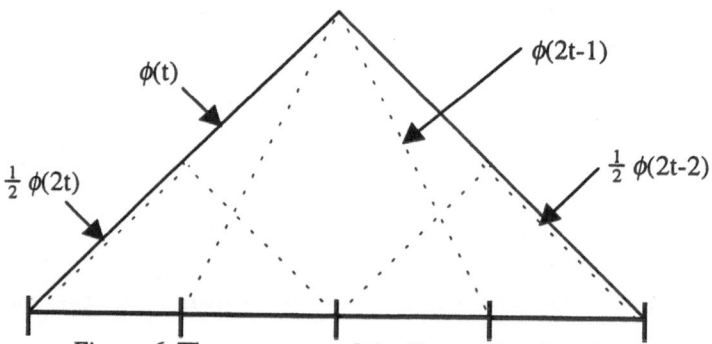

Figure 6. The geometry of the Hat scaling function

Equation (31) implies that the "averaged" inputs should be normalized by $\{(2)^{1/2}(I+A)\}^{-1}$, as indeed is done in equation (26). Equation (32) implies that the coefficients [-A, +(I+A), -I] constitute a legitimate set for a wavelet filter [49]. The averaging scheme of equation (26) constitutes a "modified Hat" averaging scheme, while the presence of A indicates that the averaging scheme is model-dependent. Note that the normal Hat-based decomposition scheme (Figure 6) uses the following filters:

scaling: $h = \frac{1}{\sqrt{2}}$ [1/2, 1, 1/2]

wavelet: $g = \frac{1}{\sqrt{2}}$ [-1/2, 1, -1/2]

However, unlike the Haar wavelet, the modified Hat (like the normal Hat) wavelet does not lead to an orthogonal basis, thus creating representations with redundancies. These redundancies lead to interaction among the wavelet coefficients on a homogeneous tree, as we will see in subsequent sections.

2. For any node, τ, of a binary tree, the corresponding value of the input, scaled through the modified Hat transform, is given by the following equation, which represents the generalization of equation (26) (see also Figure 7),

$$Bu(\tau) = \frac{1}{\sqrt{2}}(I + A^{2^m})^{-1}\left\{A^{2^m}Bu(\tau) + (I + A^{2^m})Bu(\tau\beta) + Bu(\tau\delta\alpha)\right\} \quad (33)$$

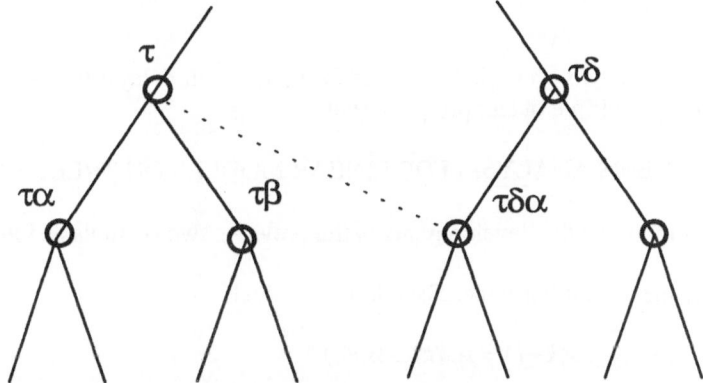

Figure 7. Transformation structure of inputs through the modified Hat transform

3. The modified Hat averaging scheme of equation (26), depends on whether the input delay is an *odd* or *even* multiple of the sampling interval. It can be easily shown that for discrete-time model of the form

$$x(0, k+1) = Ax(0,k) + Bu(0, k - p)$$

the following statements are true:

688

(a) If p = 0, or even, equation (27) takes the form

$$x(1,\tfrac{k}{2}+1) = A^2 x(1,\tfrac{k}{2}) + (I + A)Bu(1,\tfrac{k}{2}-\tfrac{p}{2})$$

where, $u(1,\dfrac{k}{2}-\dfrac{p}{2})$ is averaged as in Figure 8(a).

(b) if p = odd, equation (27) takes the form

$$x(1,\tfrac{k}{2}+1) = A^2 x(1,\tfrac{k}{2}) + (I + A)Bu^*(1,\tfrac{k}{2}-\tfrac{p-1}{2})$$

where, $u^*(1,\dfrac{k}{2}-\dfrac{p}{2})$ is averaged as in Figure 8(b).

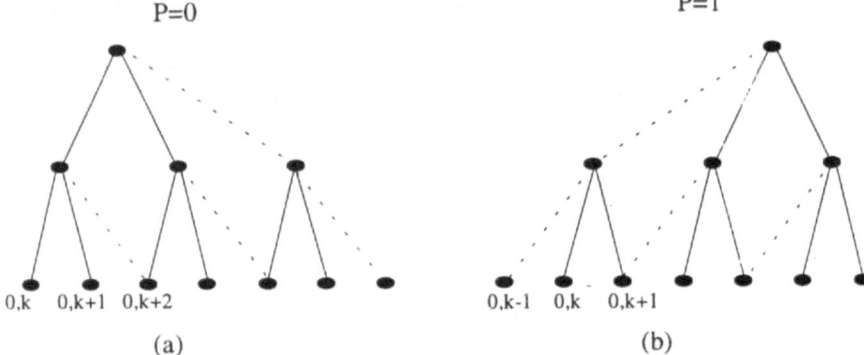

(a) (b)

Figure 8. The two averaging schemes for inputs with delays: (a) zero or even multiple, and (b) odd multiple of sampling interval.

3.4. SYSTEMS ANALYSIS FOR LINEAR MODELS ON TREES

Let us summarize the developments in the previous two sections as follows:

Given a linear, multivariable, discrete-time model,

$$x(k+1) = Ax(k) + Bu(k) \tag{21}$$

whose state and input variables are defined at discrete-time points indexed by the set of integers, we can generate an equivalent linear model, with state and input variables defined on an array of discrete points in the time-scale (or, frequency) domain, indexed by the nodes of a homogeneous tree and having the form

$$x(\tau\alpha) = A_\alpha^{(m)} x(\tau) - B^{(m)} u(\tau\alpha) \tag{28a}$$
$$x(\tau\beta) = A_\beta^{(m)} x(\tau) + B^{(m)} u(\tau\alpha) \tag{28b}$$

or equivalently,

$$x(\tau) = \left(\overline{\alpha} A_\alpha^{(m)} + \overline{\beta} A_\beta^{(m)}\right) x(\tau) - \left(\alpha\overline{\alpha} B^{(m)} - \alpha\overline{\beta} B^{(m)}\right) u(\tau) \tag{28d}$$

In terms of wavelet coefficients, the equivalent model is given by equation (30)

$$\delta x(\tau) = (I + A^{2^m})^{-1}(I - A^{2^m}) x(\tau) - \sqrt{2}(I + A^{2^m})^{-1} B^{(m)} u(\tau\alpha) \tag{30}$$

Let us augment model (21) with an output model

$$y(k) = Cx(k) \tag{34a}$$

Haar scaling of equation (34a) leads to the following output model at any node, τ, of a homogeneous tree

$$y(\tau) = Cx(\tau) \tag{34b}$$

In this section we will examine the properties of these models which link the states, measurements and inputs at various scales.

A. Input-Output Transfer Function

Substitution of $x(\tau)$ from equations (28a) into equation and (34b) yields the following input-output transfer function

$$y(\tau) = C(I - \overline{\alpha} A_\alpha^{(m)} - \overline{\beta} A_\beta^{(m)})^{-1} (\alpha\overline{\alpha} B^{(m)} - \alpha\overline{\beta} B^{(m)}) u(\tau) \tag{35}$$

or,

$$y(\tau) = Su(\tau) = S^\uparrow \vec{S} u(\tau) \tag{35a}$$

where,

$$S^\uparrow = -C(I - \overline{\alpha} A_\alpha^{(m)} - \overline{\beta} A_\beta^{(m)})^{-1} \tag{36a}$$

and

$$\vec{S} = \alpha\overline{\alpha} B^{(m)} - \alpha\overline{\beta} B^{(m)} \tag{36b}$$

The $S^\uparrow$ factor of the transfer function involves upward-shifts only (note the presence of upward-shift operators, $\overline{\alpha}$ and $\overline{\beta}$), while the $\vec{S}$ involves horizontal shifts on nodes at the same horocycle (note that the results of the $\alpha\overline{\alpha}$ and $\alpha\overline{\beta}$ operations remain at the same horocycle). Thus, $S^\uparrow$ smoothes the values along the path linking the current node, τ, to the boundary point, $-\infty$ [5]. On the other hand, $\vec{S}$ smoothes the values at a given level (horocycle) of the tree.

690

B. Stability

The fact that model (28) is equivalent to model (21) implies that the stability characteristics of the discrete-time model are passed on to the multi-scale model on the tree, and vice versa. Note that since the stability of dynamic systems is an asymptotic property as time $\rightarrow +\infty$, in any stability treatment of a multiscale system we should consider a homogeneous tree with an infinite number of nodes at each level. However, before proceeding we need to define the concept of stability for a model on a tree.

> *Definition 1.* Consider the "initial time" path (Figure 9) of an infinite tree. All nodes of the tree are considered to be to the "right" of the "initial time" path. The multi-scale system of equation (28), defined on a homogeneous tree, is ℓ_p-stable in the Lyapunov sense, if, given a bounded state at any node, τ, on the "initial time" path i.e. $\|x(\tau)\|_p \le C$, then the value of the state at any other node of the tree, computed through the recursive application of model (28), is also bounded.

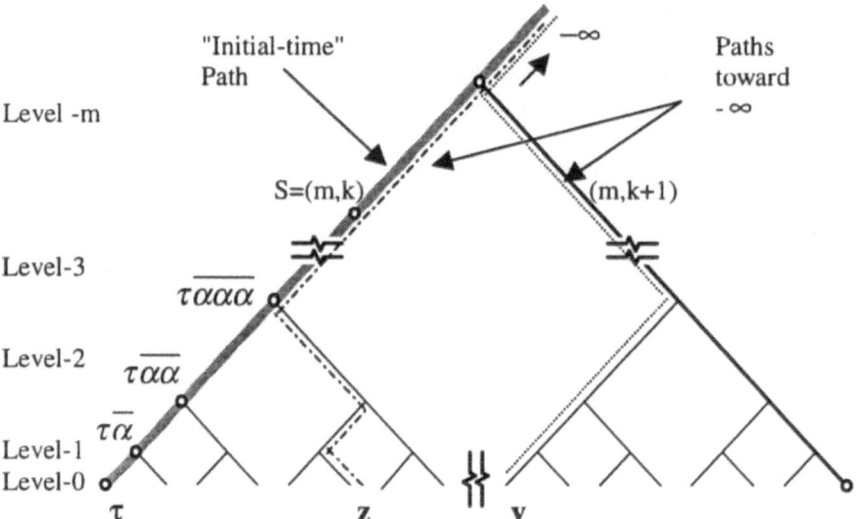

Figure 9. An infinite homogeneous tree, depicting the "Initial-Time" path. It is used in defining and proving conditions on stability, controllability and observability of multi-scale models on trees.

The following theorem is a direct result of the above definitions:

> *Theorem 2.* If the discrete-time system of equation (21) is ℓ_p-stable in the Lyapunov sense, then the multi-scale system of equation (28) is also ℓ_p-stable in the Lyapunov sense, and vice versa.

The proof of the theorem is given in Appendix A.2.

C. Steady State

The steady state in a multi-scale system corresponds to the state value at very large scales. It plays the role of "initializing conditions" in multi-scale computations on trees. Recursive application of equation (27) to the m-th level yields,

$$x(m,\frac{k}{2^m}+1) = A^{2^m} x(m,\frac{k}{2^m}) + \left(I + A^{2^{m-1}}\right)\left(I + A^{2^{m-2}}\right)\cdots(I + A)B\cdot u(m,\frac{k}{2^m}) \qquad (37)$$

As

$$m \to +\infty, \text{ for stable systems } A^{2^m} \to 0,$$

$$x(m,\frac{k}{2^m}+1) \to x(steadystate), \ u(m,\frac{k}{2^m}) \to u(steady\ state)$$

and

$$\lim_{m\to\infty} \left[\left(I + A^{2^{m-1}}\right)\cdots(I + A)\right] = \lim_{m\to\infty}\left(I + A + A^2 + A^{2^m}\right) = (I - A)^{-1}$$

Therefore,

$$x(steady\ state) = (I - A)^{-1}Bu(steady\ state)$$

as expected.

D. Controllability

Again, since systems (21) and (28) are equivalent, we expect that the controllability characteristics of the discrete-time system (21) will be possessed by the multi-scale system on a tree, and vice versa. We need though, to define the concept of controllability for a system on a tree.

> Definition 2. The multi-scale system of equation (28), defined on a homogeneous tree, is controllable, if, given an initial state at node, τ, of the "initial time" path (see Figure 9), the system can reach the desired state at any other node, s, of the tree, through a finite sequence of inputs, provided that
> - distance between s and τ is finite i.e. d(s,τ)=finite, and
> - the control inputs are at the same horocycle (level) as s or lower.

The following theorem is a direct result of the above definition (see Appendix A.2 for the proof of the theorem):

> Theorem 3. If the discrete-time system (21) is controllable, then the multi-scale model (28) is also controllable, and vice-versa.

E. Observability

The discrete-time system (21) with the output model (34a) is equivalent to the multi-scale model (28) with the output model (34b). By analogy to the previous paragraph, let's define the observability of a multi-scale system as follows:

> *Definition 3.* The multi-scale system (28) with the output model (34b), defined on a homogeneous tree, is observable, if, given a set of n measured outputs on the tree, we can compute the initial state on a node of the "initial time" path, provided that the measurements correspond to nodes of trhe tree which are at the same horocycle (level) as the initial state or lower.

The following theorem is a direct result of the above definition and its proof can be found in appendix A-3.

> *Theorem 4.* If the discrete-time system (21) with the output model (34a) is observable, then the multi-scale model (28) with output model (34b) is also observable, and vice versa.

F. Multi-Rate Control Actions and Measurements

All the developments in the previous paragraphs were limited to discrete-time systems with all inputs applied at equally distant discrete-points and all outputs measured with the same sampling intervals. In this section we will extend the discussion to account for systems whose inputs are applied with different rates and whose outputs are sampled at different discrete-time intervals, such as those discussed by Gopinath and Bequette [24] and Ohshima and Hashimoto [43].

Consider the discrete-time model

$$x(k+1) = Ax(k) + B_1 u^{(1)}(k) + B_2 u^{(2)}(k')$$
(38)

where, inputs $u^{(1)}(k)$ are applied at every discrete-time point, k, while inputs $u^{(2)}(k')$ are applied at every other discrete-time point, i.e. with half the frequency of $u^{(1)}(k)$. Consequently,

$$u^{(2)}(k') = \begin{cases} u^{(2)}(k) & \text{if} \quad k = 0 \text{ or even} \\ u^{(2)}(k-1) & \text{if} \quad k = \text{odd} \end{cases}$$
(39)

Assuming that the discretization of system (38) corresponds to the nodes at level-0 of a homogeneous tree, it is easy to generate the following discrete-time model at level-1 (see Section 3.3 and Figure 4(a)),

$$x(1,\frac{k}{2}+1) = A^2 x(1,\frac{k}{2}) + \frac{1}{\sqrt{2}}\left\{AB_1 u^{(1)}(0,k) + (I+A)B_1 u^{(1)}(0,k+1) + B_1 u^{(1)}(0,k+2)\right\}$$
$$+ \frac{1}{\sqrt{2}}\left\{AB_2 u^{(2)}(0,k) + (I+A)B_2 u^{(2)}(0,k+1) + B_2 u^{(2)}(0,k+2)\right\}$$

or,

$$x(1,\frac{k}{2}+1) = A^2 x(1,\frac{k}{2}) + (I+A)B_1 u^{(1)}(1,\frac{k}{2}) + (I+A)B_2 u^{(2)}(1,\frac{k}{2})$$

or,

$$x(1,\frac{k}{2}+1) = A^2 x(1,\frac{k}{2}) + (I+A)Bu(1,\frac{k}{2}) \qquad (40)$$

where

$$B = [B_1 B_2], \quad u = \begin{bmatrix} u^{(1)} \\ u^{(2)} \end{bmatrix}$$

and, $u^{(1)}(1,\frac{k}{2})$ and $u^{(2)}(1,\frac{k}{2})$ are given by the modified Hat averages of equation (26).
Therefore, at the discretization level-0, it is identical to what we have already generated for systems with equal rates on the application of all inputs, i.,e. equation (27).

Therefore, the properties of the multi-scale system on a tree, resulting from the dynamic system (40), remain the same, provided that level-1 represents the level of finest resolution. Thus, the state of the multi-scale system is controllable throughout the homogeneous tree from level-1 and higher, if rank [B   AB A^2B ... $A^{n-1}B$] = n. However the system is not controllable at level-0, as it can be easily deduced from the proof of Theorem 3 in Appendix A-2. The above analysis can be directly extended to cases where the interval of applying certain inputs is 2^m times the basic interval, or not a multiple of a power of 2. The latter case will be discussed below in reference to multi-rate sampled outputs.

Let us now extend the discussion to include discrete-time systems with outputs sampled at different rates. The reference discrete-time system is given by

$$x(k+1) = Ax(k) + Bu(k) \qquad (21)$$

$$y^{(1)}(k) = C_1 x(k) \qquad (41a)$$

$$y^{(2)}(2^m k) = C_2 x(2^m k) \tag{41b}$$

The discrete-time system at level-m of a homogeneous tree is given by

$$x(m, \frac{k}{2^m}+1) = A^{2^m} x(m, \frac{k}{2^m}) + \left\{ \left(I + A^{2^{m-1}}\right)\left(I + A^{2^{m-2}}\right)\cdots(I + A)B \right\} u(m, \frac{k}{2^m}) \tag{42}$$

$$y(m, \frac{k}{2^m}) = \begin{bmatrix} y^{(1)}(m,k) \\ y^{(2)}(m,k) \end{bmatrix} = \begin{bmatrix} C_1 \\ C_2 \end{bmatrix} x(m, \frac{k}{2^m}) = Cx(m, \frac{k}{2^m}) \tag{43}$$

It follows easily from the proof of Theorem 4 in Appendix A.3., that the state of system (42) with the output model (43) is observable throughout the homogeneous tree from level-m and above, if and only if rank $[C^T \quad A^T C^T \quad (A^T)^2 C^T \cdots (A^T)^{n-1} C^T]^T = n$. We cannot completely reconstruct the state of the system at any node of the tree below level-m. Partial reconstruction is attainable to the extent of the rank of $[C_1^T \quad A^T C^T \cdots (A^T)^{n-1} C_1^T]^T$. Approximate but complete reconstructions of the state are possible by propagating the values of the measurement sequence $\left\{ y(m, \frac{k}{2}); k = 0,1,2,\ldots \right\}$ down the branches of the homogeneous tree at levels of finer discretization, below level-m. The ensuing interpolation is based on the characteristics of the scaling function used on the tree.

If the sampling intervals are not 2^m, m = 1, 2, 3, ..., multiples of a basic one, the following two options are possible.

(a) Use scaling and wavelet functions which admit a dilation base different than 2.
(b) Employ an arbitrary interpolation scheme and resample, so that the sampling intervals are 2^m multiples of a basic one.

G. Generalization to Other Scaling and Wavelet Functions

Although the generation of multi-scale systems from discrete-time ones has been based on the Haar scaling and wavelet functions (for the state and measurement variables) and the corresponding modified Hat (for the inputs), the resulting multi-scale model forms are independent of the specific type of scaling and wavelet functions.

Consider a scaling filter, h(n), of support, q, and coefficients

$$h = [a_0, a_1, \ldots, a_{q-1}] \tag{44a}$$

Let the corresponding wavelet filter, g(n), also of support, q, have the following coefficients

$$g = [b_0, b_1, \ldots, b_{q-1}] \tag{44b}$$

Note that [49]

$$\sum_{i=0}^{q-1} a_i = \sqrt{2} \quad \text{and} \quad \sum_{i=0}^{q-1} b_i = 0 \tag{45}$$

and b_i's can be computed from a_i's and vice versa. The homogeneous tree, whose nodes are used to index the components of a signal scaled by h(n) or/and decomposed by g(n), is of order q + 1 and admits at each node, q down-shift opertors, i.e. α_0, α_1 , $\dots$, α_{q-1} and one up-shift operator, $\bar{\gamma}$, see Figure 10. The following results can be easily shown:

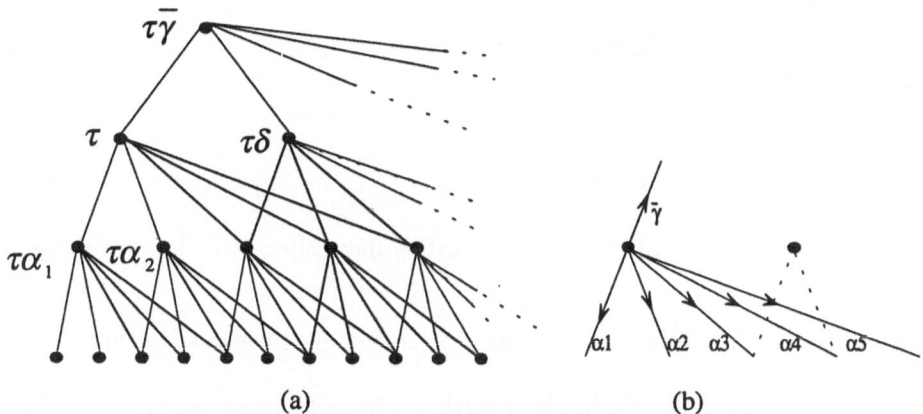

(a)   (b)

Figure 10. (a) The homogeneous tree of order q = 1 encoding the states and outputs of a homogeneous multi-scale system. (b) The associated shifts around node τ.

Generalization-1 (Homogeneous System). Given

$$x(k+1) = Ax(k) \quad \text{and} \quad y(k) = Cx(k)$$

then, at any node, τ (at level-m), of a homogeneous tree or order q+1, the state and measurements at the left-most nodes, $\tau\alpha_1$, are given by the following multi-scale model, under scaling and wavelet filters, h(n) and g(n), given by (44a) and (44b):

$$x(\tau\alpha_1) = A_1^{(m)} x(\tau) \tag{46a}$$

and

$$y(\tau\alpha_1) = Cx(\tau\alpha_1) \tag{46b}$$

where,

$$A_1^{(m)} = \sqrt{2}\left(a_o + a_1 A^{2^m \cdot 1} + a_2 A^{2^m \cdot 2} + \cdots + a_{q-1} A^{2^m \cdot (q-1)}\right)^{-1} \tag{47}$$

696

Thus, the discrete-time model defined over n time points has been transformed to a multi-scale model defined over the n left-most nodes of a tree that spans the time horizon of n points.

Generalization-2 (Forced System). Given

$$x(k+1) = ax(k) + Bu(k) \quad \text{and} \quad y(k) = Cx(k)$$

then, at any node, τ (at Level-m), of a homogeneous tree of order q + 1, the state and measurements at the left-most nodes, $\tau\alpha_1$, are given by the following multi-scale model,

$$x(\tau\alpha_1) = A_1^{(m)} x(\tau) + B_1^{(m)} u(\tau\alpha_1) \tag{48a}$$

and

$$y(\tau\alpha_1) = Cx(\tau\alpha_1) \tag{48b}$$

Sequences, $\{x(k)\}$ and $\{y(k)\}$ have been scaled by the scaling filter, h(k), of support q (given by equation 44a).

Sequence $\{Bu(k)\}$ has been scaled by a filter of support, q+1, with coefficients

$$[a_o A, \ (a_o + a_1 A), \ (a_1 + a_2 A), \ \dots, \ (a_{q-2} + a_{q-1} A), \ a_{q-1}] \tag{49}$$

and has been normalized by

$$\frac{1}{\sqrt{2}} (I + A)^{-1}.$$

H. The Issue of "Closure" Between the Time and Time-Scale Domains

In going from the time-domain to the frequency-domain, during the Fourier transformation of a signal, we need only the transformation mapping; the two domains are orthogonal to each other. This is not the case as we go from the time-domain, where the discrete-time systems are defined, to the time-scale domain where the multi-scale systems are defined. The consequences are interesting and need to be pointed out, as one contemplates the formulation of multi-scale models.

Consider the two-scale, forced linear model of equation (28a). Its variables are defined in the time-scale domain and their values are indexed by the nodes of a homogeneous tree. If we require that the model (28a) in the time-scale domain is completely equivalent to the model (27a) in the time-domain, we have imposed a requirement that constraints in a profound way any transformation from one domain to the other. This requirement can be exposed as follows:

"The values of the states and inputs on a homogeneous tree, evolving by equation (28a), must achieve "closure", i.e. be equal, with the values of the states and inputs on the discrete-time domain, evolving by equation (27a)".

As a result of the "closure requirement" the two domains, i.e. the time-domain and the time-scale-domain, cannot be independent. As we move from the 1-dimensional domain of discrete-time (indexed by the integer numbers) to the 2-dimensional domain of time-scale (indexed by the nodes of a homogeneous tree) the additional freedom is taken up by the requirement that the state evolves by e.g. Haar scaling. As a result, the input variables are not free to be scaled by Haar, but must obey the resulting constraint. This leads to the modified Hat transform for the inputs (see Remark 1 in Section 3.3.).

It is instructive to examine the consequences of lifting the "closure requirement". In such case, let us postulate the following two-scale model, on a dyadic tree

$$x(\tau\alpha) = A_\alpha^{(m)} x(\tau) + B_\alpha^{(m)} u(\tau\alpha) \tag{50a}$$

$$x(\tau\beta) = A_\beta^{(m)} x(\tau) + B_\beta^{(m)} u(\tau\beta) \tag{50b}$$

where, both the state and input variables are scaled by the Haar function.

It is rather straightforward to show that model (50a) and (50b) cannot generate a discrete-time, causal model of the form

$$x(k+1) = Ax(k) + Bu(k)$$

at any resolution. As a result, equations (50a) and (50b) cannot model adequately causal systems, where the inputs are tangible physical quantities, like those used as manipulated inputs for control. In this case the "closure requirement" is of essence and must be satisfied.

On the other hand, consider a static Markov moel with states varying over a spatial dimension e.g.

$$x(k+1) = Ax(k) + Bw(k)$$

with $w(k)$ representing a stochastic process characterizing modeling errors. For the corresponding multiscale model

$$x(\tau\alpha) = A_\alpha^{(m)} x(\tau) + B_\alpha^{(m)} w(\tau\alpha) \tag{51a}$$

$$x(\tau\beta) = A_\beta^{(m)} x(\tau) + B_\beta^{(m)} w(\tau\beta) \tag{51b}$$

no physical causality needs to be satisfied and the "closure requirement" is not necessary. In such case. The localized effect of the Markov process leads to a multi-scale model on a tree, employing common scaling on states and inputs.

4. BASIC TASKS WITH MS-MODELS ON TREES

The transformation of discrete-time models into their multi-scale counterparts offers some very attractive advantages in solving various engineering problems in process operations and control. In this section we will examine the character of algorithms which have been developed in order to address basic tasks, such as; (a) solution of sets of linear equations, (b) optimal control, and (c) optimal estimation. In subsequent sections we will examine how these algorithms can be combined to address composite issues, e.g. feedback control through Model-Predictive Controllers.

4.1 SIMULATION OF LINEAR DYNAMIC SYSTEMS

Given the inputs and initial conditions, the solution of the set of linear equations

$$x(k+1) = Ax(k) + Bu(k) \quad ; \quad k = 0, 1, \dots, 2^M - 1$$

$$x(0) \quad = c = \text{unknown}$$

$$u(k) = \text{known} \quad ; \quad k = 0, 1, \dots, 2^M - 1$$

involves algorithms requiring 2^M steps of computations. The complexity of each step depends on the structure of A.

The multi-scale counterpart of the above problem has the form (see generalizations of equations 22a and 22b):

$$x(\tau\alpha) - A_\alpha^{(m)} x(\tau) = b_\alpha \tag{52a}$$

$$x(\tau\beta) - A_\beta^{(m)} x(\tau) = b_\beta \tag{52b}$$

where, $b_\alpha = -B^{(m)} u(\tau\alpha)$ and $b_\beta = B^{(m)} u(\tau\alpha)$ are known quantities. The solution of equations (52a) and (52b), can be proceed as follows::

Step-1 (Up-shift):   Given, $x(\tau\alpha)$, compute, $x(\tau)$ from (52a)

Step-2 (Down-shift):   With, $x(\tau)$, known, compute, $x(\tau\beta)$ from (52b)

The initial condition is at a $\tau\alpha$-type node, i.e. a left-branch node, so the algorithm can "march" explicitly from the initial conditions.

The dyadic character of the tree, defining the domain for the states in (52a) and (52b), implies a complete parallelization of Step-1 and Step-2, e.g.

Step-1a: Given $x(0) \equiv x(0,0)$, compute $x(1,0)$ from (52a) with m = 0

Step-2a: Given $x(1,0)$, compute $x(0,1)$ from (52b) with m = 0

Step-1b: Given $x(1,0)$, compute $x(2,0)$ from (52a), with m = 1

Figure 11 shows the evolution of parallel computations during the simulation of a multi-scale system over 32 discrete-time intervals.

The parallelizable algorithm, described above, requires

$$K = 2 \log_2 (2^M) - p \qquad (53)$$

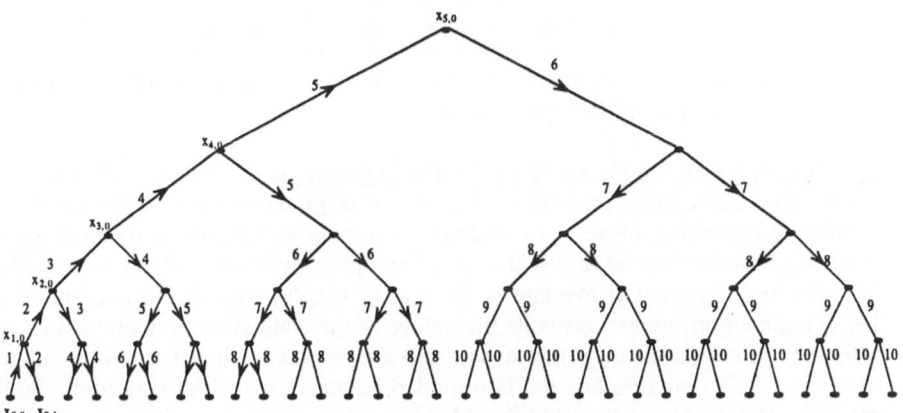

Figure 11. Groups of parallelizable computations (characterized bt the same number) generated during the simulation of a multi-scale dynamic system, allowing the simulation over $32(=2^5)$ time points in $2\log_2(2^5)=10$ steps

steps of computations, a significant reduction over the 2^M steps required for the solution of the discrete-time model, where p is the level of the tree at which the given conditions on the state are specified, i.e. p = 0 for initial conditions at t = 0. The complexity of each step depends on the structures of A, A^2, A^4, ..., A^{2M-1} and for moderately dense matrices A, remains approximately the same as in the discrete-time models. However, even for sparse A's, the dimensionality of the state vector is, normally, orders of magnitude smaller than the number of time-points in the simulation horizon.

Two additional features make the parallelizable algorithm, described above, fairly attractive:

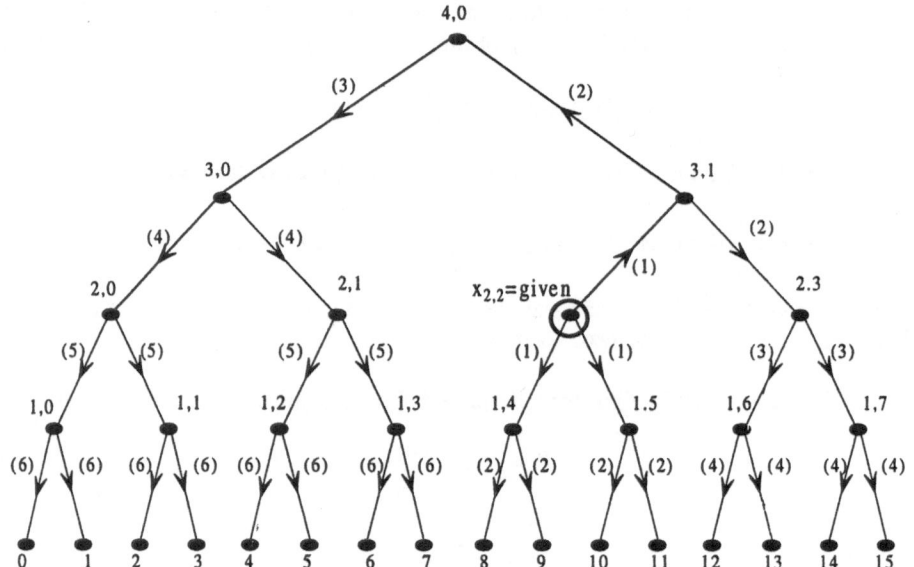

Figure 12. The parallelizable computations during the simulation of a multi-scale system with $x(2,2)$ as the "initial" given condition.

a. Handling a multitude of given "initial" conditions on states. Although state values at an initial time is the most common set of given conditions, the multi-scale simulation described above can handle in an equally straightforward manner given conditions on the state at any scale. For example (see Figure 12), if we are given the Haar (i.e. arithmetic) average of the states, $x(2,2)$, over the time-points (8, 9, 10, 11), the algorithm can generate the values of the state at all time points in $2 \log_2 (16) - 2 = 6$ steps. Clearly, the fewest-step solution is achieved when the average state over the simulation horizon is provided as the given initial condition. In this case the solution requires $\log_2 (2^M) = M$ steps.

b. Handling stiff systems with large condition numbers. Physico-chemical phenomena evolve with different time scales. Fine discretization is absolutely necessary for rapidly evolving states but inconsequential in the computation of slowly evolving ones. The multi-scale representation allows an efficient trade-off between (i) the coarseness of time-discretization, which achieves desired levels of solution accuracy, and (ii) cost of computations.

At the core of the issue is the value of the condition number, ℓ

$$\ell = \max \{\text{eigenvalue (A)}\}/\min \{\text{eigenvalue (A)}\}$$

As the condition number increases, the discretization length must decrease, thus leading to more "costly" simulations. Decompose A as follows,

$$A = P \Lambda P^{-1}$$

where, Λ is the diagonal matrix of eigenvalues of A and P its modal matrix of eigenvectors. As the multi-scale model moves up at higher levels, the dynamic evolution of scaled states is determined by the matrix, $A_\alpha^{(m)}$. (See equation 52a).

Now from equations (19)

$$A_\alpha^{(m)} = \sqrt{2}(I + A^{2^m})^{-1} = \sqrt{2}(I + P\Lambda^{2^m} P^{-1})^{-1}$$

Then, the up-shift of Step-1, yields

$$x(\tau) = \frac{1}{\sqrt{2}}\left(I + P\Lambda^{2^m} P^{-1}\right)\left(x(\tau\alpha) + b_\alpha\right)$$

For stable systems, all eigenvalues of A, satisfy the conditions, $|\lambda_i| \leq 1, i = 1,2,\cdots,n$. Therefore, $\lambda_i^{2^m}$ decreases rapidly for $|\lambda_i| \ll 1$, as m increases. Consequently, as the computations move up the tree, the size of the problems solved at higher levels decreases, thus restoring the condition number to lower values and blocking the propagation of errors.

4.2. MULTI-SCALE OPTIMAL CONTROL

Consider the following optimal control problem formulation:

Problem-P1.

$$\underset{u(k);k=0,1,2,\ldots2^M-1}{Minimize} \quad F = \sum_k \left\{(r(k) - x(k))^T (r(k) - x(k)) + u^T(k)Qu(k)\right\} \tag{54}$$

subject to

$$x(k+1) = Ax(k) + Bu(k) \qquad k = 0, 1, 2, \ldots, 2^{M-1} \tag{55}$$

$$\underline{U} \leq u(k) \leq \overline{U} \tag{56}$$

This problem has been extensively studied and presented in the time domain by numerous researchers [6], [15], [19], [20], [22], [44].

Let's generate the equivalent problem formulation, after defining the state and input variables on a homogeneous tree. First, let's transform the objective function (54) into an equivalent form involving variables on a homogeneous tree. Then, we will transform the state equations (55) and constraints (56) into equivalent expressions but with the states and inputs defined on a tree.

Consider the sequence of the i-th component, $\{Z^{(i)}\}$, of a vector, Z, over the time points, $k = 0, 1, 2, \ldots, 2^M\text{-}1$. The Haar decomposition of $Z^{(i)}$ yields the vector wavelet coefficients,

$$\delta z^{(i)} = WZ^{(i)} \tag{57}$$

where W is the orthonormal Haar wavelet decomposition matrix of the form (for $M = 3$)

$$W = \begin{bmatrix} \frac{1}{\sqrt{2}} & -\frac{1}{\sqrt{2}} & 0 & 0 & 0 & 0 & 0 & 0 \\ 0 & 0 & \frac{1}{\sqrt{2}} & -\frac{1}{\sqrt{2}} & \frac{1}{\sqrt{2}} & -\frac{1}{\sqrt{2}} & 0 & 0 \\ 0 & 0 & 0 & 0 & \frac{1}{\sqrt{2}} & -\frac{1}{\sqrt{2}} & 0 & 0 \\ 0 & 0 & 0 & 0 & 0 & 0 & \frac{1}{\sqrt{2}} & -\frac{1}{\sqrt{2}} \\ \frac{1}{2} & \frac{1}{2} & -\frac{1}{2} & -\frac{1}{2} & 0 & 0 & 0 & 0 \\ 0 & 0 & 0 & 0 & \frac{1}{2} & \frac{1}{2} & -\frac{1}{2} & -\frac{1}{2} \\ \frac{1}{\sqrt{8}} & \frac{1}{\sqrt{8}} & \frac{1}{\sqrt{8}} & \frac{1}{\sqrt{8}} & -\frac{1}{\sqrt{8}} & -\frac{1}{\sqrt{8}} & -\frac{1}{\sqrt{8}} & -\frac{1}{\sqrt{8}} \\ \frac{1}{\sqrt{8}} & \frac{1}{\sqrt{8}} & \frac{1}{\sqrt{8}} & \frac{1}{\sqrt{8}} & \frac{1}{\sqrt{8}} & \frac{1}{\sqrt{8}} & \frac{1}{\sqrt{8}} & \frac{1}{\sqrt{8}} \end{bmatrix} \tag{58}$$

From equation (57) we have

$$\delta z^{(i)^T} \delta z^{(i)} = Z^{(i)^T} W^T W Z^{(i)} = Z^{(i)^T} Z^{(i)}$$

since the Haar decomposition is orthonormal, i.e. $W^T W = I$. Thus,

$$\sum_k z^T(k)z(k) = \sum_k \sum_i (z^{(i)}(k))^2 = \sum_i Z^{(i)^T} Z^{(i)} = \sum_i \delta z^{(i)^T} \delta z^{(i)}$$

leading to

$$\sum_k z^T(k)z(k) = \sum_\tau \delta z^T(\tau)\delta z(\tau) \tag{59}$$

where the summation at the right-hand side is over all nodes of the dyadic tree, resulting from the Haar decomposition of a sequence of values over 2^M points. Equation (59) can be seen as the analog of Parseval's theorem in the time-scale domain. Applying equation (59) on the sequence, $\{r(k) - x(k)\}$, the first term of the objective function (54) yields,

$$\sum_k (r(k) - x(k))^T (r(k) - x(k)) = \sum_\tau (\delta r(\tau) - \delta x(\tau))^T (\delta r(\tau) - \delta x(\tau))$$

The second term of the objective function, i.e. $u^T(k)Qu(k)$, represents a penalty against the use of large values in the manipulated inputs, $u(k)$. In most common formulations of optimal control problems, matrix Q denotes the relative significance of the various components of the input vector, u, and there is no explicit consideration in the objective function of the time-dependent behavior of the manipulated variables.

Additional terms are usually introduced to account for such considerations, the most typical example of which is a penalty term on the magnitude of input changes, i.e.

$$\sum_k \Delta u^T(k) R \Delta u(k) \cdot$$

The multi-scale models, on the other hand , allow the formulation of objective functions which can explicitly include considerations on the time-behavior of input variables, e.g. penalize short time-scale movements more than the long time-scale ones, or the opposite. Therefore, the objective function that we will consider for the multi-scale version of optimal control has the following generalized form,

$$F = \sum_\tau (\delta r(\tau) - \delta x(\tau))^T (\delta r(\tau) - \delta x(\tau)) + \sum_{\tau\alpha} R(\tau\alpha) u^T(\tau\alpha) Q u(\tau\alpha) \qquad (60)$$

where, $u(\tau\alpha)$ is the scaled value of the input at any left-node, $\tau\alpha$, of a binary tree and $R(\tau\alpha)$ the corresponding weight.

Note 1: The scaling of the input is done through the modified Hat transform, as discussed in section 3.3., and is given by equation (33)

$$u(\tau) = \frac{1}{\sqrt{2}}(I + A^{2^m})^{-1}\left\{A^{2^m}u(\tau\alpha) + (I + A^{2^m})u(\tau\beta) + Iu(\tau\delta\alpha)\right\} \qquad (33)$$

where, m, denotes the level (or, scale) of the binary tree where node, τ, is located.

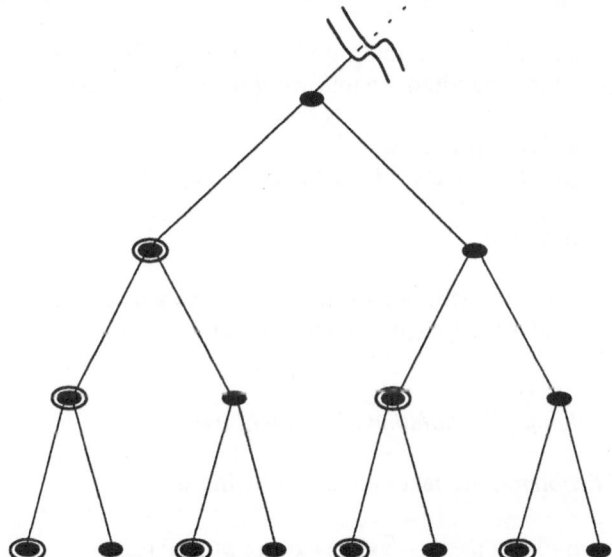

Figure 13. The left nodes of a binary tree indicating the independent values of the manipulated input variables in an optimal control formulation.

Note 2: From the state equations (3) we notice that the manipulated input is only on left-nodes. By summing up over the left-nodes, only, the second term of the objective function (60) has as many terms as the original objective (see Figure 13)

Note 3: Since $\delta x(\tau) = \sqrt{2}x(\tau\alpha) - x(\tau)$ (see Section 2.2.), the objective function (60) can be put into the following form,

$$F = \sum_{\tau\alpha}\left\{(\delta r(\tau) - \sqrt{2}x(\tau\alpha) + x(\tau))^T (\delta r(\tau) - \sqrt{2}x(\tau\alpha) + x(\tau)) + R(\tau\alpha)u^T(\tau\alpha)Qu(\tau\alpha)\right\} \quad (60a)$$

Consequently the original problem, *Problem-P1*, is transformed to the following equivalent formulation:

Problem-P1a

$$\underset{u(\tau\alpha);\,\tau\alpha\in Tree}{\text{Minimize}} F = \sum_{\tau\alpha}\left\{(\delta r(\tau) - \sqrt{2}x(\tau\alpha) + x(\tau))^T (\delta r(\tau) - \sqrt{2}x(\tau\alpha) + x(\tau)) + R(\tau\alpha)u^T(\tau\alpha)Qu(\tau\alpha)\right\} \quad (60a)$$

subject to
$$x(\tau\alpha) = A_\alpha^{(m)}x(\tau) - B^{(m)}u(\tau\alpha) \quad (28a)$$

$$\underline{U}^{(m)} \le u(\tau) \le \overline{U}^{(m)} \quad (61)$$

and the scaling relationships on the states (Haar) and inputs (modified Hat)

Problem-P1a is a quadratic programming problem and can be solved through various strategies, which can be grouped into the following three categories;

(a) identification of active constraints,
(b) interior feasible-point searches with penalizing expressions to ensure constraint satisfaction, and
(c) hybrids of the above two.

When we use the multi-scale definition of inputs and states on a tree, the identification of active constraints offers efficient implementations, as it will be shown in the following paragraph.

A Multi-Scale Algorithm for the Solution of Problem P1a.

The Lagrangian of *Problem-P1a* takes on the following form,

$$L = \sum_{\tau\alpha}\left\{(\delta r(\tau) - \sqrt{2}x(\tau\alpha) + x(\tau))^T (\delta r(\tau) - \sqrt{2}x(\tau\alpha) + x(\tau)) + R(\tau\alpha)u^T(\tau\alpha)Qu(\tau\alpha)\right\} + \quad (62)$$
$$\sum_{\tau\alpha}\lambda^T(\tau\alpha)\left\{x(\tau\alpha) - A_\alpha^{(m)}x(\tau) + B^{(m)}u(\tau\alpha)\right\} + \sum_{\tau\alpha}\mu^T(\tau\alpha)(u(\tau\alpha) - \overline{U}^{(m)}) + \sum_{\tau\alpha}\upsilon^T(\tau\alpha)(\underline{U}^{(\cdot\cdot)} - u(\tau\alpha))$$

The optimal policy must satisfy the following conditions:

$$\frac{\partial L}{\partial (x(\tau\alpha))} = 2\sqrt{2}(\delta r(\tau) - \sqrt{2}x(\tau\alpha) + x(\tau)) + \lambda(\tau\alpha) - A_\alpha^{(m)}\lambda(\tau\alpha\alpha) = 0 \tag{63a}$$

$$\frac{\partial L}{\partial \lambda(\tau\alpha)} = x(\tau\alpha) - A_\alpha^{(m)}x(\tau) + B^{(m)}u(\tau\alpha) = 0 \tag{63b}$$

$$\frac{\partial L}{\partial u(\tau\alpha)} = 2R(\tau\alpha)Qu(\tau\alpha) + B^{(m)^T}\lambda(\tau\alpha) + \mu(\tau\alpha) - \upsilon(\tau\alpha) = 0 \tag{63c}$$

$$\frac{\partial L}{\partial \mu(\tau\alpha)} = u(\tau\alpha) - \overline{U}^{(m)} \leq 0 \tag{63d}$$

$$\frac{\partial L}{\partial \upsilon(\tau\alpha)} = \underline{U}^{(m)} - u(\tau\alpha) \leq 0 \tag{63e}$$

The above conditions denote the multi-scale form of the classical two-point boundary value problem with the complementarity conditions to handle the inequality constraints on the manipulated inputs. The structure of equations (63a) and (63b) indicates that

(a) the state equations (63b) can be solved in $O(\log_2 N)$ steps of parallelizable computations (see Section 4.1), given the inputs $u(\tau\alpha)$, at all left-nodes, and

(b) the co-state equations (63a) can be solved in $O(\log_2 N)$ steps of parallelizable computations, given the inputs and the states at the nodes of the binary tree.

Although the solution of the state and co-state equations involves procedures of $O(\log_2 N)$ parallelizable computational steps, a significant reduction from the $O(N)$ steps required by serial solutions in the time-domain, the actual complexity in locating the optimum is determined by the inequality constraints and how efficiently one can identify the set of active ones.

The multi-scale formulation of *Problem-P1a* allows the implementation on the following procedure, which offers significant savings through the parallelization of many of its computational steps.

Phase-0.   Let the control horizon be $N=2^M$ discrete-time points, with M free to take the appropriate value

Phase-1.   Solve the following set of local minimizations at all left-nodes,

Minimization-at-Node-τ

$$\underset{u(\tau\alpha)}{Minimize} \quad F(\tau) = \left\{(\delta r(\tau) - \sqrt{2}x(\tau\alpha) + x(\tau))^T (\delta r(\tau) - \sqrt{2}x(\tau\alpha) + x(\tau)) + R(\tau\alpha)u^T(\tau\alpha)Qu(\tau\alpha)\right\}$$

subject to:

$$x(\tau\alpha) = A_\alpha^{(m)} x(\tau) - B^{(m)} u(\tau\alpha)$$

$$\underline{U}^{(m)} \le u(\tau) \le \overline{U}^{(m)}$$

and compute the inputs, $u(\tau\alpha)$, at all left-nodes.

Phase-2. Compute the inputs at all right-nodes using the scaling relationship (33), i.e.

$$u(\tau\beta) = \sqrt{2}u(\tau) - (I + A^{2^m})^{-1} A^{2^m} u(\tau\alpha) - (I + A^{2^m})^{-1} u(\tau\delta\alpha) \quad (64)$$

Phase-3. Identify the active input constraints and repeat *Phase-1* and *Phase-2* computations solving the reduced-size unconstrained QP.

Let us now discuss the implementational details of the above three phases.

Phase-1. Compute the inputs at all left-nodes. The local minimizations, described above, are carried out in the following order that allows extensive parallelization:

Step-1: Given the initial conditions, $x(0,0)$, solve the problem
  Minimization-at-Node-(1,0)
  Find; $u(0,0)$ and $x(1,0)$
  Compute; $x(0,1)$

Step-2: Given, $x(1,0)$, solve the problem
  Minimization-at-Node-(2,0)
  Find; $u(1,0)$ and $x(2,0)$
  Compute; $x(1,1)$

Step-3: Carry out the following in parallel:

  (a) Given, $x(2,0)$, solve the problem
  Minimization-at-Node-(3,0)
  Find; $u(2,0)$ and $x(3,0)$
  Compute; $x(2,1)$

  (b) Given, $x(1,1)$, solve the problem
  Minimization-at-Node-(1,1)
  With the additional constraint (64) i.e.
  $$\underline{U}^{(0)} \le u(0,1) = \sqrt{2}u(1,0) - (I + A)^{-1} A u(0,0) - (I + A)^{-1} u(0,2) \le \overline{U}^{(0)}$$
  Find; $u(0,2)$ and $x(0,2)$
  Compute; $x(0,3)$

Step-4:   ...

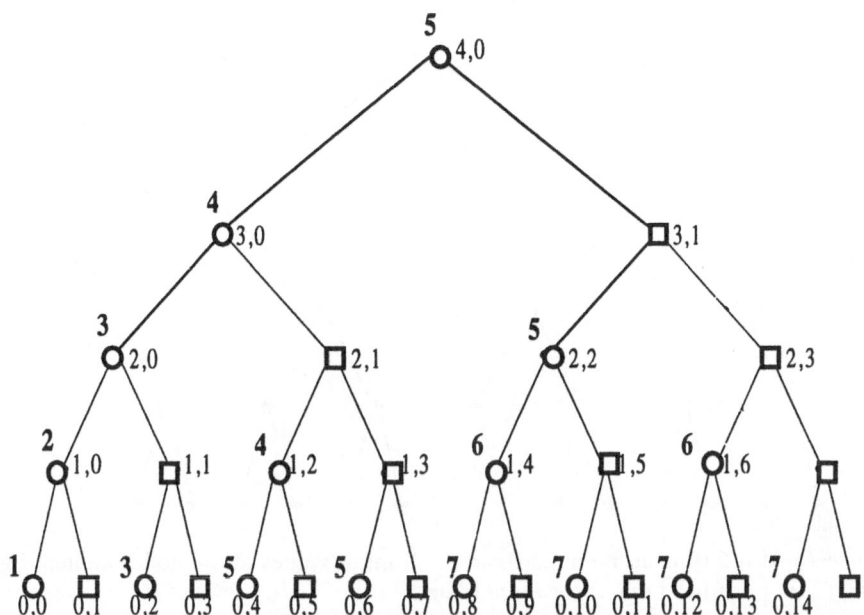

Figure 14. The sequence of parallel minimizations in *Phase-1*. The circles denote the inputs computed from the specific minimization, while the numbers next to the circles denote the step-number in the sequence of minimizations.

Figure 14 illustrates the sequencing in the computation of the inputs at all the left-nodes, over a horizon of 16 time points. It is easily shown that *Phase-1* involves $(\log_2 N - 1)$ steps of parallelizable minimizations. If the result of any of the local minimizations during *Phase-1*, leads to an input on a constraint, i.e. $u(\tau\alpha) = \overline{U}^{(m)}$ or $u(\tau\alpha) = \underline{U}^{(m)}$, then the search space for the subsequent minimizations is reduced, because all inputs on the nodes emanating from, $\tau\alpha$, must be on the same constraint. For example, if $u(2,0) = \overline{U}^{(2)}$, then the inputs, $u(0,0)$, $u(0,1)$, $u(0,2)$, $u(0,3)$, $u(0,4)$, $u(0,5)$ and $u(0,6)$, must be on the upper bound (see Figure 15).

Phase-2. Compute the inputs at all right-nodes. The computations in this phase are rather straightforward using the input scaling relationship (33) and are carried out alongside with those in *Phase-1*, as appropriate. For example;

- $u(0,1)$ can be computed after the completion of *Step-3* of *Phase-1*, when $u(0,0)$, $u(0,2)$ and $u(1,0)$ are available .

- $u(2,1)$ can be computed after the completion of *Step-5* of *Phase-1*, when $u(2,0)$, $u(2,2)$ and $u(3,0)$ are available .

708

The question arises, whether the input value on a right-node computed during *Phase-2* satisfies the bounding constraints or not. The following results guarantee the feasibility of the input values at all right-nodes:

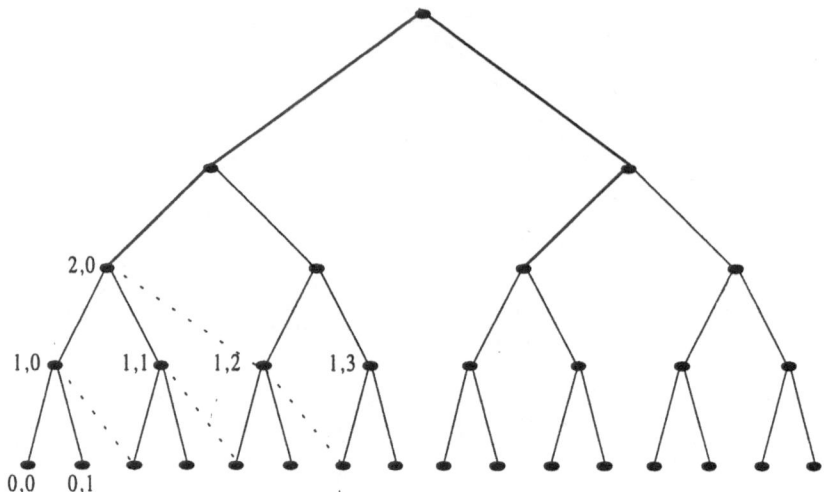

Figure 15. If $u(2,0)$ is at the upper bound, all input values at the nodes emanating from node, $(2,0)$, must be at the upper *input* bound

<u>Result-1:</u> If $u(\tau\alpha)$, $u(\tau\delta\alpha)$, and $u(\tau)$ are at a bounding value, then $u(\tau\beta)$, given by the scaling relationship (64) must be at the same bounding value.

<u>Result-2:</u> If $u(\tau\alpha)$, $u(\tau\delta\alpha)$, and $u(\tau)$ are all inside the feasible region of the upper and lower bounds, so is $u(\tau\beta)$, provided that the dynamic system is stable or stabilizable by the available manipulated inputs.

Phase-3. Identify the active input constraints. At the completion of *Phase-2* all inputs are known, as well as the value of the states at all nodes of the binary tree. Consequently, the values of co-states can be computed in $O(\log_2 N)$ steps (see equation (63a)), and from equation (63c) we can then compute the values of the Kuhn-Tucker multipliers, $\mu(\tau\alpha)$ or $\upsilon(\tau\alpha)$, at the left-nodes of the binary tree, for those inputs that were found in *Phase-1* to be at a constraint. The sign of the resulting values of the Kuhn-Tucker multipliers indicates whether a constraint should be held active or be relaxed.

After *Phase-3* we know all inputs that should be held at a bound. Consequently, in the subsequent iteration *Phase-1* and *Phase-2* computations are limited to those nodes where the input can vary within the bounds.

Remarks:

1. If the optimal control policy is an unconstrained one, the above procedure will find it at the completion of *Phase-3* computations. It would have required only $\log_2 N - 1$ local minimizations, a significant reduction over any other procedure.

2. If the optimal control policy is a fully constrained one, the above procedure will find it at the completion of *Phase-3* after only $\log_2 N - 1$ local minimizations.

3. If the optimal control policy is composed of constrained and unconstrained segments, then the above procedure will locate the optimum by solving an unconstrained QP in the remaining degrees of freedom, i.e. the inputs not on any bounding constraint.

Extension to Multi-Rate Optimal Control

The optimal control formulation of *Problem-P1a* and the associated active-constraints strategy for its solution, discussed in the previous paragraph, can be naturally extended to account for control actions applying at different rates. Specifically, consider the discrete-time model (equation 38)

$$x(k+1) = Ax(k) + B_1 u^{(1)}(k) + B_2 u^{(2)}(k) \tag{38}$$

where, inputs $u^1(k)$ are applied at every discrete-time point, k, while inputs $u^2(k)$ are applied at every other discrete-time point, i.e. with half the frequency of $u^1(k)$. Then, the multi-rate version of *Problem-P1a* remains essentially the same in form and substance.

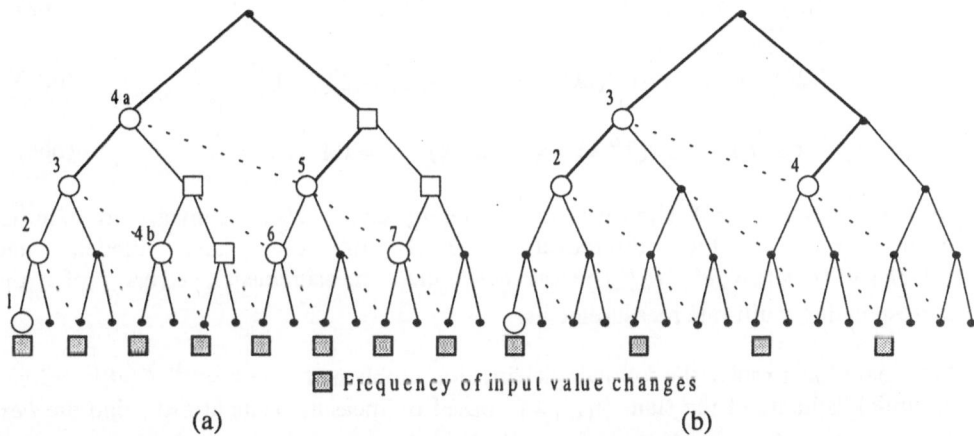

■ Frequency of input value changes

(a)   (b)

Figure 16. The circles denote the nodes with the degrees of freedom, i.e. the manipulated input values, during the solution of *Problem-P1a* when the manipulated input changes; (a) every other discrete-time point (8 degrees of freedom); (b) every fourth discrete-time point (4 degrees of freedom).

Thus, the input scaling yields

$$B_1u^{(1)}(1,\frac{k}{2}) = \frac{1}{\sqrt{2}}(I+A)^{-1}\{AB_1u^{(1)}(0,k) + (I+A)B_1u^{(1)}(0,k+1) + IB_1u^{(1)}(0,k+2)\}$$

$$B_2u^{(2)}(1,\frac{k}{2}) = \frac{1}{\sqrt{2}}(I+A)^{-1}\{AB_2u^{(2)}(0,k) + (I+A)B_2u^{(2)}(0,k) + IB_2u^{(2)}(0,k+2)\}$$

while the number of degrees of freedom

(a) remains the same for $u(1)$, and
(b) is reduced to half for $u(2)$ (see Figure 16a). Figure 16b shows the nodes on a tree with the degrees of freedom, when the manipulated inputs are applied every fourth discrete-time point.

4.3. MULTI-SCALE STATE-ESTIMATION AND OPTIMAL FUSION OF MULTI-RATE MEASUREMENTS

Consider the following discrete-time system over a horizon of 2^M discrete-time points,

$$x(k+1) = Ax(k) + w(k) \qquad\qquad k = 0,1,2,\cdots,2^M-1 \qquad\qquad (65)$$

with outputs measured at different sampling rates according to the following model

$$y^{(0)}(k) = C_o x(k) + v^{(0)}(k) \qquad\qquad k = 0,1,2,\cdots,2^M-1 \qquad\qquad (66a)$$

$$y^{(1)}(2k) = C_1 x(2k) + v^{(1)}(2k) \qquad\qquad k = 0,1,2,\cdots,2^{M-1}-1 \qquad\qquad (66b)$$

$$y^{(M)}(2^M k) = C_M x(2^M k) + v^{(M)}(2^M k) \qquad k = 0,1 \qquad\qquad (66c)$$

$w(k)$ in equation (65) signifies the modeling error or/and unmeasured external disturbances, at the discretization level of the state model, and $v^{(0)}(k), v^{(1)}(2k),\cdots,v^{(M)}(2^M k)$ represent the measurement errors of the corresponding multi-rate measurements.

We would like to solve the following estimation problem at time $k = 2^M\text{-}1$ [48]: "Given an initial estimate of the state $\hat{x}(0\mid-)$, based on measurements at $k<0$, find the *best* estimates ($\hat{w}(k)$ and $\hat{v}^{(j)}(k)$) for the values of the unknown, temporally independent sequences
$\{w(0), w(1),\cdots, w(2^M-2)\}, \{v^{(0)}(0), v^{(0)}(1),\cdots, v^{(0)}(2^M-1)\},$
$\{v^{(1)}(0), v^{(1)}(1),\cdots, v^{(1)}(2^{M-1}-1)\},\ \cdots,\ \{v^{(M)}(0), v^{(M)}(1)\}$, and the error in the initial estimate, $x^{\circ}(0) = x(0) - \hat{x}(0\mid-)$, based on the modeling relationships 65-66.

The state estimation problem for Markov static processes with zero-mean Gaussian sequences describing the modeling errors and measurement noise has been studied in [12], [21], [27] and [34].

For the case of dynamic systems, we would like to solve the following moving horizon optimal state estimation problem, by fusing measurements (possibly of the same states) available at different rates:

Problem-P2

$$
\begin{aligned}
\underset{\substack{x^e(0) \\ \hat{w}(0),\hat{w}(1),...,\hat{w}(2^M-2) \\ \hat{v}^{(0)}(0),\hat{v}^{(0)}(1),...,\hat{v}^{(0)}(2^M-1) \\ \hat{v}^{(1)}(0),\hat{v}^{(1)}(1),...,\hat{v}^{(1)}(2^{M-1}-1) \\ \vdots \\ \hat{v}^{(M)}(0),\hat{v}^{(M)}(1)}}{\text{Minimize}} \quad & \left| F = \left(x^e(0)\right)^T P\left(x^e(0)\right) + \sum_k \left(\hat{w}^T(k) R \hat{w}(k) + \hat{v}^{(0)}(k)^T Q_o \hat{v}^{(0)}(k) \right) \right. \\
& + \sum_{2k} \hat{v}^{(1)}(2k)^T Q_1 \hat{v}^{(1)}(2k) + \cdots + \sum_{2^M k} \hat{v}^{(M)}(2^M k)^T Q_M \hat{v}^{(M)}(2^M k)
\end{aligned}
$$

subject to the state and measurement equations (65) and (66)

The positive definite weighting matrices P, R, Q_0, Q_1, ..., Q_M reflect the relative confidence placed on the initial estimate, state model, and the various measurements as sources of information. Once *Problem-P2* has been solved, we can compute the state estimate at the current time, $k = 2^M$, by integrating the state equation with the corrected initial condition and the known sequence of modeling error values, $w(0), w(1), ..., w(2^M-2)$.

State Estimation with Single-Rate Measurements

If all measurements are made at the same sampling rate, the moving horizon state estimation of *Problem-P2* takes on the following form:

Problem-P2a

$$
\underset{\substack{x^e(0) \\ \hat{w}(0),\hat{w}(1),...,\hat{w}(2^M-2) \\ \hat{v}(0),\hat{v}(1),...,\hat{v}(2^M-1)}}{\text{Minimize}} \quad F = \left(x^e(0)\right)^T P\left(x^e(0)\right) + \sum_k \left\{ \hat{w}(k)^T R' \hat{w}(k) + \hat{v}(k)^T Q \hat{v}(k) \right\}
$$

subject to

$$
\begin{aligned}
\hat{x}(k+1) &= A\hat{x}(k) + \hat{w}(k) && k = 0,1,...,2^M - 2 \\
y(k) &= C\hat{x}(k) + \hat{v}(k) && k = 0,1,...,2^M - 1
\end{aligned}
$$

The multi-scale analog of *Problem-P2a* can be easily shown to take the following form:

Problem-P2b

Minimize
$x^e(0,0)$
$\hat{w}(\tau\alpha)\forall\tau\alpha$
$\hat{v}(\tau\alpha)\forall\tau\alpha$

$$F = \left(x^e(0)\right)^T P\left(x^e(0)\right) + \sum_{\tau\alpha}\left\{\hat{w}(\tau\alpha)^T R\hat{w}(\tau\alpha)\right\} + \sum_{\tau}\left\{\delta\hat{v}(\tau)^T Q\delta\hat{v}(\tau)\right\}$$

subject to

$$\hat{x}(\tau\alpha) = A_\alpha^{(m)}\hat{x}(\tau) - D^{(m)}\hat{w}(\tau\alpha)$$

$$y(\tau) = C\hat{x}(\tau) + \hat{v}(\tau)$$

where

$$D^{(m)} = \sqrt{2}\left(1 + A^{(m-1)}\right)^{-1}B^{(m-1)}$$

$$\delta v(\tau) = \sqrt{2}v(\tau\alpha) - v(\tau)$$

Problem-P2b can be solved by the following procedure with parallelizable sets of computations:

Phase-0 Consider a horizon of 2^M discrete time points and construct a homogeneous binary tree with M levels of nodes.

Phase-1 Starting from the top node, (M,0), sweep the tree downwards, solving in parallel local minimizations at the left node of the form

Minimize
$\hat{w}(\tau\alpha),\hat{v}(\tau\alpha)$

$$F = \hat{w}(\tau\alpha)^T R\hat{w}(\tau\alpha) + \left(\sqrt{2}\hat{v}(\tau\alpha) - \hat{v}(\tau)\right)^T Q\left(\sqrt{2}\hat{v}(\tau\alpha) - \hat{v}(\tau)\right)$$

subject to

$$\hat{x}(\tau\alpha) = A_\alpha^{(m)}\hat{x}(\tau) - D^{(m)}\hat{w}(\tau\alpha)$$

$$y(\tau) = C\hat{x}(\tau) + \hat{v}(\tau)$$

Phase-2 At the completion of *Phase-1*, compute the resulting error in the initial state estimate and the value of the objective function. Through an upward sweep, update the estimates of $w(\tau\alpha)$ and $v(\tau\alpha)$ at all left-nodes and iterate *Phase-1* computations.

Remarks

1. At the top node (M,0), of the binary tree, $v(M,0) = 0$ and $w(M,0) = 0$, (initial assumption). Consequently, an initial estimate of $x(M,0)$ is available to initiate the minimizations of *Phase-1*.

2. The parallel and independent minimizations of *Phase-1* are "linked" only through the impact they have on the value of the error in the initial state estimate.

3.   The Lagrangian of *Problem-P2b* is given by,

$$L = (x^e(0))^T P(x^e(0)) + \sum_{\tau\alpha} \{\hat{w}(\tau\alpha)^T R\hat{w}(\tau\alpha)\} + \sum_{\tau} (\sqrt{2}\hat{v}(\tau\alpha) - \hat{v}(\tau))^T Q(\sqrt{2}\hat{v}(\tau\alpha) - \hat{v}(\tau)) + \quad (67)$$
$$\sum_{\tau\alpha} \lambda(\tau\alpha)\{\hat{x}(\tau\alpha) - A_\alpha^{(m)}\hat{x}(\tau) + D^{(m)}\hat{w}(\tau\alpha)\} + \sum_{\tau\alpha} \pi(\tau\alpha)\{y(\tau) - C\hat{x}(\tau) - \hat{v}(\tau)\}$$

The necessary conditions for optimality are:

$$\frac{\partial L}{\partial \lambda(\tau\alpha)} = \hat{x}(\tau\alpha) - A_\alpha^{(m)}\hat{x}(\tau) + D^{(m)}\hat{w}(\tau\alpha) = 0 \qquad (67a)$$

$$\frac{\partial L}{\partial \pi(\tau\alpha)} = y(\tau) - C\hat{x}(\tau) - \hat{v}(\tau) = 0 \qquad (67b)$$

$$\frac{\partial L}{\partial \hat{x}(\tau\alpha)} = \lambda(\tau\alpha) - A_\alpha^{(m)^T}\lambda(\tau\alpha\alpha) - C^T\pi(\tau\alpha) = 0 \qquad (67c)$$

$$\frac{\partial L}{\partial x^e(0,0)} = 2P^T x^e(0,0) + \lambda(0,0) = 0 \qquad (67d)$$

$$\frac{\partial L}{\partial \hat{w}(\tau\alpha)} = 2R^T \hat{w}(\tau\alpha) + D^{(m)^T}\lambda(\tau\alpha) = 0 \qquad (67e)$$

$$\frac{\partial L}{\partial \hat{v}(\tau\alpha)} = 2Q^T \hat{v}(\tau) - \pi(\tau\alpha) = 0 \qquad (67f)$$

At the conclusion of *Phase-1*, the values of $\hat{x}(\tau\alpha)$, $\hat{w}(\tau\alpha)$ and $\hat{v}(\tau\alpha)$ are available throughout the tree. Then,

(a)   solve (67f) for $\pi(\tau\alpha)$
(b)   with $\lambda(0,0)$, given by equation (67d), as initial condition, solve costate equations (67c) upwards for all $\lambda(\tau\alpha)$
(c)   In an upward sweep, update $\hat{w}(\tau\alpha)$ at all nodes, using equation (67e).

4.   All computations in *Phase-1* and *Phase-2* are parallelizable requiring $O(\log_2 N)$ steps of computations.
5.   The formulation of *Problem-P2a* can be easily extended to include constraints on state estimates and errors, e.g.

$$\underline{g} \leq G\hat{x}(\tau\alpha) \leq \overline{g} \qquad\qquad k = 0,1,..., 2^M - 1$$
$$\underline{h} \leq \hat{w}(\tau\alpha) \leq \overline{h} \qquad\qquad k = 0,1,..., 2^M - 1$$
$$\underline{l} \leq \hat{v}(\tau\alpha) \leq \overline{l} \qquad\qquad k = 0,1,..., 2^M - 1$$

where $\hat{x}$ denotes the resulting state estimate and $\hat{w}$, the associated estimate of the modeling error or disturbance, or, $\hat{v}$, the measurement error.

The ensuing version of the multi-scale state estimation *Problem-P2b* will include the following constraints:

$$
\begin{aligned}
\underline{g}^{(m)} &\le G\hat{x}(\tau\alpha) \le \overline{g}^{(m)} \\
\underline{h}^{(m)} &\le \hat{w}(\tau\alpha) \le \overline{h}^{(m)} \\
\underline{l}^{(m)} &\le \hat{v}(\tau\alpha) \le \overline{l}^{(m)}
\end{aligned}
\tag{68}
$$

During *Phase-1*, the local minimizations must now include the corresponding constraints. If the estimate at a current node, $(\hat{x}(\tau\alpha), \hat{w}(\tau\alpha)$, or, $\hat{v}(\tau\alpha))$ reaches a bound, then the estimates at all nodes emanating from the current node will remain at the same bound, thus reducing significantly the search space.

During the upward adaptation of the estimates, only those estimates on the bounds and with Kuhn-Tucker multipliers indicating relaxation of the constraints, will be updated.

Recursive State Estimation at Multiple Scales
In this case we will assume that the modeling error or disturbance, *w(k)*, and the measurement error, *v(k)*, (and their estimates $\hat{w}(k)$ and $\hat{v}(k)$) are independent, zero mean, normally distributed random variables with covariance, R(*k*) and Q, respectively, i.e. the covariance of the modeling error or disturbance is allowed to vary with time, while the covariance of the measurement error is assumed to be constant.

The Haar transformation of the random sequence $\{v(k)\}$ leads to (see Appendix A.4)

(a)   random scaling function coefficients at any node, τ, at level-m; with $v(\tau) = \mathcal{N}(0, Q)$,

(b)   random wavelet coefficients $\delta v(\tau) = \mathcal{N}(0, Q)$.

The Haar transformation of the random sequence $\{I(k)\}$, on the other hand leads to

(a)   random scaling function coefficients at any node τ, with,
$$w(\tau) = \mathcal{N}(0, (\tfrac{1}{2})(R(\tau\alpha)+R(\tau\beta)),$$

(b)   random wavelet coefficients at any node, t, with,
$$\delta w(\tau) = \mathcal{N}(0, (\tfrac{1}{2})(R(\tau\alpha)+R(\tau\beta)).$$

The Modified-hat transformation of the random sequence $\{w(k)\}$ requires us to include the cross-correlation and the covariance matrix is tridiagonal:

(a)   at some zero-th level,

$$E[w(\tau)w(\sigma)] = R(0) \quad \text{for} \quad \tau = \sigma$$
$$= 0 \quad \text{otherwise}$$

(b)   at the first level, for the modified-hat scaling functions

$$E[w(\tau)w(\sigma)] = R(1) = \frac{a_0^2 + (a_0+1)^2 + 1}{2(a_0+1)^2} R(0) \quad \text{for} \quad \tau = \sigma$$
$$= D(1) = \frac{a_0}{2(a_0+1)^2} R(0) \quad \text{for} \quad \tau = \sigma \pm 1$$
$$= 0 \quad \text{otherwise}$$

(c)   at subsequent levels for the modified-hat scaling functions

$$E[w^{(k+1)}(\tau)w^{(k+1)}(\sigma)] = R(k+1) = \frac{a_k^2 + (a_k+1)^2 + 1}{2(a_k+1)^2} R(k) + \frac{1}{2} D(k) \quad \text{for} \quad \tau = \sigma$$
$$= D(k+1) = \frac{a_k}{2(a_k+1)^2} R(k) + \frac{1}{2} D(k) \quad \text{for} \quad \tau = \sigma \pm 1$$
$$= 0 \quad \text{otherwise}$$

(d)   at all levels for the modified-hat wavelet coefficients

$$E[\delta w^{(k+1)}(\tau)\delta w^{(k+1)}(\sigma)] = R''(k+1) = \frac{a_k^2 + (a_k-1)^2 + 1}{2(a_k-1)^2} R(k) + \frac{1}{2} D(k) \quad \text{for} \quad \tau = \sigma$$
$$= D''(k+1) = \frac{-a_k}{2(a_k-1)^2} R(k) + \frac{1}{2} D(k) \quad \text{for} \quad \tau = \sigma \pm 1$$
$$= 0 \quad \text{otherwise}$$

Note then that we have a covariance structure that varies in a predictable way, and may thus be used in the upward sweep of a Kalman filtering routine on the tree.

Let us now formulate the following multi-scale state estimation problem with modeling errors or disturbances having varying covariance.

Problem-P2c
$$\underset{\hat{x}(k) ,k=0,1,...,2^M -1}{\text{Minimize}} F = \sum_{\tau} \left\{ \delta\hat{w}(\tau)^T R'' \delta\hat{w}(\tau) + \hat{w}^T(T)R(T)\hat{w}(T) + \delta\hat{v}(\tau)^T Q\delta\hat{v}(\tau) + \hat{v}^T(T)Q\hat{v}(T) \right\}$$
subject to

$$\hat{x}(\tau\alpha) = A_\alpha^{(m)}\hat{x}(\tau) - B^{(m)}\hat{w}^*(\tau\alpha)$$
$$\hat{x}(\tau\beta) = A_\beta^{(m)}\hat{x}(\tau) + B^{(m)}\hat{w}^*(\tau\alpha)$$
$$y(\tau) = C\hat{x}(\tau) + \hat{v}(\tau)$$

716

We note that *Problem-P2c* is similar to the unconstrained version of *Problem-P2b*, where the state and measurement equations have been expressed in terms of scaling coefficients rather than wavelet coefficients. Also, the modeling error or disturbance in the state equation has been expressed in terms of the modified hat transform coefficients (see Section 3.3.).

The solution to the above problem leads to the formulation of a linear observer, which minimizes the variance of the reconstruction error as one moves up the tree, along left-nodes. The structure of the scale recursive observer is as follows (see Appendix A.5)

$$\hat{x}(\tau|\tau) = \hat{x}(\tau|\tau+) + K(\tau)\left[y(\tau) - C\hat{x}(\tau|\tau+)\right] \qquad (A.12)$$

with observer gain given by

$$K(\tau) = P(\tau|\tau\alpha)C^T\left[Q + CP(\tau|\tau\alpha)C^T\right]^{-1} \qquad (A.21)$$

The state reconstruction error $e(\tau|\tau) = x(\tau) - \hat{x}(\tau|\tau)$ evolves through the multi-scale equation,

$$e(\tau|\tau) = \left[I - K(\tau)C\right]\overline{A}_\alpha^{(m)T}e(\tau\alpha|\tau\alpha) + \left[I - K(\tau)C\right]\overline{B}_\alpha^{(m)}w^*(\tau\alpha) - K(\tau)v(\tau) \qquad (A.17)$$

where, $\overline{A}_\alpha^{(m)} = \left[A_\alpha^{(m)}\right]$ and $\overline{B}_\alpha^{(m)} = -[A_\alpha^{(m)}]^{-1}B^{(m)}$, and possesses a covariance denoted by $P(\tau)$.

Remarks
1. The observer applies only on the left-node of the dyadic tree. For a time-horizon of 2^M discrete time points, there are 2^M left-nodes. Application of a corresponding observer on the right nodes of the tree does not guarantee new information.
2. .For a time horizon of 2^M discrete-time points, the Kalman filter operating at the time-domain would require 2^M steps in order to generate state estimates at all scales. The multi-scale version of the Kalman filter defined by equations (A.12) and (A.19), requires $\log_2(2^M) = M$ steps of parallelizable state estimations.

State Estimation with Multi-Rate Measurements
The availability of measurements at various levels of discretization provides additional information, which must be incorporated in the estimation of states as the observer moves up through the various nodes of the tree. In addition, a downward propagation of the state estimates leads to an interpolated smoothing which contains the information provided by all measurements at various sampling rates. Those two observations lead to a strategy that includes

(a) an upward sweep through the Kalman filtering steps for the estimation of states at the various nodes of the tree, and

(b) a downward sweep that smoothes the estimates at all nodes of the tree.

This strategy was first proposed by Chou [12] for the optimal treatment of signals and will be further developed in the following paragraphs for the state estimation of Physico-chemical processes. The multi-scale state and measurement equations of *Problem-P2* are,

$$
\begin{aligned}
x(\tau\alpha) &= A_\alpha^{(m)} x(\tau) - B^{(m)} w^*(\tau\alpha) & \tau \in T_0 \\
x(\tau\beta) &= A_\beta^{(m)} x(\tau) + B^{(m)} w^*(\tau\beta) & \tau \in T_0 \\
y^{(0)}(\tau) &= C_1 x(\tau) + v^{(0)}(\tau) & \tau \in T_0 \\
y^{(1)}(\tau) &= C_2 x(\tau) + v^{(1)}(\tau) & \tau \in T_1 \\
&\;\;\vdots & \vdots \\
y^{(M)}(\tau) &= C_{M+1} x(\tau) + v^{(M)}(\tau) & \tau \in T_M
\end{aligned}
$$

where T_m, m=0,1,...,M, denotes the dyadic trees starting with the top node at level-M and ending with the leaf nodes at level-0, level-1, ...

Upward Sweep with State Estimation

Step 1. Begin at level-0, i.e. the finest discretization, and for any node at this level, set $\hat{x}(\tau|t+)$ =given and P(τ|τ+)=given.

Step 2. Using measurements, $y^{(0)}(\tau)$, which are available at all nodes of T_0, generate the optimal estimates of the states at any node, τ, of the tree (see Appendix 6.A).

Step 3. Using the measurements, $y^{(1)}(\tau)$, update the estimates of the state at all nodes at level-1 and higher, using the recursive filter of Appendix A.5.

Step 4. Continue in a similar spirit with measurements at higher levels, updating
- states at level-2 and higher using measurements $y^{(2)}(\tau)$,
- states at level-3 and higher using measurements $y^{(3)}(\tau)$,
- etc to higher scales.

Downward Smoothing of State Estimates.

Once the multi-scale Kalman filter, described in the previous paragraph, reached the top node of the tree, it has generated the smoothest estimate (smallest covariance), based on all available measurements. We let, $\hat{x}(T)$ act as the initial condition for a downward smoothing of the state estimates according to the following relationship,

$$
\tilde{x}(\tau) = \hat{x}(\tau|\tau) + L(\tau)\left[\tilde{x}(\tau\bar{\gamma}) - \hat{x}(\tau\bar{\gamma}|\tau)\right] \tag{69}
$$

where $\tilde{x}(\tau)$ and $\tilde{x}(\tau\bar{\gamma})$ are the smoothed estimates during the downward sweep. The gain, $L(\tau)$, is selected to be,

$$L(\tau) = P(\tau|\tau)\overline{A}_\alpha^{(m)} P^{-1}(\tau|\tau).$$ (70)

<u>Remark.</u> Both the upward and the downward sweeps are computationally fairly efficient. Assuming a horizon of 2^M points and new measurements at M levels, the upward fusion of measurements is carried out in k steps of parallelizable computations, where

$$k = M + (M-1)+\cdots+1 = \frac{M(M-1)}{2}.$$

The downward smoothing is carried out in M steps of parallelizable computations.

5. MODEL-PREDICTIVE CONTROL IN THE TIME-SCALE DOMAIN

Analysis of linear systems in the frequency domain has offered invaluable insights for the synthesis of control systems, an activity that must be carried out in the time domain. In model-predictive control (MPC), the imposition of inequality constraints on the manipulated inputs and controlled outputs, renders a nonlinear controller. Analysis in the frequency domain is inapplicable. However, a multitude of design specifications for MPC could be expressed and specified in terms of frequency; e.g. the duration or/and frequency of output constraints violation, suppression of highly oscillatory control inputs, the relative weighting of closed-loop performance characteristics. In this section we will present an alternative framework for the design of MPC. Performance index, state equations, output equations, constraints on inputs and outputs, will be expressed in terms of variables defined in the time-scale (frequency) domain. The ensuing formalism

(a)   solves essentially the same problem posed in the time domain,

(b)   offers algorithmic solutions, which are easily parallelizable and thus significantly more efficient, allowing practical solution to problems with theoretical limits, e.g. "infinite" horizon problems.

(c)   allows a natural expression of engineering specs on, for example, performance, output constraints and others,

(d)   offers distinct alternatives to a series of outstanding issues.

At time k=0, an MPC implementation solves the following open-loop optimal control problem [19], [32], [39], [40],

Problem-P3

$$\underset{u(k)}{Minimize} \quad F = \sum_k (r(k) - x(k))^T Q'(r(k) - x(k)) + \sum_k u^T(k) R' u(k) \tag{71}$$

subject to

$$x(k+1) = Ax(k) + Bu(k) + w(k) \quad k = 0,1,..., p-1 \tag{72}$$

$$\underline{U} \le u(k) \le \overline{U} \quad k = 0,1,...,m-1 \tag{73}$$

$$y(k) = Cx(k) + v(k) \quad k = 0,1,..., p \tag{74}$$

$$\underline{Y} \le y(k) \le \overline{Y} \quad k = 0,1,..., p \tag{75}$$

where, *r(k)* denotes the desired set-point trajectory, and *w(k)* and *v(k)* represent the modeling error or disturbance and measurement error at the same time point.

The prediction horizon, denoted by p is typically longer than the control horizon, denoted by m, with both being design parameters. If *u(k)*, k = 0,1,...,m is the solution to *Problem-P3*, we implement the control action, *u(0)*. At the next time point, we measure the controlled output, we formulate again *Problem-P3*, but over time horizons shifted by one sampling period, and solve the new problem. A Kalman filter can be used to process the information provided by the last measurement of outputs and generate corrected estimates of the full state vector at the current time point. This procedure yields an implicit feedback control law.

The above formulation of *Problem-P3* is an extension of the optimal control problem discussed in Section 4.2. The presence of output equations (74) and output constraints (75) does not offer any complications, and thus we can easily convert *Problem-P3* into the following equivalent form, defined in the time-scale domain:

Problem-P4

$$\underset{\delta u(\tau)}{Minimize} \quad F = \sum_\tau (\delta r(\tau) - \delta x(\tau))^T Q(\tau)(\delta r(\tau) - \delta x(\tau)) + \sum_\tau \delta u^T(\tau) R(\tau) \delta u(\tau) \tag{71}$$

subject to

$$x(\tau\alpha) = A_\alpha^{(m)} x(\tau) - B^{(m)} u(\tau\alpha) - D^{(m)} w(\tau\alpha)$$

$$x(\tau\beta) = A_\beta^{(m)} x(\tau) + B^{(m)} u(\tau\alpha) + D^{(m)} w(\tau\alpha)$$

$$\underline{U} \le u(\tau) \le \overline{U}$$

$$y(\tau) = Cx(\tau) + v(\tau)$$

$$\underline{Y} \le y(\tau) \le \overline{Y}$$

Certain clarifications in the above formulation are in order:

(1)   We will take the prediction and control horizons to be the same and of length 2^M points. Thus, the two summations in the objective function are over the same set of nodes of a dyadic tree.

(2)   The state and output variables, as well as the measured errors are expressed in terms of Haar scaling function coefficients at a particular node.

(3) $u(\tau)$ and $w(\tau)$ are scaling function coefficients resulting from the modified Hat transformation.

(4)   We note that $\delta x(\tau) = \sqrt{2}x(\tau\alpha) - x(\tau)$, $\delta r(\tau) = \sqrt{2}r(\tau\alpha) - r(\tau)$, and, $\delta u(\tau) = \sqrt{2}u(\tau\alpha) - u(\tau)$. Thus, the objective function does contain variables compatible with those contained in the constraining relationships.

In the following paragraphs, we will try to capture some of the distinct advantages offered by the time-scale formalism of *Problem-P4*.

Design Specifications
MPC requires a series of design specifications; (a) select the weighting matrices for the objective function, (b) set the control and prediction horizons, (c) specify the tolerances in the violation of output constraints, (d) select the structure for the modeling error/disturbance term in the state equation. In the time-domain formulation of MPC, no frequency characteristics are explicit both (i) over certain ranges of time and (ii) in specific ranges of frequencies (scale). Here are some examples.

Objective Function.   Relative weights can be explicitly assigned to the state behavior; $[\delta r(\tau) - \delta x(\tau)]$ is the error at specific time intervals with the state variation determined by a specific range of frequencies. Thus, one may use larger weights at higher nodes of the tree, in order to suppress longer range variations, or use larger weights at lower nodes to suppress variations of higher frequency. Furthermore, one may explicitly select the entries of R to penalize rapidly changing control efforts sustained over a large period of time. Finally, one may "encourage" high frequency movements in the control inputs at an early period of the control horizon and penalize them at later stages.

Output Constraints.   Constraints on inputs may lead to violations of output constraints (soft constraints), if feedback stability is to be maintained. The degree of "softness" in the satisfaction of output constraints can be explicitly introduced in the solution of *Problem-P4* as follows: Append the objective function with penalty terms like $k(\tau)[\underline{Y} - y(\tau)]$ and $l(\tau)[y(\tau) - \overline{Y}]$, where $k(\tau)$ and $l(\tau)$ take on large positive values when the corresponding constraint is violated, and zero otherwise. The relative values of $k(\tau)$ and $l(\tau)$ at various nodes, express the designer's relative preferences on the extent of allowed violations.

<u>Control and Prediction Horizons.</u>   These are related to the feasibility of the reachable states and the stabilizability of the feedback scheme. They will be discussed separately in a subsequent paragraph.

<u>Structure of Modeling Error/Disturbance.</u>   This is related to the state reconstruction scheme and will be discussed later in conjunction with the multi-scale state estimation.

Solving the Open-Loop Optimal Control Problem
In solving *Problem-P4* we can use the active constraint strategy developed in Section 4.2., whose main features are:
(a)   the parallelizability of the solution of the state and costate equations, resulting in $O(log_2 2^M) = M$ steps of the matrix multiplications, a significant reduction over the 2^M steps needed in the time-domain, and
(b)   the efficient multi-scale handling of the input constraints (see step-a, -b, -c in Section 4.2), which determine the complexity of the overall QP problem; the longer the constrained input segments, the more efficient the location of the optimum.

We have found that it is extremely useful to first solve *Problem-P4* without input constraints. If the optimal control is an unconstrained policy, for a given desired trajectory, then the optimal policy is found in $log_2 2^M = M$ steps. If, on the other hand, the policy is a constrained one, we can generate (i) an early indication of the extent of the unconstrained policy, and (ii) guide the iteration of Step-1, -2, -3, -4, and, -5 (see Section 4.2.) on the most likely dyadic segmentation of the input constraints. Given that the cost for solving the unconstrained problem is rather low, the computational savings in solving the constrained problem can be significant, 2 to 3 orders of magnitude shorter computational time.

Such reductions in the time for solving *Problem-P4* have allowed us to consider fairly long prediction and control horizons, reaching practical considerations of an "infinity horizon" MPC solution. Indeed, finite horizon MPC problems, with $2^6=64$ points in the time domain take longer than the solution of the equivalent multi-scale MPC problems over an "infinite horizon" of $2^{10}=1024$ points.

State Estimation and the Structure of $w(\tau)$
At each time step, the multi-scale Kalman filter of Section 4.3., for single-rate or multiple-rate measurements are used to estimate the current states at all nodes of a tree. The optimal state estimate, also produces an estimate of the modeling error/disturbance term, $w(\tau)$, at each node of the tree. Certain remarks are in order for the proper use of this procedure in the multi-scale MPC framework:

Remarks.
1.   The estimation horizon is equal to the prediction horizon i.e. of length 2^M values of past feedback error at the points $k=0,-1,-2,\ldots,-(2^M-1)$. Consequently, it provides a rich structure for the characterization of the modeling errors/disturbances over M scales.

2.   The state estimation requires $\log_2 2^M = M$ steps of computations, significantly less than the 2^M steps for a classical Kalman filter operating over the 2^M points in the time domain. Consequently, the extensive enrichment in the structure of the error does penalize the overall procedure with unduly heavy computational burden, even for large values of 2^M.

3.   The fact that $w(\tau)$ describes the modeling error disturbance over a specific time period and at a specific range of frequencies (scale), allows the introduction of logic which can weight differently these contributions in the estimation of state. Work is presently under way to establish such logic.

4.   Figure 16 provides a pictorial depiction of the multi-scale state estimation and control of the MPC strategy in the time-scale domain:
     (1)   the state estimation provides the current state at all M levels, and
     (2)   the optimal control determines the control policy at the time points through the fusion of the control actions at the various levels.

Stability, Reachability

Optimality of the open-loop finite control policy in the time-horizon, does not guarantee closed-loop stability of MPC. Infinite horizon MPC, on the other hand, does. In classical, time-domain MPC, several procedures have been established to achieve closed-loop stability of finite-horizon MPC implementations, such as (a) addition of a terminal constraint, $x(t+T)=0$ [38], (b) replacement of the finite horizon MPC by a control problem with infinite horizon cost, but finite horizon control policy [47].

In addition, when we have an MPC formulation with state constraints (equation (75)), the control policy must assure that they are satisfied for the segment of the prediction horizon beyond the end of the control horizon. Rawlings and Muske [47] have shown that there exists an N large enough, so that if the constraints within the constraint horizon m+N, then they are automatically satisfied beyond the m+N horizon. Gilbert and Tan [23] proposed an algorithm that can be used to compute N.

Finally, if the controlled system contains unstable modes, then the control horizon must be large enough to ensure the constraining of all unstable modes to return to zero, by the time MPC goes into the constant control policy segment.

All of the above indicate that one needs to establish a large enough horizon to ensure closed-loop stability under various circumstances/problem formulations. In multi-scale MPC increasing the prediction/control horizon increases moderately the computational cost. Therefore, one can employ somewhat larger horizons than those employed in time-domain MPC formulations. The practical question is how to determine the length of such a horizon. The answer lies in the solution of an auxiliary problem which is based on the idea of "blocking and condensing" [33] future inputs and predicted states, which has found wide utilization for solving practical DMC problems over large horizons.

Consider *Problem-P4* with the nodes τ constrained to be on the "initial-time path" (see Figure 9). The solution of this problem produces a "blocked and condensed" policy pictorially depicted in Figure 17. The solution of such problems involves $\log_2 2^M = M$ steps of parallelizable computation, and thus can be found with little computational cost, even for large M's, e.g. 10-20. If the modified problem admits a finite state, constrained policy for some M, then 2^M is an acceptably large horizon. The finer scale state behavior at nodes such as, (1,1) and (2,1), which is not included in the formulation of the modified problem, is bounded, as it can be easily seen. However, the state at nodes such as (1,2) and (1,3) could take on very large values of opposite signs, yielding a bounded average at node (2,1). Nevertheless, once we have established that the control/prediction horizon of length 2^M admits a finite state control policy, that uses controls at the nodes of the "initial-time path", (i.e. nodes (1,0), (2,0) and (3,0)), we can control the state variations in nodes (1,2) and (1,3) through the use of the control variables at these nodes.

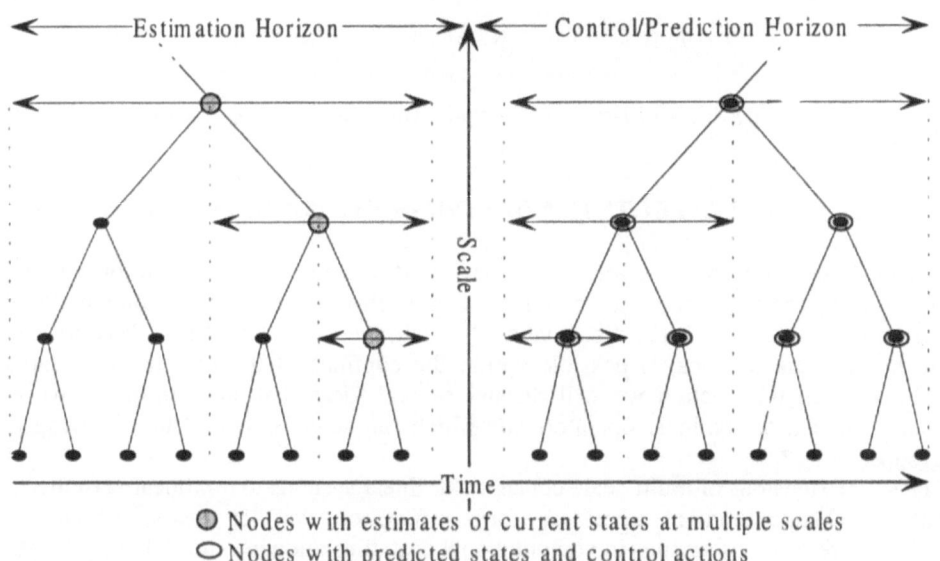

○ Nodes with estimates of current states at multiple scales
○ Nodes with predicted states and control actions

Figure 17. The merging of multi-scale state estimation and multi-scale optimal control in MPC

The auxiliary problem, solved over the nodes of the "initial-time path", can be routinely solved at each step of a multi-scale MPC in order to determine the length of the control/prediction horizon, which is necessary to ensure stability of the closed-loop and satisfaction of the state constraints

724

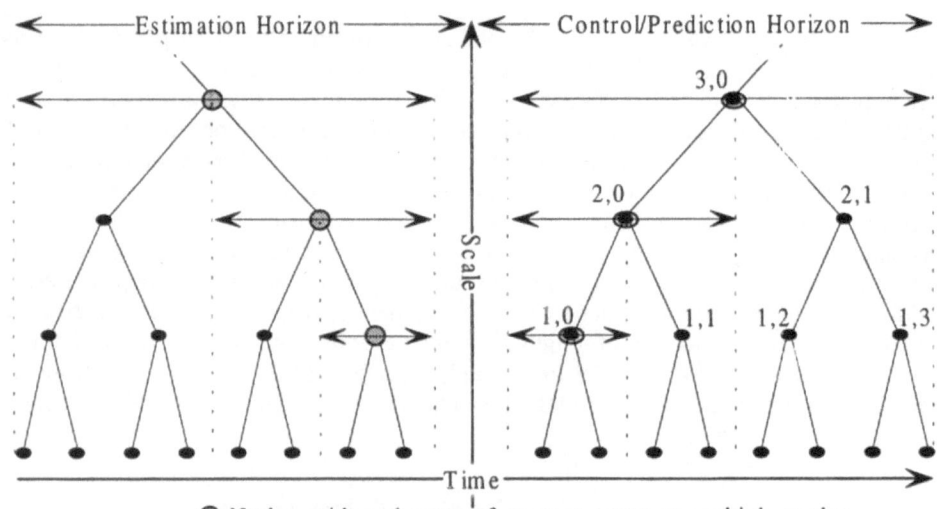

← Estimation Horizon → ← Control/Prediction Horizon →

Scale

Time

◉ Nodes with estimates of current states at multiple scales
○ Nodes with predicted states and control actions

Figure 18. Multi-scale MPC with "Blocking and Condensing"

6. MULTI-SCALE ASPECTS IN NONLINEAR SYSTEMS: TENTATIVE IDEAS

Although the concept of scale is inherent in describing the behavior of nonlinear dynamic systems, its rigorous definition and utilization is only possible under the assumption of local linearized behavior. As such, the extension of the ideas presented in the previous sections, is possible within the confines of the regions of linearized behavior. In this section we will discuss several ideas that stem from the wavelet transformation of nonlinear operators and which can be classified in the following two groups:

(1)   Extensions of multi-scale concepts for linear systems to nonlinear systems
(2)   Computation of nonlinear forms, f(x), on wavelet bases, which offer tremendous savings in computational cost, thus allowing the stable and explicit solution and inversion of nonlinear dynamics.

Time-Varying Linear Systems
Time-varying linear dynamic systems of the form

$$x(k+1) = A(k)x(k) + B(k)u(k) \qquad (76)$$
$$y(k) = C(k)x(k) \qquad (77)$$

can be converted into multi-scale systems of the form

$$x(\tau\alpha) = A_\alpha(\tau\alpha)x(\tau) + B_\alpha(\tau\alpha)u(\tau\alpha) \qquad \text{:left-nodes} \qquad (78)$$
$$x(\tau\beta) = A_\beta(\tau\beta)x(\tau) + B_\beta(\tau\beta)u(\tau\alpha) \qquad \text{:right-nodes} \qquad (79)$$
$$y(\tau) = C(\tau)x(\tau) \qquad (80)$$

i.e. with coefficients varying across the indexing nodes of the tree. The properties of such systems with time-scale varying coefficients follow very closely those of time-varying systems. Work is presently in progress [18] to explore the utility of such systems in solving optimal control and estimation problems for linear systems with time-varying coefficients.

Homogenization of Linearized Models
Given a discrete-time system with non-linear dynamics,

$$x(k+1) = f(x(k), u(k)) \tag{81}$$

we can linearize the function, $f(x(k), u(k))$, around several points. Consider two such distinct points, and the resulting linearized models given by

$$x^{(1)}(k+1) = A_1 x^{(1)}(k) + B_1 u^{(1)}(k)$$
$$x^{(2)}(k+1) = A_2 x^{(2)}(k) + B_2 u^{(2)}(k)$$

Transforming these models into linear multi-scale models (see Section III.3), we take:

$$x^{(1)}(\tau\alpha) = A_{1,\alpha}^{(m)} x^{(1)}(\tau) - B_{1,\alpha}^{(m)} u^{(1)}(\tau\alpha) \qquad \text{and}$$
$$x^{(1)}(\tau\beta) = A_{1,\beta}^{(m)} x^{(1)}(\tau) - B_{1,\beta}^{(m)} u^{(1)}(\tau\alpha)$$
$$x^{(2)}(\tau\alpha) = A_{2,\alpha}^{(m)} x^{(2)}(\tau) - B_{2,\alpha}^{(m)} u^{(2)}(\tau\alpha) \qquad \text{and}$$
$$x^{(2)}(\tau\beta) = A_{2,\beta}^{(m)} x^{(2)}(\tau) - B_{2,\beta}^{(m)} u^{(2)}(\tau\alpha)$$

If at some level-m,

$$A_{1,\alpha}^{(m)} \cong A_{2,\alpha}^{(m)} \quad \text{and} \quad B_{1,\alpha}^{(m)} \cong B_{2,\alpha}^{(m)}$$

then we say that the two equations have been numerically "homogenized" at level-m and above; they yield the same solutions at level-m and above, while their behavior is different at lower scales.

Such numerical homogenization is quite common in Physico-chemical systems where the different linearization points trigger differences, primarily in the small-scale behavior. Whenever such homogenization is possible, the coarse behavior of a nonlinear system is described by the same equation, leaving the multi-model representation to capture the small-scale differences.

Thus the multi-scale control and estimation of nonlinear systems could proceed in a downward direction, as follows:
(i)   Use the common model of the numerically homogenized linear approximations, in order to compute the control policy or optimal state estimates at coarse levels of resolution.

726

(ii) Beyond the scale of level-m, where the homogenization does not hold, continue with the specific linearized approximation that corresponds to the current level of state values. This strategy allows the smooth evolution in the description of the dynamics of a nonlinear system.

For more details on the numerical homogenization of equations describing the micro- and macro-structure of physical phenomena, see Brewster and Beylkin [9].

Representation of Nonlinear Operators in Wavelet Bases
Beylkin [8] has shown that a wide array of linear and nonlinear operators, e.g. $\frac{d}{dt}, \frac{d^n}{dt^n}, \exp(A), \sin(A), \cos(A), F(u(t))$, and others, can be represented in wavelet bases through very sparse representations, which allow the evaluation of their output through procedures of very low complexity. For example, Beylkin's non-standard form for the representation of operators in wavelet bases leads to algorithms of the following complexities:

(a) multiplication of dense matrices; $O(N)$ vs $O(N^3)$
(b) generalized inverse; $O(N\log R)$ vs $O(N^3\log R)$, where R is the condition number of the matrix
(c) square-root of an operator; $O(N\log R)$ vs $O(N^3\log R)$
(d) $\exp()$, $\sin()$, $\cos()$,; $O(N)$ vs $O(N^3)$
(e) squaring operator; $O(N)$ vs $O(N^3)$

The reductions in computational loads are remarkable and push further the boundaries of what is possible in nonlinear systems theory. For example, limitations in the tolerable computational load are at the heart of several difficulties in the deployment of nonlinear MPC, such as [37]:
1. The admissibility of a control policy may be negated, if it resulted from a local rather than the global minimum.
2. The terminal equality constraint, required to ensure feedback stability, can only be achieved asymptotically, and thus guaranteed only when the errors have become insignificant; a point in time long after the completion of a practical control horizon.

Consequently, any drastic improvements in the computational load required by certain basic tasks, e.g. solution and inversion of nonlinear equations in the presence of inequality constraints, will have a significant impact on the achievement of the theoretical requirements for a stable MPC.

Beylkin's ideas offer significant incentives in this direction and should be further explored.

References

1. Bakshi, B. R. and Stephanopoulos, G. (1994) Representation of process trends-III. multiscale extraction of trends from process data, *Computers Chem. Engng* **18**, 267-302.
2. Bakshi, B. and Stephanopoulos, G. (1993) Wave-net: a multiresolution, hierarchical neural network with localized learning, *AICHE Journal* **39**.
3. Basseville, M., Benveniste, A., Chou, K.C., Golden, S.A., Nikoukhah, R. and Willsky, A.S. (1992) Modeling and estimation of multiresolution stochastic processes, *IEEE Transactions in Information Theory*, **38**.
4. Basseville, M., Benveniste, A., and Willsky, A.S. (1992) Multiscale-autoregressive processes, part I: Schur-Levinson parametrizations, *IEEE Transactions on Signal Processing*, **40**.
5. Basseville, M., Benveniste, A., and Willsky, A.S. (1992) Multiscale-autoregressive processes, part I: Lattice structures for whitening and modeling, *IEEE Transactions on Signal Processing*, **40**.
6. Bell, M. L., Limebeer, D. J. N., and Sargent, R.W.H. (1996) Robust receding horizon optimal control, *CPSE Consortium Meeting*.
7. Benveniste, A., Nikoukhah, R., and Willsky, A. (1994) Multiscale system theory *IEEE Transactions on Circuits and Systems-1: Fundamental Theory and Applications*, **41**.
8. Beylkin, G. (1991) Wavelets, multiresolution analysis and fast numerical algorithms, Draft of INRIA lecture notes.
9. Beylkin, G. and Brewster, M.E. (1994) A multiresolution strategy for numerical homogenization, Preprint.
10. Carrier J. F. and Stephanopoulos, G. (1997) Wavelet-based modulation in control-relevant Process identification, to appear in *AIChE Journal*.
11. Cheong. C. K., Aizawa,K., Hatori, M., and Saito, T., (1995) Nonlinear noise reduction on zero-crossing representations of a wavelet transform, *Electronics and Communications in Japan*, **78**.
12. Chou, K.C., (1991) A stochastic approach to multiscale signal processing, PhD Thesis, Massachusetts Institute of Technology, Department of Electrical Engineering and Computer Science.
13. Chou, K.C., Willsky, A., and Benveniste, A. (1994) Multiscale recursive estimation, data fusion, and regularization, *IEEE Transactions on Automatic Control*, **39**.
14. Chou, K.C., Willsky,A., and Nikoukhah, R. (1994) Multiscale systems, Kalman filters, and Ricatti equations, *IEEE Transactions on Automatic Control*, **39**.
15. Cutler C. R., and Yocum, F.H., (1991) Experience with the DMC inverse for identification, Technical Report.
16. Daubechies, I., (1992) *Ten Lectures on Wavelets*, SIAM.
17. Daubechies, I, (1990) The wavelet transform, time-frequency localization and signal analysis*IEEE Trans. Inf. Theory*, **36**, 961-1005.
18. Dyer, M., (1997) Multi-scale models for time-varying linear systems, MIT-LISPE, Technical Report, under preparation.
19. Eaton, J. W., and Rawlings, J.B. (1991) Model predictive control of chemical processes *Chemical Engineering Science*.
20. Feng, W., (1996) The application of time-frequency techniques to process control and identification Ph.D. Thesis, Texas A&M Univ.
21. Fieguth, P.W., (1995) Application of multiscale estimation to large scale multidimensional imaging and remote sensing problems, PhD Thesis, Massachusetts Institute of Technology, Department of Electrical Engineering and Computer Science.
22. Genceli, H., Nikolau, M. (1996) New approach to constrained predictive control with simultaneous model identification, *AICHE Journal*, **42**.
23. Gilbert, E.G. and Tan, K.T. (1991) Linear systems with state and control constraints: The theory and application of maximal output admissable sets, *IEEE Transactions on Automatic Control*, **36**(9): 1008-1020.
24. Gopinath, R., and Bequette, B. (1991) Multirate model predictive control, an analysis for single-input single-output systems, *Proc of PSE'91* **II.5.1**.
25. Goupillaud, P., Grossman, A., and Morlet, J. (1995) Cycle octave and related transform in seismic signal analysis, *Geoexploration*, **23**, 85-102.
26. Graps, A. (1995) An introduction to wavelets, *IEEE Computational Science and Engineering*, **2**.

728

27. Irving, W. (1995) Multiscale stochastic realization and model identification with applications to large-scale estimation problems, PhD Thesis, Massachusetts Institute of Technology, Department of Electrical Engineering and Computer Science.

28. Jawerth, B., and Sweldens, W. (1991) An overview of wavelet based multiresolution analyses, Technical Report, University of South Carolina.

29. Koulouris, A., and Stephanopoulos, G. (1994) On-line empirical learning of process dynamics with Wave-nets, Technical Report, MIT-LISPE.

30. Koung, C., and Macgregor, F.J. (1994) Identification for robust multivariable control: the design of experiments, *Automatica*, **30**, 1541-1554.

31. Kreinovich, V., Sirisaengtaksin, O., and Cabrera, S. (1993) Wavelet neural networks are optimal approximators for functions of one variable, Technical Report, MIT-LISPE.

32. Lee, J., (1995) Model predictive control: A tutorial, to appear in *CRC Industrial Electronics Handbook*, Department of Chemical Engineering, Auburn University, Auburn, AL.

33. Lee, L. H., Chikkula, Y., Yu, Z., and Kantor, J.C. (1995) Improving computational efficiency of model predictive control algorithm using wavelet transformation, *Int. J. Control*, **61** 859-883.

34. Luettgen, M. (1993) Image processing with multiscale stochastic models, PhD Thesis, Massachusetts Institute of Technology, Department of Electrical Engineering and Computer Science.

35. Lindsey, A. (1994) The non-existence of a wavelet function admitting a wavelet transform convolution theorem of the Fourier type, preprint, Ohio Univeristy, Athens, OH.

36. Mallat, S. G. (1989) A theory for multiresolution signal decomposition: The wavelet representation, *IEEE Transactions on Pattern Analysis and Machine Intelligence*, **II**, 674-693.

37. Mayne, D. (1996) Nonlinear model predictive control: An assessment, Preprints of *Chemical Process Control-V*, Tahoe City, CA.

38. Mayne, D. and Michalska, H. (1990) Receding horizon control of nonlinear systems, *IEEE Transactions on Automatic Control* **AC-35**: 814-824.

39. Morari, M. (1993) Model predictive control: Multivariable control technique of choice in 1990's ? Preprint, California Institute of Technology.

40. Morari, M.,Garcia, C.E., Lee, J., and Prett, D.M., (1994) Model predictive control, Book preprint.

41. Muske K. R., Rawlings, J.B. (1992) Model predictive control with linear models", accepted to be published in *AICHE Journal*, 1992.

42. Nikolau, M. (1995) Multiscale model predictive control, Memorandum, MIT-LISPE.

43. Ohshima, M., and Hashimoto I. (1992) Multi-rate multivariable model predictive control and its application to a semi-commercial polymerization reactor, Technical Report.

44. Qin, S. J., and Badgwell, T.A. (1995) An overview of industrial model predictive control technology, Preprint, University of Texas at Austin.

45. Ramirez, W. F. (1985) Process control and identification, Academic Press, Boston.

46. Rauch, H., Tung, F., and Striebel, C. (1965) Maximum likelihood estimates of linear dynamic systems, *AIAA Journal*, **3**.

47. Rawlings, J. B., Muske, K.R. (1993) The stability of constrained receding horizon control, *IEEE Transactions on Automatic Control*, **38**.

48. Robertson, D., Lee, J.H., and Rawlings, J.B. (1994) A moving horizon-based approach for least squares state estimation, *AICHE Journal*, **42**, 2209-2223.

49. Strang, G., and Nguyen, T. (1996) Wavelets and filter banks, Wellesley.

50. Stephanopoulos, G., (1989) *The Shell Process Control Workshop*, D. Prett and C. Garcia Butterworth Publishers, Stoneham, MA.

APPENDIX

A.1 PROOF OF THEOREM 2.

If a system (21) ℓ_p-stable in a Lyapunov sense, then $0 < \|A\|_p \leq 1$, and consequently $0 < \|A^{2^m}\|_p \leq 1$. Therefore, the homogeneous discrete-time model

$$x(m, k+1) = A^{2^m} x(m, k)$$

is ℓ_p-stable in the Lyapunov sense at any level, m, of discretization, and the state of the multi-scale system is bounded at any node of the infinite tree.

To prove the reverse, consider the node, τ, on the "initial time" path and a node, s, along the same path (i.e. the "initial time" path), see Figure 8. Using equation (28),

$$x(\tau) = 2^{m/2} (I + A)^{-1} (I + A^2)^{-1} \cdots (I + A^{2^m})^{-1} x(s)$$

or

$$\|x(s)\|_p = \frac{1}{2^{m/2}} \|I + A\| \|I + A^2\| \cdots \|I + A^{2^m}\| \cdot \|x(\tau)\|_p$$

If $\|x(\tau)\|_p$ is bounded, then $\|x(s)\|_p$ is bounded for any m, as m $\to \infty$, if and only if $\|A\|_p \leq 1$, which implies the ℓ_p-stability of system (21). Since system (28) is ℓ_p-stable in the Lyapanov sense, then if $\|x(\tau)\|$ is bounded, the state is also bounded at any node on the same horocycle as τ, i.e. $\|x(z)\|$, $\|x(v)\|$, etc. are all bounded. Then, the state at any node along the path from nodes z, v, etc. towards $-\infty$ is bounded implying $\|A\|_p \leq 1$.

A.2 PROOF OF THEOREM 3

If system (21) is controllable, then

$$\text{rank } [B \ \ AB \ \ A^2B \ \dots \ A^{n-1}] = n \tag{A.1}$$

Where, n is the dimensionality of the state vector. Starting with node, τ, on the "initial time" path, equation (28) yields

$$x(\tau\alpha) = \left[A_\alpha^{(0)}\right]^{-1} x(\tau) + \left[A_\alpha^{(0)}\right]^{-1} B^{(0)} u(\tau) = (I + A)x(\tau) + Bu(\tau) \tag{A.2}$$

Take, without loss of generality, $x_\tau = 0$. Then, recursive application of equation (A.2), along the nodes of the "initial time" path, yields the state at any node of the path, e.g.

$$x(\tau\overline{\alpha}) = Bu(\tau) \tag{A.3a}$$

$$x(\tau\overline{\alpha}\,\overline{\alpha}) = (I + A)Bu(\tau) + Bu(\tau\overline{\alpha}) \tag{A.3b}$$

$$x(\tau\overline{\alpha}\,\overline{\alpha}\,\overline{\alpha}) = (I + A^2)(I + A)Bu(\tau) + (I + A)Bu(\tau\alpha) + Bu(\tau\overline{\alpha}\,\overline{\alpha}) \tag{A.3c}$$

$$x(\underbrace{\tau\overline{\alpha}\,\overline{\alpha}\cdots\overline{\alpha}}_{n}) = \prod_{j=0}^{n-1}(I + A^{2^j})Bu(\tau) + \prod_{j=0}^{n-2}(I + A^{2^j})Bu(\tau\overline{\alpha}) + \cdots + Bu(\underbrace{\tau\overline{\alpha}\,\overline{\alpha}\cdots\overline{\alpha}}_{n-1}) \tag{A.3d}$$

Equations (A.3a), (A.3b), (A.3c), (A.3d) can be solved for the unique solution in the inputs if and only if

$$\text{rank } [B \quad (I+A)B \quad (I+A)^2 B \cdots (I+A)^{n-1} B] = n \tag{A.4}$$

which is true, given (A.1). Therefore, we can bring the zero state x_τ to a desired value at any node on the "initial time" path, at a level higher than that of τ.

From equation (2.1) it follows that at any level-m, m = 1, 2, 3, $\cdots$

$$\text{rank } [B \quad A^{2m}B \quad A^{2^m+1}B \cdots A^{2^m+(n-1)}B] = n$$

which implies that the system at level-m (see Figure 8)

$$x(s) = A^{2^m}x(s\overline{\alpha}\beta) + (I + A^{2^{m-1}})(I + A^{2^{m-2}})\cdots(I + A)Bu(s)$$

is also controllable. Therefore, given an initial state at any node on the "initial time" path, we can reach the desired state value at any other node at the same horocycle (level) as the initial state. The proof of the reverse part of the theorem follows in a straightforward manner.

A.3 PROOF OF THEOREM 3

If system (21) with output model (34) is observable, then

$$\text{rank } [C^T \quad A^T C^T \quad (A^T)^2 C^T \cdots (A^{2^m+(n-1)})^T C^T]^T = n \tag{A.5}$$

leading to the conclusion that at any level-m, m = 0, 1, 2, 3, ... , the corresponding discretized model

$$x(s) = A^{2m}x(s\overline{\alpha}\beta) + (I + A^{2^{m-1}})(I + A^{2^{m-2}})\cdots(I + A)Bu(s)$$

with outputs

$$y(s) = Cx(s)$$

is also observable. Consequently, given n measurements at n nodes of the same level (horocycle), we can compute the initial state at the node on the "initial time" path and level-m.

Now, let us assume that the n measurements are at different level nodes on the "initial time" path. Then, assuming (without loss of generality) all inputs on this path to be zero, we take,

$$y(\tau) = Cx(\tau) \tag{A.6a}$$

$$y(\tau\overline{\alpha}) = Cx(\tau\overline{\alpha}) = C(I + A)x(\tau) \tag{A.6b}$$

$$y(\tau\overline{\alpha}\,\overline{\alpha}) = Cx(\tau\overline{\alpha}\,\overline{\alpha}) = C(I + A^2)(I + A)x(\tau) \tag{A.6c}$$

$$y(\underbrace{\tau\overline{\alpha}\,\overline{\alpha}\cdots\overline{\alpha}}_{n-1}) = C\prod_{j=0}^{n-1}(I + A^{2^j})x(\tau) \tag{A.6d}$$

The set of equations (A.6a), A.6b), (A.6c), ..., (A.6d) admits a unique solution, given condition (A.5).

When the measurements are distributed at various levels and are on different paths, similar reasoning, as above, allows the computation of the initial state on the "initial time" path and a level which is not lower than the level at which a measurement is available.

A.4   HAAR DECOMPOSITION OF RANDOM SEQUENCES

Let random sequence, $\{v(k); k = 0,1, ..., 2^M-1\}$ with $v(k) = \mathcal{N}(0,Q)$. If τ is a node at level-m, then the scaling coefficients resulting from the Haar transformation have a covariance,

$$\begin{aligned}
Q(\tau) = E\left[v(\tau)v(\tau)^T\right] &= E\left[\tfrac{1}{\sqrt{2}}\big(v(\tau\alpha) + v(\tau\beta)\big)\tfrac{1}{\sqrt{2}}\big(v(\tau\alpha) + v(\tau\beta)\big)^T\right] \\
&= \tfrac{1}{2}\left\{E\left[v(\tau\alpha)v(\tau\alpha)^T\right] + E[v(\tau\beta)v(\tau\beta)^T]\right\} \qquad \text{(A.7a)} \\
&= E\left[v(\tau\alpha)v(\tau\alpha)^T\right]
\end{aligned}$$

Continuing recursively at lower levels, we have,

$$
\begin{aligned}
Q(\tau\alpha) = E\!\left[v(\tau\alpha)v(\tau\alpha)^T\right] &= E\!\left[\tfrac{1}{\sqrt{2}}\!\left(v(\tau\alpha\alpha)+v(\tau\alpha\beta)\right)\tfrac{1}{\sqrt{2}}\!\left(v(\tau\alpha\alpha)+v(\tau\alpha\beta)\right)^T\right] \\
&= \tfrac{1}{2}\!\left\{E\!\left[v(\tau\alpha\alpha)v(\tau\alpha\alpha)^T\right]+E[v(\tau\alpha\beta)v(\tau\alpha\beta)^T]\right\} \\
&= E\!\left[v(\tau\alpha\alpha)v(\tau\alpha\alpha)^T\right]
\end{aligned} \tag{A.7b}
$$

Thus, at level-0 where the sequence $\{v(\mathrm{k})\}$ is defined, we have,

$$
Q(\tau) = E\!\left[v(\tau)v(\tau)^T\right] = Q \qquad \forall\, levels - m \tag{A.8}
$$

Following a similar line of proof, it can easily be shown that for the wavelet coefficients, we have,

$$
E\!\left[\delta v(\tau)\delta v(\tau)^T\right] = Q. \tag{A.9}
$$

For a sequence of random variables, $\{w(k);\ k=0,1,\ldots,2^M\text{-}1\}$, with varying covariance, i.e. $w(k) = \mathcal{N}(0,R_k)$, the covariances of the scaling and wavelet coefficients across the various nodes of the tree, vary according to the following formulae.

Covariance of scaling coefficients

$$
\begin{aligned}
R(\tau) = E\!\left[w(\tau)w(\tau)^T\right] &= E\!\left[\tfrac{1}{\sqrt{2}}\!\left(w(\tau\alpha)+w(\tau\beta)\right)\tfrac{1}{\sqrt{2}}\!\left(w(\tau\alpha)+w(\tau\beta)\right)^T\right] \\
&= \tfrac{1}{2}\!\left\{E\!\left[w(\tau\alpha)w(\tau\alpha)^T\right]+E[w(\tau\beta)w(\tau\beta)^T]\right\} \\
&= \tfrac{1}{2}\!\left(R(\tau\alpha)+R(\tau\beta)\right)
\end{aligned} \tag{A.10}
$$

Covariance of wavelet coefficients

$$
\begin{aligned}
\delta R(\tau) = E\!\left[\delta w(\tau)\delta w(\tau)^T\right] &= E\!\left[\tfrac{1}{\sqrt{2}}\!\left(w(\tau\alpha)-w(\tau\beta)\right)\tfrac{1}{\sqrt{2}}\!\left(w(\tau\alpha)-w(\tau\beta)\right)^T\right] \\
&= \tfrac{1}{2}\!\left\{E\!\left[w(\tau\alpha)w(\tau\alpha)^T\right]+E[w(\tau\beta)w(\tau\beta)^T]\right\} \\
&= \tfrac{1}{2}\!\left(R(\tau\alpha)+R(\tau\beta)\right)
\end{aligned} \tag{A.11}
$$

A.5   MULTI-SCALE RECURSIVE STATE ESTIMATOR WITH SINGLE-RATE MEASUREMENTS.

Consider the recursive estimator at node τ,

$$
\hat{x}(\tau|\tau) = \hat{x}(\tau|\tau+) + K(\tau)\!\left[y(\tau) - C\hat{x}(\tau|\tau+)\right] \tag{A.12}
$$

where,

$\hat{x}(\tau|\tau+)$ is the state estimation at node τ using the measurements at nodes descending from τ, and,

$\hat{x}(\tau|\tau)$ is the updated estimate of the state at τ after the incorporation of the measurement value $y(\tau)$.

From the left-node multi-scale state equation of *Problem-P2c*, we have,

$$\hat{x}(\tau|\tau\alpha) = \overline{A}_\alpha^{(m)}\hat{x}(\tau\alpha|\tau\alpha) \qquad (A.13)$$

where,

$$\overline{A}_\alpha^{(m)} = \left[A_\alpha^{(m)}\right].$$

Substitute $\hat{x}(\tau|\tau+)$ from equation (A.13) into equation (A.12) and take

$$\hat{x}(\tau|\tau) = \overline{A}_\alpha^{(m)}\hat{x}(\tau\alpha|\tau\alpha) + K(\tau)\left[y(\tau) - C\hat{x}(\tau|\tau+)\right].$$

Let $e(\tau|\tau) = x(\tau) - \hat{x}(\tau|\tau)$ be the reconstruction error at node τ. Then

$$e(\tau|\tau) = \left(\overline{A}_\alpha^{(m)}x(\tau\alpha) + \overline{B}_\alpha^{(m)}w^*(\tau\alpha)\right) - \left(\overline{A}_\alpha^{(m)}\hat{x}(\tau\alpha|\tau\alpha) + K(\tau)\left[y(\tau) - C\hat{x}(\tau|\tau+)\right]\right), \quad (A.14)$$

where,

$$\overline{B}_\alpha^{(m)} = -[A_\alpha^{(m)}]^{-1}B^{(m)}.$$

Now,

$$y(\tau) = Cx(\tau) + v(\tau) = C\left(\overline{A}_\alpha^{(m)}x(\tau\alpha) + \overline{B}_\alpha^{(m)}w^*(\tau\alpha)\right) + v(\tau) \qquad (A.15)$$

$$C\hat{x}(\tau|\tau+) = C\overline{A}_\alpha^{(m)}\hat{x}(\tau\alpha|\tau\alpha). \qquad (A.16)$$

Substitute (A.15) and (A.16) into (A.14) and take,

$$e(\tau|\tau) = \left[I - \dot{K}(\tau)C\right]\overline{A}_\alpha^{(m)}e(\tau\alpha|\tau\alpha) + [I - K(\tau)C]\overline{B}_\alpha^{(m)}w^*(\tau\alpha) - K(\tau)v(\tau). \quad (A.17)$$

Let $P(\tau|\tau)$ be the covariance of the reconstruction error, $e(\tau|\tau)$. Then, from equation (A.17), we have,

$$P(\tau|\tau) = \left[I - K(\tau)C\right]\overline{A}_\alpha^{(m)}P(\tau\alpha|\tau\alpha)\overline{A}_\alpha^{(m)T} + \overline{B}_\alpha^{(m)}R(\tau)\overline{B}_\alpha^{(m)T}\left[I - K(\tau)C\right]^T - K(\tau)QK(\tau) \quad (A.18)$$

Now define $e(\tau|\tau+) = x(\tau) - \hat{x}(\tau|\tau+)$ to be the updated reconstruction error at node τ.

734

A similar treatment leads to the covariance of the updated reconstruction error

$$P(\tau|\tau+) = \overline{A}_\alpha^{(m)} P(\tau\alpha|\tau\alpha) \overline{A}_\alpha^{(m)T} + \overline{B}_\alpha^{(m)} R(\tau) \overline{B}_\alpha^{(m)T} \qquad (A.19)$$

Hence, A.18 becomes

$$P(\tau|\tau) = \left[I - K(\tau)C\right]P(\tau|\tau+)\left[I - K(\tau)C\right]^T - K(\tau)QK(\tau) \qquad (A.20)$$

If $P(\tau\alpha|\tau\alpha)$ is the minimum covariance of the reconstruction error at the node τ, then the covariance $P(\tau|\tau)$ is minimized with respect to $K(\tau)$ when

$$K(\tau) = P(\tau|\tau+)C^T \left(CP(\tau|\tau+)C^T + Q\right)^{-1} \qquad (A.21)$$

Part V

INDUSTRIAL APPLICATIONS

NONLINEAR CONTROL WITH LINEAR CONTROLLERS: TRANSFORMATIONS, CALCULATED GAINS AND MODEL SCHEDULING

AYDIN KONUK
Aspen Technology Inc.
9896 Bissonnet
Houston, Texas 77036

Abstract

This paper summarizes the methods used by Aspen Technology to deal with process nonlinearities in the implementation of advanced control projects. Included is a set of DMC simulations to illustrate how these techniques work.

1. Introduction

Linear model predictive controllers have become the industry standard in the control of refinery and petrochemical processes. Since these processes are nonlinear, some more nonlinear than others, various techniques have been proposed and used to improve the control of these processes with linear controllers. A review of these methods can be found in [1]. Examples of nonlinearities are valve positions, compositions in distillation columns, product blending
and heat exchanger duties.

The current Aspen Technology linear predictive controller, called DMCplus, is based on the DMC algorithm [2]. The current DMCplus controllers can be quite large, with 20 to 30 manipulated variables and 50 to 100 controlled variables. The subject of this paper is the practical methods used by Aspen Technology to deal with process nonlinearities to be able to implement successful DMCplus applications.

The three methods that are discussed in the paper are:
a) Transformations
b) Calculated Gains
c) Model Scheduling (Separate models for different modes of operation)

Simulated Dynamic Matrix Control (DMC) responses are used to investigate and compare transformations and calculated gains techniques to each other.

R. Berber and C. Kravaris (eds.), Nonlinear Model Based Process Control, 737-748.
© 1998 *Kluwer Academic Publishers.*

2. Transformations

Transformations are used to linearize the nonlinear process responses. Steady-state functions are used for the transformations:

$$YT = F(Y) \tag{1}$$

where Y is the original nonlinear variable and YT is transformed, linearized variable. Y can be a controlled or a manipulated variable.

The most common uses of transformations are for valve positions, followed by compositions in distillation columns. When a flow measurement is available, the basis for the valve position transform is to convert the nonlinear valve position into a linear flow rate. For example, in a distillation column, where a DCS top temperature controller manipulates reflux, reflux response to reboiler duty is fairly linear, but the reflux valve position response is nonlinear. The transformed valve position looks like the flow, with linear responses. The transform is obtained by fitting flow versus valve position data to a nonlinear function describing flow as a function of valve position. When a flow measurement is not available, the choice of the transform is not that easy, and proprietary methods are used to find the best transform.

For distillation columns, various authors report that composition control is improved by using logarithmic transformations [3-6]. Our experience with the DMCplus control of high purity columns, such as those encountered in ethylene plants, is similar. With proper DCS loop configuration and tuning, step testing around the expected operating point, and transforms, the DMCplus control of these columns can be quite good.

3. Calculated Gains

This is referred to in the literature as gain scheduling. A review of gain scheduled model based control can be found in [1]. Garcia [7] proposed extending QDMC to the control of nonlinear processes as follows: integrate the nonlinear equations for the effect of past inputs on the outputs, and use local step response coefficients, obtained from a linearized model of the process, for the effect of future inputs. The linearization is updated before each control calculation.

The calculated gains method can be viewed as a simplification of this method: the linearized model is used to calculate the effect of both past and future inputs on the outputs. The advantage of this simplification for Aspen Technology is that the standard DMCplus software can be used for nonlinear problems since the model update can be done externally. The disadvantage is that the use of linear models introduces error in the effect of past inputs on the outputs. However, as will be shown by the simulations later good control is possible with this method, with the proper choice of the linearization point.

It is important to make it clear that the use of this approach at Aspen Technology has been limited to the DMCplus control of product blends. As discussed previously, transformations are used instead for other nonlinear problems such as the valve positions and compositions in distillation columns. The reason is that transforms are easier to implement and have better performance, as shown later, than calculated gains. Calculated gains are used for blend controls because this is a multivariable problem without a transform. As shown later, the nonlinear model can be obtained from first principles, without any nonlinear identification. Step testing is used to identify process dynamics, using linear identification tools. Following is a brief description of blend controls:

Most refinery products are blends of various components. For example, gasoline may be a blend of gasolines from the crude and catalytic cracking units, reformate and alkylate. For in-line blending, DMCplus is used to control blend header qualities such as octane, RVP, sulfur, etc., within specifications by manipulating the flow rate of blend components from the tanks. This is shown in Figure 1.

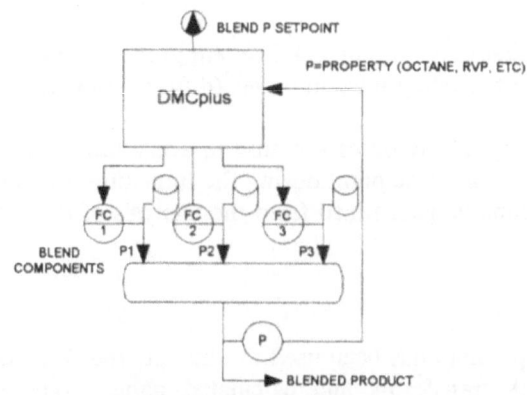

Figure 1 Blend Control

The model gain for the blend qualities with respect to component flows is not a constant, but a function of blend quality and component qualities, as shown below for three components:

$$C = (C_1F_1 + C_2F_2 + C_3F_3) / F$$
$$F = F_1 + F_2 + F_3$$

$$(2)$$

where

C = blend quality
C_1, C_2, C_3 = component qualities
F_1, F_2, F_3 = component flow rates

F = total blend flowrate

The gain of component i is then:

$$\frac{\partial C}{\partial F_i} = (C_i - C) / F \tag{3}$$

Since blend specifications are different for different grades of products (such as regular/premium gasolines, low/high sulfur fuel oils, etc.), and since component qualities in the tanks may change, the blend gains are calculated before each control calculation. This technique results in very good blend quality control when the tank qualities are known fairly well.

4. Model Scheduling

For process units which operate at different modes, the models for each mode can be sufficiently different that it may be necessary to use separate models for different modes. Examples are:

a) Delayed coker units, producing different grades of coke
b) Vacuum units feeding lube oil plants (different modes for different lube oils)

The mode may change a few times a month to every few days, and the switch to the new model takes place at some point during the transition from one mode to the other. The model switch time is determined from the analysis of the process changes during the transition.

5. Simulation

A simple example problem has been used to illustrate the DMC control of a nonlinear SISO problem with transforms and calculated gains. The first purpose of the simulation is to verify that if the transform is controlled well, the original variable will be controlled equally well. This seems to be evident for simple dynamics such as a first order model, but perhaps not so intuitive for complex dynamics, such as a second order model with a large overshoot. The second purpose of the simulation is to find the best way to calculate the gains, and then compare the performance of the best calculated gain technique with transform control.

The steady-state nonlinear function in the simulation was:

$$Y = Y^{1/2}U + Y_0 \tag{4}$$

where:
Y = output variable
U = input variable
Y_0 = constant

The transform YT is defined by:

$$YT = \frac{Y - Y0}{Y^{1/2}}$$ (5)

This transform results in the following linear equation:

$$YT = U$$ (6)

The inverse transform is given by:

$$Y = \left(YT^2 + 2Y_0 + DELTA^{1/2}\right) / 2$$ (7)

where:

$$DELTA = YT^2\left(YT^2 + 4Y_0\right)$$ (8)

5.1 DYNAMIC MODELS

Two dynamic models have been used for the simulations (both with the same steady-state nonlinear function, shown in Eq (4)):

1. First order model (Time constant = 4.48 min)

$$YT_{k+1} = .8YT_k + 0.2U_k$$ (9)

Note that this linear model is for the transformed variable YT. The nonlinear variable Y in the simulation is the inverse of YT, given in Eqs (7) and (8). The open loop responses of YT and Y are shown in Figure 2.

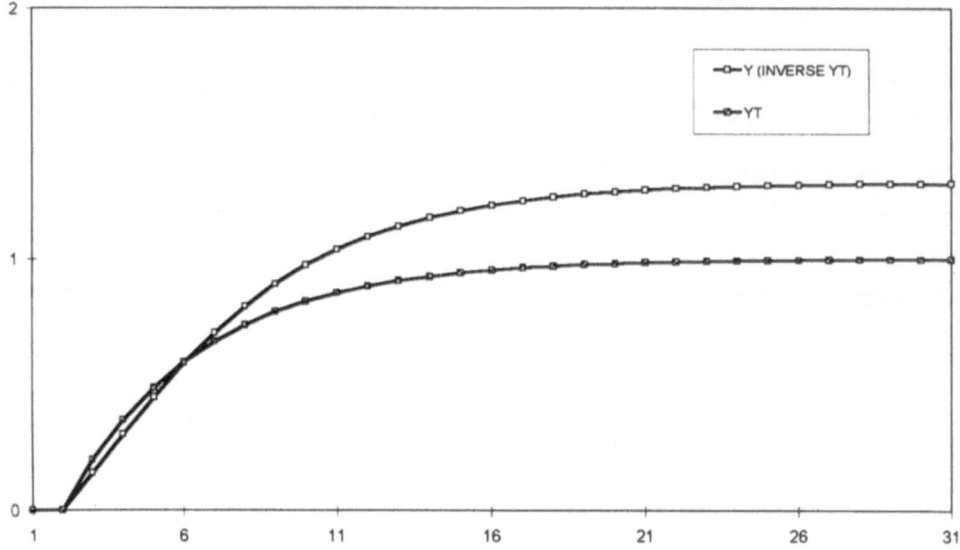

Figure 2 DMC Control of Y via Y Transform, First Order Model

2. Second order model

$$YT_{k+1} = 1.35YT_k - 0.65YT_{k-1} + 1.1U_k - 0.8U_{k-1} \qquad (10)$$

This is a second order response with 100 percent overshoot. The open loop responses of YT and Y are shown in Figure 3.

5.2 Simulation for Transforms

The simulations consisted of moving the YT setpoint from 0.0 to 1.80, and then back to 0.0. This corresponds to moving Y setpoint from 0.4 to 4.0, and then back to 0.4. The resulting gain change is 1.25 to 4.0 back to 1.25 (3.2 fold change). DMC controller has been tuned in this simulation with a move suppression of 2, where move suppression is the move penalty multiplier in the objective function of the DMC controller. Figure 4 shows the results of the simulation.

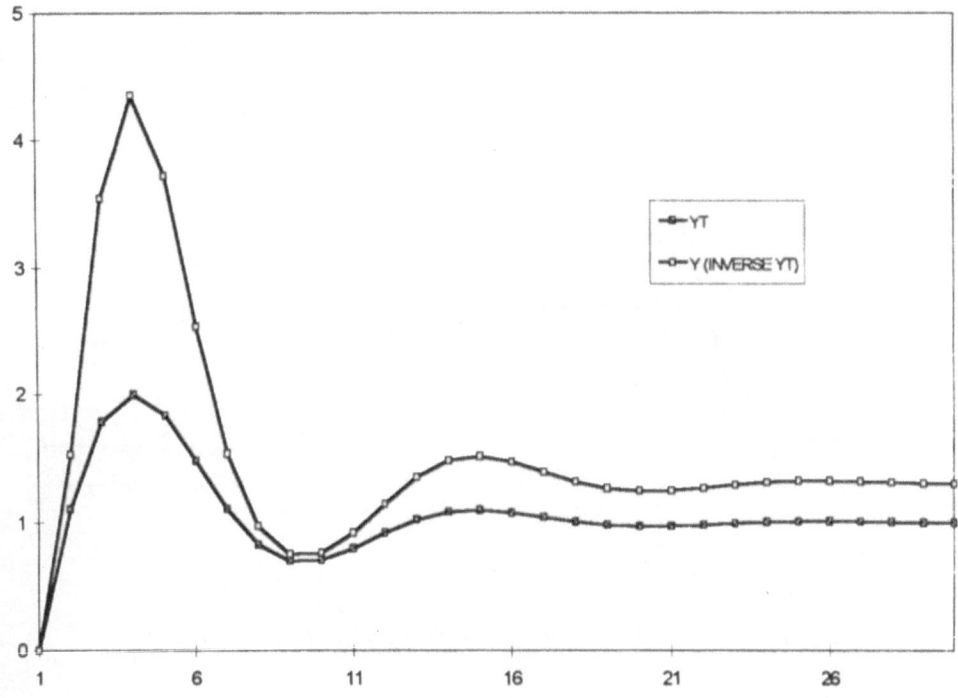

Figure 3 Y and Y Transform Models, Second Order Model

YT is controlled smoothly with very little overshoot. The original variable Y is controlled indirectly by controlling YT. The simulation shows that good control of YT results in good control of Y for this first order system.

Figure 5 shows the DMC control of the second order process. The move suppression is 2 for this case, as in the previous one. Due to large overshoot in the open loop response, YT overshoots the setpoint slightly going up and undershoots the setpoint by the same amount going down. The Y response is not symmetrical going up and down due to gain nonlinearity. Overshoot is larger going up than undershoot going down, but the Y response is still good.

744

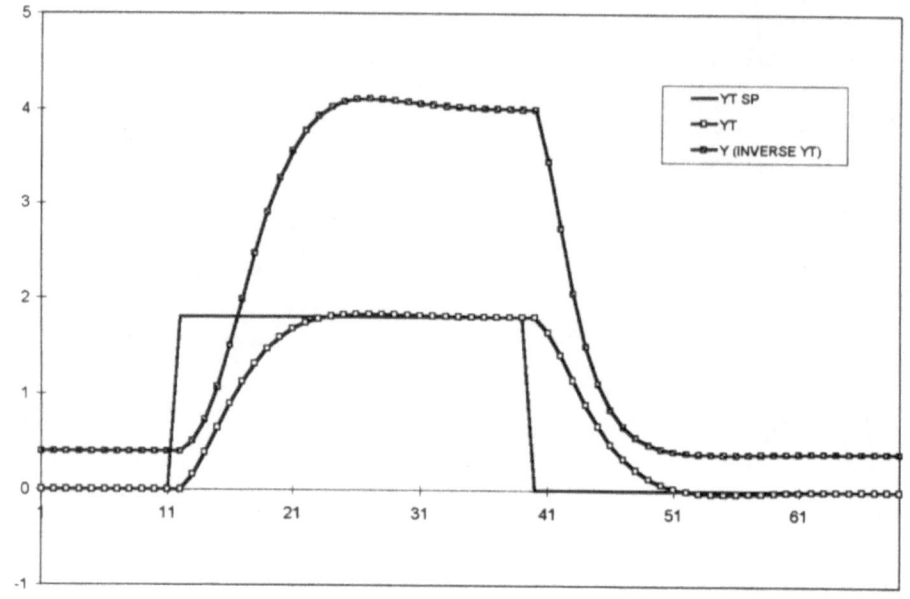

Figure 4 DMC Control of Y via Y Transform, First Order Model

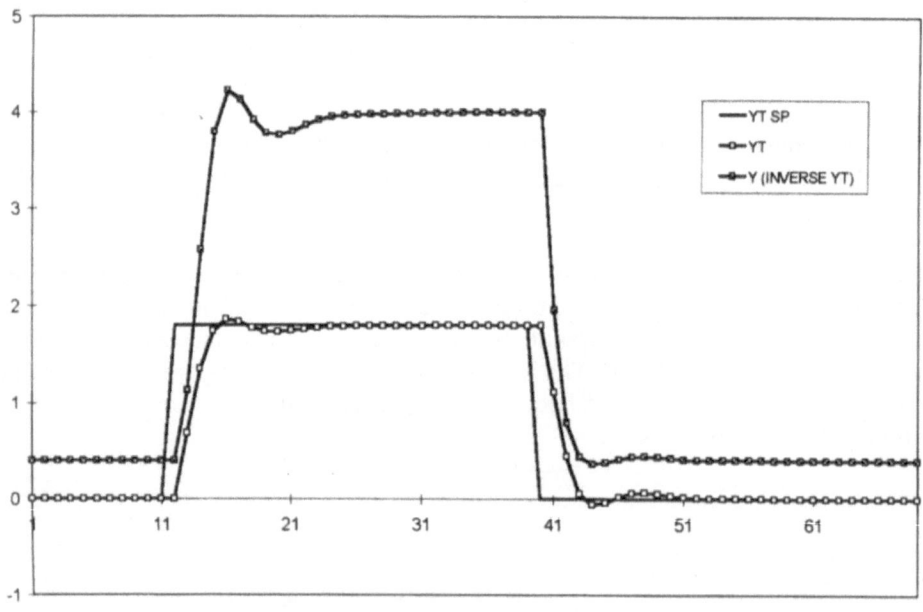

Figure 5 DMC Control of Y via Y Transform, Second Order Model

5.3 SIMULATIONS WITH CALCULATED GAINS

The nonlinear steady-state function Y is differentiated with respect to input variable(s) U to give an expression for the linear model gain in terms of Y. For the example problem, the gain is:

$$K(Y) = 2Y^{1/2} / (1 + Y_0 / Y) \tag{11}$$

This gain is used to update the local step response coefficients before each control calculation. The question which arises for implementation is what value of Y to use for the gain calculations, possibilities being Y measurement, Y setpoint or combinations.

5.4 CONTROL WITH GAIN CALCULATED AT MEASURED Y

This may seem the natural choice for linearization, but as shown in Figure 6, for the first order system, going from low to high setpoint (going from low to high gain), the control action is too aggressive, resulting in a large overshoot despite a high move suppresssion of 6. The response from high to low setpoint (from high to low gain) on the other hand is slow. Similar results are obtained for the second order system where move suppression is also set to 6 (Figure 7). Thus, the Y measurement is not a good choice for linearization.

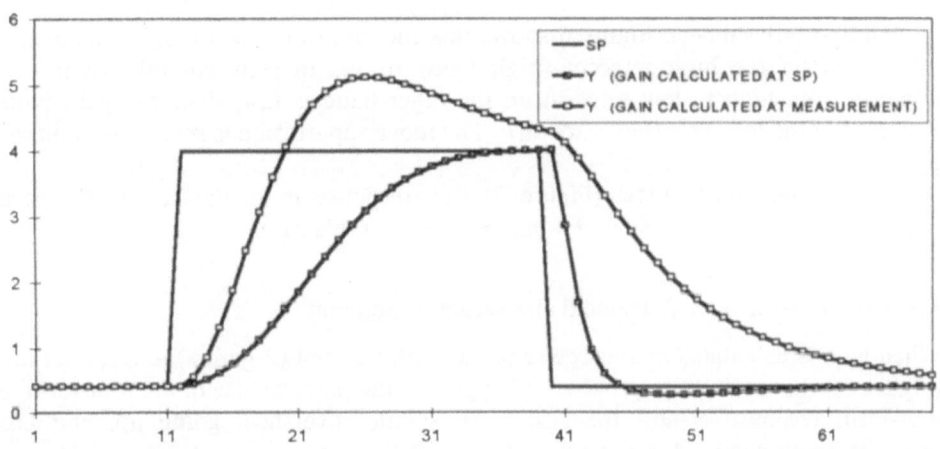

Figure 6 DMC Control of Y with Gain Calculated at Y Measurement vs. At Y Setpoint, First Order Model

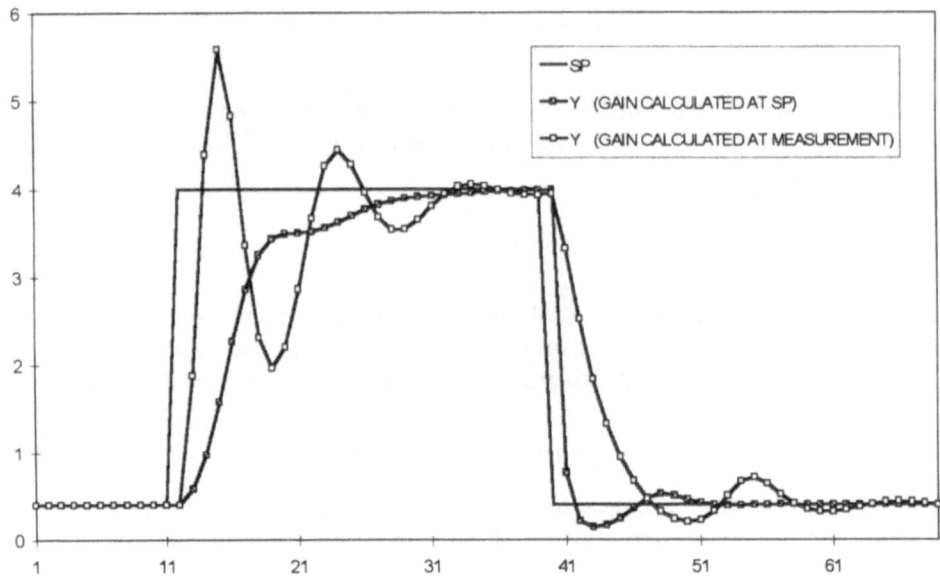

Figure 7   DMC Control of Y with Gains Calculated at Y Measurement vs. At Y Setpoint, Second Order Model

5.5   Control with Gain Calculated at Y Setpoint

For the first order model, Figure 6 shows that moving from low to high setpoints, the gain based on the high setpoint (high gain) results in slow control action. The response from high to low setpoint on the other hand is fast, since the gain is now evaluated at the low gain (low setpoint). The move suppression is set to 3 for this case.

For the second order model (Figure 7), the response is similar to the first order response. A move suppression of 4 has been used in this case.

5.6   Control with Gain Calculated at Average Y Setpoint

When gain is calculated at average setpoint, both the first (Figure 8) and second order (Figure 9) responses are improved, and approach the performance of the transform, but transform response is still the best, with smaller overshoot going up, and faster response going down. A move suppression of 2 has been used for this case. Note that better linearization reduces model error, and allows a smaller move suppression, and therefore better performance.

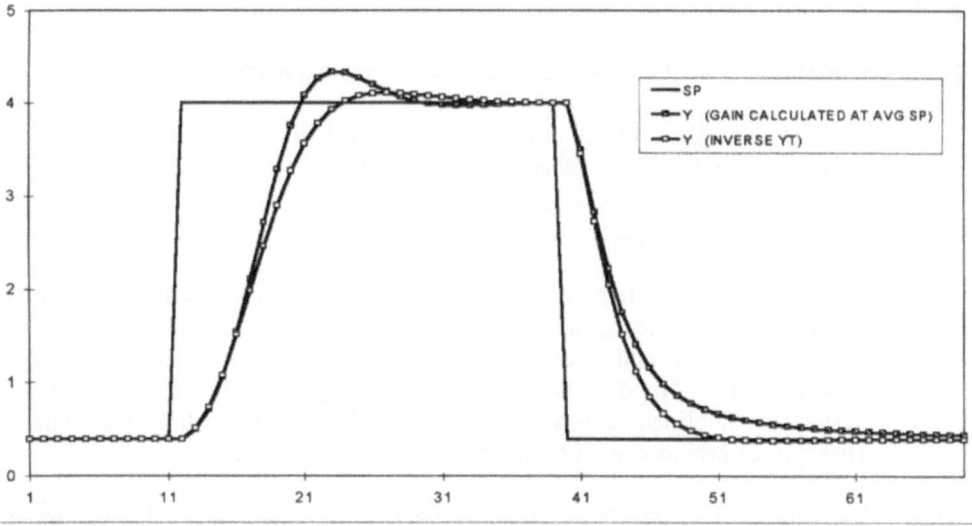

Figure 8 DMC Control of y via Transform vs. With Calculated Gain at Average Setpoint, First Order Model

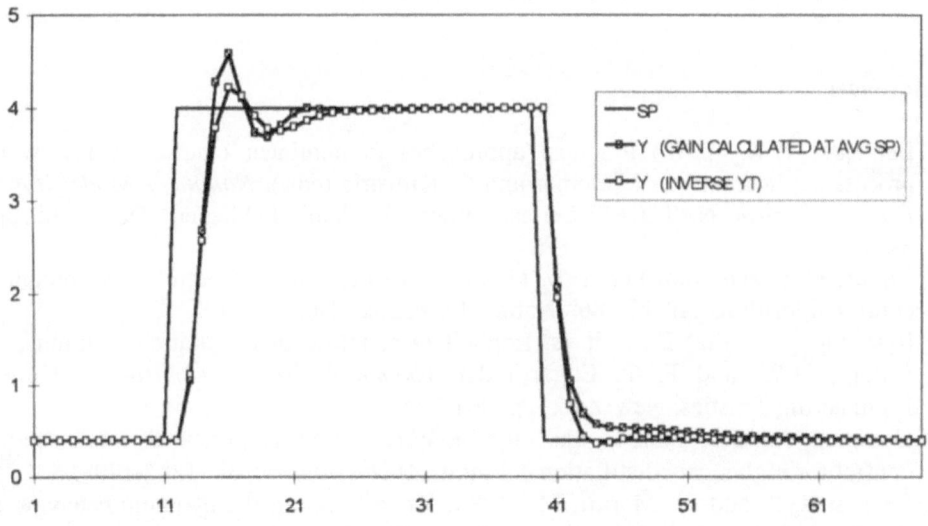

Figure 9 DMC Control of Y via Y Transform vs. With Calculated Gain at Average Setpoint, Second Order Model

6. Conclusions

Based on the simulations, the following observations can be made:

a) For the example problem, the indirect control of Y via the control of the transform YT is almost as good as the control of YT for both the first and second order dynamics. It would seem reasonable to expect that this would apply to nonlinear systems in general.

b) Again for the example problem, good control close to transform control can be obtained by linearizing the nonlinear steady-state model and evaluating the gain of the linear model at the average setpoint. This conclusion perhaps may be extended to "mildly" nonlinear systems in general. For highly nonlinear systems, a nonlinear controller would be required for better performance than can be provided with calculated gains..

These observations suggest the following practical methods for the control of nonlinear processes:

a) Use transforms and a linear controller whenever possible.
b) If transforms are not possible, or the transformed variables are still nonlinear, use calculated gains and linear controller.
c) Use a nonlinear controller for processes where the benefits of better control performance with the nonlinear controller justify the potentially large increase in the project implementation costs.

References

1. Bequette, W.B. (1998) Practical approaches to nonlinear control: A review of process applications, in R. Berber and C. Kravaris (eds.), *Nonlinear Model Based Process Control*, NATO ASI Series, Kluwer Academic Publishers, Dordrecht, pp. 3-32.
2. Cutler, C.R. and Ramaker, B.L. (1979) Dynamic Matrix Control - A computer control algorithm, AIChE 86th National Meeting, Houston, TX.
3. Ryskamp, C. (1982) Explicit vs. Implicit Decoupling in Distillation Columns, in Seborg, D.W. and F. T. Edgar (eds), *Chemical Process Control 2.*, United Engineering Trusties, New York, pp. 361-375.
4. Georgiou, A., Georgakis, C. and Luyben, W.L. (1988), Nonlinear dynamic matrix control for high purity distillation columns, *AIChE Journal*, **34**, 1287-1298.
5. Skogestad, S. and M. Morari, M. (1988) Understanding the dynamic behavior of distillation columns, *Ind. Eng. Chem. Res.*, **27**, 1848-1862.
6. Skogestad, S. and Morari, M. (1988) LV Control of high purity of distillation columns, *Chem. Eng. Sci.*, **43**, 33-48.
7. Garcia, C. E. (1984) Quadratic dynamic matrix control of nonlinear processes: An application to a batch reaction process, AIChE National Meeting, San Francisco.

CONTROL OF A STEAM BOILER BY ELEMENTARY NONLINEAR DECOUPLING (END)

JENS G. BALCHEN AND GEIR LARSEN
Department of Engineering Cybernetics
Norwegian University of Science and Technology
7034 Trondheim

Keywords: Nonlinear decoupling, multivariable control, modelbased control, steam boiler

Abstract: The paper describes the simulated application of the END-algorithm (Balchen 1998) for model-based, nonlinear, multivariable control of a steam boiler process. The choice of the tuning parameters for the optimization criterion is discussed in some detail.

1. INTRODUCTION

Elementary Nonlinear Decoupling (END) is a powerful technique for the derivation of model-based control of a nonlinear, multivariable process. The theoretical basis for END is given in Balchen (1998). The present paper demonstrates the performance of END by the simulation of a realistic physical process. The process chosen is a steam boiler for ship operations. The reason for choosing this process is that:

- it has an acceptable complexity and a good mathematical model exists.

- it is multivariable with very strong interaction between the different physical phenomena.

- it is highly nonlinear

- it has a pronounced non-minimum-phase behavior

- it is representative for many industrial processes

But these arguments in favor of using the steam boiler process as an example of END-control should not lead to the conclusion that the END-algorithm should be preferred in industrial practice for this process. In fact, well established control algorithms utilizing feed forward and feedback control have proven records of excellent performance.

R. Berber and C. Kravaris (eds.), Nonlinear Model Based Process Control, 749-780.
© 1998 *Kluwer Academic Publishers.*

750

2. A STEAM BOILER PROCESS

Figure 1 shows a simplified boiler process. The burners supply heat to vertical tubes, which are called *risers* and carry a mixture of water and steam. At the top of the risers is a *drum,* which is kept half full of water. The upper part of the drum contains saturated steam. Another set of tubes leaving the drum, the *downcomers,* carry water to a smaller drum where mud is separated from the water. The water is then fed to the risers where vaporization takes place. The steam produced in the risers passes from the drum to the *superheaters* placed in the *combustion chamber.* A control valve is located between the *primary* and *secondary superheaters* for temperature control of the superheated product steam. The product steam is sent to different consumers (heat exchangers, steam turbines) where the energy is removed and the liquid is returned to the drum. Some additional water has to be added, because of leakage losses.

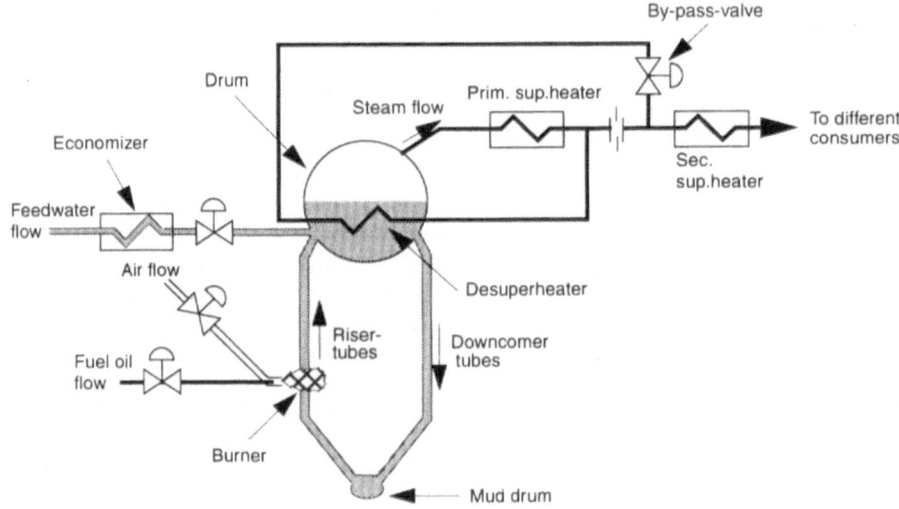

Figure 1 A simplified ship boiler process

The boiler process is multivariable with a high degree of coupling between the different state variables. The "shrink and swell" phenomena in the water level is especially important. If the pressure in the drum decreases the steam bubbles in the drum and risers will expand and this will force the water level in the drum to rise. The shrink and swell effect is a non-minimum phase phenomenon. Details regarding steam boilers in general are found in Balchen and Mumme (1988). A 7th order statespace model of a steam boiler has been developed in Tysso (1980) and is reviewed in the next section.

3. A NONLINEAR 7TH ORDER STEAM BOILER MODEL

The boiler statespace is as follow:

$$\dot{x}(t) = f(x, u, v, t)$$
$$y = h(x, u, v, t) \tag{1}$$

The state vector x is defined as (with approximate nominal values):

x_1: V_{dw} the volume of water in the drum including the air bubbles, 8.2 [m^3].
x_2: H_{dw} the enthalpy of the water in the drum, 306 [kcal/kg].
x_3: ρ_{ds} the density of saturated steam in the drum, 38 [kg/m^3].
x_4: T_{rm} the average temperature of the metal in the risers, 315 [$^\circ$C].
x_5: H_{ro} the average enthalpy of the steam-water mixture in the risers, 320 [kcal/kg].
x_6: T_{psm} the average temperature of the primary superheater metal, 463 [$^\circ$C].
x_7: T_{ssm} the average temperature of the secondary superheater metal, 558 [$^\circ$C].

The control vector u is defined as:

u_1: M_w the feed water flow, 29.2 [kg/s].
u_2: M_{oil} the oil flow, 1.94 [kg/s].
u_3: Y_{valve} the steam temperature control valve relative position (0-1).

The disturbance vector v is defined as:

v_1: M_s the mass flow of steam leaving the drum, 29.2 [kg/s].
v_2: T_{ewo} the temperature of feed-water leaving the economizer (assumd to be constant) 230 [$^\circ$C].

The measurement vector y is defined as:

y_1: h the drum liquid level, 1,1 [m].
y_2: P_d the drum steam pressure 66 [bar].
y_3: T_{sso} the steam temperature at the outlet of the secondary superheater, 488 [$^\circ$C].
y_4: P_{sso} the steam pressure at the outlet of the secondary superheater, 65 [bar].

The water steam system:

The water-steam system is divided into three components: the downcomers, the risers and the drum. Assumptions related to the water-steam system are described in Tysso (1980)

Drum system:

The mass balance of the water with steam bubbles in the drum is given by:

$$\frac{d}{dt}V_{dw} = \frac{1}{\rho_{dw}}[(1 - X_o)M_o + M_w + M_{ws} - M_{dow}]$$ (2)

V_{dw} = volume of water with steam bubbles in the drum, risers and downcomers, $[m^3]$.

ρ_{dw} = density of water in the drum (assumed to be constant), $[kg/m^3]$.

X_o = steam quality of steam-water mixture leaving the risers.

M_{dow} = mass flow of water entering the downcomers, $[kg/s]$.

M_{ws} = mass flow of steam condensing in the drum, $[kg/s]$.

M_o = mass flow of steam-water mixture leaving the risers, $[kg/s]$.

M_w = mass flow of water entering the drum, $[kg/s]$.

The drum level is given by:

$$h = \frac{V_{dw} - V_d}{A_d}$$ (3)

h = drum level, $[m]$.

A_d = area of the free water-surface in the drum is a function of h, but is assumed to be constant, $[m^2]$.

V_d = reference value of water in the drum at 100% load, $[m^3]$.

The rate of condensation is given by:

$$M_{ws} = K_{ws}(H_{ws} - H_{dw}) \tag{4}$$

H_{ws} = the enthalpy of the water in the saturation state, [kcal/kg].
H_{dw} = the enthalpy of the water in the drum, [kcal/kg].
K_{ws} = heat transfer coefficient, [kcal/kg°C].

The dynamic equation of steam density in the drum is given from the mass balance of the steam in the drum:

$$\frac{d}{dt}\rho_{ds} = \frac{1}{V_t - V_{dw}}\left[\ [M_o + M_w - M_{dow} - M_s] \right.$$
$$\left. + \frac{\rho_{ds} - \rho_{dw}}{\rho_{dw}}[(1 - X_o)M_o + M_w + M_{ws} - M_{dow}] \right] \tag{5}$$

ρ_{ds} = density of saturated steam in the drum, [kg/m^3].
M_s, = the massflow of steam leaving the drum, [kg/s].
V_t = total drum volume, [m^3].

When the density, ρ_{ds} of saturated steam in the drum is known, a steam table is used to find the corresponding value of drum pressure. The result after using a curvefitting procedure is a cubic equation given by:

$$P_d = -1.64 + 2.263\rho_{ds} - 0.007876\rho_{ds}^2 + 0.137\times10^{-4}\rho_{ds}^3 \tag{6}$$

The water in the drum is not in saturation state, and the energy balance of water in the drum becomes:

$$\frac{d}{dt}H_{dw} = \frac{1}{\rho_{dw}V_{dw}}[\ M_o(1 - X_o)(H_{rw} - H_{dw}) + M_w(H_{ewo} - II_{dw})$$
$$M_{ws}(H_{ss} - H_{dw}) + Q_{aw}] \tag{7}$$

H_{ewo} = enthalpy of the feedwater leaving the economizer, [kcal/kg].
H_{ss} = enthalpy of the steam in the saturation state, [kcal/kg].

Q_{aw} =   heat flow from attemperator, [kcal/s].

The density of the water in the drum as function of drum pressure is given by (curvefitting is used):

$$\rho_{dw} = 902.4 - 3.062 P_d + 0.014663 P_d^2 - 0.00005037 P_d^3 \qquad (8)$$

The risers:

The energy balance of the mass of riser metal is:

$$\frac{d}{dt} T_{rm} = \frac{Q_1 - Q_{rw}}{c_{rm} m_{rm}} \qquad (9)$$

Q_1 =   heat flow to the riser, [kcal/s].
Q_{rw} =   heat flow to the water in the risers, [kcal/s].
c_{rm} =   specific heat of metal of the risers (assumed to be constant), [kcal/kg°C].
m_{rm} =   mass of the metal of risers (assumed to be constant), [kg].
T_{rm} =   temperature of the metal of the risers, [°C].

$$Q_1 = k_1 M_{oil} \qquad (10)$$

The heat flow to the water in the risers can be described approximately as:

$$Q_{rw} = K_{rm}(T_{rm} - T_s)^3 \qquad (11)$$

T_s =   saturation temperature of the steam, [°C].
K_{rm} =   heat transfer coefficient, [kcal/kg°C].

If we assume that the steam water mixture in the risers is in saturation state, the saturation temperature can be calculated using steam tables.

$$T_s = 165.26 + 2829 P_d - 0.021351 P_d^2 + 0.00007537 P_d^3 \qquad (12)$$

The energy balance of steam water mixture in the risers can be written as:

$$\frac{d}{dt}H_{ro} = \frac{1}{V_r \rho_{ro}}\left[Q_{rw} - M_{dow}(H_{ro} - H_{dow}) - M_o r \frac{X_o}{2}\right] \tag{13}$$

H_{ro} = average enthalpy of the steam water mixture in the risers, [kcal/kg].
V_r = volume of the risers, [m³].
ρ_{ro} = average density of the steam water mixture in the risers, [kg/m³].
H_{dow} = enthalpy of water in downcomers, [kcal/kg].
H_{rw} = enthalpy of the water in the risers, [kcal/kg].
X_o = steam quality at top of risers.
M_o = mass flow of steam water mixture leaving risers, [kg/s].
r = latent heat of vaporization, [kcal/kg].

Assuming that the liquid phase is incompressible it is possible to express r and H_{rw} as functions of drum pressure:

$$r = 499.74 - 2.650P_d + 0.0127685P_d^2 - 0.4496 \times 10^{-4} P_d^3 \tag{14}$$

$$H_{rw} = 165.373 + 3.0991P_d - 0.0210698P_d^2 + 0.7581 \times 10^{-4} P_d^3 \tag{15}$$

It is necessary to find an explicit expression for M_o. The density of steam water mixture is given by:

$$\rho_{ro} = \rho_{rw} - \alpha(\rho_{rw} - \rho_{rs}) \tag{16}$$

ρ_{rs} = density of steam in risers, [kg/m³].
ρ_{rw} = density of water in risers, [kg/m³].
α = local void fraction.

The void fraction is given as:

$$\alpha = \frac{X\rho_{rw}}{\rho_{rs} + X(\rho_{rw} - \rho_{rs})} \qquad (17)$$

X = local steam quality.

It is shown in Tysso, (1980) that the flow of steam water mixture is given as:

$$M_o = \frac{M_{dow} - \dfrac{V_r f_2 f_6 f_{10}}{r} + f_4 f_7 f_8 + f_4 f_5 f_7 f_9}{1 + f_4 f_7 + f_4 f_5 f_7(1 - X_o) - \dfrac{V_r f_2 f_6 X_o}{2}} \qquad (18)$$

$$f_1 = \frac{X\rho_{rw}^{\,2}}{(\rho_{rs} + X(\rho_{rw} - \rho_{rs}))^2}$$

$$f_2 = \frac{\rho_{rs}\rho_{rw}(\rho_{rs} - \rho_{rw})}{(\rho_{rs} + X(\rho_{rw} - \rho_{rs}))^2}$$

$$f_3 = 5.37 e^{-0.026\rho_{ds}}$$

$$f_4 = \frac{1}{V_t - V_{dw}}$$

$$f_5 = \frac{\rho_{ds} - \rho_{dw}}{\rho_{dw}}$$

$$f_6 = \frac{1}{V_r \rho_{ro}}$$

$$f_7 = V_r f_{.1} - \frac{V_r f_2 f_3}{r}$$

$$f_8 = M_{dow} + M_s - M_w$$

$$f_9 = M_{dow} - M_w - M_{ws}$$

$$f_{10} = Q_{rw} - M_{dow}(H_{ro} - H_{dow})$$

$$M_{dow}^2 = \frac{\rho_{dw}L_{do}g - \rho_{ro}L_{ro}g}{\left[\dfrac{k_3+1}{2A_{ro}^2\rho_{ro}} + \dfrac{k_4}{\rho_{dw}} + \dfrac{k_5}{\rho_{ro}} + \dfrac{k_1+1}{2A_{do}^2\rho_{dw}} + C\left(\dfrac{X_o\rho_{dw}}{\rho_{ds}} + 1\right)\right]} \tag{19}$$

L_{do}	=	average height of downcomers, [m].
g	=	acceleration of gravity, [m/s²].
L_{ro}	=	total length of all downcomers, [m].
A_{ro}	=	total cross-section area in the riser, [m²].
A_{do}	=	total cross-section area in downcomers, [m²].
C	=	loss factor
$k_{1,4,5}$	=	loss factors

Superheater system

The primary superheater

The energy balance of the metal and the steam becomes:

$$\frac{d}{dt}T_{psm} = \frac{Q_2 + M_s(H_{psi} - H_{pso})}{c_{psm}m_{psm}} \tag{20}$$

m_{psm}	=	mass of metal of the primary superheater, [kg].
c_{psm}	=	specific heat of metal of the primary superheater, [kcal/kg°C].
T_{psm}	=	temperature of metal of primary superheater, [°C].
V_{ps}	=	volume of the steam in the primary superheater, [m³].
ρ_{ps}	=	density of the steam leaving the primary superheater, [kg/m³].
H_{pso}	=	enthalpy of the steam leaving the primary superheater, [kcal/kg].
H_{psi}	=	enthalpy of the steam entering the primary superheater, [kcal/kg].
Q_2	=	heat flow to the metal of the primary superheater, [kcal/s].

$$Q_2 = \begin{cases} k_2M_{oil} & \text{when } M_{oil} > 0.2 \\ 0 & \text{when } M_{oil} < 0.2 \end{cases} \tag{21}$$

The energy balance of the steam in the primary superheater can be described as:

$$Q_{ps} = M_s c_{ps}(T_{pso} - T_{psi}) \tag{22}$$

Q_{ps} = heat flow transferred by convection to the steam in the superheater, [kcal/s]

c_{ps} = specific heat of the steam in the primary superheater, [kcal/kg°C].

T_{pso} = temperature of the steam leaving the primary superheater.

T_{psi} = temperature of the saturated steam, [°C].

The specific heat is given as:

$$c_{ps} = f\left(P_d, \frac{T_{psm} + T_{psi}}{2}\right) \tag{23}$$

where the function $f(\ \cdot\ )$ describes the steam table over the range of operation for the boiler.

Tysso (1980) shows that the temperature of the steam leaving the primary superheater is given by solving a second order equation:

$$aT_{pso}^2 + bT_{pso} + c = 0 \tag{24}$$

where

$$a = (M_s c_{ps})^2$$

$$b = (M_s^{0.8} K_{ps})^2 (T_{psm} - T_{psi}) - 2aT_{psi}$$

$$c = (M_s^{0.8} K_{ps})^2 ((T_{psi}T_{psm} - T_{psm}^2) + aT_{psi}^2)$$

The positive square root of the second order equation is used.

The attemperator

The attemperator is located between the primary and the secondary superheaters. The steam is cooled by passing through the drum. The cooling is determined by the position of the control valve. The energy balance of the steam in the attemperators becomes:

$$Q_{aw} = M_a c_a (T_{pso} - T_{ao}) \qquad (25)$$

Q_{aw} = heat flow to the water in the drum, [kacl/s].
T_{ao} = temperature of the steam leaving the attemperator, [$^\circ$C].
T_{pso} = temperature of steam leaving primary superheater, [$^\circ$C].
c_a = specific heat of steam in the attemperator, [kcal/kg$^\circ$C].
M_a = steam flow through the attemperator, [kg/s].

The steam flow through the attemperator is determined by the control valve position:

$$M_a = M_s k_a Y_{valve} \qquad (26)$$

k_a = flow coefficient
Y_{valve} = linear control variable (Y_{valve}=0 when the valve is closed and Y_{valve}=1 when the valve is fully opened).

The drum water temperature is given by:

$$T_{dw} = \frac{H_{dw} + 7.42}{1.063} \qquad (27)$$

In the same way as the temperature out of the primary superheater is given by (24) the steam temperature after the attemperator is given by a second order equation:

$$aT_{ao}^{2} + bT_{ao} + c = 0 \qquad (28)$$

where

$$a = (M_a c_a)^2$$

$$b = (M_a^{0.8} K_{as})^2 (T_{dw} - T_{pso}) - 2aT_{pso}$$

$$c = (M_a^{0.8} K_{as})^2 ((T_{pso} T_{dw} - T_{dw}^2) + aT_{pso}^2)$$

The negative square root of the second order equation is used.

The pressure drop in the shunt valve becomes:

$$P_{pso} - P_{ssi} = k_{sv} M_{sv}^2 \tag{29}$$

$$M_{sv} = M_s - M_a \tag{30}$$

P_{ssi} = pressure of the steam entering second superheater, [bar].
k_{sv} = pressure drop constant of the shunt valve.
M_{sv} = steam flow through the shunt valve M_{sv}, [kg/s].

The secondary superheater

The steam flows from the attemperator and primary superheater are mixed and the temperature of the steam entering the second superheater will be:

$$T_{ssi} = \frac{M_a T_{ao} + M_{sv} T_{pso}}{M_s} \tag{31}$$

The temperature of the second superheater metal is described as follow:

$$\frac{d}{dt} T_{ssm} = \frac{Q_3 + M_s (H_{ssi} - H_{sso})}{c_{ssm} m_{ssm}} \tag{32}$$

Q_3 = heat flow to the metal of the secondary superheater, [kcal/s].
c_{ssm} = specific heat of the metal of the secondary superheater, [kcal/kg°C].
m_{ssm} = mass of metal of secondary superheater, [kg].

H_{ssi} = enthalpy of steam entering the secondary superheater, [kcal/kg].
H_{sso} = enthalpy of steam leaving the secondary superheater, [kcal/kg].

$$Q_3 = \begin{cases} k_3 M_{oil} & \text{when } M_{oil} > 0.2 \\ 0 & \text{when } M_{oil} < 0.2 \end{cases} \tag{33}$$

The energy balance of the steam in the secondary superheater becomes:

$$Q_{ss} = M_s c_{ss}(T_{sso} - T_{ssi}) \tag{34}$$

T_{sso} = temperature of steam leaving the secondary superheater, [°C].
c_{ss} = specific heat of the steam in the secondary superheater, [kcal/kg°C].

The specific heat is given as:

$$c_{ss} = f\left(P_{ssi}, \frac{T_{ssm} + T_{ssi}}{2}\right) \tag{35}$$

where the function $f(\cdot)$ describes the steam table over the range of operation for the boiler. Again it is shown in Tysso, (1980) that the temperature of the steam leaving the secondary superheater is given by solving a second order equation:

$$aT_{sso}^2 + bT_{sso} + c = 0 \tag{36}$$

where

$$a = (M_s c_{ss})^2$$

$$b = (M_s^{0.8} K_{ss})^2 (T_{ssm} - T_{ssi}) - 2aT_{ssi}$$

$$c = (M_s^{0.8} K_{ss})^2 ((T_{ssi}T_{ssm} - T_{psm}^2) + aT_{ssi}^2)$$

The positive square root of the second order equation is used.

K_{ss} =   heat transfer coefficient of secondary superheater, [kcal/kg°C]

The pressure drop in the secondary superheater becomes:

$$P_{ssi} - P_{sso} = k_{ss} M_s^{\ 2} \tag{37}$$

k_{ss} =   pressure drop constant of the secondary superheater
P_{sso} =   pressure of the steam leaving the secondary superheater, [bar]

Simulated step response of the nonlinear model

The dynamic response of the nonlinear model when the load (steam flow, M_s) is reduced from 100% to 90% can be studied in Figure 2. This is an open loop simulation and the control variables are set to nominal values at 100% load. The load changes after 5 sec and state variables are plotted. Of special interest is the non-minimum phase behavior of the state $x_1 = V_{dw}$ (subplot c). This inverse response is caused by the "swelling effect" in drum, downcomers and risers.

After the step change in steam flow the state changes as ramp functions. This is caused by the two eigenvalues at the origin. Table 1 shows the steady state values used during the open loop simulation with around 100% steam flow.

Table 1 Steady state values at 100% steam flow.

x_1, V_{dw} (kcal/kg)	8.19
x_2, H_{dw} (kcal/kg)	306.48
x_3, ρ_{ds} (kg/m³)	37.87
x_4, T_{trm} (°C)	314.77
x_5, H_{ro} (kcal/kg)	320.31
x_6, T_{psm} (°C)	458.60
x_7, T_{ssm} (°C)	535.29
u_1, M_w (kg/s)	29.19
u_2, M_{oil} (kg/s)	1.90
u_3, Y_{valve}	0.48
v_1, M_s (kg/s)	29.19
v_2, T_{ewo} (°C)	230.00

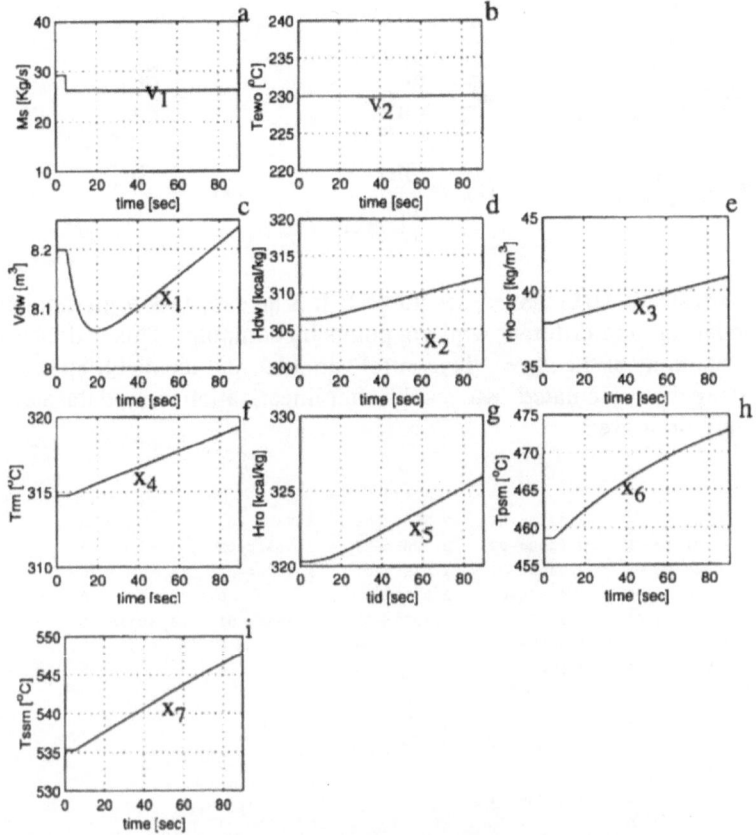

Figure 2 The nonlinear model response to a step change in steam flow rate
from 100% to 90%.

4. A LINEARIZED MODEL

Assuming small perturbations around a steady state operating point the nonlinear model
can be linearized. The linearized model my be written in matrix form as:

$$\dot{x}(t) = Ax(t) + Bu(t) + Cv(t)$$
$$y(t) = Hx(t) + H_Bu(t) + H_Cv(t)$$

(38)

764

where

$$A= \frac{\partial f}{\partial x}\bigg|_{x^0, u^0, v^0} \qquad B = \frac{\partial f}{\partial u}\bigg|_{x^0, u^0, v^0} \qquad C = \frac{\partial f}{\partial v}\bigg|_{x^0, u^0, v^0}$$

$$H = \frac{\partial h}{\partial x}\bigg|_{x^0, u^0, v^0} \qquad H_B = \frac{\partial h}{\partial u}\bigg|_{x^0, u^0, v^0} \qquad H_C = \frac{\partial h}{\partial v}\bigg|_{x^0, u^0, v^0}$$

(39)

x^0, u^0 and v^0 are steady state values, see Table 1. It is an easy task to find quite accurate linearized models around different working points by computer. This is done by generating small pertubations of the states, the control variables and the disturbances and the values of (39) are then calculated. The results after linearization around the steady state at 100% load conditions are:

A =

-5.3938e-07	-6.2219e-04	6.9331e-02	7.7565e-03	-3.7811e-02	0	0
3.4591e-05	-1.0174e-01	2.2350e-02	2.7983e-03	-1.1491e-02	2.8729e-04	0
1.6292e-06	4.6872e-02	-2.5409e-01	1.0395e-01	-9.9370e-04	0	0
0	0	4.4583e-01	-2.8378e-01	0	0	0
9.9661e-07	1.0735e-01	-2.8077e-01	2.7979e-01	-1.7538e-01	-1.3838e-02	0
0	0	1.2488e-02	0	0	8.9742e-03	0
0	1.5637e-03	-3.1141e-03	0	0	0	-1.1688e-02

B =

8.6857e-04	0	0
-1.1670e-02	0	1.0135e-01
1.4774e-03	0	0
0	1.3427e+00	0
9.0371e-04	0	0
0	1.5062e+00	0
0	8.3378e-01	-4.2148e-01

C =

9.5137e-03	0
4.9800e-03	5.1386e-03
-2.8736e-02	0
0	0
-1.7578e-02	0
-9.0748e-02	0
-5.1356e-02	0

H =

1.3333e-01	0	0	0	0	0	0
0	0	1.7254e+00	0	0	0	0
0	4.5382e-02	-1.7614e-01	0	0	2.6046e-01	6.7577e-01
0	0	1.7254e+00	0	0	0	0

H_B =

0	0	0
0	0	0
0	0	-1.2357e+01
0	0	2.4908e+00

H_C =

0	0
0	0
-3.6233e-01	0
-5.1165e-01	0

Similar matrices have been found around steady state values at 55% and 30% outlet steam flow, M_s.

The eigenvalues of the linearized system at 100% load are:

$\lambda_1 =$ -1.1688e-02   , related to the superheater system.
$\lambda_2 =$ -4.9661e-01   , related to the risers.
$\lambda_3 =$ -1.7536e-01   , related to the drum system.
$\lambda_4 =$ -1.4337e-01   , related to the risers.
$\lambda_5 =$ -1.3888e-02   , related to the superheater system.
$\lambda_6 =$ 4.0322e-04   , related to the steam pressure.
$\lambda_7 =$ -3.0839e-06   , related to the drum water volume.

There are two eigenvalues close to the origin. One of these is associated with the drum water volume. The exact eigenvalue of this state is placed at the origin. But because of inaccuracy in the linearization procedure it ends up in the left half plane, close to the origin. The second mode is unstable and this is caused by the fact that the steam outlet flow is not modeled as a function of the pressure in the drum, Tysso (1980).

Simulated step response of the linearized model

The dynamic response of the linear model when the load (steam flow, M_s) is reduced from 100% to 90% can be studied in Figure 3. This is an open loop simulation and the control variables are set to nominal values at 100% load. The load change after 5 sec and state variables are plotted. Comparing the responses of the nonlinear and the linear model shows that a linear model is very accurate around the linearization point.

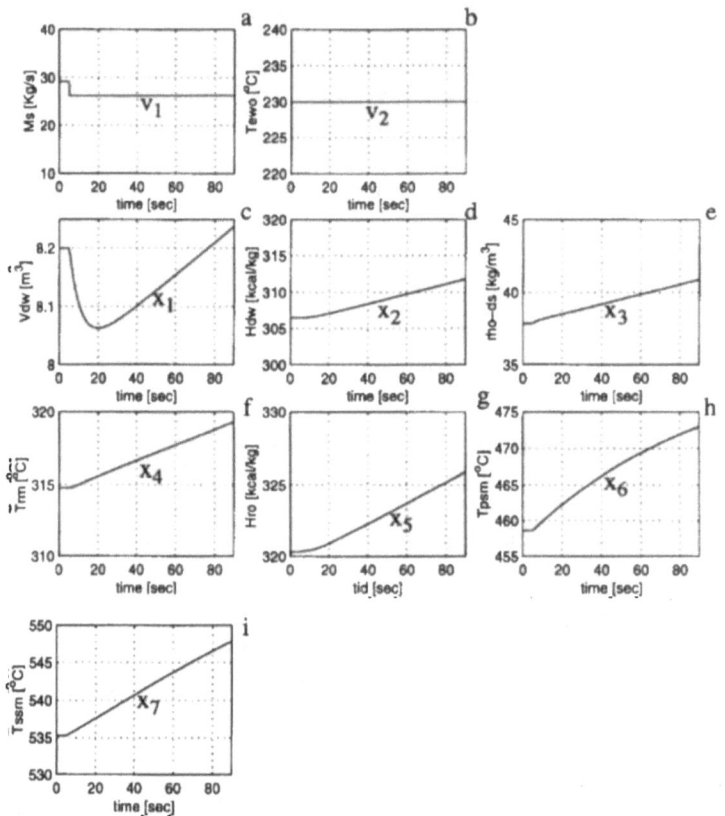

Figure 3 The linearized model response to a step change in steam flow rate from 100% to 90%.

5. ELEMENTARY NONLINEAR DECOUPLING (END) APPLIED TO CONTROL OF A STEAM BOILER

The END-algorithm is described in Balchen (1998)

There are an infinite number of property transformations that may be used for realization of the dynamic equation solver in the END control strategy. The following conditions must be satisfied:

- **DB** has to be a nonsingular matrix.
- The condition number of **DB** must be small.
- The inner system must be stable, $Re\{\lambda_{(I-B(DB)^{-1}D)A}\} < 0$.
- Δ**D** must be reasonably small.

To satisfy these four conditions the following optimization criteria may be used:

$$\mathbf{J} = \int_0^{t_1} (\Delta \mathbf{z}^T \mathbf{Q} \Delta \mathbf{z} + \alpha \Delta \mathbf{u}^T \mathbf{P} \Delta \mathbf{u}) d\tau + \beta_1 \kappa_{\mathbf{DB}} + \beta_2 \left(\sum_{i=1}^{r} \sum_{j=1}^{n} (\Delta d_{ij})^2 \right)^{1/2} \tag{40}$$

where r is the dimension of the property space $\mathbf{z}$ and n is the dimension of the statespace $\mathbf{x}$. The first part of the cost function (integral) secures stability of the innner system. The second part secures a small condition number of **DB** and the third part (Frobenius-norm) secures a small Δ**D**. The size of κ and Δ**D** calculated on $\mathbf{D}_{opt}$ are strongly dependent on the weighting factors β_1, β_2. The use of the condition number makes it possible to influence the use of the control actions more accurately compared to a criterion that only make use of the weighting factor α in the integral.

The optimization starts by generating a disturbance so that the integral can be calculated. The calculation of the integral, the condition number and the Frobenius-norm is done for each new Δ**D**. The same disturbance is used each time a new and hopefully better Δ**D** is tested. A gradient search has been used and the resulting Δ**D** matrix has proven to fullfil the four claims related to Δ**D**.

When designing a **D**-matrix of an unstable or an integrating process as the steam boiler, it is necessary to use a stabilizing feedback controller from the property vector $\mathbf{z}$. The size of **G** must be compared with the size of the eigenvalues of the dynamic equation solver. A good start is to choose a small $\mathbf{G}(\mathbf{G} = 0.1\Lambda)$. After optimization **G** can be increased to increase the closed loop system bandwidth. The bandwidth of the feedback loop is strongly related to the condition number of **DB**. A small condition number makes it possible to chose **G** close to Λ. The condition number can be forced to unity if β_1 is increased or β_2 is decreased.

A large number of testruns have been made to evaluate the optimization function. These tests have shown that optimal $(\mathbf{D}_{opt})$ is not dependent upon the start matrix $\Delta\mathbf{D}_{start}$, but the results are dependent upon the weighting parameters in the optimization function and the test signal. The signal used for optimization results in a $\Delta\mathbf{D}_{opt}$ that will satisfy perfect control of the property variables even if the disturbance is changed dramatically. Tests have shown that there is a conflict between the claim for a small Frobenius norm (a small $\Delta\mathbf{D}$) and a small $\kappa_{(DB)}$. The conflict is visualized in Figure 4, where β_2 is reduced from 1 to 0.1 to 0.01 resulting in increased values of the elements in $\Delta\mathbf{D}$. As illustrated in

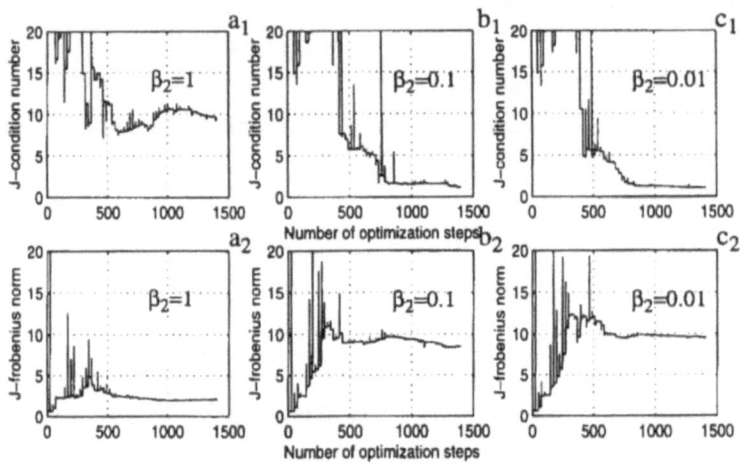

Figure 4 Scaled condition number and Frobenius-norm calculated for each optimization step as a function of β_2. ($\beta = 0.1$ and $\alpha = 0.05$)

Figures $4b_1, 4c_1, 4b_2$ and $4c_2$ the Frobenius-norm and the condition number are less dependent upon β_2 when $\beta_2 < 0.1$. This is so because the weighted Frobenius-norm will be suppressed by the integral when the total cost function is calculated.

The transients in Figure 4 are caused by the gradient search function, but will not influence the final result. Figure 4 also shows that:

The Frobenius-norm can be used to secure that the most dominant elements of $\mathbf{D}_{opt}$ are the original elements in $\mathbf{D}^0$ and therefore $\Delta z = z^0\text{-}z$ will be small.

But as explained above the claim of a small $\Delta\mathbf{D}_{opt}$ will in the case of the steam boiler result in a large condition number of $\mathbf{DB}$ and the use of the control variables will increase and saturation my occur even for small changes in the steam flow. If a larger $\Delta\mathbf{D}_{opt}$ can be used, the problem with control variable saturation will be reduced, but the cost is a larger $\Delta z=z^0\text{-}z$ and therefore the need for control of z^0 will increase. This conflict is further discussed in a following section.

769

An optimization result:

The actual property transformation is given as:

$$D^0 = \begin{bmatrix} 0.13333 & 0 & 0 & 0 & 0 & 0 & 0 \\ 0 & 0 & 1.7254 & 0 & 0 & 0 & 0 \\ 0 & 0.0454 & -0.1761 & 0 & 0 & 0.2605 & 0.6758 \end{bmatrix} \tag{41}$$

It is impossible to make use of the actual property transformation D^0 in an END-algorithm applied to the steam boiler, because the matrix $D^0 B$ is singular. Therefore a modified $D = D^0 + \Delta D$ that satisfies the claims of D must be found. This is done by using the the optimization criteria in (40). A start matrix that fullfils stability of the inner system and a reasonable condition number is found using a simple search procedure. $D_{start} = D^0 + \Delta D_{start}$ is given as:

$$D_{start} = \begin{bmatrix} 0.13333 & 0 & 0 & 0 & 0 & -0.001 & 0.001 \\ 0 & 0 & 1.7254 & 0.0001 & 0 & 0 & 0 \\ 0 & 0.0454 & -0.1761 & 0 & 0 & 0.2605 & 0.6758 \end{bmatrix} \tag{42}$$

After optimization ($\alpha=0.05$, $\beta_1=0.1$, $\beta_2=100$) the resulting D_{opt} is:

```
1.3333e-01  -2.8039e-06   2.0782e-04  -4.7020e-05   1.4150e-05  -1.9736e-05 2.0543e-05
-2.1283e-01   2.3516e-03   1.7254e+00   1.0135e-02  -3.6455e-03   4.9383e-03 -5.5870e03
1.4908e+00   4.5382e-02  -1.7614e-01  -8.0856e-02   3.4477e-02   2.6046e-01 6.7577e-01
```

$$\tag{43}$$

The boldface elements in the optimal D_{opt} were kept constant during the optimization. These elements are the desired property transformation chosen for the process. By scaling the optimal D it can be shown that the elements in ΔD are small.

After optimization the condition number DB increased slightly and the Frobenius- norm was reduced. This is a typical result when β_2 is large. It is usually easier to find a large ΔD_{start} that satisfies the claim of a stable internal system, then finding a small ΔD_{start}.

6. SIMULATION OF THE END-ALGORITHM APPLIED TO THE STEAMBOILER.

A number of test simulations have been made to evaluate the properties of the END-algorithm applied to the steam boiler. The desired property transformation is given as:

$$z^0(t) = D^0 x(t) \qquad (44)$$

z^0_1: h the level of liquid in the drum.
z^0_2: P_{sso} the steam pressure leaving the secondary superheater.
z^0_3: T_{sso} the steam temperature leaving the secondary superheater.

Responses of different D_{opt}

Different property transformations D_{opt} have been found using different weighting of the condition number and the Frobenius-norm. These optimizations were done using small changes in the steam flow M_s that prevented the control variables from saturating. Figures 5 and 6 show two simulations using the END-algorithm to control the steam boiler. In each case the load changes gradually from 100% to 105% starting at t= 0.5 sec. A soft step occurs at t=10 sec, see Figure 5d-e. The property transformation $D_{opt}=D^0+\Delta D_{opt}$ used in figure 5, was found using the optimization criteria (40) with the weighting parameters $\alpha = 0.05$, $\beta_1 = 0.1$ and $\beta_2 = 100$. The resulting D_{opt} is found to be::

```
1.3333e-01  -2.8039e-06   2.0782e-04  -4.7020e-05   1.4150e-05  -1.9736e-05 2.0543e-05
-2.1283e-01   2.3516e-03   1.7254e+00   1.0135e-02  -3.6455e-03   4.9383e-03 -5.5870e-03
 1.4908e+00   4.5382e-02  -1.7614e-01  -8.0856e-02   3.4477e-02   2.6046e-01 6.7577e-01
```

$$(45)$$

By scaling D_{opt} it can be shown that the matrix ΔD_{opt} is small compared to D^0. The scaled condition number of **DB** is as large as $\kappa_{DB} = 53$.

The property transformation $D_{opt}=D^0+\Delta D_{opt}$ used in Figure 6, was found using weighting parameters $\alpha = 0.05$, $\beta_1 = 0.1$, $\beta_2 = 0.01$. The resulting D_{opt} is found to be:

```
 1.3333e-01   1.3778e-02   5.4840e-02  -6.6503e-03  -2.5849e-02   3.1991e-03   4.8688e-03
 6.0060e-01   2.0172e-01   1.7254e+00   2.1624e-01  -1.3764e-01  -1.7508e-01  -1.9422e-02
-2.6891e+00   4.5382e-02  -1.7614e-01  -6.4780e-01   7.0562e-01   2.6046e-01   6.7577e-01
```

$$(46)$$

In this case ΔD_{opt} is large compared to D^0, but the scaled condition number of **DB** is reduced to $\kappa_{DB} = 10$.

When comparing the two simulations the following notation will be used. ΔD^L (L: large ΔD) represent the optimal ΔD found when $\beta_2 = 0.01$ and ΔD^S (S: small ΔD) represent ΔD_{opt} when $\beta_2 = 100$.

Changes of the control variables:

Using ΔD^S results in aggressive changes in the control variables despite of small changes in the disturbance, v_1, see Figure 5a-c. The reason is the large condition number of **DB** ($\kappa_{DB} = 53$).

The same figure shows that the controller does not detect the non-minimum phase behavior of the process and the control variables move in the wrong direction when a change in the steam flow rate M_s occurs. based upon the objective functional J_0

Using ΔD^L results in less changes in the control variables, see Figure 6a-c. The variables also react correctly despite the non-minimum phase behavior. This correct effect is caused by the larger elements in column four and five of ΔD^L, compared with the use of ΔD^S. These elements are actually a strong feedforward controller from the riser metal temperature T_{rm} and the enthalpy of the steam-water mixture in the risers H_{ro}.

Changes of the desired property vector z^0:

When ΔD^S is used, the desired property vector is close to constant, see Figure 5f- h. The stationary error is less than 1% and if necessary this error can be removed by a feedback controller from z^0.

The desired property vector changes more if ΔD^L is used to form the property transformation D_{opt}. The property vector z is close to perfect at 100% load conditions. When the load change occurs z^0 will no longer be constant. The reason is that a large ΔD_{opt} will result in a large Δz. Since the END-algorithm only secures perfect control of the designed property space z the desired property vector z^0 will no longer be kept constant. The stationary error of z^0 can easily be forced to zero by use of a feedback control from z^0 (Balchen 1998).

The designed property vector z is of minor interest, because the END-algorithm will always satisfy perfect control of z if the eigenvalues of the dynamic-equation solver are large and if there is no long lasting saturating control variables. z is plotted in Figures 5 and 6 subplot i-k. The state variables are plotted both figures, see subplot l-r.

The two optimal property transformations discussed above represent two extreme cases found when β_2 is chosen equal to 100 and 0.01. A better ΔD_{opt} with respect to reasonable variations in the control variables and control of the desired property space, should exist between these two extreme values. .

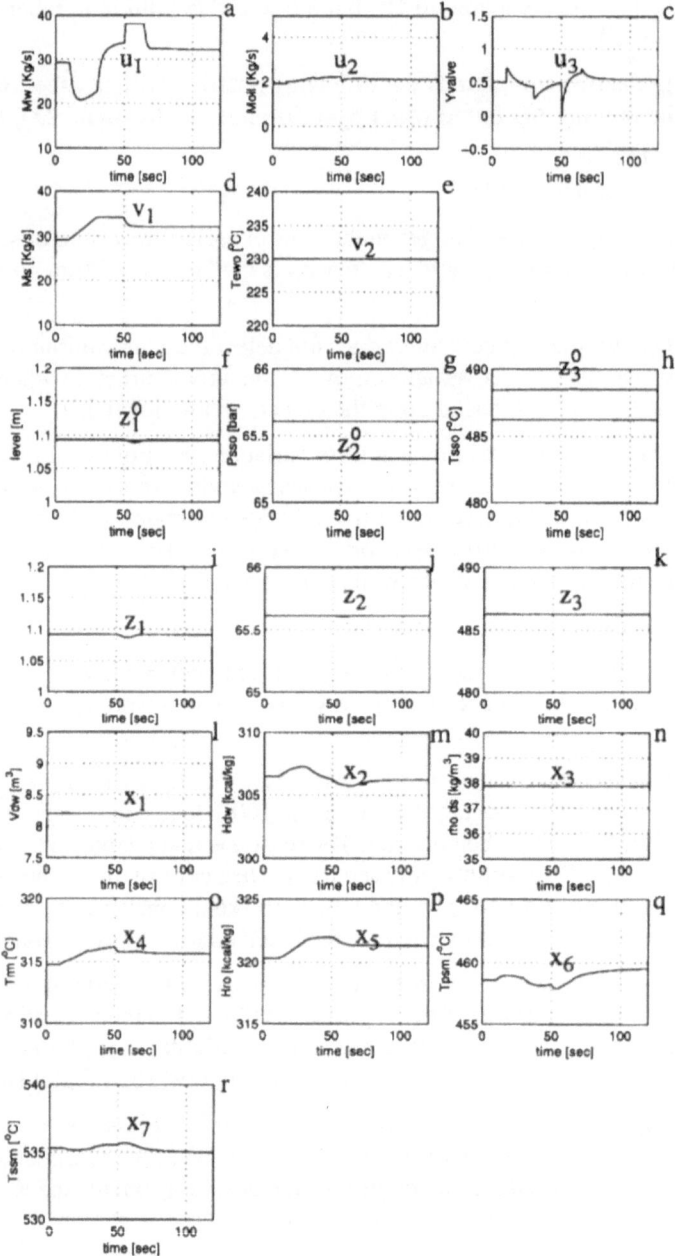

Figure 5 Responses of states and control variables when $\mathbf{D}_{opt}$ ($\Delta\mathbf{D}^S$) were found by optimization using the weighting factors $\alpha = 0.05$, $\beta_1 = 0.1$, $\beta_2 = 100$.

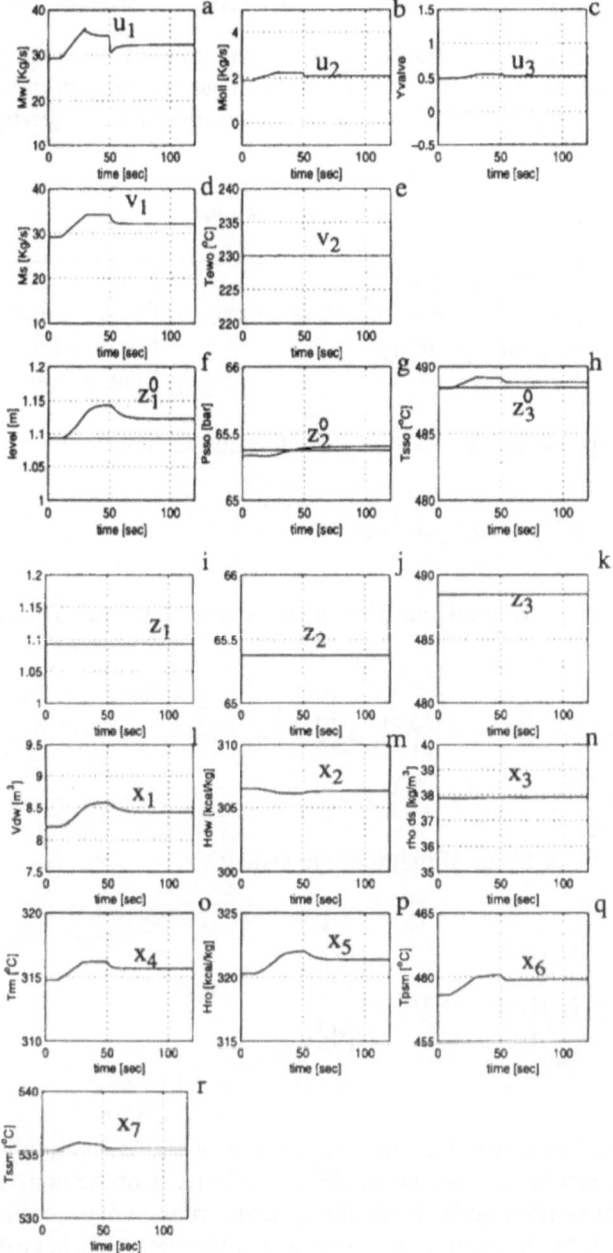

Figure 6 Responses of states and control variables when $\mathbf{D}_{opt}$ ($\Delta\mathbf{D}^L$) were found by optimization using the weighting factors $\alpha = 0.05$, $\beta_1 = 0.1$, $\beta_2 = 0.01$.

Responses of the END-algorithm with saturating controls (SEND)

When one or more of the control variables u_i reach a saturation level, the control system loses one or more degrees of freedom in control. This means that one or more of the property variables must be sacrificed (Balchen 1998). The priority of the properties chosen for the steam boiler is as follows:

Number of saturated control variables:	Saturated control variables u_i	Sacrificed properties
1	u_1 or u_2 or u_3	$z_3 = T_{sso}$
2	u_1 and u_2 or u_1 and u_3 or u_2 and u_3	$z_2 = P_{sso}$ and $z_3 = T_{sso}$
3	u_1 and u_2 and u_3	None of the properties can be controlled

For each of the seven possible combinations of saturation the control matrices are:

$$\mathbf{B}, \mathbf{B}^1, \mathbf{B}^2, \mathbf{B}^3, \mathbf{B}^{12}, \mathbf{B}^{13}, \mathbf{B}^{23}, \mathbf{B}^{123} \tag{47}$$

where $\mathbf{B}$ represents the unsaturated and $\mathbf{B}^{123}$ the fully saturated case. The resulting $\mathbf{D}$ matrices are:

$$\mathbf{D}, \mathbf{D}^1, \mathbf{D}^2, \mathbf{D}^3, \mathbf{D}^{12}, \mathbf{D}^{13}, \mathbf{D}^{23}, \mathbf{D}^{123} \tag{48}$$

As an example:

The following two pairs of $\mathbf{D}$ and $\mathbf{B}$ matrices are tested:

$$\mathbf{B} \rightarrow \mathbf{D} = \begin{bmatrix} \mathbf{d}_1^T \\ \mathbf{d}_2^T \\ \mathbf{d}_3^T \end{bmatrix}, \quad \mathbf{B}^1 \rightarrow \mathbf{D}^1 = \begin{bmatrix} \mathbf{d}_1^T \\ \mathbf{d}_2^T \end{bmatrix} \tag{49}$$

$\mathbf{B}$ represents the control matrix of the linearized system at 100% steam flow and $\mathbf{D}$ is described in (46). As can be seen the matrix $\mathbf{D}^1$ is a submatrix of the non-saturated designed property transformation $\mathbf{D}$. While the system is running it is continuously being tested to determine what combination of saturating control variables that exists. If saturation of one or more of the control variables is detected the END-algorithm with saturating control variables (SEND) is used to control the reduced property space variables.

The simulation in Figure 7 shows the system response to a change in M_s that result in saturating control variables. The upper saturation limit of u_1 has been reduced compared with the actual saturation limit of the boiler, in order that the features of the SEND-algorithm can be illustrated properly for a long lasting saturation. From t=0 to t = 15 the system operates in state $\mathbf{B} \rightarrow \mathbf{D}$ as none of the control variables are saturated. At t=15, u_1 reaches the upper saturation limit and the control system loses one degree of freedom and will now operate in state $\mathbf{B}^1 \rightarrow \mathbf{D}^1$. The property z_3 is sacrificed. This can be seen in subplot k as z_3 changes. The controlled properties z_1 and z_2 are constant. At t = 40, u_1 is no longer saturated and the system returns to the none saturated state $\mathbf{B} \rightarrow \mathbf{D}$. The result is perfect control of the three properties in z. The responses in statevariables are plotted in subplot l-r.

A simulation of the system responses using an END-algorithm (without considering input saturation) is plotted in figure 8. The disturbance is the same as in Figure 7. Saturation of u_1 occurs at t=15 and the response of the designed property vector z is shown in subplot i-k. Comparing the SEND- and END algorithm clearly indicates that the END-algorithm will not satisfy perfect control of any properties in z with saturating control variables. In the case of the SEND-algorithm the number of perfectly controlled properties is only reduced by the number of saturated control variables.

$\mathbf{D}^1$ was found as a submatrix of the non-saturated designed property matrix $\mathbf{D}$. Another approach is to use the optimization criterion to design the property transformations for each of the seven possible combinations of saturating control variables. A large number of tests using the steam boiler as an example, has shown that this is possible if the weighting factor β_1 is large and β_2 is low, claiming a small condition number of $\mathbf{DB}$ and a larger $\Delta \mathbf{D}$. Choosing a large β_2 will result in unstable modes of the inner system and cause instability of the SEND-algorithm. This is particularly apparent in non-minimum phase processes.

Responses of the END-algorithm using an integrating feedback controller from z^0.

In the case of the steam boiler which is a non-minimum phase process, it has been necessary to accept a relatively large $\Delta \mathbf{D}$ to meet the important claims:

- The condition number of $\kappa_{\mathbf{DB}}$ must be small.
- The inner system must be stable, $Re\{\lambda_{(I - B(DB)^{-1}D)A}\} < 0$.

for the seven combinations of saturating control variables. The penalty for this is an increased $\Delta z = z - z^0$. To suppress this error an integrating feedback controller from the desired property vector z^0 could be used (Balchen 1998). Figure 9 shows the responses of the END-algorithm using a feedback control from z^0. The designed property transformation $\mathbf{D}$ and changes in v_1 are the same as in the simulation plotted in Figure 6. Comparing these to responses clearly indicates that the integrating feedback control not only removes the stationary error but the transients of z^0 are reduced as well.

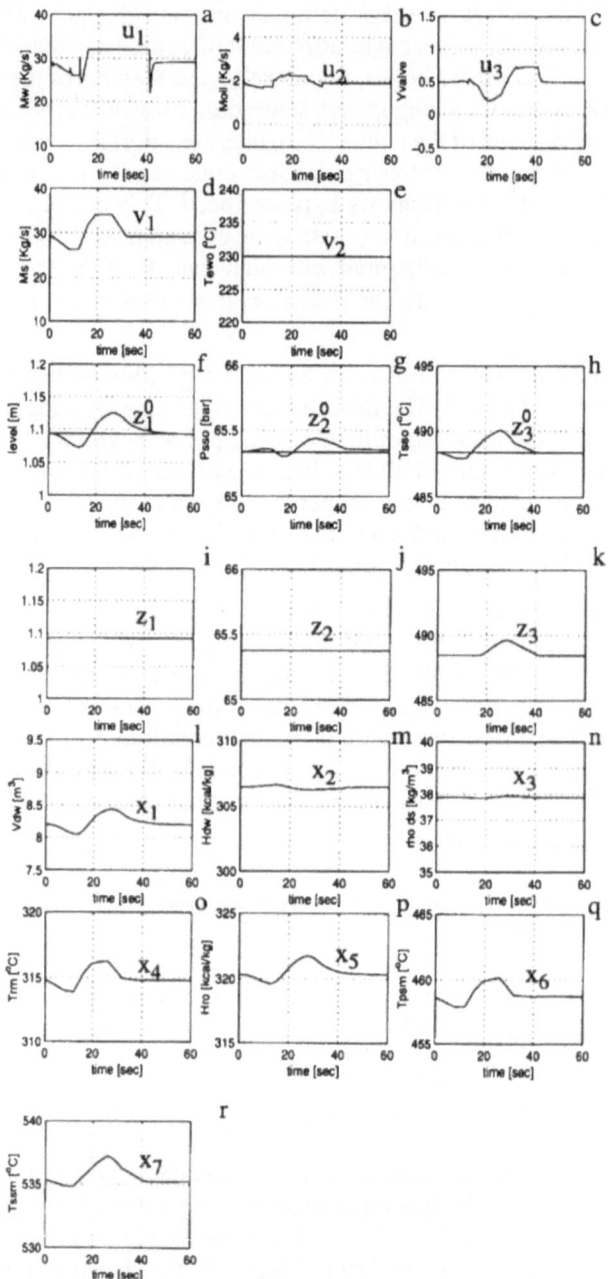

Figure 7 Responses of the SEND-algorithm to changes in M_s

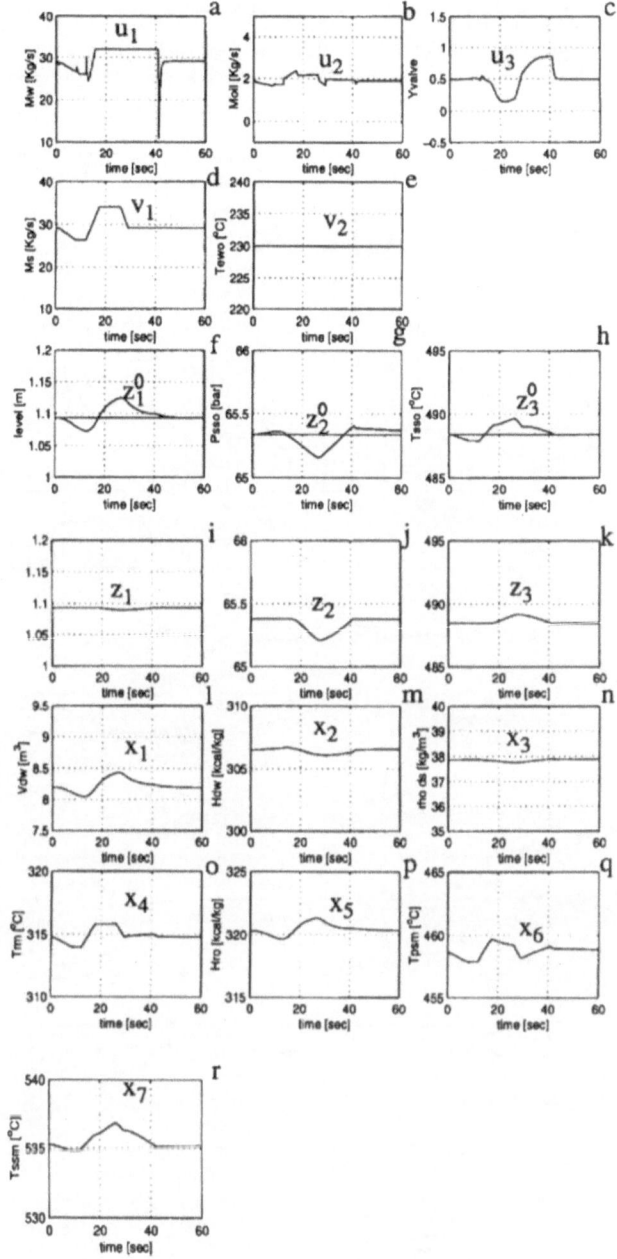

Figure 8 Responses of the END-algorithm to changes in M_s

778

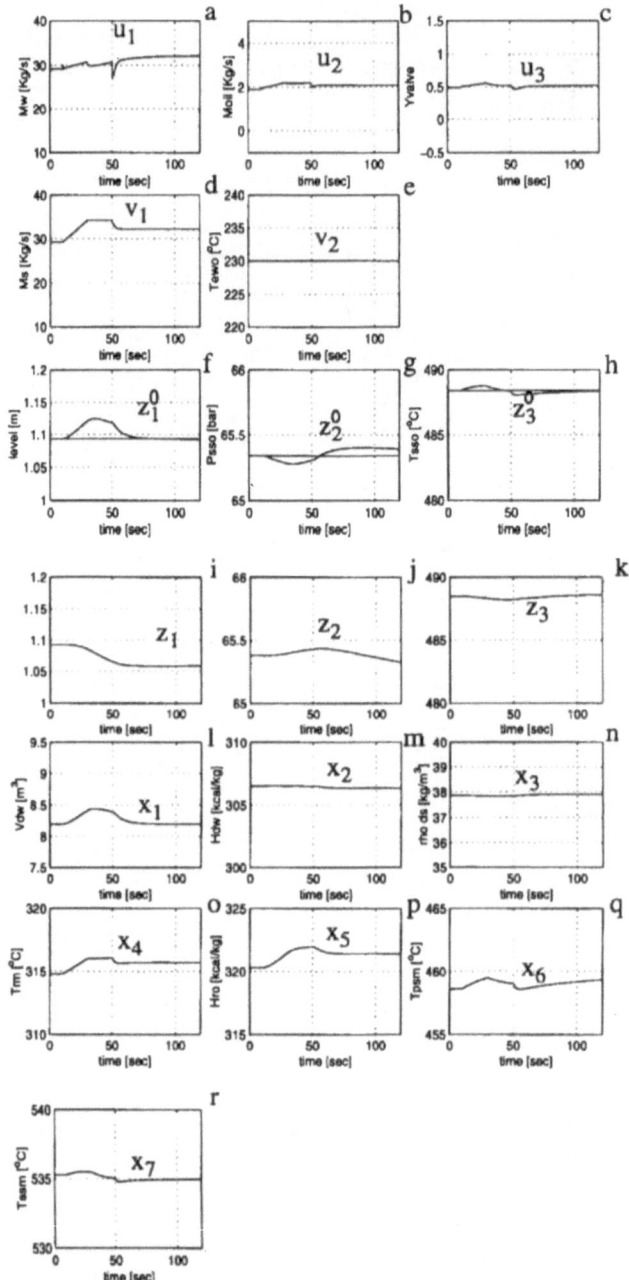

Figure 9 Responses of the END-algorithm with an integrating feedback from z^0

7. OPTIMIZATION OF ΔD USING MATLAB 5.0

The optimization of ΔD is done off-line. The results presented in the previous sections are calculated using MATLAB 5.0. The MATLAB function "fminu.m" is used to calculate new ΔD-matrices upon the results returned from the criterion function. The criterion function must include:

- A complete model of the END-algorithm for the process, with feedback from z
- A procedure for calculation of the scaled condition number of κ_{DB}.
- A procedure for calculation of the Frobenius-norm of the scaled ΔD.

These three parts are programmed as separate functions, which makes it easy to use the MATLAB code for different processes. The main program structure is shown in Figure 10

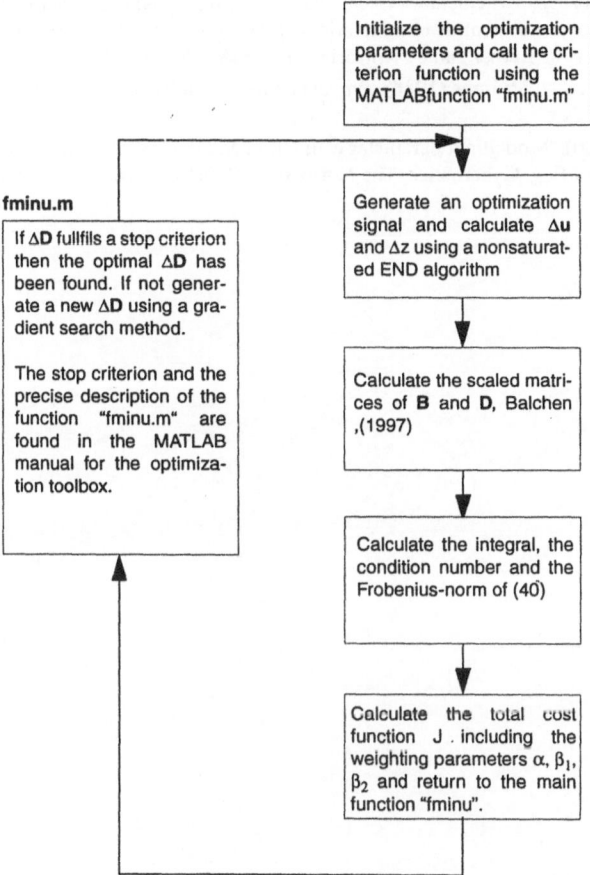

Figure 10 Flow sheet of the optimization routine implemented in MATLAB 5.0

Optimization of one **D**-matrix is time consuming and will last for 10-15 hours using a Pentium Pro 200 Mz. The most time consuming part is the calculation of Δz and Δu. The calculation time will decrease considerably if parts of the code are compiled for Borland C++.

8. ACKNOWLEDGMENT

This research has been partly supported by STATOIL

9. REFERENCES

[1]   Balchen, J.G. (1998). "Elementary Nonlinear Decoupling (END), a general approach to model based control of nonlinear multivariable prcocesses", *Nonlinear Model Based Process Control,* editors: R. Berber and C. Kravaris, Kluwer Academic Publishers, Dordrecht.

[2]   Balchen, J. G. , K. I. Mumme (1988). *Process control.* Van Nostrand Reinhold Company Inc, New York.

[3]   Tysso, A. (1980)."Modelling, parameter estimation and control of a ship boiler". *Dr. ing. Thesis*, Dept. of Eng. Cybernetics, The Norw. Inst. of Techn., Trondheim. Norway, Report ITK-80-13-W.

REDUCTION OF PVC BATCH TIME BY OPTIMAL CONTROL OF FREE RADICAL CONCENTRATION

A.J.DAMSLORA, S.SAELID
Norsk Hydro ASA
Corp. Research Centre
P.O. Box 2560, N-3901 Porsgrunn, Norway

B.LIE
Telemark College
Department of Technology
Section for Process Automation
N-3901 Porsgrunn, Norway

Abstract   In Norsk Hydro the cooling systems of many batch reactors producing PVC by emulsion polymerisation are not fully utilized. By manipulating the radical production rate it is possible to control the rate of polymerisation so that the heat generation matches the cooling capacity. The result is a significant reduction in reaction time. The desired radical production rates are found by minimizing a weighted sum of temperature variance and batch time, subject to a non-linear state space model and with bounds on the coolant temperature and average reactor temperature. The radical production rate was represented by a differentiable spline with a finite number of parameters. These parameters were used as decision variables in the optimisation problem. The framework may be used both for design and control purposes.

1. Aims and background

Polyvinyl Chloride (PVC) is an economically important industrial product. Norsk Hydro produces about 300 ktons/yr with batch polymerisation reactors. The highly exothermic polymerisation process is carried out in large vessels (autoclaves) where the heat removal represents a major limitation of the production rate. The production recipes are determined off-line and usually specifies a constant reference temperature. The cooling system must then be able to remove heat fast enough to maintain this temperature. An uncontrolled change in temperature may result in a runaway situation or cause quality deterioration. Late in the reaction, heat generation rises and may be several times larger than the average value due to the Trommsdorff effect. A common way to limit the maximum heat generation is to keep the polymerisation rate

R. Berber and C. Kravaris (eds.), Nonlinear Model Based Process Control, 781-803.
© 1998 *Kluwer Academic Publishers.*

782

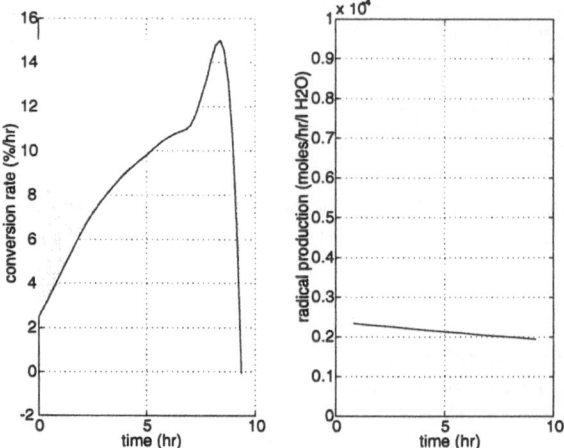

Figure 1: Experimental data from Norsk Hydro ASA. The left graph shows measured rate of polymerisation (which is proportional to the heat generation). The polymerisation was carried out at constant temperature. Assuming that the peak matches the heat removal capacity, the gray area illustrates how the capacity may be used more efficiently.
To the right the radical production profile is shown. (One single thermal decomposition initiator).

sufficiently low during the early stages of the process. But this approach leaves a substantial part of the heat transfer capacity unused for a significant part of the total batch time, see fig. 1.

What is desirable is to increase the reaction rate so that it matches the heat removal capacity during the course of the reaction. One way to obtain this is by nonisothermal operation but this strategy is complicated by the constraints imposed by quality considerations. In this work we consider a different approach. By using a sufficiently flexible initiatior system it is possible to obtain a reaction rate which is high throughout the reaction[1].

In this paper a nonlinear state space model of a Vinyl Chloride (VC) emulsion polymerisation reactor is used to calculate optimal *free radical production rates* in the polymerisation reaction. These optimal production profiles may be used for both control and design purposes.

From a design point of view the collection of optimal radical production rates provides a basis for choosing an initiation system with sufficient degrees of freedom. The variation in the profiles determines what initiator decomposition rates are necessary to realise the optimal radical production rates.

---

[1] This angle of attack was suggested, e.g. by Hamielec and MacGregor [8] .

In a control context a specific initator system is already chosen. Then it is not possible to realise all possible sets of production rates. In this case the subset of realisable radical production rates — the set of *admissible* controls — must be included as part of the optimisation problem solved by the controller.

A popular control strategy is model based predictive control [3] . This approach is by no means new. The concepts of "sliding horizon" and "shrinking horizon" control were introduced soon after Pontryagin *et al.* [16] presented their comprehensive theory of optimal control in the late 1950's. For computational reasons the early applications were LQ [2, 14] , *i.e.* with linear models, quadratic objective function. During the 70's and 80's methods were developed to handle more general constraints and model uncertainty [3, 32] , and during the 80's and 90's problems with nonlinear models have received increasing attention [10, 17, 1, 12, 29] .

According to Embiruçu *et al.* (1996) few works on predictive control of the PVC process have been reported in the open literature [5] . Some work has been done on minimum end-time initiator policies on other polymers, e.g. Ponnuswamy *et al.* (1986) on Methyl Methacrylate [25] , Hsu and Chen (1988) on styrene [13] , and Jang and Yang (1989) on Vinyl Acetate [11] . The latter system is kinetically closest to VC.

Jang and Yang parameterised the state variables by collocation methods. In this paper single shooting [19] is applied to solve the state space equations.

On a PVC system, Nagy and Agachi (1997) demonstrated that nonlinear predictive controllers are capable of following time variant temperature trajectories [21] . In addition to control the temperature of the PVC polymerisation, as Nagy and Agachi did, the controller in this paper reduces the batch time to a minimum.

2. Mathematical model

2.1 GENERAL SYSTEM DESCRIPTION

The emulsion polymerisation system studied in this work is free-radical polymerisation of VC. The physical picture given by Smith and Ewart (1948) serves to describe the system [28] . See fig. 2.

The major components are water (dispersant), monomer/polymer together with a number of additives, notably initiators and emulsifiers. The initiators, which are water soluble, decompose to produce free radical species.

The free emulsifier exceeding the critical micelle concentration limit

(CMC), tends to form micelles [~0.01μ] which in turn absorb free radicals and become polymer particles which start to grow. As the particles grow they adsorb emulsifier and eventually the emulsifier concentration drops below the CMC limit leaving no more micelles. This marks the end of the nucleation stage, or interval I, of the reaction.

After this moment the number of particles is approximately constant, but the radicals will continue to absorb radicals. When a radical is absorbed in a particle, further addition of monomer units to the radical takes place. This is the main reaction step and it is termed *propagation*. The propagation is highly exothermic and is responsible for virtually all heat generation in the polymerisation reaction.

There are essentially two fates for the free radical sitting at the end of the growing polymer chain: *i)* It may be transferred to another monomer unit and start to grow a new chain. *ii)* Another radical may be absorbed by the particle in which case the two radicals will quickly destroy each other by combination or disproportionation. These two events are referred to as *chain transfer* and *termination*, respectively. Short radicals (e.g. right after a chain transfer) may escape from the particles into the water phase. This is termed radical *desorption*.

From the end of the nucleation step and up to about 70 % conversion, the monomer exists in the form of droplets [$\lesssim 10\mu$], in micelles/particles and in the water phase. The monomer transfer between these three phases is fast enough that the monomer distribution is at thermodynamic equilibrium at all times. As long as the monomer exists as a separate phase (droplets), the water phase will be saturated with monomer and the monomer concentration in the particles will be constant. This regime is referred to as interval II of the polymerisation.

The period after the monomer droplets have been consumed is termed interval III. For this period the monomer concentration in the particles will decrease and the viscosity inside the particles will increase. The diffusivity of radicals in the particles falls by several orders of magnitude which gives rise to an accelleration due to excessive propagation with associated heat generation. The acceleration is finally quenched by the depletion of monomer. The final particle size typically ranges from $100 - 1000\mu$.

More detailed descriptions of the polymerisation system may be found in [7] and the references therein.

2.2   STATE SPACE MODEL

The model used in this work has been developed by Ugelstad and co-workers [9, 30, 31] and has proved to be valid for batch emulsion polymerisation of PVC with water soluble initiators. The differential

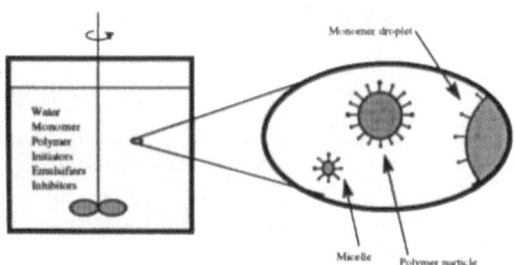

Figure 2: The major components and phases of an emulsion polymerisation system.

equations are:

$$\dot{\xi} \;=\; f_1(\xi, T)\, u f_{\text{eff}} \tag{1}$$

$$\dot{T} \;=\; \frac{(-\Delta H_p) n_{m0} f_1(\xi, T)\, u f_{\text{eff}}}{C_r} - \frac{U}{C_r}(T - T_c) \tag{2}$$

where the the states (ξ, T) are the monomer conversion and reactor temperature, u is the manipulated variable and is, for computational convenience, defined as the square root of the free radical generation rate, f_{eff} is an efficiency factor, ΔH_p is the heat of reaction, n_{m0} is the initial amount of monomer, C_r is the heat capacity of the reactor content, U is an overall heat transfer coefficient, and T_c is the temperature of the coolant in the reactor mantle. The variables and parameters of the model are summarized in the symbol list at the end of the article.

Ugelstad *et al.* [30] verified the following equation for the nonlinear function $f_1(\xi, T)$

$$f_1(\xi, T) = \frac{k_p C_M^p}{n_{m0} N_A} \sqrt{\frac{V_p}{2 k_{tp}} + \frac{N^w \frac{1}{3} V_p^{\frac{2}{3}}}{2 k_{d'}}} \tag{3}$$

where k_p, the rate constant of propagation, is given by

$$k_p = A_{cp} \exp\left[-\frac{E_{ap}}{RT}\right], \tag{4}$$

C_M^p, the monomer concentration in the particles is

$$C_M^p = \frac{\rho_m}{M_{wm}} \phi \tag{5}$$

where ϕ is the monomer volume fraction in the particles. It is given by the expression

$$\phi = \begin{cases} \phi_{\text{sat}} & \text{if} \quad \xi < \xi_{\text{crit}} \\[2mm] \frac{1-\xi}{1+\left(\frac{\rho_m}{\rho_p}-1\right)\xi} & \text{if} \quad \xi \geq \xi_{\text{crit}} \end{cases} \tag{6}$$

TABLE 1: Parameters for the polymerisation model.

Symbol	Unit	Value range	Reference
A_{cp}	$dm^3mol^{-1}hr^{-1}$	$1.19 \cdot 10^{10}$	(Brandrup and Immergut 1974)
A_{cfm}	$dm^3mol^{-1}hr^{-1}$	$3.6 \cdot 10^7$	(Odian 1991)
A_{ctp}	$dm^3mol^{-1}hr^{-1}$	$2.16 \cdot 10^{15}$	(Odian 1991, p. 275)
c_p	$kJ\,kg^{-1}K^{-1}$	0.934	(Brandrup and Immergut 1974)
c_w	$kJ\,kg^{-1}K^{-1}$	4.184	—
ΔH_p	$kJmol^{-1}$	97.6	(Nilsson, Silvegren and Törnell 1982)
E_{ap}	$kJ\,mol^{-1}$	16	(Brandrup and Immergut 1974)
E_{atp}	$kJ\,mol^{-1}$	17.6	(Brandrup and Immergut 1974, p. 275)
$k_{d'}$	dm^2hr^{-1}		(Ugelstad, Mørk, Dahl and Ragnes 1969)
M_{wm}	$kg\,mol^{-1}$	$6.25 \cdot 10^{-2}$	—
n_{m0}	mol	0.957	Recipe, Norsk Hydro
N_A	mol^{-1}	$6.023 \cdot 10^{23}$	—
N^w	dm^{-3}	$10^{17} - 10^{19}$	(Hansen and Ugelstad 1982)
R	$kJ\,mol^{-1}K^{-1}$	$8.314 \cdot 10^{-3}$	—
U	$kJK^{-1}hr^{-1}$	0.3936	Fitted to data
V_w	dm^3	$64 \cdot 10^{-3}$	Recipe, Norsk Hydro
ρ_m	$kgdm^{-3}$	0.92	(Brandrup and Immergut 1974)
ρ_p	$kgdm^{-3}$	1.38	(Brandrup and Immergut 1974)

in which ξ_{crit} is the conversion at which the monomer droplets disappear. ϕ is continuous, but not differentiable at $\xi = \xi_{crit}$.

V_p, the volume of the polymer particles is given by

$$V_p = \frac{M_{wm}n_{m0}\xi}{(1-\phi)\rho_p}. \tag{7}$$

The termination rate constant, k_{tp}, depends on both temperature and conversion. The following semi-empirical relation is used. It is a simplification of an expression proposed by Marten and Hamielec (1978) based on free volume theory [18] .

$$k_{tp} = A_{cp}\exp\left[-\frac{E_{atp}}{RT}\right]\exp\left[-\frac{\phi_{sat}}{\beta_{sat}\phi - \beta_0\left(\phi - \phi_{sat}\right)} + \frac{1}{\beta_{sat}}\right], \tag{8}$$

where $A_{cp}\exp\left[-\frac{E_{atp}}{RT}\right]$ is an ordinary Arrhenius term and β_0, β_{sat} and ϕ_{sat} are adjustable parameters that must be fitted to data. Observe that the last exponential expression is unity for $\phi = \phi_{sat}$ (interval II).

In the energy balance equation, C_r is a function of conversion

$$C_r = \rho_w V_w c_w + M_{wm}n_{m0}\left(\xi c_p + (1-\xi)c_m\right). \tag{9}$$

The parameter values used in the model are shown in table 1.

3.   Model fitting

The model described in the previous section was fitted to data obtained in a laboratory scale reactor system CPA-2 from ThermoMetric equipped with temperature control and measurement of the heat generation with a resolution of $1 \cdot 10^{-4}$ W. Detailed description of the equipment and its use is available [22] . The heat generation is proportional to the rate of change of monomer conversion.

3.1 LEAST SQUARES PARAMETER ESTIMATES

A plot of the measured rate of polymerisation is given in fig. 3. During the nucleation step, the rate rises sharply and joins the interpolation shown in the figure. Numerical integration of the data indicates that 0.9395 mol VCM was converted during the reaction period shown in figure 3. This is 98.18 % of the total monomer amount specified by the recipe. The nucleation step accounts for less than 2 % of the total conversion.

The parameters to be estimated were

$$\theta = [f_{\text{eff},0}, f_{\text{factor}}, \beta_{\text{sat}}, \beta_0, \phi_{\text{sat}}] , \qquad (10)$$

whereas the other parameters were assigned to values available in the literature. The initiator efficiency, f_{eff}, decreases as the reaction proceeds. In this work f_{eff} is assumed to decrease exponentially with conversion. $f_{\text{eff},0}$ and f_{factor} are parameters in this exponential form, $i.e.$ $f_{\text{eff}} = f_{\text{eff},0} \exp[-f_{\text{factor}}\xi]$. The three last parameters are defined in the previous section.

The fitting was done by minimising the sum of squared deviations between model predictions and measurements, $i.e.$ by minimising the following objective function

$$J(\theta) = \sum_j \left(\dot{\xi}_{\text{meas},j} - \dot{\xi}_{\text{pred},j} \right)^2 \qquad (11)$$

with respect to θ and subject to simple bounds on θ. The summation is over the set of measurements. The predicted values are calculated by single shooting, so that it is not the one step prediction error that is minimised. This might be of importance if the model structure is not completely correct [27] . The computation was carried out with Matlab and the E04UCF (SQP) routine available from NAG [20] . The predictions of the fitted model is shown in fig. 3. The parameter vector of the fitted model was

$$\theta_{\text{fitted}} = \left[8.954 \cdot 10^{-1}, 3.402 \cdot 10^{-1}, 7.541 \cdot 10^{-2}, \qquad (12) \right.$$
$$\left. 5.136 \cdot 10^{-2}, 3.913 \cdot 10^{-1} \right]$$

3.2 COMMENTS ON PARAMETER IDENTIFIABILITY

Given a model structure with a parameter θ, and assuming that there

788

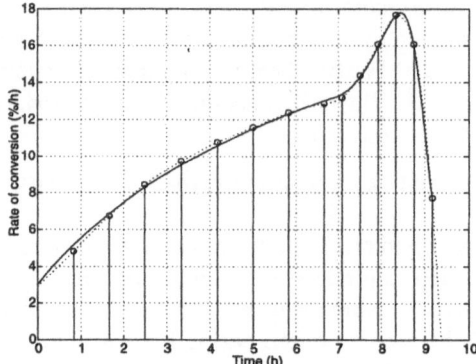

Figure 3: The measured rate of polymerisation (· · ·o· · ·) and the rate predicted by the state space model (——) are shown. The parameters were estimated by nonlinear least squares fitting.

exists a unique $\theta = \theta^*$ for which the model predicts the polymerisation system perfectly for all possible input data, the system is said to be parameter identifiable[2] [15] . From a more practical point of view, if the model fit is measured by a least squares criterion as in eqn. 11, the system is parameter identifiable provided that θ^* is the unique solution to the minimisation of $J(\theta)$.

Identifiability is thus related to the characteristics of $J(\theta)$. In practice, two problems may arise. $J(\theta)$ may have infinitely many solutions due to insufficient information in the measurements and $J(\theta)$ may have multiple minima so that θ_{fitted} may differ from θ^* in spite of convergence of the optimisation.

A suitable measure for the behaviour of J is the eigenstructure of the hessian H, whose ijth element is $H_{ij} = \left\{ \frac{\partial^2 J}{\partial \theta_i \partial \theta_j} \right\}$. The condition number of H should not be too large because this implies that there are large sets of parameter values that give approximately equally good fit to the measured data. If the hessian is close to singular the optimisation problem may be regularised by including the squared deviation of θ from a tentative value θ_0 based on *a priori* knowledge.

Concerning the possible multiple optima, it is important to have *a priori* knowledge about the parameter values. Fortunately a lot of information about the components of θ is available. f_{eff} is known to be around 0.8, so reasonable initial values are $f_{\text{eff},0} = 0.8$ and $f_{\text{factor}} = 0$ (constant efficiency f_{eff}). ϕ_{sat} is known to be about $0.3 - 0.35$. Ini-

---

[2] The notion of identifiability is not uniquely defined in the literature. E.g. Söderström and Stoica has a slightly different definition [26] .

tial values for the two last parameters, β_0 and β_{sat} are not as readily available.

No rigorous identifiability analysis is performed in this work. However, the optimisation algorithm terminated normally with no signs of a singular Hessian. Moreover, on a more heuristic basis it is clear that the β parameters are affected mainly by the data in interval III (by the shape of the peak), whereas the parameters in the initiator efficiency factor depends on the early part of the reaction rate profile. The ϕ_{sat} parameter is strongly influenced by the time at which the monomer droplets disappear, which is exhibited by the sharp rise in the rate of polymerisation[3] between $7 - 8\frac{1}{2}$ hrs in fig. 3. In other words the parameters depend on *separate* characteristics of the reaction rate profile, and this supports the hypothesis that the measurements contain enough information to estimate the parameters.

4. Control parameterisation

The manipulated variable that is to be optimised is the free radical production rate, ρ^w. In eqn. 1 the nonlinear transformation $u = \sqrt{\rho^w}$ has been introduced to make the state space equations affine in the control variable. In this work, we chose to solve the optimal control problem by using nonlinear programming techniques. In that case, u has to be represented by a finite number of parameters. This means that u must be given a form

$$u(t) = u(P_u, t), \tag{13}$$

where P_u is a parameter vector. Several parameterisations were attempted.

4.1 PIECEWISE LINEAR REPRESENTATION

In an industrial reactor initiator is typically added at a finite number of times during the polymerisation. Each injection leads to a sharp rising edge in radical production followed by a decay as the initiator is consumed. The decay is exponential. In the case when initiator is injected at a frequency comparable to the time constant of initiator decomposition or higher, a piecewise linear approximation will be acceptable. Its mathematical form is

$$u = \sum_{j=1}^{i-1}(\delta_j + \alpha_j\tau_j) + \delta_i + \alpha_i(t - \sum_{j=1}^{i-1}\tau_j) \tag{14}$$

---

[3] Another way to detect the disappearance of monomer as a seperate phase is by monitoring the pressure of the reactor.

$$\text{for } t \in [\sum_{j=1}^{i-1} \tau_j, \sum_{j=1}^{i} \tau_j), i = 1, \ldots, N,$$

or

$$P_u = \{\delta_i, \alpha_i, \tau_i\}_{i=1}^N, \tag{15}$$

where δ_j, α_j and τ_j are tuning parameters representing increments, slopes and durations between switching times, respectively. The parameterised control may be forced to be continuous by requiring the increments $\delta_j = 0 \; \forall j$. Since initiator cannot be removed, the δ_j must be nonnegative. The slopes, α_j, are related to the decay rates of the initiator system under study. Finally, τ_j are the durations between the switching times of the controller. τ_j must be positive.

4.2 A DIFFERENTIABLE REPRESENTATION

It turned out that the objective function's smoothness and continuity in the parameters had a great impact on the performance of the optimiser (cf. fig. 8). Representing the radical production with a differentiable spline function yielded better convergence properties compared to application of the nonsmooth parameterisation in eqn. 14.

The control value was assumed to be strictly decreasing with time. This assumption is justified by simulations (cf. fig. 4). The applied interpolant was a monotonicity-preserving cubic Hermite spline [6] . The monotonicity property prevents the undesirable oscillatory behavior associated with splines. Standard computational routines for calculating interpolated values, derivative values and integral of the spline are available from NAG [20] .

The spline function is uniquely determined by specifying a set of interpolation points, *i.e.*

$$P_u = \{t_i, u_i\}_{i=0}^N. \tag{16}$$

The number of interpolation points, $N + 1$, affects the resolution of the control that has been parameterised. For an explicit expression of the spline, see [6] .

If both t_i and u_i are allowed to vary, the representation is not unique. To obtain a unique representation it was chosen to fix u_i according to

$$u_i = u_0 - (i - 1)\Delta u \text{ for } i = 1, \ldots, N \tag{17}$$

$$\Delta u = (u_0 - u_{\min})/(N - 1), \tag{18}$$

which means that the ordinate interval $[u_{\min}, u_0]$ was uniformly quantised. The durations, $\tau_i = t_i - t_{i-1}$ (with $t_0 = 0$), were used as decision variables together with u_0, *i.e.*

$$P_u = \{u_0, \tau_1, \tau_2, \ldots, \tau_N\}. \tag{19}$$

Thus, a small τ_i is equivalent to a large rate of change of the control value in the time interval $[t_{i-1}, t_i]$ and *vice versa*.

This control parameterisation is suitable only if initiator can be continuously or close to continuously fed into the reactor.

4.3   THE PHYSICAL CONTROL VARIABLES

The calculation of the physical control variables, *e.g.* feed rates, from the parameterised control, $u = \sqrt{\rho^w}$, is an inverse problem. The physical control variables, here denoted w, are related to the radical production, ρ^w, through a state space model

$$\dot{x} = f(x, T) + w \tag{20}$$

$$\rho^w = g(x, T) \tag{21}$$

where T is the temperature and x is a state (vector) whose form depends on the specific initiator system. This state space model must be inverted to obtain the feed variable w [10] . As an example, consider a single thermal decomposition initiator system for which eqs. 20–21 will take the form

$$\dot{n}_i = k_i n_i + w_i \tag{22}$$

$$\rho^w = 2 k_i n_i \tag{23}$$

where n_i is the amount of initiator in the water phase, w_i is the initator feed rate and $k_i = A_{ki} \exp[-\frac{E_{ai}}{RT}]$ is the decomposition rate constant. In this case x and w in eqs. 20–21 are scalar, and w_i is obtained directly as a function of T, $\dot{T}$ and $\dot{\rho}^w$ by differentiation of the second equation and substitution for n_i and $\dot{n}_i$ in the first.

5.   Simulations

5.1   FORMULATION OF THE OPTIMISATION PROBLEM

The objective is to minimise the batch time, t_{final} , which is given by $\xi(t_{\text{final}}) \sim 0.7$ for interval II simulations and by $\xi(t_{\text{final}}) \sim 0.95$ for simulation of the entire batch time.

The constraints due to the limited cooling capacity is expressed by a lower limit on the coolant temperature: $T_c \geq T_{c,\text{min}}$. The mean temperature must not deviate significantly from a reference value, T_{ref}, which follows from the desired average molecular weight.

The states (conversion and temperature) are assumed to be measured. The following cases were considered.

Case I   The temperature was assumed to be constant and equal to the reference value, T_{ref}. A constant temperature implies that eqn. 2 is at steady state. Furthermore, when the temperature is constant, it is obvious that the batch time is minimised by adding initiator such that heat of reaction balances the heat removal with T_c at its lower limit, $T_{c,min}$. This leaves no degrees of freedom for the optimisation. The only feasible solution is (on implicit form):

$$u = \frac{U\,(T_{ref} - T_{c,min})}{\Delta H_p n_{m0} f_1\,(\xi, T_{ref}) \cdot f_{eff}} \qquad (24)$$

where T_{ref} is specified. (ξ depends on u through eq. 1.)

Case II   The temperature is *not* assumed to be constant, but the temperature mean and variance are constrained by the molecular weight specifications. As in the previous case it is obvious that the heat generation should be controlled such that T_c may be kept at its minimum. The batch time and temperature variance are sought minimised with respect to P_u and subject to eqs. 1 and 2 together with

$$\frac{1}{t_{final}} \int_0^{t_{final}} T\,dt = T_{ref} \,. \qquad (25)$$

The last constraint ensures a correct value for the average molecular weight. For PVC the average molecular weight is given by the average temperature.

5.2   RESULTS

5.2.1   Case I

The reactor temperature is assumed to be constant, say T_{ref}, and the coolant temperature is kept at its minimum, $T_{c,min}$. Then eqn. 2 implies eqn. 24 which may be substituted into eqn. 1, yielding

$$\dot{\xi} = \frac{U\,(T_{ref} - T_{c,min})}{n_{m0} \Delta H_p} = \dot{\xi}_{max} \qquad (26)$$

where $\dot{\xi}_{max}$ is the highest allowable conversion rate. If U is assumed to be constant[4], then

$$\xi = \xi_0 + \dot{\xi}_{max}(t - t_0) \text{ where } \xi(t_0) = \xi_0.$$

The optimal radical production becomes $\rho^w = u^2$ where

$$u = \left. \frac{U\,(T_{ref} - T_{c,min})}{\Delta H_p n_{m0} f_1\,(\xi, T_{ref}) \cdot f_{eff}} \right|_{\xi = \xi_0 + \dot{\xi}_{max}(t - t_0)} \qquad (27)$$

$$= \left. \frac{\dot{\xi}_{max}}{f_1\,(\xi, T_{ref}) \cdot f_{eff}} \right|_{\xi = \xi_0 + \dot{\xi}_{max}(t - t_0)} \qquad (28)$$

This profile is shown in figure 4, where the maximum allowable rate of conversion is set to the maximum in figure 3, $\dot{\xi}_{max} \approx 0.18$.

The results from *case I* may be difficult to attain in practice. Nevertheless, the calculations provide a lower bound for the attainable batch

---

[4]   Fouling is negligible for the process, due to coating of the reactor walls.

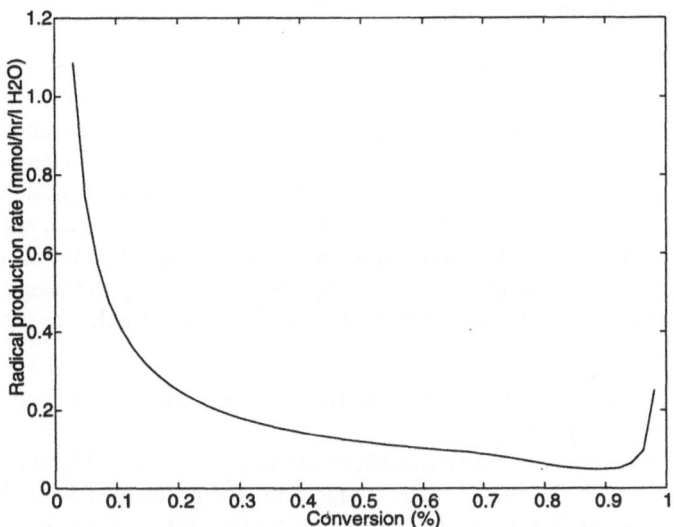

Figure 4: The optimal radical production calculated for case I is shown as a function of conversion. The rate of polymerisation is constant, so that the conversion is proportional to the reaction time. Thus, the figure gives a picture of the optimal time development of the radical production.

time and thus show the potential for improvement. This potential is a decisive factor in an industrial context. In this particular case the potential is a reduction in reaction time from 9.4 hr (cf. fig. 3) to $(0.18)^{-1}$hr ≈ 5.5 hr. Moreover, the u values from *case I* may serve as initial values in the search for optimal radical production profiles. This is especially relevant after the moment when the reactor temperature has been brought to its set point.

Finally, *case I* is interesting from a design point of view, because the radical production profile provides information on what kind of initiation system is needed to control the process.

5.2.2 Case II

Optimisation problem   Common to all optimisations carried out in this work is that the objective function was defined as

$$\Phi = \alpha_1 \cdot \frac{1}{t_{\text{final}}} \int_0^{t_{\text{final}}} (T - T_{\text{ref}})^2 \, dt + \alpha_2 \cdot t_{\text{final}} \qquad (29)$$

with t_{final} given by $\xi(t_{\text{final}}) = 0.98$ and $T_{\text{ref}} = 323$ K. Φ was minimised subject to the constraint

$$\frac{1}{t_{\text{final}}} \int_0^{t_{\text{final}}} T dt \in [T_{\text{ref}} - \epsilon, T_{\text{ref}}]. \tag{30}$$

ϵ is a feasibility parameter that was introduced for purely computational reasons. The computational time decreased when the mean temperature was allowed to depart slightly from its constraint value, T_{ref}, during the iterations. The average temperature settled at T_{ref} independent of the ϵ value because this minimises the reaction time. The initial u level and a certain number of switching times definining the shape of the radical production profile were used as decision variables, as described in section 4.

Solution method   All of the routines mentioned in this section are provided by NAG [20] .

The objective function and constraint were evaluated by integration along with the model. The integration was performed with D02CJF (Adams method). The applied optimiser was E04UCF (SQP method). The controls were evaluated by the spline interpolator E01BEF/E01BFF. The gradients were calculated by using finite differences, using the NUMJAC routine in Matlab.

Constant coolant temperature   The goal of the optimisation is to increase the reaction rate, $\dot{\xi}$ as much as possible within the limits of the cooling capacity. The ideal situation is attained when the reaction rate is exactly high enough for the reactor temperature to remain at T_{ref} and with T_c equal to its lower bound. The first optimisation strategy is to fix the coolant temperature at its minimum and let the optimisation algorithm find the radical production rate, ρ^w, that will lead to a sufficiently high reaction rate to maintain the correct reactor temperature, $T = T_{\text{ref}}$.

The results of a typical optimisation case is shown in figure 5. The SQP optimisation did not converge completely, but most of the progress had taken place in the first 20 iterations. The progression made by the optimiser after each 5th major iteration is displayed in figure 6. The weighting factors $\{\alpha_1, \alpha_2\}$ in eqn. 29 were set to $\{0.03, 0.2\}$. These two parameters adjust the trade-off between undesirable temperature oscillations and the loss associated with longer batch time. The feasibility parameter ϵ was set equal to 3 K. The initial u level was included as a decision variable in addition to the five switching times.

The final average temperature was 322.99 K, in accordance with the desired value. If the ϵ parameter was set too small (e.g. 10^{-1}) the convergence became slower. The exact value of ϵ was not critical as long as it was not too small.

The optimisation results were quite robust with respect to changes

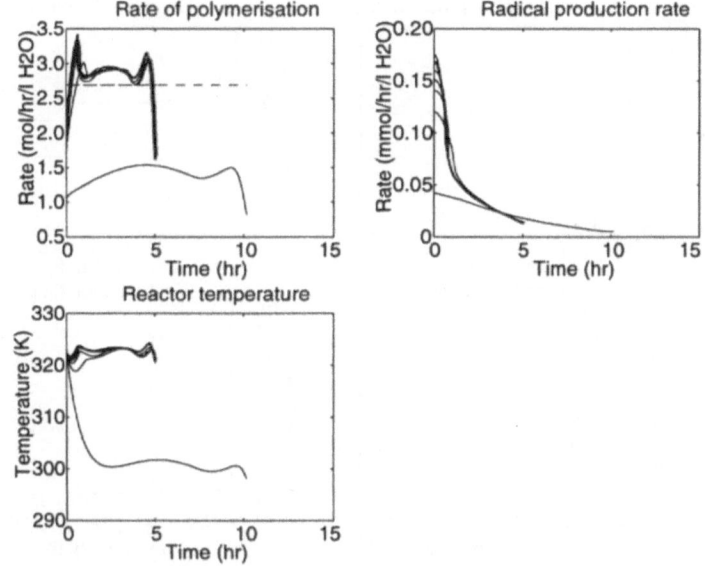

Figure 5: A typical optimisation case is shown. The batch time improvement is evident. Moreover, most of the improvement is reached within a few iterations.

in the initial value of the decision variables. The final results (after 50 major iterations) for two widely different initial sets of values are summarised in table 2[5]. The results would have been even closer if the number of iterations had been increased. The final reaction times differed by about 4 seconds, only, and the mean temperatures were also equal. The switching times were slightly different, and this may call for a regularisation, e.g., the integral of ρ^w may be included to reflect the cost of excessive use of initiator. However, in spite of the difference in switching times, the two radical production profiles are barely distinguishable.

The reaction is assumed to be preheated (i.e., the initial temperature is the correct one.) This is a reasonable assumption since the nucleation phase has already taken place. In figure 5 the temperature has a 'dip' at the start of the simulation. This has to do with the particle volume term (V_p) in eqn. 3 which is small at low conversion. To prevent this 'dip' we found it necessary to let the coolant temperature vary.

P-controlled coolant temperature An attempt was made to eliminate the temperature dip discribed above by letting the coolant tempera-

---

[5] The final reaction time is shorter than what is achieved in *Case I*. This is so since the temperature is allowed to be greater than the reference temperature.

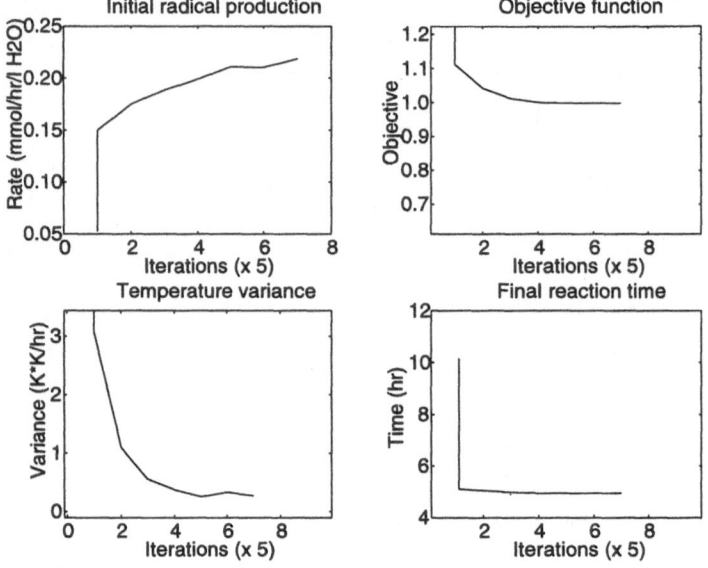

Figure 6: The progression of the optimiser is illustrated. Observe how the batch time reduction is obtained almost immediately, whereas it takes more iterations to bring down the temperature variance.

ture, T_c, move off its minimum value, $T_{c,min}$. By introducing a standard P-control law with correction for static deviation

$$T_c = \max(T_{c,min}, \frac{\Delta H_p n_{m0} f_1(\xi, T) u f_{eff} + K_p(T_{ref} - T)}{U} + T)$$
(31)

T_c is still forced to be at its minimum almost all of the time, so that no significant cooling capacity is left unused. The temperature variance is reduced at the cost of some increase in reaction time (figure 7). It is difficult, however, to set the gain factor properly. With a too small gain factor the controller will not drive T_c to its minimum, as it should. On the other hand if the gain factor is too large the optimiser is quickly stuck at a non-optimal point. The cooling temperature controller seems to create distortions in the numerical gradient calculations which destroy the convergence properties of the optimal controller.

The P-controller prevents the temperature to be lower than T_{ref}, thus making it impossible to satisfy the constraint on the mean temperature. This may be improved by introducing an integral term in the controller. Then the temperature constraint was satisfied, but the convergence properties were still lacking. When a conventional PI controller

TABLE 2: Results from to widely different sets of initial values.

Description	Case a	Case b
Initial u value	2.9830	3.0727
Switching time 1	0.4419	0.3865
Switching time 2	0.1802	0.1617
Switching time 3	0.4768	0.4566
Switching time 4	2.5052	2.3167
Switching time 5	3.2326	4.1515
Final objective function	0.9946	0.9936
Mean temperature	322.99	322.99
Total reaction time	4.9363	4.9353

$$T_c = \max(T_{c,\min}, K_p(T_{ref} - T) + \frac{K_p}{T_i} \int_0^t (T_{ref} - T)d\tau) \qquad (32)$$

was employed, together with optimisation with respect to radical production rate, the results were very similar to the simulations with T_c fixed at its lower bound (figure 5).

Including a parameterisation of the coolant temperature among the decision variable   When combining standard controllers for T_c and optimisation with respect to the radical production rate it was difficult to find tuning parameters that led to "intelligent" control of T_c. Moving one step further, the decision variables were augmented with a parameterisation of T_c to see if the optimiser would arrive at a better profile for T_c than the standard controllers did. No significant improvement was observed.

6.   Discussion

One might argue that the model of Ugelstad and co-workers was obtained on the basis of cases in which all of the initiator was added at the start of the reaction. Mechanisms like homogenous nucleation and water phase termination could significantly reduce the effect of adding initiator. However, data from Norsk Hydro's plant in Stenungsund, Sweden, confirm the effect of adding initiator throughout the entire conversion range. In Stenungsund initiator is added during the course of the reaction — with success. The addition policy is empirically based. Still it works remarkably well, and one cannot expect a dramatic reduction in the batch time at the Stenungsund plant by implementing a model based controller. However, some undesired temperature oscillations can be reduced by applying the controllers described in this paper.

Even though the control scheme in Stenungsund works quite well, the model based controller offers greater flexibility. It is simple to adapt to changing specifications on allowed temperature variations and one

798

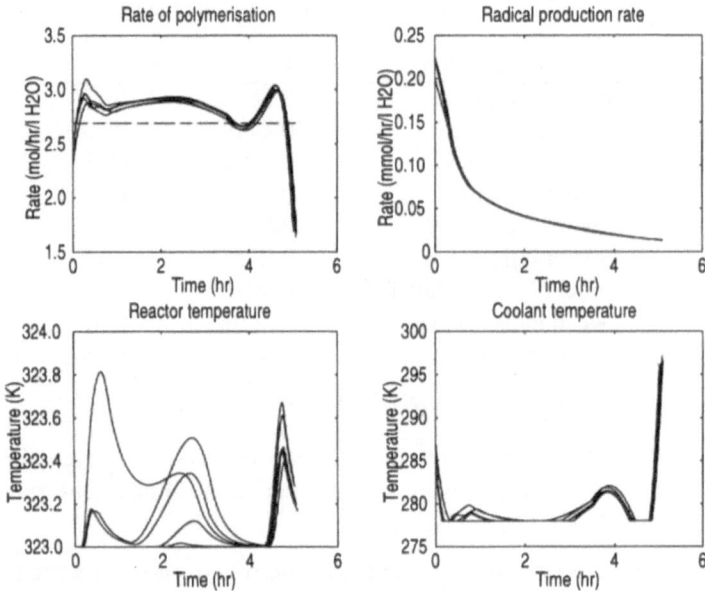

Figure 7: When a P-controller was applied to the coolant temperature, the variance of the reactor temperature was reduced. The coolant temperature is close to its lower bound (278 K) throughout the reaction.

may estimate batch-to-batch variations in parameters. In addition, the model based controller can be extended. There are control actions other than intiator additions that are carried out manually by the operators which may easily be incorporated in the model based controller.

The calculations in this paper have strictly been "optimal control" calculations rather than predictive control calculations. However, since the states are assumed to be available through measurements, predictive control just amounts to calculate the optimal control trajectory from the current measured state and to the end of the batch. The first part of this control profile is then applied. At the next sample time the procedure is repeated.

In practice, parameters like f_{eff} and U should be estimated on-line.

The parameterisation in eqn. 14 allows for fast and easy computation of the control variable. However, a lot of discontinuities are introduced, and this may destroy the smooth properties of the objective function. With insufficient integration precision the objective function may be completely intractable for optimisation. An example is shown in figure 8.

By parameterising the control variable using the spline approximation, we have avoided the problem depicted in fig. 8.

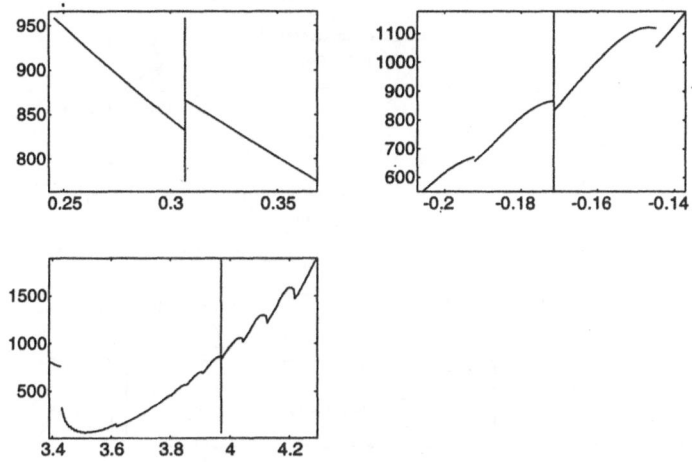

Figure 8: The objective function is illustrated with marginal plots (one decision variable is changed at a time). The point where the optimization algorithm terminated are marked by vertical lines. Perturbations show that the "solution" is far from the global optimum. (No constraints were active.) Note the discontinuities.

7. Conclusions and suggestions for further work

At Norsk Hydro it is desirable to develop an operator support system that will contribute to increased productivity and a stable product quality in the PVC production. A possible such system is depicted in fig. 9, and it builds upon the principles of model based control.

The results in this paper show how the batch time may be significantly reduced by applying a nonlinear model based controller in conjunction with a sufficiently flexible initiator system in the polymerisation process. Norsk Hydro's PVC plant in Stenungsund, Sweden, has obtained significant batch time reduction by adding initiator throughout the reaction, but they use heuristic procedures to determine when and how much to add. At this plant no dramatic reduction in batch time can be expected by introducing the model based controller outlined in this paper because the cooling capacity is almost entirely utilised with the current system. For the other E-PVC plants which do not inject initiator during the reaction, the benefit may be as big as 30 % reduction in reaction time.

The calculated optimal control actions may be used to choose an

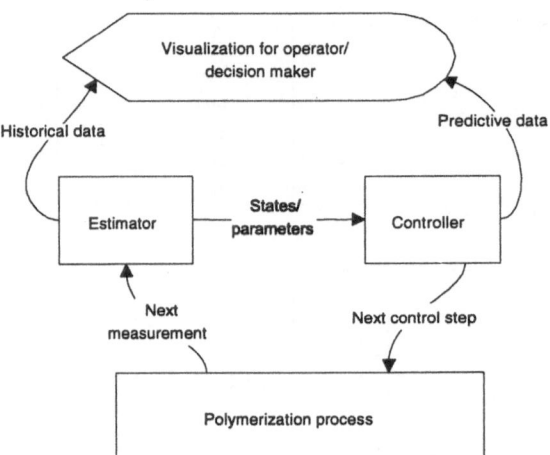

Figure 9: The controller provides the process with a set of control values for the next sample time interval, and the process generates corresponding measurements. The control action is the first step of an optimal future trajectory which may be visualized for the operator. The control horizon is to the end of the batch reaction. Thus an anticipation of the remaining part of the batch is made available to the operator. The estimator has a function dual to that of the controller. Its responsibility is to calculate the parameter and state values that optimally fits the previous measurements. In an industrial reactor it is important that the estimator also has mechanisms for detection of abnormal measurements and parameter values.

appropriate initiation system. In the design direction an obvious continuation of this work is to extend the mathematical model to full population balances so that it can be used to predict particle size distribution and other properties of the final latex. The technique applied in this paper may then be used to choose emulsifier system, monomer addition policies, etc. Experiments at the plant indicate that improvements are possible by changing the monomer addition policy.

The flexibility and extensibility of a model based controller, and its interplay with knowledge aquirement and utilisation is an argument for the introduction of model based contollers even if the short term benefits are marginal.

In a control context the optimal free radical production provided by the predictive controller must be translated into feeds of initiator or reducing agent (in case of a redox initiator system). This translation may be viewed as an actuator and the calculations are either done directly or by solving an optimisation problem.

On-line measurements of microscopic characteristics of the polymerising latex is evolving. Norsk Hydro has a strong group on soft-sensing, and the first on-line soft-sensors for polymer property measurement are about to be operative. In the near future optimisation may be applied with end-time constraints on a number of quality related properties of the latex.

References

1. J.G. Balchen, D. Ljungquist, and S. Strand. State-space predictive control. *Chemical Engineering Science*, 47(4):787–807, 1992.
2. B.D.O.Anderson and J.B.Moore. *Optimal Control – Linear Quadratic Methods*. Prentice-Hall, 1989.
3. R.R. Bitmead, M. Gevers, and V. Wertz. *Adaptive Optimal Control: The Thinking Man's GPC*. Prentice-Hall, 1990.
4. J. Brandrup and E. H. Immergut. *Polymer Handbook*. Wiley-Interscience, 2nd edition, 1974.
5. Marcelo Embirucu, Enrique L. Lima, and José Carlos Pinto. A survey of advanced control of polymerization reactors. *Polymer Engineering and Science*, 36(4):433–447, 1996.
6. F. N. Fritsch and J. Butland. A method for constructing local monotone piecewise cubic interpolants. *SIAM J. Sci Statist. Comput.*, 5:300–304, 1984.
7. Robert G. Gilbert. *Emulsion Polymerization - A Mechanistic Approach*. Academic Press Inc, San Diego, 1995.
8. A.E. Hamielec and J.F. MacGregor. Latex reactor principles: Design, operation and control. In I. Piirma, editor, *Emulsion Polymerization*, pages 319–355. Academic Press, Inc., 1982.
9. F.K. Hansen and J. Ugelstad. Particle formation mechanisms. In I. Piirma, editor, *Emulsion Polymerization*, pages 51–92. Academic Press, Inc., 1982.
10. A. Isidori. *Nonlinear Control Systems*. Springer, 3rd edition, 1995.
11. Shi-Shang Jang and Wei-Liang Yang. Dynamic optimization of batch emulsion polymerization of vinyl acetate - an orthogonal polynomial initiator policy. *Chemical Engineering Science*, 44(3):515–528, 1989.
12. C. Kravaris and C. B. Chung. Nonlinear state feedback synthesis by global input/output linearization. *AIChE J.*, 33:592–603, 1987.
13. K.Y.Hsu and S.A. Chen. *Chemical Engineering Science*, 43:1989, 1988.
14. Frank L. Lewis. *Optimal Control*. Wiley and Sons, 1986.
15. L. Ljung. *System Identification*. Prentice-Hall, 1987.
16. L.S.Pontryagin, R.V.Boltyanskii, R.V. Gamkrelidze, and E.F.Mischenko. *The Mathematical Theory of Optimal Processes*. Wiley and Sons, 1962.
17. M.A.Henson and D.E.Seborg. *Nonlinear Process Control*. Prentice – Hall, 1997.
18. L. Marten and A.E. Hamielec. In J.N. Henderson and T.C. Bouton,

editors, *Polymerization Reactors and Processes*, page 104. Am. Chem. Soc. Soc. Ser., 1978.

19. R.M.M. Mattheij and J.Molenaar. *Ordinary Differential Equations in Theory and Practice*. Wiley and Sons, 1996.

20. NAG. *NAG Foundation Toolbox - for Use with MATLAB*. The Math Works Inc., 1995.

21. Z. Nagy and S. Agachi. Model predictive control of a pvc batch reactor. *Computers and Chemical Engineering*, 21(6):571–591, 1997.

22. H. Nilsson. *Studies on the Manufacture of PVC Resins*. PhD thesis, LTH, Lund, Sweden, 1982.

23. H. Nilsson, C. Silvegren, and B. Törnell. *Chemica Scripta*, page 164, 1982.

24. George Odian. *Principles of Polymerization*. John Wiley & Sons, Inc., 3rd edition, 1991.

25. S.R. Ponnuswamy, S.L. Shah, and C.A. Kiparissides. *Journal of Applied Polymer Science*, 32:3239, 1986.

26. T. Söderström and P. Stoica. *System Identification*. Prentice-Hall, 1989.

27. Steinar Sælid. Process identification techniques. In R. Berber, editor, *Methods of Model Based Control*, pages 81–97, Dordrecht, the Netherlands, 1995. Kluwer Academic Publishers.

28. W.V. Smith and R.H. Ewart. Kinetics of emulsion polymerization. *J. Chem. Phys.*, 16:592–599, 1948.

29. Stig Strand. *Dynamic optimization in state-space predictive control schemes*. PhD thesis, Norwegian Institute of Technology, 1991.

30. J. Ugelstad, P. C. Mørk, P. Dahl, and P. Ragnes. A kinetic investigation of the emulsion polymerization of vinyl chloride. *Journal of Polymer Science, Part C*, 27:49–68, 1969.

31. J. Ugelstad, P.C. Mørk, and J.O. Aasen. Kinetics of emulsion polymerization. *Journal of Polymer Science, Part A-1*, 5:2281–2288, 1967.

32. K. Zhou, J.C. Doyle, and K. Glover. *Robust and Optimal Control*. Prentice-Hall, 1996.

List of symbols

Symbol	Unit	Description
A_{cp}	$\mathrm{dm^3mol^{-1}hr^{-1}}$	Frequency factor for the propagation reaction
A_{ctp}	$\mathrm{dm^3mol^{-1}hr^{-1}}$	Frequency factor for the termination in particles
c_m	$\mathrm{kJ\,kg^{-1}K^{-1}}$	Heat capacity of monomer
c_p	$\mathrm{kJ\,kg^{-1}K^{-1}}$	Heat capacity of polymer
c_w	$\mathrm{kJ\,kg^{-1}K^{-1}}$	Heat capacity of water
C_r	$\mathrm{kJK^{-1}}$	Heat capacity of reactor content
C_m^p	$\mathrm{mol\,dm^{-3}}$	Concentration of monomer in particles
f_{eff}	fraction	Initiator efficiency factor
$f_{\mathrm{factor}}, f_{\mathrm{eff},0}$	—	Dimensionless parameters in f_{eff}
ΔH_p	$\mathrm{kJmol^{-1}}$	Heat of reaction
E_{ap}	$\mathrm{kJ\,mol^{-1}}$	Activation energy for the propagation reaction
E_{atp}	$\mathrm{kJ\,mol^{-1}}$	Activation energy for the termination in particles
k_p	$\mathrm{dm^3mol^{-1}hr^{-1}}$	Rate constant for the propagation reaction
k_{tp}	$\mathrm{dm^3mol^{-1}hr^{-1}}$	Rate constant for the termination in particles
$k_{d'}$	$\mathrm{dm^2hr^{-1}}$	Parameter in the rate constant for the desorption of radicals from particles
M_{wm}	$\mathrm{kg\,mol^{-1}}$	Molar mass of monomer
n_{m0}	mol	Initial amount of monomer
N_A	$\mathrm{mol^{-1}}$	Avogadro's number
N^w	$\mathrm{dm^{-3}}$	Number of polymer particles per unit volume of water phase
P_u	—	Vector characterising the parameterised control u
R	$\mathrm{kJ\,mol^{-1}K^{-1}}$	Gas constant
t, t_{final}	hr	Time and final batch time
T, T_{ref}	K	Reactor temperature and its reference value
$T_c, T_{c,\mathrm{min}}$	K	Coolant temperature, and its minimum allowable value
u	—	Control variable, equals $\sqrt{\rho^w}$
U	$\mathrm{kJK^{-1}hr^{-1}}$	Overall heat transfer coefficient
V_w, V_p	$\mathrm{dm^3}$	Volume of water phase and polymer particles
α_1, α_2	—	Tunable parameters in the objective function
$\beta_0, \beta_{\mathrm{sat}}$	—	Dimensionless parameters in k_{tp}
ξ	fraction	Monomer conversion
ξ_{crit}	fraction	Monomer conversion at which the monomer droplets disappear
ξ_{meas}	fraction	Monomer conversion calculated from measured heat generation
ξ_{pred}	fraction	Monomer conversion predicted by solving state space model
ϕ	fraction	Volume fraction of monomer in polymer particles
ϕ_{sat},	fraction	*ditto* when monomer droplets are present
Φ	—	Objective function
ρ_m	$\mathrm{kgdm^{-3}}$	Density of monomer
ρ_p	$\mathrm{kgdm^{-3}}$	Density of polymer
ρ^w	$\mathrm{mol\,hr^{-3}}$	Rate of radical production in water phase

KAPPA NUMBER PROFILE CONTROL FOR CONTINUOUS DIGESTERS

FERHAN KAYIHAN
IETek - Integrated Engineering Technologies
5533 Beverly Ave NE, Tacoma WA 98422-1402, USA
e-mail : fkayihan_ietek@msn.com

ABSTRACT

Continuous digester control for improved uniformity of pulp and fiber properties is gaining more attention as strict final product quality needs and tighter environmental regulations demand much higher performance standards from strategic manufacturing processes. Removal of lignin from wood chips in digesters is a stochastic process, owing to the natural variabilities of the raw material and the physical nature of the digester vessel as a moving packed bed reactor. As additional and particular challenges for feedback control implementations, continuous digesters have long residence times and corresponding time delays, while process measurements are traditionally sparse and infrequent.

Established control approaches conventionally target the extent of reaction, measured as output Kappa number, for process regulation. In practice, on the other hand, it is common to have significant product quality variations even with good Kappa control. Fundamentally, some of the critical pulp properties, like fiber strength, depend on the reaction path in the digester as well as on final Kappa number. Therefore, for continuous digesters it is reasonable to target a nominal reaction path as the control objective rather than just the final conversion. In order to explore the possibilities of model predictive control (MPC) approach to reaction path regulation, an intuitive formulation is developed and tested for the extended modified continuous cooking (*emcc*) design. Using a simplified nonlinear process benchmark model it is shown that the digester reaction profile can be controlled effectively using five key process conditions as feedback. Three of the process measurements (upper and lower extract EA strengths and the blow line Kappa number) are traditionally measured and easily available. The other two measurements (intermediate Kappa numbers at the end of *cook* and *mcc* zones) must be estimated from available process data using a reference model. It is also shown that measurement of process disturbances like chip feed rate and moisture, and white liquor alkali strength can significantly improve controller performance.

R. Berber and C. Kravaris (eds.), Nonlinear Model Based Process Control, 805-829.
© 1998 *Kluwer Academic Publishers.*

1. Introduction

In terms of modeling and control, one of the most challenging processes in the manufacturing industries is the continuous digester. As a major capital item and a strategic process in the conversion of wood chips to pulp, digesters are getting renewed attention from systems engineers for both modeling and control improvements. Motivation for the recent interest comes mainly from (a) production scheduling needs both in processing rate and in raw material (specie) swings, (b) tight product quality commitments and (c) strict environmental requirements.

The objective of this work is simply to propose an intuitive control approach for continuous digesters and to demonstrate its relative performance compared to a conventional feedforward scheme. The method to be used is an adaptation of the established distributed parameter control approach for tubular reactors. It will be shown that controlling the reaction profile of the digester is much more effective than just controlling the blow line (exit conditions) Kappa number. Reaction profile, or the reaction path, will be defined in terms of Kappa numbers and liquor alkali strengths at a few physically reasonable locations along the vessel. Control actions will be through temperature and liquor flow adjustments. A 5x5 unconstrained MPC structure will be developed for a particular digester design and its performance will be demonstrated through simulation examples. A simplified nonlinear benchmark model will be used as the process model.

The motivation for the new control approach comes from the desire to maintain the performance of any particularly successful operating condition against changing process disturbances. Keeping the reaction history of wood chips constant is fundamentally a stronger assurance of stability for complete pulp and fiber properties, than just keeping the end-point of the reactions constant. Kappa number is defined as the fraction of lignin remaining in chips. Therefore, decreasing Kappa number trend from top to the bottom of the digester is a direct measure of the reaction path. Residual alkali strengths at exit liquor flows are used as supplementary measures, as they are sensitive indicators of possible changes in the chemical environment throughout the digester.

The organization of the paper is as follows: First, a brief description of the process will be provided for those who are not familiar with continuous digesters. Then, a recently proposed simplified benchmark mathematical model will be summarized to document the working equations of the process model, including parameters and nominal operating conditions. Following that, a conventional feedforward control policy will be described and tested against a sequence of integrated stochastic disturbances. The new approach for reaction profile control will then be introduced as an MPC problem and tested against the same disturbance sequence. The paper will be concluded with a few comments highlighting the results.

2. Process Description

Continuous digesters are very complex vertical tubular reactors, used in the pulp and paper industry to remove lignin from wood chips. Aqueous solution of sodium hydroxide and hydrosulfide, called white liquor, is used to react with porous and initially wet wood chips. Usually, continuous digesters are separated into multiple reaction and extraction zones to carry out a specific process sequence. Depending on the production needs of a pulp mill and on the state of the art of digester design at the time of installation, there may be numerous differences between digesters. In fact, almost every continuous digester is unique in some ways. However, common to all is the general sequence of transport and reaction processes that govern the overall operation. Due to the complexities of these physical and chemical process phenomena and the fact that wood chips are nonuniform in size and naturally vary in physical and chemical properties, regulating product quality in a digester is a non-trivial task.

The particular digester design chosen for this study is the dual vessel *emcc* (extended modified continuous cooking) arrangement as shown in Fig 1. Due to the limited scope of this study, only the chemical reaction problem of Kappa number and fiber properties control will be addressed here. Therefore, the focus will be on the pulping reactions in the digester. Otherwise, also important and complementary problems of production rate and chip level control will not be addressed. A brief description of the process is provided to introduce the basics. Detailed analysis and descriptions of the fundamental phenomena about digester operations are available elsewhere [1-4,8].

Wet chips are steamed to remove air in the pores and fed into the impregnation vessel (IV) together with cool white liquor. In the impregnation vessel, white liquor penetrates into the chips and equilibrates with initial moisture for about 30 minutes depending on the production rate. In the IV, both chips and liquor move in the co-current downward direction. The impregnation vessel is not always separate from the digester. In some designs, the first section at the top of the digester becomes the impregnation zone.

From the impregnation vessel, the chips are carried or sluiced into the top section of the digester with hot liquor that brings the mixture to the desired reaction temperature. The top section of the digester, referred to as the *cook* zone, is a co-current section where the initial reactions take place to remove lignin. Chips essentially react from inside out owing to the significant internal pore volume and associated surface area. Therefore, overall reaction rates depend on the concentration levels of entrapped liquor and the diffusion rates from free liquor that replenish the active ingredient holdup in the pores. Spent liquor saturated with dissolved solids at the end of the *cook* zone is extracted for chemical recovery elsewhere in the mill. Chips follow into the *mcc* (modified continuous cooking) and the *emcc* zones, now counter-current to fresh dilute white liquor which simultaneously continue mild delignification reactions and at the same time extract valuable inorganic solids from the pores of chips. Temperatures and liquor flows are adjusted to maintain desired operating conditions.

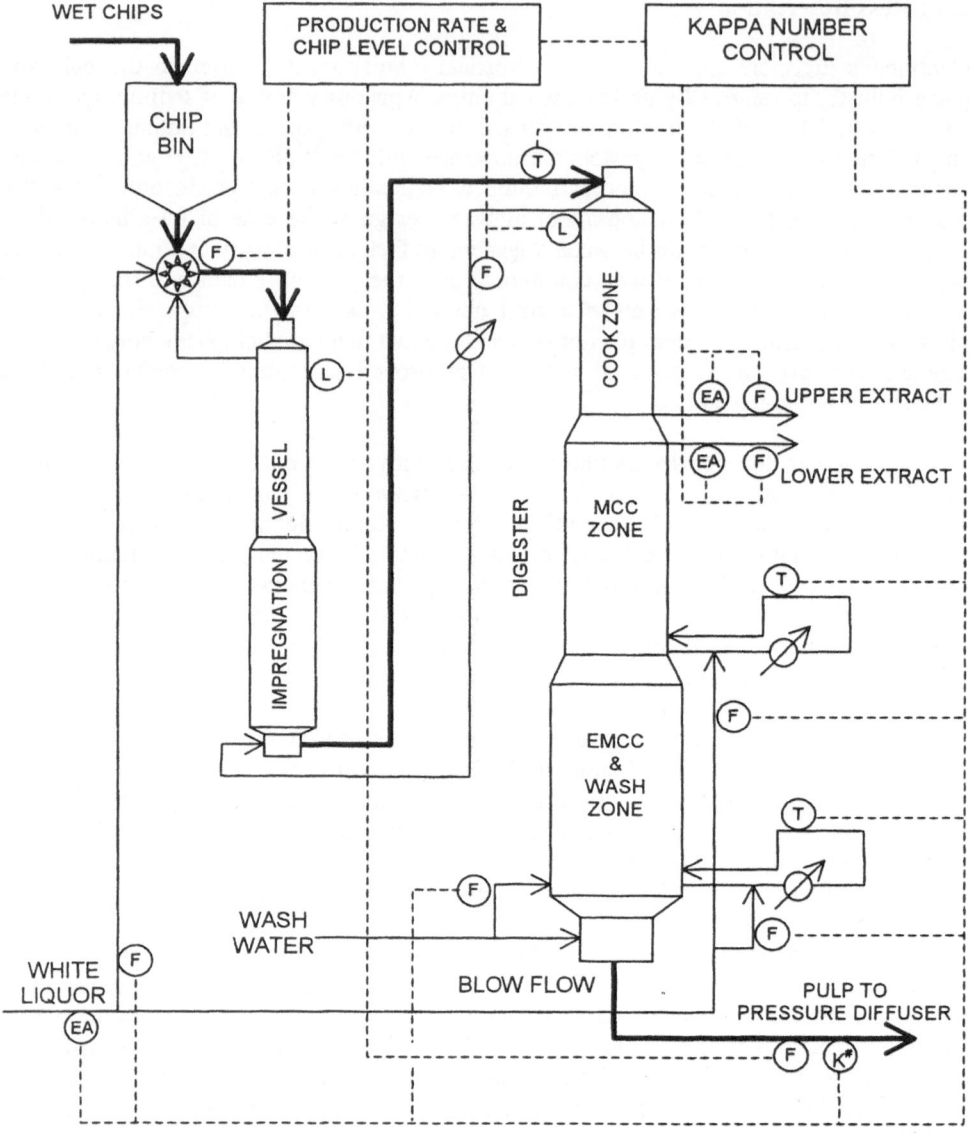

Figure 1. A simplified diagram of a dual-vessel *emcc* continuous digester showing typical control loops

As packed reactors, digesters are very unique in that the packing as the main ingredient of the process is continuously in motion, nonuniform in size and change compaction both with respect to conversion and differential head pressure. Extent of reaction, defined through the blow-line (exit) Kappa number, is the major performance measurement. Other important factors are the yield of the process and the fiber properties of the final product. Although various operating conditions may yield the same Kappa number, important fiber properties like strength are reaction path dependent. During the residence time of fiber in the vessel, rate of conversion, temperature levels and chemical compositions of residual alkali and dissolved solids are all factors in determining its properties.

3. Benchmark Model

A simplified continuous digester model is developed recently [5] in order to provide a benchmark for systems engineering research for controller design, identification, model reduction, diagnostic and monitoring. With a few minor modifications, the benchmark model will be presented here briefly. It is selected as the reference process model for simulations since it captures most of the typical process complexities and dynamic interactions without the traditional details of full-scale fundamental models. Fig 2 shows the process diagram for the benchmark model representing the equivalent of the *emcc* design in Fig 1.

Assumptions that simplify the fundamental basis of the benchmark model are:
1. Non-porous solid chips and free liquor constitute the only two phases in the system.
2. The two phases are in local thermal equilibrium.
3. There are no mass or thermal diffusion limitations within the phases.
4. The impregnation vessel is represented by instant mixing (dilution) followed by same transport delay for both phases.
5. In the digester, chip level is constant and the compaction profile is static and known.
6. Both phases in the digester move as plug flows with local space velocities based on compaction and volumetric flowrates.
7. Reaction kinetics follow the suggestions of the "Purdue Model", but the heats of reactions are ignored.
8. Reactions and mixing during the process affect solid and liquor densities but not volumes, i.e., volumes are conserved.
9. Wood extractives are ignored and initial moisture is instantly mixed with white liquor at the feed conditions.
10. Delignification reactions occur only in the digester vessel.
11. Changes in heater temperatures, trim liquor additions and other process inputs are instantaneous.

The material and enthalpy balances for the *cook* zone, where both solid and liquor phases are moving in the same direction, are

810

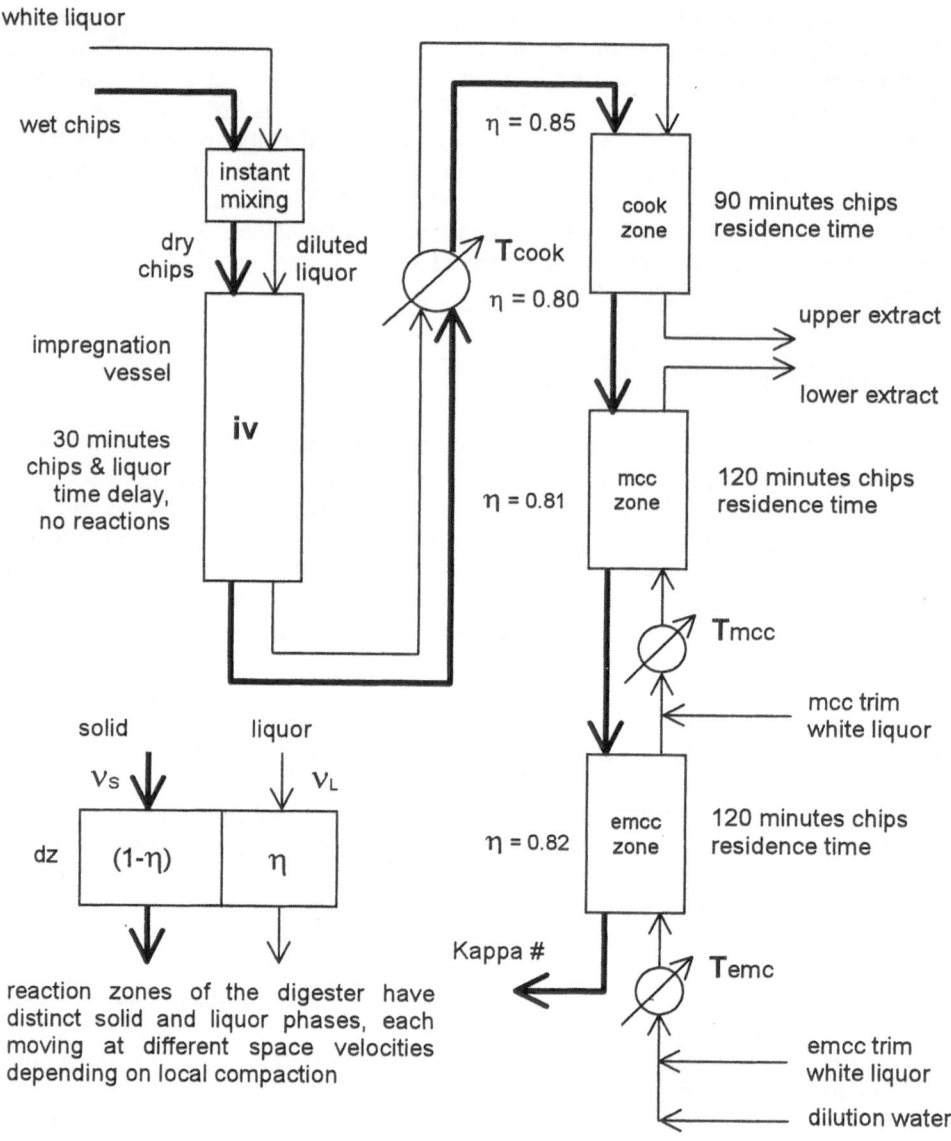

Figure 2. Simplified continuous digester benchmark model schematics showing nominal operating conditions

$$\frac{\partial \rho_{Si}}{\partial t} = -\upsilon_S \frac{\partial \rho_{Si}}{\partial z} + R_{Si} \qquad i = 1,\ldots,5 \tag{1}$$

$$\frac{\partial \rho_{Lj}}{\partial t} = -\upsilon_L \frac{\partial \rho_{Lj}}{\partial z} + R_{Lj} \qquad j = 1,\ldots,4 \tag{2}$$

$$(1+\alpha)\frac{\partial T}{\partial t} = -\upsilon_S(1 + \alpha\upsilon_L / \upsilon_S)\frac{\partial T}{\partial z} \tag{3}$$

where

$$\upsilon_S = \frac{v_S}{A(1-\eta)} \qquad \upsilon_L = \frac{v_L}{A\eta} \tag{4}$$

and

$$\alpha = \eta c_{PL}\rho_L /(1-\eta)c_{PS} \approx \eta/(1-\eta) \tag{5}$$

The corresponding equations for countercurrent flow zones are

$$\frac{\partial \rho_{Si}}{\partial t} = \upsilon_S \frac{\partial \rho_{Si}}{\partial z} + R_{Si} \qquad i = 1,\ldots,5 \tag{6}$$

$$\frac{\partial \rho_{Lj}}{\partial t} = \upsilon_L \frac{\partial \rho_{Lj}}{\partial z} + R_{Lj} \qquad j = 1,\ldots,4 \tag{7}$$

$$(1+\alpha)\frac{\partial T}{\partial t} = -\upsilon_S(1 - \alpha\upsilon_L / \upsilon_S)\frac{\partial T}{\partial z} \tag{8}$$

Reaction rates for solids are specified as

$$R_{Si} = -\theta[k_{Ai}\rho_{L1} + k_{Bi}\rho_{L1}^{1/2}\rho_{L2}^{1/2}](\rho_{Si} - \rho_{Si}^{\infty}) \qquad i = 1,\ldots,5 \tag{9}$$

with reaction rate constants $k_{Ai} = k_{Aoi}\,exp\,(-E_{Ai}/RT)$ and $k_{Bi} = k_{Boi}\,exp\,(-E_{Bi}/RT)$.

Liquor component rates are related through stoichiometric relationships as

$$R_{L1} = [(\beta_{EAL} - \tfrac{1}{2}\beta_{HSL})R_{LG} + \beta_{EAC}R_C](1-\eta)/\eta \tag{10}$$

$$R_{L2} = (\tfrac{1}{2}\beta_{HSL}R_{LG})(1-\eta)/\eta \tag{11}$$

$$R_{L3} = (-R_S)(1-\eta)/\eta \tag{12}$$

812

$$R_{L4} = \left(-R_{LG}\right)\left(1-\eta\right)/\eta \tag{13}$$

where $R_{LG} = R_{S1} + R_{S2}$, $R_C = R_{S3} + R_{S4} + R_{S5}$ and $R_S = R_{LG} + R_C$

At the mixing zone of the impregnation vessel the liquor balance and the density dilutions, due to moisture in wet chips, are

$$v_{Lo} = v_L - \frac{\rho_{So}}{\rho_w} \frac{W_o}{1-W_o} v_{So} \tag{14}$$

and

$$\rho_{Lj} = \frac{v_{Lo}}{v_L} \rho_{Loj} \quad j = 1, ..., 4 \tag{15}$$

For simulation purposes, the model equations are not solved as PDEs but as ODEs with the interpretation of the plug flow sections through a series of CSTRs as shown in Fig 3.

To develop the corresponding ODE model, let the CSTR count for any digester reaction zone be n = 1 ,..., N_{zone}, where numbering starts from the top and, for each zone, N_{zone} can be different. Then, residence times for chips and liquor are

$$\tau_S(n) = \Delta z \frac{1-\eta(n)}{v_S} \tag{16}$$

and

$$\tau_L(n) = \frac{\Delta z \eta(n)}{v_L} \tag{17}$$

For the *cook* zone, solid and liquor material and the thermal balances for any CSTR are

$$\frac{d\rho_{Si}}{dt} = \frac{\rho_{Si}(n-1) - \rho_{Si}(n)}{\tau_S(n)} + R_{Si}(n) \quad i = 1,...,5 \tag{18}$$

$$\frac{d\rho_{Lj}}{dt} = \frac{\rho_{Lj}(n-1) - \rho_{Lj}(n)}{\tau_L(n)} + R_{Lj}(n) \quad j = 1,...,4 \tag{19}$$

and

$$\frac{dT(n)}{dt} = \frac{v_S + v_L}{\Delta z}[T(n-1) - T(n)] \tag{20}$$

For the countercurrent zones the corresponding equations are

$$\frac{d\rho_{Si}}{dt} = \frac{\rho_{Si}(n-1) - \rho_{Si}(n)}{\tau_S(n)} + R_{Si}(n) \quad i = 1,...,5 \tag{21}$$

$$\frac{d\rho_{Lj}}{dt} = \frac{\rho_{Lj}(n+1) - \rho_{Lj}(n)}{\tau_L(n)} + R_{Lj}(n) \quad j = 1,...,4 \tag{22}$$

and

$$\frac{dT(n)}{dt} = \frac{v_S}{\Delta z}[T(n-1) - T(n)] + \frac{v_L}{\Delta z}[T(n+1) - T(n)] \tag{23}$$

The complete model has $10*(N_{cook} + N_{mcc} + N_{emcc})$ ODEs. For each CSTR there are 10 states: five solid densities, four liquor densities and a temperature. Initial conditions for each state are the steady-state nominal operating values. Complete nomenclature and model parameters are listed in Table 1.

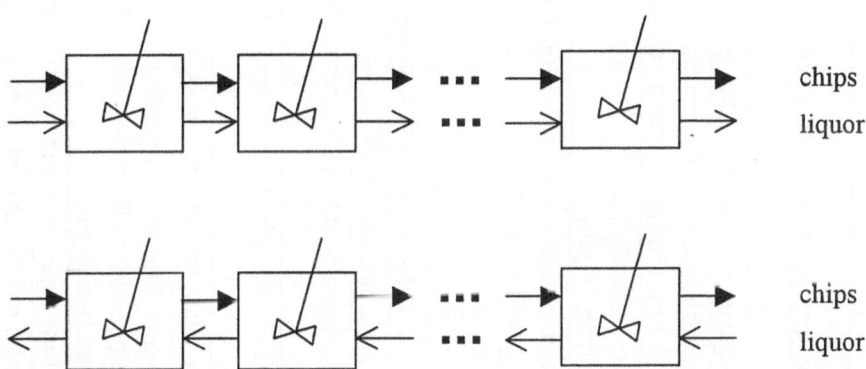

chips

liquor

chips

liquor

Figure 3. CSTR approximations for co-current and countercurrent reaction zones

Table 1. Nomenclature and model parameters

SYMBOL	DESCRIPTION AND UNITS	NOMINAL VALUES
A	Digester cross section area (same for all zones) [m^2]	1
β_{EAC}	Stoichiometric coefficient for mass of effective alkali (EA) consumed/mass of reacting carbohydrate	0.45
β_{EAL}	Stoichiometric coefficient for mass of effective alkali consumed/mass of reacting lignin	0.20
β_{HSL}	Stoichiometric coefficient for mass of hydrosulfide (SH) consumed/mass of reacting lignin	0.05
E_{Ai}	Activation energies, i = 1,…,5 [kJ/gmol]	[29.3 115 34.7 25.1 73.3]
E_{Bi}	Activation energies, i = 1,…,5 [kJ/gmol]	[31.4 37.7 41.9 37.7 167]
η	Volume fraction of liquor in digester, compaction	[0.85 0.80], 0.81, 0.82
$K^{\#}$	Kappa number = 654*lignin mass/total dry solid mass	92.3, 45.1, 29.7
k_{Aoi}	Pre-exponential kinetic factors [m^3 liq/min/kg$_{EA}$]	0.356 1.31x10^{11} 25.3 6.37 5240]
k_{Boi}	Pre-exponential kinetic factors [m^3 liq/min/(kg$_{EA}$ kg$_{EA}$)$^{1/2}$]	11.2 1.68 112 43.1 2.9x10^{16}
θ	Reaction rate effectiveness factor compensating for diffusion effects	0.65
ρ_{Lj}	Liquor component densities, j = 1,…, 4 [kg/m^3 of liquor], indices are respectively: effective alkali (EA), hydrosulfide (HS), dissolved solids and dissolved lignin	
ρ_{Loj}	White liquor component densities [kg/m^3 of liquor]	[100 30 0 0]
ρ_{Si}	Solid component densities, i = 1,…,5 [kg/m^3 of solid], indices are respectively: high reactivity lignin, low reactivity lignin, carbohydrate,	

Symbol	Description	Value
	galactoglucomman and araboxylan	
ρ_{Soi}	Solid component densities in wood chips [kg/m³ of solid]	[150 225 675 75 373]
ρ^∞_{Si}	Unreactive portion of solid component densities in wood chips [kg/m³ of solid]	[0 0 420 18 0]
ρ_w	Water density [kg/m³]	1000
R	Gas constant [kJ/gmol °K]	0.0083144
R_{Lj}	Reaction rate of liquor component j [(kg/min)/(m³ of liquor)]	
R_{Si}	Reaction rate of solid component i [(kg/min)/(m³ of solid)]	
T_C	Cook zone heater temperature [°K]	425
T_e	Emcc zone heater temperature [°K]	415
T_m	Mcc zone heater temperature [°K]	420
τ_L	CSTR liquor residence time [min]	
τ_S	CSTR solid residence time [min]	
v_L	Volumetric liquor flowrate [m³/min]	[0.090 0.090 0.080]
v_S	Volumetric solid (chips) flowrate [m³/min]	0.0267
υ_L	Liquor space velocity [m/min]	
υ_S	Chip (pulp) space velocity [m/min]	
w_o	Wet chip moisture fraction (based on total mass)	0.50
	IV time delay [minutes]	30
	cook zone height [meters]	13.71429
	mcc zone height [meters]	16.84
	emcc zone height [meters]	17.78
	upper extract flowrate [m³/min]	0.09

4. Feedforward Control

A conventional approach to regulate the Kappa number at the blow line is to maintain a constant alkali-to-wood ratio. This is essentially a feedforward policy expected to provide the necessary (stoichiometric) reaction ingredients against variations in wood feed rate. For the purposes of this study, alkali-to-wood ratio is defined in terms of wood and liquor flowrates at nominal conditions as

$$AWR = \frac{v_{Lo}\rho_{Lol}}{v_{So}\rho_{So}} \qquad (24)$$

It is important to quantify the EA density dilution effect of wood moisture and include it into the feedforward policy. This is often overlooked in practice, as on-line wood chip moisture measurement has not been as common as it should be. To formulate the necessary relationship between the manipulated variable, upper extract flowrate, and the measured deviations from the desired AWR, consider the *cook* zone liquor volumetric balance given in Eq 14. Combining it with Eq 24 gives

$$v_{L,ue} = \frac{\rho_{So}}{\rho_w} v_{So} \left(AWR \frac{\rho_w}{\rho_{Lol}} + \frac{W_o}{1 - W_o} \right) \qquad (25)$$

as the desired feedforward control policy.

In the simulation exercise, three of the five disturbances (wood flowrate, moisture and white liquor EA density) are measured at ten-minute intervals. Since there is no time delay or dynamics involved in the process between the disturbances and the necessary feedforward actions, Eq 25 was implemented as stated. A first order low-pass filter was used to prevent sudden reactions to stochastic disturbances.

Figure 4 shows the improvement of the feedforward policy in the performance of the digester. Fig 4(a) is the open loop reaction of the digester to a sequence of stochastic disturbances coming from five different sources: wood flowrate and moisture, white liquor EA density, wood lignin content and dilution (wash) water flowrate. Simulated Kappa numbers at the end of *cook* and *mcc* reaction zones are shown for reference. The main purpose of the controller is to maintain the blow line Kappa number ($K^{\#}_{emcc}$) on target. Fig 4(b) shows the difference feedforward controller makes while measuring and reacting to only the first three of the five disturbances. Although there are significant improvements in the Kappa number dynamics for each zone, regulation is not perfect due to unmeasured disturbances and process nonlinearity. In practice, the rest of the corrections will be done manually, most probably by adjusting temperatures.

Suppose we proceed with the example by trying to compensate for the deficiencies of the constant AWR policy with a simple supplemental PI control loop. The purpose of the PI controller will be to help regulate the final Kappa number. Considering the long time

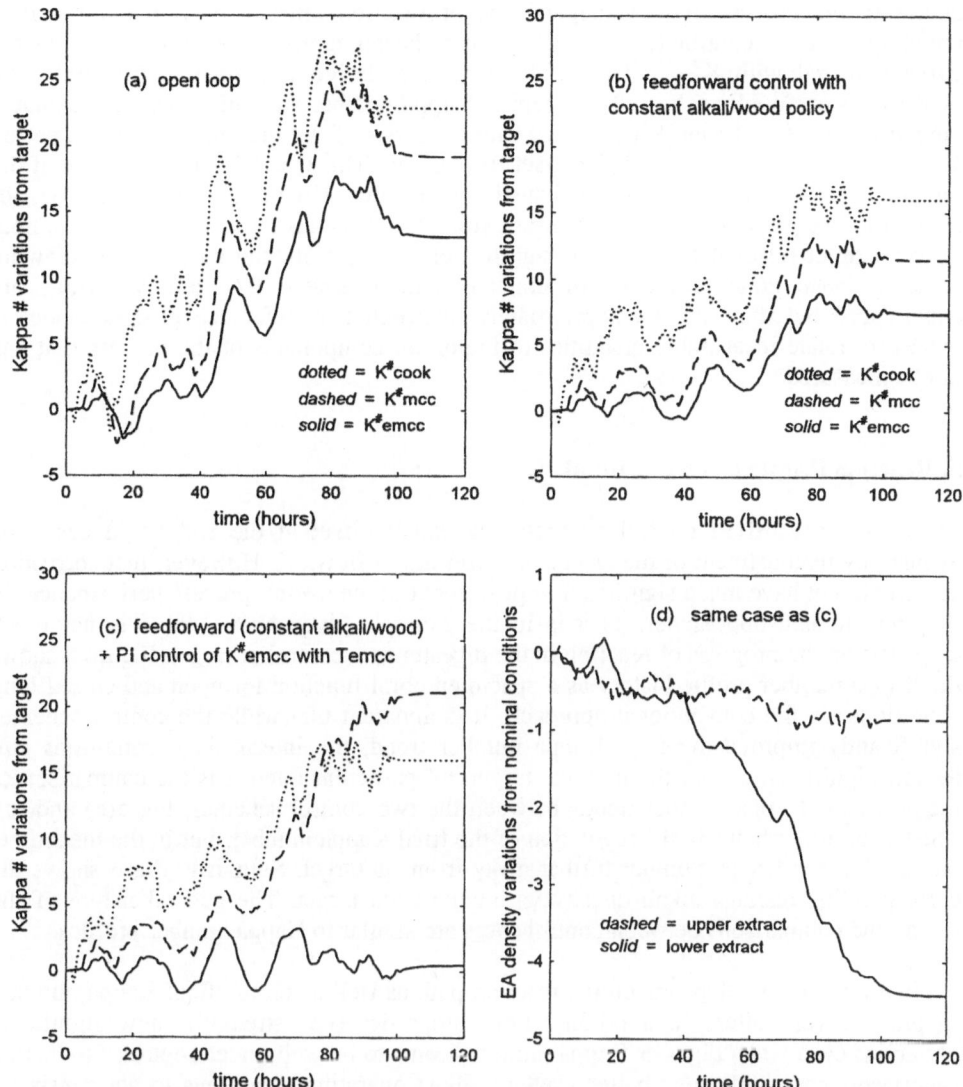

Figure 4. Performance of conventional control approaches in regulating blow line Kappa #. Measured disturbances are wood flowrate, moisture and white liquor EA density. Variations in wood lignin density and dilution water flowrate are not measured.

delays in the process and the dependence of reaction kinetics on temperature, it is tempting to use a control loop using the *emcc* heater temperature as the manipulated variable paired with $K^{\#}_{emcc}$. The result of such a design is shown in Fig 4(c). The objective is essentially achieved by keeping Kappa number variability within reasonable bounds against significant disturbances. Close analysis of the final no-disturbance period 100 to 120 hours, indicates a slight offset from target. This is not due to the failure of the integral controller, but due to the exhaustion of residual EA in the *emcc* zone. Fig 4(d) shows the changes in the upper and lower extract EA densities through the control period. Lower extract residual EA strength significantly drops from the target value showing signs of *emcc* zone problems. In practice, similar severe EA density swings are experienced but they are not appropriately interpreted. A reference process model is needed to relate measurable quantities to important components of the process that can not be measured.

5. Reaction Profile Control with MPC

In the feedforward+PI control example, the small offset at the end could easily be avoided by readjustment of the *emcc* zone trim liquor flowrate. However, that correction still would not have made significant improvement in the overall process performance. In the conventional approaches, as it is in the example given above, there is not much emphasis on the progress of reaction in the digester as a controller target. Figure 5 shows the Kappa number profile history as a spaciotemporal function for open and closed loop conditions for the conventional approach. It is apparent that while the control schemes significantly improve the final Kappa number trend, the intermediate conditions still deviate significantly from the nominal target. Of particular interest is the comparison of the *mcc* zone Kappa number trends between the two control schemes, Fig 5(b) and (c). The PI controller helps in the regulation of the final Kappa number, but in the meantime, pushes the *mcc* Kappa number further away from its target. Similarly, Fig 6 shows the corresponding residual alkali density variations from target. The general nature of the trends and comparisons between control cases are similar to Kappa number profiles.

Final pulp quality depends on the reaction path as well as on the final Kappa number. In practice very often, final product fiber properties (i.e., strength) show significant variations even when digester Kappa number seems to be well under control. Production supervisors constantly seek better ways to adjust operating conditions to get consistent fiber properties. Therefore, it seems reasonable to seek a control policy that maintains the reaction path of a successful operating condition. This would achieve the constancy of both the final Kappa number and the valuable fiber properties. Consequently, it would also minimize the variations in alkali and dissolved solids densities of the extract, and stabilize the load on downstream pulp cleaning and bleaching processes.

Controlling plug flow reactors as distributed parameter systems is an established approach. From a practical point of view, it is usually sufficient to work with a finite number of measurements and control decisions. Kappa numbers at the end of each reaction zone and the upper and lower extract EA densities are selected here as key

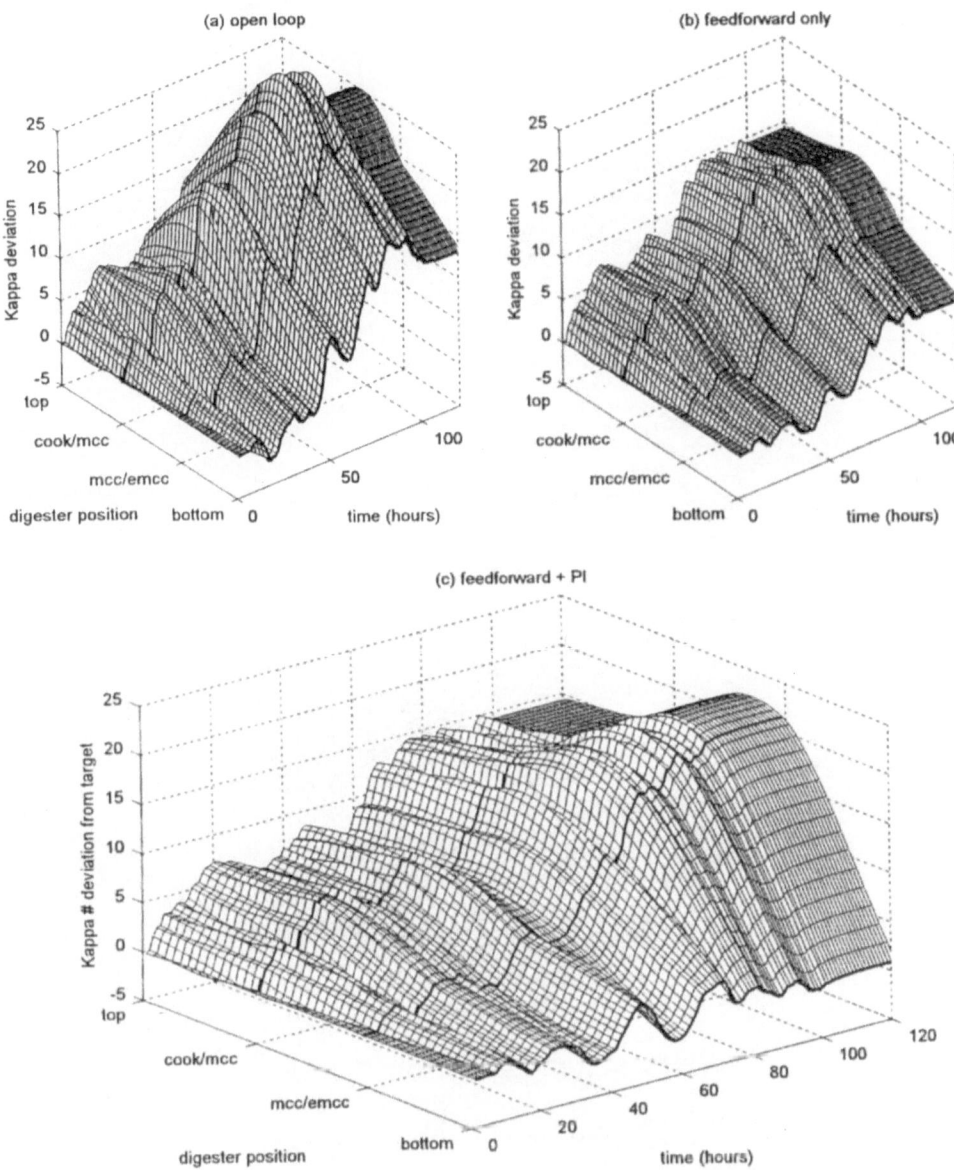

Figure 5. Changes in Kappa number profiles due to conventional control approaches. Profiles are plotted as deviations from the steady-state target.

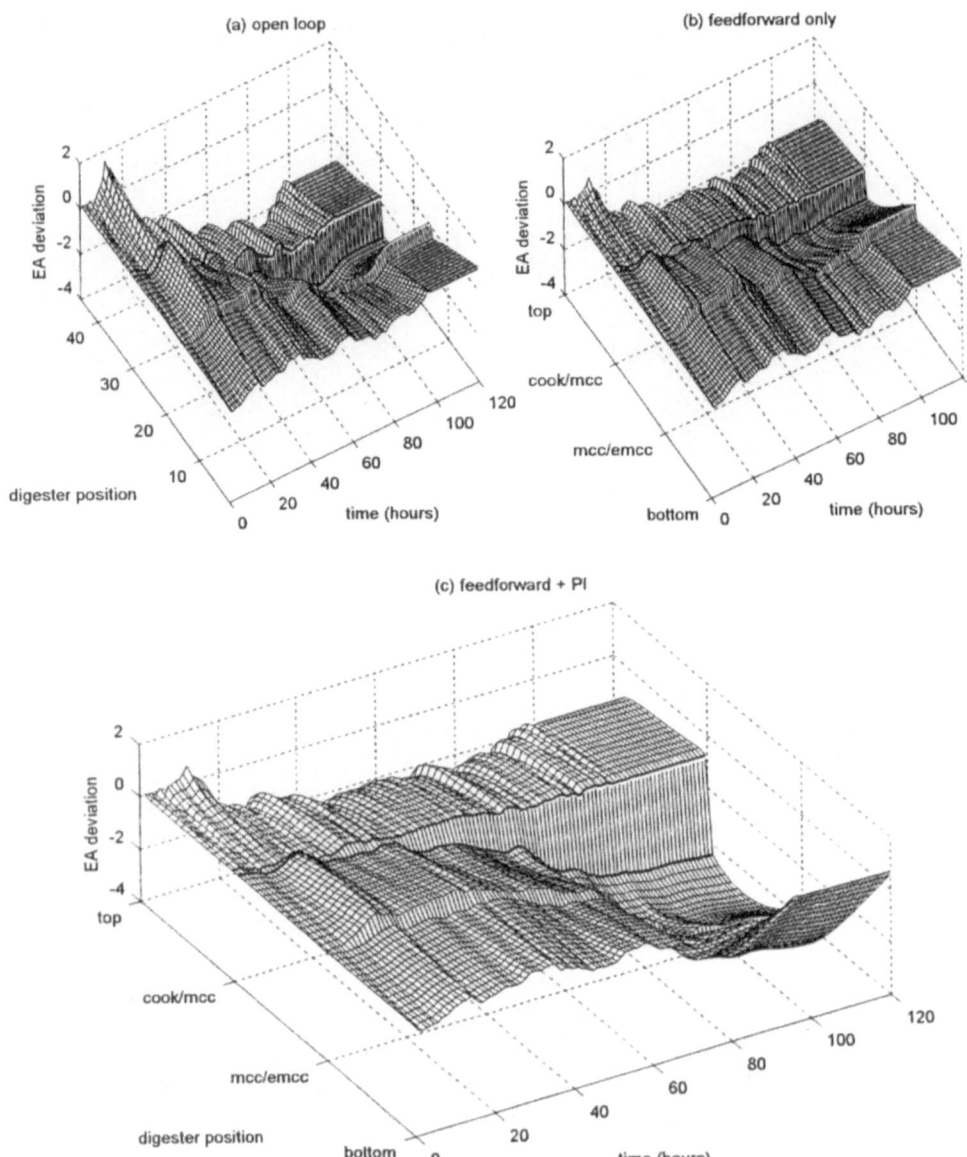

Figure 6. Changes in residual EA density profiles due to conventional control approaches. Profiles are plotted as deviations from the steady-state target.

reaction path indicators. Of these five, the two intermediate Kappa numbers can not be measured directly and must be inferred. Different ways to estimate unmeasurable digester properties for control purposes are demonstrated in [6] and [7]. In order to maintain the focus of this work towards the demonstration of potential benefits of reaction path control, we will proceed by assuming that the simulated values of $K^{\#}_{cook}$ and $K^{\#}_{mcc}$ are available as estimations for controller feedback.

In order to form a square (5x5) control system, the three reaction zone temperatures, upper extract and the *mcc* trim liquor flowrates were selected to be the manipulated variables. The *emcc* trim liquor flowrate is also available as an independent decision variable to be potentially used to relax any constraining condition, as experienced in the previous example. For this part of the study, the *emcc* trim was kept constant at its nominal flowrate.

Naturally, a linear reference model is needed for MPC implementation. Through sequences of step tests with the nonlinear benchmark model, a multi-dimensional step-response model was generated that relates both the measurable disturbances and the manipulated variables to the five process measurements. As a brief reminder, the SISO equivalent of the step-response model has the form

$$y(k) = \sum_{i=1}^{N} s^i \Delta u(k - i) + y^*(k) \tag{26}$$

where k is the time sequence, s^i is the ith step-response coefficient (i = 1,..,N) and

$$y^*(k) = y^*(k - 1) + s^N \Delta u(k - N - 1) \tag{27}$$

A simple comparison of the step-response model to the nonlinear model is shown in Fig 7. Prediction accuracy for the key Kappa numbers are displayed. The inputs to both models are the same sequence of integrated random (i.e. random walk) variations in both the disturbance and the manipulated variables. To summarize, the control relevant variables are:

U = [upper extract flowrate; temperatures of *cook*, *mcc* and *emcc* heaters; *mcc* trim white liquor flowrate]

Y = [Kappa numbers at the end of *cook*, *mcc* and *emcc* zones, upper and lower extract residual EA densities]

D – [chip flowrate; chip moisture; white liquor EA density; chip lignin density; dilution flowrate] (only first three are treated as measurable)

Fig 7 shows that for short durations, the linear model does a satisfactory job of predicting the Kappa numbers. However, for long simulation times, the differences between the linear and the nonlinear model become significant as the prediction errors accumulate. Nevertheless, the step-response model is sufficient for the purposes of this work. We can safely use a linear reference model to regulate the nonlinear process, as the controller will keep the operating conditions relatively close to the target. However, for transition control applications nonlinear methods would be necessary.

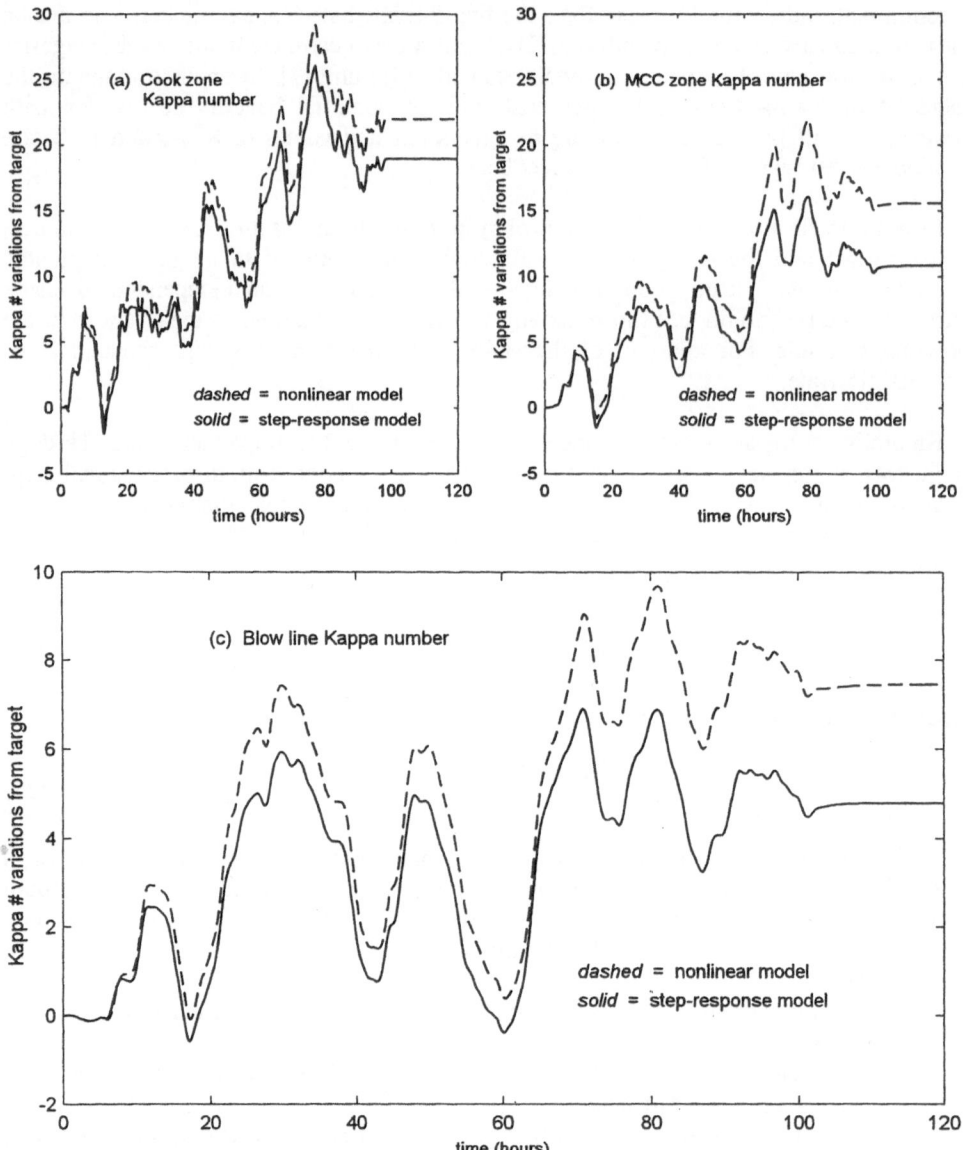

Figure 7. Comparison of step-response and nonlinear model dynamic predictions in response to a sequence of stochastic variations in manipulated and disturbance variables.

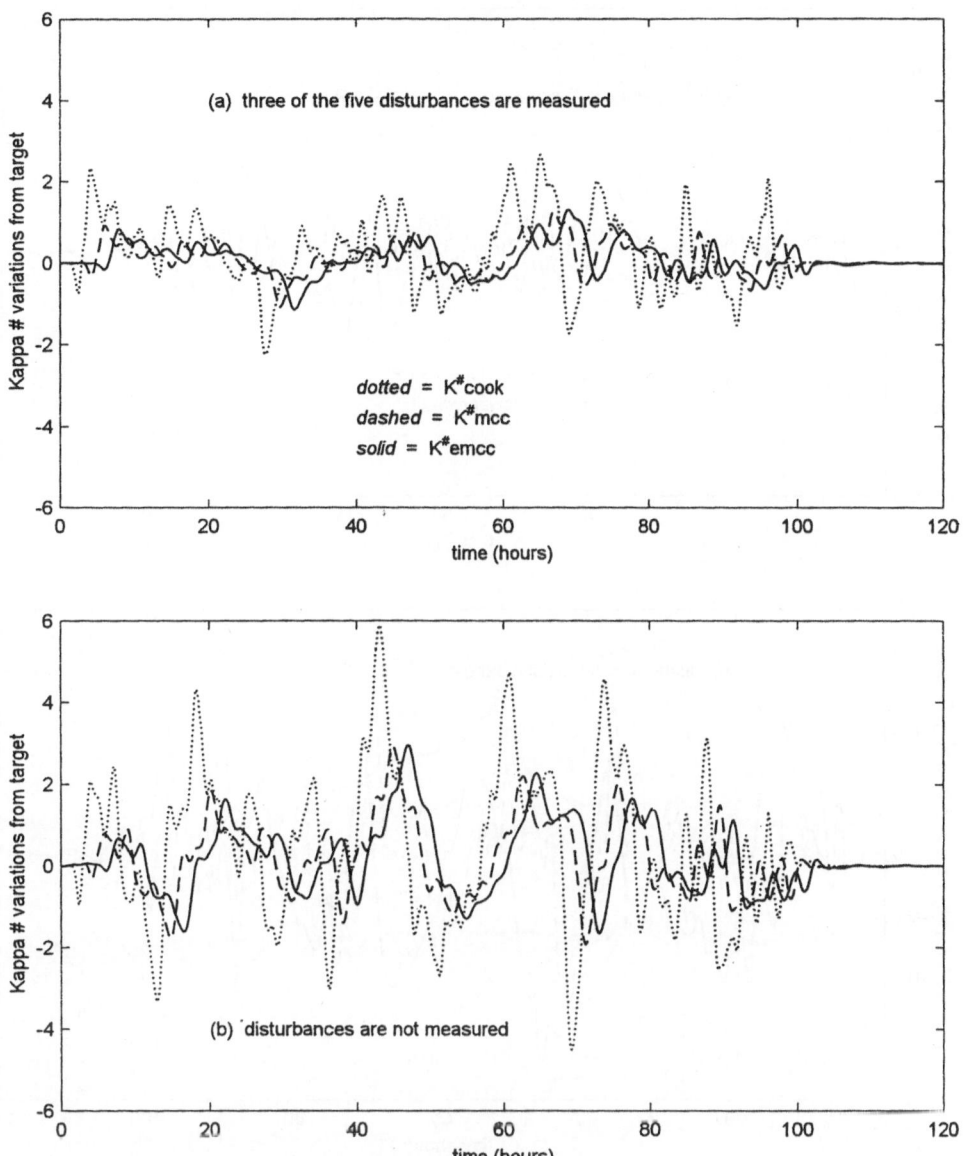

Figure 8. Regulation of reaction zone Kappa numbers along the digester with MPC, showing the improvement potential due to disturbance measurements.

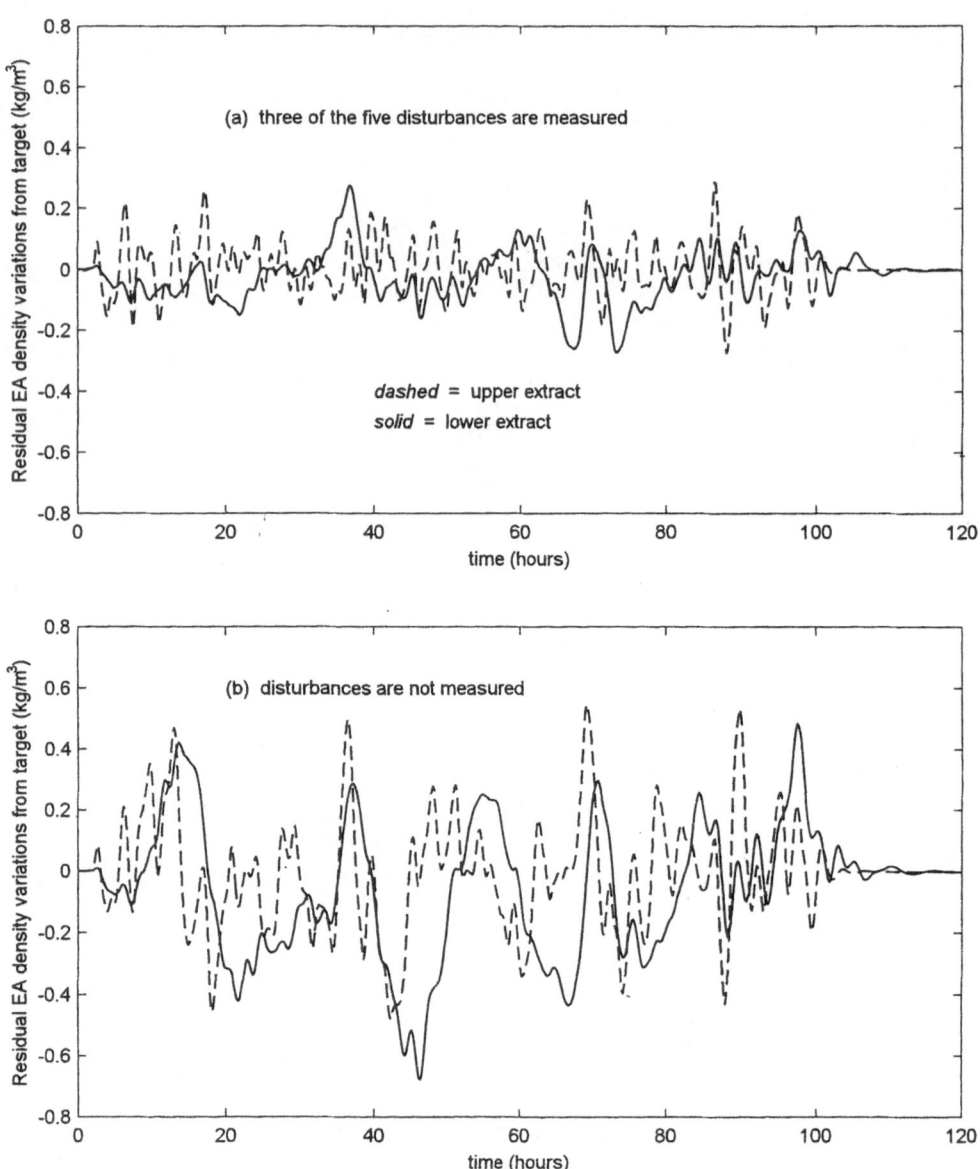

Figure 9. Regulation of upper and lower extract EA densities with MPC, as part of the control structure demonstrated in Figure 8.

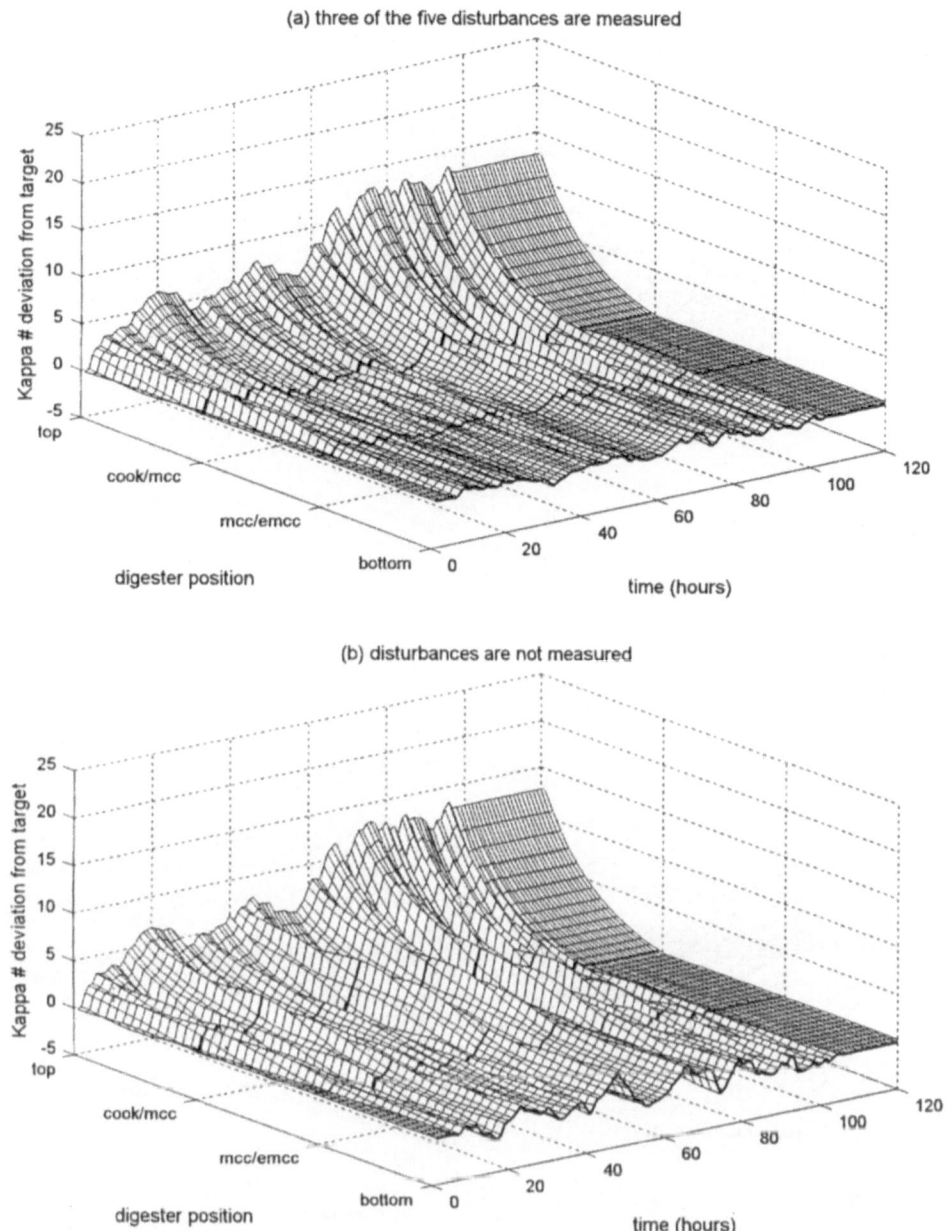

Figure 10. Kappa number profile variability during digester reaction profile control with MPC.

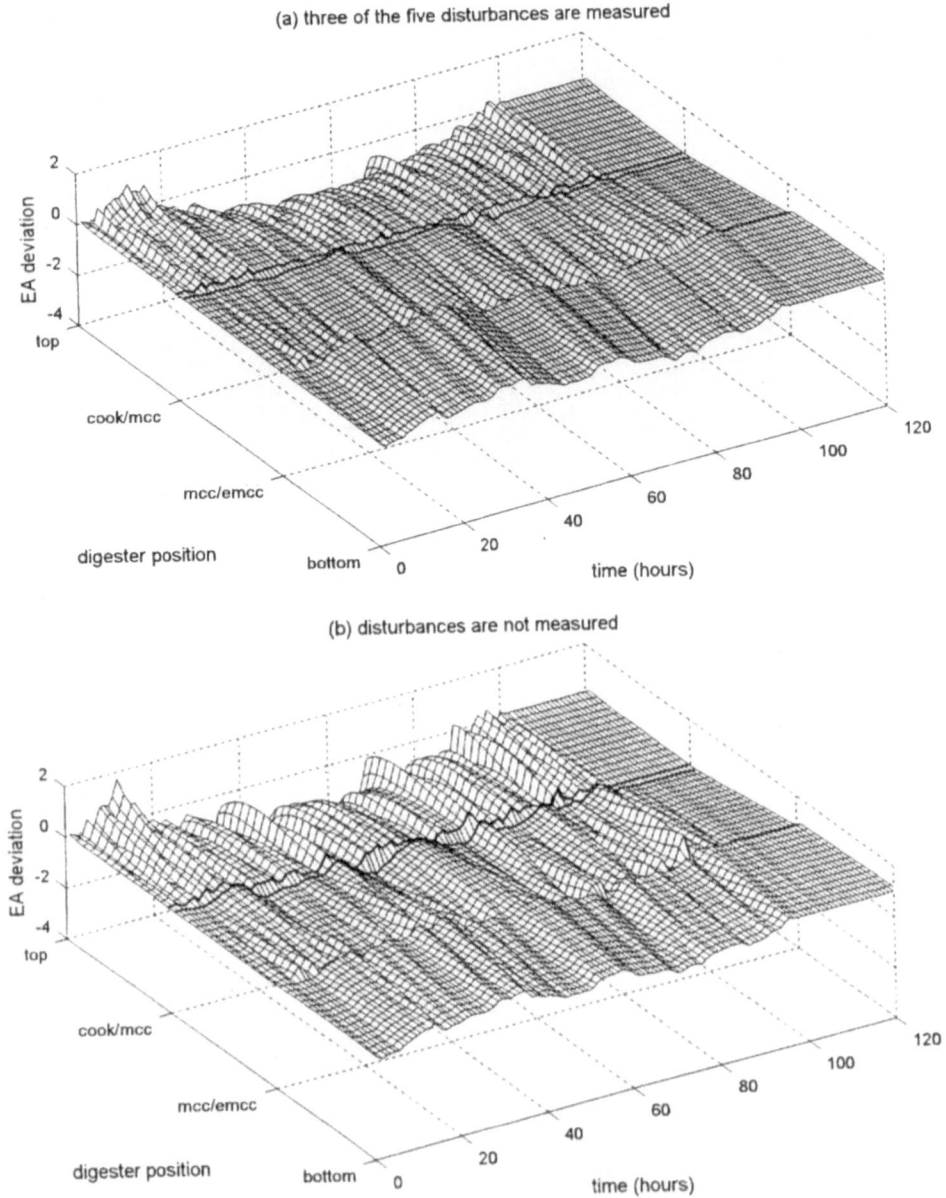

(a) three of the five disturbances are measured

(b) disturbances are not measured

Figure 11. Residual EA profile variability during digester reaction profile control with MPC.

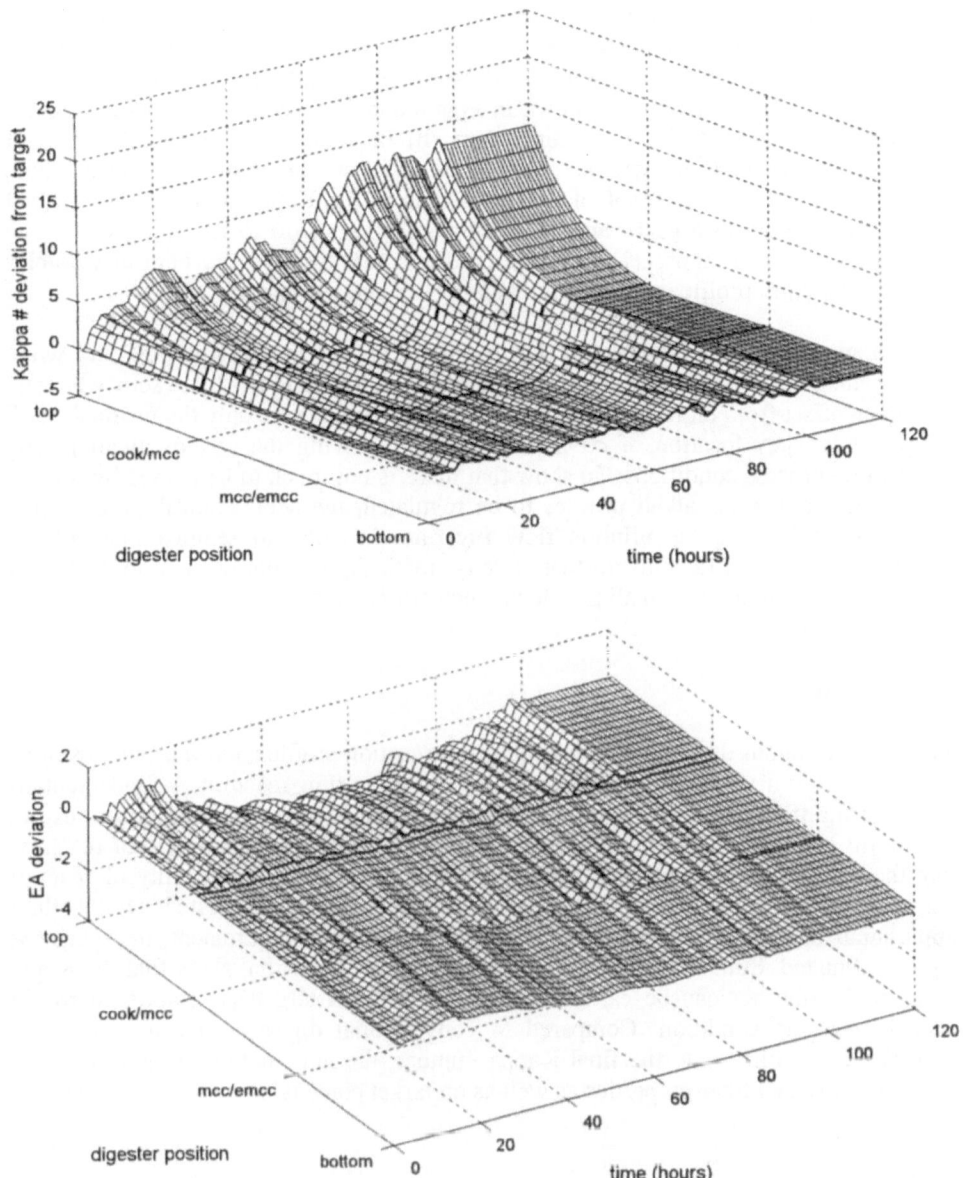

Figure 12. Kappa number and residual EA density profile variabilities with MPC. Four disturbances are active (dilution is invariant) while three are measured.

828

Using the same disturbance sequence as before, 5x5 unconstrained MPC was implemented using low pass diagonal measurement filters. As severe as the disturbances are, there was no need for constraint consideration. The results for Kappa number and residual EA density regulation are shown in Figs 8 and 9 respectively. As expected, the performance of the MPC is outstanding. Part (b) of each figure is the evidence of performance deterioration without the measurement of otherwise measurable disturbances. A convincing proof of the method is in the Kappa number and residual alkali profiles shown in Figs 10 and 11. Comparing Fig 10(a) to its equivalent in the improved conventional approach, Fig 5(c), confirms that the five key control variables selected for reaction profile regulation are appropriate and sufficient.

As a final note, in Fig 11 observe that disturbances due to dilution flow and white liquor alkali strength cause most of the *mcc* and *emcc* profiles to fluctuate somewhat and maintain an offset from target. By choice, the control variables are only the residual alkali densities at the key location of *cook/mcc* interface, leaving the rest of alkali profile dependent on process conditions. To show that there is not much to be gained by forcing all of the *mcc* and *emcc* alkali profiles to be regulated, the MPC control simulation is repeated while keeping the dilution flow invariant in order to remove most of the disturbance source. Fig 12 confirms that there is practically no improvement in the Kappa number profile although the alkali profile is much closer to target.

6. Conclusions

Control of continuous digesters is formulated as a reaction profile control problem similar to the established methods for distributed parameter plug flow reactors. A fundamentally based but simplified nonlinear benchmark process model is used as a reference to test the new control approach. Using three Kappa numbers to indicate the progress of reactions along the digester and two residual alkali densities to measure the stability of reaction conditions, a 5x5 unconstrained MPC structure is developed and tested. Of the three Kappa numbers, only one is physically available for direct measurement, the other two must be estimated. Simulation results, assuming perfect estimation, show that the Kappa profile of the digester can be easily regulated to stay on target as established by the nominal operating condition. Compared to conventional digester control approaches, which aim to stabilize only the final Kappa number, the new method is more likely to guarantee consistent fiber properties as well as on target conversion.

7. References

1. Christensen, T., Albright, L.F. and Williams, T.J. (1982) A mathematical model of the kraft pulping process, Technical Report 129, Purdue University.
2. Gustafson, R.R., Sleicher, C.A.., McKean, W.T. and Finlayson, B.A. (1983) Theoretical model of the kraft pulping process, *Ind. Eng. Chem. Process Des. Dev.*, **22**(1), pp87-96.

3. Funkquist, J. (1995) Modelling and identification of a distributed parameter process: the continuous digester, PhD thesis, Royal Institute of Technology, Stockholm.
4. Michelsen, F.A. (1995) A dynamic mechanistic model and model-based analysis of a continuous Kamyr digester, PhD thesis, University of Trondheim.
5. Kayihan, F., Gelormino, M.S., Hanczyc, E.M., Doyle, F.J. and Arkun, Y. (1996) A Kamyr continuous digester model for identification and controller design, *proc. of IFAC World Congress*, pp 37-42, San Francisco, 30 June-5 July.
6. Amirthalingam, R. and Lee, J.H. (1997) Subspace identification based inferential control of a continuous pulp digester, *proc. of PSE/ESCAPE '97*, May 26-29, Trondheim
7. Wisnewski, P.A., Doyle, F.J., Kayihan, F. and Arkun, Y. (1997) Measurement selection for model predictive control of the Weyerhaeuser benchmark digester problem, *proc. of ADCHEM '97*, June 9-11, Banff, Alberta.
8. Wisnewski, P.A., Doyle, F.J. and Kayihan, F. (in press) A fundamental continuous pulp digester model for simulation and control, *AIChE Journal*.

ARTIFICIAL NEURAL NETWORKS FOR NONLINEAR CONTROL OF INDUSTRIAL PROCESSES*

MASOUD NIKRAVESH
ESD of Lawrence Berkeley National Laboratory and
BISC of University of California at Berkeley, Berkeley, CA 94706, USA

Abstract

In this paper, the state-of-the-art methodology for model identification and control of fluid injection into petroleum reservoir is presented. The methodology uses several techniques: Dynamic Neural Network Control (DNNC) network structure, neuro-statistical (neural network, conventional statistical, and non-parametric statistical such as ACE; Alternative Conditional Expectation) techniques, Geometric Model-Based Control Strategy, and stability analysis techniques such as Liapunov theory. In this study, the DNNC model is used because it is much easier to update and adapt the network on-line. The neuro-statistical technique is used because the model can deal with uncertainty in data. The neuro-statistical model uses neural network techniques, since the functional structure of the data is unknown and uses statistical technique because the data and our requirements are imperfect. In addition, this technique in conjunction with Levenberge-Marquardt algorithm can be used as a more robust technique for network training and optimization purposes. The ACE technique is used for scaling the network's input-output data and can be used to find the input structure of the network. The result from Liapunov theory is used to find optimal neural network structure. In addition, a special neural network structure is used to insure the stability of the network for long-term prediction. In this model, the current information from the input layer is presented into a pseudo hidden layer. This model minimizes not only the conventional error in the output layer but also minimizes the filtered value of the output. This technique is a tradeoff between the accuracy of the actual and filtered prediction which will result in the stability of the long-term prediction of the network model. In this paper, an optimal neural network controller will be designed for a well-head controller for steam injection into low permeability petroleum reservoirs for oil displacement.

---

* This paper is a summary of the paper which was originally presented in the 1997 SPE Western Regional Meeting, SPE Paper 38275. Part of this paper is from a full paper submitted to Comp. & Chem. Engineering Journal. In addition, part of this paper is submitted to 1997 SPE International Operations and Heavy Oil Symposium, SPE Paper 37557. This paper is prepared as a lecture note for the NATO- ASI on Nonlinear Model Based Process Control.

831

R. Berber and C. Kravaris (eds.), Nonlinear Model Based Process Control, 831-869.
© 1998 *Kluwer Academic Publishers.*

1. Introduction

Nonlinear, time varying behavior is common in petroleum processes. For example, both isothermal and thermal oil displacement processes exhibit such behavior due to factors such as changes in permeability, changes in oil viscosity, changes in the reservoir temperature, and fracture extension. When a change in the reservoir parameters occurs, the wellhead controller frequently needs to be retuned in order to maintain satisfactory performance. Retuning the controller for nonlinear and/or non-stationary processes is usually time consuming requiring a combination of operational experience and trial-and-error. In addition, for fluid injection into low permeability rocks such as Diatomaceous field of California, little is known about the complex dynamics governing the process which makes the problem more exciting and challenging.

In the past few years, the above consideration has stimulated growing interest among control engineers and reservoir engineers [1,2] in the area of nonlinear process control especially neural network-based control strategies. Despite our incomplete knowledge, neural network models have been able to predict the complex behavior of these processes [3]. Furthermore, the neural networks have been used to design model-based controllers leading to more effective process control.

Neural networks offer the ability to generate nonlinear models for systems which are difficult to model from first principles. Neural networks with three layers can approximate any nonlinear function and generate complex decision regions for input-output mapping [4, 5]. This fact suggests that they hold great promise for modeling very complicated nonlinear systems and offer a cost-effective solution to control highly nonlinear processes. To date, the backpropagation neural network has been applied successfully in many process modeling and identification applications [6-12]. This fact suggests that neural networks, in conjunction with suitable control strategy such as Model-Based Control [11-15], State-Space and Geometric control [16-18], and Neuro-Fuzzy Control [19] can be used to control nonlinear systems.

During the past few years, several backpropagation neural network control algorithms have been proposed. Neural networks are now widely used in many nonlinear control applications [6, 7, 11, 14-16, 19-21]. Recently, neural networks have been employed in Model Predictive Control (MPC) [12-14], Internal Model Control (IMC) [15], Dynamic Matrix Control (DMC) [11, 13, 22, 23], and Adaptive Control [24, 25]. The neural network models which have been proposed for process control are complex and have several nodes in the input and hidden layers, as well as a large number of weights and bias terms. These weights and bias terms are usually initialized randomly, using no prior process knowledge to reduce network training time.

If the networks are trained using sufficient information and the parameters of the process do not change over time, neural networks give excellent prediction. However, process parameters of many systems do change over time and therefore, the performance of the network for prediction gradually deteriorates. To solve this problem, it is necessary to continuously update the parameters in the network. Continuously updating and on-line adaptation of the complex neural network structure raises a number of issues including a general approach for updating, the numerical method for recursive updating, and the speed of updating. Furthermore, in the indirect method, the neural network is trained to predict future process outputs from past and present inputs and outputs. In this approach, the inverse of the model at each sampling time must be found via an optimization routine to calculate controller outputs. Since the neural network model is frequently complex, calculation of its mathematical inverse is difficult and time consuming. In this paper, using the concept of linearization, we reduce the complexity of the mathematical expressions of the neural network models around the operating point of the process. In this case, a simpler neural network model equivalent to the more complex one is developed for any operating point of the process. It is important to mention that with linearization, we can approximate the nonlinear equations. But the more non-linear a function is, the smaller the region over which the linear approximation is accurate. However, since we update the semi-linearized model of the neural network model continuously around the process operating point, the suggested simplified neural network model will be accurate over the entire region of operating condition. As a result of the simplicity of the semi-linearized model, controller synthesis (including design, implementation, stability analysis, performance analysis, and updating) is easier using the simplified model rather than the original one. In addition, the inverse of the process is exact and does not involve approximation error. It is also easier to represent the simplified network model in the state-space model than the complex one. Once the systems are transformed into state-space models, the Liapunov theory can be used for stability analysis and in conjunction with state-space and geometrical theory can be applied to the design of reliable control systems.

In this paper, Dynamic Neural Network Control (DNNC) is presented as a control strategy which uses a neural network to model the process and then applies the mathematical inverse of the process model as the controller. DNNC falls into the large class of Model Predictive Controllers, many of which are now widely used in industry [13, 26]. The DNNC strategy differs from previous neural network controllers, neural network DMC, and MPC hybrid controllers because it offers physical meaning to the parameters. The network structure of single hidden layer DNNC is very simple, comprising limited input and hidden layer nodes and an output layer node for the SISO case. Yet in the nonlinear cases studied thus far, DNNC performs as well as, if not better than, more complex neural network control strategies. DNNC offers potential for fewer weights and bias terms, therefore, better initialization of weights and fast on-line updating of weights are possible [21]. In addition, DNNC provides tuning parameters which are analogous to those of DMC.

The structure of the paper will be as follows. First, a brief overview of neural networks with backpropagation learning [27-29], neuro-statistical model, and modified Levenberge-Marquardt algorithm will be presented. Then the linearized technique for simplifying the complex structure of the neural network will be given. The updating approach for the simplified and the complex model followed by a brief introduction on stability analysis will be discussed later. Next, a stability analysis of the neural network model will be discussed according to the Liapunov theory and based on the state-space transformation of the network model. The results from the stability analysis and the state-space model will be used to determine an optimal neural network structure, to design a reliable neural controller, and to analyze the controller

834

performance. Finally, a neural network controller will be designed for a well-head controller for steam injection into low permeability petroleum reservoirs for oil displacement.

2. Neural Network

During the last decade, application of the neural networks for modeling the complex multi-dimentional field data greatly increased. These widespread applications have been due to several attractive features of neural networks. For example, these models do not require specification of a structural relationship between input and output data and can extract and recognize underlying patterns, structures, and relationships between data. However, developing a proper neural network model that is an "accurate" representation of the data may be an arduous task which requires sufficient experience with the qualitative effects of the structural parameters of neural network models, scaling techniques for input-output data, and a minimum insight into the physical behavior of the model. In addition, neural network models are frequently complex, need a large number of precise data, and the underlying patterns and structure are not easily visible. Conventional neural networks are also not usually stable and their performances are seriously affected once subjected to long-term prediction. Also, unlike statistical methods, conventional neural network models can not deal with probability.

2. 1. THE BACKPROPAGATION NEURAL NETWORK

Details of the backpropagation neural network are available in the literature [4, 5]. Therefore, only the important characteristics of the network will be mentioned here. The typical backpropagation neural network has an input layer, an output layer, and at least one hidden layer. Each layer is fully connected to the succeeding layer with corresponding weights. In a neural network, the nonlinear elements are called nodes, neurons, or processing elements.

Consider a single neuron with a transfer function ($y1^{(i)} = f(z^{(i)})$), connection weights, w_j, and node threshold, θ. For each pattern I,

$$z^{(i)} = x_1^{(i)} w_1 + x_2^{(i)} w_2 + \bullet \bullet \bullet + x_N^{(i)} w_N + \theta$$
$$\text{for } i = 1, ..., P.$$

(1)

All patterns may be represented in matrix notation as,

835

$$\begin{bmatrix} z^{(1)} \\ z^{(2)} \\ \bullet \\ \bullet \\ \bullet \\ z^{(P)} \end{bmatrix} = \begin{bmatrix} x_1^{(1)} & x_2^{(1)} & \bullet & \bullet & \bullet & x_N^{(1)} & 1 \\ x_1^{(2)} & x_2^{(2)} & \bullet & \bullet & \bullet & x_N^{(2)} & 1 \\ \bullet & \bullet & \bullet & \bullet & \bullet & \bullet & \bullet \\ \bullet & \bullet & \bullet & \bullet & \bullet & \bullet & \bullet \\ \bullet & \bullet & \bullet & \bullet & \bullet & \bullet & \bullet \\ x_1^{(P)} & x_2^{(P)} & \bullet & \bullet & \bullet & x_N^{(P)} & 1 \end{bmatrix} \begin{bmatrix} w_1 \\ w_2 \\ \bullet \\ \bullet \\ \bullet \\ w_N \\ \theta \end{bmatrix}$$

(2)

and,

$$\underline{y1} = F(\underline{z}).$$

(3)

In more compact notation,

$$\underline{z} = \underline{\underline{X}}_1 \ \underline{w}_\theta = \underline{\underline{X}} \ \underline{w} + \underline{\theta}$$

(4)

where,

$$\underline{w}_\theta = \left[\underline{w}^T \mid \theta \right]^T$$

(5)

$$\underline{\underline{X}}_1 = \left[\underline{\underline{X}} \mid \underline{1} \right]$$

(6)

and,

$$\underline{1} = \text{column vector of ones with P rows}$$
$$\underline{\underline{X}} = \text{P x N matrix with N input and P pattern}$$
$$\underline{\theta} = \text{bias vector, vector with P rows of } \theta$$
$$\underline{w} = \text{weights, vector with N rows.}$$

During learning, the information is propagated back through the network and used to update the connection weights. The objective function for the training algorithm is usually set up as a squared error sum,

$$E = \frac{1}{2} \sum_{i=1}^{P} (y_{(observed)}^{(i)} - y_{(prediction)}^{(i)})^2.$$

(7)

This objective function defines the error for the observed value at the output layer which is propagated back through the network. During training, the weights are adjusted to minimize this sum of squared errors.

3. Dynamic Neural Network Control Strategy (DNNC), Basic Concept

This section develops the DNNC algorithm. Because DNNC is analogous to DMC, a brief explanation of the basic DMC algorithm is provided so that the similarities between DMC and DNNC can be established and so that notation is clear. Then the equations for single hidden layer DNNC will be developed. Finally the training procedure for DNNC will be presented.

3.1. DMC MODEL PREDICTION: (LINEAR INPUT-OUTPUT MODEL)

The DMC algorithm is well-established and discussed throughout the literature [30-32]. Consider a linear dynamic single-input/single-output system. A discrete representation of the process dynamics with a step-response model is given by,

$$y(k+j) = \sum_{i=1}^{j} a_i \ \Delta u(k-i+j) \ + \ y^*(k+j) \ + \ d(k+j) \tag{8}$$

where,

$$y^*(k+j) = y_0(k+j) + \sum_{i=j+1}^{N} a_i \ \Delta u(k+j-i) \tag{9}$$

$$y_0(k+j) = a_N \ u(k-N+j-1) \tag{10}$$

$$d(k) \ = \ y_m(k) \ - \ y(k) \tag{11}$$

and,

k	= discrete time
$y(k)$	= model output
$\Delta u(k)$	= change in the input (manipulated variable) defined as $u(k)- u(k-1)$
$d(k)$	= unmodelled disturbance effects on the output
a_i	= unit step response coefficients
N	= number of time intervals needed to describe the process dynamics

(Note: $a_i = a_N$ for $i \geq N$)

$y_{m(k)}$	= current feedback measurement
$y^*(k+j)$	= predicted output at $k+j$ due to input moves up to k.

In the absence of any additional information, it is assumed that

$$d(k+j) \ = \ d(k) \tag{12}$$

For N manipulated variables considered during the time interval of k-N to k+N, the equation for DMC may be represented as,

$$
\begin{bmatrix}
\Delta y(k+1) \\
\Delta y(k+2) \\
\Delta y(k+3) \\
\bullet \\
\Delta y(k+N) \\
\bullet \\
\Delta y(K+P)
\end{bmatrix}
=
\begin{bmatrix}
\Delta u(k) & \Delta u(k-1) & \Delta u(k-2) & \bullet & \Delta u(k-N+1) & \bullet & 0 & 1 \\
\Delta u(k+1) & \Delta u(k) & \Delta u(k-1) & \bullet & \Delta u(k-N+2) & \bullet & 0 & 1 \\
\Delta u(k+2) & \Delta u(k+1) & \Delta u(k) & \bullet & \Delta u(k-N+3) & \bullet & 0 & 1 \\
\bullet & \bullet & \bullet & \bullet & \bullet & \bullet & \bullet & \bullet \\
\Delta u(k+N-1) & \Delta u(k+N-2) & \Delta u(k+N-3) & \bullet & \Delta u(k) & \bullet & 0 & 1 \\
\bullet & \bullet & \bullet & \bullet & \bullet & \bullet & \bullet & \bullet \\
0 & 0 & 0 & \bullet & \Delta u(k+P-N) & \bullet & \Delta u(k-1) & 1
\end{bmatrix}
\begin{bmatrix}
a_1 \\
a_2 \\
a_3 \\
\bullet \\
a_N \\
\bullet \\
a_{P+1} \\
d(k)
\end{bmatrix}
\tag{13}
$$

where,

$$\Delta y(k + j) \;=\; y(k + j) \;-\; y_0(k + j)$$

In matrix notation,

$$\underline{\Delta y} \;=\; \underline{\underline{\Delta U}}_1 \; \underline{a}_d. \tag{14}$$

Comparing equation (14) for DMC with equation (4) for the neural network, we see that the following analogy may be drawn,

$$\begin{aligned}\underline{\Delta y} &= \underline{z}\\ \underline{\underline{\Delta U}}_1 &= \underline{\underline{X}}_1\\ \underline{w}_\theta &= \underline{a}_d.\end{aligned} \tag{15}$$

3.2. DMC CONTROLLER DESIGN

The DMC model equation is defined by equation (8). By replacing y with the desired value, y^{set} and rearranging, the DMC equation becomes,

$$\begin{bmatrix} y^{set} - y^*(k+1) - d(k) \\ \bullet \\ \bullet \\ \bullet \\ y^{set} - y^*(k+P) - d(k) \end{bmatrix} = \underline{e}(k+1) = \underline{\underline{A}} \; \underline{\Delta u}(k) \tag{16}$$

where,

$$\underline{\underline{A}} = \begin{bmatrix} a_1 & 0 & 0 & \bullet & 0 \\ a_2 & a_1 & 0 & \bullet & 0 \\ \bullet & \bullet & \bullet & \bullet & \bullet \\ a_N & a_{N-1} & a_{N-2} & \bullet & a_{N-M+1} \\ \bullet & \bullet & \bullet & \bullet & \bullet \\ a_P & a_{P-1} & a_{P-2} & \bullet & a_{P-M+1} \end{bmatrix}, \tag{17}$$

and $\underline{e}(k+1)$ is a P dimensional vector of predicted deviations from the set point.

The following objective function is used to find the M future controller moves $\Delta u(k),\dots,\Delta u(k+M-1)$,

$$\underset{\Delta u}{\text{Min}} \left\{ \left[\sum_{i=1}^{P} \gamma^2 \; [y^{set}(k+i) - y(k+i)]^2 \right] + \left[\sum_{j=1}^{M} \lambda^2 \; [\Delta u(k+M-j)]^2 \right] \right\}. \tag{18}$$

The solution of such least-squares problems is given by,

$$\underline{\Delta u}(k) = \left[\underline{\underline{A}}^T \ \underline{\underline{\Gamma}}^T \underline{\underline{\Gamma}} \ \underline{\underline{A}} + \underline{\underline{\Lambda}}^T \ \underline{\underline{\Lambda}}\right]^{-1} \ \underline{\underline{A}}^T \ \underline{\underline{\Gamma}}^T \ \underline{\underline{\Gamma}} \left[\underline{e}(k+1)\right] \tag{19}$$

where,

$$\underline{\underline{\Lambda}} = \text{diag}(\ \lambda \ \ \lambda \ \ \cdots \ \lambda \ ) \tag{20}$$
$$|\!\leftarrow \ M \ \rightarrow\!|$$
$$\lambda = \text{ move suppression parameter}$$

$$\underline{\underline{\Gamma}} = \text{diag}(\ \gamma \ \ \gamma \ \ \cdots \ \gamma \ ) \tag{21}$$
$$|\!\leftarrow \ P \ \rightarrow\!|$$

$$\gamma = \text{ output weighting parameter.}$$

By analogy, if the neural network is comprised of only a single neuron with a linear transfer function, then the controller would be given by,

$$\underline{x} = \left[\ \left[\underline{\underline{W}}^T \ \underline{\underline{\Gamma}}^T \ \underline{\underline{\Gamma}} \ \underline{\underline{W}} + \ \underline{\underline{\Lambda}}^T \ \underline{\underline{\Lambda}}\right]^{-1} \underline{\underline{W}}^T \ \underline{\underline{\Gamma}}^T \ \underline{\underline{\Gamma}} \ [\underline{z}]\right] \tag{22}$$

where,

$$\underline{\underline{W}} = \begin{bmatrix} w_1 & 0 & 0 & \bullet & 0 \\ w_2 & w_1 & 0 & \bullet & 0 \\ \bullet & \bullet & \bullet & \bullet & \bullet \\ w_N & w_{N-1} & w_{N-2} & \bullet & w_{N-M+1} \\ \bullet & \bullet & \bullet & \bullet & \bullet \\ w_P & w_{P-1} & w_{P-2} & \bullet & w_{P-M+1} \end{bmatrix} \tag{23}$$

$$\underline{x} \equiv \underline{\Delta u}$$
$$\underline{z} \equiv \underline{e}(k+1). \tag{24}$$

3.3. THE DNNC PROCESS MODEL

For a single future move in equation (8) we have,

$$\underline{\Delta u} = [\Delta u(k) \ \ 0 \ \ \cdots \ \ 0 \ ]^T,$$

Therefore,

$$y(k+1) = y^*(k+1) + a_1 \ \Delta u(k) + d(k+1). \tag{25}$$

For the nonlinear case, a_1 is a function of $\Delta u(k)$ and the system state, therefore, we can generalize equation (25) as follows,

$$y(k+1) = y^*(k+1) + a_1(k) \ \Delta u(k) + d(k+1) \tag{26}$$

or equivalently,

$$e(k+1) = a_1(k) \ \Delta u(k). \tag{27}$$

Solving equation (26) for one future move (j=1), and substituting for $y^*(k+1)$ from equations (9) and (10),

$$y(k+1) = a_1(k) \cdot \Delta u(k) + a_2(k) \cdot \Delta u(k-1) + a_3(k) \cdot \Delta u(k-2)$$
$$+ \ldots + a_N(k) \cdot \Delta u(k+1-N) + a_N(k) \cdot u(k-N) \qquad (28)$$
$$+ d(k+1)$$

with d(k+1) is defined as follows,

$$d(k+1) = y_m(k+1) - y(k+1) \qquad (29)$$

and in terms of $y_m(k)$ and y(k),

$$d(k+1) = \sigma(k) \, y_m(k) - \beta(k) \, y(k) \qquad (30)$$

where $\sigma(k)$ and $\beta(k)$ are time varying parameters. Substituting for y(k) from equation(11) into equation (30) gives,

$$d(k+1) = [\sigma(k) - \beta(k)] \, y_m(k) + \beta(k) \, d(k) \qquad (31)$$

Equation (30) clearly shows that d(k+1) is a function of $y_m(k)$, d(k), and the parameters $\beta(k)$ and $\sigma(k)$. Define $\alpha(k) = \sigma(k) - \beta(k)$ so equation (31) may be written as,

$$d(k+1) = \alpha(k) \, y_m(k) + \beta(k) \, d(k). \qquad (32)$$

It is noted that by taking $\alpha(k)=0$ and $\beta(k)=1$, equation (32) reduces to equation (11) for DMC and d(k+1) is dependent only on d(k).

Substituting equation (32) into equation (28) and simplifying one obtains,

$$y(k+1) = a_1(k) \cdot \Delta u(k) + a_2(k) \cdot \Delta u(k-1) + a_3(k) \cdot \Delta u(k-2)$$
$$+ \ldots + a_{N-1}(k) \, \Delta u(k-N+2) + a_N(k) \cdot u(k-N+1)$$
$$+ \alpha(k) \, y_m(k) + \beta(k) \, d(k)$$
$$= \underline{a}(k)^T \, \Delta \underline{uy}(k) + \beta(k) \, d(k)$$

$$\qquad (33)$$

where,

$$\Delta \underline{uy}(k) = [\Delta u(k) \quad \Delta u(k-1) \quad \cdots \quad \Delta u(k-N+2) \quad u(k-N+1) \quad y_m(k)]^T$$
$$\underline{a}(k) = [a_1(k) \quad a_2(k) \ldots \quad a_{N-1}(k) \quad a_N(k) \quad a_{N+1}(k)]^T \qquad (34)$$
$$a_{N+1}(k) = \alpha(k).$$

Equation (33) shows that y(k+1) is a function of the independent variables $\Delta \underline{uy}(k)$ and the time varying parameters $\underline{a}(k)$. (We will assume that in the training data d(k) =0.)

In a more general form, y(k+1) is given by,

$$y(k+1) = g(\Delta \underline{uy}(k), \underline{a}(k)) \qquad (35)$$

In this study, we use a neural network model for nonlinear input /output mapping given by equation (35) with the input structure of the network as defined by equation (33). In general, for nonlinear processes the single hidden layer DNNC process model is defined by,

$$y(k+1) = w_2 \ f(\underline{w_1}^T \ \underline{\Delta uy} + B_1) + B_2$$

$$\underline{\Delta uy} = [\Delta u(k) \ \ \Delta u(k-1) \ \cdots \ \Delta u(k-N+2) \ \ u(k-N+1) \ \ y_m(k)]^T$$

$$\underline{w_1} = [w1_1 \ w1_2 \ \ldots \ w1_{N-1} \ w1_N \ \ w1_{N+1}]^T$$

$$(36)$$

with the transfer function f typically defined by the hyperbolic tangent function,

$$f(z) = \frac{e^z - e^{-z}}{e^z + e^{-z}}. \tag{37}$$

It is important to note that there is no restriction on the number of hidden layers or the transfer function in the output and hidden layers.

3.3.1. *Training the DNNC Model*

With the initial guess for the network weights and biases,

$$\underline{w_1} = [\underline{a}^T \ | \ \alpha]^T$$

$$w_2 = 1 \tag{38}$$

$$B_1 = B_2 = 0$$

$$w_{1,(N+1)} = \alpha = \varepsilon,$$

where, $\quad \varepsilon \cong 0.0$, (e. g. $\varepsilon = .001$);

The DNNC model is trained with input/output data using the backpropagation algorithm. In applications, the weights and bias terms would be updated either using the backpropagation algorithm or a recursive algorithm (as discussed in Appendix A) if necessary [21, 30]. It has been shown [21, 30] that DNNC offers potential for fewer weights and bias terms, therefore, better initialization of weights and fast on-line updating of weights are possible [21].

3.4. DNNC CONTROLLER DESIGN

Before discussing the DNNC controller, we present a brief summary of current work with neural network controllers. This discussion will help to illustrate the advantages of DNNC. Neural networks can be used as a controller in either direct or indirect methods as discussed in the introduction. In the direct approach, the controller will be a neural network which represents the inverse of the process model, while in the indirect approach the controller algorithm inverts the neural network process model. In the direct approach, since the controller model is trained off line, the error in the controller prediction significantly affects the controller performance. This problem was addressed by Psichogios and Ungar [12]. In the indirect approach, the inverse of the process model at each sampling time must be calculated. Since the neural network model of the process is a nonlinear mapping between input and output, and the model is frequently complex, mathematical inversion of the process model is difficult. Therefore, the most popular method applies an optimization routine to find the controller output. To illustrate the complexity, we will outline the optimization approach.

The neural network process model for input/output mapping is defined as,

$$y(k+1) = NN(\underline{uy}(k), \underline{W\Theta}(k)), \tag{39}$$

where $\underline{uy}(k)$ includes all the inputs to the network model at time (k);

$$\underline{uy}(k)^T = [y(k), y(k-1), ..., y(k-n_y+1), u(k), u(k-1), ..., u(k-n_u+1)] \tag{40}$$

and $\underline{W\Theta}(k)$ includes all the weights and bias terms. The objective function (OF) for controller design is given by,

$$OF = E(k)^2 = \left[E(v(k), \underline{uy}(k), \underline{W\Theta}(k))\right]^2 = \left[v(k) - NN(\underline{uy}(k), \underline{W\Theta}(k))\right]^2 \tag{41}$$

where $v(k)$ is a filtered output given by,
$$v(k) = h(d(k), y^{set}(k), v(k-1)) \tag{42}$$

In the optimization routine, u(k) is calculated to minimize the objective function defined in equation (41). The backpropagation algorithm may be used, in which case the error back-propagates through the network to adjust the u(k) instead of adjusting the weights and bias terms as is done during training. Other methods such as quadratic programming also may be applied. Solving the optimization problem for conventional neural network models, which are complex and consist of several nodes and weights, is difficult and time consuming. In addition, there is approximation error in calculating u(k). Often, this error in u(k) significantly affects the controller performance. Psichogios and Ungar [12] suggest "detuning" the controller for robust performance.

Another approach to finding the controller output based on the process model employs Newton's method. Several authors have employed this technique [12, 15]. The following equations are used in this method at each iteration j,

$$u(k)^{[j]} = u(k)^{[j-1]} - \frac{E(k)^{[j-1]}}{\left[\partial E(k)^{[j-1]} \middle/ \partial u(k)^{[j-1]}\right]} \tag{43}$$

with the initial guess for u(k) as follows,

$$u(k)^{[0]} = u(k-1). \tag{44}$$

Details of the calculation of the Jacobian of $E(k)$; $\left[\dfrac{\partial E(k)^{[j-1]}}{\partial u(k)^{[j-1]}} \right]$ have been presented by Nikravesh et al. [21]. Once again there is approximation error in $u(k)$ and "detuning" is required. In addition, solving this nonlinear problem by Newton's method is not trivial. Psichogios and Ungar [12] considered some problems associated with convergence of Newton's method. In their work, consideration was given to problems including poor initial guesses, pathological functions, and singular problems with vanishing derivatives.

Since the DNNC model is very simple, its inversion does not present the difficulties encountered with other methods. The inverse of the process model is exact so, convergence and offset do not present problems. In this paper, we present the DNNC controller in an IMC framework (one step ahead prediction). The neural network IMC models are restricted to systems with stable open-loop response. However, DNNC can be employed in a more general Model Predictive Control (MPC) framework (multistep prediction). For example, DNNC can be employed in the DMC framework. Details of such a DNNC controller are given by Appendix B.

Figure 1 shows the block diagram for the IMC framework. The IMC version of the DNNC controller model will be designed using the following equations,

$$\Delta u(k) = \frac{\left[f^{-1}\left(\dfrac{v(k) - B_2}{w_2} \right) - B_1 - (\underline{w}_1^N)^T \ \underline{\Delta uy}_1^N \right]}{wl_1}$$

$$\underline{w}_1^N = [\ wl_2 \ \ wl_3 \ \ \ldots \ \ wl_{N-1} \ \ wl_N \ \ wl_{N+1} \ ]^T \tag{45}$$

$$\underline{\Delta uy}_1^N = [\ \Delta u(k-1) \ \ \Delta u(k-2) \ \ \ldots \ \ \Delta u(k-N+2) \ \ u(k-N+1) \ \ y_m(k)]^T$$

where $v(k)$ is a filtered output for $y(k+1)$. For the controller design, $y(k+1) = y^{set}(k+1)$. The transfer function f^{-1} is the inverse of equation (37),

$$f^{-1}(z) = -0.5 \ \ln\left[\frac{1-z}{1+z} \right] \tag{46}$$

The filter may be chosen to be a pulse transfer function in the form of a first-order filter. The filter equation is given by,

$$v(k) = v(k-1) + (1-\varphi) \left[y^{set}(k) - d(k) - y(k) \right] \tag{47}$$

In training the network via the backpropagation algorithm, an appropriate scaling of inputs and y(k+1) will be used so that, -1<y(k+1)<+1. In this case, $\left(\dfrac{y(k+1)-B_2}{w_2}\right)$ is always bounded between -1 and 1. It is also possible to scale the inputs and y(k+1) such that, $-1 < \left(\dfrac{v(k)-B_2}{w_2}\right) < +1$, is guaranteed. Therefore, the existence of $f^{-1}\left(\dfrac{v(k)-B_2}{w_2}\right)$ is guaranteed.

For time delay compensation, the following equation may be used,

$$v(k) = \varphi \, v(k-1) + (1-\varphi)\left[y^{set}(k) - d(k) - y(k) + v(k-\tau_d - 1)\right]. \tag{48}$$

The value for φ is bounded between zero and one. Details of filter design are available throughout the literature [15, 33-35]. In case of process/model mismatch, the filter may be used to ensure stability and robustness, but it can also compensate for certain types of disturbances. Garcia and Morari [36] and Deshpande [37, 38] provide detailed guidelines for designing IMC filters.

3.5. STABILITY ANALYSIS OF NEURAL NETWORK-BASED CONTROL SYSTEMS

The stability analysis of neural-based control systems is an important issue which must be considered for the design of a good neural-based control system [30, 39]. There are several methods which has been proposed to study the open-loop and closed-loop stability of processes and to analyze and design control systems [40]. State-space methods are best suited for analysis and synthesis of nonlinear systems and they can be applied to the design of optimal control systems [40]. Once the systems are transformed into state-space models, nonlinear model approaches such as geometric control [13-15], neuro-fuzzy control [19, 20], fuzzy logic control [41], and model-based control [11-13, 22-25] can be used to design and analysis the controller performance. In addition, the Liapunov theory [36, 38] can be used for stability analysis [7, 30, 39, 43, 44]. Liapunov stability theory plays an important role in the stability analysis of control systems described by state-space equations. The second method of Liapunov [39, 40, 42] is most commonly used and is applicable to both linear and nonlinear systems. This method is also suited for the stability of nonlinear systems for which exact solutions may be unobtainable such as neural network models. Although the second method of Liapunov is applicable to a wide class of nonlinear systems including neural network systems for stability analysis, obtaining successful results is not trivial [40]. Therefore, experience may be necessary to correctly interpret the results from the stability analysis of nonlinear systems.

3.5.1. *Dynamic Neural Network Control (DNNC); Stability Analysis*

In the following section, the stability analysis of DNNC process model and controller will be discussed.

3.5.1.1. *State-Space Representation of DNNC*

The input-output mapping of the DNNC can be represented by,

$$y(k+1) = w_2 \ \Gamma(\underline{w}_1{}^T \ \Delta\underline{uy} + B_1) + B_2 \tag{49}$$
$$\Delta\underline{uy} = [\Delta u(k) \ \ \Delta u(k-1) \ \cdots \ \Delta u(k-N+2) \ \ u(k-N+1) \ \ y_m(k)]^T$$
$$\underline{w}_1 = [wl_1 \ \ wl_2 \ \ \ldots \ \ wl_N \ \ wl_{N+1}]^T$$

$$\Gamma(z) = \frac{e^z - e^{-z}}{e^z + e^{-z}}$$

Substituting $\Delta u(k-j+2) = u(k-j+2) - u(k-j+1)$ into equation (1) and $y(k) = y_m(k)$ gives,

$$y(k+1) = w_2 \ \Gamma(\Delta\underline{w}_1{}^T \ \underline{uy} + B_1) + B_2$$
$$\underline{uy} = [u(k) \ \ u(k-1) \ \cdots \ u(k-N+2) \ \ u(k-N+1) \ \ y(k)]^T$$
$$\Delta\underline{w}_1 = [wl_1 \ \ (wl_2 - wl_1) \ \ldots \ \ (wl_N - wl_{N-1}) \ \ wl_{N+1}]^T$$
$$\tag{50}$$

Equation (50) can be written in the following discrete state-space form,

$$\underline{x}(k+1) = f(\underline{x}(k)) + g(\underline{x}(k), u(k)),$$
$$y(k) = h(z^{-1}(\underline{x}(k), u(k))),$$
$$y(k+1) = h(\underline{x}(k), u(k)) \tag{51}$$

with $\underline{x}(k)$ is given by,
$$\underline{x}(k) = [x_1(k) \ \ x_2(k) \ \cdots \ \ x_{N-1}(k) \ \ x_N(k)]^T \tag{52}$$
$$= [u(k-1) \ \ u(k-2) \ \cdots \ \ u(k-N+1) \ \ y(k)]^T$$

Therefore, $\underline{x}(k+1)$ is given by,

$$\underline{x}(k+1) = [x_1(k+1) \ \ x_2(k+1) \ \ \cdots \ \ x_{N-1}(k+1) \ \ x_N(k+1)]^T$$
$$= [u(k) \ \ \ \ \ \ u(k-1) \ \ \ \cdots \ \ u(k-N+2) \ \ y(k+1)]^T$$
$$= [u(k) \ \ \ \ \ \ x_1(k) \ \ \ \cdots \ \ x_{N-2}(k) \ \ h(\underline{x}(k), u(k))]^T$$
$$\tag{53}$$

with,

$$f(\underline{x}(k)) = [0 \ \ x_1(k) \ \ x_2(k) \ \ \cdots \ \ x_{N-2}(k) \ \ \ \ \ 0]^T$$
$$g(\underline{x}(k), u(k)) = [u(k) \ \ 0 \ \ 0 \ \ \cdots \ \ 0 \ \ 0 \ \ h(\underline{x}(k), u(k))]^T$$
$$h(\underline{x}(k), u(k)) = y(k+1) = w_2 \ \Gamma(\Delta\underline{w}_1{}^T \ \underline{uy} + B_1) + B_2 \tag{54}$$

In this study, we are interested in the stability of the overall process. In the DNNC strategy (as for any other IMC strategy (Figure 1) and with an exact model for the process, the stability of both the process and controller is sufficient for overall system stability.

3.5.1.2. *Stability of the DNNC Process Model*

The stability of the process model according to Liapunov theory is guaranteed if the magnitude of the eigenvalues of the Jacobian of $\underline{x}(k+1)$ with respect to $\underline{x}(k)$ are inside of the unit circle. In this case, the Jacobian is given by,

$$
J^Y = \begin{bmatrix}
0 & 0 & 0 & \bullet & 0 & 0 & 0 \\
1 & 0 & 0 & \bullet & 0 & 0 & 0 \\
0 & 1 & 0 & \bullet & 0 & 0 & 0 \\
\bullet & \bullet & \bullet & \bullet & \bullet & \bullet & \bullet \\
0 & 0 & 0 & \bullet & 0 & 0 & 0 \\
0 & 0 & 0 & \bullet & 1 & 0 & 0 \\
\alpha_1 & \alpha_2 & \alpha_3 & \bullet & \alpha_{N-2} & \alpha_{N-1} & w_2 \cdot wl_{N+1} \cdot \left(1 - yl(k)^2\right)
\end{bmatrix}
\tag{55}
$$

where,

$$
\begin{aligned}
yl(k) &= \Gamma(\underline{\Delta w_1}^T \ \underline{uy} + B_1) \\
\alpha_j &= w_2 \cdot \Delta w_{1,j} \cdot \left(1 - yl(k)^2\right) \\
\Delta w_{1,j} &= wl_{j+1} - wl_j \\
j &= 1, \ldots, N-1
\end{aligned}
\tag{56}
$$

The non-zero eigenvalue of the Jacobian matrix J^Y is given by,

$$
\lambda = \left(1 - yl(k)^2\right) . wl_{N+1} \ w_2
\tag{57}
$$

3.5.1.3. *Stability of the DNNC Controller*

The DNNC controller will be defined using the inverse of the DNNC process model and is given by the following equations,

$$
u(k) = \frac{\left[\Gamma^{-1}\left(\dfrac{v(k) - B_2}{w_2}\right) - B_1 - (\Delta \underline{w}_1^C)^T \ \underline{uy}^C\right]}{w_{1,1}}
\tag{58-a}
$$

with,

$$
\Delta \underline{w}_1^C = [\ \Delta w_{1,1} \ \ \Delta w_{1,2} \ \cdots \ \Delta w_{1,N-1} \ \ w_{1,N+1} \ ]^T
$$

$$\underline{uy}^C = [\; u(k-1) \quad u(k-2) \quad \ldots \quad u(k-N+2) \quad u(k-N+1) \quad y(k)]^T$$

(58-b)

$$v(k) = v(k-1) + (1-\varphi)\left[y^{set}(k) - d(k) - y(k)\right]$$
$$d(k) = y_m(k) - y(k)$$

(58-c)

$$\Gamma^{-1}(z) = -0.5 \ln\left[\frac{1-z}{1+z}\right]$$

(58-d)

In this case, the state-space representation of the controller (equation (58)) is given by,

$$\underline{x}(k+1) = \begin{bmatrix} \beta_1 & \beta_2 & \beta_3 & \bullet & \beta_{N-2} & \beta_{N-1} & \frac{-w1_{N+1}}{w1_1} \\ 1 & 0 & 0 & \bullet & 0 & 0 & 0 \\ 0 & 1 & 0 & \bullet & 0 & 0 & 0 \\ \bullet & \bullet & \bullet & \bullet & \bullet & \bullet & \bullet \\ 0 & 0 & 0 & \bullet & 0 & 0 & 0 \\ 0 & 0 & 0 & \bullet & 1 & 0 & 0 \\ 0 & 0 & 0 & \bullet & 0 & 0 & 0 \end{bmatrix} \times$$

$$\begin{bmatrix} x_1(k) \\ x_2(k) \\ x_3(k) \\ \bullet \\ x_{N-2}(k) \\ x_{N-1}(k) \\ x_N(k) \end{bmatrix} + \begin{bmatrix} R \\ 0 \\ 0 \\ \bullet \\ 0 \\ 0 \\ v^{set}(k) \end{bmatrix}$$

(59)

$$\beta_j = \frac{-\Delta w_{1,j}}{w1_1}$$

$$R = \frac{\Gamma^{-1}\left(\dfrac{v^{set}(k) - B2}{w_2}\right) - B1}{w1_1}$$

The stability of the controller according to Liapunov is guaranteed if the magnitude of the eigenvalues of the Jacobian of $\underline{x}(k+1)$ with respect to $\underline{x}(k)$ are inside of unit circle. In this case, the Jacobian is given by,

$$
J^U = \begin{bmatrix}
\beta_1 & \beta_2 & \beta_3 & \bullet & \beta_{N-2} & \beta_{N-1} & \dfrac{-wl_{N+1}}{wl_1} \\
1 & 0 & 0 & \bullet & 0 & 0 & 0 \\
0 & 1 & 0 & \bullet & 0 & 0 & 0 \\
\bullet & \bullet & \bullet & \bullet & \bullet & \bullet & \bullet \\
0 & 0 & 0 & \bullet & 0 & 0 & 0 \\
0 & 0 & 0 & \bullet & 1 & 0 & 0 \\
0 & 0 & 0 & \bullet & 0 & 0 & 0
\end{bmatrix}
\tag{60}
$$

The non-zero eigenvalues of the Jacobian matrix J^U are the solutions of the following characteristic equation,

$$
\lambda^{N-1} - \beta_1 \cdot \lambda^{N-2} - \beta_2 \cdot \lambda^{N-3} - \cdots - \beta_{N-2} \cdot \lambda - \beta_{N-1} = 0 \tag{61}
$$

3.6. OPTIMUM DNNC NETWORK STRUCTURE

The results from DNNC stability analysis will be used to define the Neural Network Stability Index (NNSI). Neural Network Stability Index (NNSI) is given in Table 1 Ref. [45]. The NNSI is a practical index which in current form can only be used with DNNC structure. The NNSI can be used to determine the optimal DNNC network structure. The optimum number of nodes required to model the process for an acceptable controller performance is given by the maximum value for N which is obtained from Guidelines 1 and 2 presented in the Ref. [45]. In addition, increasing the number of nodes in the input layer (DNNC network structure) results in the smaller value for the NNSI which implies a more stable process model and has a smoother and slower response.

4. Extension of the DNNC Model to the MIMO Case

For a MIMO system it is much more difficult to make meaningful specifications than for a SISO systems. However, the design of multivariable controllers in the DNNC control strategy is straightforward. The DNNC weights represent the process model and interaction between input/output and it is possible to determine the severity of the interaction by interpreting the network weights. The following sections will demonstrated the deign of MIMO systems in the DNNC framework.

4.1 EXTENSION OF THE DNNC MODEL TO THE MIMO CASE IN IMC FRAMEWORK

Design of multivariable controllers in the DNNC control strategy in IMC framework is straight forward. The following equations will be used for process model;

$$\underline{y}(k+1) = \underline{w}_2 \ F(\underline{\underline{W}}_1 \ \underline{\underline{uy}}(k) + \underline{B}_1) + \underline{B}_2$$

$$\underline{\underline{uy}} = [\Delta \underline{uy}^{(1)}(k) \quad \Delta \underline{uy}^{(2)}(k) \ \dots \ \Delta \underline{u}^{(j)}(k) \ ]^T$$

$$\Delta \underline{uy}^{(j)} = [\Delta \underline{u}^{(j)}(k) \quad \Delta \underline{u}^{(j)}(k) \ \dots \ \underline{u}^{(j)}(k - N_i + 1) \quad y_m^{(j)}(k) \ ]^T$$

$$\underline{y}(k+1) = [\ y^{(1)}(k+1) \quad y^{(2)}(k+1) \quad \dots \ y^{(j)}(k+1) \ ]^T \tag{62}$$

with,

$$\underline{\underline{W}}_1 = \begin{bmatrix} \underline{W}_{1,1} & \underline{W}_{1,2} & \bullet\bullet\bullet & \underline{W}_{1,j} \\ \underline{W}_{2,1} & \underline{W}_{2,2} & \bullet\bullet\bullet & \bullet \\ \bullet & \bullet & \bullet\bullet\bullet & \bullet \\ \underline{W}_{j,1} & \underline{W}_{j,2} & \bullet\bullet\bullet & \underline{W}_{j,j} \end{bmatrix} \tag{63}$$

$$\underline{W}_{i,j} = [\ W_{i,1} \quad W_{i,2} \quad \dots \quad W_{i,N_j} \quad W_{i,N_j+1} \ ]^T$$

where, $w_{i,j}$ represents the effect of input i on the output j. Therefore, the DNNC controller will be designed using the inverse of the DNNC process model as was presented in the previous chapters for SISO case. For no interaction between outputs, all the elements of matrix $\underline{W}_1$ which are not on the main diagonal will be equal to zero. In other words, if, after training the MIMO-DNNC model, all the off diagonal elements of $\underline{W}_1$ are equal to zero or a small value compared to the elements on the main diagonal, then the controller will be designed as a multi-SISO controller model.

4.2 EXTENSION OF THE DNNC MODEL TO THE MIMO CASE IN DMC FRAMEWORK

The MIMO problem formulation and equations for DNNC are the same as in the DMC case. In other words there is no difference between the form of DNNC equations and that for the DMC case, and only a change in interpretation of the variables is needed. Therefore, all properties and techniques applied to MIMO-DMC are applicable to DNNC as well. Therefore, if the DNNC controller is implemented in the DMC configuration (Appendix B as shown for SISO case), extension of SISO-DNNC to MIMO-DNNC will be straightforward and does not cause any problems [30].

5. Real-Time Controller Implementation

Nonlinear, time varying behavior is common in reservoir engineering. For example, during steam injection in the diatomite, there are several factors which cause such a behavior: (1) increase of the reservoir temperature, (2) change in the rock matrix permeability, (3) changes in the oil viscosity, (4) rock fracturing, and (5) rock dissolution. When such changes in reservoir characteristics occur, the steam injector controllers need to be retuned in order to maintain satisfactory performance. Retuning the controllers is time consuming and requires a combination of operational experience and trial-and error procedures. In addition, injection rates must be adjusted carefully, because high injection rates lead to reservoir damage, and low injection rates make oil recovery uneconomic. By using historical data for steam injectors, neural networks have been developed and used to predict the complex reservoir behavior. For instance, using historical data from Phase II steam injection in Diatomite (Kern County Field, CA), a neural network was developed to predict well head pressure as a function of injection rate, and vice versa. The resulting model provided an excellent input-output mapping in addition to pattern extraction, even though steam injection dynamics in this project were complex

6. Process and Controller Model

Using historical data from Phase III steam injection pilot in the South Belridge Diatomite (CalResources, Kern County Field, CA) two neural network models were developed.

The first model was developed to predict the wellhead pressure as a function of current and past injection flow rate and wellhead pressure. This model can be used as process model and its inverse can be used as controller. The model has 10 nodes in the input layer, 1 node in the hidden layer, and 1 node in the output layer (future wellhead pressure). As shown in Figures 2.a, b, c the resulting models provide an excellent input-output mapping. The second model was developed to predict the steam injection rate as a function of future, current, and past wellhead pressure and past injection flow rate. This model can be used in equation 5 to invert the function f to predict the manipulated variable, u(k), and be used as controller output to the process. The model has 10 nodes in the input layer, 3 nodes in the hidden layer, and 1 node in the output layer (current injection rate). As shown in Figures 2.d, e, f the resulting models provide an excellent input-output mapping.

7. Neuro-Statistical Model

In a model-based control of fluid injection into low permeability, it is of great importance to characterize how the injection pressure is related to injection flow rate based on historical data. However, data from injectors are often difficult to analyze due to their complexity, uncertainty, and the fact that a physical relationship cannot be established to show how the data are correlated. In addition, analysis of these data is laborious and human ability is limited in its understanding and use of the information. Unfortunately, only linear and simple nonlinear information can be extracted from these data by powerful tools of statistical methods such as ordinary Least-Squares (LS), Partial Least-Squares (PLS.), and nonlinear Quadratic Partial Least-Squares (QPLS). However, if priori information regarding the nonlinear input-output mapping is available, these methods become more useful. In regards to mathematical modeling, simple models may become inaccurate as several assumptions are made to simplify the models in order to solve the problem mathematically. On the other hand, complex models may become inaccurate when additional equations involving a more or less approximate description of phenomena are included in the model. In most cases, these models require a number of parameters which are not physically measurable. As a third alternative, the Neural Network-Statistical Model (NSM or Neuro-Statistical

850

Model) provide the potential to establish a model from nonlinear, complex, multi-dimentional, uncertain, and imperfect data. The model uses the advantage of the neural network in conjunction with statistical methods to analyze the data and identify the model based on data. The model uses neural network techniques, since the functional structure of the data is unknown. In addition, the model uses statistical techniques because the data and our requirements are imperfect. Statistical techniques are considered to be appropriate to deal with the nature of uncertainty in system and human error, which are not included in current neural network models. Using the nonlinear statistical techniques, we developed a neural network model in which the network parameters reflect the uncertainty in the output data. In this case, instead of one value for each network parameter, a distribution of values has been assigned to the network parameters. Therefore, the neural network prediction will be a distribution rather than a crisp value. The model has been compared with conventional neural network models. It has been concluded that the most probable parameter for the new model is similar to the crisp value for the conventional model. Using this concept the conventional Levenberge-Marquardt algorithm is modified. In this case, the final global error in the output at each sampling time is related to the network parameters and a modified version of the learning coefficient is defined. The following equations briefly shows the difference between conventional and modified technique. In the conventional technique weights can be calculated by,

$$\Delta W = \left(\underline{J}^T J + \mu^2 \, \underline{I}\right)^{-1} \, \underline{J}^T \, e \tag{64}$$

However, in modified technique the weights are given by,

$$\Delta W = \left(\underline{J}^T \underline{\Lambda}^T \underline{\Lambda} \, J + \Gamma^T \, \Gamma\right)^{-1} \, \underline{J}^T \, \underline{\Lambda}^T \underline{\Lambda} \, e \tag{65}$$

$$\underline{\Lambda}^T \underline{\Lambda} = \underline{\hat{V}}^{-1} \tag{66}$$

$$\hat{V}_{ij} = \frac{1}{2m+1} \sum_{k=-m}^{m} \hat{e}_{i+k} \hat{e}_{j+k} \tag{67}$$

$$\underline{\hat{V}} = \hat{\sigma}^2 \, \underline{I} \tag{68}$$

$$\hat{W} = W \pm k \, \hat{\sigma} \tag{69}$$

where e is error, k is gain, σ is variance, Γ is tuning parameter, and $\underline{J}$ is Jacobian matrix.

Figure 3 shows the performance of the new network model. Figure 3.a shows the predictive performance of the network model. In Figure 3, circles represent actual data, crosses represent the mean of the neural network predictions, squares represent the upper limit (max.) of the network prediction and triangles represent the lower limit (min.) of the network prediction. Figure 3a shows that the network has an excellent performance and also the actual value always lies between the upper and the lower limit predicted by neural network. Figures 3.b, 3.c, and 3.d are a magnification of Figures 3.a, 3.b, and 3.c. As one can see the trend in Figure 3.b, 3.c, and 3.d is the same as the trend in Figure 3.a. As we mentioned earlier, the output from the network is a distribution rather than a crisp value. Figure 1.e shows the distribution of the predicted values. Referring to Figure 3.e and 3.f, one can see that actual value is bounded between the one standard deviation from the most probable value. Figures 3.f , 3.g, and 3.h show the comparison between actual data, the most probable prediction based on the neural network, upper limit, lower limit, and one standard deviation from the most probable prediction. Using this technique the upper and lower bound tightened.

Even though the Levenberge-Marquardt algorithm is faster and more robust than the conventional algorithm, but it requires more memory. Therefore, to overcome this disadvantage we need to reduce the complexity of the neural network model and/or reduce the number of the data points used in each step of training. In the former case, we will use the Alternative Conditional Expectation (ACE) technique [46], a non-parametric statistical technique, to reduce the network structure. This will be done by extracting the patterns which exist in the data (Figures 4.a, 4.b, 4.d, and 4.e). In the next section, we will introduce the idea very briefly. To reduce the data points used in each step of training, we will divide the data-sets into several sub-data sets based on the pattern extracted (Figures 4.b and 4.e) using the ACE technique. In this case, we will use the recursive technique [47, 48] for network training and then use the modified Levenberge-Marquardt algorithm. Therefore, the final global error in the output at each sampling time is related to the network parameters and a modified version of the learning coefficient is defined.

8. Hybrid Neural Network-Alternative Conditional Expectation (HNACE)

Recently, application of the non-parametric statistical method such as the Alternative Conditional Expectation (ACE) scheme for modeling the complex multi-dimensional field data has greatly increased. Statistical techniques such as ACE are considered to be appropriate to deal with the nature of uncertainty. In addition, the underlying patterns and structures recognized by ACE are more visible than neural network structure (Figures 4.b and 4.e). The ACE method is a statistical technique which can be used to find optimal transformation that maximizes the correlation between the transformed variables in a reduced and normally distributed space (Figures 4.c and 4.f). Therefore, since the ACE technique transformed the variables into a scaled domain in an optimal fashion, this technique can be used for scaling the input-output data for neural network structure (Figures 4.c and 4.f). In addition, since this technique can find the correlation between the variables, it can be used to reduce the complexity of the network structure by eliminating the variables which do not introduce any new information into the network model and more knowledge into the network model (Figures 4.g, 4.h and 4.i). This can be done by examining the transformed variables (Figures 4.b, 4.e and 4.h).

9. Non-Conventional Neural Network Model

Unfortunately, conventional neural networks (Figures 5.a and 5.b) are not stable once subjected to long-term prediction. In addition, these models can not capture the dynamics of the process and their performance is seriously affected for long-term prediction. Currently, Aminzadeh et al. [49] used a special optimization algorithm and cost function [50] for training the neural network model. This technique can prevent instability and over-fitting in the training session. In this study, a different approach is used for stabilization of the long-term network prediction. In this model, the current information from the input layer is presented into the pseudo hidden layer (Figures 5.c and 5.d). During training, the network has been forced to minimize the error in the output layer in addition to the pseudo hidden layer. This technique is a tradeoff between the accuracy of individual prediction and the stability of the successive prediction and increases the stability of the network for long-term prediction.

10. Model Identification and Controller Implementation for the CalResources Steam Injectors

Recent studies of waterfloods and steamdrive projects revealed that one of the most important factors of unwanted extension of hydrofractures is caused by aggressive actions taken by PI and PID controllers during the injector start-up periods, or by operating near hydrofracturing pressure. This has resulted in reservoir damage and irreversible lost oil production. This section concerns neural network model-based control (based on DNNC strategy) of petroleum reservoirs under a steam injection process. In particular, we consider the CalResources Phase III steam injection pilot, Diatomaceous fields of California. For each injector in this pilot (Figure 6.a), a neural network model is identified by using historical data of injection-fluid flow rate and well-head pressure, depth to the top of perforation, and the length of perforation. DNNC well-head pressure controllers are synthesized by using the neural network models. The performance of the neural network model-based controller is demonstrated via numerical simulations. It is assumed that the same well-operation strategy will continue into the future. If a different well-operation strategy is chosen in the future, then one can use the approach presented by [3]. The identified model has other applications, namely, in optimal operation of the oil well (maximum production with minimum amount of injectant), in fluid injection management and in the design of recovery strategies. The neural-network model describes the behavior of an actual well more accurately than any other reservoir simulator. For this study, the behavior of each injector is considered separately.

To maintain the ``accuracy'' of the identified model within a desirable range, we update the model on-line. The model is trained via backpropagation technique. Here, we use the modified Levenberge-Marquardt algorithm for optimization.

The available data have been divided into three data sets, training, testing and validation data sets. The network model has been trained using the training data sets and continuously tested based on test data sets. The network has been trained until the prediction suffered upon continued training. Figure 6.b shows the performance of the network for training, testing and validation data sets. Figure 6.b illustrates that the network accurately predicts the wellhead pressure. Even though Figures 6.b shows that the network has a good performance for predicting the output up to 100 steps into the future, this is not always the case. Unfortunately, conventional neural networks are not stable once subjected to long-term prediction. In addition, these models can not capture the dynamics of the process. Figure 6.c shows that the performance of the network is seriously affected for long-term prediction. Currently, Aminzadeh et al. [49] used a special optimization algorithm and cost function [50] for training the neural network model. This technique can prevent instability and over-fitting in the training session. In this study, a different approach has been used for stabilization of the long-term network prediction. In this model, the current information from the input layer has been presented into the last hidden layer. During training, the network has been forced to minimize the error in the output layer in addition to the last hidden layer. This technique is a tradeoff between the accuracy of individual prediction and the stability of the successive prediction. Figure 6.d shows the structure of the network model which has been used in this study. Figure 6.d shows the performance of network model for long-term prediction.

The more effective control of well-head pressure by model based control leads not only to improvement in oil recovery but also to longer life of the reservoir. To illustrate the performance of the DNNC controller the developed controller has been applied to control the Phase III steam injectors off-line. Currently, several network models have been developed for injectors 563LR, 553AR, and 544H (Figure 6.a). In this study, the performance of the DNNC controller is demonstrated via numerical simulations. The open-loop step responses for a series of step changes in the injection flow rate is shown in Figure 6.e. It is seen that the injection process is highly nonlinear. Figure 6.f shows the performance of the smart controller for set point tracking. The smart controller was tuned with a filter constant value of 0.85. Fine tuning the DNNC controller is very easy and straightforward. In this case, increasing the filter constant will result in a slower but smoother response and decreasing this constant will result in a faster and oscillatory response.

11. Conclusions

In this study, a nonlinear controller has been developed for an oil well based on neural network and model-based techniques. The neural network model-based controller has been applied to control a steam injector through numerical simulation and real-time (off-line) implementation. The model showed promises and performed very well for set point tracking purposes. In addition, a special neural network model has been used. It has been shown that the model can stabilize the propagation of the noise and has excellent performance for long-term prediction. This technique is a tradeoff between accuracy of individual prediction and the stability of the successive prediction. However, to understand the full potential of the current network and controller model, especially in real time implementation, further investigation is required. This methodology is currently implemented at Calresources LLC off-line. Calresources LLC is working to implement the controller on-line.

12. Acknowledgements

We thank CalResources LLC, formerly Shell Western E&P, for providing the steam injector data. We thanks BISC of UC Berkeley and BISC-SIG-ES of Lawrence Berkeley National Lab for their partial support. This work was supported partially by the US Department of Energy. Partial spport was also provided by CalResources LLC.

13. References

1. Nikravesh, M. Soroush, R. M. Johnson, and T.W. Patzek (1997) Design of Smart Wellhead Controllers for Optimal Fluid Injection Policy and Producibility in Petroleum Reservoirs: A Neuro-Geometric Approach, *SPE # 37557, 1997 International Thermal perations and Heavy Oil Symposium, Bakersfield, CA.*

2. Nikravesh, M, M. Soroush, R. M. Johnson, and T.W. Patzek (1997) Design of Smart Wellhead Controllers for Optimal Fluid Injection Policy and Producibility in Petroleum Reservoirs: A Neural Network-Model Predictive Approach, *SPE # 37445, 1997 SPE Production Operations Symposium.*

3. Nikravesh, M., A. R. Kovscek, R. M. Johnson, and T. W. Patzek (1996) Prediction of Formation Damage during Fluid Injection into Fractured, Low Permeability Reservoirs via Neural Networks, *SPE Paper No. 31103, SPE 1996 Formation Damage Symposium, Lafayette, LA..*

4. Cybenko (1989) Approximation by superposition of a sigmoidal function *Math. Control Sig. System,* **2,** 303.

5. Hecht-Nielsen (1989) Theroy of bavkpropagation neural networks, *IEEE Proc., Int. Conf. Neural Network,* Washington DC, I-593.

6. Pschogios and L.H. Ungar (1992) A hybrid neural network-first principles approach to process modeling, *AIChE J,* **10,** 1499.

7. Narendra and K. Parthasarathy (1990) Identification and control of dynamical systems, using neural networks, *IEEE Tra. on Neural Networks,* **1,** 4.

8. Pollard et. al., (1992) Process identification using neural networks, *Comp. chem. Engg.,* **4,** 253.

9. Bhat et. al., (1990) Modeling chemical process systems via neural computation, *IEEE Control System Magazine,* 24.

10. Chen, S. A. Billings and P. M. Grant (1990) Non-linear systems identification using neural networks *Int. J. Control,* **51** (6), 1191.

11. Nikravesh and A.E. Farell (1993) Modeling and controlling non-linear chemical processes using hybrid Neural Network Dynamic Matrix Control algorithm (NNDMC), *AIChE annual meeting,* paper 213b8, St. Louis.

12. Pschogios and L.H. Ungar (1991) Direct and indirect model based control using artifitial neural networks, *Ind. Eng. Chem. Res,* **30,** 2564.

13. Lee and S. Park (1992) A new schem combining neural feeddforward control with model predictive control, *AIChE Journal,* **38** (2), 193-200.

14. Saint-Donat, N. Bhat, and T.J. McAvoy (1991) Neural net based model predictive control, *Int. J. Control* **54** (6), 1453.

15. Nahas, M. Henson, and D. Seborg (1992) Non-linear internal model control strategy for neural network models, *comp. and chem. engg.* **16** (12), 1039.

16. Nikravesh et al. (1996) Identification and Control of Industrial-Scale Processes via Neural

Networks, *Presented at Chemical Process Control V, Tahoe City, CA*, Jan. 5-12.

17. Soroush and C. Kravaris (1992a) Discrete-Time Nonlinear Controller Synthesis by Input/Output Linearization, *AIChE Journal*, **38**, 1923-1945.

18. Soroush and C. Kravaris (1992b) Feedforward/Feedback Control of Discrete-Time Nonlinear Systems, *1992 AIChE Annual Meeting, Nov. 1-6, Miami*, Paper 126e.

19. Gupta, M. (1996) Fuzzy Logic and Fuzzy Systems: Recent Developments and Directions, *1996 Biennial Conference of the North American Fuzzy Information Processing-NAFIPS, Berkeley, California, USA*, 155-159.

20. Gupta, M. and D.H. Rao (1992) Dynamic Neural Units in the Control of Linear and Nonlinear Systems, *International Joint Conferences on Neural Networks, Baltimore, June 7-11*, **II** 100-105.

21. Nikravesh, A. E. Farell, and T. G. Stanford (1995) Model Identification of Nonlinear Time Variant Processes Via Artificial Neural Network, *comp. chem. engg*, (in press).

22. Bhat and T.J. McAvoy (1990) Use of neural nets for dynamic modeling and control of chemical process systems, *comp. chem. engg.*, **14**, (4/5) 573.

23. Hernandez and Y. Arkun, (1990) *ACC Conf.*, 2454.

24. Kraft and D.P. Campagna (1992) A comparison between CMAC neural network control and two traditional adaptive control systems, *IEEE Control Systems Magazine*, , 36.

25. Cooper et. al. (1992) Comparing two neural networks for pattern based adaptive process control, *AIChE Journal*, **38** (1), 42.

26. Garcia, D.M. Pret, and M. Morari (1989) Model Predictive Control- A survey, *Automatica*, **25**, 335.

27. Hecht-Nielsen, R. (1989) Theory of Backpropagation Neural Networks," *presented at IEEE Proc., Int. Conf. Neural Network*, Washington, DC.

28. Rumelhart, D.E., Hinton, G.E., and Williams, R.J. (1986) Learning Internal Representations by Error Propagation, in *Parallel Data Processing*, D. Rumelhart and J. McClelland, Editor, MIT Press, Cambridge, MA,

29. Widrow, B. and Lehr, M.A. (1990) 30 Years of Adaptive Neural Networks: Perceptron, Madaline, and Backpropagation, *Proceedings of the IEEE*, **78**(9).

30. Nikravesh, M. (1994) *Dynamic Neural Network Control, Ph.D. Dissertation*, University of South Carolina, Columbia, SC.

31. Cutler, (1979) Dynamic Matrix Control- A computer control algorithm, *AIChE 86th national meeting*, Houston, TX,.

32. Cutler and B. L. Ramaker *(1980) Proc. Joint Auto. Control Conf., San Francisco, CA WP5-B, and AIChE 86th National Meeting*, Houston,TX, 1979.

33. Morari and E. Zafiiriou (1989) *Robust Process Control*, Prentice Hall, Englewood Cliffs, NJ.

34. Isidori, (1989) *Nonlinear Control Systems : An Introduction*, 2nd ed. Springer-Verlag, New York.

35. Henson and D. E. Seborg (1991) *AIChE Journal*, **37** (7), 1065.

36. Garcia and M. Morari (1982) Internal Model Control. 1. A unifying review and some new results, *Ind. Eng. Chem. Process Des.*, **21**, 308.

37. Deshpande, (1988) *Multivariable Control Methods*, ISA.

38. Deshpande, (1988) *Computer Process Control with Advanced Control Application*, ISA.

39. Jin, M.M. Gupta, and P.N. Nikiforuk (1996) 13th Triennial World Congress, San Francisco, USA, vol. **F**, 187-192.

40. Ogata, (1987) *Discrete Time Control Systems*, Prentice-Hall, Inc., Englewood Cliffs, NJ.

41. Langari and W. Li (1996) Analysis and Efficient Implementation of Fuzzy Logic Control Algorithms, *1996 Biennial Conference of the North American Fuzzy Information Processing-NAFIPS, Berkeley, California, USA*, 1-4.

42. LaSalle, (1986) *The Stability and Control of Discrete Processes,* New York, Springer Verlag.
43. Jin, P. , N. Nikiforuk, and M.M. Gupta, (1994) Absolute Stability Conditions for Discrete-Time Recurrent Neural Networks, *IEEE Trans. Neural Networks,* **5**, 945-955.
44. Jin, P., N. Nikiforuk, and M.M. Gupta (1993) Dynamics and Stability of Multilayered Recurrent Neural Networks, *Proc. of the 1993 IEEE International Conference on Neural Networks (ICNN'93),* vol. **II**, 1135-1140.
45. Nikravesh, M., A.E. Farell, and T. G. Stanford (1997) Dynamic Neural Network Control for Nonlinear Systems: Optimal Neural Network Structure and Stability Analysis, *Chemical Engineering Journal,* **68**, 41-50.
46. Breiman, L., and J. H. Friedman (1985) Estimating Optimal Transformations for Multiple Regression and Correlation, *Journal of the Americal Statistical Assoc.,* **80** (391).
47. Azimi and R. J. Liou (1992) Fast Learning Process of Multilayer Neural Networks Using Recursive Least Squares Method, *IEEE Transaction on Signal Processing,* **40**, 446-450.
48. Nikravesh, M., Farell, A.E., and Stanford, T.G. (1996) Model Identification of Nonlinear Time Variant Processes Via Artificial Neural Networks, *J. of Computers and Chemical Engineering,* **20** (11).
49. Aminzadeh, F., S. Katz, and K. Aki (1994) Adaptive Neural network for Generation of Artificial Earthquake Precursors, *IEEE Trans. On Geoscience and Remote Sensing,* **32** (6).
50. Lagarias, J.C. and F. Aminzadeh (1983) Multi-stage Planning and the Extended Linear-Quadratic-Gaussian Control Problem, *Mathematics of Operations Research,* **8** (1).
51. Jannarone, R.J. (1994) Concurrent Information Processing : *Real Time Neurocomputing for Rapidly Changing World,* book draft in review/preparation, University of South Carolina, Columbia, SC.

Appendix A: Updating of the network recursively [30, 47, 51]

The input-output mapping of the multilayer DNNC model can be represented by,

$$\underline{y}^{(Net)} = \underline{\underline{W2}} \; \underline{y}_1^{(Net)} + \underline{\theta 2} \tag{A-1}$$

with,

$$\underline{y}_1^{(Net)} = F(\underline{\underline{W1}} \; \underline{x} + \underline{\theta 1}) \tag{A-2}$$

and,

$\underline{\underline{W1}}$	:	$n_h \times n_x$; input/hidden layer weight matrix
$\underline{\underline{W2}}$	:	$n_y \times n_h$; hidden/output layer weight matrix
$\underline{\theta 1}$	:	$n_h \times 1$; hidden layer bias vector
$\underline{\theta 2}$	:	$n_y \times 1$; output layer bias vector
$\underline{y}^{(Net)}$	:	$n_y \times 1$; network outputs (predictions) vector
$\underline{x}$	:	$n_x \times 1$; network inputs vector
$\underline{y}_1^{(Net)}$	:	$n_h \times 1$; vector of outputs from hidden

layer nodes

n_x	:	number of inputs
n_h	:	number of hidden layer nodes
n_y	:	number of output layer nodes

The basic least squares method produces a parameter estimation which is the result of minimization of a weighted squared error function,

$$J = \frac{1}{2N} \sum_{t=1}^{N} \lambda(t)^2 \left(\underline{y}(t) - \underline{\hat{y}}(t) \right)^2 \tag{A-3}$$

with,

$$\underline{y}(t) = \underline{y}^{(observed)}(t) \qquad : \quad \text{actual output, observed value}$$

for output

$$\underline{y}(t) = \underline{y}^{(Net)}(t) \qquad : \quad \text{predicted value, output from}$$

network.

A column vector, $\underline{W\Theta}$ is defined such that it includes all the elements of the matrix $\underline{W1}$, vector $\underline{\Theta1}$, all the elements of the matrix $\underline{W2}$, and vector $\underline{\Theta2}$,

$$\underline{W\Theta} = \begin{bmatrix} \underline{W\Theta1} \\ \underline{W\Theta2} \end{bmatrix} \tag{A-4}$$

with,

$$\underline{W\Theta1} = \begin{bmatrix} W1_{1,j} & | & W1_{2,j} & | & \cdots & | & W1_{n_y,j} & | & \underline{\Theta1}^T \end{bmatrix}^T$$
$$j = 1, ..., n_x$$

$$\underline{W\Theta2} = \begin{bmatrix} W2_{1,k} & | & W2_{2,k} & | & \cdots & | & W2_{n_h,k} & | & \underline{\Theta2}^T \end{bmatrix}^T$$
$$k = 1, ..., n_h$$

Therefore, to minimize the cost function J, the derivative of J with respect to $\underline{W\Theta}$ should be equal to zero. The derivative of cost function (Equation (3)) is defined as follows,

$$\nabla J(t) = \frac{\partial J(t)}{\partial \underline{W\Theta}(t)} = \frac{-1}{N} \sum_{t=1}^{N} \left(\underline{\underline{\Psi}}(t, \underline{W\Theta}(t)) \underline{\underline{\Lambda}}(t) \underline{\varepsilon}(t, \underline{W\Theta}(t)) \right) \tag{A-5}$$

with,

$$\underline{\varepsilon1}(t, \underline{W\Theta1}(t)) = \underline{\varepsilon}(t) = \left(\underline{y}(t) - \underline{\hat{y}}(t) \right) \quad : \quad n_y \times 1, \text{ vector of errors}$$

$$\underline{\underline{\Psi}}(t, \underline{W\Theta}(t)) = \underline{\underline{\Psi}}(t) = \frac{\partial \underline{\hat{y}}(t)}{\partial \underline{W\Theta}(t)} \quad : \quad n_\theta \times n_y, \qquad \text{matrix of gradients}$$

$$\underline{\underline{\Lambda}}(t): \quad n_y \times n_y, \text{ diagonal matrix with } [\lambda_k(t)]^2 \text{ element on diagonal}$$
$$\text{for } k=1,..., n_y$$
$$n_\theta = (n_h \times n_x) + (n_y + n_h) + n_h + n_y : \text{ number of rows in gradient matrix.}$$

The parameters $[\lambda_k(t)]^2$ (for $k=1,..., n_y$) are scalar values ($0 < [\lambda_k(t)]^2 \leq 1$). In addition, each element of matrix $\underline{\underline{\Psi}}(t, \underline{W\Theta}(t))$ will be calculated as follows,

$$y_k(t) = \sum_{i=1}^{n_h} (W2_{ki}(t) \, yl_i(t) + \theta2_k(t)), \quad \text{for } k=1,..., n_y \tag{A-6}$$

$$yl_i(t) = f\left(\sum_{j=1}^{n_x} (W1_{ij}(t) \, x_j(t) + \theta1_i(t)) \right),$$

with transfer function, f, defined as the hyperbolic tangent,

$$f(z) = \frac{e^z - e^{-z}}{e^z + e^{-z}}.$$

Each element of $\underline{\underline{\Psi}}$ is given by,

$$\Psi_{ij}(t, \underline{W\Theta l}(t)) = \frac{\partial y_j(t)}{\partial W\Theta_i(t)}$$

$$= \begin{cases} W2_{jk}(t) \, (1+yl_k(t))((1-yl_k(t)) \, x_l(t) & \text{if } W\Theta_i(t) = W1_{kl}(t), \ 1 \leq k \leq n_h, \ 1 \leq l \leq n_x \\ W2_{jk}(t) \, (1+yl_k(t))((1-yl_k(t)) & \text{if } W\Theta_i(t) = \theta1_k(t), \ 1 \leq k \leq n_h \\ yl_k & \text{if } W\Theta_i(t) = W2_{jk}(t), \ 1 \leq k \leq n_h \\ 1 & \text{if } W\Theta_i(t) = \Theta2_k(t), \ 1 \leq k \leq n_y \end{cases}$$

$$\tag{A-7}$$

The updating routine is as follows,

$$\underline{\underline{P}}(t) = \frac{1}{\kappa_k(t)} \left\{ \underline{\underline{P}}(t-1) - \frac{\underline{\underline{P}}(t-1) \, \underline{\underline{\Psi}}(t) \, \underline{\underline{\Psi}}(t)^T \, \underline{\underline{P}}(t-1)}{\left[\kappa_k(t) \, \underline{\underline{\Lambda}}(t)^{-1} + \underline{\underline{\Psi}}(t)^T \, \underline{\underline{P}}(t-1) \, \underline{\underline{\Psi}}(t) \right]} \right\} \tag{A-8}$$

$$\underline{W\Theta}(t) = \underline{W\Theta}(t-1) + \underline{\underline{P}}(t) \, \underline{\underline{\Psi}}(t) \, \underline{\underline{\Lambda}}(t) \, \underline{\varepsilon}(t) \tag{A-9}$$

$$\kappa_k(t) = \left[\gamma_k(t-1) \, (1-\gamma_k(t)) \right] \Big/ \gamma_k(t) \tag{A-10}$$

$$\underline{\underline{\Lambda}}^{-1}(t) = \underline{\underline{\Lambda}}(t-1)^{-1} + \kappa_k(t) \left[\underline{\varepsilon}(t) \, \underline{\varepsilon}(t)^T - \underline{\underline{\Lambda}}(t-1)^{-1} \right] \tag{A-11}$$

with,

$$\kappa_k(t) : \text{ gain at time t.}$$

(Note : $n_y = n_h = 1$ for DNNC model)

Appendix B. DNNC Controller Design in DMC framework

The DNNC model in IMC framework predicts one future output at each sampling time. For predicting outputs more than one time step in the future, the iteration through the neural network would be required. The predicted output from the neural network can be used at sampling time $k+1$ as input to the neural network at sampling time k. The manipulated inputs $\underline{\Delta u}$ must be shifted accordingly. Similar iteration to obtain future prediction is also proposed by Saint-Donat et. al. [11]. In this case the controller model is given by the following equations,

$$\underline{\Delta u} = \left[\; \left[\underline{\underline{W1}}^T \; \underline{\underline{\Gamma1}}^T \; \underline{\underline{\Gamma1}} \; \underline{\underline{W1}} + \underline{\underline{\Lambda1}}^T \; \underline{\underline{\Lambda1}} \right]^{-1} \underline{\underline{W1}}^T \; \underline{\underline{\Gamma1}}^T \; \underline{\underline{\Gamma1}} \; G(\underline{e}) \right]$$

(B-1)

where,

$$G(\underline{e}) = F^{-1} \left[\frac{\underline{e} - B2 \; \underline{1}_P}{W2} \right] - B1 \; \underline{1}_P - \underline{y}^{*NN}(k+1) - \alpha \; \underline{y}_m^{pred}(k)$$

(B-2)

$$\underline{e} = \underline{y}^{set}(k+1) - \underline{d}(k)$$

(B-3)

with each element of transfer function, F^{-1}, defined by equation (46)

$$\underline{\underline{W1}} = \begin{bmatrix} w_{1,1} & 0 & 0 & \bullet & 0 \\ w_{1,2} & w_{1,1} & 0 & \bullet & 0 \\ \bullet & \bullet & \bullet & \bullet & \bullet \\ w_{1,M} & w_{1,M-1} & w_{1,N-2} & \bullet & w_{1,1} \\ \bullet & \bullet & \bullet & \bullet & \bullet \\ w_{1,P} & w_{1,P-1} & w_{1,P-2} & \bullet & w_{1,P-M+1} \end{bmatrix}$$

(B-4)

and,

$$\underline{y}^{*NN}(k+1) = \begin{bmatrix} \sum\limits_{j=2}^{N-1} w_{1,j} \; \Delta u(k-j+1) + w_{1,N} \; u(k-N+1) \\ \\ \sum\limits_{j=3}^{N-1} w_{1,j} \; \Delta u(k-j+2) + w_{1,N} \; u(k-N+2) \\ \bullet \\ \bullet \\ \bullet \\ \sum\limits_{j=P+1}^{N-1} w_{1,j} \; \Delta u(k-j+P) + w_{1,N} \; u(k-N+P) \end{bmatrix}$$

(B-5)

$\underline{\Delta u}(k)$　　　=　M x 1 vector of controller output, change in the input (manipulated variable) defined as u(k)- u(k-1)

M	=	number of manipulated move in to the future
P	=	prediction into the future
N	=	number of time intervals needed to describe the process dynamics
$y_m^{pred}(k)$	=	current feedback measurement, predicted using neural network
$y^{*NN}(k+1)$	=	output due to input moves up to the present time.
1_P	=	P x 1, vector of ones
P	=	$M \le P \le N-1$
W2	=	hidden/output weighting factor
W1	=	P x M, lower triangular matrix as defined in equation (52)
B1	=	hidden layer bias
B2	=	output layer bias
$\Gamma1$	=	P x P diagonal matrix, tuning parameter
$\Lambda1$	=	M x M diagonal matrix, tuning parameter.

DNNC in DMC framework provides the same tuning parameters as DMC parameters (N, M, P, γ, λ) for controller design.

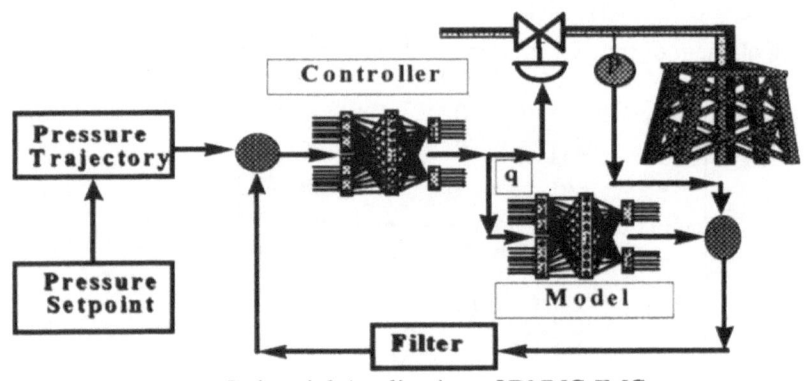

Industrial Application of DNNC-IMC

Figure 1. Block diagram of simplified DNNC strategy in IMC Framework

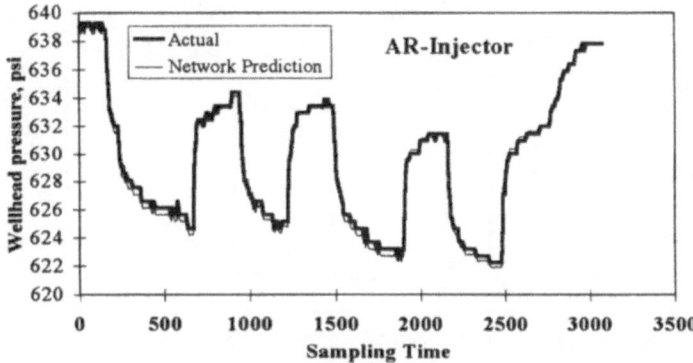

Figure 2a. Neural network prediction for wellhead pressure; Injector AR.

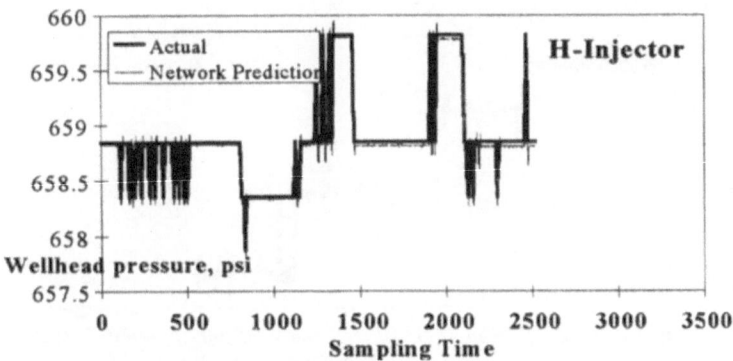

Figure 2b. Neural network prediction for wellhead pressure; Injector H.

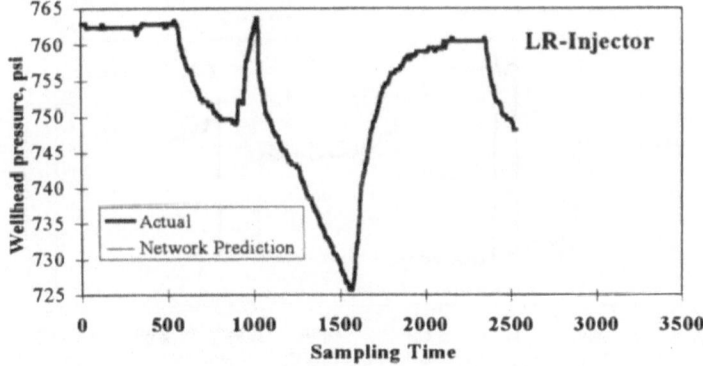

Figure 2c. Neural network prediction for wellhead pressure; Injector LR.

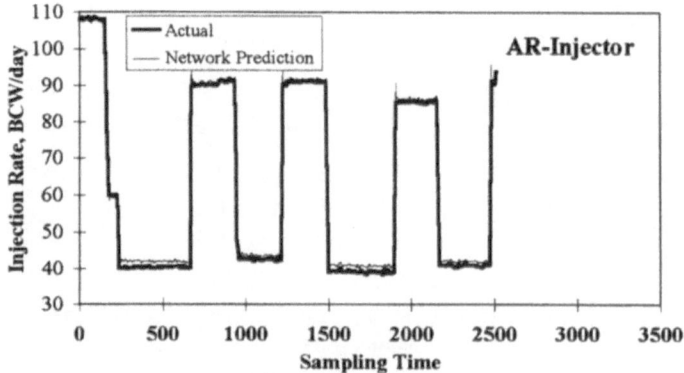

Figure 2d. Neural network prediction for injection rate; Injector AR.

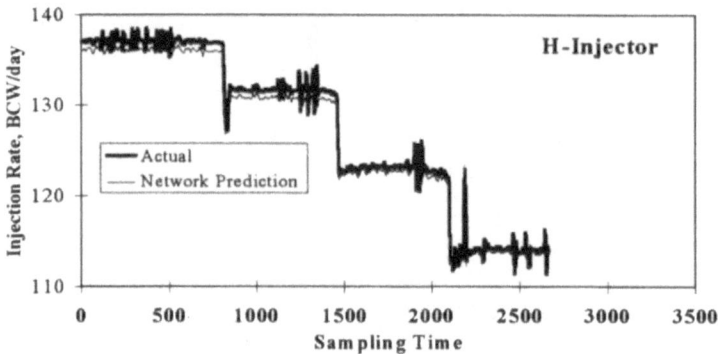

Figure 2e. Neural network prediction for injection rate;
Injector H.

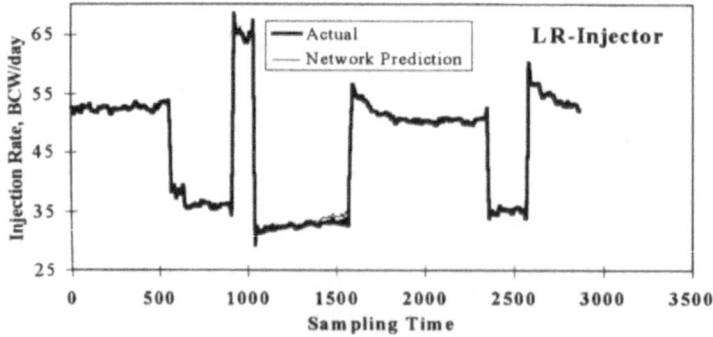

Figure 2f. Neural network prediction for injection rate;
Injector LR.

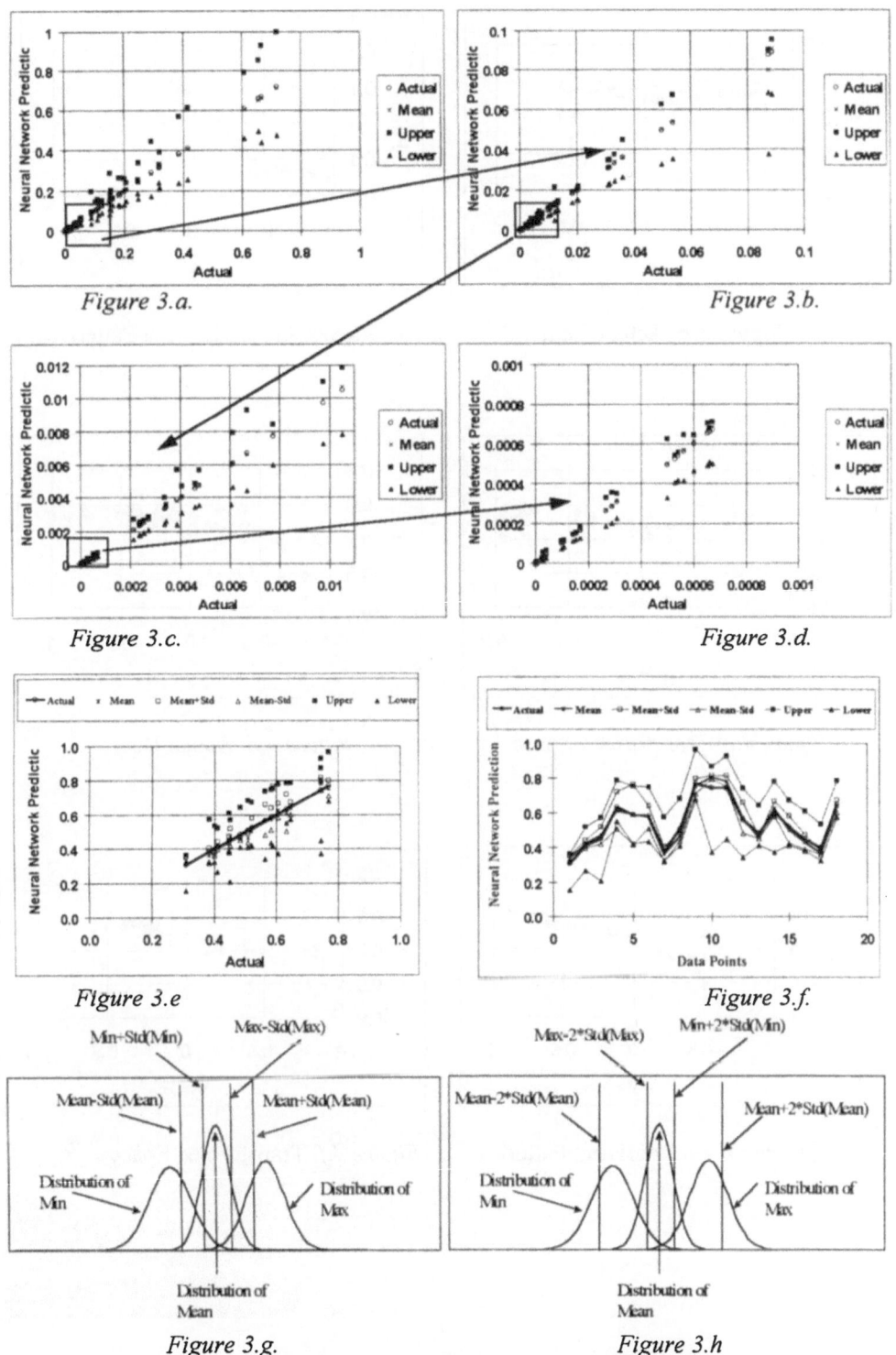

Figure 3.a.

Figure 3.b.

Figure 3.c.

Figure 3.d.

Figure 3.e

Figure 3.f.

Figure 3.g.

Figure 3.h

Figure 3. Performance of the Neuro-Statistical Model

864

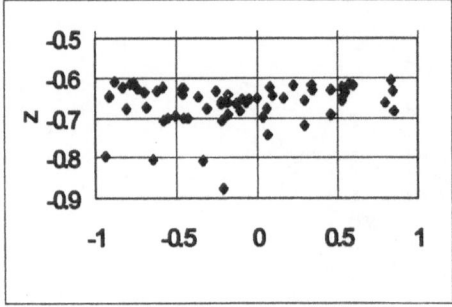

Figure 4.a. Actual Data

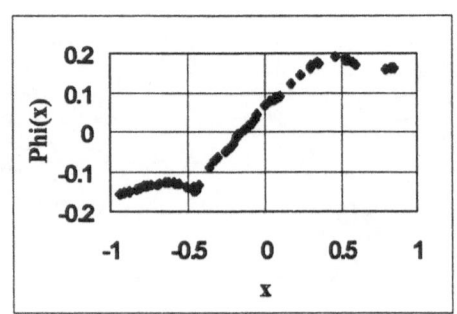

Figure 4.b. Underlying Pattern

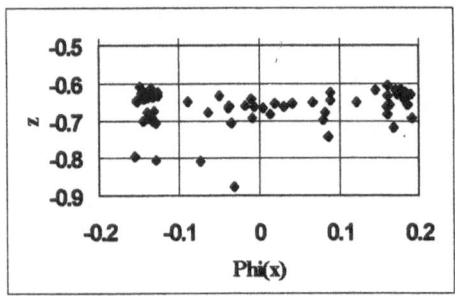

Figure 4.c. Transformed Space

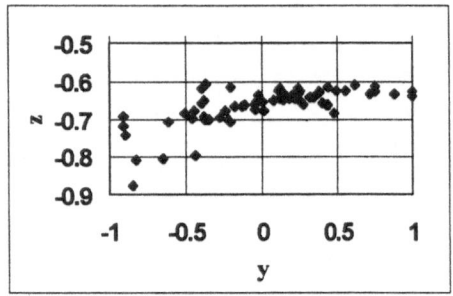

Figure 4.d. Actual Data

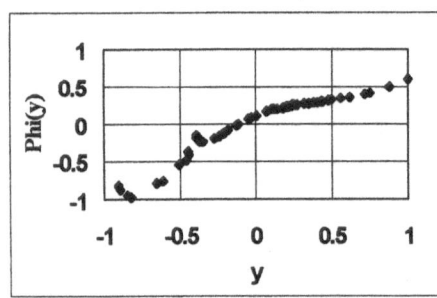

Figure 4.e. Underlying Pattern

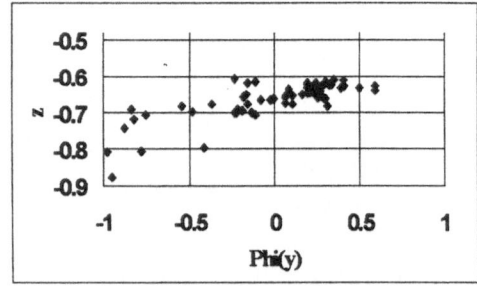

Figure 4.f. Transformed Space

NN Prediction

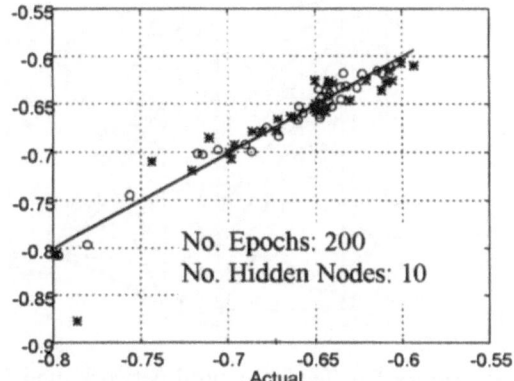

Figure 4.g. Performance of Conventional Neural Network

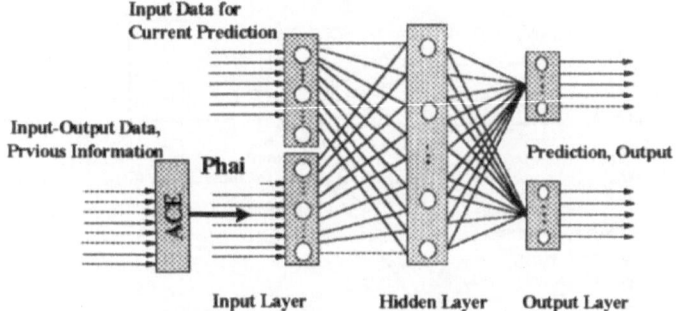

Figure 4.h. Hybrid ACE Model

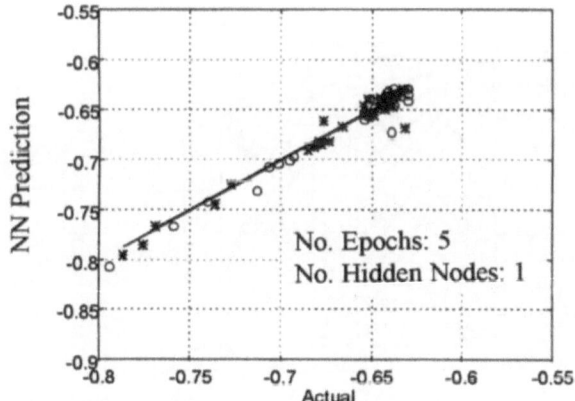

Figure 4.i. Performance of ACE-Neural Network

Figure 4. ACE, Neural Network, and ACE-Neural Network Models and Performances

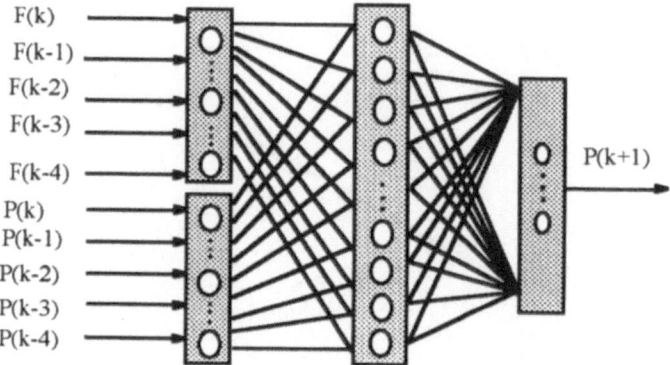

Figure 5.a. Typical neural network model

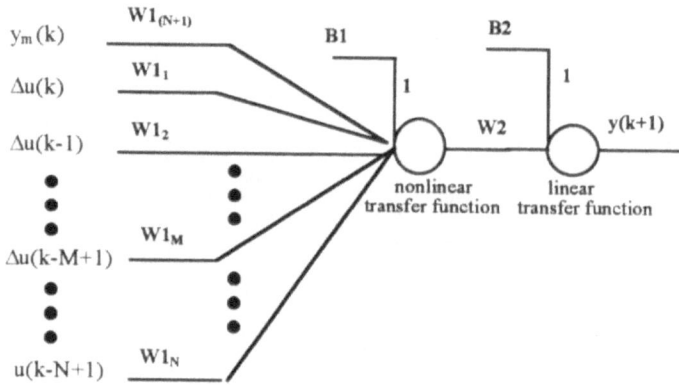

Figure 5.b. Typical single-layer DNNC process model.

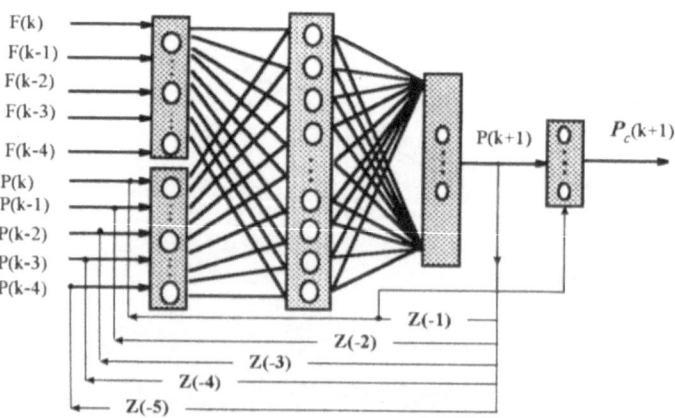

Figure 5.c. Non-conventional neural network

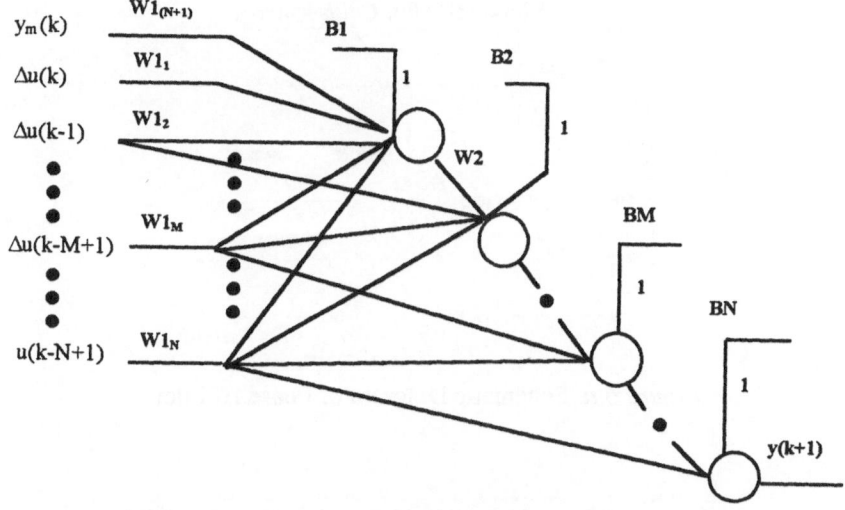

Figure 5.d. Typical multi-layer DNNC process model.

Figure 5. Comparison between conventional neural network structure and non-conventional network structure.

868

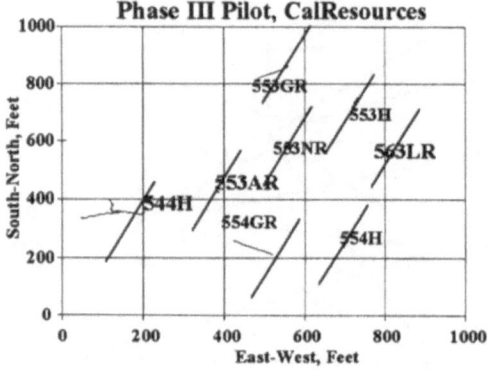

Figure 6.a. Schematic Diagram of Phase III Pilot.

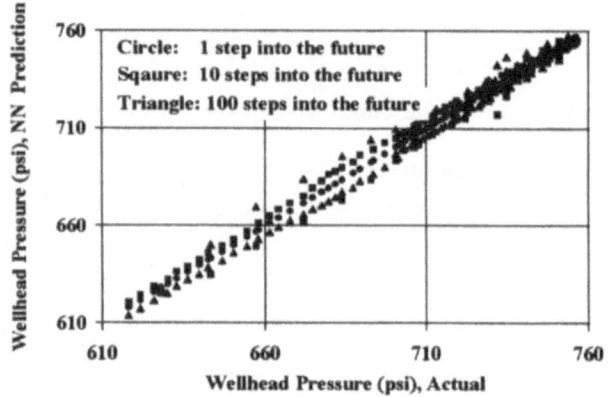

Figure 6.b. Performance of Conventional Neural Network.

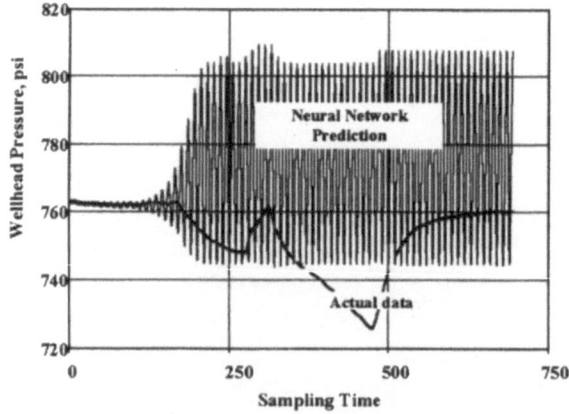

Figure 6.c. Performance of conventional neural network after 700 iterations.

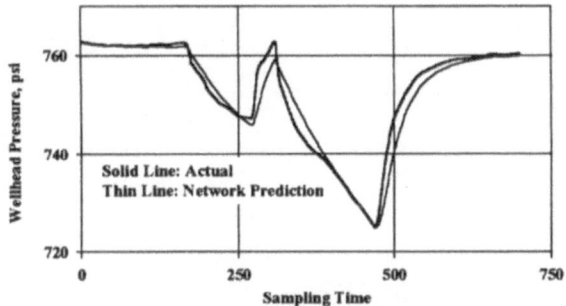

Figure 6.d. Performance of proposed neural network after 700 iterations.

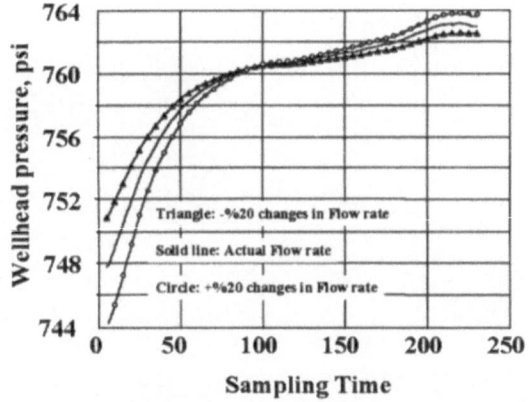

Figure 6.e. Open-loop step responses for a series of changes in injection flow rate.

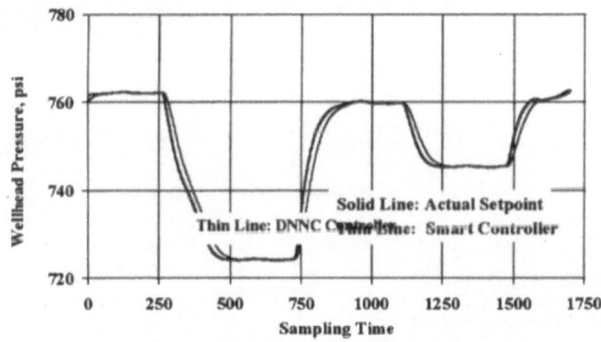

Figure 6.f. Set point tracking performance of smart controller

Figure 6. Phase III Pilot, CalResources

APPENDICES

I. Titles of Poster Presentations

II. Keeping up with Tradition:
Lectures and Events at the Workshop - A Cartoon Summary

III. List of Participants

IV. Group Picture of all Participants

V. Group Picture of Main Lecturers and Invited Contributors

TITLES OF POSTER PRESENTATIONS

☐ *M. Wellers* and *H. Rake*
Nonlinear Model Predictive Control of Simulated Chemical Processes
Using Volterra - Series

☐ *D.G. Harrell*
Supervisory Control of Nonlinear Model - Based Controllers

☐ *C.R. Edgar* and *B.E. Postlethwaite*
Using Fuzzy Relational Modelling in Nonlinear Model - Based Process
Control

☐ *U. Akman*
Nonlinear Model Predictive Control of Retrofit Heat   Exchanger
Networks

☐ *G.M. Dimirovski* and *Y. Jing*
A Decentralized Observer - Based Stabilization Control System for a Class
of Interconnected Nonlinear Plants

LECTURES AND EVENTS AT THE WORKSHOP :
A Cartoon Summary

B. W. Bequette
Rensselaer Polytechnic Institute, Troy, NY 12180-3590 USA

It has been a tradition at a number of topical control conferences to present special awards at the end of the conference. I was prepared to do the same at this workshop, when I had a vision that clearly summarized (for me, at least) the lectures and events that occurred during our 10 days together in Antalya. This vision is presented in the cartoon "Moon Over Antalya Bay" which depicts the moonlight cruise that we enjoyed early in the workshop. In the following paragraphs I present the events illustrated in the cartoon in roughly the same order as they occurred during the workshop. Footnotes are provided for the benefit of those that did not attend the conference.

Controlling a large sailboat is a complex problem and each lecturer had a different opinion about the best method to be used. In the lower left corner we notice that Yaman Arkun is attempting to use a number of model boats (sailboat, canoe, speedboat and jet-ski) to represent the behavior of the large sailboat under different operating conditions.[1] During the course of the cruise a number of people got sick because of the rocking of the boat due to wave action. We see below deck that Ahmet Palazoglu is lying on a table trying to recover from a bout with sea-sickness. Notice that Denis Dochain has come to his aid. Dr. Dochain is concerned that the illness may be due to a bacteria, and claims that Ahmet can learn all about how to control biochemical systems by "reading his book".[2] Dr. Daoutidis is recommending that Ahmet will feel better if he simply "neglects the fast modes".[3]

Going above deck we find a number of events taking place. First, we see that Andre Damslora has fallen overboard, but was saved by his tie.[4] Next, we note that Frank Allgower and Simone de Olivera are struggling to determine who can guarantee a stable control algorithm for the sailboat.[5] To the right we see that Alex Zheng is consulting a crystal ball to determine the optimal set of future control moves.[6] Wolfgang Marquardt is using a sextant to infer the position and velocity of the sailboat, and to use these for the control computation.[7] When trying to determine a model for control system design, it is important to have a proper sequence of input signals. We see that Ron Pearson is using a belly-dancer to assist him in applying forces to the boat to determine a process model.[8]

This workshop had some interested collaborations that one would not have normally expected. In the bottom center we notice that Costas Kravaris and Cole Brosilow are feverishly rowing a lifeboat in different directions. Also, we see that Ridvan Berber has fallen out of the lifeboat and is struggling to keep his cellular phone out of the water.[9]

Masoud Soroush is in a lifeboat drifting towards a rock (a hard constraint); he is busy calculating the best control moves to avoid the rock.[10] To the lower fight we see that George Stephanopolous is riding the crest of a "wave-net" to land on the sailboat and take control.[11]

On shore, we notice that Jens Balchen is preaching about his END algorithm and manipulating a number of knobs to design a nonlinear controller for the sailboat. A number of textbooks on differential geometry have fallen off the wall (recall that Lie was a Norwegian mathematician).[12] To the right we see that Masoud Nikravesh is trying to recruit a number of sailboat owners to join a consortium to study better sailboat control techniques.[13]

In the center of the shoreline we notice that Ferhan Kayihan, with ax in hand, is dreaming about how the wood from the sailboat could be used to provide chips for a Kamyr digester. Finally, we see an oven with a number of "half-baked ideas".[14] After my presentation of the cartoon summary, Costas Kravaris noted that since I did not appear I must be at the piano bar.[15]

---

[1] On the first day of the conference Yaman presented a lecture on a multiple-model-based approach for process control.

[2] Those attending the workshop will recall that Denis shamelessly held-up his book on biochemical reactor modeling and control a number of times (well, at least once) during his lecture.

[3] Prodromos presented a singular perturbation approach where a system is decomposed into slow and fast modes.

[4] During the course of his presentation Andre changed from "management mode" to "engineer mode" by taking off his tie.

[5] Recall that Frank and Simone had a heated discussion of the stability of model predictive control algorithms.

[6] During Alex's presentation there was some confusion about the C(x) function which provides an analytical solution for all control moves after the first one.

[7] Wolfgang's lecture was on dynamic data reconciliation.

[8] Ron showed us a set of semi-stochastic moves on "Turkish night" when a belly-dancer asked him to dance.

[9] Although Costas and Cole were co-authors on a paper, they did not seem to agree on much of the content of the paper. Ridvan was constantly on the move throughout the workshop, handling changes in schedules, etc. - his cell-phone never left his side in order to take care of us.

[10] Masoud gave a lecture on techniques to handle input saturation in multivariable systems.

[11] To state the obvious, George presented a lecture on wave-nets.

[12] The END algorithm that Jens presented appeared to require a tad bit more engineering insight than a number of us had. As always, Jens provided a number of provoking statements at the workshop. One was regarding the IMC "religion" so fervently practiced by many attendees.

[13] Masoud has a consortium of oil companies that are interested in oil reservoir control techniques.

[14] Cole Brosilow, during one of his lectures, claimed to have some half-baked ideas.

[15] Many of us spent an inordinate amount of the time at the piano bar, listening to music and surviving the constant verbal abuse from Barbara, the pianist.

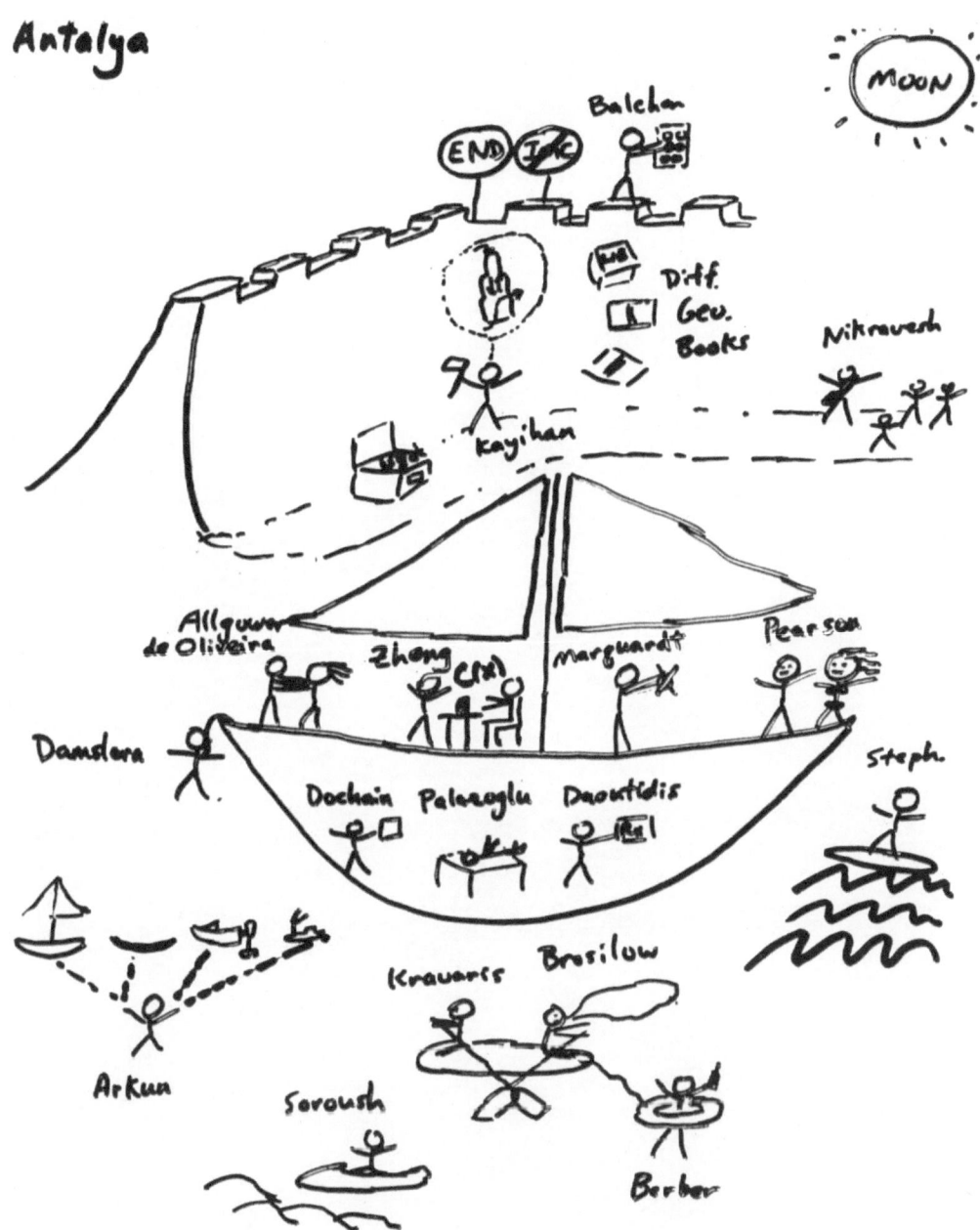

MOON OVER ANTALYA BAY

LIST OF PARTICIPANTS
NATO-ASI on Nonlinear Model Based Process Control
10-20 August 1997, Antalya - Turkey

NATO COUNTRIES

BELGIUM
Denis DOCHAIN

Université Catholique de Louvain, Bat. Euler
Av. G. Lemaitre 4-6, 1348 Louvain-la-Neuve
E-mail: dochain@auto..ucl.ac.be

Georges HEYEN

Lab. d'Analyse et Synthèse des Systèmes
Chimiques, Université de Liège
Sart Tilman, Buil. B6, B4000 Liège
E-mail: G.Heyen@ULG.ac.be

CANADA
Michel PERRIER

Dept. of Chem. Eng., École Polytechnique
P.O. B. 6079 Station Centre-ville Montreal, Quebec
E-mail: Michel.Perrier@mail.polymtl.ca

DENMARK
Torben Ravn ANDERSEN

Dept. of Chem. Eng., Building 229
Tech. Univ. of Denmark DK-2800 Lyngby
E-mail: tra@kt.dtu.dk

John BAGTERP

Dept. of Chem. Eng. Building 229
Tech. Univ. of Denmark, DK-2800 Lyngby
E-mail: bagterp@kampsax.dtu.dk

FRANCE
Ina TARALOVA

Complexe Scientifique de Rangueil GESNLA-LEISA
DGEI, INSA 31077 Toulouse, Cedex 4
E-mail: taralova@dge.insa-tlse.fr

GERMANY
Olaf ABEL

Dept. of Process Engineering, RWTH
Turmstrasse 46,D-52056 Aachen
E-mail: abel@lfpt.rwth-aachen.de

Frank ALLGÖWER

Automatic Control Laboratory, Swiss Fed. Inst. of
Technology, ETH Zentrum ETL K26
CH-8092 Zurich, Switzerland
E-mail: allgower@aut.ee.ethz.ch

Thomas BINDER   Dept. of Process Engineering, RWTH
Turmstrasse 46,D-52056 Aachen
E- mail: binder@lfpt.rwth-aachen.de

Rolf FINDEISEN   Institute fuer Automatik, ETH Zurich
CH-8092 Zurich, Switzerland
E-mail: findeise@aut.ee.ethz.ch

Felix HANISCH   Lehrstuhl für Anlagensteurungstechnik
Universitat Dortmund
E-mail: felix@ast.chemietechnik.uni-dortmund.de

Achim HELBIG   Dept. of Process Eng., RWTH
Turmstrasse 46,D-52056 Aachen
E-mail: helbig@lfpt.rwth-aachen.de

Wolfgang MARQUARDT   Dept. of Process Engineering, RWTH
Turmstrasse 46 D-5100 Aachen
E-mail: maq@lfpt.twth-aachen.de

Adel MHAMDI   Dept. of Process Eng., RWTH
Turmstrasse 46,D-52056 Aachen
E-mail: mhamdi@lfpt.rwth-aachen.de

Dirk PFESTORF   Hoechst AG, CR&T-Production Technologies
Building C584; 65926 Frankfurt/main

Matthias WELLERS   Institut für Regelungstechnik, RWTH
Steinbachstrasse 54 D-52056 Aachen
E-mail: we@irt.rwth-aachen.de

GREECE
Giorgos MOURIKAS   Dept. of Chemical Engineering, Aristotle Univ. of
Thessaloniki, PO Box 472 54006 Thessaloniki
E-mail: mourikas@alexandros.cperi.forth.gr

ITALY
Sergio M. SAVARESI   Dept. of Electronics and Inf., Milano Polytechnic
Piazza Leonarda da Vinci 32 20133 Milano
E-mail: savaresi@elet.polimi.it

NORWAY
Jens G. BALCHEN   Dept. of Engineering Cybernetics, Norwegian Univ. of
Science and Technology, N-7034 Trondheim
E-mail: jgb@itk.ntnu.no

Andre J. DAMSLORA   Norsk Hydro ASA, P.O.Box 2560

N-3901, Porsgrunn
E-mail: andre_damslora@hre.hydro.com

Ivar J. HALVORSEN

Dept. of Chem. Eng., NTNU
N-7034 Trondheim
E-mail: Ivar.J.Halvorsen@ecy.sintef.no

Truls LARSSON

Dept. of Chem. Eng., NTNU
N-7034 Trondheim
E-mail: tlarsson@kjemi.unit.no

Ingvild LOVIK

Dept. of Chem. Eng., NTNU
N-7034 Trondheim
E-mail: ingvildl@kjemi.unit.no

PORTUGAL
Carla PINHEIRO

·Dept de Eng.Quimica, Instituto Superior Tecnico
Av. Rovisco Pais, 1096, Lisboa Codex
E-mail: pinheiro@alfa.ist.utl.pt

SPAIN
Jose R. LEIZA

Faculty of Chemistry, Chem. Eng. Group
University of the Basque Country
20080 Donostia, San Sebastian
E-mail: qpplerej@sq.ehu.es

TURKEY
Uğur AKMAN

Dept. of Chem. Eng., Boğaziçi University
Bebek, 80815 İstanbul
E-mail: akman@boun.edu.tr

Levent ARABACIOĞLU

AKSA - Acrylic Chemical Ind. Inc., Yalova

Kemal ARSLAN

SEKA-Pulp and Paper Works of Turkey, Inc.,
41240 İzmit

Rıdvan BERBER

Dept. of Chem. Eng., University of Ankara
Tandoğan, 06100 Ankara
E-mail: berber@science.ankara.edu.tr

Ayla ÇALIMLI

Dept. of Chem. Eng., University of Ankara
Tandoğan, 06100 Ankara
E-mail: calimli@science.ankara.edu.tr

Mehmet C. ÇAMURDAN

Dept. of. Chem. Eng., Bogaziçi University
Bebek, 80815 İstanbul
E-mail: mccamurd@boun.edu.tr

882

Mustafa DEMİRCİOĞLU   Chemical Eng. Dept., Ege University
Bornova, 35100 İzmir
E-mail: mdemir@bornova.ege.edu.tr

Tolga EROL   Dept. of Chem. Eng., University of Ankara
Tandoğan, 06100 Ankara
E-mail: terol@science.ankara.edu.tr

Doğan GÜRBÜZ   AKSA - Acrylic Chemical Ind. Inc., Yalova

Baykan KACAR   AKSA - Acrilic Chemical Ind. Inc., Yalova

Ersan KALAFATOĞLU   TÜBİTAK- Marmara Research Center
P.O. Box 21, Gebze, 41470 Kocaeli
E-mail: ersan@mam.gov.tr

Kemal LEBLEBİCİOĞLU   Elec. Eng. Dept., METU, 06531 Ankara
E-mail: kleb@rorqual.cc.metu.edu.tr

Ertan MANTAR   SEKA-Pulp and Paper Works of Turkey, Inc.,
41240 İzmit

Canan ÖZGEN   Chem. Eng. Dept., METU, 06531 Ankara
Tel: 90-312-210 2601   Fax: 90-312- 210 1264
E-mail: cozgen@rorqual.cc.metu.edu.tr

Sibel SAIN   TÜBİTAK- Marmara Research Center
P.O. Box 21, Gebze, 41470 Kocaeli
E-mail: ssain@mam.gov.tr

Tamer TANILMIŞ   Chemical Eng. Dept., Ege University
Bornova, 35100 İzmir
E-mail: ttanil@textil.ege.edu.tr

Evren TEKOL   Dept. of Chem. Eng., University of Ankara
Tandoğan, 06100 Ankara
E-mail: tekol@science.ankara.edu.tr

Zekeriya UYKAN   Control Eng. Lab., Helsinki University of Technology
Otakaari 5 A, Fin-02150, Espoo, Finland
E-mail: zekeriya.uykan@hut.fi
(on leave from Istanbul Technical University, Turkey)

U. K.
Craig EDGAR   Dept. of Chem. Eng. University of Strathclyde
James Weir Building 75 Montrose St. Glasgow, GI 1XJ
E-mail: cedgar@chemeng.strath.ac.uk

U. S. A.

Yaman ARKUN

School of Chemical Engineering
Georgia Institute of Techology
Atlanta, GA 30332-0100
E-mail: yaman _arkun@che.gatech.edu

B. Wayne BEQUETTE

Dept. of. Chem. Eng., Rensselaer Polytechnic
Institute, Troy, NY 12180-3590
E-mail: bequeb@rpi.edu

Coleman B. BROSILOW

Department of Chemical Engineering
Case Western Reserve University
10900 Euclid Av. Cleveland OH 44106-7217
E-mail: cbb@po.cwru.edu

Prodromos DAOUTIDIS

Dept. of Chem. Eng. & Material Sciences
University of Minnesota
421 Washinton Ave. SE Minneapolis, MN 55455
Email: daoutidi@cems.umn.edu

Douglas G. HARRELL

Montell Polyolefins, R&D Center
912 Appleton Road, Elkton, MD 21921
E-mail: doug harrell@montellna.com

Costas KRAVARIS

Dept. of Chemical Engineering
The University of Michigan, Ann Arbor MI 48109
E-mail: costas@engin.umich.edu

Ferhan KAYIHAN

IETek, Integrated Engineering Technologies
5533 Beverly Ave NE, Tacoma WA 98422-1402
E-mail: fkayihan_IETek@msn.com

Aydın KONUK

Aspen Tcchnology - AC&OD
9896 Bissonnet, Houston Texas 77036
E-mail: aydin.konuk@aspentech.com

Luke MAGEE

225 Newark RJ., Mt. Vernon OH 43050
E-mail: lrm4@po.cwru.edu
(Dept. of Chem. Eng. CWRU, Cleveland OH 44106)

Masoud NIKRAVESH

Earth Sciences Division, Lawrence Berkeley
National Laboratory, Berkeley, CA 94720
E-mail: nikraves@cs.berkeley.edu

884

Luis Claudio OLIVEIRA
LOPES

Process Modeling and Control Research Center
Lehigh University
111 Research Drive Bethlehem, PA 18015
E-mail: lco2@Lehigh.Edu

Ahmet PALAZOĞLU

Dept. of Chem. Engineering, University of California
Davis - Davis, CA 95616
E-mail: anpalazoglu@ucdavis.edu

Ronald K. PEARSON

The DuPont Co.
PO Box 80101 Wilmington, DE 19880-0101
E-mail: pearsork@esvax.dnet.dupont.com

Armin SHMILOVICI

Dept. of Manufacturing Eng. Boston University
15 Saint Mary's St., Boston MA 02215
E-mail: armin1@engc.bu.edu

Masoud SOROUSH

Dept. of Chem. Eng. Drexel University
Philadelphia, PA 19104
E-mail: masoud.soroush@coe.drexel.edu

George
STEPHANOPOULOS

Dept. of Chem. Engineering, Massachusetts Inst. of
Technology, Cambridge, Massachusetts 02139
E-mail: geosteph@mit.edu

Alex ZHENG

Dept. of Chem. Eng., University of Massachusetts
Amherst, MA 01003-3110
E-mail: zzheng@spock.umass.edu

COOPERATION PARTNER COUNTRIES

ALBANIA
Petraq MARANGO

Automatic Control Department
Polytechnic University, Tirana
E-mail: marango@uptal.tirana.al

AZERBAIJAN
Abdulriza ABILOV

Ins. of Petrochem. Processes
The Azerbaijan Academy of Sciences
Telnov St. 30, Baku, 370025

BULGARIA
Alexander GEGOV

Inst. of Control and Systems Research
Bulgarian Acad. of Sciences
PO Box 79 Sofia 1113
E-mail: consys@bgcict.acad.bg

Petia KOPRINKOVA

Ins. of Control and Systems Research
Bulgarian Acad. of Sciences, PO Box 79, Sofia 1113
E-mail: apsau@bgcict.acad.bg

Marieta PETROVA

Ins. of Control and Systems Research
Bulgarian Acad. of Sciences, PO Box 79, Sofia 1113
E-mail: tdpat@bgcict.acad.bg

FORMER YUGOSLAV
REP. OF MACEDONIA
Georgi M. DIMIROVSKI

Dept. of Elect. Eng., St.Cyril and Methodius
University, PO Box 574 91000 Skopje
E-mail: dimir@cerera.etf.ukim.edu.mk

ROMANIA
Vasile MARINOIU

Dept. of Auotomation and Computer, Univ. "Oil-Gas"
Ploiesti, Bd. Bucuresti 39 2000 Ploiesti
E-mail: marv@csd.univ.ploiesti.ro

NON-NATO COUNTRIES

BRASIL
Simone L. de OLIVEIRA

Technology Center, Cidade University
Blocco G, Ilha do Fundao
Rio de Janeiro, 2194-970, RJ
E-mail: SLO@peq.coppe.ufrj.br

ISRAEL
Michal RONEN

The Prog. of Biotech., Faculty of Engineering
Ben-Gurion Univ. of the Negev
PO Box 653, Beer-sheva 84105
E-mail: michal@newton.bgu.ac.il

SWITZERLAND
Carlos CUELLAR

Measurement and Control Laboratory
Swiss Federal Institute of Technology Zurich
ETH-Zentrum CH-8092 Zürich
E-mail: cuellar@imrt.mavt.ethz.ch

APPENDIX IV - Group Picture of all Participants

887

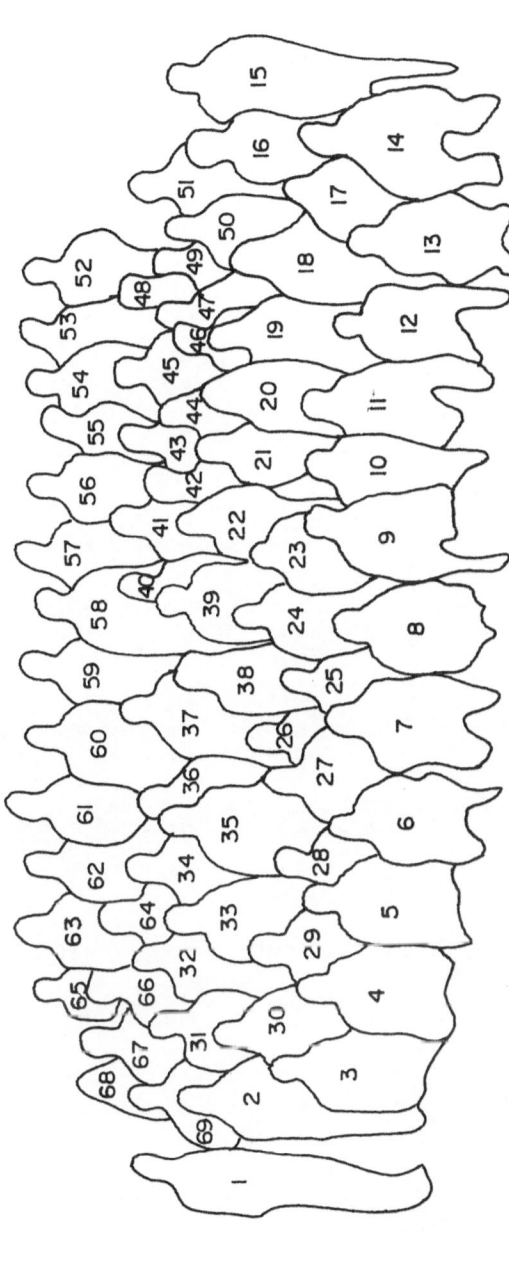

1- G. Stephanopoulos (USA); 2- A. Çalımlı (TR); 3- F. Kayıhan (USA); 4- P. Marango (AL); 5- E. Tekol (TR); 6- E. Mantar (TR); 7- B. Kacar (TR); 8- K. Leblebicioğlu (TR); 9- D. Pfestorf (D); 10- A. Gegov (BG); 11- F. Allgöwer (CH); 12- A. Zheng (USA); 13- V. Marinoiu (R); 14- L.C. Oliveira Lopes USA); 15- T. Tanılmış (TR); 16- G. Mourikas (GR); 17- M. Değirmencioğlu (TR); 18- S. Özdemir (TR); 19- S.L. de Oliveira (BR); 20– I. Loevik (N); 21- C. Pinheiro (P); 22- I. Taralova (F); 23- C. Brosilow (USA); 24- A. M'hamdi (D); 25- D. Dochain (B); 26- D. Harrell (USA); 27- M. Soroush (USA); 28- T. Erol (TR); 29- C. Cuellar (CH); 30- Y. Arkun (USA); 31- A. Konuk (USA); 32- W. Bequette (USA); 33- C. Özgen (TR); 34- M. Çamurdan (TR); 35- M. Perrier (CDN); 36- A. Abilov (Azer.); 37- M. Nikravesh (USA); 38- M. Nikravesh's sister; 39- R. Berber (TR); 40- T. Binder (D); 41- O. Abel (D); 42- W. Marquardt (D); 43- A. Helbig (D); 44- İ.E. Kalafatoğlu (TR); 45- M. Wellers (D); 46- G. Heyen (B); 47- U. Akman (TR); 48- G. M. Dimirovski (Mac.); 49- J. Bagterp (DK); 50- J. R. Leiza (E); 51- R. Pearson (USA); 52- K. Arslan (TR); 53- A. Damslora (N); 54- I.J. Halvorsen (N); 55- R. Findeisen (D); 56- S. Savaresi (I); 57- C. Edgar (UK); 58- L. Magee (USA); 59- F. Hanisch (D); 60- A. Shmilovici (USA); 61- L. Arabacioğlu (TR); 62- M. Ronen (IL); 63- D. Gürbüz (TR); 64- P. Daoutidis (USA); 65- M. Petrova (BG); 66- T. R. Andersen (DK); 67- T. Larsson (N); 68- P. Koprinkova (BG); 69-C. Kravaris (USA);
Missing in the picture: J. Balchen (N); A. Palazoğlu (USA); Z. Uykan (TR).

Main Lecturers and Invited Contributors

From left to right; Back row: C. Brosilow, D. Dochain, W. Marquardt, R. Berber, F. Allgöwer
Front Row: A. Palazoğlu, A. Damslora, A. Zheng, A. Konuk, C. Kravaris, F. Kayıhan, M. Nikravesh,
B.W. Bequette, P. Daoutidis, S.L. de Oliveira, Y. Arkun, M. Soroush, R. Pearson, G. Stephanopoulos

INDEX